REMOTE SENSING
of Natural Resources

Taylor & Francis Series in Remote Sensing Applications

Series Editor

Qihao Weng

Indiana State University
Terre Haute, Indiana, U.S.A.

Remote Sensing of Natural Resources, *edited by Guangxing Wang and Qihao Weng*

Remote Sensing of Land Use and Land Cover: Principles and Applications, *Chandra P. Giri*

Remote Sensing of Protected Lands, *edited by Yeqiao Wang*

Advances in Environmental Remote Sensing: Sensors, Algorithms, and Applications, *edited by Qihao Weng*

Remote Sensing of Coastal Environments, *edited by Qihao Weng*

Remote Sensing of Global Croplands for Food Security, *edited by Prasad S. Thenkabail, John G. Lyon, Hugh Turral, and Chandashekhar M. Biradar*

Global Mapping of Human Settlement: Experiences, Data Sets, and Prospects, *edited by Paolo Gamba and Martin Herold*

Hyperspectral Remote Sensing: Principles and Applications, *Marcus Borengasser, William S. Hungate, and Russell Watkins*

Remote Sensing of Impervious Surfaces, *edited by Qihao Weng*

Multispectral Image Analysis Using the Object-Oriented Paradigm, *Kumar Navulur*

Taylor & Francis Series in Remote Sensing Applications
Qihao Weng, Series Editor

REMOTE SENSING
of Natural Resources

Edited By **Guangxing Wang** • **Qihao Weng**

CRC Press
Taylor & Francis Group
Boca Raton London New York

CRC Press is an imprint of the
Taylor & Francis Group, an **informa** business

CRC Press
Taylor & Francis Group
6000 Broken Sound Parkway NW, Suite 300
Boca Raton, FL 33487-2742

First issued in paperback 2019

© 2014 by Taylor & Francis Group, LLC
CRC Press is an imprint of Taylor & Francis Group, an Informa business

No claim to original U.S. Government works

ISBN-13: 978-1-4665-5692-8 (hbk)
ISBN-13: 978-0-367-86745-4 (pbk)

Visit the Taylor & Francis Web site at
http://www.taylorandfrancis.com

and the CRC Press Web site at
http://www.crcpress.com

Contents

Acknowledgments ix
Editors xi
Contributors xiii
Introduction to Remote Sensing of Natural Resources xvii

Section I   Remote Sensing Systems

1. **Introduction to Remote Sensing Systems, Data, and Applications** 3
   Qihao Weng

Section II   Sampling Design and Product Quality Assessment

2. **Remote Sensing Applications for Sampling Design of Natural Resources** 23
   Guangxing Wang and George Z. Gertner

3. **Accuracy Assessment for Classification and Modeling** 45
   Suming Jin

4. **Accuracy Assessment for Soft Classification Maps** 57
   Daniel Gómez, Gregory S. Biging, and Javier Montero

5. **Spatial Uncertainty Analysis When Mapping Natural Resources Using
   Remotely Sensed Data** 87
   Guangxing Wang and George Z. Gertner

Section III   Land Use and Land Cover Classification

6. **Land Use/Land Cover Classification in the Brazilian Amazon with
   Different Sensor Data and Classification Algorithms** 111
   *Guiying Li, Dengsheng Lu, Emilio Moran, Mateus Batistella, Luciano V. Dutra,
   Corina C. Freitas, and Sidnei J. S. Sant'Anna*

7. **Vegetation Change Detection in the Brazilian Amazon with
   Multitemporal Landsat Images** 127
   Dengsheng Lu, Guiying Li, Emilio Moran, and Scott Hetrick

8. **Extraction of Impervious Surfaces from Hyperspectral Imagery:
   Linear versus Nonlinear Methods** 141
   Xuefei Hu and Qihao Weng

9. Road Extraction: A Review of LiDAR-Focused Studies 155
Lindi J. Quackenbush, Jungho Im, and Yue Zuo

Section IV   Natural Landscape, Ecosystems, and Forestry

10. Application of Remote Sensing in Ecosystem and Landscape Modeling 173
Chonggang Xu and Min Chen

11. Plant Invasion and Imaging Spectroscopy 191
Kate S. He and Duccio Rocchini

12. Assessing Military Training–Induced Landscape Fragmentation
and Dynamics of Fort Riley Installation Using Spatial Metrics
and Remotely Sensed Data 207
Steve Singer, Guangxing Wang, Heidi R. Howard, and Alan B. Anderson

13. Automated Individual Tree-Crown Delineation and Treetop Detection
with Very-High-Resolution Aerial Imagery 223
Le Wang and Chunyuan Diao

14. Tree Species Classification 239
Ruiliang Pu

15. Estimation of Forest Stock and Yield Using LiDAR Data 259
*Markus Holopainen, Mikko Vastaranta, Xinlian Liang, Juha Hyyppä, Anttoni Jaakkola,
and Ville Kankare*

16. National Forest Resource Inventory and Monitoring System 291
Erkki Tomppo, Matti Katila, and Kai Mäkisara

Section V   Agriculture

17. Remote Sensing Applications on Crop Monitoring and Prediction 315
Bingfang Wu and Jihua Meng

18. Remote Sensing Applications to Precision Farming 333
Haibo Yao and Yanbo Huang

19. Mapping and Uncertainty Analysis of Crop Residue Cover Using
Sequential Gaussian Cosimulation with QuickBird Images 353
*Cha-Chi Fan, Guangxing Wang, George Z. Gertner, Haibo Yao, Dana G. Sullivan, and
Mark Masters*

Section VI Biomass and Carbon Cycle Modeling

20. Remote Sensing of Leaf Area Index of Vegetation Covers 375
Jing M. Chen

21. LiDAR Remote Sensing of Vegetation Biomass 399
Qi Chen

22. Carbon Cycle Modeling for Terrestrial Ecosystems 421
Tinglong Zhang and Changhui Peng

23. Remote Sensing Applications to Modeling Biomass and
Carbon of Oceanic Ecosystems 443
Samantha Lavender and Wahid Moufaddal

Section VII Wetland, Soils, and Minerals

24. Wetland Classification 461
Maycira Costa, Thiago S. F. Silva, and Teresa L. Evans

25. Remote Sensing Applications to Monitoring Wetland Dynamics:
A Case Study on Qinghai Lake Ramsar Site, China 479
Hairui Duo, Linlu Shi, and Guangchun Lei

26. Hyperspectral Sensing on Acid Sulfate Soils via Mapping Iron-Bearing and
Aluminum-Bearing Minerals on the Swan Coastal Plain, Western Australia 495
Xianzhong Shi and Mehrooz Aspandiar

Index 515

Acknowledgments

More than 50 authors and coauthors were invited to contribute to this book. Many of them also offered to review the chapters. We would like to express our sincerest thanks to all the contributors and the reviewers. In addition, we wish to extend our deepest appreciation to the following reviewers for their professional services:

Xiaoyong Chen
Hannes Feihauer
Hong S. He
Xiangyun Hu
Sanna Kaasalainen
Jinxun Liu

Richard Mueller
Izaya Numata
Blanca Pérez Lapeña
Jonathan Remo
Sebastian Schmidtlein
Harini Sridharan

Conghe Song
Junmei Tang
Kelly Thorp
Zhiqiang Yang

Without the contributions and effort from the authors and the reviewers, it would have not been possible to publish this book. The acquisition editor from CRC Press, Irma Shagla-Britton, kindly offered extra support to make our collaboration extremely successful. Finally, we are indebted to our families for their love and support.

Editors

Guangxing Wang, PhD, is an associate professor of remote sensing and geographic information systems (GIS) in the Department of Geography and Environmental Resources at Southern Illinois University at Carbondale (SIUC), Illinois. He is also a guest professor at the Central South University of Forestry and Technology (CSUFT), China. He received his BS (1982) in forestry and MSc (1985) in forest biometrics at the CSUFT, China, and his PhD (1996) in remote sensing of forest resources at the University of Helsinki (UoH), Finland. On graduation, he joined the Department of Forest Resource Management, UoH, as a research scientist. As a postdoctoral research associate and academic research scientist, in 1998 Wang moved to and worked at the Department of Natural Resources and Environmental Sciences, University of Illinois at Urbana-Champaign (UIUC), Illinois, until he joined the SIUC in 2007. He has taught a forest measurement and forest management capstone course at CSUFT; remote sensing of forest resources at UoH; and introduction to remote sensing, advanced remote sensing, introduction to GIS, and cartographic design at SIUC. He has supervised more than 10 graduate students in remote sensing and GIS.

Dr. Wang's research focuses on geospatial technologies (remote sensing, GIS, global positioning system, and spatial statistics) and their applications to natural resources. The specific areas include sampling design strategies for natural resources, modeling human–environment interactions, ecosystems and environmental dynamics, human activity–induced vegetation disturbance and soil erosion, environmental quality assessment, forest carbon mapping, and quality assessment and spatial uncertainty analysis of remote sensing and GIS products. His research has been mainly funded by the U.S. Army Engineer Research and Development Center—Construction Engineering Research Laboratory; the U.S. Strategic Environmental Research and Development Program; National Science Foundation; National Aeronautics and Space Administration (NASA); Department of Commerce, Illinois Department of Commerce and Economic Opportunity; the Counties of Southern Illinois, LLC; Illinois Council on Food and Agricultural Research; and SIUC. He is an author and coauthor of more than 100 publications, including four books and more than 50 peer-reviewed journal articles in the areas of remote sensing, GIS, forestry, environment, etc.

Qihao Weng, PhD, is the director of the Center for Urban and Environmental Change and a professor of geography at Indiana State University. He was a visiting NASA Senior Fellow (2008–2009). He is also a guest/adjunct professor at the Peking University, Hong Kong Polytechnic University, Wuhan University, and Beijing Normal University, and a guest research scientist at Beijing Meteorological Bureau, China. He received his PhD in geography from the University of Georgia in 1999. In the same year, he joined the University of Alabama as an assistant professor. Since 2001, he has been a member of the faculty in the Department of Earth and Environmental Systems at Indiana State University, where he has taught five courses on remote sensing, digital image processing, remote sensing–geographic information systems (GIS) integration, and two courses on GIS and environmental modeling, and has mentored 10 doctoral and 10 master's students.

Dr. Weng's research focuses on remote sensing and GIS analysis of urban ecological and environmental systems, land use and land cover change, environmental modeling, urbanization impacts, and human–environment interactions. He is the author of more than 140 peer-reviewed journal articles and other publications and five books. He has worked extensively with optical and thermal remote sensing data, and more recently with light detection and ranging data, primarily for an urban heat island study, land cover and impervious surface mapping, urban growth detection, image analysis algorithms, and integration with socioeconomic characteristics, with financial support from U.S. funding agencies that include the National Science Foundation, NASA, U.S. Geological Survey, United States Agency for International Development, National Oceanic and Atmospheric Administration, National Geographic Society, and Indiana Department of Natural Resources. He was the recipient of the Robert E. Altenhofen Memorial Scholarship Award by the American Society for Photogrammetry and Remote Sensing (ASPRS) (1999), the Best Student-Authored Paper Award from the International Geographic Information Foundation (1998), and the 2010 Erdas Award for Best Scientific Paper in remote sensing from ASPRS (first place). At Indiana State University, he received the Theodore Dreiser Distinguished Research Award in 2006 (the university's highest research honor) and was selected as a Lilly Foundation Faculty Fellow in 2005 (one of six recipients). In May 2008, he received a prestigious NASA senior fellowship. In April 2011, he was named as the recipient of the Outstanding Contributions Award in remote sensing in 2011 sponsored by American Association of Geographers (AAG) Remote Sensing Specialty Group. He has given more than 70 invited talks (including colloquia, seminars, keynote address, and public speech) and has presented more than 100 papers at professional conferences (including copresenting).

Dr. Weng is the task leader for Group on Earth Observation (GEO) SB-04, Global Urban Observation and Information (2012–2015). In addition, he serves as an associate editor of *ISPRS Journal of Photogrammetry and Remote Sensing and Spatial Hydrology* and is the series editor for both the Taylor & Francis Series in Remote Sensing Applications and McGraw-Hill Series in GIS&T. He was a national director of ASPRS (2007–2010), chair of AAG China Geography Specialty Group (2010–2011), secretary of International Society for Photogrammetry and Remote Sensing Working Group VIII/1 (Human Settlement and Impact Analysis, 2004–2008), as well as a panel member of the U.S. Department of Energy's Cool Roofs Roadmap and Strategy in 2010.

Contributors

Alan B. Anderson
Research and Development Center
Construction Engineering Research
  Laboratory
Champaign, Illinois

Mehrooz Aspandiar
Department of Applied Geology
Curtin University of Technology
Perth, Australia

Mateus Batistella
Embrapa Satellite Monitoring
Campinas, São Paulo, Brazil

Gregory S. Biging
Department of Environmental Science,
  Policy, and Management
Center for Ecosystem Measurement,
  Monitoring and Modeling
College of Natural Resources
University of California, Berkeley
Berkeley, California

Jing M. Chen
Department of Geography and Program
  in Planning
University of Toronto
Toronto, Ontario, Canada

Min Chen
Division of Earth and Environmental
  Sciences
Los Alamos National Laboratory
Los Alamos, New Mexico

Qi Chen
Department of Geography
University of Hawai'i at Mānoa
Honolulu, Hawai'i

Maycira Costa
University of Victoria
Victoria, British Columbia, Canada

Chunyuan Diao
Department of Geography
State University of New York
Buffalo, New York

Hairui Duo
School of Nature Conservation
Beijing Forestry University
Beijing, China

Luciano V. Dutra
National Institute for Space Research
São Jose dos Campos, São Paulo, Brazil

Theresa L. Evans
University of Victoria
Victoria, British Columbia, Canada

Cha-Chi Fan
Office of Electricity, Renewables and
  Uranium Statistics
U.S. Energy Information Agency
U.S. Department of Energy
Washington, DC

Corina C. Freitas
National Institute for Space Research
São Jose dos Campos, São Paulo, Brazil

George Z. Gertner
Department of Natural Resources and
  Environmental Sciences
University of Illinois at
  Urbana-Champaign
Champaign, Illinois

Daniel Gómez
Escuela de Estadística
Complutense University
Madrid, Spain

Kate S. He
Department of Biological Sciences
Murray State University
Murray, Kentucky

Scott Hetrick
Anthropological Center for Training and
    Research on Global Environmental
    Change
Indiana University
Bloomington, Indiana

Markus Holopainen
Department of Forest Sciences
University of Helsinki
Helsinki, Finland

Heidi R. Howard
Research and Development Center
Construction Engineering Research
    Laboratory
Champaign, Illinois

Xuefei Hu
Rollins School of Public Health
Emory University
Atlanta, Georgia

Yanbo Huang
Crop Production Systems
    Research Unit
Stoneville, Mississippi

Juha Hyyppä
Finnish Geodetic Institute
Helsinki, Finland

Jungho Im
Department of Environmental Resources
    Engineering
College of Environmental Science and
    Forestry
State University of New York
Syracuse, New York

Anttoni Jaakkola
Finnish Geodetic Institute
Kirkkonummi, Finland

Suming Jin
Arctic Slope Regional Corporation
    Research and Technology Solutions
and
U.S. Geological Survey Earth Resources
    Observation and Science Center
Sioux Falls, South Dakota

Ville Kankare
Department of Forest Sciences
University of Helsinki
Helsinki, Finland

Matti Katila
Finnish Forest Research Institute
Vantaa, Finland

Samantha Lavender
Pixalytics Ltd
Tamar Science Park
Plymouth, Devon, UK

Guangchun Lei
School of Nature Conservation
Beijing Forestry University
Beijing, China

Guiying Li
Anthropological Center for Training and
    Research on Global Environmental
    Change
Indiana University
Bloomington, Indiana

Xinlian Liang
Finnish Geodetic Institute
Helsinki, Finland

Dengsheng Lu
Center for Global Change and Earth
    Observations
Michigan State University
East Lansing, Michigan

Kai Mäkisara
Finnish Forest Research Institute
Vantaa, Finland

Mark Masters
Flint River Water Planning and Policy
  Center
Albany, Georgia

Jihua Meng
Institute of Remote Sensing and Digital
  Earth
Chinese Academy of Sciences
Beijing, China

Javier Montero
Faculty of Mathematics
Complutense University of Madrid
Madrid, Spain

Emilio Moran
Center for Global Change and Earth
  Observations
Michigan State University
East Lansing, Michigan

Wahid Moufaddal
National Institute of Oceanography and
  Fisheries
Alexandria, Egypt

Changhui Peng
Department of Biology Sciences
Institute of Environment Sciences
University of Quebec at Montreal
Montreal, Quebec, Canada

Ruiliang Pu
Department of Geography, Environment,
  and Planning
University of South Florida
Tampa, Florida

Lindi J. Quackenbush
Department of Environmental Resources
  Engineering
College of Environmental Science and
  Forestry
State University of New York
Syracuse, New York

Duccio Rocchini
GIS and Remote Sensing Unit
Department of Biodiversity and Molecular
  Ecology
Research and Innovation Centre
Fondazione Edmund Mach
Trentino, Italy

Sidnei J. S. Sant'Anna
National Institute for Space Research
São Jose dos Campos
São Paulo, Brazil

Linlu Shi
School of Nature Conservation
Beijing Forestry University
Beijing, China

Xianzhong Shi
Department of Applied Geology
Curtin University of Technology
Perth, Australia

Thiago S. F. Silva
Instituto Nacional de Pesquisa Espaciais
São Paulo, Brazil

Steve Singer
Department of Geography and
  Environmental Resources
Southern Illinois University Carbondale
Carbondale, Illinois

Dana G. Sullivan
USDA-ARS Southeast Watershed Research
  Laboratory
Tifton, Georgia

Erkki Tomppo
Finnish Forest Research Institute
Vantaa, Finland

Mikko Vastaranta
Department of Forest Sciences
University of Helsinki
Helsinki, Finland

Guangxing Wang
Department of Geography and
    Environmental Resources
Southern Illinois University Carbondale
Carbondale, Illinois

Le Wang
Department of Geography
State University of New York
Buffalo, New York

Qihao Weng
Department of Earth and Environmental
    Systems
Center for Urban and Environmental
    Change
Indiana State University
Terre Haute, Indiana

Bingfang Wu
Institute of Remote Sensing and Digital
    Earth
Chinese Academy of Sciences
Beijing, China

Chonggang Xu
Division of Earth and Environmental
    Sciences
Los Alamos National Laboratory
Los Alamos, New Mexico

Haibo Yao
Geosystems Research Institute
Mississippi State University
Stennis Space Center
Starkville, Mississippi

Tinglong Zhang
Laboratory for Ecological Forecasting and
    Global Change
College of Forestry
Northwest A&F University
Yangling, Shaanxi, China

Yue Zuo
Department of Environmental Resources
    Engineering
College of Environmental Science and
    Forestry
State University of New York
Syracuse, New York

Introduction to Remote Sensing of Natural Resources

Guangxing Wang and Qihao Weng

Aims and Scope

This book focuses on remote sensing systems, algorithms, and their applications for evaluation of natural resources, especially in the areas of sampling design, land use and land cover (LULC) characterization and classification, natural landscape and ecosystem assessment, forestry and agriculture mapping, biomass and carbon cycle modeling, wetland classification and dynamics monitoring, and soils and minerals mapping. It combines review articles with case studies that demonstrate recent advances and developments of methods, techniques, and applications of remote sensing, with each chapter on a specific area of natural resources. This book aims at providing undergraduate and graduate students with a focused text or a supplementary text for those whom the principles of remote sensing, digital image processing, or remote sensing applications is a course to be taken. It may also be useful for researchers, scientists, engineers, and decision makers in the area of geospatial technology and/or its applications to natural resources as a reference text.

Synopsis of the Book

This book consists of 7 sections and a total of 26 chapters. The seven sections are remote sensing systems; sampling design and product quality assessment; land use and land cover classification; natural landscape, ecosystems, and forestry; agriculture; biomass and carbon cycle modeling; and wetland, soils, and minerals. In Section I, following an introduction to the basic principle of remote sensing, Weng presents an overall review of various remote sensors including electro-optical sensors, thermal infrared sensors, passive microwave and imaging radar sensors, and LiDAR (light detection and ranging). He further discusses the characteristics of remotely sensed data including photographs and satellite images, especially on spatial, spectral, radiometric, and temporal resolution. Finally, Weng suggests fundamental considerations for conducting a remote sensing project.

Section II consists of four chapters on advanced algorithms and applications of remote sensing for sampling design and product quality assessment. When natural resources are mapped with remotely sensed images, field observations of interest variables are collected and used for both creating spatially explicit estimates of the variables and assessing the quality of the resulting maps (Congalton and Green 2009; Campbell and Wynne 2011). Thus, selecting a cost-effective sampling design strategy is critical for both mapping and accuracy assessment of natural resources (Thompson 1992; Brus and de Gruijter 1997; Wang et al. 2008). In Chapter 2, Wang and Gertner review the applications of remote sensing for sampling design of natural resources. They discuss the recent advances in design- and model-based strategies. They highlight the roles of remote sensing in sampling design. Moreover, they present an advanced local variability–based sampling design and demonstrate a case study to explain this method.

Remote sensing data products may contain a large amount of uncertainties (Mowrer and Congalton 2000; Gertner et al. 2002; Congalton and Green 2009). Use of remote sensing–derived maps for decision-making supports may possess risks. Thus, how to quantify and reduce uncertainties in natural resource estimates is important. Chapter 3 by Jin focuses on accuracy assessment of hard-classified maps for categorical variables. She first introduces the core method (error matrix) for accuracy assessment of thematic maps. Then she discusses the factors that need to be considered to obtain an error matrix, especially when accuracy assessment of large-area land cover and change detection maps is conducted. In addition, she introduces the recent developments in map assessment of continuous variables.

Although the error matrix has been widely used for accuracy assessment, it has also been criticized owing to its inability to assess fuzzy or soft classification maps (Pontius and Cheuk 2006). In Chapter 4, Gómez, Biging, and Montero present new developments in this area. They first give a short description of different types of accuracy assessments. Then they demonstrate the main techniques that deal with the pixel accuracy assessment for soft classification maps and their applications through a real case study.

Another challenge in accuracy assessment of remote sensing data products relates to spatial uncertainty analysis, that is, how to quantify the impacts of input errors and uncertainties on the accuracy of the final products (Crosetto and Tarantola 2001; Gertner et al. 2002; Wang et al. 2005, 2009; Lilburne and Tarantola 2009). In Chapter 5, Wang and Gertner provide an assessment of the state of the art on spatial uncertainty analysis for mapping natural resources with remotely sensed data. They first discuss various error sources and existing methods for conducting spatial uncertainty analysis and the error budget. Then, they identify significant gaps that currently exist in this research area and propose a methodological framework. Moreover, they present two case studies for soil erosion prediction and aboveground forest carbon mapping.

Use of remotely sensed images for LULC classification has been substantially studied (Lu and Weng 2007). Section III, from Chapter 6 to Chapter 9, focuses on various aspects of LULC characterization and classification with remotely sensed images. In Chapter 6, Li and colleagues compare LULC classifications in the Brazilian Amazon using different remotely sensed datasets. The datasets include Landsat thematic mapper (TM) images, ALOS PALSAR L-band horizontal and horizontal polarization (HH) and horizontal and vertical polarization (HV) images, RADARSAT-2 C-band HH and HV images, and their fusions and textures. The authors further make the comparison of different classification algorithms including maximum likelihood classifier, classification tree analysis, fuzzy ARTMAP, k-nearest neighbor, object-based classification, and support vector machine.

Although many change detection techniques are available, effectively detailing "from–to" change information of LULC categories is still challenging (Singh 1989; Mas 1999; Lu et al. 2004; Im et al. 2008). In Chapter 7, Lu and colleagues present a method in which a green vegetation index is used to generate vegetation gain and loss distribution and then combined with one classified image. The green vegetation index was derived using linear spectral mixture analysis. This combination made it possible to derive detailed "from–to" change information of LULC. They then apply this method to vegetation change detection in the Brazilian Amazon with Landsat images.

Urban imperious surface is a unique type of land cover that can be extracted from various remotely sensed data (Weng 2012). In Chapter 8, Hu and Weng demonstrate a method of extraction of impervious surfaces from hyperspectral imagery. They provide an assessment of subpixel algorithms for extracting impervious surfaces. Then, they use linear spectral mixture analysis and a multilayer perceptron neural network with a back-propagation learning algorithm with a Hyperion image covering the Atlanta metropolitan area. The results indicate that the neural network method worked better but the spatial and temporal variations of impervious surfaces deserve special attention due to the diversity of urban environments. In mapping impervious surfaces, an important topic is to accurately extract roads because of their importance in community and transportation planning and natural resource management. In Chapter 9, Quackenbush and colleagues conduct a review of literature in the area of road extraction. In particular, they emphasize the use of LiDAR data and its fusion with high spatial resolution images to extract roads. They discuss various classification methods and algorithms and further analyze generation of road networks.

Traditionally, landscape and ecosystem models lack the ability to derive spatially explicit estimates of interest variables, while remotely sensed images provide the potential to obtain spatial distributions of these variables. Thus, integration of models and remotely sensed images will potentially improve predictions (Goward et al. 1994; Kimball et al. 2000; Zhao and Running 2010). Various remote sensing technologies have been developed and applied to forest resource management (Tomppo et al. 2008; Hyyppä et al. 2009; Pu 2009). Section IV, from Chapter 10 to Chapter 16, presents remote sensing algorithms and applications for natural landscapes, ecosystems, and forestry. In Chapter 10, Xu and Chen first introduce four groups of forest models that can be run at ecosystem, stand, landscape, and regional and global scale, respectively. They then discuss model initialization and validation and emphasize the integration of models with remotely sensed data using forest and lake landscapes as examples.

Although identification of invasive species in natural landscapes and ecosystems using remotely sensed data has been substantially studied, we are still facing great challenges in research mainly due to the biotic and abiotic complexities associated with the species (Sakai et al. 2001). In Chapter 11, He and Rocchini assess the applications of imaging spectroscopy for detecting and mapping the spatial spread of invasive species. They present various image-processing and classification algorithms and emphasize the need for a combination of algorithms and phenology to improve identification of invasive species.

Human activities often lead to fragmentation of landscapes and thus degrade landscape and environmental quality. In Chapter 12, Singer et al. (2012) assess military training–induced land fragmentation and dynamics for a U.S. installation where military training activities took place using a time series of Landsat TM images. They calculate several spatial metrics at both landscape and patch levels and analyze their relationships with military training intensity. Moreover, they study the effects of different spatial resolution images on landscape segmentation and LULC classification using Indian Remote Sensing images.

Remotely sensed data can be utilized in forest resource management at different scales from individual tree to forest stand and region. In Chapter 13, Wang and Diao present a two-stage approach: a Laplacian of Gaussian edge detection method followed by the marker-controlled watershed segmentation to automatically delineate individual tree crowns using high-resolution aerial imagery. In their study, a multiscale wavelet decomposition algorithm is used to enhance tree crown boundaries. In Chapter 14, Pu provides an assessment of the state of the art on remote sensing applications for tree species classification using high spatial resolution and hyperspectral images. In addition, he discusses various image-processing and information extraction techniques and advanced classification algorithms.

Forest management and strategic planning require spatially explicit estimates of forest resources. LiDAR data provide this potential. In Chapter 15, Holopainen and colleagues give a comprehensive review of LiDAR systems and their applications in estimation of forest stand stock and yield. They start with introduction of airborne, mobile, and terrestrial laser scanning and then discuss point cloud metrics and surface models, and area-based and individual tree detection methods. They also demonstrate LiDAR applications in estimation and mapping of tree and forest stand variables. In Chapter 16, Tomppo and colleagues focus on the applications of remote sensing for Finnish national forest resource inventory and monitoring. They propose a multisource method and show how to use the *k*-nearest neighbor to generate spatially explicit estimates of forest resources.

Remote sensing provides a powerful tool for agriculture (Maignan et al. 2008; Becker-Reshef et al. 2010; Wu et al. 2010), as described in Section V. In Chapter 17, Wu and Meng introduce various models for crop monitoring and prediction systems, such as crop condition monitoring models, crop acreage estimation models, and so on. They then present the widely used crop monitoring and prediction systems that are run at national, regional, and global scales. Furthermore, they discuss the role of remote sensing in crop monitoring and prediction systems.

The Global Positioning System and geographic information systems have made great advances over the past decades. When combined with remote sensing, these technologies have made it possible to manage in-field variability in agriculture, leading to precision farming (NRC 1997; Zhang et al. 2002). In Chapter 18, Yao and Huang discuss the image characteristics that are critical for precision farming and various vegetation indexes. Then, they present six aspects of remote sensing applications in precision farming, including soil property mapping, insect/pest infestation identification, crop water stress detection, in-field nitrogen stress detection, weed sensing and mapping, and herbicide drift detection.

Due to human activities, a great amount of agricultural land in the world has been classified as highly erodible, contributing to continuous degradation of soil productivity (Singer et al. 2012). Accurate assessment of conservation tillage adoption, for example, rapidly quantifying crop residue cover, becomes critical. In Chapter 19, Fan and colleagues first give a brief review of mapping crop residue cover using various remotely sensed images. They then compare an image-based sequential Gaussian cosimulation algorithm with a traditional regression modeling to generate spatially explicit estimates of crop residue cover for three sites in the Little River Experimental Watershed of Georgia by combining sample data with QuickBird images.

To mitigate the effect of global warming, it is essential to provide policy makers with accurate information on the distributions and dynamics of carbon sources and sinks (Bousquet et al. 2000; Cox et al. 2000). Remotely sensed data may be used to examine the spatial distribution of vegetation canopy and environmental characteristics, and thus provide great potential in global carbon cycle modeling (Running 1999; Cohen and Goward 2004; Turner et al. 2004). Section VI, from Chapter 20 to Chapter 23, was designed to demonstrate the remote sensing applications for biomass and carbon cycle modeling. In Chapter 20, Chen first reviews the ground-based leaf area index (LAI) measurement methods and techniques and recent developments in LAI remote sensing. He then discusses how to generate global LAI and clumping index maps using data from different satellite sensors and the challenges in mapping of LAI using remote sensing.

Vegetation, especially forest, plays an important role in carbon sink and mitigation of atmospheric concentrations of carbon dioxide (Lefsky et al. 2002; Asner 2009). Thus, accurately estimating forest biomass and carbon stock is critical (Houghton 2009; Wang et al. 2009). LiDAR provides great potential in estimation of forest biomass and carbon. In Chapter 21, Chen presents an in-depth review of the current status and research advances on biomass estimation using various LiDAR systems. Moreover, he examines critical issues and challenges in LiDAR remote sensing.

There is no doubt that both terrestrial and oceanic ecosystems play a significant role in the global biogeochemical cycles, and remote sensing is a powerful tool to model the carbon cycles (Running 1999; Bousquet et al. 2000; Cox et al. 2000). In Chapter 22, Zhang and Peng discuss remote sensing–derived vegetation characteristics and the process models used to model the carbon cycling of terrestrial ecosystems. Through three case studies, they emphasize the integration of process models and remotely sensed data for monitoring the dynamics of vegetation ecosystems. Chapter 23 shifts the focus to the ocean ecosystem. Lavender and Moufaddal introduce the widely used methods to collect *in situ* measurements and their combinations with models and remotely sensed data for estimation of oceanic ecosystem biomass. They offer an overview of algorithms and present a study to monitor human-induced fish dynamics in the Mediterranean Sea.

Section VII of this book deals with applications of remote sensing to wetland classification and dynamics monitoring and soils and minerals mapping. As transitional ecosystems between land and water, wetlands play important roles in flood control, climate regulation, carbon storage, aquifer recharge, and biodiversity management. However, accurate global coverage of wetlands is not yet available (Davidson and Finlayson 2007; Mitsch and Gosselink 2007). In Chapter 24, Costa and colleagues review various remote sensing techniques for classifying wetlands. They then demonstrate a case study in which multispectral IKONOS and airborne-based hyperspectral images were compared to map eelgrass beds in the temperate coastal waters of British Columbia, Canada. In addition, they present a classification of wetland habitats at the Nhecolândia region of the Brazilian Pantanal using object-oriented classification. Chapter 25 by Duo and colleagues examine wetland dynamics of Qinghai Lake, China, using Landsat multitemporal images acquired over a period of 34 years from 1977 to 2011. The authors investigate the areas and human-induced conversions of LULC categories including lake, grassland, sandy area, marshes, farmland, saline land, bare land, and residential areas. Moreover, they analyze the landscape dynamics using spatial metrics.

Remotely sensed data have also been utilized in classification and mapping of soils and minerals. In Chapter 26, Shi and Aspandiar present a study using hyperspectral remote sensing on acid sulfate soils via mapping iron- and aluminum-bearing minerals on the Swan Coastal Plain, Western Australia. They collected intensive soil samples and obtained their spectral data using an Analytical Spectral Devices FieldSpec 3 that was used as ground truth to match the spectrum from the HyMap sensor. The authors then study the reflectance spectral characterization of the soils and minerals in the Swan Coastal Plain and classify and map iron- and aluminum-bearing minerals, carbonates, and sulfates using spectral feature fitting, spectral indexes, and linear unmixing methods.

References

Asner, G. P. 2009. "Tropical Forest Carbon Assessment: Integrating Satellite and Airborne Mapping Approaches." *Environmental Research Letters* 4. doi:10.1088/1748-9326/1084/1083/034009.

Becker-Reshef, I., C. Justice, M. Sullivan, E. Vermote, C. Tucker, A. Anyamba, J. Small, et al. 2010. "Monitoring Global Croplands with Coarse Resolution Earth Observations: The Global Agriculture Monitoring (GLAM) Project." *Remote Sensing* 2(6):1589–609.

Bousquet, P., P. Peylin, P. Ciais, C. Le Quere, P. Friedlingstein, and P. P. Tans. 2000. "Regional Changes in Carbon Dioxide Fluxes of Land and Oceans Since 1980." *Science* 290(5495):1342–46.

Brus, D. J., and J. J. de Gruijter. 1997. "Random Sampling or Geostatistical Modeling? Choosing Between Design-Based and Model-Based Sampling Strategies for Soil (With Discussion)." *Geoderma* 80:1–44.

Campbell, J. B., and R. H. Wynne. 2011. *Introduction to Remote Sensing*. New York: The Guilford Press.

Cohen W. B., and S. N. Goward. 2004. "Landsat's Role in Ecological Applications of Remote Sensing." *BioScience* 54:535–45.

Congalton, R. G., and K. Green. 2009. *Assessing the Accuracy of Remotely Sensed Data: Principles and Practices*. Boca Raton, FL: CRC Press, Taylor and Francis Group.

Cox, P. M., R. A. Betts, C. D. Jones, S. A. Spall, and I. J. Totterdell. 2000. "Acceleration of Global Warming due to Carbon-Cycle Feedbacks in a Coupled Climate Model." *Nature* 408(6809):184–87.

Crosetto, M., and S. Tarantola. 2001. "Uncertainty and Sensitivity Analysis: Tools for GIS-Based Model Implementation." *International Journal of Geographical Information Science* 15(5):415–37.

Davidson, N. C., and C. M. Finlayson. 2007. "Earth Observation for Wetland Inventory, Assessment and Monitoring." *Aquatic Conservation: Marine and Freshwater Ecosystems* 17:219–28.

Gertner, G., G. Wang, S. Fang, and A. B. Anderson. 2002. "Mapping and Uncertainty of Predictions Based on Multiple Primary Variables from Joint Co-Simulation with TM Image." *Remote Sensing of Environment* 83:498–510.

Goward, S. N., R. H. Waring, D. G. Dye, and J. Yang. 1994. "Ecological Remote Sensing at OTTER: Satellite Macroscale Observations." *Ecological Applications* 4:322–43.

Houghton, R. A., F. Hall, and S. J. Goetz. 2009. "Importance of Biomass in the Global Carbon Cycle." *Journal of Geophysical Research-Biogeosciences* 114: G00E03. doi: 10.1029/2009JG000935.

Hyyppä, J., H. Hyyppä, X. Yu, H. Kaartinen, H. Kukko, and M. Holopainen. 2009. "Forest Inventory Using Small-Footprint Airborne LiDAR." In *Topographic Laser Ranging and Scanning: Principles and Processing*, edited by J. Shan and C. Toth. Boca Raton, FL: CRC Press, Taylor and Francis Group. 335–70.

Im, J., J. R. Jensen, and J. A. Tullis. 2008. "Object-Based Change Detection Using Correlation Image Analysis and Image Segmentation." *International Journal of Remote Sensing* 29(2):399–423.

Kimball, J. S., A. R. Keyser, S. W. Running, and S. S. Saatch. 2000. "Regional Assessment of Boreal Forest Productivity Using an Ecological Process Model and Remote Sensing Parameter Maps." *Tree Physiology* 20:761–75.

Lefsky, M. A., W. B. Cohen, D. J. Harding, G. G. Parker, S. A. Acker, and S. T. Gower. 2002. "LiDAR Remote Sensing of Above-Ground Biomass in Three Biomes." *Global Ecology and Biogeography* 11:393–99.

Lilburne, L., and S. Tarantola. 2009. "Sensitivity Analysis of Spatial Models." *International Journal of Geographical Information Science* 23(2):151–68.

Lu, D., P. Mausel, and E. Brondizio. 2004. "Change Detection Techniques." *International Journal of Remote Sensing* 25(12):2365–407.

Lu, D., and Q. Weng. 2007. "A Survey of Image Classification Methods and Techniques for Improving Classification Performance." *International Journal of Remote Sensing* 28(5):823–70.

Maignan, F., F. M. Breon, and C. Bacour. 2008. "Interannual Vegetation Phenology Estimates from Global AVHRR Measurements Comparison with In Situ Data and Applications." *Remote Sensing of Environment* 112:496–505.

Mas, J. F. 1999. "Monitoring Land-Cover Changes: A Comparison of Change Detection Techniques." *International Journal of Remote Sensing* 20(1):139–52.

Mitsch, W. J., and J. G. Gosselink. 2007. *Wetlands*. 4th ed., 600. Hoboken, NJ: John Wiley & Sons.

Mowrer, H. T., and R. G. Congalton. 2000. *Quantifying Spatial Uncertainty in Natural Resources: Theory and Applications for GIS and Remote Sensing*. Ann Arbor, MI: Sleeping Bear Press.

NRC (National Research Council). 1997. *Precision Agriculture in the 21st Century: Geospatial and Information Technologies in Crop Management*. Washington, DC: National Academy Press.

Pontius Jr., R. G., and M. L. Cheuk. 2006. "A Generalized Cross-Tabulation Matrix to Compare Soft-Classified Maps at Multiple Resolutions." *International Journal of Geographical Information Science* 20(1):1–30.

Pu, R. 2009. "Broadleaf Species Recognition with In Situ Hyperspectral Data." *International Journal of Remote Sensing* 30(11):2759–79.

Running, S. W., D. D. Baldocchi, D. P. Turner, S. T. Gower, P. S. Bakwin, and K. A. Hibbard. 1999. "A Global Terrestrial Monitoring Network Integrating Tower Fluxes, Flask Sampling, Ecosystem Modeling and EOS Satellite Data." *Remote Sensing of Environment* 70(1):108–27.

Sakai, A. K., F. W. Allendorf, J. S. Holt, D. M. Lodge, J. Molofsky, A. Kimberly, K. A. With, et al. 2001. "The Population Biology of Invasive Species." *Annual Review of Ecology and Systematics* 32:305–32.

Singer, S., G. Wang, H. Howard, and A. B. Anderson. 2012. "Comprehensive Assessment Indicator of Environmental Quality for Military Land Management." *Environmental Management* 50:529–40.

Singh, A. 1989. "Review Article: Digital Change Detection Techniques Using Remotely Sensed Data." *International Journal of Remote Sensing* 10(6):989–1003.

Thompson, S. K. 1992. *Sampling*. New York: John Wiley & Sons, Inc.

Tomppo, E., H. Olsson, G. Ståhl, M. Nilsson, O. Hagner, and M. Katila. 2008. "Combining National Forest Inventory Field Plots and Remote Sensing Data for Forest Databases." *Remote Sensing of Environment* 112:1982–99.

Turner, D. P., S. V. Ollinger, and J. S. Kimball. 2004. "Integrating Remote Sensing and Ecosystem Process Models for Landscape- to Regional-Scale Analysis of the Carbon Cycle." *BioScience* 54 (6):573–84.

Wang, G., G. Z. Gertner, A. B. Anderson, and H. R. Howard. 2008. "Repeated Measurements on Permanent Plots Using Local Variability Based Sampling for Monitoring Soil Erosion." *Catena* 73:75–88.

Wang, G., G. Z. Gertner, S. Fang, and A. B. Anderson. 2005. "A Methodology for Spatial Uncertainty Analysis of Remote Sensing Products." *Photogrammetric Engineering and Remote Sensing* 71(12): 1423–32.

Wang, G., T. Oyana, M. Zhang, S. Adu-Prah, S. Zeng, H. Lin, and J. Se. 2009. "Mapping and Spatial Uncertainty Analysis of Forest Vegetation Carbon by Combining National Forest Inventory Data and Satellite Images." *Forest Ecology and Management* 258(7):1275–83.

Weng, Q. 2012. "Remote Sensing of Impervious Surfaces in the Urban Areas: Requirements, Methods, and Trends." *Remote Sensing of Environment*, 117(2):34–49.

Wu, B., J. Meng, and Q. Li. 2010. "Review of Overseas Crop Monitoring Systems with Remote Sensing." *Advances in Earth Science* 25(10):1003–12.

Zhang, N., M. Wang, and N. Wang. 2002. "Precision Agriculture: A Worldwide Overview." *Computers and Electronics in Agriculture* 36(2–3):113–32.

Zhao, M., and S. W. Running. 2010. "Drought-Induced Reduction in Global Terrestrial Net Primary Production from 2000 through 2009." *Science* 329(5994):940–43.

Section I

Remote Sensing Systems

1

Introduction to Remote Sensing Systems, Data, and Applications

Qihao Weng

CONTENTS

1.1 Definition of Remote Sensing 3
1.2 Remote Sensors 4
    1.2.1 Electro-Optical Sensors..4
    1.2.2 Thermal Infrared Sensors...6
    1.2.3 Passive Microwave and Imaging Radar Sensors 8
    1.2.4 Light Detection and Ranging...9
1.3 Characteristics of Remotely Sensed Data...10
    1.3.1 Photographs..10
    1.3.2 Satellite Images..10
    1.3.3 Spatial Resolution 11
    1.3.4 Spectral Resolution 12
    1.3.5 Radiometric Resolution...15
    1.3.6 Temporal Resolution..15
1.4 Some Basic Considerations for Remote Sensing Applications.........................17
References...18

1.1 Definition of Remote Sensing

We perceive the surrounding world through our five senses. Some senses (e.g., touch and taste) require contact of our sensing organs with the objects. However, much information about our surroundings through the senses of sight and hearing do not require close contact between our organs and the external objects. In this sense, we are performing remote sensing all the time. Generally, remote sensing refers to the activities of recording/observing/ perceiving (sensing) objects or events at faraway (remote) places. The sensors (e.g., special types of cameras and digital scanners) can be installed in airplanes, satellites, or space shuttles to take pictures of the objects or events on the earth's surface. Therefore, the sensors are not in direct contact with the objects or events being observed. The information needs a physical carrier to travel from the objects to the sensors through an intervening medium. The electromagnetic radiation (solar radiation) is normally used as an information carrier in remote sensing. The output of a remote sensing system is usually an image (digital pictures) representing the objects/events being observed. A further step of image analysis and interpretation is often required to extract useful information from the image.

In a more restricted sense, remote sensing refers to the science and technology of acquiring information about the earth's surface (land and ocean) and atmosphere using sensors onboard airborne (aircraft and balloons) or spaceborne (satellites and space shuttles) platforms. Depending on the scope, remote sensing may be divided into (1) satellite remote sensing, when satellite platforms are used; (2) photography and photogrammetry, when photographs are used to capture visible light; (3) thermal remote sensing, when the thermal infrared (IR) portion of the spectrum is used; (4) radio detection and ranging (radar) remote sensing, when microwave wavelengths are used; and (5) light detection and ranging (LiDAR) remote sensing, when laser pulses are transmitted toward the ground and the distance between the sensor and the ground is measured based on the return time of each pulse. Remote sensing has now been integrated with other modern geospatial technologies such as geographic information system (GIS), global positioning system (GPS), and mobile mapping.

1.2 Remote Sensors

Remote sensing systems/sensors can be basically grouped into two types: passive and active sensors. The main source of energy for remote sensing comes from the sun. Remote sensors record solar radiation reflected or emitted from the earth's surface. When the source of energy comes from outside a sensor, it is called a passive sensor. Examples of passive sensors include photographic cameras, electro-optical sensors, thermal IR sensors, and antenna sensors. Because passive sensors use naturally occurring energy, they can only capture data during the daylight hours. The exception is thermal IR sensors, which can detect naturally emitted energy day or night, as long as the amount of energy is large enough to be recorded. Active sensors use the energy coming from within the sensor. They provide their own energy that is directed toward the target to be investigated. The energy scattered back/reflected from that target is then detected and recorded by the sensors. An example of active sensing is radar, which transmits a microwave signal toward the target and detects and measures the backscattered portion of the signal. Another example is LiDAR. It emits a laser pulse and precisely measures its return time to calculate the height of each target. Active sensors can be used to image the surface at any time, day or night, and in any season. Active sensors can also be used for examining wavelengths that are not sufficiently provided by the sun, such as microwaves, or to better control the way a target is illuminated (Canada Centre for Remote Sensing 2007). This section will begin our discussion with a concise comparison between across-track and along-track scanners, two commonly used electro-optical sensors for acquiring multispectral imagery; we will then discuss thermal IR sensors before turning to active sensors.

1.2.1 Electro-Optical Sensors

Satellite images are acquired by electro-optical sensors. Each sensor may consist of various amounts of detectors. The detectors receive and measure reflected or emitted radiance from the ground, as electrical signals, and convert them into numerical values, which are stored onboard or transmitted to a receiving station on the ground. The incoming radiance is often directed toward different detectors, so that the radiance within a specific range of wavelength can be recorded as a spectral band. The major types of scanning systems used to acquire multispectral image data include across-track and along-track scanners.

Instead of imaging a ground scene entirely in a given instant in time (as with a photographic camera), across-track scanners scan (i.e., image) the earth's surface in a contiguous series of narrow strips by using a rotating/oscillating mirror. These ground strips are oriented perpendicular to the flight path of the platform (e.g., satellite). As the plat moves forward, new ground strips are created. Successive ground strips build up a two-dimensional (2D) image of the earth's surface along the flight line. This technique is sometimes called "whiskbroom" scanning since the scanner "whisks" the earth's surface as the platform advances along the flight direction. The incoming reflected or emitted radiation is frequently separated into several spectral bands, as these remote sensing scanning systems are capable of multispectral imaging. The near-ultraviolet (UV), visible, reflected IR, and thermal IR radiations can all be used in a multispectral remote sensing system, which usually has 5–10 spectral bands. A set of electronic detectors, which is sensitive to a specific range of wavelengths, detects and measures the incoming radiation for each spectral band and save as digital data.

For a given altitude of the platform, the angular field of view, that is, the sweep of the mirror, determines the width of the imaged swath. Airborne scanners typically sweep large angles (between 90° and 120°), whereas satellites need only to sweep fairly small angles (10° to 20°) to cover a broad region because of their higher altitudes. The instantaneous field of view (IFOV) of the sensor determines the ground area viewed by the sensor in a given instant in time. This ground area, known as a ground resolution cell, is of course subjected to the influence of the altitude of the platform. IFOV defines the spatial resolution of a sensor. Because the distance from the sensor to the target increases toward the edges of the swath, the ground resolution cells are larger toward the edges than at the center of the ground swath (i.e., nadir). This change in the sensor–target distance gives rise to a scale distortion in the acquired images, which is often corrected during the stage of image processing before the images are provided to the users. The length of time that the IFOV "sees" a ground resolution cell as the rotating mirror scans is called the dwell time. It is generally quite short, but it influences the design of the spatial, spectral, and radiometric resolution of the sensor (Canadian Centre for Remote Sensing 2007). A small IFOV, in spite of being associated with high spatial resolution, limits the amount of radiation received by the sensor, whereas a larger IFOV would receive more radiation, allowing the detection of small variations in reflectance or emittance. Moreover, a larger IFOV generates in effect a longer dwell time over any given area and improves the radiometric resolution. A large IFOV also has a high signal-to-noise (S/N) ratio (signal greater than background noise).

Across-track scanners generally have a wide image swath. Although their mechanical system is complex, their optical system is relatively simple. Across-track scanners have been widely used in satellite remote sensing programs. Well-known examples include those scanners onboard the Landsats: multispectral scanner (MSS), thematic mapper (TM), and enhanced TM Plus (ETM+). These sensors were the prime earth-observing sensors during the 1970s into the 1980s. The Advance Very High Resolution Radiometer (AVHRR) and Defense Meteorological Satellite Program Operational Linescan Systems of the U.S. National Oceanic and Atmospheric Administration (NOAA) also use across-track scanners. However, these instruments contain moving parts, such as oscillating mirrors, which are easily worn out and often fail. Another type of sensing system was developed in the interim, namely along-track, or pushbroom, scanners, which use charge-coupled devices (CCDs) as the detector.

Like across-track scanners, along-track scanners also use the forward motion of the platform to record successive scan lines and build up a 2D image. The fundamental difference between the two types of scanning systems lies in the method with which each scan line is built. Along-track scanners detect the field of view at once by using a linear array of detectors, which are located at the focal plane of the image formed by the lens systems. Linear arrays may consist of numerous CCDs positioned end to end. The sensor records the energy of each scan line simultaneously as the platform advances along the flight track direction, which is analogous to "pushing" the bristles of a broom along a floor. For this reason, along-track scanning systems are also known as pushbroom scanners. Each individual detector measures the energy for a single ground resolution cell, and thus the IFOV of a single detector determines the spatial resolution of the system. Each spectral band requires a separate linear array. An area array can also be used for multispectral remote sensing, with which a dispersing element is often used to split light into narrow spectral bands.

Along-track scanners offer several advantages over across-track scanners. Because pushbroom scanners have no moving parts (oscillating mirrors), their mechanical reliability can be relatively high. Moreover, along-track scanners allow each detector to "see" and measure the energy from each ground resolution cell for a longer period of time (dwell time), and therefore enable a stronger signal to be recorded and a higher S/N ratio, as well as a greater range in the signal, which improves the radiometric resolution of the system. Further, the increased dwell time makes it possible for smaller IFOVs and narrower bandwidths for each detector. Therefore, along-track scanners can have finer spatial and spectral resolution without ignoring the radiometric resolution. In addition, the geometric accuracy in the across-track direction is higher owing to the fixed relationship among the detectors' elements. Because detectors are usually solid-state microelectronic devices, they are generally smaller and lighter, and require less power. These advantages make it well suited for along-track scanners to be onboard small satellites and airplanes (Avery and Berlin 1992). A disadvantage of linear array systems is the need for cross-calibrating thousands of detectors to achieve uniform sensitivity across the array. Another limitation of the pushbroom technology is that currently commercially available CCD detectors cannot operate in the mid-IR and longer wavelengths (Lillesand et al. 2008).

CCD detectors are now commonly used on air- and spaceborne sensors. The first airborne pushbroom scanner is the multispectral electro-optical imaging scanner built by the Canadian Centre for Remote Sensing. It images in eight bands from 0.39 to 1.1 µm (using optical filters to produce the narrow band intervals) and uses a mirror to collect fore-and-aft views (along track) suitable as stereo imagery. The German Aerospace Research Establishment developed the first pushbroom scanner to be flown in space. The modular optico-electronic multispectral scanner was aboard shuttle missions STS-7 and STS-11 in 1983 and 1984, respectively. It images two bands at 20 m resolution at the wavelength range of 0.575–0.625 µm and 0.825–0.975 µm. The first use of CCD-based pushbroom scanner on an earth-observing satellite was on the French SPOT-1 launched in 1986 with the sensor SPOT-HRV. Many other satellite remote sensing programs have now adopted the technology of along-track scanners, including the Indian satellite IRS, Japanese ALOS, Chinese HJ-1 series satellites, and U.S. EO-1 (ALI sensor), QuickBird, and IKONOS.

1.2.2 Thermal Infrared Sensors

Thermal IR scanners are a special type of electro-optical sensor. The thermal scanners detect and measure the thermal IR portion of the electromagnetic radiation, instead of visible, near-IR, and reflected IR radiations. Because atmospheric absorption is very strong in some portions of the electromagnetic spectrum due to the existence of water vapor and carbon dioxide, thermal IR sensing is limited to two specific regions from 3 to 5 µm and 8 to 14 µm, known as atmospheric windows. Comparing with the visible and reflected

IR spectra, thermal IR radiation is much weaker, which requires the thermal sensors to have large IFOVs sufficient to detect the weak energy to make a reliable measurement. The spatial resolution of thermal IR channels is usually coarser than that for the visible and reflected IR bands. For example, the Landsat TM sensor has a spatial resolution of 120 m in its thermal IR band and 30 m in its visible through mid-IR bands. An advantage of thermal imaging is that images can be acquired at both day- and nighttime, as long as the thermal radiation is continuously emitted from the terrain. Photodetectors sensitive to the direct contact of photons on their surface are used in the thermal sensors to detect emitted thermal IR radiation, and are made from some combinations of metallic materials. To sense thermal IR radiation in the 8–14 μm interval, the detector is usually an alloy of mercury–cadmium–tellurium (HgCdTe). Mercury-doped germanium (Ge:Hg) is also used for this thermal imaging window, which is effective over a broader spectral range of 3–14 μm. Indium–antimony (InSb) is the alloy used in detectors in the range of 3–5 μm. To maintain maximum sensitivity, these detectors must be cooled onboard to keep their temperatures between 30 and 77 K, depending on the detector type. This can be done either with cooling agents, such as liquid nitrogen or helium (contained in a vacuum bottle named "Dewar" that encloses the detector) or, for some spacecraft designs, with radiant cooling systems that take advantage of the cold vacuum of outer space (Short 2009). The high sensitivity of the detectors ensures fine differences in the emitted radiation can be detected and measured, leading to high radiometric resolution. A better understanding of this radiometric concept is via S/N ratio. This signal is an electrical current related to changes in the incoming radiant energy from the terrain, whereas noise refers to unwanted contribution from the sensor (Campbell 2007). Higher sensitivity means to improve the S/N ratio by limiting the thermal emissions of the sensor while keeping a stable, minimum magnitude of the signal from the feature imaged.

In addition to the thermal detectors, the thermal scanners have two other basic components: an optomechanical scanning system and an image recording system. For across-track scanners, an electrical motor is mounted underneath the aircraft, with the rotating shaft oriented parallel to the aircraft fuselage thus the flight direction. The total angular field of view typically ranges from 90° to 120°, depending on the type of sensor (Jensen 2007). The incoming thermal IR energy is focused from the scan mirror onto the detector, which converts the radiant energy into an appropriate level of analog electrical signal. The resulting electrical signals are sequentially amplified (because of the low level of thermal IR energy from the terrain) and recorded on magnetic tapes or other devices. These signals can be displayed in monitors and can be recorded and analyzed digitally after analog-to-digital conversion. For each scan, the detector response must be calibrated to obtain a quantitative expression of radiant temperatures. The calibrated thermal scanners usually have two internal temperature references (electrical elements, mounted at the sides of the scanning mirror), one as the cold reference and the other the warm reference. The two references provide a scale for determining the radiant temperatures of ground objects (Avery and Berlin 1992).

Recent technological advances in CCDs make it possible for thermal IR imaging based on area and linear arrays of detectors (Jensen 2007). As with other along-track scanners, thermal IR area/linear arrays permit a longer dwell time, and thus improve radiometric resolution and the accuracy of temperature measurements. Moreover, because the alignment of all detectors is fixed, the geometric fidelity of thermal IR imagery is much enhanced. Some forward-looking IR systems, used in military reconnaissance and law enforcement, adopt area-array technology. The Canadian firm, Itres (http://www.itres .com/), has built a thermal IR sensor based on linear array technology. The newest type of the sensor, TABI-1800 (thermal airborne broadband imager), measures thermal IR spectrum in the range of 8–12 μm with a swath of 320–1800 pixels. It can provide a temperature resolution up to one-tenth of a degree.

1.2.3 Passive Microwave and Imaging Radar Sensors

The electromagnetic radiation in the microwave wavelength region (1 cm–1 m in wavelength) has been widely used in remote sensing of the earth's land surface, atmosphere, and ocean. In comparison with visible, reflected, and thermal IR spectra, microwave has a longer wavelength. This property permits microwave to penetrate through cloud cover, haze, dust, and so on in the atmosphere, while not being affected by atmospheric scattering, which affects shorter optical wavelengths. The detection of microwave energy under almost all weather and environmental conditions is especially important for tropical areas, where large cloud covers occur frequently throughout the year.

Passive microwave sensors operate in a similar manner as those for thermal IR imaging. The sensors are typically radiometers, which require large collection antennas, either fixed or movable. The passive microwave sensors detect and measure naturally emitted microwave energy within their field of view, with wavelength between 0.15 and 30 cm. The emitted energy is mainly controlled by the temperature and moisture properties of the emitting object or surface. Because the wavelengths are so long, this emitted energy is fairly small compared to optical wavelengths (visible and reflected IR). As a result, the fields of view for most passive microwave sensors are large to detect enough energy to record a signal, and the data acquired by such sensors are characterized by low spatial resolution.

Passive microwave remote sensing using air- and spaceborne sensors has been used for several decades in the studies of terrestrial systems, meteorology, hydrology, and oceanography. Owing to microwave's sensitivity to moisture content, passive microwave sensing is widely used for detecting soil moisture and temperature, as well as assessing snow melt conditions. In addition, microwave radiation from below thin soil cover provides helpful information about near-surface bedrock geology. Onboard some meteorological satellites, passive microwave sensors are the devices used to measure atmospheric profiles, water and ozone content, in the atmosphere and to determine precipitation conditions. On oceans, passive microwave sensors track sea ice distribution and currents, and assess sea surface temperature and surface winds. These sensors are also useful in monitoring pollutants in the oceans and in tracking oil slicks.

Active microwave sensors provide their own source of microwave radiation to illuminate the target. Active microwave sensors can be of imaging or nonimaging type. The most common form of imaging active microwave sensors is radar. Radar typically consists of four components: a transmitter, a receiver, an antenna, and an electronics system to process and record the data. Radar measures both the strength and round-trip time of the microwave signals that are emitted by a radar antenna and reflected off a target (object or surface). The radar antenna alternately transmits and receives pulses at particular microwave wavelengths and polarizations (the orientation of the electric field) (Freeman 1996). When reaching the earth's surface, the energy in the radar pulse may be scattered in all directions, possibly with some reflected back toward the antenna. This backscattered energy then reaches the radar and is received by the antenna in a specific polarization. These radar echoes are converted to digital data and passed to a data recorder for later processing and display as an image (Freeman 1996). Because the radar pulse travels at the speed of light, it is possible to use the measured time for the round-trip of a particular pulse to calculate the distance or range to the reflecting object (Freeman 1996).

Imaging radar sensing is apparently different from optical and thermal remote sensing. As an active sensor, radar can be used to image any target at any time, day or night. This property of radar, when combined with microwave energy's ability to penetrate through clouds and most rain, makes it an anytime and all-weather sensor. Nonimaging microwave sensors include altimeters and scatterometers. Radar altimeters send out microwave pulses and measure the round-trip time delay to the targets. The height of the surface can be determined based on the time delay of the return signals. Altimeters look straight down at nadir below the platform (aircraft or satellite) to measure heights or elevations (Canada Centre for Remote Sensing 2007). Radar altimetry is commonly used on aircraft for height determination, and for topographic mapping and sea surface height estimation with the platforms of aircraft and satellites. Another type of nonimaging sensor is a scatterometer. Scatterometers are used to make reliable measurements of the amount of backscattered energy, which is largely influenced by the surface properties and the angle at which the microwave energy strikes the target (Canada Centre for Remote Sensing 2007). For example, a wind scatterometer is used to measure the wind speed and direction over the ocean surface (Liew 2001). It sends out pulses of microwaves in multiple directions and measures the magnitude of the backscattered signal, which is related to the ocean surface roughness. The information about the surface roughness can be used to derive the wind speed and direction of ocean surface (Liew 2001).

1.2.4 Light Detection and Ranging

LiDAR is a technology that dates back to the 1970s, when airborne laser scanning systems were developed. Like radar, it is an active remote sensing technology. The first system was developed by NASA, and it operated by emitting a laser pulse and precisely measuring its return time to calculate the range (height) by using the speed of light. Later, with the advent of GPS in the late 1980s and rapidly pulsing laser scanners, the necessary positioning accuracy and orientation parameters were achieved. The LiDAR system has now become comprehensive, typically with four primary components, including a rapidly pulsing laser scanner, precise kinematic GPS positioning, inertial measurement units for capturing the orientation parameters (i.e., tip, tilt, and roll angles), and a precise timing clock (Renslow et al. 2000). With all the components, along with software and computer support, assembled in a plane, and by flying along a well-defined flight plan, LiDAR data are collected in the form of records of the returned-pulse range values and other parameters (e.g., scanner position and orientation parameters). After filtering noise and correcting, laser point coordinates are calculated, and a basic ASCII file of x, y, and z values for LiDAR data is formed, which may be transformed into a local coordinate system. Since the inception of LiDAR systems, LiDAR technology has been used widely in many geospatial applications owing to its high-resolution topographic data, short acquisition time, and reasonable cost compared with traditional methods. For example, LiDAR point cloud data show an excellent potential to identify tree shape and height. In Figure 1.1a, a 6-in. resolution natural color orthophoto that was taken in March 2010 is displayed, which shows Indy Military Park, Indianapolis, IN. The photos were taken in the leaf-off season, so trees with dark shadows are conifers and those without clear shadows are deciduous (sycamores). In Figure 1.1b, various elevations from low to high are shaded with colors of blue, light blue, green, yellow, and red in the LiDAR point cloud (1 m post spacing) map. The LiDAR data were acquired in the same season, December 2009. By comparing Figure 1.1a and b, we can clearly separate conifer from deciduous trees.

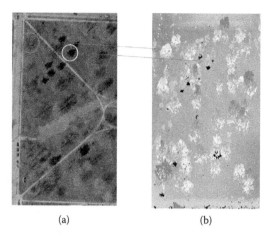

(a)   (b)

FIGURE 1.1
(See color insert.) LiDAR data used to identify tree shape, height, and life form. The red circle indicates a deciduous tree (sycamore) and the yellow circle a conifer tree. Courtesy of Jim Stout, IMGIS, IN. (a) Aerial photo and (b) LiDAR point cloud.

1.3 Characteristics of Remotely Sensed Data

Regardless of passive or active remote sensing systems, all sensing systems detect and record energy "signals" from the earth's surface features and/or from the atmosphere. Data collected by these remote sensing systems can be either in analog format, for example, hard-copy aerial photography or video data, or in digital format, such as a matrix of "brightness values" corresponded to the average radiance measured within an image pixel. The success of data collection from remotely sensed imagery requires an understanding of four basic resolution characteristics, namely, spatial, spectral, radiometric, and temporal resolution (Jensen 2005).

1.3.1 Photographs

Aerial photographs are made possible through the photographic process using chemical reactions on the surface of light-sensitive film to detect and record energy variations. Aerial photographs normally record over the wavelength range from 0.3 to 0.9 µm, that is, from UV to visible and near-IR spectra. The range of wavelengths to be detected by a camera is determined by the spectral sensitivities of the film. Sometimes, filters are used in a camera in conjunction with different film types to restrict the wavelengths being recorded or to reduce the effect of atmospheric scattering. Mapping cameras used to obtain aerial photographs are usually mounted in the nose or underbelly of an aircraft. Aerial photographs can also be taken from space shuttles, unmanned aviation vehicles, balloons, or even kites. A mapping camera that records data photographically requires that the film be brought back to the ground for processing and printing.

1.3.2 Satellite Images

Satellite imagery may be viewed as an extension of aerial photography, since satellite remote sensing relies on the same physical principles to acquire, interpret, and extract information content. Satellite images are taken at a higher altitude that allows coverage of a larger area of the earth's surface at one time, and more importantly, satellite images are acquired by electronic scanners and linear/area arrays. Satellite sensors can usually detect and record a much wider range of electromagnetic radiation, from UV radiation to microwave. The electromagnetic energy is recorded electronically as an array of numbers in digital format, resulting in digital images. In a digital image, individual picture elements, called pixels, are arranged in a 2D array in columns and rows. Each pixel has an intensity value and a location address in the 2D array, representing the electromagnetic energy received at the particular location on the earth's surface. The intensity value can have different ranges, depending on the sensitivity of a sensor to the changes in the magnitude of the electromagnetic energy. In most cases, a zero value indicating no radiation is received at a particular location and the maximum recordable number of the sensor indicates the maximum detectable radiation level. More often, satellite sensors are designed to record the electromagnetic energy at multiple channels simultaneously. Each channel records a narrow wavelength range, sometimes known as a band. The digital image from each channel can be displayed as a black-and-white image in the computer, or we can combine and display three channels/bands of a digital image as a color image by using the three primary colors (blue, green, and red) of the computer. In that case, the digital data from each channel is represented as one of the primary colors according to the relative brightness of each pixel in that channel. Three primary colors combine in different proportions to produce a variety of colors as seen in the computer.

Satellite sensors offer several advantages over photographic cameras. First, although the spectral range of photographic systems is restricted to the visible and near-IR regions, the electro-optical sensors can extend this range into the whole reflected IR and the thermal IR regions. The electro-optical sensors are capable of achieving a much higher spectral resolution. Because the electro-optical sensors acquire all spectral bands simultaneously through the same optical system, it is much easier to register multiple images (Canada Centre for Remote Sensing 2007). Second, some electro-optical sensors can detect thermal IR energy in both day- and nighttime mode (Avery and Berlin 1992). Third, since the electro-optical sensors record radiance electronically, it is easier to determine the specific amount of energy measured and to convert the incoming radiance over a greater range of values into digital format (Canada Centre for Remote Sensing 2007). Therefore, the electro-optical sensors can better utilize radiometric resolution than photographic cameras. Fourth, with photographic cameras, we have to use films as both the detector and the storage medium, whereas the electro-optical detectors can be used continuously, making the detection process renewable (Avery and Berlin 1992). Fifth, some electro-optical sensors use an in-flight display device, allowing a ground scene to be viewed in a near real-time mode (Avery and Berlin 1992). Finally, the digital recording in the electro-optical sensors facilitates transmission of data to receiving stations on the ground and immediate processing of data by computers (Canada Centre for Remote Sensing 2007). Of course, we should note that aerial photographs by cameras have the clear advantage of recording extremely fine spatial details.

1.3.3 Spatial Resolution

Spatial resolution defines the level of spatial detail depicted in an image, and it is often defined as the size of the smallest possible feature that can be detected from an image. This definition implies that only objects larger than the spatial resolution of a sensor can be picked out from an image. However, a smaller feature may sometimes be detectable if its reflectance dominates within a particular resolution cell or it has a unique shape (e.g., linear features). Another meaning of spatial resolution is that a ground feature should be distinguishable as a separate entity in the image. But the separation from neighbors or background is not always sufficient to identify the object. Therefore, spatial resolution involves ideas of both detectability and separability. For any feature to be resolvable in an image, it involves consideration of spatial resolution, spectral contrast, and feature shape. Spatial resolution is a function of sensor altitude, detector size, focal size, and system configuration (Jensen 2005). For aerial photography, spatial resolution is measured in resolvable line pairs per millimeter; whereas for other sensors, it refers to the dimensions (in meters) of the ground area that fall within the IFOV of a single detector within an array (Jensen 2005).

Spatial resolution determines the level of spatial details that can be observed on the earth's surface. Coarse spatial resolution images can detect only large features, whereas with fine resolution images, small features become visible. For a remote sensing project, image spatial resolution, however, is not the only factor needed to be considered. The relationship between the geographical scale of a study area and the spatial resolution of remote sensing image has to be studied (Quattrochi and Goodchild 1997). Generally speaking, at the local scale, high spatial-resolution imagery, such as IKONOS and QuickBird data, is more effective. At the regional scale, medium spatial resolution data, such as Landsat TM/ETM+ and Terra ASTER data, are the most frequently used. At the continental or global scale, coarse spatial resolution data, such as AVHRR and MODIS data, are most suitable. Higher resolution means the need for a larger data storage and higher cost, and may introduce difficulties in image processing for a large study area. One goal for satellite sensors since the late 1990s has been improved spatial resolution, which now is down to better than 1 m. Sub-meter satellite images have introduced a new issue of residential privacy and have resulted in some legal cases.

It is important not to confuse between the spatial resolution of a digital image and pixel size, although the two concepts are interrelated. In Figure 1.2a, a Landsat ETM+ image covering the city of Indianapolis, IN, is displayed at full resolution. The spatial resolution of Landsat reflected bands is 30 m, and in Figure 1.2a, each pixel also represents an area of 30 m × 30 m on the ground. In this case, the spatial resolution is the same as the pixel size of the image. In Figure 1.2b through f, the Landsat image is resampled to larger pixel sizes, ranging from 60 to 960 m. At the pixel size of 30 m, roads, rivers, and central business district are visible. As the pixel size becomes larger, these features and objects become indiscernible. As we discussed earlier, pixel size is primarily defined by IFOV. This "optical pixel" is generally circular or elliptical and combines radiation from all components within its ground area to produce the single pixel value in a given spectral channel (Harrison and Jupp 2000). Along each image line, optical pixel values are sampled and recorded, and the average spacing of optical pixels relative to the ground area covered is then used to determine the geometric pixel size by using various mathematical algorithms (Harrison and Jupp 2000). Coarse resolution images may include a large number of mixed pixels, where more than one surface cover type (several different features/objects) can be found within a pixel. Fine spatial resolution data considerably reduce the mixed pixel problem.

1.3.4 Spectral Resolution

Different features/materials on the earth's surface interact with the electromagnetic energy in different and often distinctive ways. A specific spectral response curve, or spectral signature, can be determined for each material type. Broad classes, such as vegetation, water, and soil, in an image can often be distinguished by comparing their responses over distinct, very broad wavelength ranges. Spectral signature, or spectral response curve, is related to the spectral resolution of a remote sensor. Each remote sensor is unique with regard to what portion(s) of the electromagnetic spectrum it detects and records. Moreover,

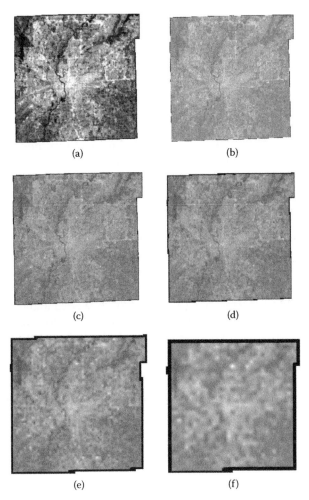

FIGURE 1.2
Effect of digitizing the same area with different pixel sizes. (a) 30 m, (b) 60 m, (c) 120 m, (d) 240 m, (e) 480 m, (f) 960 m. (From Weng, Q. et al., *Remote Sensing of Environment*, 89, 4, 467–483, 2004.)

each remote sensor records a different number of segments of the electromagnetic spectrum, or bands, which may also have different bandwidths. Spectral resolution of a sensor, by definition, refers to the number and the size of the bands it is able to record (Jensen 2005). For example, AVHRR, onboard NOAA's Polar-Orbiting Environmental Satellite platform, collects four or five broad spectral bands (depending on individual instrument) in visible (0.58–0.68 μm, red), near-IR (0.725–1.1 μm), mid-IR (3.55–3.93 μm), and thermal IR (10.3–11.3 μm and 11.5–12.5 μm) portions of the electromagnetic spectrum. AVHRR, acquiring image data at the spatial resolution of 1.1 km at nadir, has been extensively used for meteorological studies, vegetation pattern analysis, and global modeling. The Landsat TM sensor collects seven spectral bands. Its spectral resolution is higher than early instruments on board the Landsats such as MSS and Return Beam Vidicon.

Most satellite sensors are engineered to use particular electromagnetic channels to differentiate between major cover types of interest, or materials, on the earth's surface. The position, width, and number of spectral channels being sensed have been determined before a sensor is carried into the space. Spectral extent describes the range of wavelengths being sensed in all channels of an image. A remote sensing system may detect electromagnetic radiation in the visible and near-IR regions, or extend to the middle and thermal IR regions. Microwave radiation may be recorded using radar and passive microwave sensors. An increase in spectral resolution over a given spectral extent implies that a greater number of spectral channels are recorded. The finer the spectral resolution, the narrower the range of wavelength for a particular channel or band. The increased spectral resolution allows for a fine discrimination between different targets, but also increases costs in the sensing system, and data processing on board and on the ground. Therefore, the optimal spectral range and resolution for a particular cover type need to be modified with respect to practical uses of data collection and processing.

A major advance in remote sensor technology in recent years has been a significant improvement in spectral resolution, from bandwidths of tens to hundreds of nanometers (1 μm = 1000 nm), as pertains to the Landsat TM, to 10 nm or less. Hyperspectral sensors (imaging spectrometers) are the kind of instruments that acquire images in many, very narrow, contiguous spectral bands throughout the visible, near-IR, mid-IR, and thermal IR portions of the spectrum. Whereas the Landsat TM obtains only one data point corresponding to the integrated response over a spectral band 0.24 μm wide, a hyperspectral sensor, for example, is capable of obtaining 24 data points over this range using bands on the order of 0.01 μm wide. The NASA Airborne Visible/Infrared Imaging Spectrometer (AVIRIS) collects 224 contiguous bands with wavelengths from 400 to 2500 nm. A broadband system can only discriminate general differences among material types, whereas a hyperspectral sensor affords the potential for detailed identification of materials and better estimates of their abundance. Figure 1.3a shows an AVIRIS image of some circular fields in the San Luis Valley of Colorado. Vegetation or crop types in the fields, identified by different colors in the figure, are determined based on field survey and the spectral curves plotted in Figure 1.3b for the crops indicated. Another example of a hyperspectral sensor

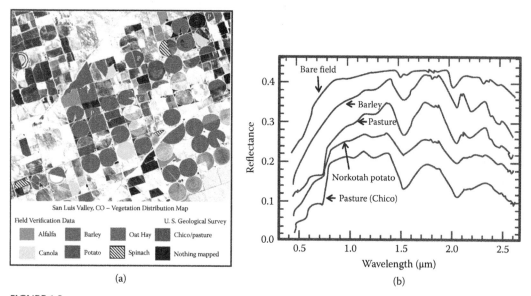

(a)   (b)

FIGURE 1.3
(See color insert.) Vegetation species identification using hyperspectral imaging technique. (a) AVIRIS image showing circular fields in San Luis Valley, CO, and (b) reference spectra used in the mapping of vegetation species. (From Clark, R.N. et al., *Proceedings: Summitville Forum '95*, edited by H.H. Posey, J.A. Pendelton, and D. Van Zyl, 64–9. Colorado Geological Survey Special Publication 38, 1995. With permission.) Available at http://speclab.cr.usgs.gov/PAPERS.veg1/vegispc2.html.

is the Moderate Resolution Imaging Spectrometer (MODIS), used on both of the NASA's Terra and Aqua missions and their follow-ons to provide comprehensive data about land, ocean, and atmospheric processes simultaneously. MODIS has a 2-day repeat global coverage with spatial resolution (250, 500, or 1000 m depending on wavelength) in 36 spectral bands.

1.3.5 Radiometric Resolution

Radiometric resolution refers to the sensitivity of a sensor to incoming radiance, that is, how much change in radiance must be there on the sensor before a change in recorded brightness value takes place (Jensen 2005). Therefore, radiometric resolution describes a sensor's ability to discriminate subtle differences in the detected energy. Coarse radiometric resolution would record a scene using only a few brightness levels, that is, at very high contrast, whereas fine radiometric resolution would record the same scene using many brightness levels. Radiometric resolution in digital imagery is comparable to the number of tones in aerial photography (Harrison and Jupp 2000). For both types of remote sensing images, radiometric resolution is related to the contrast in an image.

Radiometric resolution is frequently represented by the levels of quantization that are used to digitize the continuous intensity value of the electromagnetic energy recorded. Digital numbers in an image have a range from 0 to a selected power of 2 minus 1. This range corresponds to the number of bits (binary digits) used for coding numbers in computers. For example, Landsat-1 MSS initially records radiant energy in six bits (values ranging from 0 to 63, i.e., levels of quantization = 64 = 2^6), and was later expanded to seven bits (values ranging from 0 to 127, equal to 2^7). In contrast, Landsat TM data are recorded in eight bits, that is, its brightness levels range from 0 to 255 (equals to 2^8). Because human eyes can only perceive 20–30 different gray levels, the additional resolution levels provided by digital images are not visually discernible (Harrison and Jupp 2000). Digital image processing techniques using computers are necessary to derive maximum discrimination from the available radiometric resolution.

1.3.6 Temporal Resolution

Temporal resolution refers to the amount of time it takes for a sensor to return to a previously imaged location, commonly known as the repeat cycle or the time interval between acquisitions of two successive images. For satellite remote sensing, the length of time for a satellite to complete one entire orbital cycle defines its temporal resolution. Depending on the types of satellites, the revisit time may range from half to more than 10 days. For airborne remote sensing, temporal resolution is less pertinent since users can schedule flights for themselves.

Temporal resolution has an important implication in change detection and environmental monitoring. Many environmental phenomena constantly change over time, such as vegetation, weather, forest fire, and volcano. These changes are reflected in the spectral characteristics of the earth's surface and are recorded in remote sensing images. Temporal differences between remotely sensed imagery are not only caused by the changes in spectral properties of the earth's surface features/objects, but they can also result from atmospheric differences and changes in sun position during the course of a day and during the year. Temporal resolution is an important consideration in remote sensing of vegetation because vegetation grows according to daily, seasonal, and annual phenological cycles. It is crucial to obtain anniversary or near-anniversary images in change detection of vegetation. Anniversary images greatly minimize the effect of seasonal differences (Jensen 2005). Many weather sensors have a high temporal resolution: note the geostationary operational environmental satellite (every

15 minutes) and Meteosat first generation (every 30 minutes). Observing short-lived, time-sensitive phenomena, such as floods, fire, and oil spill, requires truly high temporal resolution imagery. In general, ideal temporal resolution varies considerably for different applications. While monitoring of crop vigor and health may require daily images in a growing season, studies of urban development only need to have an image every 1–5 years (Jensen 2007). Monitoring the earth's ecosystems, especially forest, is the main goal of many satellite missions. Forest disturbance can be caused by human activities and natural forces. Harvest, fire, and storm damage often result in abrupt changes, whereas changes due to insect and disease can last several years or longer. However, tree growth is always a slow, gradual process. Characterization of forest growth and disturbance therefore requires remote sensing data of different temporal resolutions (Huang et al. 2010).

The set overpass times of satellites may coincide with clouds or poor weather conditions. This is especially true in the tropical areas, where persistent clouds and rains in the wet season offer limit clear views of the earth's surface and thus prevent the acquisition of good quality images. Moreover, some projects require remote sensing images of particular seasons. In agriculture, phenology manifests itself through the local crop calendar—the seasonal cycle of plowing, planting, emergence, growth, maturity, harvest, and fallow (Campbell 2007). The crop calendar stipulates the acquisition satellite images in the growth season, often from late spring to fall in the mid-latitude region, but there is also a need to consider the local climate. On the contrary, detection of urban buildings and roads may well be suited in the leaf-off season in the temperate regions. In sum, the date of acquisition should be determined to ensure the extraction of maximum information content from remotely sensed data. When field work needs to coincide with image acquisition, the date of acquisition must be planned well ahead of the time.

The off-nadir imaging capability of the SPOT-5 sensor reduces the usual revisit time of 26 days to 2–3 days, depending on the latitude of the imaged areas. This feature is designed for taking stereoscopic images and for producing digital elevation models (Figure 1.4).

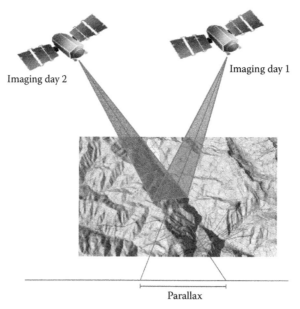

FIGURE 1.4
Off-nadir imaging for stereoscopic acquisitions.

But it obviously also allows for daily coverage of selected regions for short periods and provides another means for monitoring dynamic events such as flood or fire. The ASTER sensor has a similar capability of off-nadir imaging for its near-IR band (0.76–0.86 µm) by equipping a backward-looking telescope to acquire a stereopair image.

1.4 Some Basic Considerations for Remote Sensing Applications

The range of remote sensing applications includes archaeology, agriculture, cartography, civil engineering, meteorology and climatology, coastal studies, emergency response, forestry, geology, GISs, hazards, land use and land cover, natural disasters, oceanography, water resources, military, and so on. Most recently, with the advent of high spatial resolution imagery and more capable techniques, urban and other civic applications of remote sensing are rapidly gaining interest in the remote sensing community and beyond. A remote sensing project typically involves four steps: (1) identifying the problem and determining the objectives; (2) acquiring data based on the understanding of the characteristics of various remote sensing systems/platforms; (3) detecting, identifying, measuring, and analyzing features, objects, phenomena, and processes from remote sensing images; and (4) outputting information in the form of enhanced images, maps, tables, figures, statistics, and databases.

The knowledge on the mechanism and process through which each type of remote sensor acquires digital imagery is important for selecting an appropriate type of sensor for a specific project as well as for image interpretation and analysis. Based on the way the reflected and/or emitted energy is used and measured during the sensing, remote sensors are most commonly grouped into two categories: passive and active sensors. Passive sensors are further subgrouped according to the data recording method, and include across-track, along-track, and thermal IR scanners. Active sensors are further divided into radar and LiDAR sensors, which have distinct data collection methods and geometric properties in resultant images. Although this chapter focuses the discussion on the characteristics of remote sensors, a continued discussion on main satellite missions for the earth's observation will allow us to understand better the image collection processes and geometric characteristics of the remote sensing systems, which, in turn, provide useful information for proper use of satellite imagery.

To understand the characteristics of remotely sensed data, we may categorize remote sensors into two broad classes: analog and digital sensors. Aerial photographs, as an example of an analog image, have the clear advantage of recording extremely fine spatial details. Satellite imagery, as an example of digital imagery, tends to have better quality in spectral, radiometric, and temporal resolution. The trend over the past 40 years has been directed toward improved resolution of each type. Considerable efforts have been made to designing and constructing sensors that maximize the resolution(s) needed for their intended tasks. Superior spatial resolution permits ever smaller targets, less than a meter in size, to be seen as individuals. Greater spectral resolution means that individual entities (features, objects) can be more accurately identified because their spectral signatures are more distinguished as in hyperspectral sensors. Increased radiometric resolution offers sharper images, allowing for better separation of different target materials from their backgrounds. Finally, improved temporal resolution has made possible not only the area of monitoring ecosystems and natural hazards, but also detection of movement of ships and vehicles.

Under many situations, clear trade-offs exist between different forms of resolutions. For example, in traditional photographic emulsions, increases in spatial resolution are based on decreased size of film grain, which produces accompanying decreases in radiometric resolution, that is, the decreased sizes of grains in the emulsion portray a lower range of brightness values (Campbell 2007). In multispectral scanning systems, an increase in spatial resolution requires a smaller IFOV, thus less energy is reaching the sensor. This effect may be compensated by broadening the spectral window to pass more energy, that is, decreasing spectral resolution, or by dividing the energy into fewer brightness levels, that is, decreasing radiometric resolution (Campbell 2007). To overcome the conflict between spatial and temporal resolutions, NASA and Europe Space Agency have reportedly proposed developing remote sensing systems that are able to provide data combining both high spatial resolution and revisit capabilities, among others.

Data analysis, particularly digital image processing, is now a key step in implementing a remote sensing project. Before main image analyses take place, preprocessing of digital images is often required for detection and restoration of bad lines, geometric rectification or image registration, radiometric calibration and atmospheric correction, and topographic correction. To make interpretation and analysis easier, it may also be necessary to implement various image enhancements, such as contrast, spatial, and spectral enhancements. The core of digital image processing is thematic information extraction, aiming at detecting, identifying, measuring, and analyzing features, objects, phenomena, and processes from digital remote sensing images by using computers to process digital images. Manual interpretation and analysis is still an important method in digital image processing. However, the majority of digital image analysis methods are based on tone or color, which is represented as a digital number (brightness value) in each pixel of the digital image. The development of digital image analysis and interpretation is closely associated with availability of digital remote sensing images, advances in computer software and hardware, and the evolution of satellite systems.

Remote sensing has now been integrated with other modern geospatial technologies such as GIS, GPS, and mobile mapping. GIS has been used to enhance the functions of remote sensing image processing at various stages (Weng 2009). It also provides a flexible environment for entering, analyzing, managing, and displaying digital data from various sources necessary for remote sensing applications. Many remote sensing projects need to develop a GIS database to store, organize, and display aerial and ground photographs, satellite images, and ancillary, reference, and field data. GPS is another essential technology when the remote sensing projects need to have accurate ground control points and to collect in situ samples and observations in the fields. In conjunction with GPS and wireless communicating technologies, mobile mapping becomes a technological frontier. It uses remote sensing images and pictures taken in the field to update GIS databases regularly to support problem solving and decision making at any time and any place.

References

Avery, T.E., and G.L. Berlin. 1992. *Fundamentals of Remote Sensing and Airphoto Interpretation*. 5th ed. Upper Saddle River, NJ: Prentice Hall.

Campbell, J.B. 2007. *Introduction to Remote Sensing*. 4th ed. New York: Guilford Press.

Canada Centre for Remote Sensing. 2007. Tutorial: Fundamentals of Remote Sensing. Accessed December 3, 2010. http://www.ccrs.nrcan.gc.ca/resource/tutor/fundam/chapter2/08_e.php.

Clark, R.N., T.V.V. King, C. Ager, and G.A. Swayze. 1995. Initial vegetation species and senescence/stress mapping in the San Luis Valley, Colorado using imaging spectrometer data. In *Proceedings: Summitville Forum '95*, edited by H.H. Posey, J.A. Pendelton, and D. Van Zyl, 64–9. Colorado Geological Survey Special Publication 38.

Freeman, T. 1996. What is Imaging Radar? Accessed January 23, 2011. http://southport.jpl.nasa.gov/

Harrison, B., and D. Jupp. 2000. Introduction to remotely sensed data, a module of *the science of remote sensing*. 4th ed., prepared by Committee on Earth Observation Satellites (CNES), the French Space Agency. Accessed June 29, 2010. http://ceos.cnes.fr:8100/cdrom/ceos1/irsd/content.htm.

Huang, C., S.N. Goward, J.G. Masek, N. Thomas, Z. Zhu, and J.E. Vogelmann. 2010. "An automated approach for reconstructing recent forest disturbance history using dense Landsat time series stacks." *Remote Sensing of Environment* 114(1):183–98.

Jensen, J.R. 2005. *Introductory Digital Image Processing: A Remote Sensing Perspective*. 3rd ed. Upper Saddle River, NJ: Prentice Hall.

Jensen, J.R. 2007. *Remote Sensing of the Environment: An Earth Resource Perspective*. 2nd ed., 592. Upper Saddle River, NJ: Prentice Hall.

Liew, S.C. 2001. *Principles of Remote Sensing*. A tutorial as part of the "Space View of Asia, 2nd Edition" (CD-ROM), the Centre for Remote Imaging, Sensing and Processing (CRISP), the National University of Singapore. Accessed January 25, 2011. http://www.crisp.nus.edu.sg/~research/tutorial/rsmain.htm.

Lillesand, T.M., R.W. Kiefer, and J.W. Chipman. 2008. *Remote Sensing and Image Interpretation*. 6th ed. New York: John Wiley & Sons.

Quattrochi, D.A., and M.F. Goodchild. 1997. *Scale in Remote Sensing and GIS*. New York: Lewis Publishers.

Renslow, M., P. Greenfield, and T. Guay. 2000. *Evaluation of multi-return LIDAR for forestry applications*. Project Report for the Inventory & Monitoring Steering Committee, RSAC-2060/4810-LSP-0001-RPT1.

Short, N.M., Sr. 2009. *The Remote Sensing Tutorial*. Accessed January 25, 2011. http://rst.gsfc.nasa.gov/.

Weng, Q. 2009. *Remote Sensing and GIS Integration: Theories, Methods, and Applications*, 397. New York: McGraw-Hill.

Weng, Q., D. Lu, and J. Schubring. 2004. "Estimation of land surface temperature-vegetation abundance relationship for urban heat island studies." *Remote Sensing of Environment* 89(4):467–83.

Section II

Sampling Design and Product Quality Assessment

2

Remote Sensing Applications for Sampling Design of Natural Resources

Guangxing Wang and George Z. Gertner

CONTENTS

2.1 Introduction..23
2.2 Design- and Model-Based Sampling Design Strategies.......................24
2.3 Role of Remotely Sensed Data in Sampling Design Strategies...........28
2.4 Optimal Spatial Resolution for Sampling and Mapping....................30
2.5 Optimal Temporal Resolution for Sampling and Mapping................32
2.6 Improvement of Sampling Design Strategies.......................................34
2.7 Local Variability–Based Sampling Design Strategy............................36
2.8 Case Study...37
2.9 Conclusion..39
References...39

2.1 Introduction

When natural resources are mapped using remotely sensed images, sampling design strategies are required to collect field observations of variables of interest at the selected locations and at the same time to create spatial explicit estimates of the variables (Brus and de Gruijter 1997). Moreover, the obtained ground data can also be used to assess the quality of the resulting maps (Campbell and Wynne 2011). Thus, selection of sampling design strategies is very critical for both mapping and accuracy assessment of natural resources.

A sampling design strategy means the combination of a sampling design procedure and an estimator (Brus and de Gruijter 1997). A cost-efficient sampling design strategy means that given the obtained estimates satisfy precision requirements for a population and any subareas on its map, it can provide as much information as possible at a minimum cost. Suppose that a study area consists of square cells or pixels and field observations of an interest variable will be collected and used to create estimates of the population and the cells. Given a cost budget and the desired precision of estimates, an optimal sampling design strategy requires determining appropriate plot size and shape; optimal sampling distance, that is, sample size; and spatial distribution of plots to collect ground data (O'Regan and Arvanitis 1966; Olea 1984; Thompson 1992).

Sampling designs can be simple or very complicated. Sample plots can be square, rectangular, circular, or transect lines. Moreover, sample plots can also be fixed or variable in area. The size of sample plots can vary greatly and often range from 1 to 10,000 m^2. For example, the U.S. National Forest Inventory and Analysis (FIA) sampling design is based on an array of hexagons, each containing one sample plot that represents an area of 2404.5 ha.

Each FIA sample plot is a cluster of four 7.32 m radius subplots with a configuration of subplot 1 at the plot center and subplots 2–4 located 36.6 m from subplot 1 at azimuths of 0°, 120°, and 240° (Bechtold and Patterson 2005). Within each subplot, trees with diameter at breast height (DBH) of ≥12.7 cm are measured. Moreover, each subplot is surrounded by a 17.95 radius macroplot used to sample large trees that rarely occur (e.g., ≥101.6 cm DBH) or mortality. In addition, within each subplot a 2.07 m radius microplot offset from the subplot center by 3.66 m at an azimuth of 90° is designed and is used to measure saplings (DBH of 2.54–12.7 cm) and seedlings. The permanent sample plots are remeasured every 5 years. In contrast, the sampling design for the national forest inventory in China is relatively simple (CMF-DFRM 1996). A systematic sampling design is used. The sampling distance varies from 1 to 8 km depending on different regions. The sample plots are squares, and each has an area of 0.08 ha. The permanent plots are remeasured at 5- or 10-year intervals.

Given a budget and an accuracy requirement, the first question for a sampling design is what plot size and shape are best for sampling and mapping an interest variable in a study area? Once the plot size and shape are determined, what sampling size would be the most cost efficient? What about the spatial allocation of sample plots, randomly or systematically? What estimator should be used to interpolate the values of the variable at the unobserved locations using the sample plot data? Moreover, these decisions may be dependent on each other. Thus, the best solution may have to be searched for among a large number of combinations from different sampling design components such as plots and sample sizes.

There is a great body of literature that deals with the issues in sampling design. However, we are still facing some big challenges. For example, the existing methods that were developed based on global variation of an interest variable neglect local variation. Moreover, both global and local variations are in fact unknown before designing a sample. If the past field observations are used to estimate the variations, the estimates will definitely be associated with uncertainties. Thus, the impacts of such uncertainties on sampling design strategies are unknown. On the other hand, although remotely sensed data have been widely used in sampling design, it is still not clear how to use the data that is the most cost efficient to improve sampling design strategies. This chapter intends to provide insights into the state of the art on sampling design to collect field measurements for sampling and mapping natural resources using remotely sensed data by reviewing the existing literature and to further identify the significant gaps that currently exist and provide useful implications to improve sampling design.

2.2 Design- and Model-Based Sampling Design Strategies

Sampling design deals with two distinct frameworks used for inference: the design-based approach and the model-based approach (de Gruijter and Ter Braak 1990; Thompson 1992; Brus and de Gruijter 1997). In the design-based approach, classical sampling theory is used. It is assumed that values of a variable Z at different units or cells that compose a population are fixed, sampling means selecting a subset of cells, and the goal is to determine the probabilities used to select the sample locations to obtain the expectation of the variable (Brus and de Gruijter 1997). A simple and widely used example of the design-based approach is simple random sampling. In this method, an equal probability is used to select each of the sample plots and at a given precision level e—acceptable error of average value of an interest variable—the sample size n for a study area can be determined based on global variation of the variable (O'Regan and Arvanitis 1966; Curran and Williamson 1986):

$$n = \left(\frac{\sigma \times t}{e} \right)^2 \tag{2.1}$$

where σ is the sample standard deviation that is usually estimated by a pilot study or historical datasets and t is a value of Student's distribution with $n - 1$ degrees of freedom.

The idea behind the design-based approach is to infer population parameters using sample parameters. A sample mean and variance can be calculated and used as the estimates of the population mean and variance. The estimator for the population mean is unbiased. The estimation error—the difference between estimated and true means—can be obtained. The sample variance can be used to estimate confidence intervals given a significance level α such as 5%. For a forest inventory program, for example, if square sample plots with the size of 30 m × 30 m are used, the locations of sample plots can be randomly allocated to collect field observations of tree height and from the obtained data, a sample mean and variance of tree height can be computed and used to infer the population mean and variance of forest height.

In the design-based approach, moreover, for the estimates of individual locations, the cells of a study area are first grouped into homogeneous segments (polygons) or strata and a sample mean for each segment or stratum is calculated and assigned to each cell within the segment or stratum (Brus and de Gruijter 1997). In addition, a spatial mean of squared errors can be used as a measure of accuracy for estimation of cells (Brus et al. 1992).

The model-based approach is based on geostatistical theory. With this sampling design strategy, a value of an interest variable Z at any location u of a population, $Z(u)$, is not regarded as being fixed but random (de Gruijter and Ter Braak 1990; Brus and de Gruijter 1997). It can have more than one possible value and each is defined with a probability of occurrence, which forms a random variable. An observed value is regarded as a realization of this random variable at this location. The set of values that are associated with all the cells of the population is considered to be a realization of an underlying random process that is characterized by a joint distribution of the random variables. An infinite set of realizations can be generated from the same random model.

In a model-based approach, it is assumed that a realization at each location can be generated from a joint distribution of random variables (de Gruijter and Ter Braak 1990; Brus and de Gruijter 1997). Modeling the probabilities of the realizations of the random process thus requires determining the joint distributions of the random variables. Modeling is conducted by estimating the model mean, variance, and covariance or variogram. To make the estimation possible, first of all, the random variables are assumed to be spatially autocorrelated (de Gruijter and Ter Braak 1990); that is, their values at near locations are more similar than those farther apart. Second, an assumption of intrinsic hypothesis is made that the increments $[Z(u) - Z(u + h)]$ of this random process are second-order stationary which means the expected difference between the values $Z(u)$ and $Z(u + h)$ of the random process at two locations u and $u + h$ separated by the vector h is zero and the variance of the difference is constant (Goovaerts 1997). If the intrinsic hypothesis is met, a variogram that accounts for the spatial autocorrelation of this variable can be estimated:

$$\hat{\gamma}_{zz}(h) = \frac{1}{2N(h)} \sum_{\alpha=1}^{N(h)} \left(z(u_\alpha) - z(u_\alpha + h) \right)^2 \tag{2.2}$$

where $z(u_\alpha)$ and $z(u_\alpha + h)$ are data values of this variable at two locations u_α and $u_\alpha + h$, respectively, and $N(h)$ is the number of all the data location pairs separated by the separation vector h. Obviously the variogram is a function of h only and not a function of u. As a variance function, an experimental variogram is often obtained using sample plot data and fit using authorized models including spherical, exponential, Gaussian, and power models and their nested models (Goovaerts 1997). For example, a spherical model is as follows:

$$\hat{\gamma}(h) = \begin{cases} c_0 + c_1 \left[1.5\dfrac{h}{a} - 0.5\left(\dfrac{h}{a}\right)^3 \right] & h \le a \\ c_0 + c_1 & h > a \end{cases} \tag{2.3}$$

As the distance h increases, the variogram generally increases. In this equation, c_0 is the nugget parameter implying the spatial variability within sample plots, c_1 is the structure parameter that accounts for variance change with increasing separation distance, a is the range parameter that suggests the maximum distance of spatial dependence of this variable, and $c_0 + c_1$ is called the sill parameter that indicates the maximum variance when the distance h reaches its range parameter. Within the range, the values of the random process at different locations can be considered spatially dependent; the dependence gets weaker as the separation distance increases, and beyond the range values will be essentially independent.

Moreover, it can also be assumed that the random process is second-order stationary, which means it has both a constant mean μ and a constant variance $C(0)$ and its covariance $C_{zz}(h)$ exists and is a function of only the separation distance and direction. The covariance can be calculated using the following equation (Goovaerts 1997):

$$C_{zz}(h) = C_{zz}(0) - \gamma_{zz}(h) \tag{2.4}$$

In the model-based approach, the value at each of the cells for a population is predicted using a linear unbiased predictor such as ordinary kriging. The predicted values are obtained by weighting the sample plot data and at the same time the error variance of the predicted value is minimized (Journel and Huijbregts 1978; Goovaerts 1997). When the sample plots are consistent with the cells to be predicted in terms of size, for example, the ordinary kriging predictor and its kriging variance are as follows:

$$\hat{Z}(u) = \sum_{\alpha=1}^{n(u)} \lambda_\alpha z(u_\alpha) \tag{2.5}$$

$$\sigma_k^2(u) = 2\sum_{\alpha=1}^{n(u)} \lambda_\alpha \gamma(u_\alpha, u) - \sum_{\alpha=1}^{n(u)}\sum_{\beta=1}^{n(u)} \lambda_\alpha \lambda_\beta \gamma(u_\alpha, u_\beta) \tag{2.6}$$

where $n(u)$ is the number of sample plot data within a given neighborhood determined by the maximum distance of spatial autocorrelation for the interest variable; $\gamma(u_\alpha, u)$ and $\gamma(u_\alpha, u_\beta)$ are the values of the variogram between locations u_α and u and between u_α and u_β, respectively, and are determined by the variogram and the spatial configuration of the data locations with relation to the predicted location. The weights λ_α should be derived so that their sum is $\sum_{\alpha=1}^{n(u)} \lambda_{\alpha} = 1$. If the cells to be predicted, V, are larger than the sample plots, the block kriging predictor and its kriging variance are as follows (McBratney et al. 1981; Goovaerts 1997):

$$\hat{Z}_v(u) = \sum_{\alpha=1}^{n(u)} \lambda_{\alpha v}(u) z(u_{\alpha}) \tag{2.7}$$

$$\sigma_{kv}^2 = \sum_{\alpha=1}^{n(u)} \lambda_{\alpha v}\gamma(u_{\alpha}, V) + \mu_v - \gamma(V, V) \tag{2.8}$$

where $\lambda_{\alpha v}$ is the weight of data of a sample plot when a block V centered at location u is predicted, $\gamma(u_{\alpha}, V)$ is the average semivariance between the sample plots and the block V, $\gamma(V, V)$ is the within-block variance (average semivariance) between the cells within the block, and μ_v is a Lagrange parameter. The model-based sampling design strategy does not focus on the accuracy of the thematic map of an interest variable but on estimating the error of the random model that is used to generate the predicted value at each location.

In the model-based approach, a systematic sampling design is often assumed and the critical step is to determine a distance between sample plots—sampling grid spacing. In Equations 2.6 and 2.8, the kriging variance is minimized for any unknown location or block within a study area. Kriging variance varies depending only on the variogram that accounts for the spatial autocorrelation of the interest variable and the configuration of the sample plots in relation to the cell or block to be predicted; it does not depend on the sample data values themselves (McBratney and Webster 1981; Burgess et al. 1981; McBratney et al. 1981). Thus, if the variogram function can be estimated by a pilot survey or using a historical dataset, various estimates of the kriging variance for any sampling design can be calculated. Given an allowed maximum kriging variance, the sampling distance can be determined. Minimizing kriging variance requires determining a minimum sample size under an allowed error level. The combination of minimum kriging variance and systematic sampling leads to the neighboring sample plots being as far from one another as is practical for a fixed sample size and area.

The two aforementioned sampling design strategies have their advantages and disadvantages. For example, theoretically the design-based approach requires fewer assumptions than the model-based approach (de Gruijter and Ter Braak 1990). But the estimation variance obtained using the design-based approach is often 3.5–9 times larger than the kriging variance obtained using the model-based approach (McBratney and Webster 1983a). Van Kuilenburg et al. (1982) also implied that the model-based approach could lead to more accurate estimates than the design-based approach on the basis of equal fieldwork. After comparing a combination of systematic sampling and block kriging predictor with that of stratified simple random sampling and the Horvits–Thompson estimator, Brus and de Gruitjter (1997) concluded that the model-based approach is more accurate than the design-based one. Moreover, during the last two decades many authors (Atkinson 1991; Atkinson et al. 1992, 1994; Curran and Atkinson 1998; Wang et al. 2005; Anderson et al. 2006; Gertner et al. 2007; Wang et al. 2007, 2008a, 2009) further improved the model-based approach by introducing remotely sensed images into the sampling design strategies. The developments will be reviewed and discussed in Sections 2.3 through 2.7.

2.3 Role of Remotely Sensed Data in Sampling Design Strategies

Determining an appropriate sample size in the design-based approach requires the global variation of the interest variable, whereas determining a maximum sampling distance in the model-based approach requires the known variogram of this variable. Before sampling, however, both the global variation and the variogram are usually unknown. Remotely sensed data provide the potential to estimate the global variation and variogram of the interest variable (Wang et al. 2005).

Let an interest variable be a primary variable and a spectral variable be a secondary or an ancillary variable. Remotely sensed images are representations of the primary variables because they record the spectral signals that are reflected from objects on the earth, that is, the spatial variability of the primary variables is coded in remote sensing data. Thus, remotely sensed images are often highly correlated with the sample plot data to be collected through sampling design. Moreover, remotely sensed data are available everywhere and can indirectly reveal spatial distribution of the primary variables. The characteristics of large coverage and repeated acquisition of remotely sensed data make it possible to improve spatiotemporal sampling designs of primary variables for large areas.

In traditional design-based approaches, remotely sensed data are mainly used to conduct stratified sampling designs. In the designs, a study area is segmented into homogeneous polygons (Tomppo 1987) or the cells of a study area are classified into homogeneous strata (Wang 1996; Poso et al. 1999). Within each of the polygons or strata, random sampling is carried out. A stratified estimator is then applied to calculate a sample mean, and within each of the polygons or strata an average is computed and assigned to each cell. The idea behind the method is to minimize the within-strata variance and maximize the between-strata variance and thus increase the efficiency of sampling designs. Another alternative is to employ remotely sensed data to design multiphase sampling. For example, satellite images can be used to conduct stratification or segmentation of a study area and to draw first-phase sample plots. Aerial photos are then utilized to interpret the first-phase sample plots, and from the sample plots second-phase sample plots are drawn and measured in the field.

One example is two-phase sampling for stratification (Cochran 1977; Thompson 1992). This method has been widely used in Nordic countries, especially Finland, for forest inventory (Poso 1973, 1988, 1992; Poso and Waite 1995). A systematic set of first-phase sample plots are first generated for a study area, and remotely sensed data are acquired for each plot. These first-phase plots are then divided into homogeneous strata using remotely sensed data, and a set of plots are drawn from each stratum and measured for field observations of forest variables. Each first-phase plot is estimated by an estimator based on the combination of the first-phase plots and field information. The calculation of estimates for geographical regions or even small compartments is possible from the estimates of first-phase plots. For example, for stratification of the first-phase plots, remotely sensed data can be transformed into principal components and the k-means algorithm is then used for clustering the plots. Various estimators including mean vector, k-nearest neighbors, and regression may be applied for calculating estimates of the first-phase plots, and these may lead to quite different results (Tomppo 1992; Poso and Waite 1995; Poso et al. 1999).

One of the roles that remotely sensed images play in model-based approaches is to estimate the unknown variograms of the primary variables. One example is the study by Wang et al. (2005) in which a method to approximate the variogram of vegetation cover using the variogram that is derived from a spectral variable was proposed based on Markov models I and II (Journel 1999), and a model-based sampling design strategy for vegetation cover was then conducted. In the study by Wang et al. (2005), it was assumed that vegetation cover was significantly correlated with a spectral variable and the spatial cross variogram $\gamma_{ZY}(h)$ quantifying the joint spatial variability between these two variables was obtained (Webster et al. 1989; Goovaerts 1997; Wang et al. 2004):

$$\gamma_{ZY}(h) = \frac{1}{2N(h)} \sum_{\alpha=1}^{N(h)} \left(Z(u_\alpha) - Z(u_\alpha + h) \right) \left(Y(u_\alpha) - Y(u_\alpha + h) \right) \tag{2.9}$$

where Y is a spectral variable; $Y(u_\alpha)$ and $Y(u_\alpha + h)$ are its data values at locations u_α and $u_\alpha + h$, respectively; and the other terms are the same as in Equation 2.2. This cross variogram was approximated by the variogram of vegetation cover based on Markov model I in Equation 2.10 (Goovaerts 1997; Almeida and Journel 1994) and the variogram of the spectral variable based on Markov model II in Equation 2.11 (Journel 1999), respectively. Thus, the variogram of vegetation cover is estimated in Equation 2.12:

$$\gamma_{ZY}(h) \approx \frac{C_{ZY}(0)}{C_{ZZ}(0)} \gamma_{ZZ}(h) \tag{2.10}$$

$$\gamma_{ZY}(h) \approx \frac{C_{ZY}(0)}{C_{YY}(0)} \gamma_{YY}(h) \tag{2.11}$$

$$\gamma_{ZZ}(h) \approx \frac{C_{ZZ}(0)}{C_{YY}(0)} \gamma_{YY}(h) \tag{2.12}$$

where $C_{ZY}(0)$ is the traditional covariance between the vegetation cover and the spectral variable, $\gamma_{YY}(h)$ the variogram of the spectral variable Y and $C_{YY}(0)$ its traditional variance, and $C_{ZZ}(0)$ the traditional variance of the vegetation cover. Generalizing Equation 2.12, the traditional variance of the primary variable can be estimated using historical datasets or by a pilot study. Estimating the traditional variance of a primary variable is relatively easy in comparison with directly estimating its variogram. Traditional variance and variogram of the spectral variable can be directly computed using a remotely sensed image. If this approximation is reasonable, the sampling distance for measuring the primary variable can be worked out based on the point or block kriging variance (McBratney et al. 1981; McBratney and Webster 1983a, b; Wang et al. 2005).

It is more important that remotely sensed images can be used to improve model-based sampling design strategies. The improvements include determining optimal plot sizes and time intervals of remeasurements (Wang et al. 2001, 2008b; Gertner et al. 2006), deriving the local spatial variability (Anderson et al. 2006; Wang et al. 2008b), and reducing kriging variances of primary variables (Gertner et al. 2007; Wang et al. 2007, 2009). The use of remotely sensed data thus greatly increases the efficiency of sampling design strategies. In Sections 2.4 through 2.7, the roles of remotely sensed data in sampling designs are reviewed in detail.

2.4 Optimal Spatial Resolution for Sampling and Mapping

Given a cost budget and accuracy requirement of estimates, the basic task of a cost-efficient sampling design is determining the optimal size of sample plots used to collect field observations provided everything else is constant (Thompson 1992). On the other hand, it is also necessary to select appropriate spatial resolutions—pixel sizes of output maps—when spatially explicit estimates of natural resources are generated using remotely sensed data (Wang et al. 2005, 2009). Determining an optimal pixel size for mapping is especially required in the case of multisensor or multiresolution images used for mapping.

Traditionally, optimal sizes of sample plots used to collect field observations are determined based on the global coefficient of variation (CV) for a primary variable (Zeide 1980; Reich and Arvanitis 1992). Given a cost budget, suppose different sample plot sizes are used for the collection of field data. For each plot size, a CV for a primary variable is computed using the obtained observations. The CVs are then plotted against plot sizes. Generally, as the plot size increases the variation coefficient rapidly decreases at the beginning and then slowly and gradually becomes stable. The plot size at which the variation coefficient starts to stabilize is considered to be optimal. This method is usually applied to mapping continuous variables. In biodiversity studies, determining optimal plot sizes is similar to the aforementioned method with the difference that the CV is replaced by the number of species in such studies.

In practice, however, because of a lack of field observations before sampling, remotely sensed data provide the potential to determine optimal plot and pixel sizes for sampling and mapping natural resources. For this purpose, variance-based methods were proposed by Townshend and Justice (1988), Marceau et al. (1994), and Holopainen and Wang (1998). If the spatial resolution of a remotely sensed image is fine enough, the data can be aggregated to various coarser spatial resolutions using different size windows. The variances of the obtained spectral data from the smaller pixels can be calculated for each window size and an average value of the variances obtained. The obtained relationship between the average value of variances and the window size based on remotely sensed data is similar to the relationship between the variation coefficient and the plot size based on plot observations. That is, with an increasing window size the average variance decreases rapidly at the beginning and then slowly and eventually becomes stable. Thus, the window size at which the variance average starts to become stable is regarded as being optimal (Holopainen and Wang 1998).

Finding an optimal plot size for a collection of field observations is similar to searching for an optimal pixel size for mapping natural resources using a remotely sensed image. For this purpose, Hay et al. (1997) proposed an object-specific resampling method in which it is assumed that an image consists of image objects that are the models of ground objects; the closer the image objects in space, the more similar. Each of the image objects is regarded as an entity composed of individual pixels more similar than dissimilar to it. As the window size increases from 3 × 3 pixels to 5 × 5 pixels, 7 × 7 pixels, and so on, the variance of values from smaller pixels increases, reaches the maximum value, and then decreases. A distinct threshold in variance of pixel values can be obtained by increasing window size, and it corresponds to the optimal pixel size for mapping natural resources.

Theoretically, the aforementioned variance-based methods are, to some extent, based on spatial autocorrelation of primary and spectral variables. Woodcock and Strahler

(1987) developed a similar approach called the local variance method, which is based on the relationship between spatial resolution and spatial dependence. With this method an image or a landscape is considered to be a collection of discrete objects, and distinct boundaries exist between the objects. An average local variance of the image can be calculated at each of the different spatial resolutions, and it varies as the spatial resolution becomes coarser. The spatial resolution at which the average local variance reaches its maximum should be optimal.

The method of Woodcock and Strahler (1987) was further modified by the studies by Atkinson and Danson (1988) and Atkinson and Curran (1995) in which an average semivariance at a lag of one pixel was used to replace the aforementioned average local variance. The idea behind this modification is that the value of a variogram for a variable spatially varies and is a function of the separation distance of data locations given a direction, and as the window size increases the average semivariance at a lag of one pixel behaves in a similar way as described by Woodcock and Strahler (1987).

Based on the studies, the methods developed by Hay et al. (1997), Woodcock and Strahler (1987), Atkinson and Danson (1988), and Atkinson and Curran (1997) succeeded in finding optimal spatial resolutions for mapping the primary variables that are categorical, such as classification of land use and land cover types. If the primary variables are continuous, such as soil erosion and aboveground forest carbon, the methods might be misleading because there are no distinct boundaries between one area and another.

As Woodcock et al. (1988a, b) and Atkinson and Martin (1999) suggested, the variogram that accounts for the spatial variability of a variable varies depending on the spatial resolution of data in terms of structure of spatial variability. The structural changes of spatial variability due to different spatial resolutions can be characterized using different variogram models such as Equation 2.3 and the dynamics of their nugget, structure, and range parameters. If the assumption that as the spatial resolution becomes coarser the structure of the variogram fluctuates at the beginning and gradually becomes stable is met, then the dynamics of spatial variability at different spatial resolutions can be reproduced by calculating and fitting the experimental variograms from a remotely sensed image with different models and model parameters. The variogram obtained at an appropriate spatial resolution should capture the stable structure of spatial variability. Moreover, if the primary variable is highly correlated with the spectral variable, the spatial resolution at which the spatial structure of the variogram is captured implies the appropriate size of sample plots that should be used to collect field observations.

Based on the aforementioned idea, a so-called spatial variability–based method was developed and validated in the study by Wang et al. (2001) in which the appropriate plot sizes for sampling and mapping canopy cover percentages of different vegetation types were determined. To sample and map a soil erosion relevant ground and vegetation cover factor for a military installation, Wang et al. (2008b) further assessed this method with multiple plot sizes and multiresolution images and found that the obtained optimal spatial resolutions were 12 and 20 m for the years 1999 and 2000, respectively, and they were consistent with those using traditional methods. Their study also demonstrated that the most appropriate spatial resolutions using the high-resolution IKONOS images were consistent with those using ground sample plot data. Using ordinary kriging without images and an image-aided sequential Gaussian cosimulation algorithm to generate the maps of the cover factor, in addition, they verified that the obtained most appropriate spatial resolutions were optimal in terms of cost efficiency defined as the product of sampling cost and map error.

The aforementioned studies implied the potential of the spatial variability–based method by using high-resolution images instead of field observations to determine the optimal spatial resolutions before sampling. This method can be applied to both ground and image data. If a spectral variable is highly correlated with a primary variable, the optimal spatial resolution obtained using the remotely sensed image will be similar to that obtained using field plot data. On the other hand, the accuracy of this method will, to a great extent, be dependent on the correlation of the used spectral variables with the primary variable. When the correlation is low, the uncertainty of the optimal spatial resolutions obtained from the remotely sensed images to approximate the optimal plot sizes to collect field observations cannot be neglected. In addition, determining optimal spatial resolutions will become very complicated when a complex landscape is surveyed and mapped. This complexity will lead to multiple sizes of optimal sample plots and output map units. This requires developing a novel methodology by which multiple optimal spatial resolutions are determined using remotely sensed images.

2.5 Optimal Temporal Resolution for Sampling and Mapping

In addition to determining optimal spatial resolutions, a cost-efficient sampling design strategy to collect ground data for monitoring the dynamics of natural resources and eco-systems also needs to search for an optimal remeasurement frequency, that is, time interval to remeasure the ground plots. For example, in Nordic countries including Finland, Sweden, Norway, and Denmark, the time interval to remeasure permanent sample plots of forest resources inventory varies from 5 to 10 years (Segebaden 1992; Tomppo 1996). In Canada and the United States, the time interval for remeasuring forest resource inventory permanent plots is 5 years (Canadian Forest Service 2001; Bechtold and Patterson 2005). In China, forest inventory permanent plots are also remeasured every 5 years (CMF-DFRM 1996). The countries have different time intervals for remeasuring their forest sample plots mainly because they are located in areas ranging from tropic to subtropic, temperate, and boreal regions, and they possess different topographic and atmospheric features and soil properties, which leads to variable growth rates of forests. Moreover, the cost to collect remeasurements is another important factor that limits the use of high remeasure frequencies. In addition, accuracy requirements also lead to the selection of different time intervals. Generally, the longer the time interval, the lower the cost required for collection of ground data and the lower the accuracy of estimates. Given a budget and an accuracy requirement, determining an optimal remeasurement frequency thus becomes a big challenge.

Compared to seeking an optimal spatial resolution, there are relatively fewer reports in the area of determining an optimal time interval of remeasurements. In the area of forest resource inventory, several authors including Blight and Scott (1973), Jones (1980), Patterson (1950), and Ranneby and Rovainen (1995) investigated the determination of time interval between remeasurements by minimizing the variance of estimates in a framework of sampling with partial replacement. In this method, the weights of measurements were optimized based on their variances from permanent sample plots and the correlation between their successive measurements. The idea behind this method is that the variance of estimates increases with the increasing time interval of remeasurements.

Another method is that the time interval of remeasurements can be determined by comparing the accuracy requirement with the uncertainty of estimates obtained by extending the time interval of prediction that is made using a historical empirical model (Avery and Burkhart 1994). In this method, a dataset that is collected within a long period such as 20 years is first divided into two datasets at the middle point of the period. The first-half dataset is used to develop an empirical model, and the obtained model is then applied to make a prediction for the second half of the period. The predicted values are compared with the second-half dataset. As the time of prediction becomes longer, the accuracy of predicted values decreases. Given a precision requirement, the time interval of remeasurements can be determined.

In model-based sampling design strategies, kriging variance can be utilized for determining an optimal sampling distance (McBratney and Webster 1981; McBratney et al. 1981; McBratney and Webster 1983a). This method has been expanded to spatiotemporal sampling design in the case that a spatiotemporal variogram is available. One earlier example is the study by Rodriguez-Iturbe and Mejia (1974) in which a network design for the observations of rainfall at stations in time and space was optimized by analyzing the variance of precipitation estimates as a function of correlation in time, correlation in space, length of operation of the network, and geometry of the gauging array. The variance decreased with increasing sample size in space and with remeasurement frequency in time. This method based on spatiotemporal autocorrelation has been applied to spatiotemporal sampling design of long-term monitoring networks for changes of climate, forest, other natural resources, and environmental systems (Switzer 1979; Cressie et al. 1990; Guttorp et al. 1992). The difficulty of using this method lies in the requirement of developing spatiotemporal variograms and cross-covariance models.

Moreover, Rouhani and Wackernagel (1990) and Kyriakidis and Journel (1999) developed a multivariate time series model that allows one to analyze the time trend of a variable. Stein et al. (1994) proposed a two-step space–time kriging procedure in which spatial interpolation was first conducted and then temporal interpolation was made. This method was modified and applied to spatial and temporal sampling design for monitoring a soil erosion relevant ground and vegetation cover factor for the years 1989–1995 on a military installation (Gertner et al. 2006). The spatial sampling design was first carried out for each year. Given a spatial sample design for each year, the time interval for remeasurements was then determined so that the ordinary kriging variance of interpolation in time met the desired precision requirement. The temporal kriging variance of interpolation in time was a function of time interval between remeasurements, temporal variability, and spatial kriging variances. Given a spatial sampling design and temporal variogram, changing the remeasurement frequency could result in different kriging variances for the temporal interpolation. When the temporal kriging variance met the desired precision, an appropriate time interval of remeasurements was obtained. This method requires sufficient spatial and temporal remeasurements.

Furthermore, Gertner et al. (2006) suggested a simpler method to determine an optimal time interval of remeasurements. In this method, a temporal variogram that accounted for average temporal variability of an interest variable was calculated and modeled. The range parameter of the obtained temporal variogram model indicated the maximum length of time for the temporal autocorrelation of the variable, and it could be used to approximate the time interval between remeasurements. This method also requires a dataset consisting of remeasurements collected from permanent plots for a long period.

In both design-based and model-based sampling design strategies, determining an optimal time interval between remeasurements is often conducted together with spatial sampling. This requires sufficient remeasurements in both space and time. In the model-based approaches, if spatial and temporal measurements are sufficient, a complete spatiotemporal kriging estimator can be developed and used for simultaneously conducting spatial and temporal sampling. The model-based methods jointly model spatial and temporal variabilities of a random process, which is an obvious advantage compared with the design-based approaches. In practice, the historical field measurements often are not available or not sufficient. On the other hand, the historical datasets are associated with uncertainties. Thus, multi temporal remotely sensed images, if they are highly correlated with primary variables, provide a great potential for determining the time interval between remeasurements because they approximate both spatial and temporal variabilities of the primary variables.

2.6 Improvement of Sampling Design Strategies

When using design-based approaches such as simple random sampling it is assumed that each plot to be sampled where an interest variable is measured has an equal probability to be selected, and the number of sample plots is determined based on the variation coefficient of the variable. This method neglects the potential spatial dependency of random variables, which often leads to a larger sample size, a larger error in estimates, and the duplication of information because some sample plots may be too close together (Curran 1988; Curran and Atkinson 1998).

Model-based sampling design strategies are based on the regionalized variable theory in geostatistics. The idea behind this method is that kriging variance is determined only by the variogram of the interest variable and the spatial configuration of the plots to be sampled. Thus, various estimates of kriging variance can be obtained given different sampling distances. As long as an allowed error variance is given, the optimal sampling distance can be obtained. Curran and Atkinson (1998) reviewed the development of this method.

In its earlier stages, this method was used for the sampling design of soil properties by Burgess et al. (1981), McBratney et al. (1981), McBratney and Webster (1981), and McBratney and Webster (1983a, b). In these studies, kriging variances were plotted against sampling distance–grid spacing for a given plot size and the optimal sampling strategy was read from the graph. More than one optimal sampling strategy might exist, and the final choice was dependent on the cost of measurement. Sometimes, a larger plot size might be more efficient than a larger sample size (Atkinson and Curran 1995). This method led to systematic samples.

Based on this idea, Xiao et al. (2005) developed an optimization of sampling design for mapping vegetation cover percentage, which was used in the management of soil erosion for an army installation. Their study focused on searching for optimal solutions for sampling designs in which both plot size and sample size were taken into account given a cost budget and a desirable precision requirement for regional estimation and local mapping of vegetation cover. Their results showed that the model-based method was more cost efficient than the design-based methods. They also found that when local estimates were created, increasing the plot size was more cost efficient than increasing the sample size. But cost efficiency was reversed when global estimation was made.

Atkinson (1991) introduced remotely sensed data with the model-based approach with an unbiased cokriging estimator. In this estimator, cokriging variance is used to determine the sampling distance of a primary variable because it varies depending only on the variogram of a primary variable, the cross variograms between the primary variable and spectral variables, and the spatial configuration of data locations to be sampled and image data locations. Due to the spatial cross correlation between the primary variable and secondary variables, the sample size required is often smaller compared with that of the previous method without the use of remotely sensed images. In this method, the coregionalization consisting of one variogram for the primary variable, one variable for each spectral variable, and cross variograms between all pairs of variables has to be formalized. Given a set of variograms, cokriging variances can be computed for any configuration of sampling and can be used to design an optimal scheme that meets the given precision for least effort. Atkinson et al. (1992) used this method for sampling design in intensive agriculture in Britain and found that the sampling designs that incorporated cokriging using remotely sensed data were nine times as efficient as the ones that entailed kriging using the primary variable alone. This modified method was further applied by Atkinson et al. (1994, 2000) for sampling and mapping the green leaf area index of barley, the biomass of pasture, and tropical forest biophysical properties.

Using image data can greatly reduce the number of ground sample plots and sampling cost required for the collection of data. But the efficiency of using remotely sensed data varies depending on the ratio of ground sample plot data to remotely sensed data. Wang et al. (2009) investigated the efficiencies of remotely sensed data and the sensitivity of sampling distance for sampling and mapping a ground and vegetation cover factor in a monitoring system of soil erosion dynamics by cokriging with Landsat Thematic Mapper (TM) images. They found that under the same precision requirement the efficiency gain is significant as the ratio of ground to image data used varies from 1:1 to 1:16. They also pointed out that directly using image data from neighboring pixels in sampling design and mapping was more efficient in increasing the accuracy of maps than using image data from sampled pixels.

Compared with design-based methods, model-based sampling designs can lead to the need for a smaller number of sample plots given a precision level for a study area. However, these methods require variograms that are derived from ground data. In fact, ground data are often not available before sampling. Based on a method developed by Wang et al. (2005), the variograms of primary variables can be estimated using remotely sensed images. This method was very promising for determining the optimal sampling distance for estimating regional averages. However, a high correlation between primary variables and spectral variables was required when determining the sampling distance for local estimation (Wang et al. 2005).

In model-based methods, moreover, maximum kriging variance and cokriging variance are calculated based on variograms that measure global spatial variability of a primary variable and spatial configuration of spatial data locations to be systematically sampled. This implies that the sample designs are globally optimal, but they are not optimal for local areas. Wang et al. (2007) and Gertner et al. (2007) proposed a novel method that integrated the cokriging variance–based sampling design with traditional stratification in which the within-strata variances were minimized and the between-strata variances were maximized. Thus, this method can, to some extent, achieve optimal sampling designs at both global and local levels. Wang et al. (2007) indicated that this method could greatly increase the cost efficiency of sampling and mapping a soil erosion relevant ground and vegetation cover factor using Landsat TM images compared with traditional sample designs. Gertner et al. (2007) further combined the block cokriging variance–based sampling design with stratification.

The aforementioned integration of cokriging and stratification greatly improved the cost efficiency of sampling designs by taking into account the spatial variability of a primary variable based on homogeneous strata. In other words, the optimization of sampling designs can be achieved at global and stratum levels. Within each of the homogeneous strata, the sampling distances may not be optimal from location to location. The significant gap was overcome by a novel method that was developed by Anderson et al. (2006) and Wang et al. (2008a).

2.7 Local Variability–Based Sampling Design Strategy

Sampling design strategies have been widely studied, including various design-based and model-based approaches. But the existing methods cannot lead to truly optimal sampling designs when neglecting the local spatial variability of primary variables. For this purpose, Anderson et al. (2006) and Wang et al. (2008a) proposed a local variability–based sampling design strategy. In this method, an image-aided sequential Gaussian cosimulation that combines remotely sensed and ground data was applied to simulate local spatial variability. This method theoretically can lead to a sampling design with variable sampling distances, that is, grid spacing varies from place to place and is optimal at both local and global levels.

In practice, the theoretical assumption of second-order stationary, even the weaker intrinsic hypothesis, is rarely met, that is, often the spatial variability of a primary variable is not constant but varies from place to place and is dependent on both the local spatial configurations of data locations and the data values themselves. In a cost-efficient sampling design strategy, in addition to global spatial variability and the spatial configuration of data locations, the local spatial variability of the primary variable should thus be taken into consideration. The development of the method focuses on obtaining the estimates of local spatial variability that can be used to design a reasonable allocation of sample plots at local areas. This can be realized using an image-aided sequential Gaussian cosimulation algorithm (Goovaerts 1997; Wang et al. 2002, 2008a; Anderson et al. 2006).

In this cosimulation algorithm, a value of a primary variable at a location is considered as a realization of a random process at this location (Goovaerts 1997). At each location, more than one realization can be generated by randomly drawing values from a multivariate joint distribution. This distribution can be approximated by a conditional cumulative distribution function (CCDF) that is determined using sampled data, previously simulated values if any, and image data within a given neighborhood. The realizations can then be used to calculate a sample mean, a sample variance, and thus a sample CV that quantifies the local spatial variability. Thus, the local uncertainty about a predicted value of the variable at any particular location is modeled through the set of possible realizations of the random process at that location (Anderson et al. 2006). This multivariate joint distribution is assumed to be Gaussian. If this assumption is not met, a normal score transformation of data should be conducted.

In this cosimulation algorithm, it is supposed that a study area consists of many cells. To generate a realization of the random process at each cell u, a random path to visit each cell is first set up and a computer program is used to follow the order of visiting each cell and calculate a conditional mean and a conditional variance using a simple unbiased cokriging estimator:

$$z^{\text{sck}}(u) = \sum_{\alpha=1}^{n(u)} \lambda_\alpha^{\text{sck}}(u) \left[z(u_\alpha) - m_z \right] + \lambda_y^{\text{sck}}(u) \left[y(u) - m_y \right] + m_z \qquad (2.13)$$

$$\sigma^{2(\text{sck})}(u) = C_{zz}(0) - \sum_{\alpha=1}^{n(u)} \lambda_\alpha^{\text{sck}}(u) C_{zz}(u_\alpha - u) - \lambda_y^{\text{sck}}(u) C_{zy}(0) \qquad (2.14)$$

where $z^{sck}(u)$ and $\sigma^{2(sck)}(u)$ are the simple cokriging estimate and variance, respectively; $n(u)$ is the number of data used to predict location u, which varies from location to location; $y(u)$ is the data of a spectral variable at a cell u; m_z and m_y are the means of sample plot and image data, respectively; and $\lambda_\alpha^{sck}(u)$ and $\lambda_y^{sck}(u)$ are the weights for sampled and image data, respectively.

The conditional mean and conditional variance obtained earlier are used to determine the CCDF from which a value is randomly drawn and used as a realization of the random process at this location. This value is also used as a conditional datum for the next cosimulation. Once a cell is predicted, the algorithm moves the cosimulation to the next cell by following a random path. When all the cells are visited, a realization of spatial distribution—a map of the primary variable—is achieved. The aforementioned cosimulation process can be repeated many times by setting up different random paths to visit the cells of the study area. Thus, many realizations of the spatial distribution of the primary variable can be obtained. The optimal number of realizations should be the one at which the variance of predicted values becomes stable.

From the earlier multiple realizations, a sample mean and a sample variance are calculated and further a sample CV is derived at each cell. Based on the CV, grid spacing can be computed given an allowed error, E, and a significance level, α:

$$\text{Grid spacing} = \sqrt{\frac{A \times E^2}{t_\alpha \times CV^2}} \tag{2.15}$$

In Equation 2.15, the study area A is given and t_α is the Student's distribution at the significance level α. The variable CV leads to a variable grid spacing. Thus, the grid spacing reflects the local optimal sampling distance and varies from place to place. This method has, to some extent, bridged the significant gaps mentioned earlier in the area of advancing sampling design strategies for sampling and mapping natural resources. The disadvantage of this method is that it still requires field observations to derive the variograms of primary variables. Future studies may focus on how to use the variograms from remotely sensed data to approximate the variograms of primary variables and conduct spatial uncertainty analysis for this method.

2.8 Case Study

The local variability–based sampling design discussed in Section 2.7 was assessed and compared with simple random sampling in a case study (Wang et al. 2008a). The case study area is located at Fort Riley, a military installation in which a variety of military training activities have taken place since the 1950s. This study area has an area of 41,154 ha. It is located in the Bluestem Prairie region, Kansas, and is characterized by rolling plains, dissected by stream valleys, and dominated by tall grasslands. These activities seriously disturbed the vegetation canopy cover and led to an increase in soil erosion. Thus, an optimal sampling design was needed to monitor the dynamics of soil erosion.

In the study, a soil erosion relevant ground and vegetation cover factor that is a function of ground cover, vegetation canopy cover, and minimum vegetation canopy height for raindrops was sampled and mapped using Landsat TM images. The field data of the cover factor from a total of 154 permanent plots were available for the years from 1989 to 2001. The size of the permanent plots was 6 m × 100 m with a 100 m line transect located in the center of each plot. Ground cover, vegetation canopy cover, and average minimum canopy height were recorded at points of 1 m intervals (0.5 m, 1.5 m, …, 99.5 m) along the transect. The ground and vegetation cover percentages were then calculated and the values of plot cover factor were derived based on a set of empirical models (Wischmeier and Smith 1978; Wang et al. 2008a). Moreover, a set of Landsat 5 TM images at a spatial resolution of 30 m × 30 m were also acquired for these years and the image data were aggregated to a spatial resolution of 90 m × 90 m using a window average.

These data were utilized to simulate the optimal sampling design using the aforementioned local variability–based sampling design strategy. The maps of local variation coefficient of the cover factor were derived using the aforementioned spatial cosimulation algorithm. Based on local variability and Equation 2.15, the sampling distance at each pixel was then calculated. As examples, the local variability maps for the years 1992 and 1998 are shown in Figure 2.1 (Wang et al. 2008a). If sampling efficiency is defined as the ratio of coefficients of correlation between the predicted and observed values of the cover factor to the corresponding sample size used, Figure 2.2 presents that out of 13 years there were 8 years at which the sampling efficiency per unit from the local variability sampling design was higher than that from the random samples (Wang et al. 2008a). This means that, statistically, in comparison with simple random sampling, the local variability–based sampling design strategy greatly reduced the number of sampled plots and increased the cost efficiency for sampling.

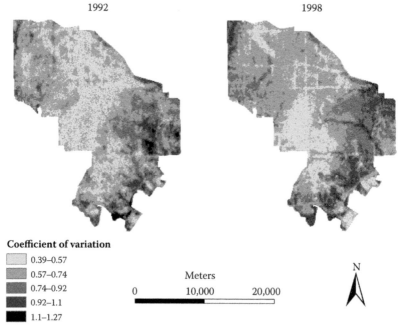

FIGURE 2.1
Coefficients of variation for soil erosion relevant cover factor variation for the years 1992 and 1998.

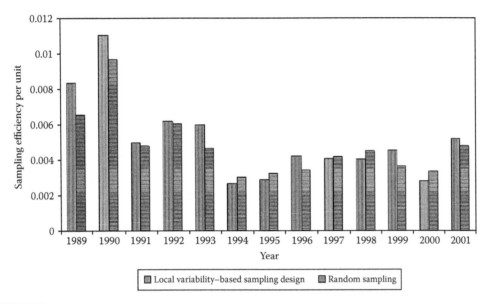

FIGURE 2.2
Comparison of sampling efficiency per unit between local variability–based sampling and random sampling. (From Wang, et al., *Catena*, 73, 75–88, 2008a.)

2.9 Conclusion

Mapping natural resources using remotely sensed images requires cost-efficient sampling design strategies to collect field observations of an interest variable, generate its spatially explicit estimates, and further assess the quality of the resulting maps. Thus, selection of sampling design strategies is very critical. Moreover, a cost-efficient sampling design strategy means the combination of an optimal sampling design procedure and an unbiased estimator for both global and local estimates, and it includes determining appropriate plot size and shape, optimal sample size, and spatial distribution of plots given a cost budget and the desired precision of the estimates. In practice, the development of a cost-efficient sampling design strategy is very complicated. Although a great body of literature is available in this area, we are still facing some big challenges. In this chapter, a review of literature on the development and advancement of sampling design strategies that are used to collect field observations for sampling and mapping natural resources is conducted. Moreover, the significant gaps that currently exist in this area are identified and discussed.

References

Almeida, A. S., and A. G. Journel. 1994. "Joint Simulation of Multiple Variables with a Markov-Type Coregionalization Model." *Mathematical Geology* 26:565–88.

Anderson, A. B., G. Wang, and G. Z. Gertner. 2006. "Local Variability Based Sampling for Mapping a Soil Erosion Cover Factor by Co-Simulation with Landsat TM Images." *International Journal of Remote Sensing* 27(12):2423–47.

Atkinson, P. M. 1991. "Optimal Ground-Based Sampling for Remote Sensing Investigations: Estimating the Regional Mean." *International Journal of Remote Sensing* 12:559–67.

Atkinson, P. M., and P. J. Curran. 1995. "Defining an Optimal Size of Support for Remote Sensing Investigations." *IEEE Transactions of Geoscience and Remote Sensing* 33:768–76.

Atkinson, P. M., and P. J. Curran. 1997. "Choosing an Appropriate Spatial Resolution for Remote Sensing Investigations." *Photogrammetric Engineering and Remote Sensing* 63(12):1345–51.

Atkinson, P. M., and F. M. Danson. 1988. "Spatial Resolution for Remote Sensing of Forest Plantations." In *Proceedings of IGARSS'88 Symposium*, September 13–16, 1988, Edinburgh, Scotland.

Atkinson, P. M., G. M. Foody, P. J. Curran, and D. S. Body. 2000. "Assessing the Ground Data Requirements for Regional Scale Remote Sensing of Tropical Forest Biophysical Properties." *International Journal of Remote Sensing* 21:2571–87.

Atkinson, P. M., and D. Martin. 1999. "Investigating the Effect of Support Size on Population Surface Models." *Geographical & Environmental Modeling* 3:101–19.

Atkinson, P. M., R. Webster, and P. J. Curran. 1992. "Cokriging with Ground-Based Radiometry." *Remote Sensing of Environment* 41:45–60.

Atkinson, P. M., R. Webster, and P. J. Curran. 1994. "Cokriging with Airborne MASS Imagery." *Remote Sensing of Environment* 50:335–45.

Avery, T. E., and H. E. Burkhart. 1994. *Forest Measurements*. New York: McGraw-Hill.

Bechtold, W. A., and P. L. Patterson. 2005. "The Enhanced Forest Inventory and Analysis Program—National Sampling Design and Estimation Procedures." *General Technical Report SRS-80*, 85, U.S. Department of Agriculture, Forest Service, Southern Research Station, Asheville, NC.

Blight, B. J. N., and A. J. Scott. 1973. "A Stochastic Model for Repeated Surveys." *Journal of the Royal Statistics Society B* 35:61–66.

Brus, D. J., and J. J. de Gruijter. 1997. "Random Sampling or Geostatistical Modeling? Choosing Between Design-Based and Model-Based Sampling Strategies for Soil (with Discussion)." *Geoderma* 80:1–44.

Brus, D. J., J. J. de Gruijter, and A. Breeuwsma. 1992. "Strategies for Updating Soil Survey Information: A Case Study to Estimate Phosphate Sorption Characteristics." *Journal of Soil Science* 43:567–81.

Burgess, T. M., R. Webster, and A. B. McBratney. 1981. "Optimal Interpolation and Isarithmic Mapping of Soil Properties. IV Sampling Strategy." *Journal of Soil Science* 32:643–59.

Campbell, J. B., and R. H. Wynne. 2011. *Introduction to Remote Sensing*. New York: The Guilford Press.

Canadian Forest Service. 2001. "Canada's National Forest Inventory." Accessed on April 14, 2013, https://nfi.nfis.org/index.php.

Chinese Ministry of Forestry–Department of Forest Resource and Management (CMF-DFRM). 1996. *Forest Resources of China 1949 to 93*. Chinese Forestry Press, Liu-hai-hu-tong 7, Beijing, People's Republic of China.

Cochran, W. G. 1977. *Sampling Techniques*. 3rd ed. New York: John Wiley & Sons.

Cressie, N., C. A. Gotway, and M. O. Grondona. 1990. "Spatial Prediction from Networks." *Chemometrics and Intelligent Laboratory Systems* 7:251–71.

Curran, P. J. 1988. "The Semivariogram in Remote Sensing: An Introduction." *Remote Sensing of Environment* 24:493–507.

Curran, P. J., and P. M. Atkinson. 1998. "Geostatistics and Remote Sensing." *Progress in Physical Geography* 22:61–78.

Curran, P. J., and H. D. Williamson. 1986. "Sample Size for Ground and Remotely Sensed Data." *Remote Sensing of Environment* 20:31–41.

de Gruijter, J. J., and C. J. F. Ter Braak. 1990. "Model-Free Estimation from Spatial Samples: A Reappraisal of Classical Sampling Theory." *Mathematical Geology* 22:407–15.

Gertner, G. Z., G. Wang, and A. B. Anderson. 2006. "Determination of Frequency for Re-Measuring Ground and Vegetation Cover Factor for Monitoring of Soil Erosion." *Environmental Management* 37(1):84–97.

Gertner, G. Z., G. Wang, A. B. Anderson, and H. R. Howard. 2007. "Combining Stratification and Up-Scaling Method—Block Cokriging with Remote Sensing Imagery for Sampling and Mapping an Erosion Cover Factor." *Ecological Informatics* 2:373–86.

Goovaerts, P. 1997. *Geostatistics for Natural Resources Evaluation.* New York: Oxford University Press.

Guttorp, P., P. D. Sampson, and K. Newman. 1992. "Non-Parametric Estimation of Spatial Covariance with Application to Monitoring Network Evaluation." In *Statistics in Environmental and Earth Sciences,* edited by A. T. Walden and P. Guttorp, 39–51. London: Edward Arnold.

Hay, G. J., K. O. Niemann, and D. G. Goodenough. 1997. "Spatial Thresholds, Image-Objects, and Up-Scaling: A Multi-Scale Evaluation." *Remote Sensing of Environment* 62:1–19.

Holopainen, M., and G. Wang. 1998. "The Calibration of Digitized Aerial Photographs for Forest Stratification." *International Journal of Remote Sensing* 19:677–96.

Jones, R. G. 1980. "Best Linear Unbiased Estimators for Repeated Surveys." *Journal of the Royal Statistics Society B* 42:221–26.

Journel, A. G. 1999. "Markov Models for Cross-Covariances." *Mathematical Geology* 31:955–64.

Journel, A. G., and C. J. Huijbregts. 1978. *Mining Geostatistics.* London: Academic Press.

Kyriakidis, P. C., and A. G. Journel. 1999. "Geostatistical Space-Time Models: A Review." *Mathematical Geology* 31(6):651–84.

Marceau, D. J., D. J. Gratton, R. A. Fournier, and J. Fortin. 1994. "Remote Sensing and the Measurement of Geographical Entities in a Forested Environment. 2. The Optimal Spatial Resolution." *Remote Sensing of Environment* 49:105–17.

McBratney, A. B., and R. Webster. 1981. "The Design of Optimal Sampling Schemes for Local Estimation and Mapping of Regionalized Variables—II Program and Examples." *Computers & Geosciences* 7:335–65.

McBratney, A. B., and R. Webster. 1983a. "How Many Observations Are Needed for Regional Estimation of Soil Properties?" *Soil Science* 135(3):177–83.

McBratney, A. B., and R. Webster. 1983b. "Optimal Interpolation and Isarithmic Mapping of Soil Properties V. Co-Regionalization and Multiple Sampling Strategy." *Journal of Soil Science* 34:137–62.

McBratney, A. B., R. Webster, and T. M. Burgess. 1981. "The Design of Optimal Sampling Schemes for Local Estimation and Mapping of Regionalized Variables—I Theory and Method." *Computers & Geosciences* 7:331–34.

Olea, R. A. 1984. "Sampling Design Optimization for Spatial Functions." *Mathematical Geology* 16(4):369–92.

O'Regan, W. G., and L. G. Arvanitis. 1966. "Cost Effectiveness in Forest Sampling." *Forest Science* 12(4):406–14.

Patterson, H. D. 1950. "Sampling in Successive Occasions with Partial Replacement of Units." *Journal of the Royal Statistics Society B* 12:241–55.

Poso, S. 1973. "A Method of Combining Photo and Field Samples in Forest Inventory." *Communication Institute Forestry Fennica* 76(1):1–133.

Poso, S. 1988. "Seeking for an Optimal Path for Using Satellite Imageries for Forest Inventory and Monitoring." In *Proceedings of the IUFRO subject group 4.02.05 meeting in Finland,* August 29–September 2.

Poso, S. 1992. "Establishment and Analysis of Permanent Sample Plots. Remote Sensing and Permanent Plot Techniques for World Forest Monitoring." In *Proceedings of the IUFRO S4.02.05 Wacharakitti International Workshop,* January 13–17, Pattaya, Thailand.

Poso, S., and M.-L. Waite. 1995. "Sample Based Forest Inventory and Monitoring Using Remote Sensing." In *Invited papers of IUFRO XX World Congress,* August 6–12, Tampere, Finland.

Poso, S., G. Wang, and T. Sakkari. 1999. "Weighting Alternative Estimates When Using Multi-Source Auxiliary Data for Forest Inventory." *Silva Fennica* 33(1):41–50.

Ranneby, B., and E. Rovainen. 1995. "On the Determination of Time Intervals between Remeasurements of Permanent Plots." *Forest Ecology and Management* 71:195–202.

Reich, R. M., and L. G. Arvanitis. 1992. "Sampling Unit, Spatial Distribution of Trees, and Precision." *North Journal of Applied Forest* 9:3–6.

Rodriguez-Iturbe, I., and J. M. Mejia. 1974. "The Design of Rainfall Networks in Time and Space." *Water Resources Research* 10(4):713–28.

Rouhani, S., and H. Wackernagel. 1990. "Multivariate Geostatistical Approach to Space-Time Data Analysis." *Water Resources Research* 26:585–91.

Segebaden, G. V. 1992. "The Swedish National Forest Inventory—a Review of Aims and Methods." In *Proceedings of Ilvessalo Symposium on National Forest Inventories, August 17–21*, edited by A. Nyyssönen, S. Poso, and J. Rautala, 41–46, Helsinki, Finland: Finnish Forest Research Institute.

Stein, A., C. G. Kocks, J. C. Zadoks, H. D. Frinking, M. A. Ruissen, and D. E. Myers. 1994. "A Geostatistical Analysis of the Spatio-Temporal Development of Downy Mildew Epidemics in Cabbage." *The American Phytopathological Society* 84(10):1227–39.

Switzer, P. 1979. "Statistical Considerations in Network Design." *Water Resources Research* 15(6):1712–16.

Thompson, S. K. 1992. *Sampling*. New York: John Wiley & Sons.

Tomppo, E. 1987. "Stand Delineation and Estimation of Stand Variates by Means of Satellite Images." University of Helsinki, Department of Forest Mensuration and Management Research Notes 19.

Tomppo, E. 1992. "Multi-Source National Forest Inventory of Finland." In *Proceedings of Ilvessalo Symposium on National Forest Inventories in Finland*, edited by A. Nyyssönen, S. Poso, and J. Rautala, August 17–21, 1992, 16–25, Helsinki, Finland: Finnish Forest Research Institute.

Tomppo, E. 1996. "Multi-Source National Forest Inventory of Finland." In *New Thrusts in Forest Inventory, Proceedings of the Subject Group S4.02-00 "Forest Resource Inventory and Monitoring" and Subject Group S4.12-00 "Remote Sensing Technology" Volume I IUFRO XX World Congress*, edited by R. Paivinen, J. Vanclay, and S. Miina, August 6–12, 1995, Vol. 7, 27–42, Tampere, Finland: EFI Proceedings.

Townshend, J. R. G., and C. O. Justice. 1988. "Selecting the Spatial Resolution of Satellite Sensors Required for Global Monitoring of Land Transformations." *International Journal of Remote Sensing* 9:187–236.

Van Kuilenburg, J., J. J. De Gruijter, B. A. Marsman, and J. Bouma. 1982. "Accuracy of spatial interpolation between point data on soil moisture supply capacity, compared with estimates from mapping units." *Geoderma* 27:311–325.

Wang, G. 1996. *An Expert System for Forest Resource Inventory and Monitoring Using Multi-Source Data*. University of Helsinki, Department of Forest Resource Management, Helsinki, Finland, Ph.D. dissertation, ISBN 951-45-7289-0.

Wang, G., G. Z. Gertner, and A. B. Anderson. 2004. "Spatial-Variability-Based Algorithms for Scaling-Up Spatial Data and Uncertainties." *IEEE Transactions in Geoscience and Remote Sensing* 42:2004–15.

Wang, G., G. Z. Gertner, and A. B. Anderson. 2005. "Sampling Design and Uncertainty Based on Spatial Variability of Spectral Reflectance for Mapping Vegetation Cover." *International Journal of Remote Sensing* 26(15):3255–74.

Wang, G., G. Z. Gertner, and A. B. Anderson. 2007. "Sampling and Mapping a Soil Erosion Relevant Cover Factor by Integrating Stratification, Model Updating and Cokriging with Images." *Environmental Management* 39(1):84–97.

Wang, G., G. Z. Gertner, and A. B. Anderson. 2009. "Efficiencies of Remotely Sensed Data and Sensitivity of Grid Spacing in Sampling and Mapping a Soil Erosion Relevant Cover Factor by Cokriging." *International Journal of Remote Sensing* 30(17):4457–77.

Wang, G., G. Z. Gertner, A. B. Anderson, and H. R. Howard. 2008a. "Repeated Measurements on Permanent Plots Using Local Variability Based Sampling for Monitoring Soil Erosion." *Catena* 73:75–88.

Wang, G., G. Z. Gertner, H. Howard, and A. B. Anderson. 2008b. "Optimal Spatial Resolution for Collection of Ground Data and Multi-Sensor Image Mapping of a Soil Erosion Cover Factor." *Journal of Environmental Management* 88:1088–98.

Wang, G., G. Z. Gertner, X. Xiao, S. Wente, and A. B. Anderson. 2001. "Appropriate Plot Size and Spatial Resolution for Mapping Multiple Vegetation Cover Types." *Photogrammetric Engineering & Remote Sensing* 67(5):575–84.

Wang, G., S. Wente, G. Z. Gertner, and A. B. Anderson. 2002. "Improvement in Mapping Vegetation Cover Factor for Universal Soil Loss Equation by Geo-Statistical Methods with Landsat TM Images." *International Journal of Remote Sensing* 23:3649–67.

Webster, R., P. J. Curran, and J. W. Munden. 1989. "Spatial Correlation in Reflected Radiation from the Ground and Its Implications for Sampling and Mapping by Ground-Based Radiometry." *Remote Sensing of Environment* 29:67–78.

Wischmeier, W. H., and D. D. Smith. 1978. *Predicting Rainfall-Erosion Losses from Cropland East of the Rock Mountains: Guide for Selection of Practices for Soil and Water Conservation. Agriculture Handbook*, 1–58. No. 282, US Government Printing Office, SSOP Washington, DC: USDA.

Woodcock, C. E., and A. H. Strahler. 1987. "The Factor of Scale in Remote Sensing." *Remote Sensing of Environment* 21:311–22.

Woodcock, C. E., A. H. Strahler, and D. L. B. Jupp. 1988a. "The Use of Variogram in Remote Sensing: I. Scene Models and Simulated Images." *Remote Sensing of Environment* 25:323–48.

Woodcock, C. E., A. H. Strahler, and D. L. B. Jupp. 1988b. "The Use of Variogram in Remote Sensing: II. Real Digital Images." *Remote Sensing of Environment* 25:349–79.

Xiao, X., G. Z. Gertner, G. Wang, and A. B. Anderson. 2005. "Optimal Sampling Scheme for Remote Sensing Mapping of Vegetation Cover." *Landscape Ecology* 20(4):375–87.

Zeide, B. 1980. "Plot Size Optimization." *Forest Science* 26(2):251–57.

3

Accuracy Assessment for Classification and Modeling

Suming Jin

CONTENTS

3.1 Introduction 45
3.2 Error Matrix and Analysis 46
3.3 Essential Components and Issues of Accuracy Assessment 49
3.4 Special Considerations for Accuracy Assessment 51
    3.4.1 Large-Area Land Cover 51
    3.4.2 Change Detection Maps 52
    3.4.3 Maps of Continuous Variables 52
References 54

3.1 Introduction

The importance of characterizing, quantifying, and monitoring land cover and land use and their changes has been widely recognized by global and environmental change studies (Matthews et al. 2004; Foley et al. 2005; Turner et al. 2007). Remote sensing is an attractive source of thematic maps that are available at a range of spatial and temporal scales (Foody 2002). Assessing the quality of thematic maps derived from remotely sensed data provides information about the quality of a map and its fitness for a particular purpose (Janssen and van der Wel 1994; Foody 2002; Congalton and Green 2009). Quality assessment also provides a better understanding of map error and its likely implications, which helps guide effective decisions. Researchers and users of remotely sensed data must have a strong knowledge of the factors that need to be considered and the techniques used in performing any accuracy assessment. Failure to know these techniques and considerations can severely limit one's ability to effectively use remotely sensed data (Congalton 1991).

The accuracy of a map or spatial dataset is a function of both positional accuracy and thematic accuracy. Positional accuracy measures how close a spatial feature on a map is to its true location on the ground both in horizontal and vertical positions (Bolstad 2005). Thematic accuracy measures whether the mapped feature at a particular time is different from what was actually on the ground at the same time (Congalton and Green 2009). Positional and thematic errors typically are not differentiated in accuracy assessment reports of land cover maps (Foody 2002). Congalton (1994) categorized four definite historical stages of development of accuracy assessment techniques: (1) stage of visual appraisal, during which no real accuracy assessment was performed; (2) stage of aspatial area summaries, where accuracy assessment was based on comparisons of the areal extent of the classes in the derived thematic map relative to their extent in some ground or other reference dataset; (3) stage of primitive error matrix analysis, which was based on a comparison of the class labels in the thematic map and ground data for a set of specific locations (however, only an overall map accuracy was obtained); and (4) the age of the error matrix, a refinement of the third stage in which greater use of the information on the correspondence of the thematic map labels to those from the reference data is made. This stage has the confusion or error matrix at its core and uses this to describe the pattern of class allocation relative to the reference data (Foody 2002).

3.2 Error Matrix and Analysis

An error matrix, alternatively called a confusion matrix, is a simple cross-tabulation of mapped class labels against that observed on the ground or the reference data for a sample of cases at specified locations (Congalton 1991) (see Table 3.1 for a hypothetical example). The columns usually represent the reference data that are considered to be correct or at least more accurate than the map data, whereas the rows indicate the classification generated from the remotely sensed data (i.e., the map). The confusion matrix provides the information of describing classification accuracy and characterizing errors and lies at the core of much work on accuracy assessment. Based on Table 3.1, various accuracies can be calculated.

Producer's Accuracy	User's Accuracy	Overall Accuracy
A = 90/101 = 89%	A = 90/96 = 94%	(90 + 81 + 75)/275 = 89%
B = 81/90 = 90%	B = 81/98 = 83%	
C = 75/84 = 89%	C = 75/81 = 93%	
Omission Error	Commission Error	
A = (10 + 1)/101 = 11%	A = (4 + 2)/96 = 6%	
B = (4 + 5)/90 = 10%	B = (10 + 7)/98 = 17%	
C = (2 + 7)/84 = 11%	C = (1 + 5)/81 = 7%	

An error matrix is a very effective way to represent map accuracy because the accuracies of each category are plainly described along with both the errors of inclusion (commission errors) and the errors of exclusion (omission errors) present in the classification (Congalton 1991). A "commission error" is simply defined as including an area in a category when it does not belong to that category. An "omission error" is excluding an area from the category to which it belongs. In addition to clearly showing errors of omission and commission, the error matrix can be used to compute other accuracy measures such as overall accuracy, producer's accuracy, and user's accuracy (Story and Congalton 1986). "Overall accuracy" is simply the

TABLE 3.1

Hypothetical Example of an Error Matrix

		Reference Data			Row Total (No. of Classified Sample Units)
		A	B	C	
Thematic map classes	A	90	4	2	96
	B	10	81	7	98
	C	1	5	75	81
Column total (No. of reference sample units)		101	90	84	275

sum of the major diagonal (i.e., the correctly classified sample units) divided by the total number of sample units in the entire error matrix (Table 3.1). "Producer's accuracy" is a measure of how much of the area in each class was classified correctly, which is calculated by dividing the number in the diagonal cell of the error matrix (i.e., number of correct classifications) by the number in the column total (i.e., total number of the reference data in the class). "User's accuracy" is a measure of the proportion of each map class that is correct, which is calculated by dividing the number in the diagonal cell of the error matrix (i.e., number of correct classifications) by the number in the row total (i.e., total number of classifications in the category). Producer's and user's accuracies indicate individual category accuracies instead of just the overall classification accuracy. A simple example involving only three classes (Table 3.1) shows the calculation of all those accuracy and error measures. Studying the error matrix in Table 3.1 reflects an overall accuracy of 89%. For example, the producer's accuracy for class A is 89% and the user's accuracy is 94%; the omission error for class A is 11% and the commission error is 6%. It indicates that 89% of the areas called class A on the ground have been correctly identified as A, and 94% of the areas called class A on the map are actually A on the ground. The sum of producer's accuracy and omission error is equal to 100%, and the sum of user's accuracy and commission error is equal to 100%.

Table 3.2 shows an error matrix in mathematical terms, assuming that n samples are distributed into k^2 cells. n_{ij} denotes the number of samples classified into category i ($i = 1, 2, ..., k$) in the map and category j ($j = 1, 2, ..., k$) in the reference dataset; n_{i+} denotes the total number of classifications in category i; and n_{+j} is the total number of the reference data in category j.

Overall, producer's and user's accuracies are then computed as follows:

$$\text{overall accuracy} = \frac{\sum_{i=1}^{k} n_{ii}}{n} \tag{3.1}$$

$$\text{producer's accuracy } j = \frac{n_{jj}}{n_{+j}} \tag{3.2}$$

$$\text{user's accuracy } i = \frac{n_{ii}}{n_{i+}} \tag{3.3}$$

The error matrix is used mainly to provide a basic description of thematic map accuracy and for comparison of accuracies. Several basic analysis techniques can be performed to help interpret and/or compare the accuracy assessment results. The Kappa analysis is the most commonly used measurement along with overall accuracy, producer's accuracy, and user's accuracy. It measures how well the remotely sensed classification agrees with the

TABLE 3.2

Error Matrix in Mathematical Terms

		Reference Data			Row Total (n_{i+})
		1	**2**	**K**	
Thematic map	1	n_{11}	n_{12}	n_{1k}	n_{1+}
classes	2	n_{21}	n_{22}	n_{2k}	n_{2+}
	k	n_{k1}	n_{k2}	n_{k3}	n_{k+}
Column total (n_{+j})		n_{+1}	n_{+2}	n_{+3}	n

reference data. A KHAT value can be computed based on error matrix using the following mathematical equation:

$$\hat{K} = \frac{n \sum_{i=1}^{k} n_{ii} - \sum_{i=1}^{k} n_{i+} \, n_{+i}}{n^2 - \sum_{i=1}^{k} n_{i+} \, n_{+i}}$$

(3.4)

For the example in Table 3.1, the KHAT value is calculated as follows:

$$\text{KHAT} = \frac{(275 \times (90 + 81 + 75)) - (96 \times 101 + 98 \times 90 + 81 \times 84))}{(275 \times 275 - (96 \times 101 + 98 \times 90 + 81 \times 4))} = 0.84$$

The Kappa analysis provides a means for testing if the remote sensing classification is significantly better than a random classification (Congalton and Green 2009). Professional image-processing software generally provides the Kappa analysis in its accuracy assessment package. A test can also be performed to check if two independent KHAT values, and therefore two error matrices, are significantly different. However, some researchers and scientists have objected to use of the Kappa coefficient for assessing the accuracy of remotely sensed classifications because the degree of chance agreement may be overestimated and therefore, the Kappa coefficient may underestimate the classification agreement (Foody 1992). The Kappa analysis is appropriate when all the errors in the matrix can be considered of equal importance. A weighted Kappa analysis can be conducted and is powerful if appropriate weights can be selected for all the errors (Congalton and Green 2009).

In addition to the Kappa analysis, a second technique called Margfit can be applied to "normalize" or standardize the error matrices for comparison purposes (Congalton and Green 2009). Margfit uses an iterative proportional fitting procedure that forces each row and column (i.e., marginal) in the matrix to sum to a predetermined value, hence the name Margfit (marginal fitting). Therefore, the normalization process provides a convenient way of comparing individual cell values between error matrices regardless of the number of samples used to derive the matrix. It is noted that some researchers and scientists have objected to the use of the Margfit to normalize or standardize the error matrices. The main concern is that normalizing an error matrix can lead to large bias in the accuracy estimates (Stehman 2004). Table 3.3 shows a normalized error matrix for the example in Table 3.1.

$$\text{Normalized accuracy} = \frac{(0.8911 + 0.9000 + 0.8929)}{3} = 89\%$$

Normalized accuracy directly includes the off-diagonal elements (omission and commission errors) because of the iterative proportional fitting procedure. KHAT accuracy

TABLE 3.3

Normalized Error Matrix from Table 3.1

		Reference Data		
		A	**B**	**C**
Thematic map classes	A	0.8911	0.0444	0.0238
	B	0.0990	0.9000	0.0833
	C	0.0099	0.0556	0.8929
Column total		1	1	1

indirectly incorporates the off-diagonal elements as a product of the row and column marginals. Overall accuracy only incorporates the major diagonal and excludes omission and commission errors. These three measures may not agree with each other because each measure incorporates different information about the error matrix. In each accuracy assessment, the original error matrix, the derived overall accuracy, the producer's and user's accuracies, and the Kappa coefficient are typically recommended to be present in an accuracy assessment report because they are easily computed and understood, and also capture the essential quality of the map product (Congalton and Green 2009). In addition to that, confidence intervals are expected for any statistical estimate.

3.3 Essential Components and Issues of Accuracy Assessment

Since the error matrix is at the core of the accuracy assessment, it is imperative to know how the error matrix is generated and populated and to understand the related issues. The three essential components of an accuracy assessment are sampling design, response design, and analysis (Stehman and Czaplewski 1998).

Sampling design will determine both the cost and the statistical rigor of the assessment and is one of the most challenging and important components of accuracy assessment (Wulder et al. 2006). Sampling design requires knowledge of the distribution of thematic classes across the landscape, determination of sampling unit and numbers to be taken, and choice of a sampling scheme for selecting samples (Congalton and Green 2009). Sample units are the portions of the map that are selected for accuracy assessment. There are four possible choices for the sampling unit: (1) a single pixel, (2) a cluster of pixels, (3) a polygon (object), and (4) a cluster of polygons. The number of samples per map class must be sufficiently large to be statistically representative of the map area. The appropriate sample size can and should be computed for each project using the multinomial distribution. Many researchers have published equations and guidelines for choosing the appropriate sample size (Congalton 1988; Stehman 2001; Foody 2008). Congalton (1988) presented a general guideline that suggests a minimum of 50 samples for each map class should be collected for maps of less than 1 million acres in size and fewer than 12 classes. Stehman (2001) stated that a sample size of 100 per land cover class would ensure that accuracy can be estimated with a standard error of no greater than 0.05. Foody (2008) included several equations that could be used to determine sample sizes for common applications in remote sensing, using both independent and related samples. However, the sampling number varies depending on the method used to estimate. Five common sampling schemes have been applied for collecting the reference data: (1) simple random sampling; (2) systematic sampling; (3) stratified random sampling; (4) cluster sampling; and (5) stratified, systematic, unaligned sampling. Congalton (1988) performed sampling simulations on three spatially diverse areas using all five of these sampling schemes and concluded that in all cases, simple random and stratified random samplings provided satisfactory results.

The response design is the protocol for determining the reference classification recorded at each sampling unit (Stehman 2001). There are two distinct components in a response design: the evaluation protocol, which is the procedure used to collect the reference information, and the labeling protocol, which specifies the land cover label that will be assigned to the sampling unit. Congalton (2001) suggested that the reference data must be classified using the same classification scheme as the map data to reduce confusion. However, a scenario where the reference data have a more detailed classification schema than the map data may be advantageous for a statistical estimation (Czaplewski 2003). A classification scheme should be mutually exclusive and totally exhaustive. Mutual exclusivity requires that each mapped area falls into one and only one category or class. A totally exhaustive classification scheme results in every area on the mapped landscape receiving a map label; no area can be left unlabeled.

During the accuracy assessment analysis, several issues may be kept in mind. Foody (2002) summarized them as follows: First, the measurement and meaning of classification accuracy depend considerably on one's particular viewpoint and requirements. Second, a number of fundamental assumptions are typically made in the derivation of indices from the error matrix. For example, it is generally assumed implicitly that each class (e.g., pixel) to be classified belongs fully to one of the classes in an exhaustively defined set of discrete and mutually exclusive classes (Congalton et al. 1998; Townsend 2000; Congalton and Green 2009). However, there is a mixed-pixel problem that will get more serious when the scale of thematic maps gets coarser. In some cases, soft or fuzzy classification may be more appropriate than traditional hard classification. Gómez et al. (2008) extended standard accuracy measures (e.g., overall, producer's, user's, or Kappa statistic) for soft classification. Their method can be applied to accuracy assessment of classification maps in any framework: errors with different importance, soft classifier and crisp reference data (expert), or with a fuzzy expert. Pontius and Cheuk (2006) and Pontius and Connors (2009) formulated the cross-tabulation matrix for accuracy assessment of soft or fuzzy classification. This approach can be also used to quantify and compare the accuracy of soft classification maps at multiple resolutions.

Third, a variety of errors are encountered in an image classification. Nonthematic errors may result in misrepresentation, typically underestimation, of the actual accuracy. For example, positional uncertainty can have a major detrimental effect on thematic mapping studies. Pontius (2000, 2002) advanced the assessment by dividing and quantifying positional and thematic errors, respectively. Fourth, the reference data often contain errors. The accuracy of the ground data is rarely known and the level of effort needed to collect the appropriate data is not clearly understood. The methods and protocols of the reference data acquisition may influence their accuracy and suitability for relation to the thematic map to assess its accuracy (Czaplewski 1992; Scepan 1999; Yang et al. 2000; Powell et al. 2004). The size of support of sampling units used in ground data collection is often different from that of the units mapped from the imagery, leading to difficulties in analyzing the datasets (Atkinson et al. 2000; Yang et al. 2000; Wulder et al. 2006; Riemann et al. 2010). Sampling is often consciously constrained to large homogeneous regions of the classes, and regions in and around the vicinity of complexities such as boundaries are excluded. As a result of this type of strategy, the accuracy statement derived may be optimistically biased.

Fifth, the confusion matrix and the accuracy metrics derived from it provide no information on the spatial distribution of error. Thematic map accuracy may vary across a landscape related to terrain, landscape complexity, land cover types, and land use patterns (Steele et al. 1998). Incomplete information about the pattern and variability of map accuracy may lead to negative outcomes and reluctance on the part of decision makers to apply information derived from remote sensing and geographic information system analysis. Spatially identifying the sources of uncertainties, modeling their accumulation and propagation, and finally, quantifying them will be critical to control the quality of spatial data (Wang et al. 2005). Confidence-based quality assessment may provide valuable and complementary information to the conventional error matrix method because it provides an estimate of the map quality at pixel level and does not require additional reference data (Strahler et al. 2006). Sixth, equal-weighted error is usually assumed in error matrix accuracy assessment.

3.4 Special Considerations for Accuracy Assessment

3.4.1 Large-Area Land Cover

Large-area land cover products are increasingly prevalent because of increasing demand for regional and global environment monitoring and assessment; however, standard operational protocols for their validation do not exist (Justice et al. 2000). The issues that need to be addressed for large-area land cover products assessment include logistically feasible and statistically valid sampling strategies, the capability to assess the accuracy of the reference data, and stable and informative metrics of accuracy (Franklin and Wulder 2002). Wulder et al. (2006) developed an accuracy assessment framework for large-area land cover classification products derived from medium-resolution satellite data and showed the framework with an example of the land cover map of the forested area of Canada produced by the Earth Observation for Sustainable Development program. First, they defined their objective of the accuracy assessment, then determined an appropriate sampling frame and unit, sample size, and design, and then selected an evaluation protocol and labeling protocol. The example showed the compromise between the theoretical aspects of accuracy assessment and the practical realities of implementation over a specific jurisdiction. Stehman et al. (2003) presented an accuracy assessment protocol for the 1992 National Land Cover Database for the eastern United States. They developed a probability sampling design incorporating three levels of stratification and two stages of selection. Agreement between the map and the reference land cover labels is defined as a match between the primary or alternate reference label determined for a sample pixel and a mode class of the mapped 3 × 3 block of pixels centered on the sample pixel. In their report, the estimated error matrices, user's and producer's accuracies, and standard errors are provided to document the regional, class-specific accuracy for Anderson Levels I and II. Scepan (1999) first validated a 1 km global land cover database developed from Advanced Very High Resolution Radiometer data using a stratified random sample with the interpretation of high-resolution satellite imagery (i.e., Landsat and SPOT images) to determine reference land cover.

Sometimes, cross-validation was conducted to obtain information on the quality of large land cover products when independent validation data are not available. To conduct an N-fold cross-validation, the testing data will be divided into N equal-sized subsets. A cross-validation accuracy estimate can be derived by using each subset to evaluate the classification results developed using the remaining testing data. The process can be repeated N times. The average and standard deviations of the N estimates provide an accuracy estimate of the classification. It should be noted that the validity of any accuracy estimate depends primarily on whether the testing data are independent from the training data used for mapping and if the data are collected based on a probability sampling.

3.4.2 Change Detection Maps

Change detection maps involve individual from/to class scenarios. Each dimension of the change detection standard error matrix is the square of the number of classes involved. This new matrix has the same characteristics of the single-date classification error matrix, but it assesses errors in changes between two periods and not simply a single classification. Obtaining the sample of data to use in the construction of the change detection confusion matrix can be difficult. Perhaps a more significant problem, however, is that these approaches are appropriate only for use with conventional hard classifications. This, therefore, limits the change detection indicating where a conversion of land cover appears to have occurred.

A two-step approach is suggested when it is not possible to use the change detection error matrix approach (Congalton and Green 2009). The first step is to assess the accuracy of just the areas that changed between the two periods in question; in other words, conduct a single-date accuracy assessment only on the areas that changed between time 1 and time 2. The accuracy assessment only needs to be conducted for the changed areas for time 2 because the rest of the map has the same accuracy as the map in time 1 for all the areas that did not change. The second step is simply a change/no-change validation. The binomial distribution can be used to calculate the sample size. Computing the sample size for the binomial approach requires the use of a lookup table that presents the required sample size for a given minimum error and a desired level of confidence. In general, the use of hard classifications within a postclassification comparison-based approach would be expected to underestimate the area of land undergoing a change and, where a change is detected, to overestimate the magnitude of change as it is a simple binary technique (e.g., Foody 2001). The second step can also aggregate change types into few strata and calculate the sample size accordingly. Li and Zhou (2009) developed a practical methodology to assess the accuracy of multitemporal change detection using a trajectory error matrix. In this error matrix, one axis represents the land cover change trajectory categories derived from single-date classified images, and the other represents the land cover change trajectories identified from the reference data. The overall accuracies of change trajectories and states of change/no-change are used as indices for accuracy assessment. They practically aggregate all possible change trajectories into six subgroups. Their method provides a realistic and detailed assessment of the results of multitemporal change detection using postclassification comparison-based methods.

3.4.3 Maps of Continuous Variables

Geospatial modeling with remotely sensed data is being used to produce geospatial datasets of continuous variables, such as proportion of forest cover, biomass, leaf area index, forest composition variables, and impervious surface (e.g., Ohmann and Gregory 2002; Yang et al. 2003; Cohen and Goward 2004; Xian and Homer 2010). Many accuracy assessments of continuous geospatial datasets are frequently provided with only a single measure of agreement (e.g., a coefficient of determination [r^2] or root-mean-square error [RMSE] value). However, r^2 merely indicates the linear covariation between two datasets rather than the actual difference; RMSE is a dimensional measure of disagreement, so it is not independent of data scale and unit (Ji and Gallo 2006). To circumvent the problems associated with r, r^2, and RMSE, Willmott (1981, 1982) developed the index of agreement, which is used especially for validating prediction models. The index of agreement (d) is expressed as follows:

$$d = 1 - \frac{\sum_{i=1}^{n}(X_i - Y_i)^2}{\sum_{i=1}^{n}\left(\left|X_i - \bar{X}\right| + \left|Y_i - \bar{X}\right|\right)^2} \tag{3.5}$$

where X_i is the observed value, Y_i is the modeled or simulated value, and $\bar{X}$ is the mean of observed values. Willmott (1981, 1982) also defined two error indices: the systematic and unsystematic MSE. Mielke (1984, 1991) developed a measure of agreement (ρ) to validate prediction models as well.

$$\rho = 1 - \frac{1/n\sum_{i=1}^{n}(X_i - Y_i)^2}{1/n^2\sum_{i=1}^{n}\sum_{j=1}^{n}(X_i - Y_j)^2} \tag{3.6}$$

Ji and Gallo (2006) summarized the advantages and disadvantages of the measures of agreement as mentioned earlier in a table. They too developed an agreement coefficient (AC) that provides a symmetric, bounded, and nondimensional measure of agreement. AC can quantify actual difference between different datasets and separate systematic and unsystematic errors. AC between two datasets X and Y is defined as

$$AC = 1 - \frac{\sum_{i=1}^{n}(X_i - Y_i)^2}{\sum_{i=1}^{n}\left(\left(|\bar{X} - \bar{Y}| + |X_i - \bar{X}|\right)\left(|\bar{X} - \bar{Y}| + |Y_i - \bar{Y}|\right)\right)} \tag{3.7}$$

where $\bar{X}$ and $\bar{Y}$ are the mean values of X and Y, respectively. The numerator of the main term of Equation 3.7 is the sum of square difference (SSD) of X and Y, which indicates the degree of disagreement between X and Y. The denominator is the sum of potential difference (SPOD) used to standardize SSD. The AC equation can also be expressed in a simplified form:

$$AC = 1 - \frac{SSD}{SPOD} \tag{3.8}$$

They also developed the systematic agreement coefficient (AC_s) and unsystematic agreement coefficient (AC_u) to measure the systematic agreement and unsystematic agreement, respectively. AC_s and AC_u are calculated as follows:

$$AC_s = 1 - \frac{SPD_s}{SPOD} \tag{3.9}$$

$$AC_u = 1 - \frac{SPD_u}{SPOD} \tag{3.10}$$

$$SPD_u = \sum_{i=1}^{n}\left(|X_i - X_i| + |Y_i - Y_i|\right) \tag{3.11}$$

$$SPD_s = SSD - SPD_u \tag{3.12}$$

Ji and Gallo (2006) recommended that AC, AC_s, AC_u, and some associated parameters (e.g., root-mean-square difference, which is known as the sum of RMSE) should be used to indicate data agreement or disagreement.

The challenges to effectively assess the accuracy of geospatial datasets of modeled continuous variables are primarily from the error and uncertainty in the reference data and mismatches in spatial support between the reference data and the map data. Riemann et al. (2010) developed an effective assessment protocol for continuous geospatial datasets of forest characteristics using the U.S. Forest Service Forest Inventory and Analysis data. They stated several characteristics of effective assessments. The first one includes multiple types of assessment for continuous geospatial datasets. They summarized a suite of statistics (such as r^2, RMSE, AC, AC_s, and AC_u) at four different scales for their study example. The second one describes characteristics relevant to how the dataset will be used. The third one describes the location of errors because error varies across the landscape or by subpopulations. The fourth one is assessing across a range of scales. The fifth one is timely and consistent application of assessment. The assessments and measures they proposed provide information on the location, magnitude, frequency, and type of error in geospatial datasets of continuous variables. Their assessment framework should help researchers and managers improve and more effectively use modeled map products.

References

Atkinson, P. M., G. M. Foody, P. J. Curran, and D. S. Boyd. 2000. "Assessing the Ground Data Requirements for Regional-Scale Remote Sensing of Tropical Forest Biophysical Properties." *International Journal of Remote Sensing* 21:2571–87.

Bolstad, P. 2005. *GIS Fundamentals*. 2nd ed. White Bear Lake, MN: Eider Press.

Cohen,W., and S. Goward. 2004. "Landsat's Role in Ecological Applications of Remote Sensing." *BioScience* 54:535–45.

Congalton, R. G. 1988. "Using Spatial Autocorrelation Analysis to Explore the Errors in Maps Generated from Remotely Sensed Data." *Photogrammetric Engineering and Remote Sensing* 54:587–92.

Congalton, R. G. 1991. "A Review of Assessing the Accuracy of Classifications of Remotely Sensed Data." *Remote Sensing of Environment* 37:35–46.

Congalton, R. G. 1994. "Accuracy Assessment of Remotely Sensed Data: Future Needs and Directions." In *Proceedings of Pecora 12 Land Information from Space-Based Systems*, 383–8. Bethesda, MD: ASPRS.

Congalton, R. G. 2001. "Accuracy Assessment and Validation of Remotely Sensed and Other Spatial Information." *International Journal of Wildland Fire* 10:321–28.

Congalton, R. G., M. Balogh, C. Bell, K. Green, J. A. Milliken, and R. Ottman. 1998. "Mapping and Monitoring Agricultural Crops and Other Land Cover in the Lower Colorado River Basin." *Photogrammetric Engineering and Remote Sensing* 64:1107–13.

Congalton, R. G. and K. Green. 2009. *Assessing the Accuracy of Remotely Sensed Data: Principles and Practices*. 2nd ed. Boca Raton, FL: CRC Press.

Czaplewski, R. L. 1992. "Misclassification Bias in Areal Estimates." *Photogrammetric Engineering and Remote Sensing* 58:189–92.

Czaplewski, R. L. 2003. "Statistical Design and Methodological Considerations for the Accuracy Assessment of Maps of Forest Condition." In *Remote Sensing of Forest Environments: Concepts and Case Studies*, edited by M. A. Wulder and S.E. Franklin, 115–41. Boston, MA: Kluwer Academic.

Foley, J. A., R. DeFries, G. P. Asner, C. Barford, G. Bonan, S. R. Carpenter, et al. 2005. "Global Consequences of Land Use." *Science* 309:570–4.

Foody, G. M. 1992. "On the Compensation for Chance Agreement in Image Classification Accuracy Assessment." *Photogrammetric Engineering and Remote Sensing* 58:1459–60.

Foody, G. M. 2001. "Monitoring the Magnitude of Land-Cover Change Around the Southern Limits of the Sahara." *Photogrammetric Engineering and Remote Sensing* 67:841–7.

Foody, G. M. 2002. "Status of Land Cover Classification Accuracy Assessment." *Remote Sensing of Environment* 80:185–201.

Foody, G. M. 2008. "Harshness in Image Classification Accuracy Assessment." *International Journal of Remote Sensing* 29:3137–58.

Franklin, S. E., and M. A. Wulder. 2002. "Remote Sensing Methods in Large Area Land Cover Classification Using Satellite Data." *Progress in Physical Geography* 26:173–205.

Gómez, D., G. Biging, and J. Montero. 2008. "Accuracy Statistics for Judging Soft Classification." *International Journal of Remote Sensing* 29(3):693–709.

Janssen, L. L. F., and F. J. M. van der Wel. 1994. "Accuracy Assessment of Satellite Derived Land-Cover Data: A Review." *Photogrammetric Engineering and Remote Sensing* 60:419–26.

Ji, L., and K. Gallo. 2006. "An Agreement Coefficient for Image Comparison." *Photogrammetric Engineering and Remote Sensing* 72:823–33.

Justice, C., A. Belward, J. Morisette, P. Lewis, J. Privette, and F. Baret. 2000. "Developments in the 'Validation' of Satellite Sensor Products for the Study of the Land Surface." *International Journal of Remote Sensing* 21:3383–90.

Li, B., and Q. Zhou. 2009. "Accuracy Assessment on Multi-Temporal Land-Cover Change Detection Using a Trajectory Error Matrix." *International Journal of Remote Sensing* 30:1283–96.

Matthews, H. D., A. J. Weaver, K. J. Meissner, N. P. Gillett, and M. Eby. 2004. "Natural and Anthropogenic Climate Change: Incorporating Historical Land Cover Change, Vegetation Dynamics and the Global Carbon Cycle." *Climate Dynamics* 22:461–79.

Mielke, P. W., Jr. 1984. "Meteorological Applications of Permutation Techniques Based on Distance Functions." In *Handbook of Statistics*, edited by P. R. Krishnaiah and P. K. Sen, Vol. 4, 813–30. New York: Elsevier.

Mielke, P. W., Jr. 1991. "The Application of Multivariate Permutation Methods Based on Distance Functions in the Earth Science." *Earth-Science Review* 31:55–71.

Ohmann, J. L., and M. J. Gregory. 2002. "Predictive Mapping of Forest Composition and Structure with Direct Gradient Analysis and Nearest-Neighbor Imputation in Coastal Oregon, U.S.A." *Canadian Journal of Forest Research* 32:725–41.

Pontius, R. G., Jr. 2000. "Quantification Error Versus Location Error in Comparison of Categorical Maps." *Photogrammetric Engineering & Remote Sensing* 66(8):1011–6.

Pontius, R. G., Jr. 2002. "Statistical Methods to Partition Effects of Quantity and Location during Comparison of Categorical Maps at Multiple Resolutions." *Photogrammetric Engineering and Remote Sensing* 68(10):1041–9.

Pontius, R. G., Jr. and M. L. Cheuk. 2006. "A Generalized Cross-Tabulation Matrix to Compare Soft-Classified Maps at Multiple Resolutions." *International Journal of Geographical Information Science* 20(1):1–30.

Pontius, R. G., Jr. and J. Connors. 2009. "Range of Categorical Associations for Comparison of Maps with Mixed Pixels." *Photogrammetric Engineering & Remote Sensing* 75(8):963–9.

Powell, R. L., N. Matzke, C. de Souza Jr., M. Clark, I. Numata, L. L. Hess, and D. A. Roberts. 2004. "Sources of Error in Accuracy Assessment of Thematic Land-Cover Maps in the Brazilian Amazon." *Remote Sensing of Environment* 90:221–34.

Riemann, R., B. T. Wilson, A. Lister, and S. Parks. 2010. "An Effective Assessment Protocol for Continuous Geospatial Datasets of Forest Characteristics Using USFS Forest Inventory and Analysis (FIA) Data." *Remote Sensing of Environment* 114:2337–52.

Scepan, J. 1999. "Thematic Validation of High-Resolution Global Land-Cover Data Sets." *Photogrammetric Engineering and Remote Sensing* 65:1051–60.

Steele, B. M., J. C. Winne, and R. L. Redmond. 1998. "Estimation and Mapping of Misclassification Probabilities for Thematic Land Cover Maps." *Remote Sensing of Environment* 66: 192–202.

Stehman, S. V. 2001. "Statistical Rigor and Practical Utility in Thematic Map Accuracy Assessment." *Photogrammetric Engineering and Remote Sensing* 67:727–34.

Stehman, S. V. 2004. "A Critical Evaluation of the Normalized Error Matrix in Map Accuracy Assessment." *Photogrammetric Engineering and Remote Sensing* 70: 743–51.

Stehman, S. V., and R. L. Czaplewski. 1998. "Design and Analysis for Thematic Map Accuracy Assessment: Fundamental Principles." *Remote Sensing of Environment* 64: 331–44.

Stehman, S. V., J. D. Wickham, J. H. Smith, and L. Yang. 2003. "Thematic Accuracy of the 1992 National Land-Cover Data for the Eastern United States: Statistical Methodology and Regional Results." *Remote Sensing of Environment* 86: 500–16.

Story, M., and R. G. Congalton. 1986. "Accuracy Assessment: A User's Perspective." *Photogrammetric Engineering and Remote Sensing* 52: 397–9.

Strahler, A. H., L. Boschetti, G. M. Foody, M. A. Friedl, M. C. Hansen, M. Herold, P. Mayaux, Morisette J. T., S. V. Stehman, and C. E. Woodcock. 2006. *Global Land Cover Validation: Recommendations for Evaluation and Accuracy Assessment of Global Land Cover Maps*, EUR 22156 EN. Luxemburg: Office for Official Publications of the European Communities.

Townsend, P. A. 2000. "A Quantitative Fuzzy Approach to Assess Mapped Vegetation Classification for Ecological Applications." *Remote Sensing of Environment* 72: 253–67.

Turner, B. L., II, E. Lambin, and A. Reenberg. 2007. "The Emergence of Land Change Science for Global Environmental Change and Sustainability." *Proceedings of the National Academy of Sciences of the United States of America* 104: 20666–71.

Wang, G., G. Z. Gertner, S. Fang, and A. B. Anderson. 2005. "A Methodology for Spatial Uncertainty Analysis of Remote Sensing Products." *Photogrammetric Engineering and Remote Sensing* 71: 1423–32.

Willmott, C. J. 1981. "On the Validation of Models." *Physical Geography* 2: 184–94.

Willmott, C. J. 1982. "Some Comments on the Evaluation of Model Performance." *Bulletin American Meteorology Society* 63: 1309–13.

Wulder, M. A., S. E. Franklin, J. C. White, J. Linke, and S. Magnussen. 2006. "An Accuracy Assessment Framework for Large-Area Land Cover Classification Products Derived from Medium-Resolution Satellite Data." *International Journal of Remote Sensing* 27: 663–83.

Xian, G., and C. Homer. 2010. "Updating the 2001 National Land Cover Database Impervious Surface Products to 2006 Using Landsat Imagery Change Detection Methods." *Remote Sensing of Environment* 114: 1676–86.

Yang, L. M., G. Xian, J. M. Klaver, and B. Deal. 2003. "Urban land-cover change detection through sub-pixel imperviousness mapping using remotely sensed data." *Photogrammetric Engineering and Remote Sensing* 69:1003–10.

Yang, L., S. V. Stehman, J. D. Wickham, J. H. Smith, and N. J. Van Driel. 2000. "Thematic Validation of Land Cover Data of the Eastern United States Using Aerial Photography: Feasibility and Challenges." *Proceedings of the 4th International Symposium on Spatial Accuracy Assessment in Natural Resources and Environmental Sciences*, 747–54. The Netherlands: Delft University Press.

4

Accuracy Assessment for Soft Classification Maps

Daniel Gómez, Gregory S. Biging, and Javier Montero

CONTENTS

4.1   Introduction 57
4.2   Accuracy Assessment Types 59
      4.2.1   Accuracy Based on Matrix Error 59
      4.2.2   Choice of Sampling Unit for Accuracy Assessment 60
      4.2.3   Object-Based Classification 60
      4.2.4   Change Detection 60
      4.2.5   Criticisms of Kappa 61
      4.2.6   Other Measures of Accuracy 61
4.3   Pixel Accuracy Assessment Measures for Soft Classification Problems 62
      4.3.1   Definition 3.1: Overall Accuracy of Binaghi et al. (1999) 64
      4.3.2   Definition 3.2: Overall Accuracy of Gómez et al. (2008) 65
      4.3.3   Definition 3.3: Gómez et al. (2008) 66
4.4   Accuracy Assessment Example 66
4.5   Real Application 68
      4.5.1   Image Description 69
      4.5.2   Classification Maps and Reference Data 73
      4.5.3   Accuracy Measures 73
4.6   Final Conclusions and Remarks 82
Acknowledgments 84
References 84

4.1 Introduction

An important topic in using maps derived from a statistical classifier is the accuracy assessment of the classification. Analysts usually need to compare various techniques, algorithms, or different approaches. As pointed out by Stehman and Czaplewski (1998), the accuracy assessment of classification maps generally involves three different steps: the sampling design, the response or measurement design to obtain the true classes for each sampling (usually requiring an expert), and the analysis of the data obtained.

In the sampling design step, the objects of the study (subregions [pixels] in the pixel classification, objects in object-oriented problems, or areas in change detection problems) are the sampling units for conducting the accuracy assessment. In this chapter, we focus on pixels as the basic sampling unit, but more complex situations could be considered (see Section 4.2 for a general view of the different methods of accuracy assessment). Choosing a sampling design is a difficult step and an open problem in classification maps. One can find many papers in the literature (see, e.g., Congalton and Green 2008) that deal with this problem.

In the second step, the "true" or reference values for each sampling unit are determined. Moreover, this step is very complex and there are several possible approaches. Most of these approaches require the final decision of an expert to match each sample to a well-known class. Although most of the approaches force the expert to be crisp in his or her classification, recent research (Binaghi et al. 1999; Woodcock and Gopal 2000; Laba et al. 2002; Driese et al. 2004; Gómez et al. 2008) necessarily considers a generalization of this idea into a soft framework.

Finally, the last step of conducting the accuracy assessment compares the results obtained by the classifier with the reference dataset.

Nevertheless, classical classification models typically assume the existence of underlying crisp classes that decision makers are willing to fit, either because those classes have been already precisely defined or because we can assume that those classes exist and that with experience they can be extensively determined. That is, we can utilize a family of classes allowing a crisp partition of all the objects under study, in such a way that each object can be associated to one and only one class. Under such a classical approach, "accuracy" estimates the probability of misclassification. Misclassification occurs when we assign a class that, due to the natural uncertainty of our observational system, becomes a wrong assignment. This approach might be considered appropriate for certain kinds of problems, but certainly it is unrealistic when dealing with the analysis of the natural environment like land cover classes. In fact, most of the concepts we consider in analyzing land cover do not allow a crisp definition necessitating a clear boundary between them. A forest has gradations, so we find it difficult to place an exact boundary between two ecotones. But human knowledge representation has the ability to manage these complex concepts that are naturally subject to graduation from extreme situations that rarely appear in a pure stage. Under these circumstances, classical accuracy measures fail in trying to identify the class that each object belongs to, since objects belong *to some extent* to several classes. As pointed out in Amo et al. (2000), a family of classes defining a crisp partition might simply not exist. And if such a family of classes exists, they might be too complex to be practical. Even in the case that a manageable family of classes exists, they can most probably be reached only after a long learning experience.

This work focuses on the accuracy measurement when dealing with soft classification systems where each object can simultaneously verify *to some extent* several fuzzy properties, to be distinguished from crisp properties that either hold or not. It is interesting to point out that the only feasible uncertainty for such crisp properties is the probabilistic one: with the available information, we do not know which class is correct, but we are sure that only one of the classes is considered correct. In a soft classification, we allow for intermediate conditions that occur in the gradations between the classes. In this sense, fuzzy approaches (see Zadeh 1965) look for models that are closer to the standard uncertainty in human knowledge, not to be confused with the uncertainty in human decision making, as pointed out in Montero et al. (2007). Even though our knowledge is most often fuzzy, in the end we have to choose an action that is always crisp (e.g., choosing a class for a certain object). Crispness (and probability) belongs to a decision-making framework, but we should not force reality into crispness just because we need a crisp decision. Knowledge should first fit reality, and whenever we require fuzzy concepts to understand reality, we should be looking for fuzzy models. The key issue we address in this work is how the probabilistic accuracy measures associated with crisp classification systems can be translated to fuzzy classification systems to evaluate their quality.

This chapter is organized as follows: In Section 4.2, we give a short description of different types of accuracy assessments. In Section 4.3, we describe some of the main techniques that deal with the pixel accuracy assessment for soft classification problems. In Section 4.4, we present an accuracy assessment example. In Section 4.5, we present a real case in which we analyze the importance of a correct soft assessment. Finally, some conclusions and remarks are presented in Section 4.6.

4.2 Accuracy Assessment Types

In this section, we discuss various issues related to accuracy assessment. These topics include measures of accuracy, criticisms of Kappa, the choice of sampling unit for accuracy assessment, accuracy assessment of object-based classification, and change detection accuracy assessment.

4.2.1 Accuracy Based on Matrix Error

The most widely used approaches for image classification accuracy assessment are site-specific methods based on the analysis of the entries in a confusion or error matrix (Congalton and Green 1999; Foody 2002). The confusion or error matrix represents the differences between the reference data and the classifier taking into account the classes that are the target of the classification. In binary accuracy assessment, the error matrix associated with a pair (i, j) represents the estimation of the number of pixels that have been classified into the class i, but in the reference data are assigned into class j. Obviously, if $i = j$, the expert and classifier coincide in their opinion and the rest of the matrix represents errors in the classification. As pointed out by Foody (2002), deciding how to build a confusion matrix and interpret its contents is sometimes one of the hardest parts of the accuracy assessment process. And this difficulty is often compounded by the use of inappropriate measures to quantify classification accuracy.

Most of the relevant studies use basic aggregation measures of accuracy as the proportion of correctly allocated cases. This measure is obtained as a function of the sample design used in acquiring the testing set (Stehman 1995). Thus, the estimates of classification accuracy derived from confusion matrices are constructed from testing sets drawn by simple random and stratified random sampling from the same map, without any allowance for the difference in the sample design, which may differ substantially if the classes vary in abundance and spectral separability.

Other measures, such as the Kappa coefficient of agreement in the assessment of classification accuracy (Congalton and Mead 1983; Congalton and Green 1999; Smits et al. 1999; Wilkinson 2005), have gained use, but its real utility is being debated in the remote sensing community. The arguments made for the adoption of the Kappa coefficient are typically based on reasons such as its calculation corrects for chance agreement and utilizes the entire confusion matrix as well as that a variance term can be calculated for it. Despite its limitations and some strong criticisms (see, e.g., Pontius and Millones 2011), the use of the Kappa coefficient and related approaches over the past 20 years has encouraged an increasingly rigorous and quantitative evaluation of classification accuracy, which should be regarded as a useful, if somewhat incorrect, step in the direction toward an appropriate evaluation method.

Another important problem, which is the topic of this chapter, is the fact that the confusion matrix is based implicitly on the assumption that the pixels are pure, and the ground dataset is perfectly colocated with the image classification. Both these assumptions are rarely satisfied. The proportion of mixed pixels in an image is a function of the spatial resolution of the imagery and the land cover mosaic, but it is often substantial. In addition to this, the classification output could be obtained in a soft way since fuzzy classifiers offer a more realistic characterization of a pixel. For all these reasons, it is necessary to develop a different methodology and accuracy measures that permit us to measure the accuracy when the information given by the reference data or by the classifier is soft.

4.2.2 Choice of Sampling Unit for Accuracy Assessment

There is no agreement in the literature on the sampling unit to use for accuracy assessment. For example, Stehman and Czaplewski (1998) list sampling units used in over 30 selected accuracy assessment projects. The sample units included points, pixels, pixel blocks, and mapped polygons. Any of these sampling unit types are admissible as sampling units and can be used in accuracy assessment, although there may be locational inaccuracies particularly for points and pixels. The choice of the sampling unit for accuracy assessment need not be associated with the way the map is constructed. For example, if the classified map uses polygons to represent classes, the sampling units need not be polygons; likewise, if the map is classified by individual pixels, the sampling units need not be pixels. The investigator has much leeway in deciding what the sampling unit will be in a probability sampling design and she or he can take into consideration factors such as cost and locational accuracy of sampling units. For simplicity, in this chapter, we assume that pixels are the sampling units for accuracy assessment and that they are correctly located in the image.

4.2.3 Object-Based Classification

Recently, there has been interest in object-based classification for land use land cover (LULC) classification (see Yu et al. 2006; Ke et al. 2010; Zhang et al. 2010). In this scenario, objects, which may be of varying size, are the map units (Clinton et al. 2010). Accuracy assessment of segmented objects involves comparison of the object-based image segmentation with training objects. Persello and Bruzzone (2009) proposed a set of geometric indices that characterize geometric errors in the classification map. Clinton et al. (2010) measure the shape of the extracted and reference objects and determine the distance between boundary pixels in the segments and reference objects. These methods are not true measures of accuracy, but are related to them. Because work on object-based image processing and accuracy assessment is inchoate, we do not include it in this chapter.

4.2.4 Change Detection

Change detection is a special case in remote sensing because it involves detection of objects, pixels, and groups of pixels or polygons that have changed over time. There are numerous methods for detecting change (e.g., Khorram 1999; Chen et al. 2003; Gong and Xu 2003; Liu et al. 2008; Wang et al. 2009). Regardless of the method for detecting change, sampling to determine the accuracy of a change map is inherently different from sampling for accuracy of a one-point-in-time thematic map. This is because the change categories usually represent a very small fraction of the change image. For example, in Dobson and Bright's (1993) study of the wetlands, submerged vegetated habitat and adjacent upland cover of the

Chesapeake Bay, around 3% of the area had experienced change over a 5-year period. In this case, the change pixels or polygons can be considered rare classes. To capitalize on this situation, Biging et al. (1999) proposed using disproportionate stratified sampling that increases the sampling fraction in the stratum in which the rare population is concentrated. Fuzzy accuracy assessment can be applied to this problem even though there are k LULC classes, in which case there would be k^2 classes in the change error matrix. Applying fuzzy accuracy assessment to this problem would likely be cost prohibitive. For example, Laba et al. (2002) found that field observations for fuzzy accuracy assessment were more complex and time-consuming than those made with conventional accuracy assessment. We do not include accuracy assessment of change detection in this chapter, although the methods needed parallel those of a one-point-in-time accuracy assessment as presented in this chapter.

4.2.5 Criticisms of Kappa

Kappa is widely used in remote sensing studies and has been recommended by Rosenfield and Fitzpatrick-Lins (1986), Congalton and Mead (1983), and Congalton (1991), among many others. Conversely, Kappa has also been extensively criticized. Some authors (see, e.g., Stehman 1997; Pontius 2000; Foody 2002; Liu et al. 2007) point out that the Kappa coefficient is not the only way to compensate for chance agreement or to test the significance of differences in accuracy among classifiers. Recent extensions of Kappa based on the error matrix are the Kappa location (Pontius 2000) and the Kappa histogram (Hagen 2002). These two statistics are sensitive to respective differences in location and in the histogram shape of all the categories. However, Pontius and Millones (2011) now believe that Kappa indices including the Kappa location and Kappa histogram are invalid. They posit some reasons for this, one of them is that the Kappa index attempts to compare observed accuracy relative to a baseline of accuracy expected due to randomness. But, they believe that the baseline is irrelevant and misleading. They offer two new measures, quantity disagreement and allocation disagreement, to replace the Kappa index.

Another important criticism of the Kappa coefficient is the assumption of the independence of samples. Rarely is the independence assumption met (see Foody 2004 or McKenzie et al. 1996 for more details). To solve this problem, some authors present alternative ways to assess values. For example, the method presented by Donner et al. (2000), in which the significance of the difference between two Kappa coefficients derived from a related sample includes the covariance between the related samples.

We acknowledge the limitations of Kappa expressed by these authors, but we also concur with Congalton and Green (2008) that Kappa "must still be considered a vital accuracy assessment measure." We say this because the most common method of comparing different classifiers is to use the Kappa statistic. This method allows for a statistical comparison between two classifiers, when both classifiers and the data reference set are crisp. Now that Kappa has been extended (Gómez et al. 2008) to work with a soft classifier and crisp reference data or with a fuzzy expert, there is additional incentive to use Kappa in conjunction with other overall measures.

4.2.6 Other Measures of Accuracy

Chen et al. (2010), using simulated data, created a 400 × 400 pixel image with three reference classes, and investigated root-mean-square error (RMSE) as an index of disagreement useful in accuracy assessment. They found that as a map-level index RMSE had a high consistency with Kappa and overall accuracy, as judged by Kendall's tau, regardless of the random sampling intensity used for the accuracy assessment sample. At the category level, the correlations were weaker than at the map level. If sample size was greater than 200 pixels, then the consistencies among category-level accuracy indices were almost the same level as for the complete sampling case. Their specific findings would be expected to change as the number of references classes and the image size increase.

Sarmento et al. (2009) proposed using an accuracy measure for each land cover class that incorporates uncertainty inherent in the reference labeling protocol. Two forms of uncertainty are modeled: the first is the confidence that the interpreter has in the reference class label and the second allows for a primary and secondary reference land cover label for defining a match with the map class at an accuracy assessment location. They show that this measure is a compromise between agreement provided by the MAX (= 1 when the membership function is maximum among all map categories) and RIGHT (= 1 when the membership function exceeds a threshold) operators that have wide use in fuzzy accuracy assessment (Gopal and Woodcock 1994).

Pontius and Millones (2011) eschew Kappa in favor of two measures of disagreement: quantity and allocation. The quantity disagreement q_g measures the absolute differences in proportions between the reference map and the comparison map for an arbitrary category g. The allocation disagreement a_g calculates the minimum of the proportion of category g omitted in the reference map and the proportion of category g committed in the comparison map and is multiplied by 2. Based on these two quantities, the authors propose four different versions of Kappa (Kappa quantity, Kappa allocation, Kappa histo, and Kappa no). While the concept of quantity disagreement and allocation disagreement make intuitive sense to us, it is too early to know if these measures will be adopted by the remote sensing community. Recently, in Gómez et al. (2008) and in Gómez and Montero (2011), some of these new concepts of Kappa were presented and extended into a soft framework.

In this chapter, we concentrate our analysis on the use of pixels or pixel blocks (plots) as the sampling units for accuracy assessment. We do so because pixels and pixel blocks retain their identity through time and they are not affected by revisions and updates that may be made to land cover maps. Map polygons do not share this property. We utilize the error matrix and associated measures: overall, producer's, and user's accuracies (as recommended, e.g., by Liu et al. 2007 and Congalton and Green 2008). We also utilize Kappa for the case of soft classification and soft reference data for reasons previously discussed in Section 4.2.5.

4.3 Pixel Accuracy Assessment Measures for Soft Classification Problems

As was pointed out in Section 4.1, the pixel accuracy assessment measures for any classification map try to compare the results obtained by the classifier with the reference dataset. This comparison can be done in different ways that can be divided, following Congalton and Green (1999), between *nonsite-specific assessments* and *site-specific assessments*. In a nonsite-specific assessment, only total areas for each category are computed and compared. This approach is less expensive but has been criticized because it does not take into account the correct localization of the classifier. Given the limitations of nonsite-specific assessment, the site-specific assessment is usually preferred. For the evaluation of soft classifications in general, various suggestions have been made (Congalton 1991; Gopal and Woodcock 1994; Foody 1995; Binaghi et al. 1999; Lewis and Brown 2001; Green and Congalton 2004; Pontius and Cheuk 2006; Gómez et al. 2008), among which, the fuzzy-soft error matrix is one of the most appealing approaches, as it represents a generalization (grounded in fuzzy set theory) of the traditional confusion matrix.

Also, the RMSE, also called the root-mean-square deviation (RMSD), is a frequently used index to reflect the differences between values predicted by a model and the values actually observed or measured. RMSE serves to aggregate individual differences from the comparison of each data pair into a single index of disagreement. It is often used to assess the accuracy of each class of soft classification (category-level accuracy) (e.g., Carpenter et al. 1999; Silván-Cárdenas and Wang 2008).

In nonsite-specific assessment, there exist some approaches to measure the accuracy in a soft framework (see, e.g., Finn 1993), but this presents similar problems as with classical nonsite-specific assessment. As is pointed out by Foody (1999), these accuracy measures only take into account the percentage or area of pixels that belong to each class and its relation with the true reference data. So you can have a perfect agreement between classifier and reference data because both coincide in the number of pixels that belong to each class, but not in the location.

In site-specific assessment, most of the accuracy measures (overall accuracy, user accuracy, producer accuracy, Kappa statistic, Kappa location, Kappa histo, etc.) for classification maps can be obtained in a natural way based on the matrix error. In the crisp framework, the error matrix M is defined for each i and j (M_{ij}) as the number of sampling units that have been placed into class i by the classifier and into class j by the expert. Obviously, the diagonal of the matrix represents the correct matching between classifier and expert/true value. In hard/crisp classification models, it is clear how to build this matrix. In a fuzzy or soft framework, this problem is still open. This problem appears since in a crisp framework it is clear if the classifier output coincides with the true reference data, but in a soft framework, it is necessary to build an agreement or disagreement function for each sampling unit. Let us suppose that we have three classes, and for one sampling unit p, we have the values $(0.4, 0.6, 0)$ and $(0.3, 0.7, 0)$ that correspond to the output of the classifier and the true/reference data. What should be the agreement or error between these two values?

Some approaches propose that the error function between classifier and expert be measured by an expert based on a linguistic scale of accuracy (see Woodcock and Gopal 2000; Laba et al. 2002). Unfortunately, as pointed in Woodcock and Gopal (2000), this approach does not take full advantage of the potential allowed by fuzzy sets.

Binaghi et al. (1999) present the first work that builds an error matrix using a soft classifier and expert for the reference. According to their approach, the degree to which the sampling unit (pixel) has been classified into class i by the classifier and into class j by the expert is calculated based on the MIN operator. Once the fuzzy error (fuzzy confusion matrix M) has been obtained, the overall, producer's, and user's accuracies can be computed taking into account that the number of pixels n used in these expressions is now the sum in all pixels of all fuzzy reference's coordinates. Based on the assumption that the fuzzy classification is a Ruspini partition, Pontius and Batchu (2003) and Pontius and Cheuck (2006) include some new operators (multiplication operator and composite operator) to build a new fuzzy confusion matrix in the same way as Binaghi et al. (1999). From these fuzzy error matrices, one can compute an overall accuracy measure in a soft framework.

An alternative approach of how to build the error matrix is presented in Gómez et al. (2008) that improves the overestimation of accuracy that can occur using the Binaghi approach. In Gómez et al.'s work, a disagreement measure (error between expert and classifier) was defined, which did not assume that all errors are equally important. From their disagreement measure, a new family of accuracy measures is built (overall accuracy, producer accuracy, user accuracy, and Kappa statistic) in a soft framework that can be viewed as an extension of the classical measures since these approaches coincide in the case in which both the classification and the expert are crisp. Recently, Gómez and Montero (2011) extended the crisp accuracy measures of Kappa histo and Kappa allocation into a soft framework.

Now, we introduce some notation to formally define the main accuracy measures that deal with soft classification maps. From a mathematical point of view, a sampling unit that has been classified by the expert/reference dataset (R) or by a classifier (C) can be modeled as two functions assigning to each sampling unit a vector **P**.

$$R \text{ or } C: \quad \mathbf{P} \to \left\{ x \in [0,1]^k \right\}$$

4.3.1 Definition 3.1: Overall Accuracy of Binaghi et al. (1999)

Let $A_1, \ldots, A_k$ be the classes under consideration, let R be the expert/reference data function, let C be the classifier function with k classes, and let T be the sampling accuracy set composed of T pixels. Then the overall Binaghi accuracy is defined as

$$\text{Overall}_B = \frac{\sum_{p \in T} \sum_{i=1}^{k} \min\{R(p)_i, C(p)_i\}}{\sum_{p \in T} \sum_{i=1}^{k} \min\{R(p)_i, R(p)_i\}} = \frac{\sum_{p \in T} \sum_{i=1}^{k} \min\{R(p)_i, C(p)_i\}}{\sum_{p \in T} \sum_{i=1}^{k} R(p)_i} \tag{4.1}$$

Note that when we have a Ruspini (1969) partition in the reference data (i.e., for each pixel, the sum of the reference values $R(p)_i$ in all the classes is equal to 1), the denominator is equal to the number of pixels in T. On the other hand, in the numerator, we have a sum for each pixel of the agreement between the classifier and the reference data. So, in the end, the overall formula is the average agreement in all pixels, as it happens in the crisp case. In Section 4.5, we analyze not only the average but also the agreement distribution to show the real differences between different ways of measuring the accuracy of a classification map.

$$\text{Overall}_B = \frac{1}{|T|} \sum_{p \in T} \left(\sum_{i=1}^{k} \min\{R(p)_i, C(p)_i\} \right) \tag{4.2}$$

Let us also observe that the overall measure can be viewed (as in the crisp case) as the sum of the diagonal of a fuzzy error matrix X divided by the sum of the pixels in T of the coordinates of the reference data (that for Ruspini partitions is equal to $|T|$). Formally the fuzzy error matrix is obtained as

$$X_{ij} = \sum_{p \in T} \min\{C(p)_i, R(p)_j\} \tag{4.3}$$

and then

$$\begin{aligned}
\text{Overall}_B &= \frac{\sum_{p \in T} \sum_{i=1}^{k} \min\{R(p)_i, C(p)_i\}}{\sum_{p \in T} \sum_{i=1}^{k} R(p)_i} = \frac{\sum_{i=1}^{k} \sum_{p \in T} \min\{R(p)_i, C(p)_i\}}{\sum_{p \in T} \sum_{i=1}^{k} R(p)_i} \\
&= \frac{\sum_{i=1}^{k} X_{ii}}{\sum_{p \in T} \sum_{i=1}^{k} R(p)_i} = \frac{\sum_{i=1}^{k} X_{ii}}{\sum_{p \in T} 1} = \frac{\sum_{i=1}^{k} X_{ii}}{|T|}
\end{aligned} \tag{4.4}$$

The last two equalities are only true when the reference data satisfies the Ruspini condition.

So let us note that the overall accuracy measure defined by Binaghi can be viewed or presented in different formats: as the original formula, as an average of the different agreements in each pixel, or as the sum of the diagonal of a normalized error matrix.

A variant of the MIN operator is sometimes used as a similarity index $\left(SI_{i,j} = 1 - \dfrac{|R(p)_i - C(p)_j|}{R(p)_i + C(p)_j} \right)$ for comparing soft classifications (see, e.g., Townsend 2000). This variant results after normalizing the MIN operator by the sum of the grade values. The SI operator is also meaningful for subpixel comparison, as it corresponds to a normalized maximum subpixel overlap.

Another alternative is the use of the product operator (PROD) that arises from a pure probabilistic view of the pixel–class relationship. In the traditional probabilistic ontology, the pixel–class relationship represents the probability that a pixel (entirely) belongs to a class, and the PROD operator gives the joint probability that the reference and assessed pixels belong to two given classes, provided that the pixels have been independently classified. Pontius and Batchu (2003) and Pontius and Cheuck (2006) present more variants of the previous approach using other operators (multiplication operator and composite operator) instead of the minimum aggregation operator. It is important to note that most of these operators require that the soft classification imposes the constraint that $\sum_{i=1,k} R(p)_i = \sum_{i=1,k} C(p)_i = 1$, for all sampling units p of the test set.

Another way to assess the accuracy of soft classification maps can be found in Gómez et al. (2008). This approach is based on a dissimilarity measure between expert or reference data and classifier. Their approach corrects for a slight overestimation produced by Binaghi's method, which is discussed in detail Section 4.4.

4.3.2 Definition 3.2: Overall Accuracy of Gómez et al. (2008)

Let $A_1, ..., A_k$ be the classes under consideration, let R be the expert/reference data function, let C be the classifier function, and let T be the sampling accuracy set with $|T|$ unique pixels. Then the overall Gómez accuracy is defined as

$$\left(\text{Overall}_G = \sum_{p \in T} \frac{1 - D(R(p), C(p))}{|T|} \right) \tag{4.5}$$

where $D(R(p), C(p))$ represents the disagreement between reference and classifier.

Formally, this distance is calculated as $D(R(p), C(p)) = \min \left\{ 1, \sum_{j=1}^{k} w_{ij} |R(p)_j - C(p)_j| \right\}$ and w_{ij} is the error importance of classifying a pixel into class j when the true reference data belongs to class i.

Let us observe that, as happened with Definition 3.1, the overall measure is an average of the agreement between reference and classifier for each pixel in T. Also note that the disagreement measure D between expert and classifier depends on the number of weights that represent the importance of the errors. If you believe that all errors are equally important, then two weighting schemes have been used in literature to reflect this equality. The first weighting scheme is $W = 1 - I$ (i.e., $w_{ij} = 1$ if $i \neq j$ and 0 otherwise). We have denoted the overall measure that uses $W = 1 - I$ as Gómez 1 (Overall$_{G1}$). The other possibility is to consider a (Manhattan) distance function between vectors and thus we have $W = \frac{1}{2}$ to have distances bounded into the interval [0, 1] We have denoted the overall accuracy measure that uses $W = \frac{1}{2}$ as Gómez 2 (Overall$_{G2}$).

Let us note that if the classifier produces a Ruspini's partition and the expert/reference is crisp, then the overall accuracy measure defined previously as Gómez 1 coincides with the overall accuracy defined by Binaghi et al. (1999). Or if the classifier C is crisp and the reference data is a Ruspini partition, the overall accuracy defined by Binaghi et al. (1999) is the same as the overall accuracy measure denoted as Gómez 2. For a more general case, these approaches do not coincide. Taking into account this idea of measuring the distance between points, Chen et al. (2010) proposed an accuracy measure based on the RMSE of the kT values that correspond to each coordinate of each reference pixel. This approach takes all the kT numbers between [0] and [1] and computes the differences between expert or reference and the classifier. This approach follows the same idea of measuring the difference between expert and classifier as a distance, but it loses the interpretability of the result and it is strongly dependent on the number of classes. For example, the error associated with the pair $C = (1, 0, 0)$ and $R = (0, 1, 0)$ is different from the error associated with the pair $C = (1, 0, 0, 0)$ and $R = (0, 1, 0, 0)$. Nevertheless, it is an interesting alternative that could provide more information. Formally, the RMSE is calculated as

$$\text{RSME} = \frac{\sum_{p \in T} \sum_{i=1}^{k} (C(p)_i - R(p)_i)^{\frac{1}{2}}}{kT} \tag{4.6}$$

Accuracy measures such as the Kappa statistic, among others, have been extended from Definition 3.2 of the disagreement measure.

4.3.3 Definition 3.3: Gómez et al. (2008)

Let $A_1, \ldots, A_k$ be the classes under consideration, let R be the expert/reference data function, let C be the classifier function, and let T be the sampling accuracy set with $|T| = t$ unique pixels. Then the extended Kappa statistic K_E is defined as

$$K_E = \frac{\hat{p}_0 - \hat{p}_c}{1 - \hat{p}_c} \tag{4.7}$$

where

$$\hat{p}_c = \sum_{p \in T} \sum_{q \in T} \frac{1 - D(R(p), C(q))}{t^2} \tag{4.8}$$

It is important to note that this new definition is an extension of the standard Kappa measure for two raters as shown in Gómez et al. (2008). Other accuracy measures can be extended in a similar way (e.g., see Gómez and Montero 2011) from the agreement/error definition for soft reference and classification data.

4.4 Accuracy Assessment Example

In this section, we show how it is possible to obtain an accuracy assessment in a soft framework. Given a classification map problem with three classes, let us suppose that we have only two sampling units in the reference dataset $T = \{p_1, p_2\}$. Additionally, suppose that we have two classifiers C_a and C_b for which we want to measure the performance and also let us suppose that the expert/true reference set (R) produces soft information in the following set (as shown in the following table). Please note that when using a soft classifier, the sum of the coordinates can exceed 1 as with $C_a(p_1)$ in the following table. What should be the overall accuracy for these two classifiers?

$$R(p_1) = [0.3, 0.4, 0.3] \quad R(p_2) = [0.6, 0.1, 0.3]$$
$$C_a(p_1) = [0.5, 0.5, 0.5] \quad C_a(p_2) = [0.3, 0.1, 0.6]$$
$$C_b(p_1) = [0.5, 0.4, 0.1] \quad C_b(p_2) = [0.6, 0.1, 0.3]$$

1. Following the overall accuracy defined in Binaghi et al. (1999) and the fuzzy error matrix formula described in Definition 3.1, the fuzzy error matrix X (in the format reference data columns-classifier rows) for C_a is calculated as

$C_a /R \rightarrow$	A_1	A_2	A_3
A_1	$0.3 + 0.3 = 0.6$	$0.4 + 0.1 = 0.5$	$0.3 + 0.3 = 0.6$
A_2	$0.3 + 0.1 = 0.4$	$0.4 + 0.1 = 0.5$	$0.3 + 0.1 = 0.4$
A_3	$0.3 + 0.6 = 0.9$	$0.4 + 0.1 = 0.5$	$0.3 + 0.3 = 0.6$

For example, the entry X_{12} of the above table (that corresponds to the classifier for class A_1 and the reference [R] for class A_2) is obtained as $\min\{C_a(p_1)_1, R(p_1)_2\} + \min\{C_a(p_2)_1, R(p_2)_2\}$ since we have only two pixels, and this is equal to $\min\{0.5, 0.4\} + \min\{0.3, 0.1\} = 0.4 + 0.1 = 0.5$. The denominator for $\text{Overall}_B = \sum_{p \in T} R(p)_i$ for $i = 1, 2, 3$, which equals 0.9, 0.5, and 0.6, respectively, and sums to 2. Then we have that $\text{Overall}_B = \dfrac{0.6 + 0.5 + 0.6}{0.9 + 0.5 + 0.6} = \dfrac{1.7}{2} = 0.85$. Let us observe that as the reference data R satisfies the Ruspini condition (i.e., the sum of coordinates for each pixel is equal to 1), the overall measure is obtained also as the sum of the diagonal values divided by the number of sampling units that for this case is 2.

2. In a similar way, the fuzzy error matrix for C_b is calculated as

$C_b /R \rightarrow$	A_1	A_2	A_3
A_1	$0.3 + 0.6 = 0.9$	$0.4 + 0.1 = 0.5$	$0.3 + 0.3 = 0.6$
A_2	$0.3 + 0.1 = 0.4$	$0.4 + 0.1 = 0.5$	$0.3 + 0.1 = 0.4$
A_3	$0.1 + 0.3 = 0.4$	$0.1 + 0.1 = 0.2$	$0.3 + 0.1 = 0.4$

and $\sum_{p \in T} R(p)_i$ for $i = 1, 2, 3$ are 0.9, 0.5, and 0.6, respectively. Thus, we have that

$\text{Overall}_B = \dfrac{0.9 + 0.5 + 0.4}{0.9 + 0.5 + 0.6} = \dfrac{1.8}{2} = 0.9$.

Both classifiers produce a similar accuracy performance. This problem is due to the overestimation of agreement measure made by Binaghi that appears in some situations (e.g., when the soft classification or reference is not a Ruspini partition). For the classifier C_a, the Binaghi approach calculates the overall agreement between R and C_a to be perfect, that is, equal to 1, even though the values for R and C_a are (0.3, 0.4, 0.3) and (0.5, 0.5, 0.5), respectively.

3. Following Gómez et al. (2008), if we assume that all the errors are equally important (i.e., weights are specified as $W = 1 - I$), the overall accuracy (Overall_{G1}) is

calculated as the agreement between expert and classifier in the two sampling units. The disagreement between expert and C_a in p_1 is obtained as $|C_a(p_1)_1 - R(p_1)_1|$ + $|C_a(p_1)_3 - R(p_1)_3| = |0.5 - 0.3| + |0.5 - 0.3| = 0.4$. And for p_2, it is $|C_a(p_2)_2 - R(p_2)_2|$ + $|C_a(p_2)_3 - R(p_2)_3| = |0.1 - 0.1| + |0.6 - 0.3| = 0.3$. This is because the maximum in the reference data for p_1 is reached in the second coordinate, so we only consider the differences between reference data and classifier in the first and third coordinates (since $w_{21} = 1$, $w_{22} = 0$, $w_{23} = 1$). In a similar way, as the maximum in the reference data is reached in the first coordinate ($w_{11} = 0$, $w_{12} = 1$, $w_{13} = 1$), we only consider the differences between both vectors in the second and third coordinates. So,

$$\text{Overall}_{G1} = A(C_a, R) = \sum_{i=1,2} \frac{\left(1 - D\left(C_a(p_i), R(p_i)\right)\right)}{2} = \frac{0.6 + 0.7}{2} = 0.65$$

In a similar way, it is possible to obtain

$$\text{Overall}_{G1} = A(C_b, R) = \sum_{i=1,2} \frac{\left(1 - D\left(C_b(p_i), R(p_i)\right)\right)}{2} = \frac{0.6 + 1}{2} = 0.8$$

since the agreement between C_b and R is 1 in p_2.

The other possibility is to consider the weights $W = \frac{1}{2}$ (which is a simplification of previous measure, but could result in a small overestimation of the agreement). For these weights, the disagreement between C_a and R for pixel p_1 is as follows:

$\frac{1}{2}(|0.5 - 0.3| + |0.5 - 0.4| + |0.5 - 0.3|) = 0.3$, so agreement is 0.7, and in a similar way, we have that the disagreement between C_a and R for pixel p_2 is $\frac{1}{2}(|0.3 - 0.6| + |0.1 - 0.1| + |0.6 - 0.3|) = 0.3$, so agreement is 0.7. And thus, the overall accuracy for C_a is 0.7.

4. In a nonsite-specific assessment, we only compare the total area in the different classes estimated by the reference data and by the classifier. So only this information is relevant. To make this calculation, we first need to aggregate the information of R and C for all p, and after, a similarity function (usually an entropy measure) between these two vectors represents the accuracy measure. A possible aggregation of R could be $\frac{1}{2}(0.3 + 0.6) = 0.45$ for the first class, $\frac{1}{2}(0.4 + 0.1) = 0.25$ for the second class, and $\frac{1}{2}(0.3 + 0.3) = 0.3$ for the third class, so $R = (0.45, 0.25, 0.3)$.

Following similar guidelines, we can aggregate C_a and C_b over all reference pixels. Thus, $C_a = (0.4, 0.3, 0.55)$ and $C_b = (0.55, 0.25, 0.2)$. Now, we can apply any similarity measure between vectors $\mathbf{R}$ and $\mathbf{C_b}$. Note that most of the nonsite-specific assessment measures model the uncertainly in a probabilistic or Ruspini way. This means that they require the sum of the coordinates to be equal to 1. Clearly C_a does not meet this requirement, since the sum of the coordinates is greater than 1.

4.5 Real Application

In this section, our main aim is to demonstrate, through a realistic application dataset, the applicability of the fuzzy accuracy measures previously presented and to evaluate the differences between fuzzy measures and hard/classical measures. In this section, we

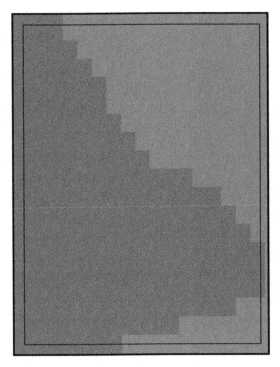

FIGURE 4.1
(See color insert.) One pixel in which we can observe more than one class. The red color represents the mountainous arid class and the tan-orange color represents the sand class.

obtain different accuracy measures from a real image when the information that we have from the reference data and the classifier is given in a soft way. Also, we calculate classical accuracy measures to compare and show the main differences between classical and soft approaches. It is important to note that the term "fuzzy" or "soft" appears in different frameworks in remote sensing classification. Situations exist in which fuzziness or softness appears as an intermediate step in obtaining a final crisp classification of the image. If this is the case, there is not any problem in using classical accuracy measures in assessing the performance of a classifier or a classification map. Nevertheless, there exist other situations, in which the classification or the reference for a specific pixel is given in a soft way. For example, we could have a pixel that belongs to a *forest* class with a value of 0.65 and also belongs to the *sand* class with a value of 0.35 (see Figure 4.1).

4.5.1 Image Description

For our study site, we utilize a MODIS image of size 300 columns and 200 rows with pixel size having a nominal spatial resolution of 500 m. This image is from a place close to Shijiazhuang on the North China Plain. The study area is approximately 345 km southwest of Beijing (see Figure 4.2). The study area center coordinates are approximately 37.46° latitude and 114.145° longitude. The study area (Figure 4.3) has an elevation gradient whose high point is approximately 1400 m in the southwest, 460 m in the central image, and around 30 m in the eastern portion of the image. It was selected because of the diversity of LULC types and also because there is very good reference data. The reference data was

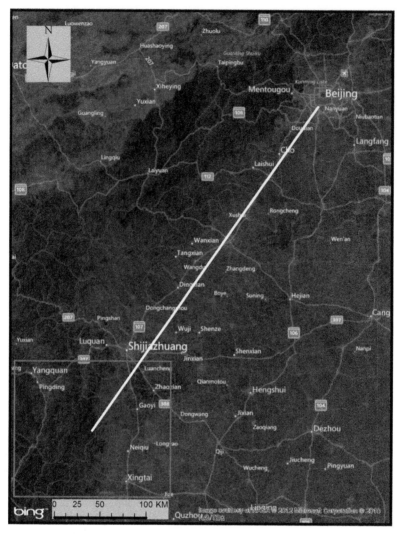

FIGURE 4.2
(See color insert.) The location of our study area, southwest of Beijing in the North China Plain. Source: Bing Maps.

produced by Liu et al. (2003). Liu et al. interpreted 30-m resolution Landsat data to create reference data to study land use change dynamics in China during the 1990s. The LULC classes for this area were classified for this study into nine classes (Table 4.1).

Our training data consisted of usually 40–100 polygons per class typically of size 3–4 MODIS pixels. For very small classes such as unused land (class 61), the training set was necessarily much smaller than for the larger LULC classes. The training sets were well distributed throughout the study area. The training dataset was used to create a supervised classification of a composite MODIS image (bands 5, 4, 3 corresponding to NIR, green, and blue) using the nine LULC categories in Table 4.1 (see Figure 4.4b). The classified image was then smoothed using the four nearest neighbors in a 3 × 3 majority filter to help remove the "pepper and salt effect." The resulting classification map is presented in Figure 4.4a. The focus of this chapter is on the accuracy assessment of soft classification

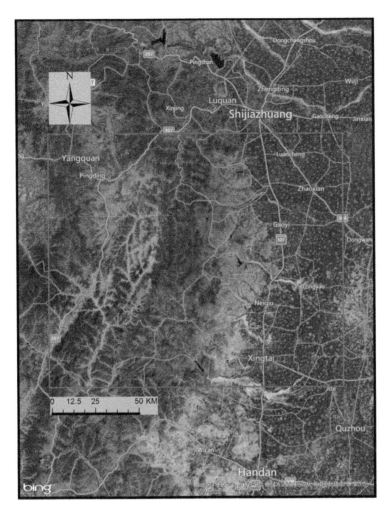

FIGURE 4.3
(**See color insert.**) Close-up of our study area near Shijiazhuang, China. (Courtesy of NASA and Bing Maps.)

TABLE 4.1

Aggregated Land Cover Land Use Classes Used in This Study

Class	Class Number in Figure 4.4	Class Name
21	1	Forest (natural and planted with canopy cover >30%)
22	2	Forest with shrub (canopy cover >40% and shrub height <2 m)
23	3	Forest with low cover (10%–30%) or other forest types (fruit garden)
32	4	Grassland with medium cover (20%–50%)
33	5	Grassland with low cover (5%–20%)
41	6	Water (channel, pond, and water bottomland)
51	8	Impervious land (urban, rural residential, and other construction)
61	9	Unused land (sand)
121–123	10	Arid land (mountainous, hilly, and plain area)

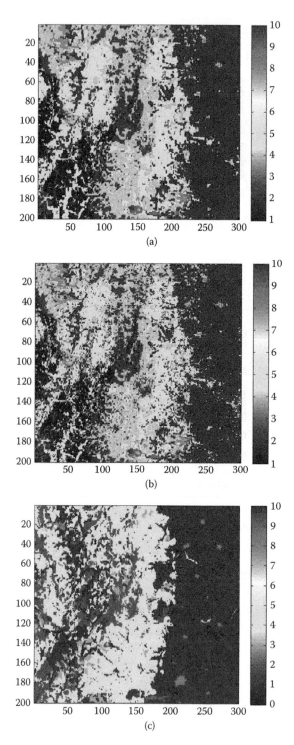

FIGURE 4.4
(**See color insert.**) (a, b, c) Classification maps made by the classifiers C_1, C_2, and C_3, respectively. The color numbers (right-hand vertical scale) represent the class of the crisp classifier. Color 1 corresponds to class 21, color 2 corresponds to class 22, and so on (see Table 4.1 for definition of classes).

maps. To demonstrate our techniques for accuracy assessment, we needed a reasonable, not optimal, classification that was achieved with this procedure.

We can compare the classification based on 500-m MODIS pixels to the reference data based on 30-m Landsat data. We can compute class membership percentages within each MODIS pixel. This allows us to construct (pseudo) fuzzy class membership for each MODIS pixel.

The dominant reference class type as determined by the thematic mapper (TM) data was arid land (classes 121–123), followed by grassland (classes 33 and 32) and forest (classes 21, 22, and 23) in order of frequency on the landscape. Water and impervious land were minor components on the landscape, as was unused land.

4.5.2 Classification Maps and Reference Data

In this real application, we measure the accuracy of a classification map given by three classifications (see the following) when the reference data is given in a soft way. The three classification maps are obtained in the following way:

1. Let us denote by C_1 the classifier that gives us the first classification map of the image (see Figure 4.4a). This classification uses the maximum likelihood classifier (MLC) with the MODIS composite image (bands: 5, 4, 3) that has been smoothed using the four nearest neighbors in a 3×3 majority filter. We also refer to C_1 as MLC + majority.

2. By C_2 we denote the classifier that gives us the second classification map of the image (see Figure 4.4b). This classification uses the MLC with the MODIS composite image (bands: 5, 4, 3) without any postclassification smoothing. We refer to C_2 as MLC.

3. Finally, let us denote by C_3 as an artificial (and unreal) crisp classifier that takes the maximum of the reference data (see Figure 4.4c). This classifier will give us the third classification map of the image. Let us observe that this classifier knows the classification made by the expert and defuzzyficates this soft information. We also refer to C_3 as Crisp max. The defuzzyfication process in this framework transforms the soft information into crisp taking the criteria of maximum degree of membership. For example, a pixel with soft information in three classes (0.85, 0.1, 0.05) will be transformed into (1, 0, 0) (i.e., this pixel belongs to the class 1).

The three classifiers are crisp, but as we have pointed out in the following, the reference data is soft because each pixel may belong to more than one class and with varying degrees of membership. In the following nine images (see Figure 4.5), we show the degree of membership of each pixel to each class.

4.5.3 Accuracy Measures

To measure the accuracy of these classification maps, we concentrate on the overall accuracy measure and the agreement distribution between expert or reference and the classifier. Other accuracy measures such as user's and producer's accuracy, rather than overall accuracy, could be obtained for this real case. In addition, we can compute Kappa values as an accuracy measure for the Gómez approach. However, there is no equivalent measure for the Binaghi approach. Nevertheless, for simplicity and to be able to compare different ways of measuring the accuracy of classification maps, we have focused just on the overall accuracy. And we also focus on the agreement distribution.

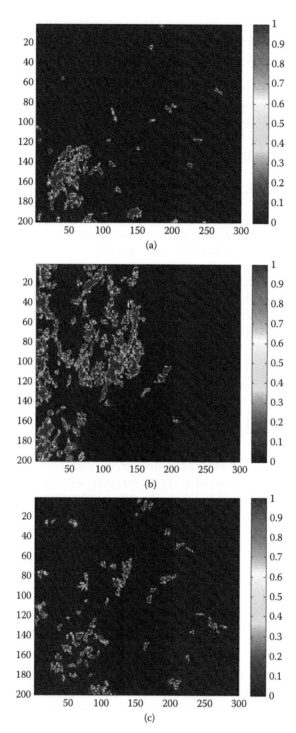

FIGURE 4.5
(**See color insert.**) Visualization of the soft reference data. In each image, we show the degree of membership (on a scale of 0–1) of each pixel to each of the nine classes (a) forest, (b) forest with shrub, (c) forest with low cover, (d) grassland with medium cover, (e) grassland with low cover, (f) water, (g) impervious land, (h) unused land, and (i) arid land.

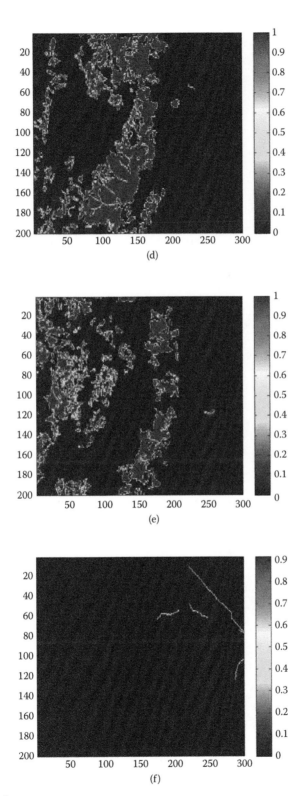

FIGURE 4.5 (*Continued*)

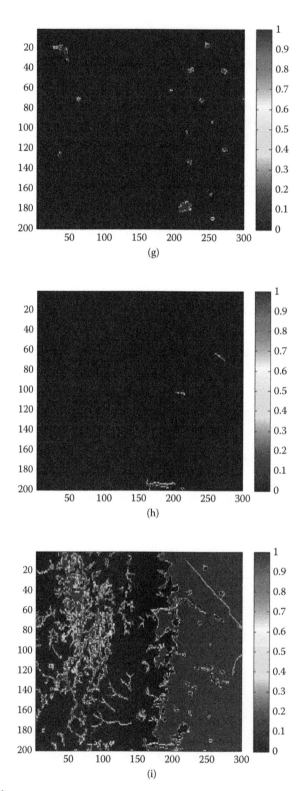

FIGURE 4.5 (*Continued*)

We compare the results obtained by using various overall accuracy measures: classical overall accuracy, Binaghi et al. (1999) overall accuracy (with MIN and PROD product aggregation operator), and Gómez et al. (2008) overall accuracy with different weights (see Definition 3.2 of Section 4.3.2).

In Table 4.2, we can see the classical error matrix (and its corresponding overall accuracy measure) of the classification map made by classifier C_1. Because the reference data is soft, it is necessary to convert this information into hard/crisp reference data. We use the classical operator MAX assigning each pixel to the class in which the maximum is reached. After the transformation, the error matrix, the overall accuracy, and a histogram of the agreement distribution are shown in Table 4.2 and Figure 4.6.

In Table 4.3, we show the error matrix given by the Binaghi approach with the MIN operator aggregation of the classification map obtained with C_1. From the similarity of these two tables and their nearly equivalent overall accuracy measures, one could form an opinion that both measures present similar results, but this first impression is wrong. These results are also repeated for this particular data when you are obtaining the error matrix using other approaches mentioned in Section 4.3. As discussed, the agreement for each pixel between reference and classifier in the crisp case can only take the values 0 or 1. In this situation, the overall accuracy measure coincides with the average of the different agreements in all pixels (i.e., the percentage of pixels in which the agreement is 1). Let us note that in the crisp case, the agreement only takes values of 0 or 1 (Bernoulli distribution), and thus from the average of the agreement (i.e., overall), it is possible to recover all the original information. In the soft case, since the agreement can take values in the interval [0, 1], it is necessary to include an agreement distribution analysis (e.g., using a histogram) to have a clear idea of what is really happening.

In Figure 4.6, we present a histogram showing the agreement distribution using the Binaghi approach for the classification map made with C_1. We also show the distribution of the differences between Binaghi agreement with a soft approach and a classical agreement using a hard approach.

In Figure 4.6a, we have the histogram of the Binaghi agreement measure for each pixel. We can see that there are around 22,000 pixels with an agreement lower than 0.1. On the other side of the scale, we see that there are close to 30,000 pixels with a very high agreement (between 0.9 and 1). For the rest of the pixels (around 8000 pixels), C_1 has an agreement with

TABLE 4.2

Error Matrix for the Classification Map Obtained with the C_1 Classifier Using a Hard Approach

C_1 R→	21	22	23	32	33	41	51	61	121–123
21	*1222*	1904	515	771	742	0	0	0	153
22	230	*2480*	414	1603	227	0	0	0	98
23	112	165	*30*	392	158	0	9	0	198
32	180	1239	336	*6418*	2244	3	44	2	2096
33	173	773	308	2191	*5331*	18	33	13	5158
41	4	9	5	24	30	*0*	34	0	81
51	1	23	10	36	135	0	*211*	2	354
61	1	0	1	1	2	4	0	*55*	23
121–123	108	331	174	807	916	20	51	63	*18501*

Notes: The overall accuracy obtained is 34,248/60,000 = 0.5708. The left-hand column is the class predicted with the classifier C_1 and the reference classes (R) are the nine right-most columns after the defuzzyfication into crisp reference data.

The diagonal values are in italics.

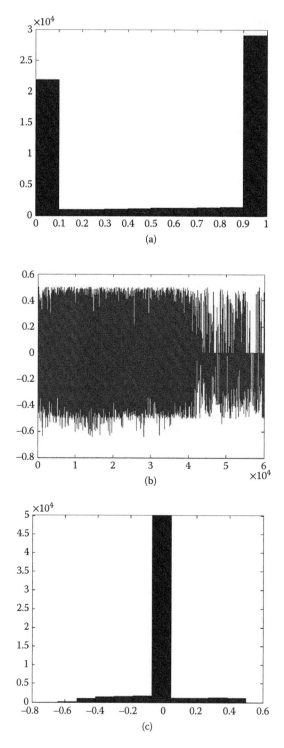

FIGURE 4.6
(a) Binaghi agreement histogram for the classification map made with C_1, (b) agreement using the Binaghi approach minus the agreement using a classical approach, and (c) histogram of the center distribution.

TABLE 4.3

Error Matrix for the C_1 Classifier Using the Binaghi Approach with the MIN Operator

$C_1 R\rightarrow$	21	22	23	32	33	41	51	61	121–123
21	*1193.23*	181.47	516.79	797.79	744.90	0.19	0	0	213.21
22	224.65	*2411.26*	416.38	1629.10	252.75	0	0.86	0	116.96
23	110.23	166.87	*30.34*	387.71	159.44	0	9.69	0.02	199.67
32	179.90	1251.17	353.64	*6319.31*	2213.20	9.67	45.82	1.45	2187.79
33	182.62	805.10	325.79	2174.990	*5235.01*	17.95	33.71	17.84	5205.03
41	3.80	9.07	5.59	24.68	30.48	*0*	32.23	0	81.12
51	0.80	24.23	10.98	39.18	141.04	0	*207.41*	1.25	347.07
61	1.56	0	1.17	2.06	2.12	2.64	0	*52.92*	24.49
121–123	121.03	343.95	183.58	798.59	931.10	68.95	59.39	64.64	*18399.72*

Notes: The overall accuracy obtained is 0.567. The left-hand column is the class predicted with the classifier C_1 and the reference classes (R) are the nine right-most columns. See Binaghi et al. (1999) for additional details on how to form this error matrix and also see the example in Section 4.4.
The diagonal values are in italics.

TABLE 4.4

Overall Accuracy for the Classification Maps Done by C_1, C_2, and C_3 Using Five Accuracy Assessment Measures/Approaches

	Overall Accuracy				
Classifier	Classical	Binaghi MIN	Binaghi PROD	Gómez 1 $W = 1 - I$	Gómez 2 $W = 1/2$
C_1 (MLC + majority)	0.5708	0.5642	0.5641	0.5631	0.5652
C_2 (MLC)	0.5539	0.5470	0.5470	0.5459	0.5470
C_3 (Crisp max)	0.9923	0.8935	0.8935	0.8925	0.8935

the reference data between 0.1 and 0.9, which is homogeneously distributed. This analysis cannot be obtained by a crisp approach, since the agreement or errors are solely 0 or 1. In Figure 4.6b, we see that the crisp agreement is very different from the Binaghi agreement in each pixel (since all the agreements are forced to be 0 or 1). In Figure 4.6c, we show the reason why these two approaches (even if they are very different) present a similar overall measure. The agreement measure with Binaghi minus the agreement measure using a classical approach has an average close to zero, so, on average, both agreement distributions are equal. We emphasize here that the *distributional analysis of the agreement measure is very informative* in understanding the accuracy of a classification map and has heretofore not been utilized.

The rest of the classification maps follow the same pattern as shown in Figure 4.6. From Table 4.4, we can see that the overall accuracy measure obtained by soft approaches (e.g., Binaghi et al. 1999; Gómez et al. 2008) and a classical approach (hard/crisp) produces similar overall accuracy values for most of the classification maps. These approaches only coincide in the average of the agreement (i.e., overall), but they have big differences in the agreement distribution (see Figures 4.6 and 4.7). The overall accuracy is the average of the agreement in all pixels. So, from Table 4.4, we can conclude that the average of the agreement in all pixels with the different accuracy measures is very similar for each classifier studied. The average overall accuracy for these three classifiers (C_1, C_2, C_3) coincides by chance since the percentage of the situations in which the classical approach produces an overestimation of the agreement between expert or reference and classifier (e.g., $R = (0.7, 0.2, 0.1)$, $C_1 = (1, 0, 0)$

produces an agreement of 1, which is greater than the real agreement) is very similar to the percentage of pixels in which the classical approach produces an underestimation of the agreement (e.g., $R = (0.4, 0.5, 0.1)$, $C_1 = (1, 0, 0)$ produces an agreement of 0, which is lower than the real agreement).

With more details, it can be seen in the agreement distribution of both approaches that the classical approach produces large errors in each agreement assignment and a classical overall measure is not ideal for measuring the overall accuracy of the classified map (even if ultimately the final overall accuracy values are very similar). We will see later, with the classified map produced by C_3, that we do not always have this situation (Figure 4.8).

In Figure 4.7a, we can observe a relatively minor overestimation (<0.02) of the Binaghi et al. (1999) approach in estimating overall accuracy when it is compared with the overall

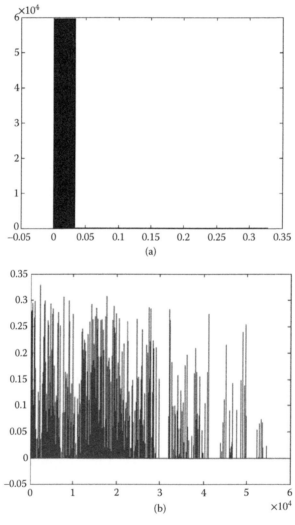

FIGURE 4.7
In (a), we have the Binaghi agreement minus the Gómez 1 agreement histogram for the classification map made by C_1. In (b), we have the value for each pixel resulting from the agreement using the Binaghi approach minus the agreement using the Gómez 1 approach for the classification map made by C_1 with weights $W = 1 - I$.

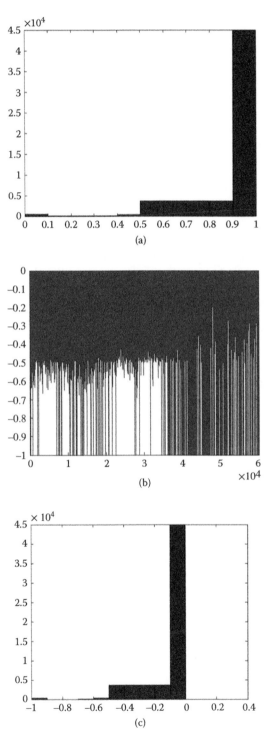

FIGURE 4.8
(a) Displays the Binaghi agreement histogram for the classification map made by C_3, (b) shows using the Binaghi approach minus the agreement using the classical approach for the classification map made by C_3, (c) shows histogram of the center distribution.

accuracy defined in Gómez et al. (2008). The histogram in Figure 4.7b is informative because it displays the distribution of these differences. From the agreement analysis of Binaghi and Gómez 1, we have to say that for the 99% of the pixels, the differences in the agreement measure between the Binaghi and Gómez approaches are in the range [0–0.02]. So, most of these differences are very small. Only 0.02% (around 15 pixels) of the pixels reach larger differences in the range of 0.25–0.30. For this subset of pixels, these two approaches could lead to different pixel classifications. Clearly, conducting a distributional analysis of the agreement measures (or differences) is quite informative in understanding the accuracy of the classification map. From an agreement analysis distribution (not just the average), we can conclude that the Binaghi and Gómez 1 (with weights $W = 1 - I$) approaches present very similar results in terms of measuring the accuracy of the classification map made by C_1.

In terms of comparison between the different measures used, the results associated with the accuracy assessment of the classification maps done by C_2 present the same conclusions as for C_1 (see Table 4.4). Both the Binaghi and the classical methods overall produce similar values for overall accuracy, but when a deeper analysis is done, it can be observed that this is by chance. This is because the performance of the classical/crisp accuracy measures is poor. Because we are comparing a crisp classification versus soft reference data, the errors measured by Binaghi with the minimum or the product aggregation operator (MIN or PROD) are the same, since MIN $\{1, a\} = 1 \times a$. The same situation happens with the agreement measure with Gómez et al. (2008). In the more general situation in which both (classifier and reference) are soft, these measures will not coincide.

The analysis of the overall accuracy performance for the classification map created by C_3 produces markedly different results than with the previous two classifiers. For example, in Figure 4.8a, we see that 45,000 pixels using the Binaghi approach have an agreement in the range of 0.9–1.0. And the differences in agreement/error between the Binaghi approach and the classical approach in each pixel produces a histogram with many large negative values in the range of −0.5 to −0.6 (Figure 4.8b). This happens because now the differences between the classical and the soft accuracy measures are not compensating, as happened in classification maps C_1 and C_2.

So, in Figure 4.8, we see that there are large differences between the Binaghi and the classical agreements. In fact for all three classifiers (C_1, C_2, and C_3), there exist large differences between the crisp/soft approaches. This provides evidence that a fuzzy accuracy approach, such as Binaghi or Gómez, is worth pursuing. Using a classical accuracy approach for our study area image from the North China Plain is not recommended because of the implications from our histogram analysis and also because a classical approach overestimates overall accuracy in relation to any of the Binaghi or Gómez measures.

4.6 Final Conclusions and Remarks

Classification and accuracy assessment are fundamental to analysis of remotely sensed data. The use of hard/crisp classifiers has a long history going back at least to Landsat 1 introduced in July 1972. By the early 1980s, extending to the present, many authors have proposed and advanced techniques for hard/crisp accuracy assessment (see, e.g., Cohen 1960; Congalton and Mead 1983; Rosenfield and Fitzpatrick-Lins 1986; Congalton 1991; Stehman 1995, 1997; Biging et al. 1999; Khorram 1999; Pontius 2000; Foody 2002). While highly valuable, the use of a hard/crisp classification and hard/crisp accuracy assessment has its limitations. Those limitations arise when objects (pixels, block of pixels, and polygons) are associated with more than one reference class. This is to be expected in many LULC studies. To overcome the limitations of the classical approach, Binaghi and her coauthors introduced the use of fuzzy mathematics for accuracy assessment in 1999 followed by the work of Woodcock and Gopal in 2000.

What has been occurring, and is the subject of this chapter, is that there are recent efforts to extend the standard accuracy measures (overall, producer's, user's, Kappa, and other measures as well) so that they can be used in a flexible framework: hard or soft classifier with crisp or fuzzy reference data.

We present two overall accuracy measures that use soft reference data (Overall$_B$ and Overall$_G$, B = Binaghi et al. 1999 and G = Gómez et al. 2008). We used a MODIS image from the North China Plain to demonstrate these methods. This area has excellent reference data derived from TM data and it has various LULC classes including forest, grasslands, arid land, and to a lesser extent water. We constructed fuzzy class membership for each MODIS pixel of nominal 500 m size by intersecting it with a comprehensive reference map based on 30 m TM data. We then used three crisp classifiers and calculated the overall accuracy with both soft and hard reference data. It was instructive that two of three classifiers had very similar overall performance as measured with Overall$_B$, Overall$_G$, or the classical overall accuracy measure. However, the overall accuracy measure alone is not sufficient for understanding the full implication of this aggregate measure.

In crisp classification models, the agreement or error between expert and classifier for a given pixel is a value that is forced to be 0 or 1. It is important to note that when a variable has a Bernoulli distribution (it only takes values in the {0,1} set), then the average of these values represents the percentage of pixels correctly classified and recovers all the information of the variable. So in crisp classification problems, when the errors or agreements are 0 or 1, the overall accuracy and error matrix (that are obtained from the average of the agreement values) summarize and capture the real accuracy of the classification. Nevertheless, this situation is not the same when the agreement (or errors) for a given pixel can take values in the interval [0,1]. As we have shown in this chapter, by introducing an agreement distribution, rather than just using the average of the agreement values in the sampling unit, we can generate a deeper analysis that leads to a more complete understanding of the real accuracy. We have shown how two very similar error matrices or overall accuracy measures can be found to be very different when the agreement distribution is utilized.

In fact, when we produce histograms of the Binaghi, Gómez, or classical agreement assignment for the classification map, we can see some important differences between the approaches. The classical approach produces large errors in each agreement assignment, which is not captured in the overall metric. While the Binaghi and Gómez agreement histograms are quite comparable (Figure 4.7), both the Binaghi accuracy measure (Figure 4.6) and, by inference (from Figure 4.7), the Gómez accuracy measure have radically different agreement histograms in comparison to the classical approach. This is because of the fundamental differences between a crisp and soft approach to characterizing class membership for each pixel. Because the natural environment contains ecotones and gradations, it is natural to consider that a pixel can have class membership in several reference classes. Conversely, with crisp classification, class membership can be only 0 or 1. This last fact leads to the poor performance of the classical overall accuracy measure that is further elucidated by studying agreement histograms.

We focused on overall accuracy because it is the most widely recognized measure of accuracy. However, our main techniques of analysis and conclusions extend naturally to other standard accuracy measures in additional to overall accuracy: producer's, user's, Kappa, Kappa histo, and Kappa no. Both the Binaghi and Gómez approaches to accuracy measures have been extended to be applicable using relevant combinations of soft/hard reference data and soft/hard classification. These extensions include both producer's and user's accuracies as well as overall accuracy. However, an advantage of the Gómez approach is that it has been extended to Kappa, which gives us the means for a statistical comparison of two classifiers, whereas Binaghi's approach does not.

Acknowledgments

We thank Dr. Peng Gong, UC Berkeley, for locating and making accessible the remote sensing and reference data used in this chapter and providing a valuable review. We are also grateful to Ms. Lu Liang, a PhD student in remote sensing at UC Berkeley, for her able assistance in readying the digital data (MODIS and reference) for analysis. This work has been partially supported by the Government of Spain, research grant TIN2012-32482, and by USDA McIntire Stennis project 7671 MS.

References

Biging, G. S., D. Colby, and R. G. Congalton. 1999. "Sampling Systems for Change Detection Accuracy Assessment." In *Remote Sensing Change Detection Environmental Monitoring Methods and Applications*, edited by R. S. Lunetta and C. D. Elvidge, 281–308. Chelsea, MI: Ann Arbor Press.

Binaghi, E., P. Brivio, P. Ghezzi, and A. Rampini. 1999. "A Fuzzy Set Based Accuracy Assessment of Soft Classification." *Pattern Recognition Letters* 20:935–48.

Carpenter, G. A., S. Gopal, S. Martens, and C. E. Woodcock. 1999. "A Neural Network Method for Mixture Estimation for Vegetation Mapping." *Remote Sensing of Environment* 70:138–52.

Chen, J., P. Gong, C. He, R. Pu, and P. Shi. 2003. "Land-Use/Land-Cover Change Detection Using Improved Change-Vector Analysis." *Photogrammetric Engineering & Remote Sensing* 69(4):369–79.

Chen, J., X. Zhu, H. Imura, and X. Chen. 2010. "Consistency of Accuracy Assessment Indices for Soft Classification: Simulation Analysis." *ISPRS Journal of Photogrammetry and Remote Sensing* 65(2):156–64.

Clinton, N., A. Holt, J. Scarborough, L. Yan, and P. Gong. 2010. "Accuracy Assessment Measure for Object-Based Image Segmentation Goodness." *Photogrammetric Engineering and Remote Sensing* 76(3):289–99.

Cohen, J. 1960. "A Coefficient of Agreement for Nominal Scales." *Educational and Psychological Measurement* 20:37–46.

Congalton, R. G. 1991. "A Review of Assessing the Accuracy of Classifications of Remotely Sensed Data." *Remote Sensing of Environment* 37:35–46.

Congalton, R. G., and K. Green. 1999. *Assessing the Accuracy of Remotely Sensed Data: Principles and Practices*. Boca Raton, FL: Lewis Publications.

Congalton, R. G., and K. Green. 2008. *Assessing the Accuracy of Remotely Sensed Data: Principles and Practices*. 2nd ed. Boca Raton, FL: CRC Press.

Congalton, R. G., and R. A. Mead. 1983. "A Quantitative Method to Test for Consistency and Correctness in Photointerpretation." *Photogrammetric Engineering and Remote Sensing* 49(1):69–74.

del Amo, A., J. Montero, and G.S. Biging. 2000. "Classifying pixels by means of fuzzy relations." *International Journal General Systems* 29: 605–21.

Dobson, J., and E. Bright. 1993. "Large-Area Change Analysis: The Coastwatch Change Analysis Project (C-CAP)." *Proceedings of Peccary 12*. August 24–26, 1993, Sioux Falls, SD. American Society for Photogrammetry and Remote Sensing, 1994, Bethesda, MD, 73–81.

Donner, A., M. Shoukri, N. Klar, and E. Bartfay. 2000. "Testing the Equality of Two Dependent Kappa Statistics." *Statistics in Medicine* 19:373–87.

Driese, K., W. Reiners, G. Lovett, and S. Simkin. 2004. "A Vegetation Map for the Catskill Park, NY, Derived From Multi-Temporal Landsat Imagery and GIS Data." *Northeastern Naturalist* 11:421–42.

Finn, J. T. 1993. "Use of the Average Mutual Information Index in Evaluating Classification Error and Consistency." *International Journal of Geographical Information Systems* 7(4):349–66.

Foody, G. M. 1995. "Cross-Entropy for the Evaluation of the Accuracy of a Fuzzy Land Cover Classification with Fuzzy Ground Data." *ISPRS Journal of Photogrammetry and Remote Sensing* 50(5):2–12.

Foody, G. M. 1999. "The Continuum of Classification Fuzziness in Thematics Mapping." *Photogrammetric Engineering and Remote Sensing* 65:443–51.

Foody, G. M. 2002. "Status of Land Cover Classification Accuracy Assessment." *Remote Sensing of Environment* 80:185–201.

Foody, G. M. 2004. "Thematic Map Comparison: Evaluating the Statistical Significance of Differences in Classification Accuracy." *Photogrammetric Engineering and Remote Sensing* 70:627–33.

Gong, P., and B. Xu. 2003. "Remote Sensing of Forests Over Time: Change Types, Method and Opportunities." In *Remote Sensing of Forest Environments: Concepts and Case studies*, edited by M. A. Wulder and S. E. Franklin, 301–34. Dordrecht/Boston/London: Kluwer Academic Publishers.

Gopal, S., and C. Woodcock. 1994. "Theory and Methods for Accuracy Assessment of Thematic Maps Using Fuzzy Sets." *Photogrammetric Engineering and Remote Sensing* 60:181–88.

Gómez, D., G. S. Biging, and J. Montero. 2008. "Accuracy Statistics for Judging Soft Classification." *International Journal of Remote Sensing* 29:693–709.

Gómez, D., and J. Montero. 2011. "Determining the Accuracy in Image Supervised Classification Problems." *Advances in Intelligent Systems Research* 1:342–49.

Green, K., and R. G. Congalton. 2004. "An Error Matrix Approach to Fuzzy Accuracy Assessment: The NIMA Geocover Project." In *Remote Sensing and GIS Accuracy Assessment*, edited by R. S. Lunetta and J. G. Lyon, 163–72. Boca Raton, FL: CRC Press.

Hagen, A. 2002. "Multi-Method Assessment of Map Similarity." *Proceedings of the 5th AGILE Conference on Geographic Information Science*, April 25–27. Palma, Spain, 171–82.

Ke, Y., L. Quackenbush, and J. Im. 2010. "Synergistic Use of QuickBird Multispectral Imagery and LIDAR Data for Object-Based Forest Species Classification." *Remote Sensing of Environment* 114(6):1141–54.

Khorram, S., ed. 1999. *Accuracy Assessment of Remote Sensing-Derived Change Detection*. Bethesda, MD: American Society for Photogrammetry and Remote Sensing.

Laba, M., S. Gregory, J. Braden, D. Ogurcak, E. Hill, E. Fegraus, J. Fiore, and S. D. DeGloria. 2002. "Conventional and Fuzzy Accuracy Assessment of the New York Gap Analysis Project Land Cover Map." *Remote Sensing of Environment* 81:443–55.

Lewis, H. G., and M. Brown. 2001. "A Generalized Confusion Matrix for Assessing Area Estimates from Remotely Sensed Data." *International Journal of Remote Sensing* 22:3223–35.

Liu, C., P. Frazier, and L. Kumar. 2007. "Comparative Assessment of the Measures of Thematic Classification Accuracy." *Remote Sensing of Environment* 108:606–16.

Liu, D., K. Song, J. R. G. Townshend, and P. Gong. 2008. "Using Local Transition Probability Models in Markov Random Fields for Forest Change Detection." *Remote Sensing of Environment* 112(5):2222–31.

Liu, J., Z. Zhang, D. Zhuan, Y. Wang, W. Zhou, S. Zhang, R. Li, N. Jiang, and S. Wu. 2003. "A Study on the Spatial-Temporal Dynamic Changes of Land-Use and Driving Forces Analyses of China in the 1990s." *Geographical Research* 22:1–12.

McKenzie, D. P., A. J. Mackinnon, N. Péladeau, P. Onghena, P. C. Bruce, D. M. Clarke, S. Harrigan, and P. D. McGorry. 1996. "Comparing Correlated Kappas by Resampling: Is One Level of Agreement Significantly Different from Another?" *Journal of Psychiatric Research* 30:483–92.

Montero, J., V. López, and D. Gómez. 2007. "The role of fuzziness in decision making." *In Fuzzy Logic.* Vol. 215, 337–49. Studies in Fuzziness and Soft computing.

Persello, C., and L. Bruzzone. 2009. "A Novel Protocol for Accuracy Assessment in Classification of Very High Resolution Images." *IEE Transactions on Geoscience and Remote Sensing* 48(3):1232–44.

Pontius, R. 2000. "Quantification Error versus Location Error in Comparison of Categorical Maps." *Photogrammetric Engineering and Remote Sensing* 66:1011–16.

Pontius, R., and K. Batchu. 2003. "Using the Relative Operating Characteristic to Quantify Certainty in Prediction of Location of Land Cover Change in India." *Transactions in GIS* 7:467–84.

Pontius, R., and M. Cheuck. 2006. "A Generalized Cross-Tabulation Matrix to Compare Soft-Classified Maps at Multiple Resolutions." *International Journal of Geographical Information Science* 20:1–30.

Pontius, R., and M. Millones. 2011. "Death to Kappa: Birth of Quantity Disagreement and Allocation Disagreement for Accuracy Assessment." *International Journal of Remote Sensing* 32(15):4407–29.

Rosenfield, G., and K. Fitzpatrick-Lins. 1986. "A Coefficient of Agreement As a Measure of Thematic Classification Accuracy." *Photogrammetric Engineering and Remote Sensing* 52:223–27.

Ruspini, E. 1969. "A New Approach to Clustering." *Information and Control* 15:22–32.

Sarmento, P., H. Carrão, M. Caetano, and S. V. Stehman. 2009. "Incorporating Reference Classification Uncertainty into the Analysis of Land Cover Accuracy." *International Journal of Remote Sensing* 30(20):5309–21.

Silván-Cárdenas, J., and L. Wang. 2008. "Sub-Pixel Confusion–Uncertainty Matrix for Assessing Soft Classifications." *Remote Sensing of Environment* 112(3):1081–95.

Smits, P. C., S. G. Dellepiane, and R. A. Schowengerdt. 1999. "Quality Assessment of Image Classification Algorithms for Land-Cover Mapping." *International Journal of Remote Sensing* 20:1461–86.

Stehman, S. 1995. "Thematic Map Accuracy Assessment from the Perspective of Finite Population Sampling." *International Journal of Remote Sensing* 16(3):589–93.

Stehman, S. 1997. "Selecting and Interpreting Measures of Thematic Classification Accuracy." *Remote Sensing of Environment* 62:77–89.

Stehman, S., and R. Czaplewski. 1998. "Design and Analysis for Thematic Map Accuracy Assessment: Fundamental Principles." *Remote Sensing of Environment* 64:331–44.

Townsend, P. A. 2000. "A Quantitative Fuzzy Approach to Assess Mapped Vegetation Classifications for Ecological Applications." *Remote Sensing of Environment* 72:253–67.

Wang, L., J. Chen, P. Gong, H. Shimazaki, and M. Tamura. 2009. "Land Cover Change Detection with a Cross-Correlogram Spectral Matching Algorithm." *International Journal of Remote Sensing* 30(2):3259–73.

Wilkinson, G. G. 2005. "Results of Implications of a Study of Fifteen Years of Satellite Classification Experiments." *IEEE Transaction on Geoscience and Remote Sensing* 43(3):433–40.

Woodcock, C., and S. Gopal. 2000. "Fuzzy Set Theory and Thematic Maps: Accuracy Assessment and Area Estimation." *International Journal Geographical Information Science* 14:153–72.

Yu, Q., P. Gong, N. Clinton, G. S. Biging, M. Kelly, and D. Schirokauer. 2006. "Object-Based Detailed Vegetation Classification with Airborne High Spatial Resolution Remote Sensing Imagery." *Photogrammetric Engineering and Remote Sensing* 72(7):799–811.

Zadeh, L. 1965. "Fuzzy Sets." *Information Sciences* 8:338–53.

Zhang, X., F. Xuezhi, and J. Hong. 2010. "Object-Oriented Method for Urban Vegetation Mapping Using IKONOS Imagery." *International Journal of Remote Sensing* 31(1):177–96.

5

Spatial Uncertainty Analysis When Mapping Natural Resources Using Remotely Sensed Data

Guangxing Wang and George Z. Gertner

CONTENTS

5.1 Introduction 87
5.2 Sources of Uncertainties and Errors 88
5.3 Existing Methods for Uncertainty and Sensitivity Analysis................................. 92
5.4 A General Methodology for Uncertainty and Sensitivity Analysis..................... 96
5.5 Case Studies.. 99
5.6 Conclusions... 102
References... 102

5.1 Introduction

Combining field plot observations and remotely sensed data has been widely used to generate spatially explicit estimates of natural resources such as leaf area index, forest biomass/carbon, biodiversity, soil types, and soil erosion (Wang et al. 2007, 2009b; Mascaro et al. 2011a; Lu et al. 2012). The used data and information possess a large amount of uncertainties and errors (Wang et al. 2009b, 2011). The procedures to generate maps of natural resources also lead to uncertainties (Wang et al. 2005). Therefore, natural resource estimates are associated with uncertainties (Gertner et al. 1995, 1996, 2002a, 2002b; Fang et al. 2002; Sierra et al. 2007; Larocque et al. 2008; Nabuurs et al. 2008; Mascaro et al. 2011b). Using the obtained maps in decision supports will definitely result in numerous risks (Heath and Smith 2000). Thus, how to quantify and reduce uncertainties of natural resource estimates is becoming very important.

There are many examples of creating natural resource maps by combining field plot observations and remotely sensed images. One kind of example is that forest biomass/carbon has often been mapped by combining forest inventory plot and remotely sensed data (Lefsky et al. 2002; Chave et al. 2003; Chen et al. 2003; Lu 2006; Saatchi et al. 2007; Tang et al. 2007a, 2007b; Asner et al. 2009; Wang et al. 2009b). Another kind of example is that human activity–induced soil erosion is usually estimated by mapping disturbance of vegetation canopy using field plot observations and remotely sensed images (Gertner 2002a, 2002b; Wang et al. 2002b, 2003, 2009a; Parysow et al. 2003; Anderson et al. 2005). Decisions can be made from the digital maps on which spatial patterns, distributions, processes, and relationships are clearly visualized and easily updated. Because of the potential of this technology, the use of these products for natural resource management such as forest management planning, global carbon cycle modeling, and land management has skyrocketed in the past decade. The key is to provide accurate spatial data products. However, most of the current systems produce products that do not meet the requirements in accuracy by natural resource management. The main reason may be partly due to the complexity and difficulties in spatially identifying uncertainties and errors and partly because the current systems lack the capabilities for spatially assessing the accuracy and validity of maps or ignore uncertainties and errors involved in the maps (Heuvelink et al. 1989; Lunetta et al. 1991; Goodchild and Gopal 1992; Lanter and Veregin 1992; Fang 2000; Crosetto and Tarantola 2001; Gertner et al. 2002a; Chave et al. 2004; Wang et al. 2005; Lilburne and Tarantola 2009; Mascaro et al. 2011b).

Natural resource managers often implicitly assume that the values that characterize model entities are true or error-free. This is usually known as the deterministic assumption. However, spatial data and maps are usually produced in geographic information systems (GIS) and image analysis systems using field measurements of thematic variables and remote sensing images by various interpolation methods from traditional regression modeling (Lefsky et al. 2002; Chave et al. 2003; Chen et al. 2003; Lu 2006; Saatchi et al. 2007; Asner et al. 2009) to nonparametric methods such as neural network and k-nearest neighbors (Foody et al. 2001, 2003; Muukkonen and Heiskanen 2005; Saatchi et al. 2007; Tomppo et al. 2008), and cokriging and sequential Gaussian cosimulation in geostatistics (Wang et al. 2002b, 2007, 2009b, 2011). Most values employed in spatial modeling are estimates of the true parameters and, therefore, have associated uncertainties (Gertner et al. 1995, 1996, 2002a, 2002b). Even the field measurements contain errors due to inaccurate sampling and measuring (Wang et al. 2005). At the same time, analyzing and processing remotely sensed images and modeling their relationships with thematic variables leads to inaccuracy of maps (Wang et al. 2009b). The uncertainties can be also due to nonsampling errors such as measurement errors, prediction errors from inappropriate modeling methods, and expert knowledge uncertainty. Obviously, when there are uncertainties in the inputs to a system, there must be uncertainties in the maps as well. Moreover, the sensitivity of predictions to these uncertainties can vary considerably in both time and space. Thus, remote sensing products possess a large amount of uncertainties and errors. To increase accuracy of the data products and information, there is a strong need to develop a systematic methodology for identifying sources of the uncertainties and errors, quantifying them, and assessing their contributions to the uncertainty of the final products, and potential impact on the risks in decision making.

This chapter intends to provide insight on the state of the art on spatial uncertainty analysis when mapping natural resources using remotely sensed data, to identify the significant gaps that currently exist in this area, further to suggest a methodological framework for this purpose, and demonstrate its application in two case studies. Through the case studies, the spatial effects of different sources of error in uncertainty of prediction maps generated by modeling and simulation will be accounted for.

5.2 Sources of Uncertainties and Errors

Error is defined as the deviation of a measured or computed value from its true or theoretically correct value of a variable (Longley et al. 2011). In practice, the correct value is often unknown and thus a value that has higher accuracy can be used as its reference. Uncertainty means the lack of certainty, is often defined as a measure of the difference between the contents of a dataset and the real phenomena, and can be represented using a set of possible states with probabilities assigned to each possible outcome. Uncertainty of a computed or measured value exists due to various errors, inaccuracy, ambiguity, vagueness, knowledge gap, and so on.

The products created using remotely sensed data possess a large amount of uncertainties and errors. They can be primarily divided into two groups: position and thematic errors (Lanter and Veregin 1992; Pontius 2000, 2002; Asner et al. 2009; Mascaro et al. 2011b). Position errors are mainly caused by inaccuracy from ground control and geometric rectification of map coordination (Wang et al. 2009b, 2011). First of all, position errors occur when the field plots used to collect ground data and the ground control points used to georeference remote sensing images are located using the global positioning system (GPS). For example, dense forest canopy will lead to difficulties for GPS to receive signals of satellites and will further result in location errors of field plots. Secondly, position errors also come from a digital elevation model (DEM) when it is used for georeferencing of images. Georeferencing images to projected coordinate systems such as the Universal Transverse Mercator (UTM) and registering one image to another also lead to geometric errors. Thirdly, position errors can come from transformations of map projection and coordinate systems. Position errors are also associated with the uncertainties due to mismatch of sample plots with image pixels and the inconsistency in size of sample plots and image pixels. Moreover, position errors are associated with disagreement between remotely sensed data and plot measurements because parts of trees along the boundaries are out of the plots even though both sample plots and pixels have the same spatial resolutions. Another source of position errors is inaccurate digitizing of thematic maps. Position errors lead not only to shifting of pixels on the resulting maps but also to inaccurate estimation of variables of interest (Wang et al. 2011). Unfortunately, position errors are usually neglected because it seems true that a position error of 100 m does not affect locating a city such as New York. In fact, however, a position error of 50 m may result in a significant decrease of accuracy of forest biomass/carbon estimates (Wang et al. 2011).

Compared to position errors, there are more sources of uncertainties due to thematic errors. All the processes from sampling and measuring data, acquiring and rectifying image data, analyzing data, and modeling to final generation of a map produce thematic errors and uncertainties. First of all, a nonrepresentative sample leads to sampling errors of interest variables. Inaccurately measuring an interest variable in the field results in measurement errors. There may also be variation of the variables and recording and grouping errors when they are measured. Secondly, some thematic errors are related to conversion coefficients, models or equations, and their parameters. For example, values of forest carbon at sample plots are often obtained by converting tree volumes to biomass and then to carbon, but variation of the conversion factors by tree species and even within a group of the same species will lead to uncertainties of forest carbon. Inappropriate allometric equations used to account for the relationship of tree volume with diameter and height and incorrect regression models used to explain the relationship of forest biomass/carbon with spectral variables will result in uncertainties of forest carbon.

Moreover, thematic errors are related to spectral variables. The uncertainties of spectral values can be caused by unbalance of platforms, motion of scanners, poor atmospheric conditions, slope, and shadow. The uncertainties of spectral values can also be due to the use of inappropriate spatial interpolation methods used when geometrical and radiometric corrections are conducted. In addition, selecting correct image transformation and enhancement methods and using optimal spatial and temporal resolutions can greatly increase accuracy of product maps. Other image and GIS operations such as data conversion and overlapping also create thematic errors in the spatial data. Finally, thematic errors can also come from gaps of knowledge.

Some of the errors and uncertainties may be cancelled out, while others will accumulate and be propagated into the product maps. Identifying the sources of the errors and uncertainties, modeling their accumulation and propagation, and finally quantifying their contributions to the uncertainty of outputs will be critical to control the quality of spatial data or maps and can be used to improve their accuracy (Gertner et al. 1995, 1996).

Many authors have analyzed the potential sources, amount, accumulation, and propagation of position errors in map production. For example, Amaud and Flori (1998) studied bias and precision of different sampling methods for GPS positions. GPS position accuracy varies with GPS receiver configuration, location, and surrounding objects possibly blocking reception or causing multipath reception, satellite constellation status, and ionosphere conditions (Misra and Enge 2011). In general, currently, GPS horizontal position errors vary from 0 to 15 m. Ortí (1981) presented optimal distribution of control points to minimize Landsat image geometric correction errors. Ford and Zanelli (1985) developed a method for quantifying position errors for geometric correction of images. The geometric errors to georeference images with UTM vary depending on spatial resolution. Chrisman (1982a, 1982b) analyzed potential errors from digitizing existing maps. Campbell and Mortenson (1989) presented a procedure for controlling quality of digitizing maps. Shi and Liu (2000) proposed a stochastic process–based model for position errors.

Mascaro et al. (2011a, 2011b) investigated the impacts of plot location errors on thematic errors when GPS was used to locate sample plots for mapping forest carbon using Light Detection and Ranging (LiDAR) and found that a plot location error of less than 10 m caused an error of 5 Mg C/ha. Wang et al. (2011) combined forest inventory sample plot data and Landsat Thematic Mapper (TM) images and studied the impacts of plot location errors on accuracy of mapping aboveground forest carbon. They concluded that increased plot location errors weakened both the spatial autocorrelation of aboveground forest carbon and its traditional correlation with spectral variables and decreased the accuracy of the forest carbon estimates.

Many authors have also studied potential sources, amount, accumulation, and propagation of thematic errors and uncertainties. Due to the limitation of space, as examples of thematic errors, this chapter only reviews the studies in the areas of mapping forest biomass and carbon and estimates human activity–induced vegetation disturbance and soil erosion.

In the United States, soil erosion is often predicted by a widely used Revised Universal Soil Loss Equation in which a soil erosion value is defined as a product of five input factors including rainfall-runoff erosivity R, soil erodibility K, slope length L, slope steepness S, and ground and vegetation canopy cover factor C (Wischmeier and Smith 1978; Renard et al. 1997). Each of these factors is dependent on several primary variables. The estimates of soil erosion are thus associated with many sources of uncertainty. Many authors have investigated the uncertainties of predicting soil erosion based on this equation. In a study of estimating soil erosion for an U.S. army installation, Parysow et al. (2001) found that the values of soil erodibility K factor obtained from soil samples for each soil type had a variation of 20% and significantly differed from that obtained using the National Cooperative Soil Survey. Moreover, Parysow et al. (2003) investigated the impacts of input uncertainties from five soil properties including very fine sand and silt, sand percentage, organic matter, soil permeability, and soil structure on the accuracy of estimates for this K factor for the same study area and pointed out that the major and minor sources of uncertainty were very fine sand and silt, and soil structure, respectively. In mapping military training–induced disturbance, Fang et al. (2002) grouped the uncertainties into four general categories: modeling, mapping, decision, and measurement errors. The modeling errors were related to the uncertainties of the model parameter estimates. The mapping errors were classified into road mapping errors, slope mapping errors, and vegetation mapping errors. They learned that the output uncertainties mainly came from the mapping errors, especially vegetation classification errors, and the model parameter errors. In another study of predicting soil erosion, Wang et al. (2002a) concluded that the variation from slope steepness had the largest impact on predicting topographic factor LS defined as a product of slope steepness and slope length L, while the model parameters and measurement errors had the least impacts. Wang et al. (2005) conducted an error and uncertainty budget for the aforementioned soil erosion equation and concluded that the topographic factor LS was the main source of uncertainty, followed by cover factor, soil erodibility factor, and rainfall runoff factor.

Another important topic for uncertainty analysis when mapping natural resources is related to mapping forest carbon. Forests are first sampled and tree variables such as tree height and tree diameter at breast height (DBH) are measured within each sample plot. Based on the measurements of tree diameter and height, tree volumes are calculated using allometric equations. The tree volumes are then summed and converted to forest biomass and carbon using the conversion coefficients. Finally, the forest biomass and carbon is interpolated from the sample plots to unobserved locations using spatial interpolation methods such as regression modeling. This process is complicated and the obtained estimates are associated with many sources of uncertainty. The uncertainty also varies spatially and temporally.

Saatchi et al. (2007) mapped aboveground live and dead tree biomass and belowground biomass in the Amazon basin and reported an uncertainty of 20%. In estimation of forest biomass in tropical forests by combining field plot and LiDAR data, Asner et al. (2010, 2011) and Mascaro et al. (2011a, 2011b) concluded that the errors of the estimates varied from 17 to 40 Mg C/ha (1 Mg = 1000 kg). Moreover, Gonzalez et al. (2010) found that LiDAR data–derived estimates of aboveground forest carbon had smaller errors than the corresponding estimates by QuickBird images and the reason was because the former was more correlated with tree height.

Based on Phillips et al. (2000), sampling error for sample plot selection and measurement error for tree height and diameter and regression error for tree volume were the most important sources of uncertainty for estimation of forest biomass/carbon. Keller et al. (2001) studied the effects of sampling errors and uncertainties due to allometric equations and ratios used to estimate biomass of tree roots, lianas and epiphytes, and necromass, on the quality of forest biomass estimates in the Tapajos National Forest of Brazil and pointed out that inappropriate allometric equations were the major source of uncertainty. Chave et al. (2004) also found that the allometric equations led to greater uncertainty than sampling and measurement errors of tree variables when forest biomass was estimated in a vast tropical forest landscape of Panama. Nabuurs et al. (2008) implied that it was often difficult to estimate the change of carbon sequestration due to forest management and planning because of great uncertainty in estimating forest carbon and that the stem model parameters mainly determined the uncertainty of aboveground forest carbon estimates. Wang et al. (2009b) suggested that the variation of the conversion coefficients from tree volume to biomass for the same tree species was up to 15%. They also indicated that the variation of image data had more impact on the accuracy of mapping aboveground forest carbon than the sample plot data.

The spatial resolutions of the used sample plots and images will also affect the quality of forest biomass/carbon estimates. As the spatial resolution becomes coarser, often the uncertainty of natural resource estimates decreases. One example is the study by Keller et al. (2001) in which the uncertainty of forest biomass estimates due to sampling error decreased by 10% if the size of sample plots was increased from 0.25 to 1 ha. This finding was also supported by the Chave et al. (2004) study. By increasing plot size from 0.36 to 1 ha, moreover, Mascaro et al. (2011a, 2011b) decreased the error of forest carbon stock by 38%.

The studies mentioned earlier are only examples of the research in the area of uncertainty analyses for natural resource mapping by combining field measurements and remotely sensed data. Most of these studies focused on analyzing individual sources of uncertainty and only a few of them dealt with a systematic investigation and exploration of spatial uncertainty analysis in this area. For example, Lunetta et al. (1991) and Walsh et al. (1987) systematically discussed the sources and classification of the errors and uncertainties for remote sensing and GIS products. Veregin (1992) and Chrisman (1992) presented concepts and methods for modeling position and thematic errors in GIS and remote sensing mapping. Friedl et al. (2001) summarized three sources of uncertainty in terms of image acquisition process, data processing techniques, and differences between spatial resolutions of instrument and scale of an ecological process on the ground. Saatchi et al. (2007) mentioned two sources of uncertainty in mapping forest biomass/carbon, including uncertainties associated with measurements at individual plots and those in extrapolating data from plots to the entire area.

Other examples of systematically investigating sources of uncertainty in natural resource mapping include the studies by Asner et al. (2009), Mascaro et al. (2011b), and Wang et al. (2009b). The authors summarized the uncertainties in generation of forest carbon maps into six categories: (1) errors related to tree variables such as sampling, measurement, recording, and grouping errors; (2) errors related to conversion coefficients, equations, and models, such as variation of conversion factors and parameters of equations and models; (3) errors related to spectral variables such as image data errors due to unbalance of platforms and motion of scanners; (4) errors associated with locations such as sample plot location errors and geometric correction errors; (5) errors associated with inconsistency in sizes of sample plots and image pixels and in borders of plots and tree crowns; and (6) errors associated with inconsistency in time of field observations and image data.

We have previously discussed potential sources of uncertainty in natural resource mapping. However, identifying the uncertainty sources is only the first step of spatial uncertainty analysis for mapping natural resources using remotely sensed images. The more important issue is how to quantify the uncertainties and model their propagation from inputs to outputs.

5.3 Existing Methods for Uncertainty and Sensitivity Analysis

The accuracy of natural resource maps obtained by combining field plot and remotely sensed data is traditionally assessed by calculating a Pearson product moment correlation coefficient, root-mean-square error between estimated and observed values of a continuous variable, or an error matrix for a categorical variable. These traditional measures are simple and focus on global accuracy of maps. On the other hand, quality assessment of natural resource maps can also be conducted using uncertainty analysis and sensitivity analysis (Heuvelink et al. 1989; Lodwick et al. 1990; Lanter and Veregin 1992; Heuvelink and Burrough 1993; Arbia et al. 1998; Heuvelink 1998; Crosetto and Tarantola 2001; Lilburne and Tarantola 2009;). The former requires modeling and assessing the uncertainty propagation from inputs to outputs through a mapping system. The latter requires apportioning the output uncertainties of the mapping system into different components of the input uncertainties.

Furthermore, error and uncertainty budgets can be used to assess the quality of an overall mapping system (Gertner and Guan 1991). An error and uncertainty budget can be considered as a catalog of the different error sources (Gelb et al. 1974) that allows the partitioning of the projection variance and bias according to their origins. As a specialized form of sensitivity analysis, an error and uncertainty budget shows the effects of individual errors and uncertainties and their groups on the quality of a multicomponent model's predictions. The goal in developing the error and uncertainty budget is to account for all major sources of errors that can be expected in a mapping system. By doing this, the sources of errors and uncertainties can be examined and partitioned in different ways. Furthermore, an error and uncertainty budget can be generated for different time steps and spatial scales.

Because of the way an error and uncertainty budget is generated, the components that cause the most uncertainty can be identified immediately. These components will be the ones that contribute the most toward final prediction variance and/or bias. Additionally, if the model is modified, the newly created uncertainty contributions can be assessed quickly. Accounting for error and uncertainty may have management implications as well (Gertner and Guan 1991). For example, management decisions may take the error and uncertainty analysis into consideration so as to achieve objectives in spite of the uncertainty in the mapping system.

There is a great body of literature in the area of quality assessment of natural resource maps. However, only a few reports have dealt with a systematical identification and analysis of uncertainties. For example, Lodwick et al. (1990) developed attribute error and sensitivity analysis methods associated with map-based suitability analysis. Heuvelink et al. (1989) introduced modeling propagation of errors in spatial modeling with GIS. Lanter and Veregin (1992) suggested a research paradigm for error and uncertainty propagation in layer-based GIS. Crosetto and Tarantola (2001) proposed a general procedure for uncertainty and sensitivity analysis of GIS-based products. Lilburne and Tarantola (2009) systematically described the procedure of sensitivity analysis and corresponding methods.

There are several widely used methods for conducting global uncertainty and sensitivity analysis of remote sensing–based mapping systems. They include the Monte Carlo method (Openshaw 1992; Helton 1993; Heuvelink 1998; Jansen 1999; Crosetto and Tarantola 2001), Fourier amplitude sensitivity test (FAST) (Cukier et al. 1973; Collins and Avissar 1994; Saltelli et al. 1999; Wang et al. 2002a; Fang et al. 2003; Xu and Gertner 2007, 2008a, 2008b, 2011; Lilburne and Tarantola 2009; Xu et al. 2009), Taylor series (Gertner et al. 1995; Parysow et al. 2003; Fang et al. 2004), polynomial regression (Gertner et al. 1996, 2002a; Wang et al. 2009b; Lu et al. 2012), Sobol's (1993) method, and response surface modeling (Downing et al. 1985; Iman and Helton 1988). Moreover, several widely used sampling designs are applied in global uncertainty and sensitivity analysis, including simple random sampling, Latin hypercube sampling (LHS), replicated LHS, quasi-random sequences, search curve–based sampling, and random balance design (Sobol 1967; Cukier et al. 1978; McKay 1995; Saltelli et al. 1999; Tarantola et al. 2006; Lilburne and Tarantola 2009; Xu and Gertner 2011).

In uncertainty and sensitivity analysis, screening experiments are often used to identify the input components that do not have significant effects on the output variability (Morris 1991; Campolongo et al. 2007). Variance-based methods that decompose the variance of a model prediction into partial variances to be explained by the model inputs are usually employed to calculate sensitivity indices (Patil and Frey 2004). Regression-based methods are applied to link the input uncertainties to the output uncertainty (Saltelli et al. 2000; Gertner et al. 2002a; Wang et al. 2009b; Lu et al. 2012).

Helton (1993), Saltelli et al. (2000), Helton and Davis (2003), and Lilburne and Tarantola (2009) summarized the advantages and disadvantages of these uncertainty and sensitivity analysis methods. The Monte Carlo method and Sobol's method are computationally intensive when the number of input parameters increases, although they can be used to deal with interactions among the input parameters. The Taylor series expansion–based methods can handle interactions among input parameters, but require that the model functions be continuous and differentiable. The FAST method is computationally efficient, but assumes that all the input parameters are independent, and there is also another problem about aliasing effect among parameters by using integer characteristic frequencies.

To overcome the shortcomings, some authors developed improvements of the existing methods. For example, Saltelli et al. (1999) and Xu and Gertner (2007, 2008a) extended FAST to models with correlated parameters based on the reordering of the independent sample in the traditional FAST. Xu and Gertner (2008b) conducted uncertainty and sensitivity analysis for models with correlated parameters and then proposed a regression-based method to quantitatively decompose the total uncertainty in model output into partial variances contributed by the correlated variations and partial variances contributed by the uncorrelated variations. Furthermore, Xu and Gertner (2011) compared different sampling approaches for FAST to reduce the effect of sampling errors on the estimation of partial variances and suggested that variance-based sensitivity indices estimated by search curve–based sampling had higher precision but larger underestimations than simple random sampling and random balance design sampling.

On the other hand, the accuracy of natural resource maps obtained using remotely sensed images often varies spatially depending on the complexity of landscape, soil properties, topographical features, density of sample data, and accuracy of remotely sensed data used (Congalton 1988; Steele et al. 1998; Gertner et al. 2002a; Wang et al. 2009b, 2011). That is, one kind of error is the largest source to the uncertainty of the output at a location, but not at another. Therefore, uncertainty and sensitivity analysis needs to be done on a pixel-by-pixel basis to account for spatial variation of uncertainty. However, a common feature of the existing uncertainty and sensitivity analysis methods is that they were originally developed for aspatial model systems and cannot be directly used to conduct spatial uncertainty and sensitivity analysis and make spatial error and uncertainty budgets for remote sensing–derived maps of natural resources. At the same time, there is abundant evidence to support the use of spatial information from neighboring locations to improve estimation at an unknown location (Gertner et al. 2002a). However, there are no existing methods available to assess the effect of the interactions and spatial information from neighbors in mapping. A systematic and practical methodology of spatial uncertainty and sensitivity analysis for natural resource mapping is needed.

After some improvement, the Taylor series–based method and FAST have been implemented to conduct spatial uncertainty and sensitivity analysis of remote sensing and GIS-derived natural resource maps (Openshaw 1992; Crosetto and Tarantola 2001; Fang et al. 2002; Wang et al. 2002a; Parysow et al. 2003; Xu and Gertner 2007, 2008a, 2011; Xu et al. 2009). For spatially correlated model parameters and interactions between the inputs, Gertner et al. (2002a) and Wang et al. (2009b) combined an image-based spatial cosimulation algorithm and polynomial regression method to spatially partition uncertainties of soil erosion and forest carbon estimates into various input components of errors.

Particularly, Gertner et al. (2002a) integrated a jointly spatial cosimulation algorithm and polynomial regression to map multiple variables, identify and quantify various resource errors, including the effect of spatial information from neighboring pixels, and further calculate their contributions to the output uncertainty.

Lilburne and Tarantola (2009) grouped spatial sensitivity analysis methods into (1) a one-at-a-time approach in which one input parameter at a time is varied by a certain amount, for example, 10%, to determine its effect on the output variation of a mapping system given other parameters are fixed; (2) a global spatial sensitivity analysis method (Crosetto and Tarantola 2001) in which extended FAST is used; (3) Sobol's method–based spatial sensitivity analysis in which a variance-based method is used to calculate sensitivity indices (Tang et al. 2007a, 2007b); (4) subregion-based spatial sensitivity analysis in which spatial variation of uncertainty is simplified, that is, an area is first divided into subzones, and within each subzone that is characterized by a set of scalar inputs, traditional sensitivity analysis is conducted without spatial uncertainty modeling (Hall et al. 2005); (5) the method of Lilburne et al. (2003) in which 1000 realizations of soil map were generated and the sequences were then sorted according to the objective function so that the inputs had some meaning; and (6) the method of Crosetto and Tarantola (2001) in which a binary input to switch uncertainties on and off at the same rate is used to determine their relative importance. Lilburne and Tarantola (2009) summarized the limitations of these methods and further enhanced the method of Tarantola (2006) by combining the methods of Sobol (1993) and Saltelli (2002) in spatial sensitivity analysis of spatially dependent models.

Although there has been substantial research in spatial uncertainty and sensitivity analysis of natural resource mapping, there is still a long way to go for one to clearly understand how errors and uncertainties from the input field plot and image data and the used methods, models, parameters, and knowledge gaps accumulate and are propagated to a map product of natural resources. For example, most of the existing methods neglect spatial autocorrelation of variables to be mapped and their spatial cross-correlation with remotely sensed images. Until a recent study (Wang et al. 2011), we did not know the characteristics of error in its estimation due to an error of 2 m to locate sample plots on the ground using GPS when a variable such as forest carbon is mapped. We also do not know the amount of error in classification of categorical variables when an error of 5% in remotely sensed data occurs. Furthermore, error varies spatially and temporally. There has been only one study that reported how the quality of estimates at a pixel is affected by use of information from neighboring pixels. It is also not clear as to how uncertainties accumulate and are propagated from the input components to the output when spatial data are scaled up from a finer spatial resolution such as 30 × 30 m to a coarser one such as 1 × 1 km. More examples can be easily given. In a word, we are lacking a systematic study on error accumulation and propagation in remote sensing–based mapping systems. When multiple variables that are spatially correlated with each other are jointly mapped, to account for the accumulation and propagation of error will become more complicated. These might be the reasons why the accuracy of a map is still often low, although we spend a large amount of money to develop various advanced sensors for image collection.

In an attempt to develop a methodological framework for spatial uncertainty and sensitivity analysis of remote sensing–based mapping for natural resources, Wang et al. (2005) suggested that an ideal and general methodology for spatial uncertainty and sensitivity analysis should have the following characteristics: (1) it is able to simulate all the interest variables and input components that vary spatially and temporally; (2) it takes into account the spatial autocorrelation of the variables to be mapped and their spatial cross-correlation with remotely sensed images that provide the linkage of the interest variables from sample locations to unobserved locations; (3) it is based on the realizations of the entire input and output maps, not on the individual pixels or subareas, so that the uncertainties of resulting estimates are not dependent on the order at which the realizations are generated and it is possible to account for the entire spatial structure of error and uncertainty; (4) when the interest variables are correlated with each other, their realizations can be generated using a spatially joint simulation; (5) it takes into consideration the effect of spatial information from neighboring locations; (6) it can model uncertainty propagation from various input components to output and the modeling is scale-independent, that is, given spatial resolutions of input spatial data, the output maps can have any spatial resolutions that are coarser than the pixel sizes of the input data; and (7) it has the capability to calculate the uncertainty contributions of all the input components to the output uncertainty.

5.4 A General Methodology for Uncertainty and Sensitivity Analysis

Based on Wang et al. (2005, 2009b), a methodological framework is suggested in this study and it meets the above-mentioned important characteristics for a general procedure of spatial uncertainty and sensitivity analysis. The framework integrates an image-based spatial cosimulation algorithm and an error budget method and can be used to identify various sources of errors and uncertainties; model their accumulation and propagation from input plot and image data, image processing and analysis, methods, models, and parameters to map products; develop an error and uncertainty budget; and quantify relative contributions of the input errors and uncertainties to the output uncertainty. Let the following model be a remote sensing–based mapping system for natural resources:

$$Y = f(X_1, X_2, \ldots, X_i, \ldots, X_M) \tag{5.1}$$

where Y is an interest variable and its spatially explicit estimates are generated, and X_i with $i = 1, 2, \ldots, M$ are the components that contain various input uncertainties. The methodological framework consists of the following steps:

1. Given a mapping system in which sample plot data are available only at sample locations and remotely sensed images are available everywhere, identify and calculate all sources of errors and uncertainties.
2. Make an assumption of probability distribution function for each input component and test the assumption.
3. Based on the above-mentioned conditional data and assumptions, an image-based spatial cosimulation algorithm is used to generate realizations of the expected map for the interest variable, and from the realizations, the variances of spatially explicit estimates as a measure of the output uncertainty are calculated.
4. Develop the relationships of the input uncertainties with the output uncertainty using uncertainty and sensitivity methods such as polynomial regression.
5. Calculate relative contributions of the input uncertainties to the output uncertainty.

First of all, a mapping system can be linear or nonlinear and associated with many sources of uncertainty. For example, aboveground forest carbon is usually obtained through conversions of tree volumes to forest biomass and then to carbon based on conversion coefficients or factors (Wang et al. 2009b). But, tree volume often has a nonlinear relationship with tree height and DBH. Moreover, spatially explicit estimates of aboveground forest carbon are often created by combining forest inventory plot data and remotely sensed images. The values of forest carbon only exist at the sample plots, while remotely sensed data are available everywhere and provide the association of forest carbon from sample locations to unobserved locations. Thus, the potential sources of uncertainty for the estimates of aboveground forest carbon will include variation of tree variables, their sampling and measurement errors; errors due to the conversion coefficients and inappropriate allometric equations; errors related to spectral variables; errors associated with locations of sample plots, their mismatch with image pixels, and their inconsistency in size with output map units; and so on.

Secondly, a probability distribution can be assumed for an interest variable and each of its input components. The widely used assumption is normal distribution. If normal distribution is not met, normal score transformation of data can be conducted. Moreover, in the methodological framework, it is supposed that a study area consists of square cells or pixels used for sampling and collecting ground plot and remotely sensed data. At the locations of sample plots, the image data are colocated. The method for map generation is based on a theory in geostatistics that states the spatial autocorrelation of variables. That is, an interest variable can be regarded as a random process and its values at different locations are the realizations of this random process. The values are spatially similar when the separation distance of data is small, and the similarity becomes weaker as the separation distance increases up to a threshold distance at which the data values become independent. If this variable is characterized by normal distribution, the realizations can be generated by integrating plot data and remotely sensed images using a sequential Gaussian cosimulation algorithm (Wang et al. 2005, 2009b).

In this cosimulation algorithm, a random function is first used to determine a random path to generate a realization of the interest variable at each location. Following this path, a conditional mean of the variable is derived at each location using an unbiased simple collocated cokriging estimator by weighting plot and remotely sensed data and the previously simulated values, if any, within a given neighborhood in which the variables are spatially autocorrelated. The weight of the data varies depending on the spatial autocorrelation and spatial configuration of the data locations in relation to the cell to be estimated. The closer the data location from the estimated cell, the more weight the data has. At the same time, a conditional variance of the mean is calculated based on the spatial autocorrelation and spatial configuration. Using the obtained mean and variance, a conditional distribution is then determined and used to randomly draw a value that is considered as a realization of this variable at this location. The value can be also used as a conditional data for the next simulation. After all the cells are visited, a realization of the expected map for the variable is yielded.

The process described previously can be repeated many times by creating and using different random paths, which will lead to many realizations at each location. From the realizations, a sample mean and sample variance can be calculated for each location and regarded as a predicted value of this variable and its measure of uncertainty, respectively. How many realizations should be generated? Generally, the optimal number of realizations should be the one at which the sample variance becomes stable. When more than one variable is mapped, a joint sequential cosimulation should be employed (Gertner et al. 2002a; Wang et al. 2004). That is, the interest variable that has the highest correlation with remotely sensed images is first simulated and then the variable that has the second highest correlation. When the second variable is simulated, the realization of the first variable can be used as conditional data for the second variable.

When the output map units are larger than the used plots and image pixels, a block cosimulation upscaling algorithm can be used for mapping natural resource (Wang et al. 2004, 2009b). The major advantage of this method is that it directly scales up sample plot data from finer to coarser spatial resolutions and at the same time produces variances of estimates at larger map units. In this upscaling method, it is assumed that a larger map unit can be divided into multiple smaller pixels that are similar to the sample plots in size. An unbiased cokriging estimator can be used to produce the estimate of each smaller pixel by weighting plot data, remotely sensed data, and previously simulated values, if any, within a given neighborhood. The estimates for the smaller pixels within the larger map unit are then used to determine a conditional distribution for the larger map unit. From the distribution, a value can be randomly drawn and regarded as a realization of a random process at this larger map unit. The realization will become conditional data for sequential simulation of neighboring larger map units. The rest of the simulation process is the same as mentioned earlier when the output map units are similar to the sample plots in size. The obtained variance maps will be input into the following spatial uncertainty analysis procedure.

The spatial uncertainty analysis procedure deals with how to explicitly model the relationship between prediction and input uncertainty. The relationship will provide a means to estimate uncertainty and error sensitivity of the predictions as a function of the uncertainty in the inputs. Different approaches including deterministic and stochastic can be used to build the relationships and further generate the spatial uncertainty and error budgets. The approach used depends on the structure of component models and the characteristics of errors. The deterministic approaches are based on analytical statistical estimators (expected mean square error models) and Taylor series approximations based on component models that are mathematically differentiable (Parysow et al. 2003). In terms of the stochastic approaches, they are Monte Carlo techniques based on simple random and LHS and on FAST (Wang et al. 2002a). Another alternative is polynomial regression modeling of uncertainty propagation (Gertner et al. 2002a; Wang et al. 2009b). An advantage of the regression is its ability to partition the uncertainty contribution of different model components, their interactions, and the effect of the used spatial information from neighbors on a pixel-by-pixel basis. The form of the polynomial regression is

$$\mathrm{Var}[Y(u)] = \sum_{h=0}^{H} \sum_{i=1}^{M} \sum_{j \geq i}^{M} b_{ijh} \mathrm{Cov}[X_i(u), X_j(u+h)] + \varepsilon \tag{5.2}$$

where $X_i(u)\,(i=1,2,\ldots,M)$ are the input uncertainty components including the independent variables, cross-covariances between them, and the components with respect to the effect of spatial information from neighboring pixels on the variance of the predicted variable. u is a location to be estimated. $h\,(=1,2,\ldots,H)$ is defined as the distance of a neighbor to the estimated location u. $h=0$ indicates a center location without neighbors to be considered and $h=H$ is the distance of the farthest neighbor. b_{ijh} is the coefficient of the regression model, and ε is the error term that is assumed to be independent and has a Gaussian distribution. When $i=j$ and $h=0$, the $\mathrm{Cov}[Z_i(u), Z_j(u+h)]$ represents the traditional variance of the input component i. $\mathrm{Var}[Y(u)]$ is the variance of estimate for the interest variable Y at the location u and is the sample variance calculated based on the realizations. The regression model can be used to spatially partition the variance of the mapped variable into the input components. The relative variance contribution $\mathrm{RVC}_i(u)$ for each input uncertainty source $Z_i(u)$ at location u can be calculated as follows:

$$\text{RVC}_i(u) = \frac{\sum_{j=1}^{M}\sum_{h=0}^{H} b_{ijh}\text{Cov}[Z_i(u), Z_j(u+h)]}{\text{Var}[Y(u)]} \tag{5.3}$$

In Section 5.5, we demonstrate the applications of this framework for two case studies.

5.5 Case Studies

The methodology has been applied to several studies related to mapping and spatial uncertainty analysis of natural resources. One of the case studies was conducted by Wang et al. (2009b) at Wuyuan County located in the northeast of Jiangxi Province, East China, and in a subtropical zone. In this case study, evergreen broad-leaf forests, deciduous and evergreen broad-leaf mixed forests, coniferous forests, and bamboo dominated the area. Aboveground forest carbon for the year 2001 was mapped by combining forest inventory plot data of 46 sample plots and Landsat Thematic Mapper™ images. In the plots, tree heights (H) and DBH were measured and used to predict tree volumes (V) based on an empirical equation $V = aD^b\text{DBH}^c$ (where a, b, and c were regression coefficients). The tree volumes were further converted to values of forest biomass and aboveground forest carbon using conversion factors. The spatially explicit estimates of aboveground forest carbon were then generated using the cosimulation algorithm mentioned in Section 5.4 that combined the plot and image data and at the same time the values of the forest carbon were scaled up from a spatial resolution of 28.5 × 28.5 m to a map unit of 969 × 969 m. A total of 400 simulations were conducted, which led to 400 predicted values at each pixel. From the predicted values, a sample mean and a sample variance were calculated for each location.

The sample variances mentioned previously were regarded as a measure of uncertainty of the sample mean estimates. A polynomial regression model was developed to account for the relationship of the sample variances of the forest carbon estimates with the variances of the used plot data and TM images and their interaction. The relative variance contributions (RVC) of the input plot data and TM images to the uncertainties of the forest carbon estimates were calculated using Equation 5.3. Figure 5.1 shows the results of RVCs. The plot data had larger values of RVC at the southwest and northwest and smaller values at the east, southeast, and northeast parts (Figure 5.1a), while a reverse spatial pattern of RVC was obtained for the band ratio (Figure 5.1b). Overall, the variation of the band ratio contributed more uncertainties to the estimates of forest carbon than the variation of the plot data. Moreover, the values of RVC from the product of the input plot forest carbon variance and band ratio variance, implying the interaction between these two variables, were negative throughout the study area with especially larger absolute values at the southwest parts (Figure 5.1c). That is, the product reduced the uncertainties of the output forest carbon estimates.

Another case study was conducted at Fort Hood, Texas, where soil erosion was of a major concern because military training activities damaged vegetation and degraded land condition (Gertner et al. 2002a). The annual soil erosion of the study area was mapped using DEMs, Landsat TM images, and field measurements of variables related to soil erosion. The basic model for soil erosion is the Revised Universal Soil Loss Equation consisting of the five input factors mentioned in Section 5.2. Each input factor is dependent on

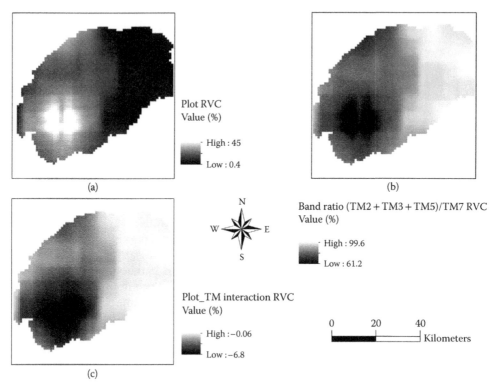

FIGURE 5.1
Relative variance contributions (RVC) of (a) plot tree carbon variance, (b) band ratio—(TM2 + TM3 + TM5)/ TM7, and (c) plot tree carbon–TM image interaction to the variances of upscaled forest carbon estimates. (From Wang, G. et al., *Forest Ecol. Manag.*, 258, 1275–83, 2009b.)

more than one primary variable. Thus, multiple variables that were spatially correlated with each other need to be mapped to derive the soil erosion map, and spatial error budgets must be implemented for all the factors and the entire prediction system.

Given the study area, soil erosion was mainly controlled by the vegetation cover factor that was a function of ground cover, canopy cover, and minimum vegetation height. These three variables were jointly mapped using the joint sequential cosimulation with field measurements of these variables and Landsat TM images of Fort Hood landscape (Gertner et al. 2002a). According to the resulting maps, the vegetation cover factor was then derived. In addition to prediction maps, the cosimulation also led to variance maps of the predicted variables. These variance maps were inputs into the spatial uncertainty and error budget procedure mentioned in Section 5.4 and a spatial partitioning of variances was conducted. Figure 5.2 presented the results of spatial uncertainty analysis and an error budget for the predicted map of vegetation cover factor, including variance maps of predicted ground cover (top left), canopy cover (top right), vegetation height (middle left), vegetation cover factor (middle right), and RVCs (bottom) of the input variables to the uncertainty of predicted vegetation cover factor values for pixels along a transect line indicated at the study area (middle right). RVC varies over space and the largest uncertainty source differs from place to place. Overall, the ground cover contributed more uncertainties than the canopy cover and vegetation height. But, at pixels 1–30 and

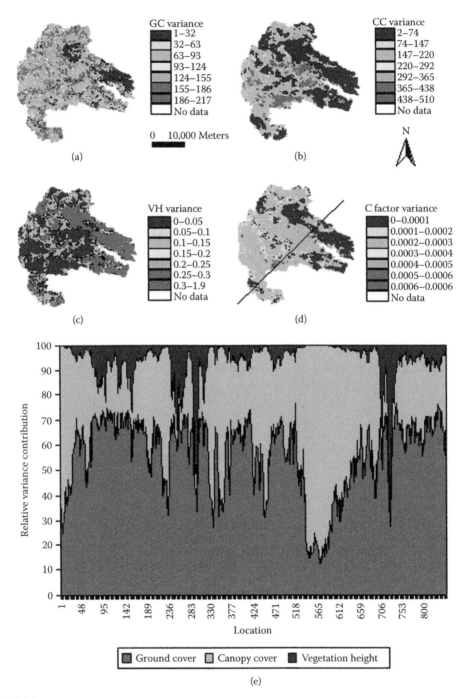

FIGURE 5.2
(**See color insert.**) Variance maps of predicted ground cover (a), canopy cover (b), vegetation height (c), and vegetation cover factor (d) related to soil erosion at Fort Hood, Texas, and relative variance contributions (e) of the input variables to uncertainty of the predicted cover factor for the pixels at a transect line marked at the vegetation cover factor variance (d). (From Gertner, G. et al., *Remote Sens. Environ.*, 83, 498–510, 2002a.)

pixels 530–610 along the transect line, the main uncertainties came from the canopy cover. This case study also found that there were significant variance contributions from the interactions between the ground cover and canopy cover and between the canopy cover and vegetation height and the spatial information from neighbors obviously reduced the uncertainties of the maps.

5.6 Conclusions

Combining field plot observations and remotely sensed data has been widely used to generate spatially explicit estimates of natural resources. The obtained map products are associated with uncertainties. Using the obtained maps in decision supports will definitely result in multiple risks. Thus, how to quantify and reduce uncertainties of natural resource estimates is very important. Although there has been a great body of literature in this area, a systematical study in which various sources of uncertainties that affect quality of natural resource maps are explored and spatial uncertainty analysis is conducted is still lacking. This chapter reviewed the existing studies and intended to provide an insight of the state of the art on spatial uncertainty analysis when mapping natural resources using remotely sensed data. Furthermore, this chapter identified significant gaps that currently exist in this area and suggested a methodological framework for the purpose for filling these gaps. Through two case studies, this chapter also demonstrated the applications of this framework to mapping and spatial uncertainty analysis of natural resources.

References

Amaud, M., and A. Flori. 1998. "Bias and Precision of Different Sampling Methods for GPS Positions." *Photogrammetric Engineering & Remote Sensing* 64(6):597–600.

Anderson, A. B., G. Wang, S. Fang, G. Z. Gertner, B. Güneralp, and D. Jones. 2005. "Assessing and Predicting Changes in Vegetation Cover Associated with Military Land Use Activities Using Field Monitoring Data at Fort Hood, Texas." *Journal of Terramechanics* 42(3–4):207–29.

Arbia, G., D. Griffith, and R. Haining. 1998. "Error Propagation Modelling in Raster GIS: Overlay Operations." *International Journal of Geographical Information Science* 12:145–67.

Asner, G. P., R. F. Hughes, J. Mascaro, A. Uowolo, D. E. Knapp, J. Jacobson, T. Kennedy-Bowdoin, J. K. Clark, and A. Balaji. 2011. "High-Resolution Carbon Mapping on the Million-Hectare Island of Hawai'i." *Frontiers in Ecology and the Environment* 9(8):434–439. doi:10.1890/100179.

Asner, G. P., R. F. Hughes, T. A. Varga, D. E. Knapp, and T. Kennedy-Bowdoin. 2009. "Environmental and Biotic Controls over Aboveground Biomass throughout a Tropical Rain Forest." *Ecosystems* 12:261–78.

Asner, G. P., G. V. N. Powell, J. Mascaro, D. E. Knapp, J. K. Clark, J. Jacobson, T. Kennedy-Bowdoin, et al. 2010. "High-Resolution Forest Carbon Stocks and Emissions in the Amazon." *Proceedings of the National Academy of Sciences of the United States of America* 107:16738–42.

Campbell, W. G., and D. C. Mortenson. 1989. "Ensuring the Quality of Geographic Information System Data: A Practical Application of Quality Control." *Photogrammetric Engineering and Remote Sensing* 55(11):1613–18.

Campolongo, F., J. Cariboni, and A. Saltelli. 2007. "An Effective Screening Design for Sensitivity Analysis of Large Models." *Environmental Modelling and Software* 22:1509–18.

Chave, J., R. Condit, S. Aguilar, A. Hernandez, S. Lao, and R. Perez. 2004. "Error Propagation and Scaling for Tropical Forest Biomass Estimates." *Philosophical Transactions of the Royal Society B: Biological Sciences* 359:409–20.

Chave, J., R. Condit, S. Lao, J. P. Casperson, R. B. Foster, and S. P. Hubbell. 2003. "Spatial and Temporal Variation of Biomass in a Tropical Forest: Results from a Large Census Plot in Panama." *Journal of Ecology* 91:240–52.

Chen, J. M., W. Ju, J. Cihlar, D. Price, J. Liu, W. J. Chen, J. Pan, T. A. Black, and A. Barr. 2003. "Spatial Distribution of Carbon Sources and Sinks in Canada's Forests Based on Remote Sensing." *Tellus B* 55:622–42.

Chrisman, N. R. 1982a. "Methods of Spatial Analysis Based on Error in Categorical Maps." Ph.D. thesis, University of Bristol, United Kingdom.

Chrisman, N. R. 1982b. "A Theory of Cartographic Error and Its Measurement in Digital Data Bases." *Proceedings of Auto-Carto* 5:159–68.

Chrisman, N. R. 1992. "Modeling Error in Overlaid Categorical Maps." In *Accuracy of Spatial Databases*, edited by M. Goodchild and S. Gopal, 21–34. Bristol, PA: Taylor & Francis.

Collins, C., and R. Avissar. 1994. "An Evaluation with the Fourier Amplitude Sensitivity Test (FAST) of Which the Land Surface Parameters Are of Greatest Importance in Atmospheric Modeling." *Journal of Climate* 7:681–703.

Congalton, R. G. 1988. "Using Spatial Autocorrelation Analysis to Explore the Errors in Maps Generated from Remotely Sensed Data." *Photogrammetric Engineering and Remote Sensing* 54:587–92.

Crosetto, M., and S. Tarantola. 2001. "Uncertainty and Sensitivity Analysis: Tools for GIS-Based Model Implementation." *International Journal of Geographical Information Science* 15(5):415–37.

Cukier, R. I., C. M. Fortuin, K. E. Shuler, A. G. Petschek, and J. H. Schaibly. 1973. "Study of the Sensitivity of Coupled Reaction Systems to Uncertainties in Rate Coefficients. I. Theory." *Journal of Chemical Physics* 59:3873–78.

Cukier, R. I., H. B. Levine, and K. E. Schuler. 1978. "Nonlinear Sensitivity Analysis of Multiparameter Model Systems." *Journal of Computational Physics* 26:1–42.

Downing, D. J., R. H. Garder, and F. O. Hoffman. 1985. "An Examination of Response-Surface Methodologies for Uncertainty Analysis in Assessment Models." *Technometrics* 27(2):151–63.

Fang, S. 2000. "Uncertainty Analysis of Biological Nonlinear Models Based on Bayesian Estimation." PhD diss., University of Illinois at Urbana-Champaign, IL.

Fang, S., G. Z. Gertner, and A. B. Anderson. 2004. "Estimation of Sensitivity Coefficients of Nonlinear Model Input Parameters Which Have a Multinormal Distribution." *Computer Physics Communications* 157:9–16.

Fang, S., G. Z. Gertner, S. Shinkareva, G. Wang, and A. Anderson. 2003. "Improved Generalized Fourier Amplitude Sensitivity Test (FAST) for Model Assessment." *Statistics and Computing* 13(3):221–26.

Fang, S., S. Wente, G. Z. Gertner, G. Wang, and A. B. Anderson. 2002. "Uncertainty Analysis of Predicted Disturbance from Off-Road Vehicular Traffic in Complex Landscapes." *Environmental Management* 30(2):199–208.

Foody, G. M., D. S. Boyd, and M. E. Cutler. 2003. "Predictive Relations of Tropical Forest Biomass from Landsat TM Data and Their Transferability between Regions." *Remote Sensing of Environment* 85:463–74.

Foody, G. M., M. E. Cutler, J. McMorrow, D. Pelz, H. Tangki, D. S. Boyd, and I. Douglas. 2001. "Mapping the Biomass of Bornean Tropical Rain Forest from Remotely Sensed Data." *Global Ecology and Biogeography* 10:379–87.

Ford, G. E., and C. I. Zanelli. 1985. "Analysis and Quantification of Errors in the Geometric Correction of Satellite Images." *Photometric Engineering and Remote Sensing* 51(11):1725–34.

Friedl, M. A., K. C. McGwire, and D. K. McIver. 2001. "An Overview of Uncertainty in Optical Remotely Sensed Data for Ecological Applications." In *Spatial Uncertainty in Ecology: Implications for Remote Sensing and GIS Applications*, edited by C. T. Hunsaker, M. F. Goodchild, M. A. Friedl, and T. J. Case, 258–83. New York: Springer-Verlag.

Gelb, A., J. Kasper Jr., R. Nash Jr., C. Price, and A. Sutherland Jr. 1974. *Applied Optimal Estimation*. Cambridge, MA: The MIT Press.

Gertner, G. Z., X. Cao, and H. Zhu. 1995. "A Quality Assessment of a Weibull Based Growth Projection System." *Forest Ecology and Management* 71:235–50.

Gertner, G. Z., and B. T. Guan. 1991. "Using an Error Budget to Evaluate the Importance of Component Models within a Large Scale Simulation Model." *Proceedings of Conference on Mathematical Modelling of Forest Ecosystems*, 62–74. Frankfurt am Main, Germany: J.D. Sauerländer's Verlag.

Gertner, G. Z., P. Parysow, and B. Guan. 1996. "Projection Variance Partitioning of a Conceptual Forest Growth Model with Orthogonal Polynomials." *Forest Science* 42(4):474–86.

Gertner, G., G. Wang, S. Fang, and A. B. Anderson. 2002a. "Mapping and Uncertainty of Predictions Based on Multiple Primary Variables from Joint Co-Simulation with TM Image." *Remote Sensing of Environment* 83:498–510.

Gertner, G. Z., G. Wang, S. Fang, and A. B. Anderson. 2002b. "Error Budget Assessment of the Effect of DEM Spatial Resolution in Predicting Topographical Factor for Soil Loss Estimation." *Journal of Soil and Water Conservation* 57(3):164–74.

Gonzalez, P., G. P. Asner, J. J. Battles, M. A. Lefsky, K. M. Waring, and M. Palace. 2010. "Aboveground Forest Carbon Densities and Uncertainties from Lidar, QuickBird, and Field Measurements in California." *Remote Sensing of Environment* 114(7):1561–75.

Goodchild, M., and S. Gopal. 1992. *Accuracy of Spatial Databases*. Bristol, PA: Taylor & Francis.

Hall, J.W., S. Tarantola, P.D. Bates, and M.S. Horritt. 2005. Distributed sensitivity analysis of flood inundation model calibration. *Journal of Hydraulic Engineering*, ASCE 131:117–126.

Heath, L. S., and J. E. Smith. 2000. "An Assessment of Uncertainty in Aboveground Forest Carbon Budget Projections." *Environmental Science and Policy* 3:73–82.

Helton, J. C. 1993. "Uncertainty and Sensitivity Analysis Techniques for Use in Performance Assessment for Radioactive Waste Disposal." *Reliability Engineering and System Safety* 42:327–67.

Helton, J. C., and F. J. Davis. 2003. "Latin Hypercube Sampling and the Propagation of Uncertainty in Analysis of Complex Systems." *Reliability Engineering and System Safety* 81:23–69.

Heuvelink, G. B. M. 1998. *Error Propagation in Environmental Modelling with GIS*. London: Taylor & Francis.

Heuvelink, G. B. M., and P. A. Burrough. 1993. "Error Propagation in Cartographic Modelling Using Boolean Logic and Continuous Classification." *International Journal of Geographical Information Systems* 7:231–46.

Heuvelink, G. B. M., P. A. Burrough, and A. Stein. 1989. "Propagation of Errors in Spatial Modeling." *International Journal of Geographical Information Systems* 3(4):323–34.

Iman, R. L., and J. C. Helton. 1988. "An Investigation of Uncertainty and Sensitivity Analysis Techniques for Computer Models." *Risk Analysis* 8(1):71–90.

Jansen, M. J. W. 1999. "Analysis of Variance Designs for Model Output." *Computer Physics Communications* 117(1):35–43.

Keller, M., M. Palace, and G. C. Hurtt. 2001. "Biomass Estimation in the Tapajos National Forest, Brazil: Examination of Sampling and Allometric Uncertainties." *Forest Ecology and Management* 154:371–82.

Lanter D. P., and H. Veregin. 1992. "A Research Paradigm for Propagating Error in Layer-Based GIS." *Photogrammetric Engineering & Remote Sensing* 58(6):825–33.

Larocque, G. R., J. S. Bhatti, R. Boutin, and O. Chertov. 2008. "Uncertainty Analysis in Carbon Cycle Models of Forest Ecosystems: Research Needs and Development of a Theoretical Framework to Estimate Error Propagation." *Ecological Modelling* 219:400–12.

Lefsky, M. A., W. B. Cohen, G. G. Parker, and D. J. Harding. 2002. "Lidar Remote System for Ecosystem Studies." *American Institute of Biological Sciences* 52:19–30.

Liburne, L., and S. Tarantola. 2009. "Sensitivity Analysis of Spatial Models." *International Journal of Geographical Information Science* 23(2):151–68.

Lilburne, L.R., T.H. Webb, and G.S. Francis. 2003. Relative effect of climate, soil, and management of nitrate leaching under wheat production in Canterbury, New Zealand. Australian Journal of Soil Research 41:699–709.

Lodwick, W. A., W. Monson, and L. Svoboda. 1990. "Attribute Error and Sensitivity Analysis of Map Operations in Geographical Information Systems: Suitability Analysis." *International Journal of Geographical Information Systems* 4(4):413–28.

Longley, P. A., M. F. Goodchild, D. J. Maguire, and D. W. Rhind. 2011. *Geographic Information Systems and Science.* New Jersey: John Wiley.

Lu, D. 2006. "The Potential and Challenge of Remote Sensing–Based Biomass Estimation. *International Journal of Remote Sensing* 27(7):1297–328.

Lu, D., Q. Chen, G. Wang, E. Moran, M. Batistella, M. Zhang, G. V. Laurin, and D. Saah. 2012. "Estimation and Uncertainty Analysis of Aboveground Forest Biomass with Landsat and LiDAR data: Brief Overview and Case Studies." *International Journal of Forestry Research* 1:1–16.

Lunetta, R. S., R. G. Congalton, L. K. Fenstermaker, J. R. Jensen, K. C. McGwire, and L. R. Tinney. 1991. "Remote Sensing and Geographic Information System Data Integration: Error Sources and Research Issues." *Photogrammetric Engineering & Remote Sensing* 57(6):677–87.

Mascaro, J., G. P. Asner, H. C. Muller-Landau, M. van Breugel, J. Hall, and K. Dahlin. 2011a. "Control over Aboveground Forest Carbon Density on Barro Colorado Island, Panama." *Biogeosciences* 8:1615–29.

Mascaro, J., M. Detto, G. P. Asner, and H. C. Muller-Landau. 2011b. "Evaluating Uncertainty in Mapping Forest Carbon with Airborne LiDAR." *Remote Sensing of Environment* 115:3770–74.

McCkay, M. D. 1995. Technical Report NUREG/CR-6311, LA-12915-MS. Washington, DC: US Nuclear Regulatory Commission and Los Alamos National Laboratory.

Misra, P., and P. Enge. 2011. *Global Positioning System.* Lincoln, MA: Ganga-Jamuna Press.

Morris, M. D. 1991. "Factorial Sampling Plans for Preliminary Computational Experiments." *Technometrics* 33:161–74.

Muukkonen, P., and J. Heiskanen. 2005. "Estimating Biomass for Boreal Forests Using ASTER Satellite Data Combined with Standwise Forest Inventory Data." *Remote Sensing of Environment* 99:434–47.

Nabuurs, G. J., B. van Putten, T. S. Knippers, and G. M. J. Mohren. 2008. "Comparison of Uncertainties in Carbon Sequestration Estimates for a Tropical and a Temperate Forest." *Forest Ecology and Management* 256:237–45.

Openshaw, S. 1992. "Learning to Live with Errors in Spatial Databases." In *Accuracy of Spatial Databases,* edited by M. Goodchild and S. Gopal, 263–76. New York: Taylor & Francis.

Ortí, F. 1981. "Optimal Distribution of Control Points to Minimize Landsat Image Registration Error." *Photogrammetric Engineering and Remote Sensing* 47(1):101–10.

Parysow, P., G. Wang, G. Z. Gertner, and A. B. Anderson. 2001. "Assessing Uncertainty of Soil Erodibility Factor in the National Cooperative Soil Survey: A Case Study at Fort Hood, Texas." *Journal of Soil and Water Conservation* 56(3):206–10.

Parysow, P., G. Wang, G. Z. Gertner, and A. B. Anderson. 2003. "Spatial Uncertainty Analysis for Mapping Soil Erodibility Based on Joint Sequential Simulation." *Catena* 53(1):65–78.

Patil, S. R., and H. C. Frey. 2004. "Comparison of Sensitivity Analysis Methods Based on Applications to a Food Safety Risk Assessment Model." *Risk Analysis* 24:573–85.

Phillips, D. L., S. L. Brown, P. E. Schroeder, and R. A. Birdsey. 2000. "Toward Error Analysis of Large-Scale Aboveground Forest Carbon Budgets." *Global Ecology & Biogeography* 9(4):305–13.

Pontius, R. G. 2000. "Quantification Error versus Location Error in Comparison of Categorical Maps." *Photogrammetric Engineering and Remote Sensing* 66(8):1011–16.

Pontius, R. G. 2002. "Statistical Methods to Partition Effects of Quantity and Location during Comparison of Categorical Maps at Multiple Resolutions." *Photogrammetric Engineering and Remote Sensing* 68(10):1041–49.

Renard, K. G., G. R. Foster, G. A. Weesies, D. K. McCool, and D. C. Yoder. 1997. *Predicting Soil Erosion by Water: A Guide to Conservation Planning with the Revised Universal Soil Loss Equation (RUSLE)*. Washington, DC: USDA, Agriculture Handbook Number 703, U.S. Government Printing Office, SSOP.

Saatchi, S. S., R. A. Houghton, R. Alvalá, J. V. Soares, and Y. Yu. 2007. "Distribution of Aboveground Live Biomass in the Amazon Basin." *Global Change Biology* 13:816–37.

Saltelli, A. 2002. "Making Best Use of Model Evaluations to Compute Sensitivity Indices." *Computer Physics Communications* 145:280–97.

Saltelli, A., K. Chan, and M. Scott. 2000. *Sensitivity Analysis*. Chichester: John Wiley.

Saltelli, A., S. Tarantola, and K. Chan. 1999. "A Quantitative Model-Independent Method for Global Sensitivity Analysis of Model Output." *Technometrics* 41:39–56.

Shi, W., and W. Liu. 2000. A stochastic process-based model for the positional error of line segments in GIS. *International Journal of Geographical Information Science* 14:51–66.

Sierra, C. A., J. I. Valle, S. A. Orrego, F. H. Moreno, M. E. Harmon, M. Zapata, G. J. Colorado, et al. 2007. "Total Carbon Stocks in a Tropical Forest Landscape of the Porce Region, Colombia." *Forest Ecology and Management* 243:299–309.

Sobol', I. M. 1967. "On the Distribution of Points in a Cube and the Approximate Evaluation of Integrals." *USSR Computational Mathematics and Mathematical Physics* 7:86–112.

Sobol', I. M. 1993. "Sensitivity Analysis for Non-Linear Mathematical Models." *Mathematical Modelling and Computational Experiment* 1:407–14.

Steele, B. M., J. C. Winne, and R. L. Redmond. 1998. "Estimation and Mapping of Misclassification Probabilities for Thematic Land Cover Maps." *Remote Sensing of Environment* 66:192–202.

Tang, Y., P. Reed, K. van Werkhoven, and T. Wagener. 2007a. "Advancing the Identification and Evaluation of Distributed Rainfall-Runoff Models Using Global Sensitivity Analysis." *Water Resources Research* 43:W06415.

Tang, Y., P. Reed, T. Wagener, and K. van Werkhoven. 2007b. "Comparing Sensitivity Analysis Methods to Advance Lumped Watershed Model Identification and Evaluation." *Hydrology and Earth System Science* 11:793–817.

Tarantola, S., D. Gatelli, and T. Mara. 2006. "Random Balance Designs for the Estimation of First Order Global Sensitivity Indices." *Reliability Engineering and System Safety* 91:717–27.

Tomppo, E., H. Olsson, G. Ståhl, M. Nilsson, O. Hagner, and M. Katila. 2008. "Combining National Forest Inventory Field Plots and Remote Sensing Data for Forest Databases." *Remote Sensing of Environment* 112:1982–99.

Veregin, H. 1992. "Error Modeling for the Map Overlay Operation." In *Accuracy of Spatial Databases*, edited by M. Goodchild and S. Gopal, 3–18. Bristol, PA: Taylor & Francis Ltd.

Walsh S. J., D. R. Lightfoot, and D. R. Butler. 1987. "Recognition and Assessment of Error in Geographic Information Systems." *Photogrammetric Engineering and Remote Sensing* 53(10):1423–30.

Wang, G., S. Fang, S. Shinkareva, G. Z. Gertner, and A. B. Anderson. 2002a. "Spatial Uncertainty in Prediction of Topographical Factor for the Revised Universal Soil Loss Equation (RUSLE)." *Transactions of American Society of Agricultural Engineer* 45(1):109–18.

Wang, G., G. Z. Gertner, and A. B. Anderson. 2004. "Mapping Vegetation Cover Change Using Geostatistical Methods and Bi-Temporal Landsat TM Images." *IEEE Transactions on Geoscience and Remote Sensing* 42(3):632–43.

Wang, G., G. Z. Gertner, A. B. Anderson, and H. R. Howard. 2009a. "Simulating Spatial Pattern and Dynamics of Military Training Impacts for Allocation of Land Repair Using Images." *Environmental Management* 44:810–23.

Wang, G., G. Z. Gertner, A. B. Anderson, H. R. Howard, D. Gebhard, D. P. Althoff, T. Davis, and P. B. Woodford. 2007. "Spatial Variability and Temporal Dynamics Analysis of Soil Erosion due to Military Land Use Activities: Uncertainty and Implications for Land Management. *Land Degradation and Development* 18:519–42.

Wang, G., G. Gertner, S. Fang, and A. B. Anderson. 2003. "Mapping Multiple Variables for Predicting Soil Loss by Joint Sequential Co-Simulation with TM Images and Slope Map." *Photogrammetric Engineering & Remote Sensing* 69(8):889–98.

Wang, G., G. Z. Gertner, S. Fang, and A. B. Anderson. 2005. "A Methodology for Spatial Uncertainty Analysis of Remote Sensing Products." *Photogrammetric Engineering and Remote Sensing* 71(12):1423–32.

Wang, G., T. Oyana, M. Zhang, S. Adu-Prah, S. Zeng, H. Lin, and J. Se. 2009b. "Mapping and Spatial Uncertainty Analysis of Forest Vegetation Carbon by Combining National Forest Inventory Data and Satellite Images." *Forest Ecology and Management* 258(7):1275–83.

Wang, G., S. Wente, G. Gertner, and A. B. Anderson. 2002b. "Improvement in Mapping Vegetation Cover Factor for Universal Soil Loss Equation by Geo-Statistical Methods with Landsat TM Images." *International Journal of Remote Sensing* 23(10):3649–67.

Wang, G., M. Zhang, G. Z. Gertner, T. Oyana, R. E. McRoberts, and H. Ge. 2011. "Uncertainties of Mapping Forest Carbon due to Plot Locations Using National Forest Inventory Plot and Remotely Sensed Data." *Scandinavia Journal of Forest Research* 26:360–73.

Wischmeier W. H., and D. D. Smith. 1978. *Predicting Rainfall-Erosion Losses from Cropland East of the Rocky Mountains: Guide for Selection of Practices for Soil and Water Conservation.* Washington, DC: USDA, Agriculture Handbook. No. 282, U.S. Government Printing Office, SSOP.

Xu, C., and G. Z. Gertner. 2007. "Extending a Global Sensitivity Analysis Technique to Models with Correlated Parameters." *Computational Statistics & Data Analysis* 51(12):5579–90.

Xu, C., and G. Z. Gertner. 2008a. "A General First-Order Global Sensitivity Analysis Method." *Reliability Engineering & System Safety* 93(7):1060–71.

Xu, C., and G. Z. Gertner. 2008b. "Uncertainty and Sensitivity Analysis for Models with Correlated Parameters." *Reliability Engineering & System Safety* 93(10):1563–73.

Xu, C., G. Z. Gertner, and R. M. Scheller. 2009. Uncertainties in the response of a forest landscape to global climatic change. *Global Change Biology* 15:116–131.

Xu, C., and G. Z. Gertner 2011. "Understanding and Comparisons of Different Sampling Approaches for the Fourier Amplitudes Sensitivity Test (FAST)." *Computational Statistics and Data Analysis* 55:184–98.

Section III

Land Use and Land Cover Classification

6

Land Use/Land Cover Classification in the Brazilian Amazon with Different Sensor Data and Classification Algorithms

Guiying Li, Dengsheng Lu, Emilio Moran, Mateus Batistella,
Luciano V. Dutra, Corina C. Freitas, and Sidnei J. S. Sant'Anna

CONTENTS

6.1 Introduction 111
6.2 Study Area 112
6.3 Methods 112
    6.3.1 Data Collection and Preprocessing 113
    6.3.2 Selection of Suitable Variables from Remotely Sensed Data 114
        6.3.2.1 Identification of Suitable Textural Images 114
        6.3.3.2 Integration of Multisensor Data 114
        6.3.2.3 Comparison of Different Datasets for LULC Classification 115
    6.3.3 Comparison of Different Classification Algorithms 116
    6.3.4 LULC Classification and Evaluation of Results 117
6.4 Resultant Analysis 117
    6.4.1 Comparison of Accuracy Assessment Results from Different Datasets 117
    6.4.2 Comparison of Accuracy Assessment Results from Different Algorithms 119
6.5 Discussion 121
    6.5.1 Roles of Radar Data in Improving Vegetation Classification 121
    6.5.2 Selection of Suitable Nonstatistical-Based Algorithms 122
6.6 Conclusions 123
Acknowledgments 123
References 123

6.1 Introduction

Land use/land cover (LULC) classification from remote sensing data is a complex procedure. Many factors such as the spatial resolution of the remotely sensed data, availability of different data sources, a suitable land cover classification system, availability of image-processing software, use of a suitable classification algorithm, and the analyst's experience may all affect the classification results (Lu and Weng 2007). The major steps for LULC classification from remotely sensed data have been discussed by Lu and Weng (2007). For a specific LULC classification, two critical questions need to be answered: What kinds of remote sensing data and variables are used? What algorithm is suitable for this kind of dataset for LULC classification?

As different kinds of sensor data are available, how to effectively incorporate these different features inherent in remote sensing data into a classification procedure for improving LULC classification becomes a new challenge. In particular, optical and radar data have different discrimination capabilities, thus combination of these features may improve LULC classification (Zhang 2010; Lu et al. 2011). Meanwhile, many classification algorithms such as neural network, decision tree, support vector machine, object-based algorithms, and subpixel-based algorithms have been developed in the past four decades (Lu and Weng 2007; Tso and Mather 2009; Blaschke 2010). It is important to better understand the capability of different classification algorithms and to identify a suitable algorithm for a specific dataset. In practice, it is still unclear which remote sensing variables and which classification algorithm can provide the best LULC classification in a study area, especially in the moist tropical region. Therefore, the objectives of this chapter are to examine different discrimination capabilities of Landsat Thematic Mapper (TM) image and radar data (i.e., Advanced Land Observing Satellite [ALOS] Phased Array type L-band Synthetic Aperture Radar [PALSAR] and RADARSAT C-band data in this research) and their integration in LULC classification, and to compare different classification algorithms for identifying a suitable algorithm for classifying a Landsat multispectral image in the moist tropical region of the Brazilian Amazon.

6.2 Study Area

Altamira is located along the Trans-Amazon Highway (BR-230) in the northern Brazilian state of Pará. The city lies on the Xingu River at the eastern edge of the study area. Major deforestation in this region began in the early 1970s, coincident with the construction of the Trans-Amazon Highway, and extensive deforestation has occurred since the 1980s, forming fishbone deforestation patterns. The dominant native types of vegetation are mature moist forest and liana forest. Deforestation has led to a complex landscape consisting of different stages of secondary succession, pasture, and agricultural lands (Moran et al. 1994; Moran and Brondizio 1998; Lu et al. 2011). Various stages of successional vegetation are mainly distributed along the Trans-Amazon Highway and feeder roads.

6.3 Methods

The framework of LULC classification based on different datasets is shown in Figure 6.1. The major steps include image preprocessing (radiometric and atmospheric calibration for a Landsat TM image, image-to-image registration between a Landsat TM image and radar data, and speckle reduction for radar data), development of textural images from original radar data, principal component analysis of radar radiometric and textural images, data fusion with the wavelet-merging technique, LULC classification with maximum likelihood classifier, and evaluation of the classification results. Meanwhile, different classification algorithms were also used to examine LULC classification based on a Landsat TM multispectral image for identifying the best algorithm.

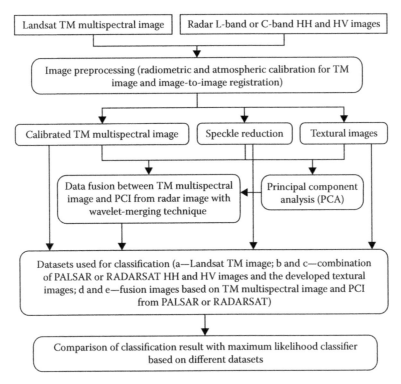

FIGURE 6.1
Framework of land use/land cover classification with multisensor data.

6.3.1 Data Collection and Preprocessing

Field data collection was conducted during July through August 2009, and a total of 432 plots were collected from the field survey and the 2008 QuickBird image. Field work was mainly to collect different successional stages and pastures in the rural regions, and the QuickBird image was used to collect sample plots in urban and urban–rural frontiers. Of the sampled plots, 220 plots were randomly selected for use as training plots during image classification procedure, and 212 plots were used as test plots for accuracy assessment. The polygons of these sample plots were created by identifying areas of uniform pixel reflectance in window sizes from approximately 3 × 3 pixels to 9 × 9 pixels on the Landsat TM imagery, depending on the patch sizes of different LULC classes. Based on the research objectives and field surveys, three forest classes (upland, flooding, and liana), three succession stages (initial, intermediate, and advanced), agropasture, and three non-vegetated classes (water, wetland, and urban) were designed and used for the land cover classification system. A detailed description of field data collection and LULC classification system was provided by Li et al. (2011).

Landsat 5 TM (Path/row: 226/62; image acquisition date: July 2, 2008), ALOS PALSAR L-band (fine beam dual level 1.5 products: L-band HH and HV polarization options with 12.5 m pixel spacing; image acquisition date: July 2, 2009), RADARSAT-2 C-band (C-band standard beam mode SGX with HH and HV polarization options with 8 m pixel spacing; image acquisition date: August 30, 2009), and QuickBird images (image acquisition date: June 20, 2008) were used in this research. The preprocessing of the satellite images included radiometric and atmospheric calibration for Landsat multispectral images with an image-based dark object subtraction model (Chavez 1996; Lu et al. 2002; Chander et al. 2009), and image-to-image registration between Landsat TM images and radar data (Universal Transverse Mercator coordinate system, zone 22, south). Both radar images were resampled to a pixel size of 10 m × 10 m using the nearest-neighbor technique during image-to-image registration. The Lee-Sigma filter with a window size of 5 × 5 pixels was used to reduce the speckle problem in radar data (Li et al. 2012).

6.3.2 Selection of Suitable Variables from Remotely Sensed Data

Remotely sensed data have different characteristics in spectral, spatial, radiometric, and temporal resolutions, as well as polarization options and others. Understanding the strengths and weaknesses of different types of sensor data is essential for selecting suitable remotely sensed data for LULC classification (Lu and Weng 2007). For example, optical sensor data often contain multispectral bands that cover visible, near-infrared, and shortwave infrared wavelengths, and mainly capture land surface features. Radar data often contain a single wavelength such as L-band and C-band but have different polarization options such as HH, HV, and VV. A radar system can penetrate forest canopy or bare soil into a certain depth to capture the undersurface information, depending on the use of different wavelengths and polarization options (Kasischke et al. 1997). It is important to make full use of different remote sensing features for improving LULC classification.

6.3.2.1 Identification of Suitable Textural Images

Our previous research in this study area has examined gray-level co-occurrence matrix-based texture measures (e.g., variance, VAR; homogeneity, HOM; contrast, CON; dissimilarity, DIS; entropy, ENT; and second moment, SM) with various window sizes (e.g., 5 × 5, 9 × 9, 15 × 15, 19 × 19, 25 × 25, and 31 × 31) on radar HH and HV images separately (Li et al. 2012). Based on training sample data, the transformed divergence (Mausel et al. 1990; Landgrebe 2003) was used to identify potential combinations of textural images. For the selected candidates of textural images, standard deviation of each textural image and correlation coefficients between the textural images were used to identify the best combination of two textural images, based on the highest value that was calculated from the sum of standard deviations divided by the sum of absolute correlation coefficients between the two textural images (Li et al. 2011). We found that the best combination of textural images for an ALOS PALSAR L-band HH image was the textures SM25 (second moment with a window size of 25 × 25 pixels) and CON31 (contrast with a window size of 31 × 31 pixels), and the best combination for an L-band HV image was the textures CON25 (contrast with a window size of 25 × 25 pixels) and SM19 (second moment with a window size of 19 × 19 pixels). For RADARSAT-2 C-band data, DIS25 (dissimilarity with a window size of 31 × 31 pixels) and HOM31 (homogeneity with a window size of 31 × 31 pixels) were the best combination for a C-band HH image, and CON25 and HOM31 were the best combination for a C-band HV image (Li et al. 2012). This chapter directly used the identified textural images. Figure 6.2 provides parts of radar radiometric and textural images for showing their different characteristics in reflecting land surfaces.

6.3.3.2 Integration of Multisensor Data

Data fusion is often used for the integration of multisensor or multiresolution data to enhance visual interpretation and/or to improve quantitative analysis performance (Pohl and van Genderen 1998). Many data fusion methods, such as principal component

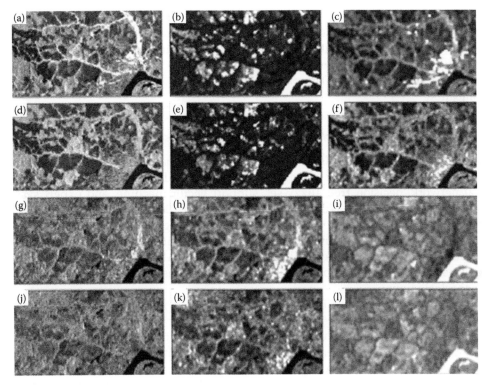

FIGURE 6.2
Comparison of ALOS PALSAR L-band and RADARSAT-2 C-band images and their textural images (a, b, and c are an ALOS PALSAR L-band HH image and HH-derived SM25 and CON31 textural images; d, e, and f are an ALOS PALSAR L-band HV image and HV-derived CON25 and SM19 textural images; g, h, and i are a RADARSAT-2 C-band HH image and HH-derived DIS25 and HOM31 textural images; j, k, and l are a RADARSAT-2 C-band HV image and HV-derived CON25 and HOM31 textural images).

analysis, wavelet-merging technique, intensity–hue–saturation, and Ehlers fusion, have been developed to integrate spectral and spatial information (Pohl and van Genderen 1998; Ehlers et al. 2010; Zhang 2010). Our previous research has indicated that the wavelet-based fusion method provided the best LULC classification in the Brazilian Amazon (Lu et al. 2011). Because radar HH and HV images and corresponding textural images may represent different land surface characteristics, it is important to make full use of their different discrimination capabilities. Therefore, principal component analysis was used to convert the HH, HV, and textural images from ALOS PALSAR L-band or from RADARSAT C-band data into a new dataset. Since the first principal component (PC1) contained the largest information, the PC1 was used in data fusion method. In this research, the PC1 from radar data (e.g., HH and HV images and textural images from ALOS PALSAR or from RADARSAT data) and the Landsat TM multispectral image were used to generate a new dataset with the wavelet-merging technique (Lu et al. 2011).

6.3.2.3 Comparison of Different Datasets for LULC Classification

To identify which dataset can provide better LULC classification, different datasets were examined: (1) Landsat TM multispectral image, (2) ALOS PALSAR L-band HH and HV images and textural images, (3) RADARSAT-2 C-band HH and HV images and textural images, (4) fusion image based on TM multispectral bands and PC1 from ALOS PALSAR data, and (5) fusion image based on TM multispectral bands and PC1 from RADARSAT data. A maximum likelihood algorithm was used to classify each dataset into a thematic map by using training samples. The classification results were then evaluated by using test sample plots and were compared for identifying a suitable dataset for LULC classification.

6.3.3 Comparison of Different Classification Algorithms

Many classification methods, from traditional statistical-based algorithms such as maximum likelihood classification to advanced nonstatistical algorithms such as artificial neural network, decision tree, fuzzy-set, support vector machine, and expert systems, are available (Franklin and Wulder 2002; Lu and Weng 2007; Tso and Mather 2009). One critical issue is how to select a suitable classification algorithm for a specific dataset. Therefore, in addition to the maximum likelihood classifier, another five classification algorithms—classification tree analysis, fuzzy ARTMAP, k-nearest neighbor, object-based classification, and support vector machine—were used to classify the Landsat TM multispectral image.

Maximum likelihood classification assumes normal or near-normal spectral distribution for each feature of interest (Jensen 2005). This classifier is based on the probability that a pixel belongs to a particular class by taking the variability of classes into account with the covariance matrix. However, the statistical-based algorithms are often criticized on the requirement of normal distribution because this assumption is often violated, especially when multisource data are used. Nonstatistical-based algorithms do not have this requirement and are regarded as having more advantages than traditional statistical-based algorithms (Pal and Mather 2003; Lu et al. 2004).

Classification tree analysis has the characteristics of distribution-free and easy interpretation over traditional supervised classifiers and has received increasing attention in remote sensing classification (Friedl and Brodley 1997; Zambon et al. 2006). The basic concept of a classification tree is to split a dataset into homogeneous subgroups based on measured attributes. The key is to identify a proper splitting criterion and auto-pruning value. Fuzzy ARTMAP is one of the neural network classification methods, which synthesizes fuzzy logic and adaptive resonance theory (ART) models (Carpenter et al. 1992; Mannan and Roy 1998). Fuzzy ARTMAP network consists of four layers of neurons: input, category, mapfield, and output. A choice parameter, learning rate parameters, and vigilance parameters are critical variables in this algorithm. The k-nearest-neighbor algorithm is based on the minimum distance from image pixels to the training samples to determine the k-nearest neighbors (Franco-Lopez et al. 2001; McRoberts and Tomppo 2007). The key for this approach is to identify a suitable k value because a large k value reduces the effect of noise on the classification but makes unclear boundaries between classes, whereas a small k value may result in overfit and reduce classification accuracy (McRoberts et al. 2002).

Object-based classification provides an alternative for classifying remotely sensed images into a thematic map based on segments comparing to the traditional per-pixel-based classification methods (Blaschke 2010). Segmentation is the process partitioning an image into isolated objects so that each object shares a homogeneous spectral similarity. These objects have better representative in the landscape than do the original pixels. These homogeneous objects are then analyzed using traditional classification methods such as minimum distance and maximum likelihood classification (Jensen 2005). The whole classification process consists of three steps: (1) image segmentation, where groups of neighboring pixels that grow in a systematic way assess spectral similarity across space and over all input bands, and segments are defined based on a user-specified similarity threshold; (2) creation of training sites and signature classes based on image segments; and (3) classification of the segments.

The support vector machine is a relatively new supervised classifier for remote sensing image classification but has gained great attention in recent years (Camps-Valls et al. 2008; Perumal and Bhaskaran 2009). It is a classification system derived from statistical learning theory developed by Vapnik (1998). It separates the classes with a decision surface that maximizes the margin between the classes. This surface for the linear case is called the optimal hyperplane, and the data points closest to the surface are called support vectors. The support vectors are the critical elements of the training set. The optimal surface solution is achieved by different functions called kernels. The common kernel types include linear, polynomial, radial basis function, and sigmoid. A recent paper by Mountrakis et al. (2011) provided a detailed review of a support vector machine used in the remote sensing field.

6.3.4 LULC Classification and Evaluation of Results

In this research, a maximum likelihood classifier was used to conduct LULC classification based on different datasets. Based on the field survey and QuickBird image, a total of 220 sample plots (over 3500 pixels) covering 10 LULC types, each having 15–30 plots, were used for each dataset. Meanwhile, the same sample plots were used to conduct Landsat TM multispectral image classification with different nonstatistical-based algorithms. All the classification results were evaluated with test sample plots.

An error matrix provides the detailed assessment of the agreement between the classified result and reference data and provides the information of how the misclassification happened (Congalton and Green 2008). Overall classification accuracy, producer's accuracy, user's accuracy, and kappa coefficient are then calculated from the error matrix (Congalton 1991; Foody 2002; Congalton and Green 2008). Both overall accuracy and kappa coefficient reflect the overall classification situation but cannot provide the reliability of each LULC class. Producer's accuracy and user's accuracy provide the complementary analysis of the accuracy assessment for each LULC class. In this study, a total of 212 test sample plots from the field survey and the QuickBird image were used for accuracy assessment. For each LULC class, 12–33 plots were collected. An error matrix was developed for each classification image and then producer's accuracy and user's accuracy for each class, and overall classification accuracy and kappa coefficient for each classification result were calculated from the corresponding error matrix.

6.4 Resultant Analysis

6.4.1 Comparison of Accuracy Assessment Results from Different Datasets

The accuracy assessment results from different datasets are summarized in Table 6.1, indicating that the TM image provided relatively good classification for different LULC classes, except SS1, and provided much higher accuracy than radar data, that is, 81.1% for the TM

TABLE 6.1

Comparison of LULC Classification Accuracy Assessment Results among Different Datasets with Maximum Likelihood Classifier

LULC Type	TM Image		Radar Data				TM and Radar Fusion			
	TM		L-band		C-band		TM-L-PC1		TM-C-PC1	
	PA	UA	PA	UA	PA	UA	PA	UA	PA	UA
Upland forest	69.7	88.5	51.5	39.5	21.2	30.4	78.8	86.7	78.8	89.7
Flooding forest	93.3	73.7	73.3	61.1	13.3	8.0	93.3	77.8	86.7	68.4
Liana forest	83.3	71.4	25.0	15.8	25.0	13.6	91.7	84.6	83.3	90.9
SS1	57.9	57.9	42.1	50.0	31.6	21.4	79.0	71.4	79.0	75.0
SS2	87.5	75.0	66.7	64.0	20.8	45.5	87.5	91.3	91.7	88.0
SS3	85.7	85.7	23.8	38.5	14.3	13.0	90.5	86.4	90.5	82.6
Agropasture	73.1	82.6	76.9	62.5	73.1	57.6	80.8	91.3	80.8	91.3
Water	87.5	100.0	83.3	95.2	91.7	100.0	87.5	100.0	87.5	100.0
Wetland	80.0	92.3	33.3	55.6	13.3	22.2	80.0	100.0	80.0	100.0
Urban	100.0	82.1	60.9	87.5	56.5	81.3	100.0	79.3	100.0	79.3
Overall accuracy	81.1		56.1		38.7		86.3		85.9	
Kappa coefficient	0.79		0.51		0.32		0.85		0.84	

Note: SS1, SS2, and SS3 represent initial, intermediate, and advanced succession vegetation; PA and UA represent producer's and user's accuracy.

TM-L-PC1, fusion image based on TM multispectral bands and PC1 from PALSAR L-band data; TM-C-PC1, fusion image based on TM multispectral bands and RADARSAT-2 C-band data.

image, 56.1% for the PALSAR L-band, and 38.7% for the RADARSAT C-band. Both radar data had poor classification performance for vegetation types. In particular, C-band data cannot effectively separate different vegetation types, while L-band data had relatively good performance for flooding forest and SS2. Both L-band and C-band data had good classification performance for agropasture and water. Overall, L-band data have much better performance than C-band data. Integration of radar and TM multispectral data improved overall accuracy by 4.8%–5.2%, that is, almost all LULC classes, especially vegetation classes, were improved. Figure 6.3 provides a comparison of LULC classification results, indicating the poor classification for radar data and satisfactory classification among the TM-based datasets.

The error matrix summarized in Table 6.2 provides detailed information about misclassification among different LULC classes. For example, in the classification results based on Landsat TM multispectral bands, the major misclassification errors were among upland forest, flooding forest, and liana forest, among initial succession, intermediate succession, and agropasture, and among water, wetland, and urban. The misclassification from radar data was much problematic, especially in the vegetation types such as within the three primary forest classes, within the three successional classes, between successional vegetation and primary forest, and between initial succession and agropasture. In the radar data classification results, urban was confused with upland forest because of their rough surfaces having similar radiometric values. The misclassification problem was reduced

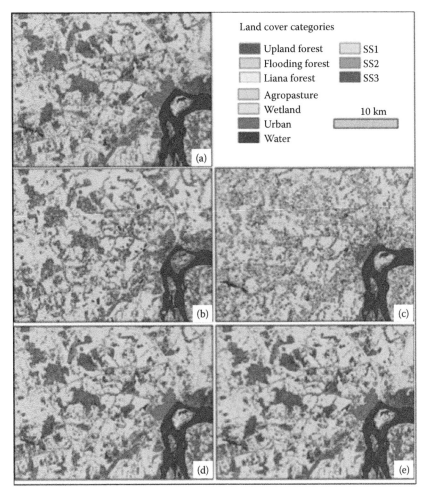

FIGURE 6.3
(See color insert.) Comparison of classification results based on (a) Landsat Thematic Mapper (TM) multispectral image, (b, c) HH, HV, and their corresponding textural images from ALOS PALSAR L-band and RADARSAT-2 C-band, respectively, (d) multisensor wavelet-based fusion images from TM multispectral bands and PALSAR L-band, and (e) multisensor wavelet-based fusion images from TM multispectral bands and RADARSAT C-band.

by using the TM and radar fusion image, especially for upland forest, liana forest, SS1, and agropasture.

6.4.2 Comparison of Accuracy Assessment Results from Different Algorithms

Selection of a suitable classification algorithm for LULC classification is important, but no single algorithm is best for each LULC class, as shown in Table 6.3. Overall, classification tree analysis provided the highest accuracy of 84.9%, followed by k-nearest neighbor, object-based classification, and maximum likelihood classification with accuracy of 81.1%–82.1%. The fuzzy ARTMAP and support vector machine had relatively low

TABLE 6.2

Comparison of Error Matrices among Different Datasets

	UPF	FLF	LIF	SS1	SS2	SS3	AGP	WAT	WET	URB	PA	UA
	Landsat TM Multispectral Image											
UPF	23	1	1	0	0	1	0	0	0	0	69.7	88.5
FLF	4	14	1	0	0	0	0	0	0	0	93.3	73.7
LIF	4	0	10	0	0	0	0	0	0	0	83.3	71.4
SS1	0	0	0	11	1	0	7	0	0	0	57.9	57.9
SS2	1	0	0	4	21	2	0	0	0	0	87.5	75.0
SS3	1	0	0	0	2	18	0	0	0	0	85.7	85.7
AGP	0	0	0	4	0	0	19	0	0	0	73.1	82.6
WAT	0	0	0	0	0	0	0	21	0	0	87.5	100.0
WET	0	0	0	0	0	0	0	1	12	0	80.0	92.3
URB	0	0	0	0	0	0	0	2	3	23	100.0	82.1
	Combination of PALSAR L-band HH, HV, and their textural images											
UPF	17	2	6	1	2	8	0	0	0	7	51.5	39.5
FLF	2	11	2	0	1	0	0	0	1	1	73.3	61.1
LIF	10	1	3	0	2	3	0	0	0	0	25.0	15.8
SS1	1	0	0	8	1	1	2	0	3	0	42.1	50.0
SS2	0	0	0	2	16	4	0	1	2	0	66.7	64.0
SS3	3	1	1	1	2	5	0	0	0	0	23.8	38.5
AGP	0	0	0	6	0	0	20	2	3	1	76.9	62.5
WAT	0	0	0	0	0	0	1	20	0	0	83.3	95.2
WET	0	0	0	0	0	0	3	1	5	0	33.3	55.6
URB	0	0	0	1	0	0	0	0	1	14	60.9	87.5
	Fusion image based on TM and PC1 from ALOS PALSAR data											
UPF	26	1	1	0	0	2	0	0	0	0	78.8	86.7
FLF	4	14	0	0	0	0	0	0	0	0	93.3	77.8
LIF	2	0	11	0	0	0	0	0	0	0	91.7	84.6
SS1	0	0	0	15	1	0	5	0	0	0	79.0	71.4
SS2	0	0	0	2	21	0	0	0	0	0	87.5	91.3
SS3	1	0	0	0	2	19	0	0	0	0	90.5	86.4
AGP	0	0	0	2	0	0	21	0	0	0	80.8	91.3
WAT	0	0	0	0	0	0	0	21	0	0	87.5	100.0
WET	0	0	0	0	0	0	0	0	12	0	80.0	100.0
URB	0	0	0	0	0	0	0	3	3	23	100.0	79.3

Note: UPF, FLF, and LIF represent upland, flooding, and liana forest; SS1, SS2, and SS3 represent initial, intermediate, and advanced succession vegetation; AGP represents agropasture; WAT, WET, and URB represent
water, nonvegetation wetland, and urban; PA and UA represent producer's and user's accuracy.

accuracy of 79.7%–78.3%. Considering individual LULC classification results, if the classification result with maximum likelihood classifier was used as a benchmark, classification tree analysis mainly improved upland forest, initial succession, and agropasture
classes; k-nearest-neighbor algorithm improved upland forest and initial succession;
and object-based classification improved intermediate and advanced succession stages.

TABLE 6.3

Comparison of Accuracy Assessment Results from Different Classification Algorithms Based on Landsat TM Multispectral Data

	MLC		CTA		ARTMAP		KNN		OBC		SVM	
	PA (%)	UA (%)	PA (%)	UA (%)	PA (%)	UA (%)	PA (%)	UA (%)	PA (%)	UA (%)	PA (%)	UA (%)
UPF	69.7	88.5	90.9	85.7	90.9	65.2	72.7	92.3	66.7	88.0	87.9	65.9
FLF	93.3	73.7	86.7	72.2	73.3	64.7	80.0	70.6	86.7	68.4	86.7	81.3
LIF	83.3	71.4	75.0	81.8	58.3	100.0	83.3	58.8	83.3	71.4	50.0	100.0
SS1	57.9	57.9	68.4	59.1	52.6	71.4	63.2	57.1	57.9	55.0	47.4	53.0
SS2	87.5	75.0	75.0	78.3	79.2	70.4	83.3	83.3	91.7	81.5	62.5	83.3
SS3	85.7	85.7	81.0	94.4	33.3	63.6	95.2	74.1	90.5	86.4	61.9	56.5
AGP	73.1	82.6	76.9	87.0	88.5	82.1	69.2	81.8	69.2	81.8	73.1	73.1
WAT	87.5	100.0	100.0	100.0	100.0	100.0	100.0	100.0	95.8	100.0	100.0	100.0
WET	80.0	92.3	86.7	86.7	100.0	100.0	73.3	100.0	80.0	92.3	100.0	100.0
URB	100.0	82.1	100.0	100.0	100.0	100.0	100.0	100.0	100.0	85.2	100.0	100.0
OCA	81.1		84.9		79.7		82.1		81.6		78.3	
KC	0.79		0.83		0.77		0.80		0.79		0.76	

Note: UPF, FLF, and LIF represent upland, flooding, and liana forest; SS1, SS2, and SS3 represent initial, intermediate, and advanced succession vegetation; AGP represents agropasture; WAT, WET, and URB represent water, nonvegetation wetland, and urban; PA and UA represent producer's and user's accuracy in percentage. OCA is the overall classification accuracy in percentage, and KC is the kappa coefficient. MLC, CTA, ARTMAP, KNN, OBC, and SVM represent maximum likelihood classification, classification tree analysis, fuzzy ARTMAP, k-nearest neighbor, object-based classification, and support vector machine, respectively.

Although overall accuracy from fuzzy ARTMAP and support vector machine slightly decreased, comparing with maximum likelihood, both algorithms indeed improved nonvegetation classes.

6.5 Discussion

6.5.1 Roles of Radar Data in Improving Vegetation Classification

Optical sensor data have distinct discrimination capabilities as compared with radar data. Optical sensor data mainly capture land surface features, thus, they are often used for LULC classification, and have proven their importance. However, the complex landscape often makes optical sensor data difficult in classifying certain LULC classes, such as some vegetation types, because of their species composition and complex forest stand structures resulting in data saturation in optical sensor data. Radar data, especially L-band data, can penetrate forest canopy to a certain depth to capture understory information or penetrate soil to a certain depth to capture different soil moisture conditions.

Therefore, use of radar data may capture some new information that optical sensor data do not have. This research indicated that radar, especially L-band, provided reasonably good classification for flooding forest, intermediate succession, and agropasture. Incorporation of radar data into optical multispectral image has proven valuable to improve some LULC types in this research. In general, data fusion is often applied to improve spatial resolution, and thus may reduce the mixed pixel problem. This spatial improvement is helpful for sites having relatively small patches of LULC such as different stages of succession vegetation. However, because of the complex stand structure in vegetation types, especially for primary forest and advanced succession, increased spatial resolution may enlarge the spectral variation within the same LULC, and thus reduce the classification accuracy. Therefore, there is a trade-off between patch size of LULC classes and the spectral variation caused by improved spatial resolution. Future research should examine how different spatial resolutions affect the selection of data fusion methods and what classification algorithm is suitable for land cover classification corresponding to the fusion images.

6.5.2 Selection of Suitable Nonstatistical-Based Algorithms

Considering the time for completing one classification procedure, fuzzy ARTMAP requires the most time, followed by object-based classification. The k-nearest neighbor and classification tree analysis require much less time than fuzzy ARTMAP and object-based classification but need more time than maximum likelihood classification. Fuzzy ARTMAP requires a long time during the training process, and object-based classification requires a relatively long time during image segmentation. For the nonstatistical classification algorithms, one critical step is to optimize the parameters used in corresponding algorithms. Depending on the complexity of algorithms, the time for identifying optimal parameters is significantly different. For example, fuzzy ARTMAP requires optimizing more parameters than other nonstatistical algorithms and requires much more time to identify suitable parameters. In object-based classification, much time is required for identifying parameters suitable for the development of a segmentation image. Classification tree analysis requires selecting a suitable split rule, and k-nearest neighbor requires identifying a suitable k value, but both algorithms require much less time for the optimization of parameters than fuzzy ARTMAP and object-based classification. On the other hand, object-based classification can considerably reduce the salt-and-pepper effect compared to the per-pixel-based classification methods. One disadvantage of this method is that the accuracy for some LULC classes may improve but others may not, depending on the complexity of LULC classes under investigation and the size of segments.

The support vector machine uses kernel functions to generate nonlinear decision boundaries and introduces a cost parameter to quantify the penalty of misclassification errors to handle the nonseparable classification problem. Therefore, the support vector machine is designed to search for an optimal solution to a classification problem, while other learning machine algorithms just provide a solution for classification. However, the complex landscape, especially vegetation types in this research and a relatively large study area, requires a huge amount of time and computer resources to implement the support vector machine classification, thus resulting in the difficulty in determining the optimal parameters such as the penalty parameter.

6.6 Conclusions

Compared with radar data, the Landsat TM multispectral image provided much better LULC classification because of its suitable spatial and spectral resolutions. However, radar data are sensitive to forest stand structure and water content; thus, incorporation of radar data into a TM multispectral image with a proper data fusion technique is valuable for improving LULC classification. Since single-date radar data have proved difficult for LULC, especially vegetation classification, much research has explored the use of multi-temporal radar data for LULC classification. Because a radar system can capture LULC information almost unaffected by atmospheric interference, radar data have become an important data source for land cover classification in the moist tropical regions because of the cloud problem, but more research should be conducted on the exploration of methods to better extract information from radar radiometric data.

This research indicated that maximum likelihood classification provided reasonably good LULC classification based on a Landsat TM multispectral image. The time and labor required in maximum likelihood classification were much less than in nonstatistical-based classification algorithms. The classification tree analysis provided better performance than the maximum likelihood algorithm. Other nonstatistical-based algorithms provided similar or slightly lower classification accuracy than maximum likelihood classification. One critical step in the nonstatistical-based algorithm is to optimize the parameters used in the corresponding algorithms, which are often time consuming. More research should be conducted on the development of methods to more automatically optimize the parameters used in the nonstatistical-based algorithm.

Acknowledgments

We acknowledge the funding support from the National Science Foundation (Grant No. BCS 0850615) for this research. Corina C. Freitas, Luciano V. Dutra, and Sidnei J.S. Sant'Anna thank JAXA (AO 108) for providing the ALOS PALSAR data through its Science Program. Mateus Batistella thanks the Canadian SOAR Program (SOAR Project No. 1957) for the RADARSAT-2 data used in this research. We also thank Anthony Cak for his assistance with fieldwork and Scott Hetrick for his assistance in organizing the field data.

References

Blaschke, T. 2010. "Object Based Image Analysis for Remote Sensing." *ISPRS Journal of Photogrammetry and Remote Sensing* 65:2–16.

Camps-Valls, G., L. Gómez-Chova, J. Muñoz-Marí, J. L. Rojo-Álvarez, and M. Martínez-Ramón. 2008. "Kernel-Based Framework for Multi-Temporal and Multi-Source Remote Sensing Data Classification and Change Detection." *IEEE Transactions on Geoscience and Remote Sensing* 46(6):1822–35.

Carpenter, G. A., S. Crossberg, N. Markuzon, J. H. Reynolds, and D. B. Rosen. 1992. "Fuzzy ARTMAP: A Neural Network Architecture for Incremental Supervised Learning of Analog Multidimentional Maps." *IEEE Transactions on Neural Networks* 3:698–713.

Chander, G., B. L. Markham, and D. L. Helder. 2009. "Summary of Current Radiometric Calibration Coefficients for Landsat MSS, TM, ETM+, and EO-1 ALI Sensors." *Remote Sensing of Environment* 113:893–903.

Chavez, P. S. Jr. 1996. "Image-Based Atmospheric Corrections—Revisited and Improved." *Photogrammetric Engineering & Remote Sensing* 62:1025–36.

Congalton, R. G. 1991. "A Review of Assessing the Accuracy of Classification of Remotely Sensed Data." *Remote Sensing of Environment* 37:35–46.

Congalton, R. G. and K. Green 2008. *Assessing the Accuracy of Remotely Sensed Data: Principles and Practices*, 2nd ed., 183. Boca Raton, FL: CRC Press, Taylor & Francis.

Ehlers, M., S. Klonus, P. J. Astrand, and P. Rosso. 2010. "Multisensor Image Fusion for Pansharpening in Remote Sensing." *International Journal of Image and Data Fusion* 1:25–45.

Foody, G. M. 2002. "Status of Land Cover Classification Accuracy Assessment." *Remote Sensing of Environment* 80:185–201.

Franco-Lopez, H., A. Ek, and M. Bauer. 2001. "Estimation and Mapping of Forest Stand Density, Volume, and Cover Type Using the k-nearest Neighbors Method." *Remote Sensing of Environment* 77:251–74.

Franklin, S. E. and M. A. Wulder. 2002. "Remote Sensing Methods in Medium Spatial Resolution Satellite Data Land Cover Classification of Large Areas." *Progress in Physical Geography* 26:173–205.

Friedl, M. A. and C. E. Brodley. 1997. "Decision Tree Classification of Land Cover from Remotely Sensed Data." *Remote Sensing of Environment* 61:399–409.

Jensen, J. R. 2005. *Introductory Digital Image Processing: A Remote Sensing Perspective*. 3rd ed., 526. Upper Saddle River, NJ: Prentice Hall.

Kasischke, E. S., J. M. Melack, and M. C. Dobson. 1997. "The Use of Imaging Radars for Ecological Applications: A Review." *Remote Sensing of Environment* 59(2):141–56.

Landgrebe, D. A. 2003. *Signal Theory Methods in Multispectral Remote Sensing*, 508. Hoboken, NJ: John Wiley and Sons.

Li, G., D. Lu, E. Moran, and S. Hetrick. 2011. "Land-Cover Classification in a Moist Tropical Region of Brazil with Landsat TM Imagery." *International Journal of Remote Sensing* 32(23):8207–30.

Li, G., D. Lu, E. Moran, L. Dutra, and M. Batistella. 2012. "A Comparative Analysis of ALOS PALSAR L-Band and RADARSAT-2 C-Band Data for Land-Cover Classification in a Tropical Moist Region." *ISPRS Journal of Photogrammetry and Remote Sensing* 70:26–38.

Lu, D., P. Mausel, E. S. Brondízio, and E. Moran. 2002. "Assessment of Atmospheric Correction Methods for Landsat TM Data Applicable to Amazon Basin LBA Research." *International Journal of Remote Sensing* 23:2651–71.

Lu, D., P. Mausel, M. Batistella, and E. Moran. 2004. "Comparison of Land-Cover Classification Methods in the Brazilian Amazon Basin." *Photogrammetric Engineering & Remote Sensing* 70:723–31.

Lu, D., and Q. Weng. 2007. "A Survey of Image Classification Methods and Techniques for Improving Classification Performance." *International Journal of Remote Sensing* 28:823–70.

Lu, D., G. Li, E. Moran, L. Dutra, and M. Batistella. 2011. "A Comparison of Multisensor Integration Methods for Land-Cover Classification in the Brazilian Amazon." *GIScience & Remote Sensing* 48(3):345–70.

Mannan, B. and J. Roy. 1998. "Fuzzy ARTMAP Supervised Classification of Multi-Spectral Remotely-Sensed Images." *International Journal of Remote Sensing* 19:767–74.

Mausel, P. W., W. J. Kramber, and J. K. Lee. 1990. "Optimum Band Selection for Supervised Classification of Multispectral Data." *Photogrammetric Engineering & Remote Sensing* 56:55–60.

McRoberts, R. E., M. D. Nelson, and D. G. Wendt. 2002. "Stratified Estimation of Forest Area Using Satellite Imagery, Inventory Data, and the k-nearest Neighbors Technique." *Remote Sensing of Environment* 82:457–68.

McRoberts, R. R. and E. O. Tomppo. 2007. "Remote Sensing Support for National Forest Inventories." *Remote Sensing of Environment* 110:412–9.

Moran, E. F., and E. S. Brondízio. 1998. Land-use change after deforestation in Amazônia. In *People and Pixels: Linking Remote Sensing and Social Science*, edited by D. Liverman, E. F. Moran, R. R. Rindfuss, and P. C. Stern, 94–120. Washington, DC: National Academy Press.

Moran, E. F., E. S. Brondízio, P. Mausel, and Y. Wu. 1994. "Integrating Amazonian Vegetation, Land Use, and Satellite Data." *Bioscience* 44:329–38.

Mountrakis, G., J. Im, and C. Ogole. 2011. "Support Vector Machines in Remote Sensing: A Review." *ISPRS Journal of Photogrammetry and Remote Sensing* 66(3):247–59.

Pal, M. and P. M. Mather. 2003. "An Assessment of the Effectiveness of Decision Tree Methods for Land Cover Classification." *Remote Sensing of Environment* 86:554–65.

Perumal, K. and R. Bhaskaran. 2009. "SVM-Based Effective Land Use Classification System for Multispectral Remote Sensing Images." *International Journal of Computer Science and Information Security* 6(2):97–105.

Pohl, C. and J. L. van Genderen. 1998. "Multisensor Image Fusion in Remote Sensing: Concepts, Methods, and Applications." *International Journal of Remote Sensing* 19:823–54.

Tso, B. and P. M. Mather. 2009. *Classification Methods for Remotely Sensed Data*, 356. London: Taylor & Francis.

Vapnik, V. N. 1998. *Statistical Learning Theory*. New York: Wiley.

Zambon, M., R. Lawrence, A. Bunn, and S. Powell. 2006. "Effect of Alternative Splitting Rules on Image Processing Using Classification Tree Analysis." *Photogrammetric Engineering & Remote Sensing* 72:25–30.

Zhang, J. 2010. "Multisource Remote Sensing Data Fusion: Status and Trends." *International Journal of Image and Data Fusion* 1:5–24.

7

Vegetation Change Detection in the Brazilian Amazon with Multitemporal Landsat Images

Dengsheng Lu, Guiying Li, Emilio Moran, and Scott Hetrick

CONTENTS

7.1 Introduction 127
7.2 Methods 128
    7.2.1 Data Collection and Preprocessing 129
    7.2.2 Development of Fractional Images with Linear Spectral
        Mixture Analysis 130
    7.2.3 Vegetation Gain/Loss Detection 130
    7.2.4 Land Use/Cover Classification and Evaluation 132
    7.2.5 Detection of Detailed "From–To" Vegetation Change Trajectories 133
7.3 Resultant Analysis and Discussion 133
    7.3.1 Analysis of Vegetation Gain/Loss Results 133
    7.3.2 Analysis of Detailed Vegetation Change Trajectories 135
7.4 Conclusions 138
Acknowledgments 139
References 139

7.1 Introduction

Since the 1970s, the Brazilian Amazon has experienced high deforestation rates largely because of colonization projects initiated in the 1970s and 1980s, road construction, and land use change (Moran 1981; Laurance et al. 2004). According to a report prepared by the National Institute for Space Research in Brazil, a total of 392,020 km^2 of deforestation occurred in the Brazilian Amazon between 1988 and 2011 (http://www.mongabay.com/brazil.html). Deforestation has converted large areas of primary forest and cerrado/savanna to agricultural lands, pasture, successional vegetation, and agroforestry (Lucas et al. 2000; Roberts et al. 2002; Sano et al. 2010). Because of the impacts of deforestation on climate change, biodiversity, and other environmental processes, monitoring of forest and savanna deforestation in the Brazilian Amazon has become an important task and has received great attention in the past two decades. For example, Brazil has developed two systems—that is, PRODES (Program for the Estimation of Deforestation in the Brazilian Amazon) (http://www.obt.inpe.br/prodes/) and DETER (Real Time Deforestation Monitoring System) (http://www.obt.inpe.br/deter/)—to monitor annual deforestation by using Landsat and MODIS data, respectively.

Remotely sensed data, with its unique characteristics in repetitive data acquisition, synoptic view, and digital format suitable for computer processing, make it the primary data source for examining land use/cover change at different scales (Lu et al. 2004). Since the first earth observation satellite was launched in the early 1970s, different kinds of sensor data have been readily available for land use/cover change detection (Althausen 2002; Lefsky and Cohen 2003). The optical sensors aboard the Landsat satellite platform series may be the most common data source for change detection because they offer a relatively long history of data collection (from the early 1970s to 2011), with suitable spectral and spatial resolutions. In particular, because of recent free public access to these images, the use of time-series Landsat images for detecting land use/cover change has recently attracted great interest (Cohen et al. 2010; Huang et al. 2010; Kennedy et al. 2010; Thomas et al. 2011; Hansen and Loveland 2012).

In the past four decades, a large number of change detection techniques have been developed (see review papers by Singh [1989], Coppin et al. [2004], Lu et al. [2004], and Kennedy et al. [2009]). The techniques may be based on per-pixel-, subpixel-, and object-based methods (Chen et al. 2012). The change detection methods can be conducted with spectral signatures, textures, and biophysical variables such as biomass. Lu et al. (2004) grouped the change detection methods into seven categories: (1) algebra, (2) transformation, (3) classification, (4) advanced models, (5) geographic information system approaches, (6) visual analysis, and (7) other approaches, and briefly introduced each category. In change detection, the critical issues to consider include (1) image preprocessing, including radiometric and atmospheric calibration and image-to-image registration; (2) selection of suitable variables from remotely sensed data; and (3) selection of suitable algorithms. The major steps for change detection analysis have been described by Lu et al. (2004).

The change detection contents are often grouped into two categories: (1) binary change and nonchange detection, and (2) detailed "from–to" change trajectories. In practice, change and nonchange detection is not informative enough for understanding land cover change, and knowing the specific changes such as deforestation, agricultural expansion, and urbanization is often required (Lu et al. 2012). For the "from–to" change trajectories, much previous research focused on the land use/cover conversion, such as from forest to agriculture or from agriculture to urban area (Lu et al. 2012). For vegetation change detection, the information of both vegetation conversion and modification is often required (Lu et al. 2008). This chapter introduces a new method to detect vegetation gain and loss based on the fraction images, which were developed from Landsat multispectral images using linear spectral mixture analysis (LSMA) in Altamira, Pará State, Brazil. A combination of a vegetation gain/loss image and a classification image was then used to detect detailed "from–to" vegetation change trajectories according to the designed system. The objective of this research is to examine vegetation modification such as degradation and growth in addition to examining vegetation conversion such as deforestation and regeneration.

7.2 Methods

The same study area—Altamira, Pará State, Brazil, as described in Chapter 6 in this book—was selected for examining vegetation change by using multitemporal Landsat Thematic Mapper (TM) images. The framework is shown in Figure 7.1. The major steps include (1) image preprocessing, including atmospheric calibration and image-to-image

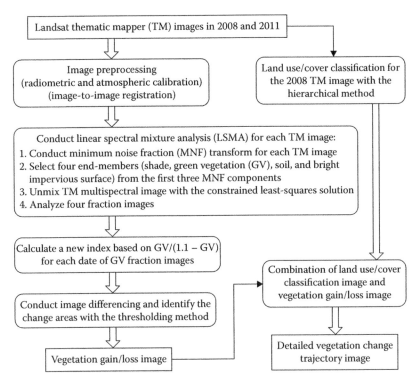

FIGURE 7.1
Framework of multitemporal Landsat Thematic Mapper images for vegetation change detection.

registration, (2) fraction image development with LSMA, (3) development of a new index from the green vegetation (GV) fraction image, (4) image differencing of the new index images and identification of thresholds for detection of vegetation gain and loss, (5) land use/cover classification of the 2008 TM image, and (6) combination of the 2008 classification image and vegetation gain/loss image to generate detailed "from–to" vegetation change trajectories.

7.2.1 Data Collection and Preprocessing

Landsat 5 TM images (path/row: 226/062), which were acquired on July 2, 2008 and July 27, 2011, were used in this research. Both TM images were geometrically registered into Universal Transverse Mercator coordinate system (zone 22, south) with root-mean-square error of 0.49 pixels. Meanwhile, radiometric and atmospheric calibration for both images was conducted with the dark-object-based subtraction method. This is an image-based procedure for reducing the effects caused by solar zenith angle, solar radiance, and atmospheric scattering (Lu et al. 2002; Chander et al. 2009). The equations used for Landsat TM image calibration are as follows:

$$R_\lambda = \mathrm{PI} \times D \times \frac{\left(L_\lambda - L_{\lambda.\mathrm{haze}}\right)}{\left(\mathrm{Esun}_\lambda \times \cos(\theta)\right)} \tag{7.1}$$

$$L_\lambda = \mathrm{gain}_\lambda \times \mathrm{DN}_\lambda + \mathrm{bias}_\lambda \tag{7.2}$$

where L_λ is the apparent at-satellite radiance for spectral band λ, DN_λ is the digital number of spectral band λ, R_λ is the calibrated reflectance, $L_{\lambda.haze}$ is path radiance, $Esun_\lambda$ is exoatmospheric solar irradiance, D is the distance between the earth and sun, and θ is the sun zenith angle. PI is a constant, which equals to 3.14159. The path radiance for each band is identified based on the analysis of water bodies in the images. The $gain_\lambda$ and $bias_\lambda$ are radiometric gain and bias corresponding to spectral band λ, and they are often provided in an image header file or a metadata file, or calculated from maximal and minimal spectral radiance values (Lu et al. 2002).

7.2.2 Development of Fractional Images with Linear Spectral Mixture Analysis

LSMA is a physically based image-processing tool based on the assumption that the spectrum measured by a sensor is a linear combination of the spectra of all pure materials (called end-members) within the pixel (Roberts et al. 1998). The mathematic model of LSMA can be expressed as

$$R_i = \sum_{k=1}^{n} f_k R_{ik} + \varepsilon_i \tag{7.3}$$

where i is the number of spectral bands used; $k = 1, \ldots, n$ (number of end-members); R_i is the spectral reflectance of band i of a pixel that contains one or more end-members; f_k is the proportion of end-member k within the pixel; R_{ik} is the known spectral reflectance of end-member k within the pixel on band i; and ε_i is the error for band i.

In the LSMA approach, one critical step is to identify suitable end-members. Although different end-member selection methods are available (Lu et al. 2003), an image-based end-member selection method based on a scatterplot of two images is often used. To better identify end-members, image transformation methods such as minimum noise fraction (MNF) transform can be used to convert the TM multispectral bands into a new dataset without correlation between the new components (van der Meer and de Jong 2000). In this research, four end-members—GV, shade, soil, and high-albedo object (mainly impervious surface)—were selected from the scatterplots of the first three transformed MNF components (Lu et al. 2011). After four end-members were identified, a constrained least-squares solution was used to unmix the TM multispectral image into four fraction images.

7.2.3 Vegetation Gain/Loss Detection

Since the developed fraction images have physical meanings (Roberts et al. 1998; Lu et al. 2003), they can be directly used for examining land cover change. Considering vegetation types, shade and GV are regarded as two important parameters for describing the status of a vegetation stand structure (Lu et al. 2005a). As vegetation grows, for example, from initial succession to advanced succession, vegetation stand structure becomes complex; thus, shade amount in a unit increases but GV amount decreases. For a primary forest, the stand structure is relatively stable in different sites or in the same site over time if no disturbance has occurred, thus shade and GV components are stable. However, when a forest site is disturbed because of factors such as selective logging or extreme weather, shade and GV components in a forest site may be changed. Because the shade component represents shade and water information in a site, it is sensitive to external factors such as the atmospheric conditions; however, the GV component is a more reliable variable than the shade component, but it is indeed sensitive to the change of vegetation amount such as leaf area index (Lu et al. 2003). Therefore, GV may be a better variable than shade and soil variables for examining vegetation change conditions and is therefore used in this research. To further enlarge the difference between vegetation change categories, a new index called the GV index, which is based on the GV image, is used and expressed as Equation 7.4:

$$f_{index} = \frac{GV}{1.1 - GV} \qquad (7.4)$$

The GV data range is within 0 and 1. Use of 1.1 is to avoid 0 in the denominator in case the GV value equals 1 for some pixels. As Figure 7.2 shows, the GV index has an exponential relationship with GV, thus the GV index enlarges in value when GV has a relatively small change. Our previous research has indicated that a transform of GV and/or shade variable can better reflect a forest stand structure status than single GV or shade variables (Lu et al. 2003). Therefore, use of the GV index may be better for examining small vegetation changes because of disturbance or natural growth.

The vegetation gain and loss are then calculated with the following image differencing based on GV index images at two dates:

$$f_{change} = f_{index}\left(t_1\right) - f_{index}\left(t_2\right) \qquad (7.5)$$

where t_1 and t_2 are two change detection dates. Here they are the years of 2008 and 2011. High positive f_{change} value indicates vegetation loss, high negative f_{change} value indicates vegetation gain, and a value close to zero indicates no change.

To identify vegetation gain or loss, one critical step is to select appropriate thresholds in both tails of the histogram representing the changed areas (Singh 1989). Two methods are often used for the selection of thresholds (Singh 1989; Yool et al. 1997; Lu et al. 2005b): (1) interactive procedure or manual trial-and-error procedure, where an analyst interactively adjusts the thresholds and evaluates the resulting image until satisfied; and (2) statistical measures, which involved selection of a suitable standard deviation from the mean. In this research, a comparative analysis of both color composites (i.e., 2008 and 2011 TM images) and the f_{change} image is used to identify multiple thresholds for determining vegetation gain and loss. Figure 7.3 shows the concept of identifying multiple thresholds based on the GV index differencing

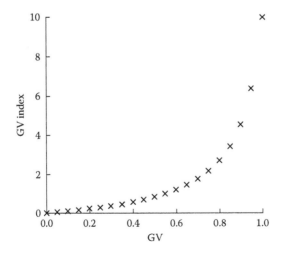

FIGURE 7.2
Relationship between green vegetation (GV) and GV index showing the exponential effect.

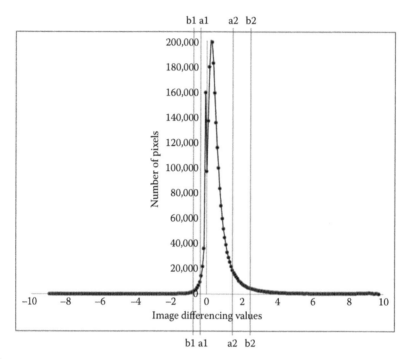

FIGURE 7.3
Concept of determining vegetation change by using multiple thresholds.

image. Usually, unchanged areas account for the majority of the study area, and changed areas are mainly located in the two tails of the histogram. According to this concept, the following rules can be established when the GV-index data are within the following ranges:

[a1, a2] indicates a unchanged area

[b1, a1] or [a2, b2] indicates a small change in vegetation amount or small disturbance

[<b1] or [>b2] indicates a large change in vegetation amounts or large disturbance

Based on identified thresholds, we can develop vegetation gain or loss at different levels. Here, we separate gain or loss into two categories each, that is, large gain or large loss and small gain or small loss. Large gain or large loss means a large change in GV proportion, such as the conversion from forest to agropasture (large loss) and the conversion from agropasture to secondary succession (large gain). Small gain or small loss means a relatively small change in GV, such as vegetation growth (small gain) or forest degradation because of selective logging (small loss).

7.2.4 Land Use/Cover Classification and Evaluation

A classification system with five land use/cover classes—forest, nonforest vegetation (e.g., secondary succession, plantations), agropasture (e.g., agricultural fields, pasture), impervious surfaces, and water—was designed for this study. The hierarchical classification method has been proven to provide reliable land use/cover classification (Lu et al. 2012), and thus, this method is further modified for use in this research for classifying the 2008 Landsat TM image into a thematic map.

TABLE 7.1

Combination of a Prior Land Use/Cover Classification Image and Vegetation Gain/Loss Image
for Generating Detailed Vegetation Change Trajectories

Land Use/Cover Class in Prior Date	Vegetation Change from Prior to Posterior Date			
	Large Gain	Small Gain	Small Loss	Large Loss
Primary forest	Forest restoration	Forest growth	Forest degradation	Forest deforestation
Nonforest vegetation	Vegetation restoration	Vegetation growth	Vegetation degradation	Vegetation deforestation
Agropasture	Regeneration	Regeneration		

Evaluation of the classification image is often required for understanding the quality of classification results. Overall classification accuracy and kappa coefficient are used to assess the overall performance in a classification, whereas producer's accuracy and user's accuracy are used to evaluate the performance of each land cover class. These parameters are calculated from the error matrix (Congalton 1991; Foody 2002; Congalton and Green 2008). In this research, a total of 413 sample plots were collected from field work conducted in 2009 and a 2008 QuickBird image over the study area (Li et al. 2011). These data were used to evaluate the 2008 classification image.

7.2.5 Detection of Detailed "From–To" Vegetation Change Trajectories

Detailed vegetation change trajectory information is often required for forest management and planning. Postclassification comparison is a common method for examining land use/cover change trajectories (Lu et al. 2004). However, this method requires accurate classification images for both dates, and image classification is especially difficult when training sample data or high spatial resolution images are not available (Lu et al. 2012). To avoid this challenge, we adopted a new method based on the combination of one classification image and one vegetation gain/loss image for generating the vegetation change trajectories. Table 7.1 summarizes the potential change trajectories, including forest restoration/growth, forest degradation/deforestation, nonforest vegetation restoration/growth, nonforest vegetation degradation/deforestation, and regeneration.

Quantitative evaluation of the change detection results is required, but it is a challenge because of the lack of reference data for both dates. In this research, no detailed quantitative analysis of the change detection result was implemented, but the result was overlaid on the color composites of both TM images to visually evaluate the quality of the change detection result.

7.3 Resultant Analysis and Discussion

7.3.1 Analysis of Vegetation Gain/Loss Results

The LSMA approach provides fractional images having physical meanings. As an example, Figure 7.4 showed four fraction images—shade, GV, soil, and impervious surface—that were developed from the 2008 and 2011 Landsat TM images. Considering forest or nonforest vegetation types, shade and GV are two important parameters representing the differences of forest stand structures. In the shade fraction image, water has the highest

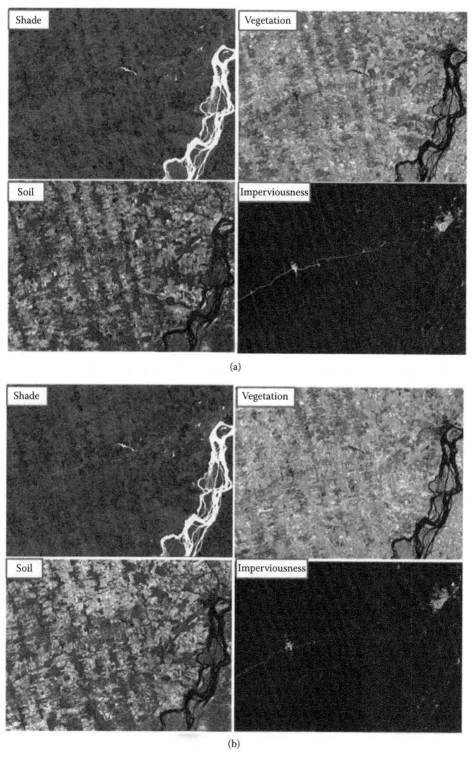

FIGURE 7.4
Comparison of fraction images from Landsat images. Fraction images in (a) 2008 and (b) 2011.

FIGURE 7.5
Differencing image between the 2008 and 2011 green vegetation index images.

value, followed by primary forest. Urban, roads, and agricultural fields have very low shade values; thus, they appear black in this image. In the GV fraction image, nonforest vegetation types such as different successional vegetation stages have high values and nonvegetation types (e.g., water and urban) have very low values. A comparison of the fraction images from the 2008 and 2011 TM images indicated the reliable results from their LSMA approach.

The GV-index differencing image, as shown in Figure 7.5, indicates that the majority of the study area is unchanged land cover (gray), and the pixels having high vegetation loss (white in the figure) are highlighted. By using the multiple thresholds, vegetation gain and loss distribution can be developed and is shown in Figure 7.6. This figure indicates that some areas are not the truly changed vegetation, they are only the changes within nonvegetation types, especially within the agropasture class. Therefore, caution should be taken when discussing vegetation gain or loss because of the greenness change in agricultural lands caused by different crop phenology. It is necessary to remove the noninteresting changed areas in this study. Incorporation of a classification image and vegetation gain/loss image is one way to solve this problem.

7.3.2 Analysis of Detailed Vegetation Change Trajectories

The classification accuracy assessment result as summarized in Table 7.2 indicates that an overall accuracy of 84% was obtained for the 2008 classification image (Figure 7.7). The results indicate that agropasture accounts for a large proportion of the total land cover area, followed by primary forest. The error matrix shows that the nonforest vegetation class has relatively lower accuracy than primary forest and agropasture. This is because nonforest vegetation has wider spectral variation, for example, advanced succession vegetation has similar spectral signatures with primary forest because of the complex forest stand structure resulting in data saturation in Landsat TM spectral signatures; and initial succession has similar spectral signatures with dirty pasture (agropasture) without clear boundaries between them (Li et al. 2011). Meanwhile, impervious surface areas are often confused with bare soils in agriculture fields during the dry season because of their similar spectral signatures (Lu et al. 2011).

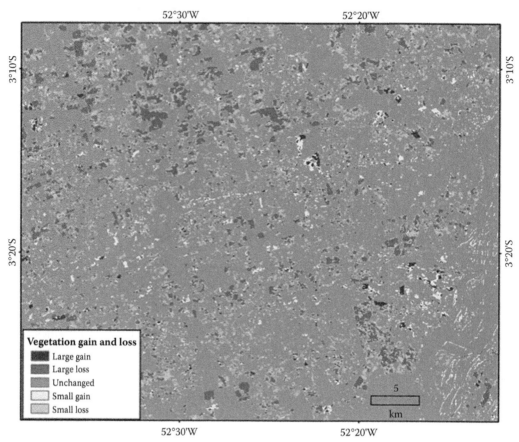

FIGURE 7.6
Distribution of vegetation gain and loss between 2008 and 2011.

TABLE 7.2

Error Matrix of the 2008 TM Image Classification Result

	Forest	Nonforest	Agropasture	Impervious	Water	RT	CT	PA	UA
Forest	126	31	1		2	160	131	96.2	78.8
Nonforest	5	79	5			89	122	64.8	88.8
Agropasture		12	88	5	2	107	96	91.7	82.2
Impervious			2	31	1	34	36	86.1	91.2
Water					23	23	28	82.1	100.0

Note: RT and CT represent row total and column total; PA and UA represent producer's accuracy and user's accuracy.
Overall accuracy, 84.0%; kappa, 0.78.

Based on the definition of vegetation change trajectories as summarized in Table 7.1, the combination of the 2008 classification image and the vegetation gain/loss image provided nine potential vegetation change trajectories. In addition to vegetation conversion (forest deforestation, from primary forest to nonvegetation land cover [e.g., agropasture, impervious surface]; vegetation deforestation, from nonforest vegetation [successional vegetation] to nonvegetation land cover; and regeneration, from agropasture to nonforest vegetation), the

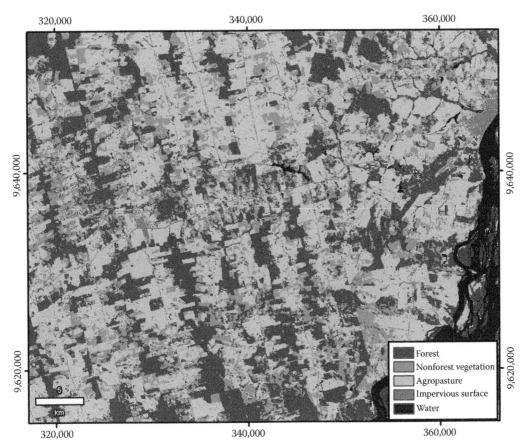

FIGURE 7.7
(**See color insert.**) Land use/cover distribution from the 2008 Landsat Thematic Mapper image.

change detection result also provides vegetation modification: forest restoration or growth, nonforest vegetation restoration or growth, and forest degradation or nonforest vegetation degradation. The result shown in Figure 7.8 indicates that the major vegetation changes include nonforest vegetation deforestation and degradation as well as regeneration.

Although no accuracy assessment for the vegetation change detection result was conducted in this research because of the lack of field measurements related to vegetation dynamics, a qualitative analysis based on the comparison of change detection results and the color composites has proven the value of this method in detecting vegetation conversion and modification. If field measurement data about vegetation modification are available, multiple thresholds can be more accurately identified. Thus, vegetation gain/loss distribution can be better determined.

In previous research, forest disturbance is often detected by using vegetation indices, such as the Normalized Difference Vegetation Index and tasseled cap transform (Healey et al. 2005; Jin and Sader 2005). The external factors such as moisture and atmospheric conditions may affect the detection results. The advantage of this GV-index-based change detection overcomes this problem, and proper use of this method by identifying multiple thresholds may provide better detection performance. The disadvantage of this new method is the difficulty in automatically separating the changes caused by agricultural change from the forest disturbance because of the impacts of crop phenology. In the moist

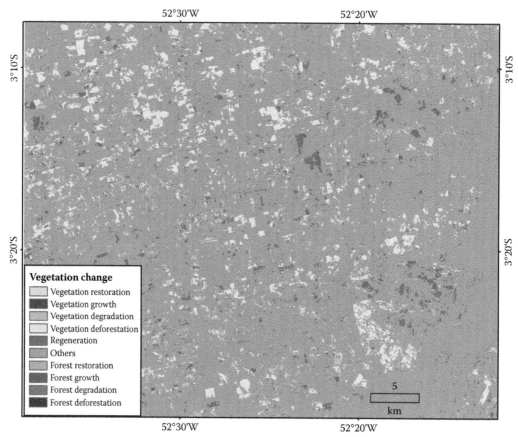

FIGURE 7.8
(See color insert.) Detailed vegetation change trajectories developed from the combination of the 2008 land use/ cover classification and vegetation gain/loss images.

tropical regions such as in this study area, cloud-free TM images are only available in the dry season. Crops are harvested, and thus the agricultural lands have similar spectral features as bare soils. When the GV-index-based method is used for vegetation change detection in other study areas, it is important to take the crop phenology into account. More research is needed to modify this method to minimize the effects of crop phenology.

7.4 Conclusions

Accurate detection of land use/cover change, especially forest disturbance, is still a challenge and has been an active research topic for a long time. Although many change detection techniques are available, most of them cannot effectively detect detailed "from–to" change trajectories, especially quantitative detection of vegetation modification. This research proposes the GV-index-based method to generate vegetation gain and loss distribution and a combination of one classification image and the vegetation gain/ loss data to generate detailed vegetation change trajectories. This research shows that this method is promising for detecting vegetation growth and degradation, in addition to deforestation and regeneration. More research is needed to identify optimal multiple thresholds based on field measurements. Caution should be taken to minimize the effects of crop phenology when the GV-index-based method is transferred to other study areas.

Acknowledgments

We acknowledge the support from the National Institute of Child Health and Human Development at NIH (grant no. R01 HD035811) for this research, addressing population and environment reciprocal interactions in several regions of the Brazilian Amazon. Any errors are solely the responsibility of the authors and not of the funding agencies.

References

Althausen, J. D. 2002. "What Remote Sensing System Should Be Used to Collect the Data?" In *Manual of Geospatial Science and Technology*, edited by J. D. Bossler, J. R. Jensen, R. B. McMaster, and C. Rizos, 276–97. New York: Taylor and Francis.

Chander, G., B. L. Markham, and D. L. Helder. 2009. "Summary of Current Radiometric Calibration Coefficients for Landsat MSS, TM, ETM+, and EO-1 ALI Sensors." *Remote Sensing of Environment* 113:893–903.

Chen, G., G. J. Hay, L. M. T. Carvalho, and M. A. Wulder. 2012. "Object-Based Change Detection." *International Journal of Remote Sensing* 33(14):4434–57.

Cohen, W. B., Z. Yang, and R. E. Kennedy. 2010. "Detecting Trends in Forest Disturbance and Recovery Using Yearly Landsat Time Series: 2. TimeSync—Tools for Calibration and Validation." *Remote Sensing of Environment* 114:2911–24.

Congalton, R. G. 1991. "A Review of Assessing the Accuracy of Classification of Remotely Sensed Data." *Remote Sensing of Environment* 37:35–46.

Congalton, R. G., and K. Green. 2008. *Assessing the Accuracy of Remotely Sensed Data: Principles and Practices.* 2nd ed., 183. Boca Raton, FL: CRC Press, Taylor & Francis.

Coppin, P., I. Jonckheere, K. Nackaerts, B. Muys, and E. Lambin. 2004. "Digital Change Detection Methods in Ecosystem Monitoring: A Review." *International Journal of Remote Sensing* 25:1565–96.

Foody, G. M. 2002. "Status of Land Cover Classification Accuracy Assessment." *Remote Sensing of Environment* 80:185–201.

Hansen, M. C., and T. R. Loveland. 2012. "A Review of Large Area Monitoring of Land Cover Change Using Landsat Data." *Remote Sensing of Environment* 122:66–74.

Healey, S. P., W. B. Cohen, Z. Yang, and O. N. Krankina. 2005. "Comparison of Tasseled Cap-Based Landsat Data Structures for Use in Forest Disturbance Detection." *Remote Sensing of Environment* 97:301–10.

Huang, C., S. N. Goward, J. G. Masek, N. Thomas, Z. Zhu, and J. E. Vogelmann. 2010. "An Automated Approach for Reconstructing Recent Forest Disturbance History Using Dense Landsat Time Series Stacks." *Remote Sensing of Environment* 114:183–98.

Jin, S., and S. A. Sader. 2005. "Comparison of Time Series Tasseled Cap Wetness and the Normalized Difference Moisture Index in Detecting Forest Disturbances." *Remote Sensing of Environment* 94:364–72.

Kennedy, R. E., P. A. Townsend, J. E. Gross, W. B. Cohen, P. Bolstad, Y. Q. Wang, and P. Adams. 2009. "Remote Sensing Change Detection Tools for Natural Resource Managers: Understanding Concepts and Tradeoffs in the Design of Landscape Monitoring Projects." *Remote Sensing of Environment* 113:1382–96.

Kennedy, R. E., Z. Yang, and W. B. Cohen. 2010. "Detecting Trends in Forest Disturbance and Recovery Using Yearly Landsat Time Series: 1. LandTrendr—Temporal Segmentation Algorithms." *Remote Sensing of Environment* 114:2897–910.

Laurance, W., A. K. M. Albernaz, P. M. Fearnside, H. L. Vasconcelos, and L. V. Ferreira. 2004. "Deforestation in Amazonia." *Science* 304:1109.

Lefsky, M. A., and W. B. Cohen. 2003. "Selection of Remotely Sensed Data." In *Remote Sensing of Forest Environments: Concepts and Case Studies*, edited by M. A. Wulder and S. E. Franklin, 13–46. Boston, MA: Kluwer Academic.

Li, G., D. Lu, E. Moran, and S. Hetrick. 2011. "Land-Cover Classification in a Moist Tropical Region of Brazil with Landsat TM Imagery." *International Journal of Remote Sensing* 32(23):8207–30.

Lu, D., M. Batistella, and E. Moran. 2005a. "Satellite Estimation of Aboveground Biomass and Impacts of Forest Stand Structure." *Photogrammetric Engineering and Remote Sensing* 71(8):967–74.

Lu, D., M. Batistella, and E. Moran. 2008. "Integration of Landsat TM and SPOT HRG Images for Vegetation Change Detection in the Brazilian Amazon." *Photogrammetric Engineering & Remote Sensing* 74(4):421–30.

Lu, D., S. Hetrick, E. Moran, and G. Li. 2012. "Application of Time Series Landsat Images to Examining Land Use/Cover Dynamic Change." *Photogrammetric Engineering & Remote Sensing* 78(7):747–55.

Lu, D., P. Mausel, M. Batistella, and E. Moran. 2005b. "Land-Cover Binary Change Detection Methods for Use in the Moist Tropical Region of the Amazon: A Comparative Study." *International Journal of Remote Sensing* 26(1):101–14.

Lu, D., P. Mausel, E. Brondízio, and E. Moran. 2002. "Assessment of Atmospheric Correction Methods for Landsat TM Data Applicable to Amazon Basin LBA Research." *International Journal of Remote Sensing* 23:2651–71.

Lu, D., P. Mausel, E. Brondízio, and E. Moran. 2004. "Change Detection Techniques." *International Journal of Remote Sensing* 25(12):2365–407.

Lu, D., E. Moran, and M. Batistella. 2003. "Linear Mixture Model Applied to Amazônian Vegetation Classification." *Remote Sensing of Environment* 87(4):456–69.

Lu, D., E. Moran, and S. Hetrick. 2011. "Detection of Impervious Surface Change with Multitemporal Landsat Images in an Urban-Rural Frontier." *ISPRS Journal of Photogrammetry and Remote Sensing* 66(3):298–306.

Lucas, R. M., M. Honzák, P. J. Curran, G. M. Foody, R. Mline, T. Brown, and S. Amaral. 2000. "The Regeneration of Tropical Forests Within the Legal Amazon." *International Journal of Remote Sensing* 21:2855–81.

Moran, E. F. 1981. *Developing the Amazon*. Bloomington, IN: Indiana University Press.

Roberts, D. A., G. T. Batista, J. L. G. Pereira, E. K. Waller, and B. W. Nelson. 1998. "Change Identification Using Multitemporal Spectral Mixture Analysis: Applications in Eastern Amazônia." In *Remote Sensing Change Detection: Environmental Monitoring Methods and Applications*, edited by R. S. Lunetta and C. D. Elvidge, 137–61. Ann Arbor, MI: Ann Arbor Press.

Roberts, D. A., I. Numata, K. Holmes, G. Batista, T. Krug, A. Monteiro, B. Powell, and O. A. Chadwick. 2002. "Large Area Mapping of Land-Cover Change in Rondonia Using Decision Tree Classifiers." *Journal of Geophysical Research* 107 (D20):8073, LBA 40-1 to 40-18.

Sano, E. E., R. Rosa, J. L.S. Brito, and L. G. Ferreira. 2010. "Land Cover Mapping of the Tropical Savanna Region in Brazil." *Environmental Monitoring and Assessment* 166:113–24.

Singh, A. 1989. "Digital Change Detection Techniques Using Remotely Sensed Data." *International Journal of Remote Sensing* 10:989–1003.

Thomas, N. E., C. Huang, S. N. Goward, S. Powell, K. Rishmawi, K. Schleeweis, and A. Hinds. 2011. "Validation of North American Forest Disturbance Dynamics Derived from Landsat Time Series Stacks." *Remote Sensing of Environment* 115:19–32.

Van der Meer, F., and S. M. de Jong. 2000. "Improving the Results of Spectral Unmixing of Landsat Thematic Mapper Imagery by Enhancing the Orthogonality of End-Members." *International Journal of Remote Sensing* 21:2781–97.

Yool, S. R., M. J. Makaio, and J. M. Watts. 1997. "Techniques for Computer-Assisted Mapping of Rangeland Change." *Journal of Range Management* 50:307–14.

8

Extraction of Impervious Surfaces from Hyperspectral Imagery: Linear versus Nonlinear Methods

Xuefei Hu and Qihao Weng

CONTENTS

8.1 Impervious Surfaces and Their Impact 141
8.2 Methods for Impervious Surface Extraction... 142
8.3 Case Study 144
    8.3.1 Study Area and Data 144
    8.3.2 Linear Spectral Unmixing Method 144
    8.3.3 Multilayer Perceptron Neural Network Method 146
8.4 Conclusions and Future Directions 148
References.. 150

8.1 Impervious Surfaces and Their Impact

Impervious surface is the anthropogenic surface that prevents water from infiltrating into soil (Arnold and Gibbons 1996), and it is widely distributed in urban areas. Impervious surface can be categorized into two primary types: rooftops and transportation systems (e.g., roads, sidewalks, and parking lots) (Schueler 1994). Impervious surfaces are made of materials impenetrable for water, such as asphalt, concrete, bricks, or stones. Compacted soils are also highly impervious and considered another type of impervious surface (Arnold and Gibbons 1996).

Impervious surface is associated with various adverse environmental outcomes and is a crucial environmental indicator (Arnold and Gibbons 1996). Impervious surface decreases the recharge of underground water, increases the velocity and volume of the surface runoff, and as a result, increases the risk of flooding. Moreover, increased runoff further erodes construction sites and river banks (Arnold and Gibbons 1996). In addition, impervious surface leads to nonpoint source pollution and threatens surface water quality (Civico and Hurd 1997; Sleavin et al. 2000). Nonpoint source pollutants include pathogens, nutrients, toxic contaminants, and sediments, which degrade water qualities and are harmful to both animals and humans. Impervious surface also affects the energy balance in urban areas and is one of the important factors that cause the urban heat island effect. Previous studies showed that impervious surface was positively related to increased surface temperatures in urban areas (Lu and Weng 2006a; Yuan and Bauer 2007). Likewise, the increase of impervious surface inevitably results in the decrease of vegetation cover, which reduces ecological productivity, interrupts atmospheric carbon cycling, and degrades air quality.

Hence, impervious surface is crucial for urban environmental management (Arnold and Gibbons 1996; Flanagan and Civco 2001; Wu and Murray 2003).

8.2 Methods for Impervious Surface Extraction

To extract impervious surface cover, traditional methods include ground survey, global positioning system (GPS), aerial photo interpretation, and satellite remote sensing interpretation (Stocker 1998). Ground measurement is both time and cost inefficient, whereas GPS is not feasible for mapping large areas. Extraction of impervious surfaces from aerial photos is expensive. Satellite remote sensing has been widely used in impervious surface estimation studies because of its relatively low cost and capability for mapping large areas (Bauer et al. 2004). A challenge in extracting impervious surfaces from medium spatial resolution satellite imagery is to tackle mixed pixels. A mixed pixel is a pixel that contains multiple land cover types, as compared to pure pixels that contain only one land cover class. Traditional supervised and unsupervised classification techniques can only identify land cover features at the pixel level and cannot effectively deal with mixed pixels. As a result, subpixel techniques need to be applied. To date, numerous subpixel classification methods have been developed for extracting impervious surface from remote sensing imagery (medium spatial resolution images), including linear spectral mixture analysis (LSMA), regression tree, artificial neural networks (ANNs), and multiple regression (Civico and Hurd 1997; Wu and Murray 2003; Yang et al. 2003a, 2003b; Bauer et al. 2004; Lee and Lathrop 2005; Lu and Weng 2006b). Object-based classification techniques have also been used for impervious surface estimation from remote sensing imagery, especially from high spatial resolution images. As the spatial resolution increases, the proportion of pure pixels increases and mixed pixels are reduced (Hsieh et al. 2001). Thus, the subpixel methods might not be appropriate. Object-based classification methods incorporate not only the color and tone of the pixels, but also other crucial characteristics such as shape, texture, and context, and thus can extract impervious surfaces with higher accuracy.

Weng (2012) provided a comprehensive review of remote sensing methods for estimating and mapping impervious surfaces in the urban areas. We focus on an examination of subpixel estimation techniques, including LSMA, ANNs, and fuzzy classifiers. A linear spectral mixture model is based on the assumption that each photon interacts with only one land cover type on the ground before being reflected back to the sensor, and as a result, the mixed spectra can be modeled as a linear combination of the spectra of land cover features weighted by the proportion of each feature within the instantaneous field of view (Singer and McCord 1979; Roberts et al. 1998; Small 2001). The spectra of these land cover features are called end-members. End-member selection is a key step to LSMA, and end-members can be selected from image data themselves, spectra libraries, or reference spectra collected from field (Roberts et al. 1998; Small 2001). Selecting end-members directly from the image's feature space is relatively simple, and thus has been used by numerous previous studies. The results of LSMA are fraction images for end-members, in which the pixel values indicate the percentage of the end-member within that pixel (Small 2001). LSMA may provide a reasonable approximation for complex landscapes (Haglund 2000), but also has several limitations (Foody et al. 1997). First, the linear assumption is not necessarily true. When scattered photons interact with multiple components, the mixture becomes nonlinear (Roberts et al. 1993; Gilabert et al. 2000). When the nonlinearity is significant, it should not be neglected (Roberts et al. 1993; Ray and Murray 1996). The difficulty of LSMA also comes from end-member selection. End-member selection is a challenge because of within-class spectral variability. The number of end-members is limited by the dimensionality of the image and the correlation between bands. Limited number of end-members reduces the capability of unmixing because of the image spectra being undersampled (Small 2001). The spectra of land cover types can be very diverse. For example, there may be various degrees of bright/dark impervious surfaces, which are located in different locations within scatterplots. Thus, the selection of end-members that can utterly represent the spectra of specific land cover classes is difficult.

Another approach for impervious surface estimation is fuzzy classification. Fuzzy classification is based on the fuzzy set theory that defines the strength of membership of a pixel to land cover class by using a fuzzy membership grade within a range from 0% to 100%. In general, the fuzzy membership grade is calculated by a fuzzy membership function. Different types of membership function have been developed, which include sigmoidal (S-shaped), J-shaped, and linear. Fuzzy classification has also been widely applied in land use and land cover classifications, vegetation classifications, change analysis, cloud cover classifications, and flooded area mapping (Fisher and Pathirana 1990; Foody and Cox 1994; Foody 1996, 1998; Zhang and Foody 1998, 2001; Mohan et al. 2000; Townsend 2000; Lee and Lathrop 2002; Amici et al. 2004; Tapia et al. 2005; Filippi and Jensen 2006; Ghosh et al. 2006; Okeke and Karnieli 2006a,b; Tang et al. 2007). So far, the use of fuzzy classification in impervious surface estimation is limited. To date, Lee and Lathrop (2002) used a supervised fuzzy c-means clustering to extract impervious surface fractions from Landsat thematic mapper (TM) images. Hu and Weng (2011) compared the performance of fuzzy classification and LSMA in impervious surface extraction at the subpixel level. Their results showed that fuzzy classification outperformed LSMA in terms of extraction accuracy.

The third subpixel method for extracting impervious surfaces is ANN. The neural networks mimic the functions of the human brain and can learn through trial and error. The ANNs can generate more accurate results, perform more rapidly, incorporate a prior knowledge in the calibration, and incorporate different types of data. There are many types of neural networks, and a commonly used algorithm is the multilayer perceptron (MLP) neural network (Atkinson and Tatnall 1997; Foody et al. 1997). MLP has been used in land use and land cover classification, change detection, and water properties estimation (Foody et al. 1997; Schiller and Doerffer 1999; Zhang and Foody 2001; Li and Yeh 2002; Corsini et al. 2003; Kavzoglu and Mather 2003). MLP has also been used in impervious surface estimation. Chormanski et al. (2008) conducted MLP to map the fractions of four major land cover classes (impervious surfaces, vegetation, bare soil, and water/shade) with both high spatial resolution and medium spatial resolution imagery. Weng and Hu (2008) used LSMA and an MLP neural network for impervious surface extraction at the subpixel level and compared the performance. Hu and Weng (2009) compared the performance of the MLP neural network and the self-organizing map (SOM) neural network for estimating impervious surfaces. In addition, Hu and Weng (2010) applied MLP to spectrally normalized images to extract impervious surfaces. Van de Voorde et al. (2011) used an MLP-based supervised classification to extract impervious surfaces. Limitations of MLP are also obvious. First, it is a challenge to determine how many nodes are needed in each layer and how to design the number of the hidden layers. Second, the training of MLP requires both presence and absence data. However, in many cases, the absence data are not available.

The SOM neural network has also been used for impervious surface estimation, although it has not been widely applied in other remote sensing applications. SOMs were previously used for both supervised and unsupervised classifications. Ito and Omatu (1997) applied an SOM neural network and a k-nearest-neighbor method on Landsat TM data. Moreover, Ito (1998) used the SOM method on synthetic aperture radar data for image classification. Ji (2000) conducted a Kohonen self-organizing feature map for land use and land cover classification. Lee and Lathrop (2006) combined SOM, learning vector quantization, and a Gaussian mixture model to estimate the percentage of impervious surface coverage.

8.3 Case Study

8.3.1 Study Area and Data

The study area is the Atlanta metropolitan area. Atlanta is the capital and the most populous city in Georgia. According to the U.S. Census Bureau, the population of the Atlanta metropolitan area has increased from 4.11 million in 2000 to 5.27 million in 2010, which is the ninth largest in the United States. Atlanta serves as the major transportation hub in the southeastern United States, and is a center for services, finance, information technology, government, and higher education. Atlanta is the world headquarters of the Coca-Cola company, the Home Depot, AT&T Mobility, UPS, and Delta Airlines.

A Hyperion image covering the study area was downloaded. The Hyperion image has 242 bands covering 400–2500 nm with a spatial resolution of 30 m. The image was georectified to a Universal Transverse Mercator coordinate system using the nearest-neighbor resampling method. A root-mean-square error (RMSE) of less than 0.2 pixels was obtained in geometric correction.

8.3.2 Linear Spectral Unmixing Method

The LSMA model can be expressed mathematically by the following equation:

$$R_b = \sum_{i=1}^{N} f_i R_{i,b} + e_b \tag{8.1}$$

where R_b is the reflectance for each band b, N is the number of end-members, f_i is the fraction of end-member i, $R_{i,b}$ is the reflectance of end-member i in band b, and e_b is the error for band b. For a fully constrained least-square unmixing solution, the following conditions are required to be satisfied:

$$\sum_{i=1}^{N} f_i = 1, \quad f_i \geq 0 \tag{8.2}$$

Model fitness is assessed by the RMSE. The mathematical expression of RMSE is shown as follows:

$$\text{RMSE} = \sqrt{\sum_{b=1}^{M} \frac{e_b^2}{M}} \tag{8.3}$$

where e_b is the unmodeled residual and M is the number of the bands. The end-member can be selected at the vertexes of the triangles in the scatterplots (Figure 8.1). In this study, four end-members were identified from the image plots, including high albedo, low albedo, soil, and vegetation.

The relationship between impervious surface, high albedo, and low albedo can be modeled in the following mathematical equation:

$$R_{\mathrm{imp,b}} = f_{\mathrm{low}} R_{\mathrm{low,b}} + f_{\mathrm{high}} R_{\mathrm{high,b}} + e_{\mathrm{b}} \tag{8.4}$$

where $R_{\mathrm{imp,b}}$ is the spectra of impervious surfaces for band b, $R_{\mathrm{high,b}}$ and $R_{\mathrm{low,b}}$ are the spectra of high albedo and low albedo end-members, f_{low} and f_{high} are the fractions of low albedo and high albedo end-members, and e_{b} is the unmodeled residual. The impervious surface fractions can be calculated by adding high albedo and low albedo together. Nevertheless, some low-albedo materials (e.g., water and shade) and high-albedo materials (e.g., cloud and sand) need to be removed because they tended to be confused with impervious surfaces spectrally.

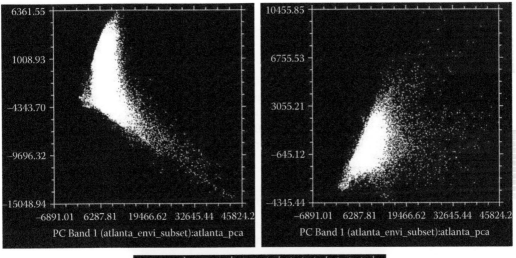

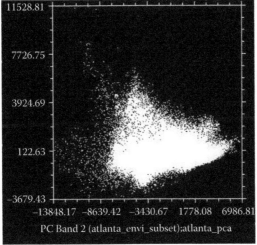

FIGURE 8.1
Scatterplots of selected end–members.

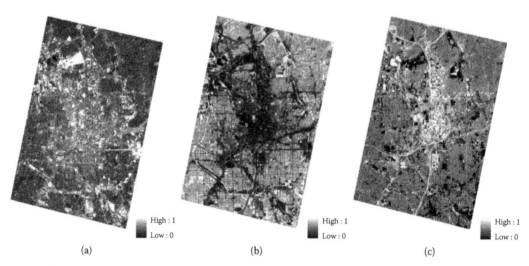

High : 1
Low : 0

High : 1
Low : 0

High : 1
Low : 0

(a)   (b)   (c)

FIGURE 8.2
Fraction images of the end-members derived from linear spectral mixture analysis. (a) High albedo,
(b) vegetation, and (c) low albedo.

Figure 8.2 shows the unmixed results, including high albedo, vegetation, and low albedo.
The brighter the pixel, the higher the percentage of the end-member within a particular pixel.
The impervious surface fractions were calculated using Equation 8.4 and shown in
Figure 8.3. The high percentages of impervious surfaces are located in the central business
district area, whereas the low percentages are within rural areas. Most residential areas
had medium percentages.

8.3.3 Multilayer Perceptron Neural Network Method

An MLP neural network with a back-propagation learning algorithm was also implemented. The MLP classifier used the following algorithm to calculate the input that a single node *j* received:

$$\text{net}_j = \sum_i w_{ij} I_i \tag{8.5}$$

where net_j refers to the input that a single node *j* receives; w_{ij} represents the weights
between node *i* and node *j*; and I_i is the output from node *i* of a sender layer (input or hidden layer). Output from a node *j* was calculated as follows:

$$O_j = f(\text{net}_j) \tag{8.6}$$

The function *f* usually is a nonlinear sigmoidal function. In the study, an input layer with
76 nodes corresponding to 76 selected Hyperion image bands (visible and near-infrared
and shortwave infrared), and one output layer with 4 nodes corresponding to 4 land cover
classes, high albedo, low albedo, vegetation, and soil were used. The number of hidden
layer nodes was calculated by the following formula:

$$N_h = \text{INT}\sqrt{N_i \times N_o} \tag{8.7}$$

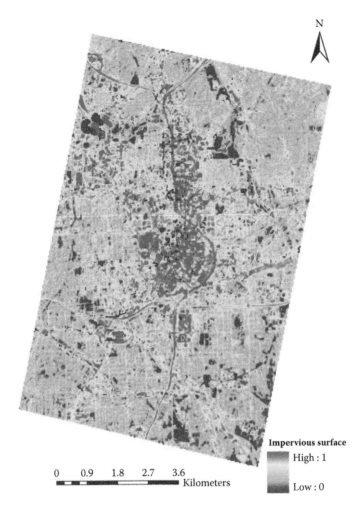

FIGURE 8.3
(**See color insert.**) Fraction image of impervious surface derived from linear spectral mixture analysis.

where N_h is the number of hidden layer nodes; N_i is the number of input layer nodes, and N_o is the number of output layer nodes.

An accuracy assessment was performed to evaluate the final results using an aerial photo covering the study area. One hundred testing samples were selected using a stratified random sampling method based on a land use and land cover classification map. To avoid geometric error, the size of each sample was set to 3×3 pixels, which equaled 90 m × 90 m in ground dimension for Hyperion images. Impervious surface were digitized in each sample on an aerial photo. The percentage of impervious surface in each sample was calculated by dividing the area of impervious surface by the total sampling area. RMSE and R^2 were calculated to quantify the accuracy of impervious surface estimation. Some of the samples are shown in Figure 8.4.

The fraction images of high albedo, low albedo, vegetation, and soil are shown in Figure 8.5. The impervious surface map is shown in Figure 8.6, which illustrates a similar spatial pattern as that shown in Figure 8.3. The accuracy assessment result indicates that an RMSE of 16.3% and R^2 of 0.75 were achieved for the whole study area.

FIGURE 8.4
(See color insert.) Testing samples in downtown Atlanta.

8.4 Conclusions and Future Directions

This chapter provides a summary of the current research on urban impervious surface estimation and mapping. In addition, a case study was conducted to demonstrate the capability of two conventional methods (LSMA and MLP) for impervious surface estimation using Hyperion imagery. Satellite remote sensing provides a cost-effective and time-efficient way for impervious surface mapping. Medium spatial resolution imagery (e.g., Landsat TM, enhanced TM Plus, and ASTER) has been used for large-area mapping, and high spatial resolution imagery (e.g., IKONOS and QuickBird), air photos, and LiDAR data for extracting urban features (e.g., roads and buildings). Numerous methods have been developed and applied in previous studies based on per-pixel-, subpixel-, and object-based algorithms. However, fewer studies have examined the spectral diversity of impervious surfaces. Hyperspectral imagery with rich spectral information is suitable for spectral analysis and should be extensively used in future studies. The spatial and temporal variations of impervious surfaces are the two issues deserving more attention because of the diversity of urban environments and the changes caused by urbanization (Weng and Lu 2009).

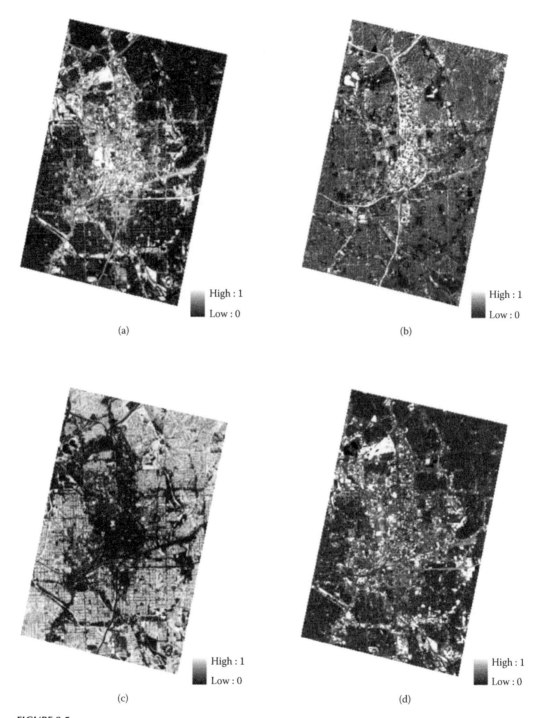

FIGURE 8.5
Fraction images of end-members derived from multilayer perceptron. (a) High albedo, (b) low albedo, (c) vegetation, and (d) soil.

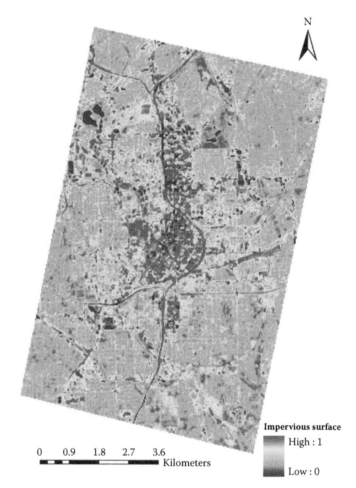

FIGURE 8.6
(**See color insert.**) Fraction image of impervious surface derived from multilayer perceptron.

References

Amici, G., F. Dell'Acqua, P. Gamba, and G. Pulina. 2004. "A Comparison of Fuzzy and Neuro-Fuzzy Data Fusion for Flooded Area Mapping Using SAR Images." *International Journal of Remote Sensing* 25(20):4425–30.

Arnold, C. L., and C. J. Gibbons. 1996. "Impervious Surface Coverage: The Emergence of a Key Environmental Indicator." *Journal of the American Planning Association* 62:243–58.

Atkinson, P. M., and A. R. L. Tatnall. 1997. "Introduction: Neural Networks in Remote Sensing." *International Journal of Remote Sensing* 18:699–709.

Bauer, M. E., N. J. Heinert, J. K. Doyle, and F. Yuan. 2004. "Impervious Surface Mapping and Change Monitoring Using Landsat Remote Sensing." In *Proceedings of ASPRS Annual Conference* May 24–28, Denver, Colorado.

Chormanski, J., T. V. d. Voorde, T. D. Roeck, O. Batelaan, and F. Canters. 2008. "Improving Distributed Runoff Prediction in Urbanized Catchments with Remote Sensing Based Estimates of Impervious Surface Cover." *Sensors* 8:910–32.

Civico, D. L., and J. D. Hurd. 1997. "Impervious Surface Mapping for the State of Connecticut." In *Proceedings of ASPRS/ACSM Annual Convention*, April 7–10, Vol. 3, 124–35. Seattle, Washington.

Corsini, G., M. Diani, R. Grasso, M. De Martino, P. Mantero, and S. Serpico. 2003. "Radial Basis Function and Multilayer Perceptron Neural Networks for Sea Water Optically Active Parameter Estimation in Case II Waters: A Comparison." *International Journal of Remote Sensing* 24(20):3917.

Filippi, A. M., and J. R. Jensen. 2006. "Fuzzy Learning Vector Quantization for Hyperspectral Coastal Vegetation Classification." *Remote Sensing of Environment* 100(4):512–30.

Fisher, P. F., and S. Pathirana. 1990. "The Evaluation of Fuzzy Membership of Land Cover Classes in the Suburban Zone." *Remote Sensing of Environment* 34(2):121–32.

Flanagan, M., and D. L. Civco. 2001. "Subpixel Impervious Surface Mapping." In *Proceedings of 2001 ASPRS Annual Convention*, April 23–27. St. Louis, MO.

Foody, G. M. 1996. "Approaches for the Production and Evaluation of Fuzzy Land Cover Classifications from Remotely-Sensed Data." *International Journal of Remote Sensing* 17(7):1317–40.

Foody, G. M. 1998. "Sharpening Fuzzy Classification Output to Refine the Representation of Sub-Pixel Land Cover Distribution." *International Journal of Remote Sensing* 19(13):2593–99.

Foody, G. M., and D. P. Cox. 1994. "Sub-Pixel Land Cover Composition Estimation Using a Linear Mixture Model and Fuzzy Membership Functions." *International Journal of Remote Sensing* 15(3):619–31.

Foody, G. M., R. M. Lucas, P. J. Curran, and M. Honzak. 1997. "Non-Linear Mixture Modelling Without End-Members Using an Artificial Neural Network." *International Journal of Remote Sensing* 18:937–53.

Ghosh, A., N. R. Pal, and J. Das. 2006. "A Fuzzy Rule Based Approach to Cloud Cover Estimation." *Remote Sensing of Environment* 100 (4):531–49.

Gilabert, M. A., F. J. Garcia-Haro, and J. Meli. 2000. "A Mixture Modeling Approach to Estimate Vegetation Parameters for Heterogeneous Canopies in Remote Sensing." *Remote Sensing of Environment* 72(3):328–45.

Haglund, A. 2000. *Towards Soft Classification of Satellite Data: A Case Study Based Upon Resurs MSU-SK Satellite Data and Land Cover Classification Within the Baltic Sea Region*. Project Report, Royal Institute of Technology, Department of Geodesy and Photogrammetry, Stockholm, Sweden.

Hsieh, P.-F., L. C. Lee, and N.-Y. Chen. 2001. "Effect of Spatial Resolution on Classification Errors of Pure and Mixed Pixels in Remote Sensing." *IEEE Transactions on Geoscience and Remote Sensing* 39:2657–63.

Hu, X., and Q. Weng. 2009. "Estimating Impervious Surfaces from Medium Spatial Resolution Imagery Using the Self-Organizing Map and Multi-Layer Perceptron Neural Networks." *Remote Sensing of Environment* 113(10):2089–102.

Hu, X., and Q Weng. 2010. "Estimation of Impervious Surfaces of Beijing, China, with Spectral Normalized Images Using Linear Spectral Mixture Analysis and Artificial Neural Network." *Geocarto International* 25(3):231–53.

Hu, X., and Q Weng. 2011. "Estimating Impervious Surfaces from Medium Spatial Resolution Imagery: A Comparison between Fuzzy Classification and LSMA." *International Journal of Remote Sensing* 32(20):5645–63.

Ito, Y. 1998. "Polarimetric SAR Data Classification Using Competitive Neural Networks." *International Journal of Remote Sensing* 19(14):2665–84.

Ito, Y., and S. Omatu. 1997. "Category Classification Method Using a Self-Organizing Neural Network." *International Journal of Remote Sensing* 18(4):829–45.

Ji, C. Y. 2000. "Land-Use Classification of Remotely Sensed Data Using Kohonen Self-Organizing Feature Map Neural Networks." *Photogrammetric Engineering & Remote Sensing* 66:1451–60.

Kavzoglu, T., and P. M. Mather. 2003. "The Use of Backpropagating Artificial Neural Networks in Land Cover Classification." *International Journal of Remote Sensing* 24(23):4907–38.

Lee, S., and R. G. Lathrop. 2002. "Sub-Pixel Estimation of Urban Land Cover Intensity Using Fuzzy C-Means Clustering." In *Proceedings of 2002 ASPRS Annual Convention*, April 19–26. Washington, DC.

Lee, S., and R. G. Lathrop. 2005. "Sub-Pixel Estimation of Urban Land Cover Components with Linear Mixture Model Analysis and Landsat Thematic Mapper Imagery." *International Journal of Remote Sensing* 26(22):4885–905.

Lee, S., and R. G. Lathrop. 2006. "Subpixel Analysis of Landsat ETM+ Using Self-Organizing Map (SOM) Neural Networks for Urban Land Cover Characterization." *IEEE Transactions on Geoscience and Remote Sensing* 44(6):1642–54.

Li, X., and A. G-O. Yeh. 2002. "Neural-Network-Based Cellular Automata for Simulating Multiple Land Use Changes Using GIS." *International Journal of Geographical Information Science* 16(4):323–43.

Lu, D., and Q. Weng. 2006a. "Spectral Mixture Analysis of ASTER Images for Examining the Relationship between Urban Thermal Features and Biophysical Descriptors in Indianapolis, Indiana, USA." *Remote Sensing of Environment* 104(2):157–67.

Lu, D., and Q. Weng. 2006b. "Use of Impervious Surface in Urban Land-Use Classification." *Remote Sensing of Environment* 102(1–2):146–60.

Mohan, B. K., B. B. Madhavan, and U. M. D. Gupta. 2000. "Integration of IRS-1A L2 Data by Fuzzy Logic Approaches for Land Use Classification." *International Journal of Remote Sensing* 21(8):1709–23.

Okeke, F., and A. Karnieli. 2006a. "Methods for Fuzzy Classification and Accuracy Assessment of Historical Aerial Photographs for Vegetation Change Analyses. Part I: Algorithm Development." *International Journal of Remote Sensing* 27(1/2):153–76.

Okeke, F., and A. Karnieli. 2006b. "Methods for Fuzzy Classification and Accuracy Assessment of Historical Aerial Photographs for Vegetation Change Analyses. Part II: Practical Application." *International Journal of Remote Sensing* 27(9):1825–38.

Ray, T. W., and B. C. Murray. 1996. "Nonlinear Spectral Mixing in Desert Vegetation." *Remote Sensing of Environment* 55:59–64.

Roberts, D. A., M. Gardner, R. Church, S. Ustin, G. Scheer, and R. O. Green. 1998. "Mapping Chaparral in the Santa Monica Mountains Using Multiple Endmember Spectral Mixture Models." *Remote Sensing of Environment* 65:267–79.

Roberts, D. A., M. O. Smith, and J. B. Adams. 1993. "Green Vegetation, Nonphotosynthetic Vegetation, and Soils in AVIRIS Data." *Remote Sensing of Environment* 44(2–3):255–69.

Schiller, H., and R. Doerffer. 1999. "Neural Network for Emulation of an Inverse Model Operational Derivation of Case II Water Properties from MERIS Data." *International Journal of Remote Sensing* 20(9):1735–46.

Schueler, T. 1994. "The Importance of Imperviousness." *Watershed Protection Techniques* 1(3):100–11.

Singer, R. B., and T. B. McCord. 1979. "Mars: Large Scale Mixing of Bright and Dark Surface Materials and Implications for Analysis of Spectral Reflectance." In *Proceedings of 10th Lunar and Planetary Science Conference, American Geophysical Union*, March 19–23, 1835–48. Washington, DC.

Sleavin, W. J., D. L. Civco, S. Prisloe, and L. Giannotti. 2000. "Measuring Impervious Surfaces for Non-Point Source Pollution Modeling." In *Proceedings of 2000 ASPRS Annual Convention*, May 22–26. Washington, DC.

Small, C. 2001. "Estimation of Urban Vegetation Abundance by Spectral Mixture Analysis." *International Journal of Remote Sensing* 22(7):1305–34.

Stocker, J. 1998. "Methods for Measuring and Estimating Impervious Surface Coverage." NEMO Technical Paper No. 3, University of Connecticut, Haddam Cooperative Extension Center.

Tang, J., L. Wang, and S. W. Myint. 2007. "Improving Urban Classification through Fuzzy Supervised Classification and Spectral Mixture Analysis." *International Journal of Remote Sensing* 28(18):4047–63.

Tapia, R., A. Stein., and W. Bijker. 2005. "Optimization of Sampling Schemes for Vegetation Mapping Using Fuzzy Classification." *Remote Sensing of Environment* 99(4):425–33.

Townsend, P. A. 2000. "A Quantitative Fuzzy Approach to Assess Mapped Vegetation Classifications for Ecological Applications." *Remote Sensing of Environment* 72(3):253–67.

Van de Voorde, T., W. Jacquet, and F. Canters. 2011. "Mapping Form and Function in Urban Areas: An Approach Based on Urban Metrics and Continuous Impervious Surface Data." *Landscape and Urban Planning* 102(3):143–55.

Weng, Q. 2012. "Remote Sensing of Impervious Surfaces in the Urban Areas: Requirements, Methods, and Trends." *Remote Sensing of Environment* 117(2):34–49.

Weng, Q., and X. Hu. 2008. "Medium Spatial Resolution Satellite Imagery for Estimating and Mapping Urban Impervious Surfaces Using LSMA and ANN." *IEEE Transactions on Geoscience and Remote Sensing* 46(8):2397–406.

Weng, Q., and D. Lu. 2009. "Landscape as a Continuum: An Examination of the Urban Landscape Structures and Dynamics of Indianapolis City, 1991–2000." *International Journal of Remote Sensing* 30(10):2547–77.

Wu, C., and A. T. Murray. 2003. "Estimating Impervious Surface Distribution by Spectral Mixture Analysis." *Remote Sensing of Environment* 84:493–505.

Yang, L., X. George, J. M. Klaver, and B. Deal. 2003a. "Urban Land-Cover Change Detection through Sub-Pixel Imperviousness Mapping Using Remotely Sensed Data." *Photogrammetric Engineering and Remote Sensing* 69(9):1003–10.

Yang, L., C. Huang, C. G. Homer, K. Bruce Wylie, and M. J. Coan. 2003b. "An Approach for Mapping Large-Area Impervious Surfaces: Synergistic Use of Landsat-7 ETM+ and High Spatial Resolution Imagery." *Canadian Journal of Remote Sensing* 29(2):230–40.

Yuan, F., and M. E. Bauer. 2007. "Comparison of Impervious Surface Area and Normalized Difference Vegetation Index as Indicators of Surface Urban Heat Island Effects in Landsat Imagery." *Remote Sensing of Environment* 106(3):375–86.

Zhang, J., and G. M. Foody. 1998. "A Fuzzy Classification of Sub-Urban Land Cover from Remotely Sensed Imagery." *International Journal of Remote Sensing* 19(14):2721–38.

Zhang, J., and G. M. Foody. 2001. "Fully-Fuzzy Supervised Classification of Sub-Urban Land Cover from Remotely Sensed Imagery: Statistical and Artificial Neural Network Approaches." *International Journal of Remote Sensing* 22(4):615–28.

9

Road Extraction: A Review of LiDAR-Focused Studies

Lindi J. Quackenbush, Jungho Im, and Yue Zuo

CONTENTS

9.1 Introduction.. 155
9.2 Overview of Recent LiDAR Applications in Road Extraction.................... 156
9.3 Road Extraction Using LiDAR Remote Sensing... 158
    9.3.1 Defining Road Clusters.. 158
        9.3.1.1 Classification Framework.. 158
        9.3.1.2 Algorithms Used for Road Identification 160
    9.3.2 Generating Road Networks.. 162
        9.3.2.1 Road Classification Refinement 162
        9.3.2.2 Centerline Extraction... 163
9.4 Discussion and Conclusion 164
References.. 165

9.1 Introduction

Accurately mapped roads are essential for a wide variety of applications including studies related to community and transportation planning, and water resource and wildlife management. Road extraction techniques have been applied to many different types of remote sensing data over recent decades and include both automated and semiautomated approaches. Development during this period has gone from low to high spatial resolution image data and more recently has extended into data fusion. Mena (2003) and Quackenbush (2004) reported many different types of remote sensed data used in linear feature extraction including multi- and hyperspectral imagery, synthetic aperture radar (SAR), and light detection and ranging (LiDAR) sources. Studies seeking to define roads from a single data source frequently face both spectral and spatial challenges. Road extraction is complicated by geometric noise—for example, because of cars or shadows—that increase spectral variance and decrease homogeneity in radiometry along a road (Weng 2012). Another problematic area is the similarity in radiometry between roads and other impervious features such as buildings and parking lots (Péteri and Ranchin 2007). Mena (2003) and Jung et al. (2006) reported semiautomatic algorithms that mitigate some of these challenges by guiding road extraction using existing geographic information system (GIS) databases or manual inputs; however, researchers have also explored factors related to successfully automating road extraction. Gecen and Sarp (2008) extracted roads automatically in an urban area from four satellite images with different ground sample distance—that is, IKONOS, QuickBird, ASTER, and Landsat—to study the impact of spatial resolution on

155

automatic road extraction. Although accuracy tended to increase with spatial resolution, Gecen and Sarp (2008) found that objects with spectral property similar to roads, such as house roofs, were frequently misclassified into road clusters at all resolutions. Baumgartner et al. (1999) extracted roads from digital aerial imagery and found road edges contained significant noise caused by adjacent features such as trees and roofs. Amini et al. (2002) proposed an object-based approach for automatic extraction of major roads in a large-scale image, generating a binary road map through image segmentation. The project presented by Amini et al. (2002) suffered a problem that is common when working with impervious land cover types, in that their map characterized some elevated objects as roads.

Because of the limitations in studies that rely exclusively on passive imagery, many researchers have moved to explore the benefits of LiDAR datasets in road extraction. Because LiDAR data provide a variety of geometric information, it has been increasingly used to detect roads (Weng 2012). LiDAR data contain accurate elevation information and this component has proven useful to separate roads from other impervious objects such as buildings. In addition, LiDAR intensity data provide good separability for ground materials, such as grass, that may have similar elevation to roads (Song et al. 2002). Prior research has shown that LiDAR datasets are useful for generating vector road networks that are suitable for incorporation in GIS databases as well as a variety of other end goals. This chapter summarizes some of the recent approaches used to extract roads from remotely sensed data, with a particular focus on applications that consider LiDAR data.

9.2 Overview of Recent LiDAR Applications in Road Extraction

LiDAR is frequently used to acquire terrain elevation data because of relatively short data acquisition and processing times, the potential for high accuracy and point density, and reduced acquisition costs (Brennan and Webster 2006). LiDAR is an active remote sensing technique that transmits pulses of laser light toward the ground and measures the time of pulse return to calculate the distance between the sensor and features on (or above) the ground (Lillesand et al. 2008). LiDAR is a useful technology for automated acquisition of the elevation information, with high accuracy that is comparable to traditional land surveys and photogrammetry, but faster, denser, and more economical data collection (Hu 2003). The use of LiDAR for accurate determination of terrain elevations began in the late 1970s. Initial systems were profiling devices that obtained elevation data only directly under the path of an aircraft (Lillesand et al. 2008). Modern LiDAR systems are capable of quickly scanning large areas and producing accurate point clouds. Many LiDAR systems collect the intensity of the reflected signal along with multiple returns for each pulse (Hu 2003). Newer systems allow full waveform data collection (Mallet and Bretar 2009). These improvements in LiDAR data collection have increased the availability of high-quality elevation information for road extraction projects.

Rottensteiner (2009, 2010) highlights the state of the art in terms of LiDAR-based urban object extraction and trends in the field. In contrast to other remote sensing data sources, the application of LiDAR to road extraction is relatively undeveloped (Clode et al. 2007). There have been several attempts to extract roads from LiDAR data (Choi et al. 2008; Rottensteiner and Clode 2008; Samadzadegan et al. 2009) but many studies use some form of data fusion—either using passive imagery or input from GIS databases. Depending on the characteristics of the input data, the processing approach, and the algorithms used, data fusion may be at the point or pixel level, or it may be at a decision level (Hall and Llinas 1997). In Hatger and Brenner (2003), estimation of road geometry parameters was performed by combining high-resolution LiDAR data (4 points per m^2) and an existing digital road database to derive the height, slope, curvature, and width of roads. Boyko and Funkhouser (2011) also used an initial road map to extract roads from dense LiDAR point clouds. Zhu et al. (2004) combined high-resolution digital images and laser scanner data to automatically extract roads in an urban area, primarily extracting roads from the digital images but using the laser data to assist the extraction process by identifying and removing high objects to recover hidden road edges. Hu et al. (2004) extracted urban roads automatically from high spatial resolution imagery (0.5 m ground sample distance) and LiDAR data (1.1 points per m^2). Both intensity and height information from the LiDAR data were used for road detection, while the high spatial resolution imagery was applied to separate grassland and trees from open areas. Wang et al. (2011) also fused LiDAR data with aerial imagery to perform road extraction.

Many of the studies that fuse imagery with LiDAR data have found that the elevation information inherently supplied by the LiDAR data is particularly beneficial for separating roads from other impervious objects such as buildings. However, because roads have relatively homogeneous reflectivity in terms of LiDAR intensity, and are typically at the same elevation as the bare surface, researchers are now exploring road extraction from LiDAR data without reliance on other data sources. Using a single data source has the advantage of decreasing costs as well as mitigating the challenges associated with data registration. In addition, because LiDAR is both an active sensor and an explicit three-dimensional (3D) data source, several benefits can be realized from a LiDAR-only approach. In particular, data acquisition is not limited to daylight hours as with passive sensors and accurate height information is contained in the data (Clode et al. 2007). While the majority of the LiDAR-focused road extraction studies described in this chapter target the urban environment, some researchers, for example Rieger et al. (1999), have used LiDAR data to detect roads within a forest area. Rieger et al. (1999) initially detected roads to generate break lines, which enhanced the quality of digital terrain model production.

Alharthy and Bethel (2003) detected roads in an urban area from LiDAR data. They used intensity and height information to filter the raw LiDAR data and remove unrelated noise. Filtering was performed by locating penetrable, that is, nonroad, objects using differences in the first and last pulse observations. Clode et al. (2004) used both height and intensity of a LiDAR point cloud with density of 1 point per 1.3 m^2 to extract roads using a hierarchical classification technique. Clode et al. (2004) introduced the idea of a local point density for noise reduction; roads were assumed to be consistent in nature so that most of the discrete noise regions, which had low local point density, were eliminated. Clode et al. (2005) applied a region growing algorithm to classify roads from LiDAR data. They tested their approach on two LiDAR datasets with 1 point per 1.3 m^2 and 1 point per 0.5 m^2, attaining better results for the study site with higher point density. Choi et al. (2007) proposed a method to extract urban road networks from LiDAR data using height and reflectance of LiDAR data and clustered road point information. Choi et al. (2007) applied their algorithm to one study site with 0.5 LiDAR points per m^2 and a second site with 2.8 points per m^2 and found that their algorithm was not sensitive to LiDAR point density within this range. Li et al. (2008) developed a sequential road extraction algorithm using smart area partitioning to extract road points based on both the intensity and height information of LiDAR data points.

Choi et al. (2008) clustered potential road points in LiDAR data using height information, and then reclustered the points using average-intensity metrics. They used a buffered clustering approach, which struggled to separate roads from parking lots. Recognizing that height and intensity measures are not sufficient to mitigate this confusion, Rottensteiner and Clode (2008) removed large parking lots by using a threshold of the maximum width of roads extracted from LiDAR data. Poullis and You (2010) developed a novel vision-based system to automatically extract complex road networks from various sensor datasets including satellite images and LiDAR data. Their system integrated perceptual grouping theory and optimized segmentation techniques for linear feature extraction.

9.3 Road Extraction Using LiDAR Remote Sensing

An important consideration in developing a road extraction algorithm is the spatial resolution of the data used for extraction. In lower spatial resolution images, roads appear as thin elongated lines, so classic line extraction algorithms are appropriate; however, as ground sample distance decreases, roads appear as homogeneous areas, necessitating the application of different methods (Grote et al. 2012). Because of these structural considerations, automatic road extraction methods in higher spatial resolution datasets commonly divide the process into two main steps: (1) define road clusters and (2) generate road networks. However, depending on the available data resolution and the project goals, either of these steps may be the primary focus of analysis. Sections 9.3.1 and 9.3.2 explore techniques used for these two stages of road extraction, with a focus on studies that use LiDAR data either independently or integrated with other data sources. There are a variety of cues that aid in the extraction of roads from LiDAR data. One characteristic of use is that roads tend to have specific reflectance properties in the wavelength ranges typically used by LiDAR systems. In addition, the noise common to LiDAR intensity is minimized by the consistency of road material along a section of a road (Rottensteiner and Clode 2008). Topological and shape properties are also useful, since there are usually no isolated road points, and in dense datasets, roads appear as relatively thick lines (Rottensteiner and Clode 2008).

9.3.1 Defining Road Clusters

Successful road extraction in suburban areas often applies region-based methods to high spatial resolution data (Grote et al. 2012). For many studies, identification of road clusters is the primary research goal; however, it is often a necessary preprocessing step before performing vectorization, particularly in complex urban scenes, and may also be a component of a broader land cover classification. The characterization of methods within this section is somewhat arbitrary and includes division based on the processing framework as well as the algorithms used. It should be noted that many studies use multiple algorithms, for example, using one approach to create a binary layer that identifies road candidates, and then using another method to clean up errors.

9.3.1.1 Classification Framework

LiDAR analysis may be based on points within the LiDAR cloud or based on raster surfaces that are derived from the LiDAR points—for example, bare earth, intensity, and first return height. These points or pixels may be treated individually or as homogeneous groups using object-based approaches. This section summarizes some of the studies that use each of these fundamental frameworks to identify roads using LiDAR data.

9.3.1.1.1 *Point- and Pixel-Based Classification*

Like pixels in passive data sources, both LiDAR point clouds and pixels within LiDAR-derived surfaces can be classified to distinguish fundamental land cover types, such as buildings, roads, trees, and other urban features (Alharthy and Bethel 2003). Working at the point and pixel level, many studies use hierarchical classification techniques to extract roads from LiDAR data (Alharthy and Bethel 2003; Hu 2003; Clode et al. 2004). This may include preprocessing steps, such as high-pass filters to enhance road edges prior to extraction (Gecen and Sarp 2008), or selective application of various LiDAR components. Alharthy and Bethel (2003) first used the intensity information of LiDAR data to detect roads based on their surface reflectivity, and then used inferences from the LiDAR elevation information to filter out misclassified pixels that did not belong to the road category. Hu (2003) and Clode et al. (2004) performed a similar classification in the reverse direction, using the elevation and then the intensity information in a hierarchical framework. Samadzadegan et al. (2009) used a classifier fusion approach to extract roads from LiDAR data using both the intensity and the range information. They generated raster layers with 1 m pixels from the intensity and range information, including first and last return data, and then explored two methods for fusing maximum likelihood and minimum distance classifiers: weighted majority voting and selected naïve bays. Samadzadegan et al. (2009) found that both fusion methods produced better results than the single classifiers, with selected naïve bays producing the best overall result.

Jiangui and Guang (2011) used elevation and intensity information to identify roads in airborne LiDAR data by first classifying the LiDAR point cloud into ground and non-ground points, and then using the LiDAR intensity to classify the ground points as either road or nonroad. Rottensteiner and Clode (2008) also directly classified the LiDAR points. They used a rule-based approach based on assumptions related to height—that is, roads are generally on or near the digital surface model—and intensity—that is, roads tend to appear as dark features—to identify potential road points within the LiDAR cloud. Rottensteiner and Clode (2008) also used the continuous nature of roads—that is, assuming that road points are typically surrounded by other road points—to use a local point density to filter out isolated pixels that may have met the height and intensity thresholds.

9.3.1.1.2 *Object-Based Approaches*

Increased access to high spatial resolution data sources—from both passive and active sensors—has stimulated a swell in object-based analysis for road extraction. Tiwari et al. (2009) used a multiresolution segmentation approach to include spectral, shape, context, and texture features in identifying roads in an urban environment. Tiwari et al. (2009) performed an initial segmentation based on QuickBird imagery, and then used a LiDAR point cloud to separate ground and elevated roads. Chen et al. (2009) also used a hierarchical object-based approach to extract roads from QuickBird and LiDAR data sources. Chen et al. (2009) used a variety of object characteristics to decrease confusion, for example, roads were separated from spectrally similar classes using object compactness measures. Another common challenge in road extraction studies that Chen et al. (2009) sought to solve through the object-based analysis was bridge identification. They used object symmetry measures to separate bridges from other tall, impervious objects, such as buildings. Im et al. (2008) used an object-based approach to classify land cover in an urban setting using decision trees based only on high-posting-density LiDAR data. They evaluated shape and texture-based metrics but found object characteristics based on height and intensity produced the best classification results. Zhu and Mordohai (2009) used object-based analysis after first creating an initial ground plane mask using LiDAR height data. They then used image segmentation of the masked LiDAR intensity data to create both boundary and interior features of image objects to support road classification using a minimum cover approach.

9.3.1.2 Algorithms Used for Road Identification

Road clusters can be defined using a variety of image processing procedures (Clode et al. 2007) working from both point or pixel and object-based frames of reference. Mena (2003) summarizes many techniques used to identify roads from a variety of data sources. This section extends Mena's work by exploring methods presented in recent studies, particulary those that incorporate LiDAR components.

9.3.1.2.1 Clustering and Calibration-Based Methods

Numerous studies have explored the utility of partition and grid-based clustering methods for road extraction. Gong et al. (2010) applied a k-means clustering approach to extract roads from LiDAR intensity data. Their two-class study—roads versus vegetation—showed improved results when LiDAR data were fused with multispectral imagery. Zhang (2006) also used a simple k-means clustering algorithm for performing image segmentation, and then used a fuzzy logic classifier to automatically identify road clusters from the clustering results. Agouris et al. (2001) first applied an unsupervised classification, and then determined roads from the defined classes through a k-medians algorithm. Choi et al. (2008) used a clustering algorithm to automatically extract road networks from 3D LiDAR data in an urban environment. The clustering implemented by Choi et al. (2008) integrated height, reflectance, and geometric information to extract road points and used the shape and size of point clusters to remove nonroad features such as cars and trees. In addition to statistical approaches, there is also a range of clustering techniques used for spatial analysis—such as genetic algorithms—that have been inspired by natural systems. Saeedi et al. (2009) extracted urban features from LiDAR data using one such clustering algorithm, which was based around the foraging habits of a colony of honeybees.

Zuo and Quackenbush (2010) considered two components in delineating potential raster road strips: (1) identifying road-level features and (2) selecting impervious features. The first component was based around the commonly used assumption that roads generally lie on or near the bare earth surface, with the exception of elevated roads, bridges, and tunnels. Zuo and Quackenbush (2010) sought to establish an optimal height threshold to select all the road-level features and remove all features with nonzero height (e.g., buildings and trees) and an optimum intensity for detecting impervious features. While thresholds can be established manually, Zuo and Quackenbush (2010) applied an automatic calibration model (Im et al. 2007) to delineate potential road strips. The automatic calibration model uses an exhaustive search technique with a stratified sampling design of the research space to determine optimal thresholds for each input layer. Zuo and Quackenbush (2010) used this automatic calibration model to find optimal thresholds for LiDAR-derived height and intensity to identify a preliminary raster road network.

9.3.1.2.2 Machine-Learning Approaches

Researchers have explored machine-learning algorithms for road extraction within both object- and pixel-based frameworks. Im et al. (2008) used decision trees to classify high-posting-density LiDAR data. Huang and Zhang (2009) used support vector machines (SVMs) to identify roads from multiscale object features. The SVM approach seeks to identify hyperplanes in the feature space to best separate the desired classes by maximizing the distance of the training point nearest to the decision boundary (Rottensteiner 2010). Zhan and Yu (2011) also applied SVMs to classify objects within a LiDAR point cloud. Jin (2012) compared a random forest (RF) approach to SVMs and the more traditional maximum likelihood classifier to categorize land cover classes, including roads, in an urban environment. RFs combine multiple decision trees that are based on a random vector sampled independently and with the same distribution for all trees (Rottensteiner 2010). Jin (2012) integrated LiDAR data with optical imagery and found the RF-based feature selection and classification approach attained the best results in a pixel-based classification. The research presented by Carlson and Danner (2010) focused on bridge extraction to support hydrologic studies using the high-resolution digital elevation models (DEMs) that LiDAR can produce. Carlson and Danner (2010) used an adaptive boosting machine learning approach to identify bridges. The adaptive boosting approach systematically combines sets of classifiers, which may individually have high error, to form a single classifier with low overall error (Carlson and Danner 2010).

9.3.1.2.3 Template-Based Methods

Researchers have also used template-based approaches to perform road extraction. Although some studies (Lin et al. 2011) have used template-based methods to extract roads from very high-resolution imagery, other researchers have explored the benefit of templates in working with LiDAR datasets (Zhao and You 2012). After creating a ground plane dataset, Zhao and You (2012) used elongated structure templates to fit local intensity distributions to identify road candidates within LiDAR data. Zhao and You (2012) allowed the template center to shift within the road, to find the best local fit.

9.3.1.2.4 Mathematical Morphology

Mathematical morphology has proven useful in linear feature extraction (Quackenbush 2004) and has been successfully applied by many researchers in road extraction. Since mathematical morphology methods are useful for segmentation and image enhancement, such approaches are commonly applied to enhancing road network extraction from a variety of data sources. Zhang et al. (1999) performed mathematical morphology to mitigate errors from objects that had spectral characteristics similar to road surfaces. Amini et al. (2002) found that mathematical morphology operations simplified and eliminated errors in the image while maintaining shape characteristics. Chanussot et al. (1999) applied a mathematical morphological approach with fuzzy fusion techniques to extract roads from SAR satellite data.

The core operators in mathematical morphology are dilation and erosion. Dilation is used for expanding features in an image and closing any gaps, while erosion shrinks image features and eliminates small features. Additional operators, such as closing and opening, build from these fundamentals; the closing operation is dilation followed by erosion, and opening is erosion followed by dilation. Like many researchers, Zuo and Quackenbush (2010) also used multiple morphology operations. They applied morphological closing twice to a binary road image using small structural elements, initially an element with a three-pixel radius and then again with an element with a two-pixel radius, to remove small gaps in road strips and connect neighboring road pixels. Zuo and Quackenbush (2010) also applied morphological opening with a one-pixel radius to remove small or narrow clusters without affecting large ones. Wan et al. (2009) also used a variety of morphological tools to extract road networks from multispectral imagery. After applying a minimum distance classifier to create a binary road candidate image, Wan et al. (2009) used morphological opening to remove isolated elements and also applied a skeleton extraction algorithm to obtain the centerline of the road network. Jin et al. (2012) also used morphological closing and opening to remove small holes and noise and to eliminate small pathways from their road network.

9.3.2 Generating Road Networks

Many road extraction studies end with a classified land cover layer, generated using one of the approaches described earlier. However, other studies seek to perform the refinement needed to convert raster road clusters to a vector network. This section summarizes some of the techniques used in refinement and creation of a skeletonized road framework. Because of the proprietary nature of the algorithms, this chapter does not discuss road extraction that uses commercial software (Gecen and Sarp 2008).

9.3.2.1 Road Classification Refinement

Optimizing the accuracy of the raster road clusters usually includes refinement to remove noise and correct errors with the potential road strips (Mena 2003; Quackenbush 2004; Zhang 2006). Because of blockage caused by features such as vehicles and overhanging trees, LiDAR points are frequently unequally distributed; however, road points are generally clustered and the local point density for a pixel in the middle of a road is generally expected to be higher than that of a pixel lying on the edge of a road. Building from the basic binary classification of roads versus nonroads, Clode et al. (2004) developed a local density method that preserved LiDAR points that were expected to be road points and eliminated noise. Clode et al. (2004) used this approach to examine pixels within a local neighborhood (e.g., 3×3 window) to compute the local density of potential road pixels within a binary road layer. Rottensteiner and Clode (2008) used a circular local density measure to reduce noise in a set of potential road points that met intensity and height thresholds. Zuo and Quackenbush (2010) applied a kernel density function with a search radius of 5 m (half the average road width) to calculate the local density of LiDAR points within potential road strips derived from initial processing. Through appropriate threshold selection, points along the road edge were maintained as road elements, while points that did not satisfy the minimum local density were removed. Wan et al. (2009) also used a distribution density measure to separate roads and buildings based on their differences in compactness and symmetry of pixel distribution.

Researchers have also used the expected size and shape of road clusters to remove erroneous clusters. Wan et al. (2009) used the eccentricity of image objects to remove buildings based on the assumption that their shape would be less elongated than roads. Zuo and Quackenbush (2010) applied a morphological area operator, using multiple cluster size thresholds in sequential applications with the area threshold decreasing with each iteration, to remove isolated nonroad elements. Jin et al. (2012) used a connected component analysis to remove small clusters of pixels that were initially labeled as being road candidates. Connected component analysis groups pixels based on connectivity, and then removes components with area below a selected threshold. Jiangui and Guang (2011) used a Delaunay triangulated irregular network (TIN) to decrease confusion between roads and other features with similar LiDAR intensity and elevation values such as parking lots. Their approach constructs a nonconstrained Delaunay TIN from the candidate road points and then uses the length of the edge of the triangles to segment the nonconstrained TIN. Through this segmentation, Jiangui and Guang (2011) were able to remove small constrained TINs that corresponded to nonroad features.

Road refinement may also consider local complexity measures, for example, Haverkamp (2002) and Gibson (2003) describe the angular texture signature (ATS) developed for road extraction from high-resolution panchromatic imagery. To apply the ATS methods on a classified binary image, where pixels are identified as road or nonroad, Zhang and Couloigner (2006) defined three different ATS descriptors—ATS-mean, ATS-compactness, and ATS-eccentricity—and applied these methods to reduce parking lots misclassified as roads in a binary road map derived from high-resolution multispectral imagery. Given the nonlinear nature of a parking lot, pixels within a parking lot typically have a larger ATS-mean than road pixels. ATS-compactness considers the area and perimeter of the ATS polygon, and points along linear features, for example, roads tend to generate lower ATS-compactness values compared to points in large blocks, for example, parking lots. Zuo and Quackenbush (2010) computed ATS-mean and ATS-compactness based on 18 rectangles with a size of 10 × 40 m to reduce errors related to parking lots in a binary road map generated from LiDAR data. They found several differences in the two indicators, such as moderate ATS values for the linear roads compared to high ATS values in larger parking lots. While most of the parking lots were successfully removed using the ATS-mean, this measure was sensitive to neighboring pixels, which made it difficult to threshold. As a result, Zuo and Quackenbush (2010) used ATS-compactness to remove the remaining parking lots and a few large driveways from a binary road map.

9.3.2.2 Centerline Extraction

Road centerline extraction is an essential procedure to produce the vector road network required for most GIS databases; the quality of the extracted road centerlines usually determines the positional accuracy of the extracted road network (Zhang 2006). While simplified by the creation of a binary raster road layer, this procedure is challenging and many methods have been considered. In most cases, these techniques apply to any binary road layer; hence at this point in the analysis, the underlying source data become less of a consideration. Niblack et al. (1992) created an algorithm for generating connected skeletons and centerlines of objects in a binary image based on a distance transform. Huang et al. (2003) developed a thinning algorithm for binary images that was very robust to noise and eliminated some spurious branches. Karathanassi et al. (1999) and Amini et al. (2002) used mathematical morphology operations to find line skeletons in binary images. Amini et al. (2002) also applied a wavelet transform to reduce the resolution of a simplified image to extract road skeletons. While these methods all aim to skeletonize road clusters, a common limitation is that they do not produce the straight-line segments typical of a road network.

Extracting straight centerlines from road clusters is challenging and many methods have been considered. Many researchers have used the Hough transform (Duda and Hart 1972) for linear feature extraction. Hu et al. (2004) used the Hough transform to extract roads from clusters derived from imagery and LiDAR fusion. However, the Hough transform is limited in identifying road centerlines in high spatial resolution images because unless road clusters have been reduced to single-pixel skeletons, the peak of the Hough transform corresponds to the longest line, that is, the diagonal of a road segment (Clode et al. 2004). The Radon transform, first introduced by Radon (1917), provides another alternate method for line detection (Zhang and Couloigner 2007). For a 2D image, the computation of its Radon transform considers projections across the image at varying orientations and offsets (Murphy 1986). The Radon transform contains bright peaks corresponding to each line in the image thus the problem of line detection is reduced to detecting the peaks in the transform domain (Zhang and Couloigner 2007). While the transform works well in simple scenarios, it struggles with larger image sizes and complex road networks that include segments with a variety of lengths. When an image contains multiple road lengths, the Radon transform detects long road components and ignores shorter road segments. To improve centerline detection accuracy, researchers have used an iterative and localized

Radon transform (Zhang 2006; Zuo and Quackenbush 2010). Zhang and Couloigner (2006) detected road centerlines and estimated line width using a gliding-box approach to localize the Radon transform. Zuo and Quackenbush (2010) also implemented the gliding-box approach using a window with a size of 100 × 100 pixels and a 100 pixel increment for road networks based on LiDAR-derived inputs with a 1 m pixel size. Based on this image subdivision, long roads were split by the moving window, thus reducing complexity and generating road components in each subset with more consistent length.

A variety of other approaches have also been considered for road centerline extraction. Zhao and You (2012) used a multistep marching algorithm to create centerlines and then modeled the road network using a Markov graph. By varying the step length, the marching algorithm can represent straight (large steps) and curved (short steps) roads. The Markov graph approach used by Zhao and You (2012) completes missing roads and removes false road candidates as it creates a connected network. Clode et al. (2007) completed the vectorized road model by obtaining the centerline and width of the road based on a phase-coded disk (PCD) approach. They were able to detect road centerlines from the position of the peak in the magnitude image resulting from the complex convolution. However, a limitation of the PCD method is that it is relatively complex and time-consuming. Naouai et al. (2010) used a constrained Delaunay TIN approach for line extraction from raster imagery. They created a skeletonized road network and found their approach to have relatively low computational demand while producing acceptable accuracy.

9.4 Discussion and Conclusion

The road extraction studies presented in this chapter and in earlier reviews (Mena 2003; Quackenbush 2004) show that researchers have harnessed algorithms from many different fields. Future studies in road extraction will likely continue with this discipline overlap, with separate exploration of the techniques used to extract raster road clusters and approaches suitable for road network linearization. For many LiDAR-based road extraction studies, for example, urban mapping and hydrologic applications, the delineation of raster clusters may be a sufficient goal, and there is value in exploring the wide variety of machine learning approaches developed for passive imagery (Lu and Weng 2009; Simler 2011). In many cases, road extraction through adaptation from other fields is not straightforward. For example, the output from the Radon analysis used for line extraction is a series of continuous, and infinitely long, lines, which is not a true representation of most segment-based road networks. Since the Radon transform generates continuous lines, it does not provide any information about the endpoints of a given line segment, and can extend lines across places where there should be a break. Zuo and Quackenbush (2010) identified road segment endpoints by overlaying the detected centerlines on the binary raster road image the localized Radon transform had been applied to. The final centerlines were identified by fusing gaps and removing short segments that were likely to be nonroad components. Future studies must continue to explore the unique adaptations required for the road extraction scenario and the impact that these adaptations have on the resultant output quality.

Researchers must also consider the changing role of different input data sources for road extraction. Data from airborne LiDAR sensors has recently played a useful role in road extraction for projects with a wide variety of goals. However, low-cost data acquisition alternatives, coupled with significant improvement in wireless communication and global positioning system technologies, may lead to a decrease in the role of airborne LiDAR for generalized updates of GIS road databases. Increasing interest in terrestrial or mobile LiDAR (Boyko and Funkhouser 2011; Tongtong et al. 2011) and floating car technology (Li et al. 2012) may offer new avenues for road data collection, particularly as related to autonomous vehicle navigation systems (Peterson et al. 2008; Malik et al. 2011). Li et al. (2012) proposed an incremental road network extraction method to update digital road maps using floating car data. Pu et al. (2011) used a hierarchical approach to process data from a mobile laser scanned point cloud for data acquisition to support road safety inspection. They first classified points into three broad categories—ground surface, objects on the ground, and objects off the ground—and then used point cloud segment characteristics—such as size, shape, orientation, and topological relationship—to assign more detailed classes. While algorithms developed for airborne LiDAR processing do have applicability for mobile and terrestrial systems, the data collection process for ground-based collection is often much more specifically targeted, facilitating higher detail, for example, individual lane, curb, or road edge detection (Peterson et al. 2008; Zhang 2010; Kang et al. 2012).

As road extraction approaches have borrowed methodology from other disciplines, so too can the techniques described in this chapter be extended to other linear extraction fields. Researchers are already exploring the applicability of algorithms developed for road extraction to extract other linear features. D'Orazio et al. (2012) used a modified region-based active contour approach for delineating archaeological trace and found that their proposed method overcame the limits of standard methods of region uniformity and different consistencies with respect to the background. Beger et al. (2011) integrated high spatial resolution orthoimagery and dense LiDAR points to reconstruct railroad track centerlines using object-based analysis and an adapted random sample consensus algorithm.

As technology evolves, the need for extraction and update of GIS road network databases from airborne or satellite data may be decreasing. However, LiDAR systems still hold great promise for enhanced collection of roads for urban mapping and impervious surface extraction and the increasing pressure on the urban environment means that research into remote sensing–based road extraction should continue. This may consider new sensors and algorithms but should also address other important considerations. For example, while many studies have focused on the extraction of roads from LiDAR data, little research has been performed on investigating uncertainty in extracted road networks. Zhou and Stein (2012) used a Markov chain Monte Carlo method to simulate positional errors in laser points and found random sets are well suited to model the uncertainty of road polygons extracted from LiDAR points. Future research needs to continue to address this important component of road extraction research to ensure the products generated are suitable for their targeted application.

References

Agouris, P., P. Doucette, and A. Stefanidis. 2001. "Spatiospectral Cluster Analysis of Elongated Regions in Aerial Imagery." In *Proceedings of the IEEE International Conference on Image Processing*, 7–10 October, 789–792. Thessaloniki, Greece: IEEE.

Alharthy, A., and J. Bethel. 2003. "Automated Road Extraction from LIDAR Data." In *Proceedings of American Society of Photogrammetry and Remote Sensing*, 7–9 May. Anchorage, Alaska: American Society for Photogrammetry and Remote Sensing.

Amini, J., M. R. Saradjian, J. A. R. Blais, C. Lucas, and A. Azizi. 2002. "Automatic Road-Side Extraction from Large Scale Image Maps." *International Journal of Applied Earth Observation and Geoinformation* 4:95–107.

Baumgartner, A., C. Steger, H. Mayer, W. Eckstein, and H. Ebner. 1999. "Automatic Road Extraction Based on Multi-Scale, Grouping, and Context." *Photogrammetric Engineering & Remote Sensing* 65:777–85.

Beger, R., C. Gedrange, R. Hecht, and M. Neubert. 2011. "Data Fusion of Extremely High Resolution Aerial Imagery and LiDAR Data for Automated Railroad Centre Line Reconstruction." *ISPRS Journal of Photogrammetry and Remote Sensing* 66:S40–51.

Boyko, A., and T. Funkhouser. 2011. "Extracting Roads from Dense Point Clouds in Large Scale Urban Environment." *ISPRS Journal of Photogrammetry and Remote Sensing* 66:S2–12

Brennan, R., and T. L. Webster. 2006. "Object-Oriented Land Cover Classification of Lidar-Derived Surfaces." *Canadian Journal of Remote Sensing* 32:162–72.

Carlson, R., and A. Danner. 2010. "Bridge Detection in Grid Terrains and Improved Drainage Enforcement." In *Proceedings of the 18th SIGSPATIAL International Conference on Advances in Geographic Information Systems*, 2–5 November, 250–259. San Jose, CA:ACM.

Chanussot, J., G. Mauris, and P. Lambert. 1999. "Fuzzy Fusion Techniques for Linear Features Detection in Multitemporal SAR Images." *IEEE Transactions on Geoscience and Remote Sensing* 37:1292–305.

Chen, Y., W. Su, J. Li, and Z. Sun. 2009. "Hierarchical Object Oriented Classification Using Very High Resolution Imagery and LIDAR Data Over Urban Areas." *Advances in Space Research* 43: 1101–10.

Choi, Y. W., Y. W. Jang, H. J. Lee, and G. S. Cho. 2007. "Heuristic Road Extraction." In *Proceedings of International Symposium on Information Technology Convergence*, 23–24 November, 338–42. Jeonju, Korea: IEEE.

Choi, Y. W., Y. W. Jang, H. J. Lee, and G. S. Cho. 2008. "Three-Dimensional LiDAR Data Classifying to Extract Road Point in Urban Area." *IEEE Geoscience and Remote Sensing Letters* 5:725–29.

Clode, S., P. Kootsookos, and F. Rottensteiner. 2004. "The Automatic Extraction of Roads from LiDAR Data." In *Proceedings of the International Society for Photogrammetry and Remote Sensing*, July 12–23, Vol. 35, 231–36. Istanbul, Turkey: ISPRS.

Clode, S., F. Rottensteiner, and P. Kootsookos. 2005. "Improving City Model Determination by Using Road Detection from LIDAR Data." In *International Archives of Photogrammetry, Remote Sensing and Spatial Information Sciences*, 29–30 August, Vol. XXXVI, 159–64. Vienna, Austria: School of Information Technology and Electrical Engineering Publications.

Clode, S., F. Rottensteiner, P. Kootsookos, and E. Zelniker. 2007. "Detection and Vectorisation of Roads from LIDAR Data." *Photogrammetric Engineering and Remote Sensing* 73:517–35.

D'Orazio, T., F. Palumbo, and C. Guaragnella. 2012. "Archaeological Trace Extraction by a Local Directional Active Contour Approach." *Pattern Recognition* 45:3427–38.

Duda, R., and P. E. Hart. 1972. "Use of the Hough Transform to Detect Lines and Curves in Pictures." *Communications of the Association of Computing Machines* 15:11–15.

Gecen, R., and G. Sarp. 2008. "Road Detection from High and Low Resolution Satellite Images." In *International Archives of the Photogrammetry, Remote Sensing and Spatial Information Sciences*, 3–11 July, Vol. XXXVII, 355–58. Beijing, China: ISPRS.

Gibson, L. 2003. "Finding Road Networks in IKONOS Satellite Imagery." In *Proceedings of the Annual Conference of the American Society of Photogrammetry and Remote Sensing*, 7–9 May. Anchorage, Alaska: ASPRS.

Gong, L., Y. Zhang, Z. Li, and Q. Bao. 2010. "Automated Road Extraction from LiDAR Data Based on Intensity and Aerial Photo." In *IEEE 3rd International Congress on Image and Signal Processing*, 16–18 October, 2130–2133. Yantai, China: IEEE.

Grote, A., C. Heipke, and F. Rottensteiner. 2012. "Road Network Extraction in Suburban Areas." *The Photogrammetric Record* 27:8–28.

Hall, D. L., and J. Llinas. 1997. "An Introduction to Multisensor Data Fusion." *Proceedings of the IEEE* 85:6–23.

Hatger, C., and C. Brenner. 2003. "Extraction of Road Geometry Parameters from Laser Scanning and Existing Databases." In *International Archives of Photogrammetry, Remote Sensing and Spatial Information Sciences*, 8–10 October, Vol. XXXIV, 225–230. Dresden, Germany: ISPRS.

Haverkamp, D. 2002. "Extracting Straight Road Structure in Urban Environments Using IKONOS Satellite Imagery." *Optical Engineering* 41:2107–10.

Hu, Y. 2003. "Automated Extraction of Digital Terrain Models, Roads and Buildings Using Airborne Lidar Data." Ph.D. Thesis, University of Calgary, Alberta, CA.

Hu, X., C. V. Tao, and Y. Hu. 2004. "Automatic Road Extraction from Dense Urban Area by Integrated Processing of High Resolution Imagery and LIDAR Data." *International Archives of Photogrammetry, Remote Sensing and Spatial Information Sciences* 35:288–92.

Huang, X., and L. Zhang. 2009. "Road Centreline Extraction from High-Resolution Imagery Based on Multiscale Structural Features and Support Vector Machines." *International Journal of Remote Sensing* 30:1977–87.

Huang, L., G. Wan, and C. Liu. 2003. "An Improved Parallel Thinning Algorithm." In *Proceedings of International Conference on Document Analysis and Recognition*, 3–6 August, Vol. II, 780–83. Edinburgh, Scotland: IEEE Computer Society.

Im, J., J. R. Jensen, and M. E. Hodgson. 2008. "Object-Based Land Cover Classification Using High-Posting-Density LiDAR Data." *GIScience & Remote Sensing* 45:209–28.

Im, J., J. Rhee, J. R. Jensen, and M. E. Hodgson. 2007. "An Automated Binary Change Detection Model Using a Calibration Approach." *Remote Sensing of Environment* 106:89–105.

Jiangui, P., and G. Guang. 2011. "A Method for Main Road Extraction from Airborne LiDAR Data in Urban Area." In *International Conference on Electronics, Communications and Control*, 9–11 September, 2425–2428. Ningbo, China: IEEE Xplore Digital Library.

Jin, J. 2012. "Random Forest Based Method for Urban Land Cover Classification Using LiDAR Data and Aerial Imagery." M.S. Thesis, University of Waterloo, Canada.

Jin, H., M. Miska, E. Chung, M. Li, and Y. Feng. 2012. "Road Feature Extraction from High Resolution Aerial Images Upon Rural Regions Based on Multi-Resolution Image Analysis and Gabor Filters." In *Remote Sensing—Advanced Techniques and Platforms*, edited by B. Escalante-Ramirez, 387–414. InTech. Available at http://www.intechopen.com/books/remote-sensing-advanced-techniques-and-platforms (accessed 26 February 2013).

Jung, W., Kim, W., Youn, J., and J. Bethel. 2006. "Automatic Urban Road Extraction from Digital Surface Model and Aerial Imagery." In *Proceedings of the Annual Conference of the American Society for Photogrammetry and Remote Sensing*, 1–5 May. Reno, NV: ASPRS.

Kang, Y., C. Roh, S.-B. Suh, and B. Song. 2012. "A Lidar-Based Decision-Making Method for Road Boundary Detection Using Multiple Kalman Filters." *IEEE Transactions on Industrial Electronics* 11:4360–68.

Karathanassi, V., C. Iossifidis, and D. Rokos. 1999. "A Thinning-Based Method for Recognizing and Extracting Peri-Urban Road Networks from SPOT Panchromatic Images." *International Journal of Remote Sensing* 20:153–68.

Li, J., H. J. Lee, and G. S. Cho. 2008. "Parallel Algorithm for Road Points Extraction from Massive LiDAR Data." In *Proceedings of IEEE International Symposium on Parallel and Distributed Processing with Applications*, 10–12 December, 308–315. Sydney, NSW: IEEE.

Li, J., Q. Qin, C. Xie, and Y. Zhao. 2012. "Integrated Use of Spatial and Semantic Relationships for Extracting Road Networks from Floating Car Data." *International Journal of Applied Earth Observation and Geoinformation* 19:238–47.

Lillesand, T. M., R. W. Kiefer, and J. W. Chipman. 2008. *Remote Sensing and Image Interpretation*. 6th ed. Hoboken, NJ: John Wiley.

Lin, X., J. Zhang, Z. Liu, J. Shen, and M. Duan. 2011. "Semi-Automatic Extraction of Road Networks by Least Squares Interlaced Template Matching in Urban Areas." *International Journal of Remote Sensing* 32:4943–59.

Lu, D., and Q. Weng. 2009. "Extraction of Urban Impervious Surfaces from an IKONOS Image." *International Journal of Remote Sensing* 30:1297–311.

Malik, U. A., S. U. Ahmed, and F. Kunwar. 2011. "A Self-Organizing Neural Scheme for Road Detection in Varied Environments." In *IEEE International Joint Conference on Neural Networks*, 31 July–5 August, 3049–3054. San Jose, CA: IEEEXplore.

Mallet, C., and F. Bretar. 2009. "Full-Waveform Topographic Lidar: State-of-the-Art." *ISPRS Journal of Photogrammetry and Remote Sensing* 64:1–16.

Mena, J. B. 2003. "State of the Art on Automatic Road Extraction for GIS Update: A Novel Classification." *Pattern Recognition Letters* 24:3037–58.

Murphy, L. M. 1986. "Linear Feature Detection and Enhancement in Noisy Images Via the Radon Transform." *Pattern Recognition Letters* 4:279–84.

Naouai, M., M. Narjess, and A. Hamouda. 2010. "Line Extraction Algorithm Based on Image Vectorization." In *IEEE International Conference on Mechatronics and Automation* 4–7 August, 470–476. Xi'an China: IEEE ICMA.

Niblack, C. W., P. B. Gibbons, and D. W. Capson. 1992. "Generating Skeletons and Centerlines from the Distance Transform." *Graphical Models and Image Processing* 54:420–37.

Péteri, R., and T. Ranchin. 2007. "Road Networks Derived from High Spatial Resolution Satellite Remote Sensing Data." In *Remote Sensing of Impervious Surfaces*, edited by Q. Weng, 215–36. Boca Raton, FL: CRC Press.

Peterson, K., J. Ziglar, and P. E. Rybski. 2008. "Fast Feature Detection and Stochastic Parameter Estimation of Road Shape Using Multiple LIDAR." In *IEEE International Conference on Intelligent Robots and Systems*, 22–26 September, 612–619. Nice, France: IEEE Robotics and Automation Society.

Poullis, C., and S. You. 2010. "Delineation and Geometric Modeling of Road Networks." *ISPRS Journal of Photogrammetry and Remote Sensing* 65:165–81.

Pu, S., M. Rutzinger, G. Vosselman, and S. Oude Elberink. 2011. "Recognizing Basic Structures from Mobile Laser Scanning Data for Road Inventory Studies." *ISPRS Journal of Photogrammetry and Remote Sensing* 66:S28–39.

Quackenbush, L. J. 2004. "A Review of Techniques for Extracting Linear Features from Imagery." *Photogrammetric Engineering & Remote Sensing* 70:1383–92.

Radon, J. H. 1917. Ueber die Bestimmng von Funktionen durch Ihre Integralwerte laengs gewisser Mannigfaltigkeiten [On the determination of functions from their integral values along certain manifolds]. *Reports on the proceedings of the Saxony Academy of Science* 69:262–77.

Rieger, W., M. Kerschner, T. Reiter, and F. Rottensteiner. 1999. "Roads and Buildings from Laser Scanner Data within a Forest Enterprise." *International Archives of Photogrammetry and Remote Sensing Spatial*, Vol. 32, 185–191. La Jolla, CA: International Archives of Photogrammetry and Remote Sensing.

Rottensteiner, F. 2009. "Status and Further Prospects of Object Extraction from Image and Laser Data." Invited paper at IEEE Urban Remote Sensing Joint Event, 20–22 May, 1–10. Shanghai, China: IEEE.

Rottensteiner, F. 2010. "Automation of Object Extraction from LiDAR in Urban Areas." In *IEEE International Geoscience and Remote Sensing Symposium*, Honolulu, HI: IEEE: 25–30 July, 1343–1346.

Rottensteiner, F., and S. Clode. 2008. "Building and Road Extraction by LiDAR and Imagery." In *Topographic Laser Ranging and Scanning: Principles and Processing*, edited by J. Shan and C. K. Toth, 445–78. Boca Raton, FL: CRC Press.

Saeedi, S., F. Samadzadegan, and N. El-Sheimy. 2009. "Object Extraction from LIDAR Data Using an Artificial Swarm Bee Colony Clustering Algorithm." In *ISPRS Workshop: City Models, Roads and Traffic*, 3–4 September, Vol. XXXVIII-3/W4, 133–138. Paris, France: ISPRS.

Samadzadegan, F., M. Hahn, and B. Bigdeli. 2009. "Automatic Road Extraction from LIDAR Data Based on Classifier Fusion." In *IEEE Urban Remotes Sensing Joint Event*, 20–22 May, 1–6. Shanghai, China: IEEE.

Simler, C. 2011. "An Improved Road and Building Detector on VHR Images." In *IEEE International Geoscience and Remote Sensing Symposium*, 24–29 July, 507–510. Vol. XXXIV, 3/B Vancouver, CA IEEE.

Song, J. H., S. H. Han, K. Yu, and Y. Kim. 2002. "Assessing the Possibility of Land-Cover Classification Using LiDAR Intensity Data." In *International Archives of Photogrammetry, Remote Sensing and Spatial Information Sciences*, 9–13 September, Vol. XXXIV, 259–62. Graz, Austria: International Society for Photogrammetry and Remote Sensing.

Tiwari, P. S., H. Pande, and A. K. Pandey. 2009. "Automatic Urban Road Extraction Using Airborne Laser Scanning/Altimetry and High Resolution Satellite Data." *Journal of the Indian Society of Remote Sensing* 37:223–31.

Tongtong, C., D. Bin, L. Daxue, and L. Zhao. 2011. "LIDAR-Based Long Range Road Intersection Detection." In *IEEE Sixth International Conference on Image and Graphics (ICIG)*, 12–15 August, 754–759. Hefei, China: IEEE.

Wan, Y., K. Wang, and D. Ming. 2009. "Road Extraction from High-Resolution Remote Sensing Images Based on Spectral and Shape Features." In *Proceedings of the International Society for Optical Engineering 7495 MIPPR: Automatic Target Recognition and Image Analysis*, 30 October, 74953R-1–74953R-6. Yichang, China: SPIE.

Wang, G., Y. Zhang, J. Li, and P. Song. 2011. "3D Road Information Extraction from LiDAR Data Fused with Aerial-Images." In *IEEE International Conference on Spatial Data Mining and Geographical Knowledge Services*, 29 June–1 July, 362–366. Fuzhou, China, IEEE.

Weng, Q. 2012. "Remote Sensing of Impervious Surfaces in the Urban Areas: Requirements, Methods, and Trends." *Remote Sensing of Environment* 117:24–49.

Zhan, Q., and L. Yu. 2011. "Objects Classification from Laser Scanning Data Based on Multi-Class Support Vector Machine." In *IEEE International Conference on Remote Sensing, Environment and Transportation Engineering (RSETE)*, 24–26 June, 520–523. Nanjing, China: IEEE.

Zhang, Q. 2006. "Automated Road Network Extraction from High Spatial Resolution Multi-Spectral Imagery." Ph.D. Thesis, University of Calgary, Alberta, CA.

Zhang, W. 2010. "LIDAR-Based Road and Road-Edge Detection." In *IEEE Intelligent Vehicles Symposium (IV)* 21–24 June, 845–848. San Diego, CA: IEEE.

Zhang, Q., and I. Couloigner. 2006. "Benefit of the Angular Texture Signature for the Separation of Parking Lots and Roads on High Resolution Multi-Spectral Imagery." *Pattern Recognition Letters* 27:937–46.

Zhang, Q., and I. Couloigner. 2007. "Accurate Centerline Detection and Line Width Estimation of Thick Lines Using the Radon Transform." *IEEE Transactions on Image Processing* 16:310–16.

Zhang, C., S. Murai, and E. Baltsavias. 1999. "Road Network Detection by Mathematical Morphology." In *Proceedings of ISPRS Workshop: 3D Geospatial Data Production: Meeting Application Requirements*, 7–9 April, 185–200. Paris, France: International Society for Photogrammetry and Remote Sensing.

Zhao, J., and S. You. 2012. "Road Network Extraction from Airborne LiDAR Data Using Scene Context." In *IEEE Computer Society Conference on Computer Vision and Pattern Recognition Workshops*, 16–21 June, 9–16. Providence, RI, IEEE.

Zhou, L., and A. Stein. 2012. "Application of Random Sets to Model Uncertainty of Road Polygons Extracted from Airborne Laser Points." *Computers, Environment and Urban Systems* (in press). Available at http://dx.doi.org/10.1016/j.compenvurbsys.2012.06.006.

Zhu, P., Z. Lu, X. Chen, K. Honda, and A. Eiumnoh. 2004. "Extraction of City Roads through Shadow Path Reconstruction Using Laser Data." *Photogrammetric Engineering & Remote Sensing* 70:1433–40.

Zhu, Q., and P. Mordohai. 2009. "A Minimum Cover Approach for Extracting the Road Network from Airborne LIDAR Data." In *IEEE International Conference on Computer Vision Workshops*, 27 September–4 October, 1582–1589. Kyoto, Japan: IEEE.

Zuo, Y., and L. J. Quackenbush. 2010. "Road Extraction from LiDAR Data in Residential and Commercial Areas of Oneida County, New York." In *Proceedings of the Annual Conference of the American Society for Photogrammetry and Remote Sensing*, 26–30 April. San Diego, CA: American Society for Photogrammetry and Remote Sensing.

Section IV

Natural Landscape, Ecosystems, and Forestry

10

Application of Remote Sensing in Ecosystem and Landscape Modeling

Chonggang Xu and Min Chen

CONTENTS

10.1 Introduction 173
10.2 Forest Landscape Dynamics 175
    10.2.1 Model Initialization 176
    10.2.2 Model Validation/Benchmarking 178
    10.2.3 Model-Data Integration.. 178
10.3 Lake Landscape Dynamics 181
    10.3.1 Extraction of Lakes from Remote Sensing Images 181
    10.3.2 Use of Lake Data for Model Initialization..................................... 184
    10.3.3 Potential Use of Lake Data for Model Validation/Benchmarking 184
10.4 Conclusion 185
Acknowledgments.. 185
References... 185

10.1 Introduction

A landscape is an area of land at a variety of scales (Wiens and Milne 1989), containing patterns that affect and are affected by different ecological processes of interest (Turner 1989). The landscape can be composed of different physical and living elements including topography, vegetation, water, and human activities (Green et al. 1996). An ecological process is an expenditure of thermal, kinetic, and biochemical energy that results in a change in state of an ecosystem (Forman and Godron 1986). Important ecological processes may include different biogeochemical cycles of carbon, water, and nitrogen and different types of disturbances including fire, insect, and harvesting that regulate ecosystem states and landscape patterns. The interaction between landscape patterns and ecological processes represents one of the key topics in ecological research (Forman 1995; Turner 1989; Wu and Hobbs 2002).

Current global climatic changes such as air temperature increase, carbon dioxide (CO_2) enrichment, and precipitation reduction have already resulted in substantial changes in ecological processes and landscape patterns at different scales (Hansen et al. 2001; Walther et al. 2002). At small scales, the increase in temperature and CO_2 enrichment may change only the growth of individual plants by regulating photosynthesis (Long et al. 2004); however, at large scales, the warming may cause catastrophic landscape changes such as substantial forest dieback resulting from intensified drought (Allen et al. 2010). For small-scale processes (e.g., photosynthesis and respiration), field experiments in open top chamber or the free air CO_2 experiment (FACE) can be used to study the effects of future climatic changes on ecosystems (Ainsworth and Long 2005); however, the information generated at small scale may not be sufficient to understand the large-scale process due to the system complexity resulting from spatial heterogeneity and interdependence of different driving factors (Wiens 1989). Furthermore, it would be infeasible to conduct large-scale field experiments to understand mechanisms causing large-scale landscape changes (e.g., fire disturbances). Therefore, numerical models and computer simulation are important tools to help us better understand and predict the effects of climatic change on large-scale landscapes (He 2008). Even for small-scale processes, models are helpful for understanding and predicting ecosystem changes under different environmental conditions (Farquhar et al. 1980; Ball et al. 1987).

Since ecosystem response to climatic change is composed of very complex processes at multiple scales, it is necessary that we build different models to help better understand the change in chemical, hydrological, biological, and ecological processes under future climatic change. Two key components of modeling are model initialization and model validation (or benchmarking). Model initialization refers to the process in which necessary data is fed into a model to provide the initial condition for the model to begin simulation. The model validation/benchmarking process refers to comparing the model output with observations to evaluate the predictability of models. For small-scale process models, measurement data are sufficient for model initialization and validation; however, for large-scale models (e.g., landscape models or global vegetation models), the biggest challenge is that they generally are difficult to initialize and validate due to the lack of large-scale datasets. Remote sensing is an essential tool that can provide the key components for large-scale model initialization and validation.

Remote sensing is useful not only for model initialization and validation but also in model-data integration targeted at monitoring ecosystems or landscape dynamics. The model-data integration is generally based on the following two steps (see Figure 10.1): first,

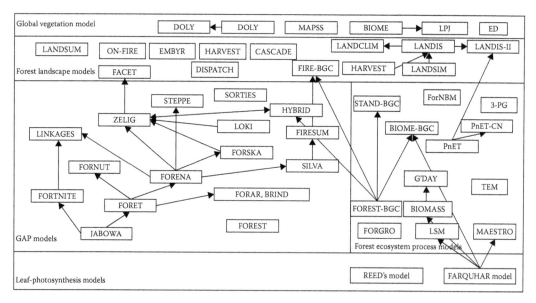

FIGURE 10.1
Linkages between leaf-photosynthesis models, forest growth and succession models, GAP models, forest landscape models, and global vegetation models: the arrows indicate the lineage among models. The lineage for GAP models is based on Urban and Shugart (1992). See text for the explanation of the models.

the model is used to predict the ecosystem status of interest (e.g., tree density) based on an initial condition. Second, predicted ecosystem status is adjusted so that the simulated value of variable of interest is close to that derived from remote sensing. This model-data integration approach is able to merge prior knowledge about the ecosystem built in the model with the signal from remote sensing and thus provide an appealing approach to gain further insights into ecosystem dynamics.

In this chapter, we will use forest and lake landscapes as two examples to demonstrate how remote sensing can be applied in landscape and ecosystem modeling studies. In particular, a forest ecosystem case study will show the application of remote sensing in model initialization, validation, and model-data integration for forest dynamics models. A lake landscape case study will show how remote sensing is used in both detection of lake landscape changes and lake model initializations.

10.2 Forest Landscape Dynamics

Based on the processes or objects being simulated at different scales, we can classify all forest models into four major groups: (1) forest ecosystem process models at the ecosystem level, (2) forest growth and succession models at the stand level, (3) landscape models at the landscape/watershed level, and (4) global vegetation models at regional and global scales (Figure 10.2). Forest ecosystem process models simulate carbon and water dynamics and/ or geochemical cycles in forest ecosystems, which are driven by different climate variables (e.g., temperature, CO_2, precipitation, radiation, and/or pollutants). The model generally does not distinguish individual trees (but see MAESTRO [Wang and Jarvis 1990]). The main outputs of the ecosystem process models include (net) primary production and biomass allocation for different components (e.g., foliage, stem, and root production) of the ecosystem. The application scale can be at plot, regional, and global levels (Running and Hunt 1993).

Forest growth and succession models simulate seedling establishment, individual tree growth, mortality, and competition among trees for nutrients and light. Forest growth and succession is normally simulated at forest plot or stand levels (100–1000 m^2). One popular simulation effort for forest growth and succession is the "GAP model" that originated from the first GAP model known as "JABOWA" developed by Botkin et al. (1972). The GAP refers to the gap in a canopy caused by a disturbance resulting in increased light on the forest floor. In the JABOWA model, succession dynamics is determined by species life history (e.g., maximum age, maximum height, and maximum diameter), empirical relationship of diameter growth to environmental factors (e.g., soil moisture, temperature, and radiation), and the competition for light.

Landscape models simulate landscape composition and pattern change as a result of forest succession, disturbances (e.g., fire, insects, and windthrow), and management (e.g., harvest and prescribed fire). Landscape models can incorporate ecosystem processes (e.g., water and carbon dynamics) (Figure 10.2); however, to simulate large-scale processes (e.g., 10^3–10^6 ha), the ecosystem processes generally are greatly simplified. For example, the FACET model (Urban et al. 1999) is based on a simple scale-up of the GAP model ZELIG. The FIRE-BGC model (Keane 1996) was evolved from the GAP model FIRESUM and the mechanistic fire model FARSITE (Finney 1998). Due to its simplicity in process representation, the landscape model may rely on ecosystem process models to provide estimates on the physiological

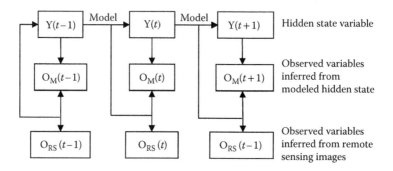

FIGURE 10.2
Integration of remote sensing data with an ecological model to detect the hidden state [Y(t), e.g., tree density]: at each time step, the model predicts the next time step hidden state [Y(t + 1)]. Each hidden state [Y(t)] corresponds to an observed value [O(t), e.g., reflectance] with a known linear/nonlinear function. The observed value can be inferred from modeled hidden states [$O_M(t)$] or from remote sensing images [$O_{RS}(t)$]. If $O_M(t)$ is very different from $O_{RS}(t)$, then we change the hidden state Y(t) to achieve a good fit. By adjusting $O_M(t)$ so that $O_M(t)$ is in the range of specified error of $O_{RS}(t)$, a good estimation of the hidden state [i.e., Y(t)] is obtained.

response of trees to climatic change. For example, the PnET-II process model (Aber and Federer 1992; Aber et al. 1995) is used to provide the estimates of establishment probability and growth capacity for different species (Xu et al. 2009), which are two key input parameters for the LANDIS-II forest landscape model (Scheller and Mladenoff 2004).

Global vegetation models are used to predict the distribution and growth of plant life forms at regional and global scales. They are the key components for the Earth System Model that is targeted to predict future climate with greenhouse gas emissions. Global vegetation models generally are composed of a biogeographical model that predicts vegetation distribution and a vegetation model that simulates vegetation growth and mortality. Biogeographical models generally are based on two important limiting factors: temperature and moisture (Neilson and Running 1996). Most biogeographical models use the growing degree days or thermal thresholds as temperature constraints, which are calibrated to the physiological cold-hardiness limit. The moisture constraint is generally more difficult to define since water balance generally depends on vegetation types, which may affect transpiration. There are two approaches to predict moisture constraints (Neilson and Running 1996): (1) iterative fitting for best leaf area index (e.g., MAPSS [Neilson 1995] and DOLY [Woodward et al. 1995]) and (2) the decoupling of water balance and site vegetation by estimating transpiration as a function of soil moisture (e.g., BIOME [Prentice et al. 1992]). The traditional vegetation model treats vegetation as "big leaf layers" with no explicit representation of trees. Therefore, it is difficult to realistically assess how trees will dynamically grow and compete for light after disturbances. Today, more advanced global vegetation models specifically are being developed based on fundamental scaling up from the GAP model (Moorcroft et al. 2001; Strigul et al. 2008; Fisher et al. 2010). By tracking disturbance histories and canopy structures, these models are expected to have better predictive power of biosphere–atmosphere interactions.

10.2.1 Model Initialization

In the past two decades, many spatially explicit forest models have been developed to simulate forest landscape dynamics (Scheller and Mladenoff 2007; He 2008). Most forest landscape models conceptualized the landscape into a grid of equal-sized cells or sites. For each cell, such a model requires the input of dominant canopy tree species, secondary tree species, and/or other stand-related parameters (e.g., age). Since the study generally comprises millions of cells, it is often infeasible to obtain cell-level information through ground surveys. Thus, methods have been developed to derive such information from the combination of forest inventory and remote sensing. Generally, remote sensing images are used to detect the dominant forest types (see Figure 10.3a) (Wolter et al. 1995). Then the dominant forest types are linked with forest inventory data to get more detailed information of canopy and understory species and their ages (see Figure 10.3b) (He et al. 1998). One key challenge of the methods using a combination of ground survey and remote sensing is that they cannot guarantee the accuracy at each cell, and this raises a question: Can the spatial information generated by such methods produce robust model results? To address this question, uncertainty analysis is necessary. Xu et al. (2004) conducted an uncertainty analysis to quantify how uncertainty in the input data of tree species and age information can affect model simulation results. Their results showed that, at the individual cell level, seed dispersal, seedling establishment, mortality, and fire disturbance caused uncertainty to increase with simulation year. The uncertainty finally reached an equilibrium state where input errors in original species age cohorts had little effect on simulation outcomes. At the landscape level, percentage area of species and their spatial patterns were not substantially affected by the uncertainties in species age structure at the cell level (Figure 10.4). Since the typical use of LANDIS is to predict long-term landscape pattern change, the combination of remote sensing and forest inventory data can be used to parameterize species age cohorts for individual cells.

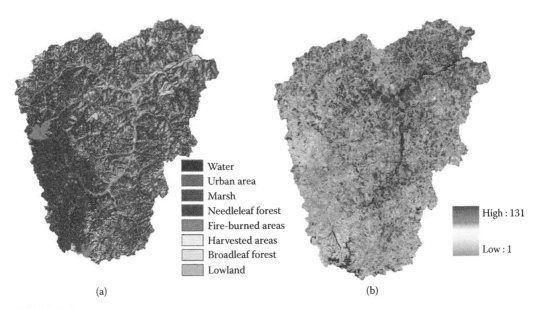

	Water
	Urban area
	Marsh
	Needleleaf forest
	Fire-burned areas
	Harvested areas
	Broadleaf forest
	Lowland

High : 131

Low : 1

(a)   (b)

FIGURE 10.3
(See color insert.) The use of remote sensing images to initialize the LANDIS model: (a) shows the landscape types classified from Landsat imagery and (b) shows 131 unique forest stands that have different species compositions and age structures, by merging the remote sensing products (a) with an elevation map and forest inventory data at different landscape locations. See Xu et al. (2004) for details.

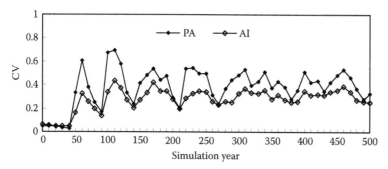

FIGURE 10.4

Uncertainty in the predicted percentage area (PA) and aggregation index (AI) for the distribution of larch (*Larix gmelinii*) in a forest landscape in northeastern China, simulated by LANDIS: uncertainty is measured by the coefficient of variation (CV) from 20 Monte Carlo simulations. See Xu et al. (2004) for details.

10.2.2 Model Validation/Benchmarking

Model validation/benchmarking is an essential component of model development. Before a model can be used for prediction and management purposes, it must be validated or benchmarked against independent datasets. Here, we show how data derived from remote sensing images could be useful for the development of an important global vegetation model. The ecosystem demography (ED) model (Moorcroft et al. 2001; Fisher et al. 2010), an advanced dynamic vegetation model, has recently been incorporated into the Community Land Model (CLM; the land surface component of the Community Earth System Model [CESM]), through a collaboration between the National Center for Atmospheric Research (NCAR), Colorado, and Los Alamos National Laboratory (LANL), New Mexico. Compared to the big-leaf model in current Earth System Models, ED more realistically represents successional stage, forest canopy structure, growth, and competition by tracking individual cohorts and land disturbance histories (Figure 10.5). Furthermore, it employs a mechanistic tree mortality function based on the amount of carbon storage (sugar and starch) for each individual cohort (Fisher et al. 2010).

The incorporation of ED into CLM represents a substantial improvement of vegetation dynamic modeling and makes it possible to realistically quantify vegetation dynamics under conditions of deforestation, fire, and drought. To evaluate model performance, we compared simulated leaf area index with that estimated from reflectance data obtained by the National Aeronautics and Space Administration's Moderate-Resolution Imaging Spectroradiometer (MODIS). CLM(ED) has been fully developed, and it has now been tested in the Amazon to simulate vegetation dynamics. The model prediction of gross primary production and leaf area index is in good agreement with observed values that are derived from MODIS products (Figure 10.6).

Model validation/benchmarking is not limited to model evaluation. They can be very useful for model development. If model prediction fails to benchmark against data from remote sensing, the model could be further developed to incorporate mechanisms that are missing in the model. For example, a new nitrogen limitation model (Xu et al. 2012) is being incorporated into the CLM(ED) model for better model performance in the Arctic, where nitrogen limitation is important (Shaver et al. 2001).

10.2.3 Model-Data Integration

Model-data integration is a promising approach to detect unknown ecosystem state variables by merging empirical knowledge of a simulated ecosystem and signals from

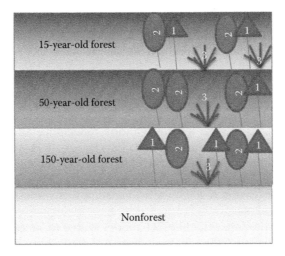

FIGURE 10.5
Ecosystem demography (ED) model structure: the ED model conceptualized the forest landscape into different succession stages. For each successional stage, the canopy is divided into different canopy layers. For each layer, the model tracks individual cohorts (represented by numbers) based on tree heights and crown areas.

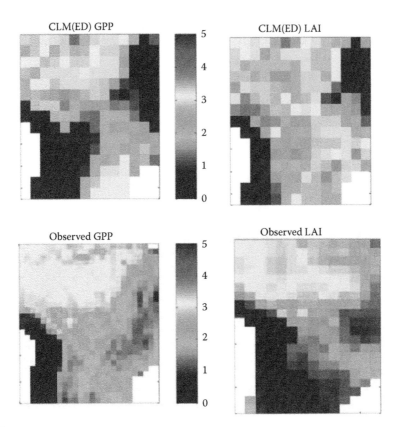

FIGURE 10.6
(See color insert.) Benchmarking of CLM(ED) model simulation against observed gross primary production (GPP) (in kg C/m²/year) and leaf area indices (LAIs) (m² leaf/m² ground) in the Amazon derived from MODIS products. Courtesy of Rosie Fisher from the National Center for Atmospheric Research, Colorado.

remote sensing. Here, we will show how model-data integration can be used to detect a key ecosystem state, tree mortality. Global vegetation models are currently unable to simulate mortality reliably in large extent due to the lack of knowledge on mortality mechanisms and the lack of data on when and where drought mortality occurs globally (McDowell et al. 2011). The development of a global dataset of tree mortality is of critical importance for the development and validation of dynamic global vegetation models. To detect tree mortality from MODIS datasets, we first coupled the ED model (Moorcroft et al. 2001; Fisher et al. 2010) with a forest canopy reflectance and transmittance (FRT) model (Kuusk and Nilson 2000; Kuusk et al. 2008) (Figure 10.7). The FRT model is composed of a model that simulates the light travel in the air (6S), a leaf-level spectrum model (PROSPECT), and a light travel model in the canopy with different cohorts of height. An ensemble Kalman filter (Evensen 2003) with 10 ensemble members is used to fit ED-FRT to MODIS reflectance values.

Preliminary results show that the ED-FRT model is able to successfully identify piñon tree mortality in a single site (500 × 500 m²) close to LANL (Figure 10.8). The site mainly is a piñon pine–juniper woodland (*Pinus edulis–Juniperus monosperma*). The sampling transect in this site showed that 80% of piñon pine died due to the 2002–2003 drought at this site. By fitting the ED-FRT model to the MODIS NDVI dataset (Figure 10.8a), it is shown that the simulated mortality of piñon pine is about 90% (Figure 10.8a and b), which is reasonably close to the observation. This approach integrates the empirical knowledge of vegetation dynamics and tree mortality from ED, a three-dimensional representation of cohort-based light travel in canopy, and different sources of reflectance signals from

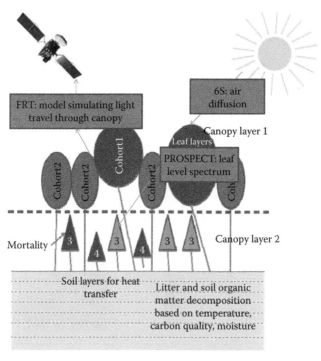

FIGURE 10.7
Coupling of the ED model and the forest reflectance and transmittance (FRT) models.

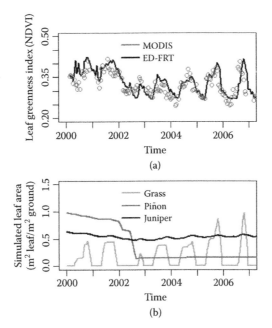

FIGURE 10.8
Evaluation of simulated normalized difference vegetation index (NDVI) by ED-FRT against MODIS NDVI using the mean values of 10 ensemble members (a); (b) shows the estimated LAIs for different vegetation types in ED.

MODIS. Thus, it has the potential to develop a global tree mortality map that can be used for global vegetation model validation/benchmarking.

10.3 Lake Landscape Dynamics

Lakes are important at different scales including local, regional, and global scales. They are an essential component of the Earth not only because of their large quantity (307 million lakes) and area coverage (4.2 million km²) (Downing et al. 2006) but also because they have major impacts on hydrological and biogeochemical cycles and energy balance (Lehner and Döll 2004; Prigent et al. 2007), which are important for the accurate prediction of climatic models on various spatial scales (Bonan 1995; Krinner 2003; Krinner and Boike 2010). Remote sensing is an essential tool for monitoring global lake dynamics. Here, we show how remote sensing can be used to detect lake changes in the Arctic and for the initialization of lake representation in Earth System Models.

10.3.1 Extraction of Lakes from Remote Sensing Images

Two types of remote sensing products that were used widely to extract lake areas are aerial photographs and satellite images. Aerial photographs were mainly used during the period from 1950s to 1980s, whereas Landsat images have been used more widely by scientists since Landsats were launched in 1972 because of their wide spatial coverage, high temporal resolution, as well as the ease and efficiency in data processing due to their digital format, as seen from existing Arctic lake studies (Frohn et al. 2005; Hinkel et al. 2005; Smith et al. 2005; Riordan et al. 2006; Plug et al. 2008; Corcoran et al. 2009; Jones et al. 2009; Labrecque et al. 2009; Arp et al. 2011; Jones et al. 2011; Roach et al. 2011; Rover et al. 2012).

Lakes can be delineated from Landsat images manually (Riordan et al. 2006; Corcoran et al. 2009; Labrecque et al. 2009); however, this manual delineation process is time and labor intensive and lake delineation results are subjective. The automated and supervised classification of lakes by computer software is more preferable, especially when lakes in large regions and/or from multiple time periods are needed. Density slicing (Frazier and Page 2000), decision trees (Rover et al. 2012), and feature extraction (Frohn et al. 2005; Hinkel et al. 2005; Jones et al. 2009; Jones et al. 2011; Chen et al. In revision) are three popular supervised classification methods used for the extraction of lakes from Landsat images, and the feature extraction method performed best in terms of the accuracy of lake basin area estimates with the use of specialized software (Roach 2011).

Here, we demonstrate how Genie Pro (Brumby et al. 1999; Perkins et al. 2005), an automated feature detection/classification system that is an example of feature extraction software, was used to extract lakes from Landsat images for a lowland Arctic region within Yukon Flats, Alaska (Chen et al. In revision). Genie Pro uses an evolutionary algorithm to generate textural–spectral image processing pipelines to classify multispectral imagery. In particular, Genie Pro integrates spectral information and spatial cues such as texture, local morphology, and large-scale shape information and allows human experts to interact efficiently with the software through an interactive and iterative "training dialog." Detailed descriptions of classification algorithms of Genie Pro can be found in the works of Perkins et al. (2005). Genie Pro was trained using human markups of about 20 lakes out of thousands of lakes (or ~0.5% of all pixels) in a study area as an initial step, and a map of lake pixels was produced by the Genie Pro algorithm. The lake map was then manually inspected to determine classification accuracy, and the training class was revised until Genie Pro generated a solution that best classified the lakes. An example lake extraction result is shown in Figure 10.9, with the rivers manually removed from the lake result map.

Lakes in the Arctic are very important for two reasons: on the one hand, they cover 15%–50% of total land area in many lowland Arctic landscapes (Mackay 1988; Rampton 1988; Burn 2002; Frohn et al. 2005; Grosse et al. 2005; Emmerton et al. 2007). On the other hand, their expansion, shrinkage, and disappearance are sensitive to global warming (Marsh et al. 2009) and associated with the storage and release of greenhouse gases such as methane and CO_2 (Kling et al. 1992; Repo et al. 2007), since most of them are thermokarst lakes, which formed in response to ground subsidence resulting from the thawing of permafrost (soil at or below the freezing point of water for 2 or more years). Permafrost degradation due to global warming is difficult to measure directly at the regional scale; therefore, thermokarst lake area change has been used to quantify the temporal rate and spatial extent of permafrost degradation (Yoshikawa and Hinzman 2003; Smith et al. 2005; Riordan et al. 2006; Labrecque et al. 2009; Chen et al. In revision).

Depending on the availability of cloud-free Landsat images for the same region, total lake areas from multiple dates of different years can be obtained, as shown in Figure 10.10. Then the temporal and spatial patterns of lake area changes can be identified and associated with possible driving factors such as precipitation, evaporation, river stages,

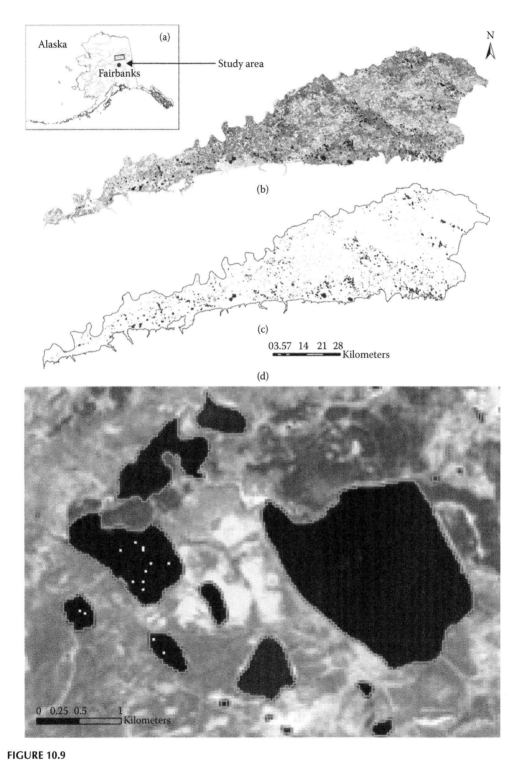

FIGURE 10.9
Landsat images and lake extraction results: (a) study area in Alaska; (b) Landsat images on August 16, 2000; (c) lake extracted by Genie Pro; and (d) zoomed-in area showing lake boundaries (light gray lines) generated by Genie Pro.

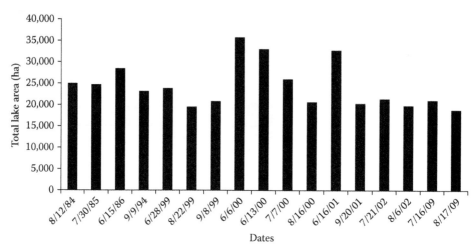

FIGURE 10.10

Time series of total lake areas.

connection to river channels, mean temperature since snowmelt (representing seasonal thaw depth), ice-jam flooding, and permafrost degradation (Chen et al. in review). The relationship between lake area and important driving factors can also be useful for the parameterization of models targeted to predict dynamic lake areas.

10.3.2 Use of Lake Data for Model Initialization

In climatic models, lakes affect the flux exchange of moisture, heat, and momentum with the overlying atmosphere, whereas in hydrological models lakes affect overland flow, river discharge, flooding, and groundwater recharge/discharge. To quantify the effects of lakes on climate and hydrology in such models, one of the fundamental input data required for model simulation is the spatial distribution of lakes (lake area percentage at the grid level). By merging lake data from existing large-scale land cover maps and newly available remote sensing images, it became possible to create global-scale inland water datasets such as the Global Lakes and Wetlands Database (GLWD) (Lehner and Döll 2004). GLWD then served as an estimate of lake and wetland extents to initialize global hydrology and climatology models, such as CLM (the land component of CESM) (http://www.cesm.ucar. edu/models/cesm1.0/clm/index.shtml), the global hydrological model PCR-GLOBWB (Van Beek et al. 2011; Wada et al. 2012; Weiland et al. 2012), and the ISBA-TRIP hydrological model (Decharme et al. 2012).

10.3.3 Potential Use of Lake Data for Model Validation/Benchmarking

Modeling studies can examine the effect of lakes on regional and global climates; they should also be able to examine the influence of climate variability and change on lakes. However, existing climate models only have fixed lake areas as initial input data and do not allow the lake areas to change in the following simulation steps in a response to climate change. To improve model performance, generate more realistic simulations, and make more accurate predictions, it is necessary that Earth System Models include processes that generate new lake areal coverage corresponding to changes in important driving factors such as precipitation and evaporation at each simulation step and reflect intra- and inter-annual lake area variabilities. For the models including dynamic lake areas in their output, remote sensing data can be used for model validation and benchmarking because model parameters can be adjusted according to the results of comparison between lake areas predicted by the models and those extracted from remote sensing images.

10.4 Conclusion

We have shown that remote sensing is an important tool for ecological model initialization, validation, and model-data integration. With improved computational power resulting from the use of supercomputers, an increased number of global datasets from remote sensing (e.g., MODIS NPP, GPP, LAI, and NDVI) (Zhao and Running 2010; Mu et al. 2011), improvements of modern model-data integration approaches (Vrugt et al. 2009), and development of more complex models of key Earth system processes (e.g., CESM) (Moorcroft 2006), it is possible to monitor important ecosystem and landscape dynamics that previously were difficult to detect. The marriage among supercomputers, models, remote sensing, and modern model-data fitting approaches has opened a new horizon for future ecological research targeted to better understand mechanisms underlying observed ecosystem and landscape dynamics.

Acknowledgments

This work is supported by the Laboratory Directed Research and Development (LDRD) Program at Los Alamos National Laboratory (LANL), and the Office of Biological and Environmental Research within DOE Office of Science. This chapter is published under the public release code of LA-UR-12-25632.

References

Aber, J. D., and C. A. Federer. 1992. "A Generalized, Lumped-Parameter Model of Photosynthesis, Evapotranspiration and Net Primary Production in Temperate and Boreal Forest Ecosystems." *Oecologia* 92(4):463–74.

Aber, J. D., S. V. Ollinger, C. A. Federer, P. B. Reich, M. L. Goulden, D. W. Kicklighter, J. M. Melillo, and R. G. Lathrop. 1995. "Predicting the Effects of Climate Change on Water Yield and Forest Production in the Northeastern U.S." *Climate Research* 5:207–22.

Ainsworth, E. A., and S. P. Long. 2005. "What Have We Learned from 15 years of Free-Air CO_2 Enrichment (FACE)? A Meta-Analytic Review of the Responses of Photosynthesis, Canopy Properties and Plant Production to Rising CO_2." *New Phytologist* 165(2):351–71.

Allen, C. D., A. K. Macalady, H. Chenchouni, D. Bachelet, N. McDowell, M. Vennetier, T. Kitzberger, et al. 2010. "A Global Overview of Drought and Heat-Induced Tree Mortality Reveals Emerging Climate Change Risks for Forests." *Forest Ecology and Management* 259(4):660–84.

Arp, C. D., B. M. Jones, F. E. Urban, and G. Grosse. 2011. "Hydrogeomorphic Processes of Thermokarst Lakes with Grounded-Ice and Floating-Ice Regimes on the Arctic Coastal Plain, Alaska." *Hydrological Processes* 25:2422–38.

Ball, J. T., I. E. Woodrow, and J. A. Berry. 1987. "A Model Predicting Stomatal Conductance and Its Contribution to the Control of Photosynthesis under Different Environmental Conditions." In *Progress in Photosynthesis Research*, edited by J. Biggins. The Netherlands: Martinus Nijhoff Publishers.

Bonan, G. B. 1995. "Sensitivity of a GCM Simulation to Inclusion of Inland Water Surfaces." *Journal of Climate* 8:2691–704.

Botkin, D. B., J. F. Janak, and J. R. Wallis. 1972. "Some Ecological Consequences of a Computer Model of Forest Growth." *Journal of Ecology* 60:849–73.

Brumby, S. P., J. Theiler, S. J. Perkins, N. R. Harvey, J. J. Szymanski, J. J. Bloch, and M. Mitchell. 1999. "Investigation of Image Feature Extraction by a Genetic Algorithm." *Proceedings of SPIE* 3812:24–31.

Burn, C. R. 2002. "Tundra Lakes and Permafrost, Richards Island, Western Arctic Coast, Canada." *Canadian Journal of Earth Sciences* 39:1281–98.

Chen, M., J. C. Rowland, C. J. Wilson, G. L. Altmann, and S. P. Brumby. In revision. "Understanding intra-annual and inter-annual lake area change dynamics in Yukon Flats, Alaska." *Permafrost and Periglacial Processes*.

Chen, M., J. C. Rowland, C. J. Wilson, G. L. Altmann, and S. P. Brumby. In revision. "Temporal and spatial pattern of thermokarst lake area changes at Yukon Flats, Alaska." *Hydrological Processes*. doi:10.1002/hyp.9642.

Corcoran, R. M., J. R. Lovvorn, and P. J. Heglund. 2009. "Long-Term Change in Limnology and Invertebrates in Alaskan Boreal Wetlands." *Hydrobiologia* 620(1):77–89.

Decharme, B., R. Alkama, F. Papa, S. Faroux, H. Douville, and C. Prigent. 2012. "Global Off-Line Evaluation of the ISBA-TRIP Flood Model." *Climate Dynamics* 38(7–8):1389–412.

Downing, J. A., Y. T. Prairie, J. J. Cole, C. M. Duarte, L. J. Tranvik, R. G. Striegl, W. H. McDowell, et al. 2006. "The Global Abundance and Size Distribution of Lakes, Ponds, and Impoundments." *Limnology and Oceanography* 51(5):2388–97.

Emmerton, C. A., L. F. W. Lesack, and P. Marsh. 2007. "Lake Abundance, Potential Water Storage, and Habitat Distribution in the Mackenzie River Delta, Western Canadian Arctic." *Water Resource Research* 43(5):W05419.

Evensen, G. 2003. "The Ensemble Kalman Filter: Theoretical Formulation and Practical Implementation." *Ocean Dynamics* 53(4):343–67.

Farquhar, G. D., S. von Caemmerer, and J. A. Berry. 1980. "A Biochemical Model of Photosynthetic CO_2 Assimilation in Leaves of C_3 Species." *Planta* 149(1):78–90.

Finney, M. A. 1998. *FARSITE: Fire Area Simulator-Model Development and Evaluation*. USDA Forest Service Rocky Mountain Research Station. Fort Collins, CO, USA.

Fisher, R., N. McDowell, D. Purves, P. Moorcroft, S. Sitch, P. Cox, C. Huntingford, P. Meir, and F. I. Woodward. 2010. "Assessing Uncertainties in a Second-Generation Dynamic Vegetation Model Caused by Ecological Scale Limitations." *New Phytologist* 187(3):666–81.

Forman, R. T. T. 1995. "Some General Principles of Landscape and Regional Ecology." *Landscape Ecology* 10:133–42.

Forman, R. T. T., and M. Godron. 1986. *Landscape Ecology*. New York: Wiley.

Frazier, P. S., and K. J. Page. 2000. "Water Body Detection and Delineation with Landsat TM Data." *Photogrammetric Engineering & Remote Sensing* 66(12):1461–7.

Frohn, R. C., K. M. Hinkel, and W. R. Eisner. 2005. "Satellite Remote Sensing Classification of Thaw Lakes and Drained Thaw Lake Basins on the North Slope of Alaska." *Remote Sensing of Environment* 97(1):116–26.

Green, B., E. A. Simmons, and I. Woltjer. 1996. *Landscape Conservation: Some Steps Towards Developing a New Conservation Dimension: A Report of The IUCN-CESP Landscape Conservation Working Group*. Kent, UK: Department of Agriculture, Horticulture and the Environment, University of London.

Grosse, G., L. Schirrmeister, V. V. Kunitsky, and H.-W. Hubberten. 2005. "The Use of CORONA Images in Remote Sensing of Periglacial Geomorphology: An Illustration from the NE Siberian Coast." *Permafrost and Periglacial Processes* 16(2):163–72.

Hansen, A. J., R. R. Neilson, V. H. Dale, C. H. Flather, L. R. Iverson, D. J. Currie, S. Shafer, R. Cook, and P. J. Bartlein. 2001. "Global Change in Forests: Responses of Species, Communities, and Biomes." *Bioscience* 51(9):765–79.

He, H. S. 2008. "Forest Landscape Models: Definitions, Characterization, and Classification." *Forest Ecology and Management* 254(3):484–98.

He, H. S., D. J. Mladenoff, V. C. Radeloff, and T. R. Crow. 1998. "Integration of GIS Data and Classified Satellite Imagery for Regional Forest Assessment." *Ecological Application* 8:1072–83.

Hinkel, K. M., R. C. Frohn, F. E. Nelson, W. R. Eisner, and R. A. Beck. 2005. "Morphometric and Spatial Analysis of Thaw Lakes and Draind Thaw Lake Basins in the Western Arctic Coastal Plain, Alaska." *Permafrost and Periglacial Processes* 16:327–41.

Jones, B., C. Arp, K. Hinkel, R. Beck, J. Schmutz, and B. Winston. 2009. "Arctic Lake Physical Processes and Regimes with Implications for Winter Water Availability and Management in the National Petroleum Reserve Alaska." *Environmental Management* 43(6):1071–84.

Jones, B. M., G. Grosse, C. D. Arp, M. C. Jones, K. W. Anthony, and V. E. Romanovsky. 2011. "Modern Thermokarst Lake Dynamics in the Continuous Permafrost Zone, Northern Seward Peninsula, Alaska." *Journal of Geophysical Research* 116:G00M03.

Keane, R. 1996. *FIRE-BGC—A Mechanistic Ecological Process Model for Simulating Fire Succession on Coniferous Forest Landscapes of the Northern Rocky Mountains.* Ogden, UT: United States Department of Agriculture, Forest Service Intermountain Forest and Range Experiment Station.

Kling, G., G. Kipphut, and M. Miller. 1992. "The Flux of CO_2 and CH_4 from Lakes and Rivers in Arctic Alaska." *Hydrobiologia* 240(1):23–36.

Krinner, G. 2003. "Impact of Lakes and Wetlands on Boreal Climate." *Journal of Geophysical Research* 108:4520.

Krinner, G., and J. Boike. 2010. "A Study of the Large-Scale Climatic Effects of a Possible Disappearance of High-Latitude Inland Water Surfaces During the 21st Century." *Boreal Environment Research* 15:203–17.

Kuusk, A., and T. Nilson. 2000. "A Directional Multispectral Forest Reflectance Model." *Remote Sensing of Environment* 72(2):244–52.

Kuusk, A., T. Nilson, M. Paas, M. Lang, and J. Kuusk. 2008. "Validation of the Forest Radiative Transfer Model FRT." *Remote Sensing of Environment* 112(1):51–8.

Labrecque, S., D. Lacelle, C. R. Duguay, B. Lauriol, and J. Hawkings. 2009. "Contemporary (1951–2001) Evolution of Lakes in the Old Crow Basin, Northern Yukon, Canada: Remote Sensing, Numerical Modeling, and Stable Isotope Analysis." *Arctic* 62(2):225–38.

Lehner, B., and P. Döll. 2004. "Development and Validation of a Global Database of Lakes, Reservoirs and Wetlands." *Journal of Hydrology* 296(1–4):1–22.

Long, S. P., E. A. Ainsworth, A. Rogers, and D. R. Ort. 2004. "Rising Atmospheric Carbon Dioxide: Plants Face the Future." *Annual Review of Plant Biology* 55:591–628.

Mackay, R. J., ed. 1988. *Catastrophic Lake Drainage, Tuktoyaktuk Peninsula Area, District of Mackenzie, Current Research, Part D.* Ottawa, Canada: Geological Survey of Canada.

Marsh, P., M. Russell, S. Pohl, H. Haywood, and C. Onclin. 2009. "Changes in Thaw Lake Drainage in the Western Canadian Arctic from 1950 to 2000." *Hydrological Processes* 23(1):145–58.

McDowell, N. G., D. J. Beerling, D. D. Breshears, R. A. Fisher, K. F. Raffa, and M. Stitt. 2011. "The Interdependence of Mechanisms Underlying Climate-Driven Vegetation Mortality." *Trends in Ecology & Evolution* 26(10):523–32.

Moorcroft, P. R. 2006. "How Close Are We to a Predictive Science of the Biosphere?" *Trends in Ecology & Evolution* 21(7):400–7.

Moorcroft, P. R., G. C. Hurtt, and S. W. Pacala. 2001. "A Method for Scaling Vegetation Dynamics: The Ecosystem Demography Model (ED)." *Ecological Monographs* 71(4):557–86.

Mu, Q., M. Zhao, and S. W. Running. 2011. "Improvements to a MODIS Global Terrestrial Evapotranspiration Algorithm." *Remote Sensing of Environment* 115(8):1781–800.

Neilson, R. P. 1995. "A Model for Predicting Continental-Scale Vegetation Distribution and Water Balance." *Ecological Applications* 5(2):362–85.

Neilson, R. P., and S. W. Running. 1996. "Global Dynamic Vegetation Modelling: Coupling Biogeochemistry and Biogeography Models." In *Global Change and Terrestrial Ecosystems*, edited by B. Walker and W. Steffen. Cambridge: Cambridge University Press.

Perkins, S. J., K. Edlund, D. Esch-Mosher, D. Eads, N. Harvey, and S. Brumby. 2005. "Genie Pro: Robust Image Classification Using Shape, Texture, and Spectral Information." *Proceedings of SPIE* 5806:139–48.

Plug, L. J., C. Walls, and B. M. Scott. 2008. "Tundra Lake Changes from 1978 to 2001 on the Tuktoyaktuk Peninsula, Western Canadian Arctic." *Geophysical Research Letters* 35(3):L03502.

Prentice, I. C., W. Cramer, S. P. Harrison, R. Leemans, R. A. Monserud, and A. M. Solomon. 1992. "A Global Biome Model Based on Plant Physiology and Dominance, Soil Properties and Climate." *Journal of Biogeography* 19(2):117–34.

Prigent, C., F. Papa, F. Aires, W. B. Rossow, and E. Matthews. 2007. "Global Inundation Dynamics Inferred from Multiple Satellite Observations, 1993–2000." *Journal of Geophysical Research* 112:D12107.

Rampton, V. N., ed. 1988. *Quaternary Geology of the Tuktoyaktuk Coastlands, Northwest Territories, Geological Survey of Canada Memoir*. Canada: Energy, Mines and Resources.

Repo, M. E., J. T. Huttunen, A. V. Naumov, A. V. Chichulin, E. D. Lapshina, W. Bleuten, and P. J. Martikainen. 2007. "Release of CO_2 and CH_4 from Small Wetland Lakes in Western Siberia." *Tellus B* 59(5):788–96.

Riordan, B., D. Verbyla, and A. D. McGuire. 2006. "Shrinking Ponds in Subarctic Alaska based on 1950–2002 Remotely Sensed Images." *Journal of Geophysical Research* 111:G04002.

Roach, J. 2011. *Lake Area Change in Alaskan National Wildlife Refuges: Magnitude, Mechanisms, and Heterogeneity*. Fairbanks, AK: University of Alaska.

Roach, J., B. Griffith, D. Verbyla, and J. Jones. 2011. "Mechanisms Influencing Changes in Lake Area in Alaskan Boreal Forest." *Global Change Biology* 17:2567–83.

Rover, J., L. Ji, B. K. Wylie, and L. L. Tieszen. 2012. "Establishing Water Body Areal Extent Trends in Interior Alaska from Multi-Temporal Landsat Data." *Remote Sensing Letters* 3(7):595–604.

Running, S. W., and E. R. J. Hunt. 1993. "Generalization of a Forest Ecosystem Process Model for Other Biomes, BIOME-BGC, and an Application for Global-Scale Models." In *Scaling Physiological Processes: Leaf to Globe*, edited by J. R. Ehleringer and C. B. Field. San Diego, CA: Academic Press.

Scheller, R. M., and D. J. Mladenoff. 2004. "A Forest Growth and Biomass Module for a Landscape Simulation Model, LANDIS: Design, Validation, and Application." *Ecological Modelling* 180:211–29.

Scheller, R., and D. Mladenoff. 2007. "An Ecological Classification of Forest Landscape Simulation Models: Tools and Strategies for Understanding Broad-Scale Forested Ecosystems." *Landscape Ecology* 22(4):491–505.

Shaver, G. R., S. M. Bret-Harte, M. H. Jones, J. Johnstone, L. Gough, J. Laundre, and F. S. Chapin. 2001. "Species Composition Interacts with Fertilizer to Control Long-Term Change in Tundra Productivity." *Ecology* 82(11):3163–81.

Smith, L. C., Y. Sheng, G. M. MacDonald, and L. D. Hinzman. 2005. "Disappearing Arctic Lakes." *Science* 308(5727):1429.

Strigul, N., D. Pristinski, D. Purves, J. Dushoff, and S. Pacala. 2008. "Scaling from Trees to Forests: Tractable Macroscopic Equations for Forest Dynamics." *Ecological Monographs* 78(4):523–45.

Turner, M. G. 1989. "Landscape Ecology: The Effect of Pattern on Process." *Annual Review of Ecology and Systematics* 20:171–97.

Urban, D. L., M. F. Acevedo, and S. L. Garman. 1999. "Scaling Fine-Scale Processes to Large Scale Patterns Using Models Derived from Models: Meta-Models." In *Spatial Modeling of Forest Landscape Change: Approaches and Applications*, edited by D. J. Mladenoff and W. L. Baker. Cambridge: Cambridge University Press.

Urban, D. L., and H. H. Shugart. 1992. "Individual-Based Models of Forest Succession." In *Plant Succession: Theory and Prediction*, edited by D. C. Glenn-Lewin, R. K. Peet, and T. T. Veblen. Vol. 249–92, London: Chapman & Hall.

Van Beek, L. P. H., Y. Wada, and M. F. P. Bierkens. 2011. "Global Monthly Water Stress: 1. Water Balance and Water Availability." *Water Resources Research* 47:w07517. doi:10.1029/2010WR009791.

Vrugt, J., C. ter Braak, H. Gupta, and B. Robinson. 2009. "Equifinality of Formal (DREAM) and Informal (GLUE) Bayesian Approaches in Hydrologic Modeling?" *Stochastic Environmental Research and Risk Assessment* 23(7):1011–26.

Wada, Y., L. P. H. van Beek, F. C. S. Weiland, B. F. Chao, Y.-H. Wu, and M. F. P. Bierkens. 2012. "Past and Future Contribution of Global Groundwater Depletion to Sea-Level Rise." *Geophysical Research Letters* 39:L09402.

Walther, G. R., E. Post, P. Convey, A. Menzel, C. Parmesan, T. J. C. Beebee, J. M. Fromentin, O. Hoegh-Guldberg, and F. Bairlein. 2002. "Ecological Responses to Recent Climate Change." *Nature* 416(6879):389–95.

Wang, Y. P., and P. G. Jarvis. 1990. "Description and Validation of an Array Model—MAESTRO." *Agricultural and Forest Meteorology* 51(3–4):257–80.

Weiland, F. C. S., L. P. H. Van Beek, J. C. J. Kwadijk, and M. F. P. Bierkens. 2012. "Global Patterns of Change in Discharge Regimes for 2100." *Hydrology and Earth System Sciences* 16(4):1047–62.

Wiens, J. 1989. "Spatial Scaling in Ecology." *Functional Ecology* 3:385–97.

Wiens, J. A., and T. T. Milne. 1989. "Scaling of "Landscapes" in Ecology, or Landscape Ecology from a Beetle's Perspective." *Landscape Ecology* 3:87–96.

Wolter, P. T., D. J. Mladenoff, G. E. Host, and T. R. Crow. 1995. "Improved Forest Classification in the Northern Ladke States Using Multi-Temporal Landsat Imagery." *Photogrametric Engineering and Remote Sensing* 61:1129–43.

Woodward, F. I., T. M. Smith, and W. R. Emanuel. 1995. "A Global Land Primary Productivity and Phytogeography Model." *Global Biogeochemical Cycles* 9:471–490.

Wu, J., and R. Hobbs. 2002. "Key Issues and Research Priorities in Landscape Ecology: An Idiosyncratic Synthesis." *Landscape Ecology* 17:355–65.

Xu, C., R. Fisher, S. D. Wullschleger, C. J. Wilson, M. Cai, and N. G. McDowell. 2012. "Toward a Mechanistic Modeling of Nitrogen Limitation on Vegetation Dynamics." *Plos One* 7(5):e37914.

Xu, C., G. Z. Gertner, and R. M. Scheller. 2009. "Uncertainties in the Response of a Forest Landscape to Global Climatic Change." *Global Change Biology* 15:116–31.

Xu, C., H. S. He, Y. Hu, Y. Chang, D. R. Larsen, X. Li, and R. Bu. 2004. "Assessing the Effect of Cell-Level Uncertainty on a Forest Landscape Model Simulation in Northeastern China." *Ecological Modelling* 180:57–72.

Yoshikawa, K., and L. D. Hinzman. 2003. "Shrinking Thermokarst Ponds and Groundwater Dynamics in Discontinuous Permafrost Near Council, Alaska." *Permafrost and Periglacial Processes* 14(2): 151–60.

Zhao, M., and S. W. Running. 2010. "Drought-Induced Reduction in Global Terrestrial Net Primary Production from 2000 through 2009." *Science* 329(5994):940–43.

11

Plant Invasion and Imaging Spectroscopy

Kate S. He and Duccio Rocchini

CONTENTS

11.1 Introduction 191
11.2 Role of Imaging Spectroscopy in Invasion Research 195
    11.2.1 Image Processing and Classification Algorithms 195
    11.2.2 Advantages of Incorporating Phenology in Invasion Research 198
11.3 Summary and Future Directions 201
Acknowledgments 201
References 201

11.1 Introduction

In the past two decades, a considerable amount of research on invasive species has been carried out globally due to the concerns for the implications and consequences of successful invasions. Invasion biologists search for answers to questions such as what types of biophysical traits make a superior invader and what types of communities are more susceptible to invasion. Ultimately, the goal of invasion research is to develop a unified and comprehensive framework that allows ecologists, conservation biologists, and land managers to make accurate predictions on the potential invaders even before their introductions and on the types of communities that are more vulnerable to invasions.

Despite the fact that a large amount of research has been carried out and published over the past decade, factors determining the invasion success of introduced species, regardless of their native origins, are still not yet fully identified. This is mainly because of the biotic and abiotic complexities associated with invasive species (Sakai et al. 2001).

Currently, the key hypotheses explaining species' invasion success and spatial range expansion include propagule pressure (Lonsdale 1999; Lockwood et al. 2005; Pyšek and Richardson 2006; Richardson and Pyšek 2006; Wilson et al. 2009), residence time (Rejmánek and Richardson 1996; Rejmánek 2000; Pyšek and Vojtech 2005; Richardson and Pyšek 2006; Wilson et al. 2007; Pyšek et al. 2009; Williamson et al. 2009; Ahern et al. 2010), enemy release (Maron and Vila 2001; Keane and Crawley 2002; Colautti et al. 2004; Joshi and Vrieling 2005), empty niche (Elton 1958; MacArthur 1970; Levine and D'Antonio 1999; Davis et al. 2000; Mack et al. 2000; Hierro et al. 2005), evolution of increased competitive ability (Blossey and Notzold 1995; Callaway and Ridenour 2004), and the novel weapons hypothesis (Callaway and Ridenour 2004; Hierro et al. 2005). Detailed information on all invasion hypotheses can be found in recent review articles by Henderson et al. (2006), Richardson and Pyšek (2006), Catford et al. (2009), Pyšek and Richardson (2010), and Vitousek et al. (2011). Significant developments in invasion ecology have been made during recent years; however, these key hypotheses still need to be further tested, and reproducible results using different invasive species at various spatial and temporal scales could shed more light on the mechanisms of invasion success.

More recently, new approaches and technologies have been applied in invasion research to test existing hypotheses and generate new ones. In particular, advances in remote sensing (Turner et al. 2003; Underwood et al. 2003; Lass et al. 2005; Asner et al. 2008a,b; Andrew and Ustin 2009; Ustin and Gamon 2010, Vitousek et al. 2011; Khanna et al. 2012; Somers and Asner 2012) have allowed invasion ecologists to obtain new knowledge that aids the understanding of the invasion success of introduced species. These technologies have made a suite of tools available to study the historic spreading, current spatial distribution, and future dispersal of invasive plants at local, regional, and global scales. In this chapter, we draw attention to remote sensing investigations that have resulted in new ecological insights for plant invasion that would not otherwise have been possible.

Remotely sensed data acquired by sensors on board satellites or aircrafts have provided usable and repeatable measurements for mapping vegetative covers and analyzing ecosystem changes (Townshend et al. 1993). The underlying mechanism of remote sensing is that certain key environmental parameters such as light, water, and temperature, with remotely detectable biophysical properties, drive the distribution and abundance of species across landscapes and determine how they occupy habitats (Turner et al. 2003). In particular, spectral information provided by hyperspectral sensors such as Hyperion, Airborne Visible/Infrared Imaging Spectrometer (AVIRIS), Compact Airborne Spectrographic Imager (CASI), Airborne Imaging Spectroradiometer for Applications (AISA), and HyMap (from HyVista) allows for the detection of invaders at the species level (Clark et al. 2005; Andrew and Ustin 2006, 2008, 2009; Lawrence et al. 2006; Miao et al. 2006; Pengra et al. 2007; Underwood and Ustin 2007; Hestir et al. 2008; Pu et al. 2008; Asner et al. 2008a,b; Narumalani et al. 2009; Somers and Anser 2012). The utility and limitations of imaging spectroscopy or hyperspectral remote sensing in invasion research have been discussed in a comprehensive review by He et al. (2011). In this chapter, we pay close attention to two key aspects in remote sensing: (1) the innovative classification approaches used in image processing and (2) the advantages of using phenology in mapping species distribution. To facilitate the readers, we provide summary information in terms of the type of sensors used, sensor characteristics, habitat type, invader studied, study area, main model used, and reference citation information in Table 11.1. A flowchart is also provided to show the role of imaging spectroscopy in invasion research (Figure 11.1). Although we mainly focus on hyperspectral sensors in this chapter, a few studies using multispectral sensors in invasion studies have also been discussed briefly.

We plan not to take too much space to address the types of spaceborne sensors commonly used in remote sensing in this chapter, but see Lass et al. (2005), Gillespie et al. (2008), Nagendra and Rocchini (2008), Schaepman et al. (2009), and Boyd and Foody (2010) for a detailed introduction on satellites with passive or active sensors as well as airborne sensors commonly used in remote sensing research for ecological and biodiversity studies.

TABLE 11.1

Summary of the Type of Sensors Used, Sensor Characteristics, Habitat Type, Invader Studied, Study Area, Main Model Used, and Article Citation Information

Satellite Sensor	Satellite Sensor Characteristics	Habitat Type	Species	Study Area	Main Model Used	Reference
AISA hyperspectral imager	492 bands, spectral range: 395–2503 nm, spatial resolution: 75 cm to 4 m	Crops	Canada thistle (*Cirsium arvense*), Russian olive (*Elaeagnus angustifolia*)	North Platte River, Nebraska	Spectral angle mapping	Narumalani et al. (2009)
AVIRIS	224 bands, spectral range: 400–2500 nm, spatial resolution: 3.5 m	Mixed forest-suburban areas	24 Introduced tree species	Hawaiian Islands	Canopy spectral signatures profiling	Asner et al. (2008a)
AVIRIS	224 bands, spectral range: 400–2500 nm, spatial resolution: 3.5 m	Schrubland, chaparral, grassland	Iceplant (*Carpobrotus edulis*), jubata grass (*Cortaderia jubata*), and blue gum (*Eucalyptus globulus*)	Vandenberg Air Force Base, California	Quality assurance analysis	Underwood and Ustin (2007)
CASI	288 bands, spectral range: 430–870 nm, spatial resolution: 3 m	River, riparian vegetation	Salt cedars (*Tamarix chinensis, T. ramosissima, and T. Parviflora*)	Humboldt River, Nevada	Classification by ANNs and LDA	Pu et al. (2008)
CASI	288 bands, spectral range: 430–870 nm, spatial resolution: 3 m	Grasslands	Weed yellow starthistle (*Centaurea solstitialis*)	California's Central Valley	PCA, unconstrained linear spectral mixture models (LSMMs)	Miao et al. (2006)
EO-1 Hyperion hyperspectral sensor	220 bands, spectral range: 357–2576 nm, spatial resolution: 30 m	Pastures, grasslands, natural forests	Guava (*Psidium guajava*)	Galapagos Islands, Ecuador	Spectral unmixing	Walsh et al. (2008)
EO-1 Hyperion hyperspectral sensor	220 bands, spectral range: 357–2576 nm, spatial resolution: 30 m	Wetlands	*Phragmites australis*	Great Lakes, Wisconsin	Spectral correlation mapper algorithm	Pengra et al. (2007)
EO-1 Hyperion hyperspectral sensor	220 bands, spectral range: 357–2576 nm, spatial resolution: 30 m	Mountain rainforests	*Myrica faya*	Hawaii Volcanoes National Park	Remotely sensed photochemical and carotenoid reflectance indexes (PRI, CRI)	Asner et al. (2006)
HyMap hyperspectral sensor	128 bands, spectral range: 450–2500 nm, spatial resolution: 3 m	Wetlands	Perennial weed (*Lepidium latifolium*), water hyacinth (*Eichhornia crassipes*), Brazilian waterweed (*Egeria densa*)	Sacramento-San Joaquin Delta	Binary decision tree, spectral angle mapping	Hestir et al. (2008)
HyMap hyperspectral sensor	126 bands, spectral range: 450–2500 nm, spatial resolution: 3.5 m	Mixed riparian zones and sagebrush—steppe vegetation	Leafy spurge (*Euphorbia esula*)	Swan Valley, Idaho	Classification by mixture-tuned matched filtering algorithm	Glenn et al. (2005)

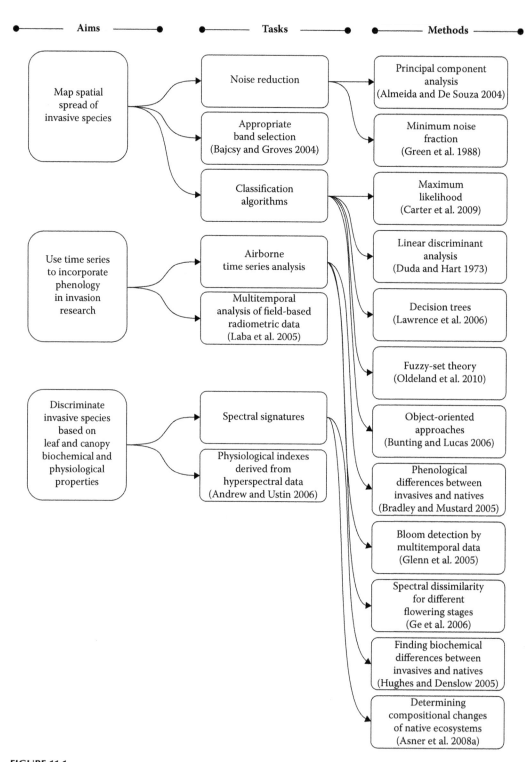

FIGURE 11.1
A flowchart outlining the role of imaging spectroscopy in invasion research.

11.2 Role of Imaging Spectroscopy in Invasion Research

11.2.1 Image Processing and Classification Algorithms

Hyperspectral data, with their ability to collect information at a high spectral resolution using a series of contiguous spectral bands, each with a narrow spectral range, can be used to record information pertaining to a range of critical plant properties including leaf pigment, water content, and chemical composition (Curran 1989; Martin and Aber 1997; Townsend et al. 2008). To cope with the large number of intercorrelated spectral bands, a number of different algorithms have been adopted for extracting those bands carrying significant information avoiding spectral noise such as principal component analysis (PCA) (Almeida and De Souza 2004), minimum noise fraction (MNF) (Green et al. 1988), and semi-automated band selection (Bajcsy and Groves 2004). Further, once specific spectral signatures of single vascular plant species (mostly trees and shrubs) (Carlson et al. 2007) have been identified, one can apply different classification algorithms. When using hyperspectral data, the most commonly used algorithms are based on (1) unsupervised classification that aggregates pixels into classes based on spectral values with numerous clustering algorithms including ISODATA and K-means (Shanmugam et al. 2006); (2) maximum likelihood (Carter et al. 2009), which simply estimates each class mean and variation from training data; and (3) linear discriminant analysis (LDA) (Duda and Hart 1973) using discriminant functions for each class based on the within-class covariance matrix, or spectral angle mapper (SAM) (Clark et al. 2005), which is a spectral matching technique comparing each sample spectrum with several reference spectra (Kruse et al. 1993). Together with these algorithms, recent techniques have been developed based on (1) decision trees (such as random forest) (Lawrence et al. 2006), (2) support vector machines (Melgani and Bruzzone 2004), (3) fuzzy-set theory (Oldeland et al. 2010), or (4) object-oriented approaches (Bunting and Lucas 2006; Plaza et al. 2009). We discuss a few case studies that use these classification algorithms in invasion research in the following paragraphs.

Images acquired by hyperspectral sensors have been used to map invasive plants in various regions in the United States and other parts of the globe. Underwood et al. (2003) successfully used the AVIRIS imagery with 4 m resolution to detect iceplant (*Carpobrotus edulis*) and jubata grass (*Cortaderia jubata*) in California's mediterranean-type ecosystems using three image-processing techniques (MNF, continuum removal, and band ratio indexes). Their study concluded that the continuum removal is a reliable method for depicting presence/absence of iceplant within the scrub community. However, the MNF and band ratio methods were most accurate in delineating the spatial distribution and density of iceplant and jubata grass (Figure 11.2).

The distribution of four dominant invasive plant species along the floodplain of the North Platte River corridor, Nebraska, is mapped by Narumalani et al. (2009). The four species include saltcedar (*Tamarix* spp.), Russian olive (*Elaeagnus angustifolia*), Canada thistle (*Cirsium arvense*), and musk thistle (*Carduus nutans*). The authors used the AISA hyperspectral imager and evaluated the spectral angle classification algorithm for detecting invasive species distribution. They confirmed an overall mapping accuracy of 74%. A higher overall accuracy was achieved from another study by Lawrence et al. (2006). They mapped two common invasive species, leafy spurge (*Euphorbia esula*) and spotted knapweed (*Centaurea maculosa*), using hyperspectral imagery through a random forest classification algorithm in Madison County, Montana. The overall mapping accuracy was 84% and 86% for spotted knapweed and leafy spurge, respectively.

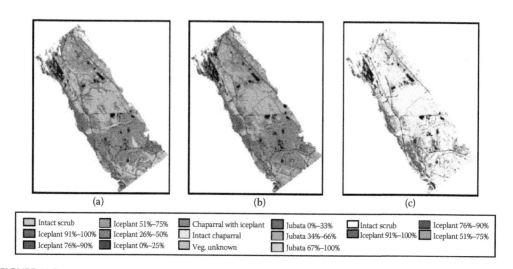

Intact scrub | Iceplant 51%–75% | Chaparral with iceplant | Jubata 0%–33% | Intact scrub | Iceplant 76%–90%
Iceplant 91%–100% | Iceplant 26%–50% | Intact chaparral | Jubata 34%–66% | Iceplant 91%–100% | Iceplant 51%–75%
Iceplant 76%–90% | Iceplant 0%–25% | Veg. unknown | Jubata 67%–100% | |

FIGURE 11.2
(See color insert.) Mapping of iceplant (*Carpobrotus edulis*) and jubata grass (*Cortaderia jubata*) using (a) minimum noise fraction, (b) band ratio indexes, and (c) continuum removal processing algorithms at Vandenberg Air Force Base in California. (Reproduced from Ustin et al., *2002 IEEE International Geoscience and Remote Sensing Symposium, June 24–28, Vol 3*, 1658–60. Toronto, Canada: IEEE International. With permission.)

Furthermore, multiple classification algorithms have been applied for comparison purposes to assess accuracy in mapping the spatial extent of invasive species. For example, DiPietro (2002) investigated the potential for mapping giant reed (*Arundo donax*), a problematic invasive species in the western part of the United States with AVIRIS hyperspectral imagery. Classification methods used in the study included unsupervised classifications (ISODATA and K-means), continuum removal, supervised classification (maximum likelihood), and SAM. The authors concluded that the maximum likelihood classification was the best method for mapping giant reed with an overall accuracy rate of 95%. SAM achieved 85% accuracy rate and was considered a good method to provide repeatable results in different images. Unsupervised classifications were not successful in mapping the spatial distribution of the targeted species and vegetation types.

More recently, an extended and comparative study was carried out by Hamada et al. (2007) to detect tamarisk in the riparian habitat of Southern California using very high spatial (0.5 m) and spectral (4 nm) resolution imagery acquired using a 700 hyperspectral sensor (Surface Optics Corporation, San Diego, CA). Several classification approaches were compared in their study including stepwise discriminant analysis and hierarchical clustering (HC). They concluded that HC was a particularly effective and efficient statistical method for identifying wave bands and spectral transforms having the greatest discriminatory power for detecting tamarisk infestation. The highest correct detection rate reached 90% with a pixel size of 25 m^2.

The same invasive species, tamarisk, was also studied by Pu et al. (2008) using CASI hyperspectral imagery with two different classification methods in Lovelock, Nevada. A newer classifier, the artificial neural network (ANN) was used and also compared with a traditional classification method, the LDA. The authors concluded that the ANN outperformed the LDA method in detecting and mapping the tamarisk distribution in the study area. They suggested that the ANN algorithm's nonlinear property might catch more information from the same input data than the LDA method, thus resulting in a higher overall classification accuracy.

Besides using multiple classification algorithms, an integrated approach involving the use of diverse remote sensing systems, data types, spatial and spectral resolutions, and analytical and image-processing algorithms can be effectively applied in invasion studies. Walsh et al. (2008) assessed an important invasive species, guava (*Psidium guajava*), in the Galapagos Islands of Ecuador. The authors used both QuickBird (with a spatial resolution of 2.44 m × 2.44 m) and Hyperion data (with a spatial resolution of 30 m × 30 m) in their study. The QuickBird data were examined through a pixel-based classification approach (i.e., an unsupervised classification using the ISODATA algorithm) and an object-based image analysis (OBIA) approach. Hyperion data were assessed using the MNF and pure pixel index. The Hyperion hyperspectral data were assessed to construct spectral end-members from QuickBird data using linear and nonlinear mixture modeling approaches. Multiple image classifications were performed using neural network and linear mixture modeling. Their results indicated two different representations of guava's spatial distribution. The OBIA tends to smooth the map of guava relative to the pixel-based classification method. They suggested that the spatial extent and rate of spread of the guava invasion is likely affected by the age of establishment of the guava source areas and land management strategies.

Imaging spectroscopy with innovative processing and classification methods is also very effective in mapping and detecting invasive plant species in aquatic and riparian habitats. This is more promising since these types of habitats bring more challenges in image processing, including meteorological, physical, and biological heterogeneity. Hestir et al. (2008) presented three case studies that address these issues using airborne imaging spectroscopy to develop regional-scale monitoring of invasive aquatic and wetland weeds in the Sacramento–San Joaquin Delta: a terrestrial riparian weed, perennial pepperweed (*Lepidium latifolium*); a floating aquatic weed, water hyacinth (*Eichhornia crassipes*); and a submerged aquatic weed, Brazilian waterweed (*Egeria densa*). They showed a range of techniques (a binary decision tree that incorporated spectral mixture analysis, spectral angle mapping, band indexes, and continuum removal products) that can be used to produce species distribution maps for wetland and estuarine resource managers. More specifically, the HyMap, an airborne hyperspectral imager that collects 128, 126, and 125 bands in the visible and near-infrared (VNIR) (0.45–1.50 μm) through the shortwave infrared (SWIR) (1.50–2.5 μm), at bandwidths from 10 nm in the VNIR to 15–20 nm in the SWIR, was used in their study. The spatial resolution of the data is 3 m, with a swath width of 1.5 km. They achieved the user's and producer's accuracies for perennial pepperweed detection of 75.8% and 63.0%, respectively; for water hyacinth detection of 89.8% and 69.1%; and for Brazilian waterweed detection of 92.1% and 59.2%. The authors suggested that perennial pepperweed and water hyacinth both exhibited significant spectral variation related to plant phenology, which will be discussed in Section 11.2.2.

In a separate study, Laba et al. (2008) mapped invasive wetland plants in the Hudson River National Estuarine Research Reserve using QuickBird satellite imagery. Three invasive plants, purple loosestrife (*Lythrum salicaria*), common reed (*Phragmites australis*), and water chestnut (*Trapa natans*), were assessed and mapped using maximum-likelihood classification technique. Mapping accuracy was evaluated with conventional contingency tables and a fuzzy-set analysis. The overall accuracies ranged from 68.4% to 83%. These encouraging accuracies suggest that high-resolution satellite imagery offers significant potential for the mapping of invasive plant species in estuarine environments. Further, the common reed species was also mapped in coastal wetlands of the Great Lakes using the EO-1 Hyperion hyperspectral sensor by Pengra et al. (2007). The authors achieved a user's accuracy for predicted *Phragmites* stands of 61.1% and suggested that the modest accuracy compares favorably with previous attempts at remote sensing of *Phragmites* with airborne hyperspectral sensors such as the HyMap airborne sensor (Bachmann et al. 2002) and PROBE-1 airborne hyperspectral sensor (Lopez et al. 2004), despite Hyperion's much coarser spatial resolution and increased signal-to-noise ratio.

11.2.2 Advantages of Incorporating Phenology in Invasion Research

Phenology refers to particular life-cycle events of living organisms, such as flowering, leaf green-up, or senescence for plants and migration for animals. Phenological changes allow ecologists to understand more about species life-cycle events and seasonal dynamics of individuals and assemblages. Phenology also plays a significant role in detecting and mapping the spatial distribution of invasive species in remote sensing applications (He et al. 2011). Multidate remotely sensed images have become more useful in invasion studies. In particular, the uniqueness in phenological development of some invasive species provides a sound basis for spectral differences between targeted species and co-occurring native vegetation (Williams and Hunt 2004; Peterson 2005; Ge et al. 2006; Evangelista et al. 2009; Singh and Glenn 2009). Examples of invaders in this regard include downy brome (*Bromus tectorum*), leafy spurge (*E. esula*), yellow starthistle (*Centaurea solstitialis*), perennial pepperweed (*L. latifolium*), and Amur honeysuckle (*Lonicera maackii*).

Bradley and Mustard (2005) showed how interannual data collected from AVHRR can effectively detect downy brome populations in the Great Basin. They were able to identify the phenological differences between the invaders and associated native flora within a single growing season. The same invasive species was also studied by Noujdina and Ustin (2008) using multidate AVIRIS data in south-central Washington, D.C. The authors compared detectability of downy brome from single-date and multidate AVIRIS data using a mixture-tuned matched filtering algorithm for image classification. They concluded that the use of multidate data increased the accuracy of downy brome detection in the semiarid rangeland ecosystems. The accuracy is a direct result of clear spectral differences controlled by phenological dissimilarities between downy brome and surrounding vegetation (Figure 11.3).

Glenn et al. (2005) used HyMap hyperspectral data collected over 2 years to detect the infestation of leafy spurge in Idaho. A slight difference in leafy spurge reflectance was found between the 2002 and 2003 images by the authors and this was likely as a result of slight changes in leafy spurge bloom and time of the image acquisition. The authors also performed accuracy assessments for each year's classification data and found that user's accuracies were all above 70%, suggesting image-processing methods were repeatable between years.

As we have discussed earlier, a study carried by Hestir et al. (2008) used imaging spectroscopy to identify invasive aquatic and wetland weeds in the Sacramento–San Joaquin Delta. They found significant differences in phenological development among the three species, including perennial pepperweed, water hyacinth, and Brazilian waterweed (Figure 11.4). For example, Brazilian waterweed has two growth peaks, an early one during the beginning of the summer and the other in the later summer. They suggested that a later acquisition in summer may reduce omission errors as plants may be given more time to grow to the water surface. They concluded that the classification accuracy can be improved by mapping each phenological stage individually for all three invasive species.

In the tropics, time series of spaceborne Hyperion data have been used to study the dynamic changes and invasive species of Hawaiian rainforests (Asner et al. 2006). The authors compared the structural, biochemical, and physiological characteristics of an invasive nitrogen-fixing tree *Myrica faya* and native *Metrosideros polymorpha* in humid montane forests. By using nine scenes spanning from July 2004 to June 2005, including a transition

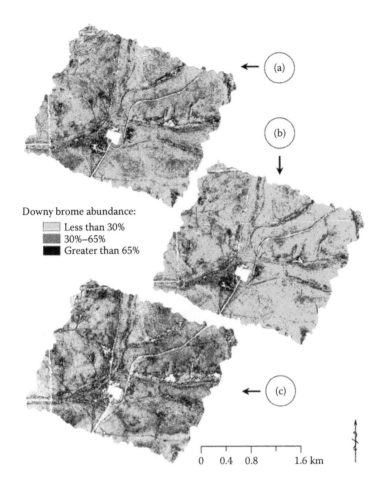

FIGURE 11.3
The maps of downy brome (*Bromus tectorum*) abundance predicted by the analysis of three different datasets: (a) multitemporal spectral stack, (b) July 2000 spectral data, and (c) May 2003 data. The overall accuracy coefficients for the three downy brome occurrence maps were 0.81 for the multitemporal dataset, and 0.70 and 0.72, respectively, for the 2000 and 2003 datasets. (Reproduced from Noujdina, N.V., and S.L. Ustin, *Weed Science*, 56, 173, 2008. With permission.)

from drier/warmer to wetter/cooler conditions, the authors successfully identified the basic biological mechanisms favoring the spread of an invasive tree species and provided a better understanding of how vegetation–climate interactions affect plant growth during the invasion process.

It is generally known that most understory invasive species are hard to detect and mapped by remote sensing since they could be completely hidden by overstory canopy. However, there might be a temporal window when a clear phenological difference exists between native overstory species and understory invaders. Wilfong et al. (2009) successfully detected the distribution of an understory invasive shrub, Amur honeysuckle (*L. maackii*), in the deciduous forest in southwestern Ohio using phenological difference between Amur honeysuckle and co-occurring native tree species in the canopy. In this case, the invading shrub leafs out earlier in the spring and retains leaves longer in fall than native deciduous species. Therefore, the best acquisition window for remote sensing is the early spring and late fall when native deciduous species are leafless.

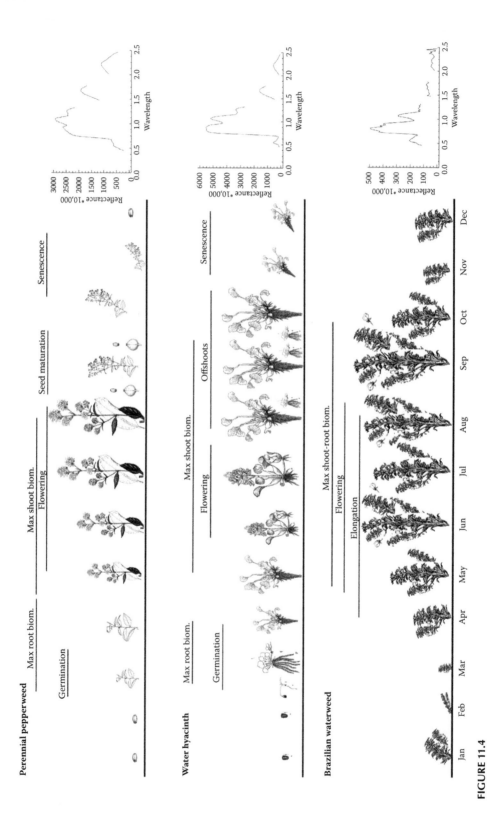

FIGURE 11.4

The annual life cycle of three invasive species and respective June spectra extracted from HyMap imagery using ground reference. (Reproduced from Hestir et al., *Remote Sensing of Environment*, 112, 4034, 2008. With permission.)

All these case studies discussed earlier allow us to conclude that using time series to incorporate phenological events in species invasion research is very effective and more accurate in detecting and mapping the spatial extent of invasive species despite its higher cost of image acquisition at times.

11.3 Summary and Future Directions

We have discussed the strength and opportunities of using imaging spectroscopy in detecting and mapping the spatial spread of invasive species in this chapter. We provided detailed case studies that show the utility of imaging spectroscopy with different operational approaches applied to diverse ecosystems in plant invasion research. It is evident that imaging spectroscopy holds great promise for invasion research in various spatial and temporal scales. Based on current study results, it is important and critical that future interdisciplinary research among ecologists and geographers should focus on developing a systematic way of understanding the spectral signature of individual species across all types of ecosystems. As suggested by Asner and Martin (2009), a fundamental understanding of the linkage between species biochemistry and spectral diversity can facilitate the mapping of species distribution even in the tropics. For this purpose, optimal spectral and spatial resolutions and their relation to classification algorithms should be thoroughly tested and identified. Furthermore, species' chemical composition (such as leaf pigments and water content) and leaf biophysical properties should be thoroughly studied and well understood at the same time. Recent development in satellite and sensor technology can further improve spectral and spatial resolutions for detecting invasive species distribution remotely. In addition, optimal seasons or dates for image acquisition should be determined by using multitemporal spectra of targeted areas incorporating information provided by species phenological characteristics. Finally, image classification methods can be further improved by using more intricate statistical methods and algorithms.

Acknowledgments

Kate S. He is supported in part by a grant from the National Science Foundation (DUE 1028125). Duccio Rocchini is supported by the EU BON (Building the European Biodiversity Observation Network) project, funded by the European Union under the 7th Framework programme, Contract No. 308454.

References

Ahern, R. G., D. A. Landis, A. A. Reznicek, and D. W. Schemske. 2010. "Spread of Exotic Plants in the Landscape: The Role of Time, Growth Habit, and History of Invasiveness." *Biological Invasions* 12:3157–69.

Almeida, T., and F. De Souza. 2004. "Principal Component Analysis Applied to Feature-Oriented Band Ratios of Hyperspectral Data: A Tool for Vegetation Studies." *International Journal of Remote Sensing* 25:5005–23.

Andrew, M. E., and S. L. Ustin. 2006. "Spectral and Physiological Uniqueness of Perennial Pepperweed (*Lepidium Latifolium*)." *Weed Science* 54:1051–62.

Andrew, M. E., and S. L. Ustin. 2008. "The Role of Environmental Context in Mapping Invasive Plants with Hyperspectral Image Data." *Remote Sensing of Environment* 112:4301–17.

Andrew, M. E., and S. L. Ustin. 2009. "Habitat Suitability Modeling of an Invasive Plant with Advanced Remote Sensing Data." *Diversity and Distribution* 15(4):1–14.

Asner, G. P., M. O., Jones, R. E. Martin, D. E. Knapp, and R. F. Hughes. 2008a. "Remote Sensing of Native and Invasive Species in Hawaiian Forests." *Remote Sensing of Environment* 112:1912–26.

Asner, G. P., D. E. Knapp, T. Kennedy-Bowdoin, M. O. Jones, R. E. Martin, J. Boardman, and R. F. Hughes. 2008b. "Invasive Species Detection in Hawaiian Rainforests Using Airborne Imaging Spectroscopy and LiDAR." *Remote Sensing of Environment* 112:1942–55.

Asner, G. P., and R. E. Martin. 2009. "Airborne Spectranomics: Mapping Canopy Chemical and Taxonomic Diversity in Tropical Forests." *Frontiers in Ecology and the Environment* 7(5):269–76.

Asner, G. P., R. E. Martin, K. M. Carlson, U. Rascher, and M. Vitousek. 2006. "Vegetation-Climate Interactions among Native and Invasive Species in Hawaiian Rainforest." *Ecosystems* 9:1106–17.

Bachmann, C. M., T. F. Donato, G. M. Lamella, W. J. Rhea, M. H. Bettenhausen, and R. A. Fusina. 2002. "Automatic Classification of Land Cover on Smith Island, VA, Using HyMAP Imagery." *IEEE Transactions on Geoscience and Remote Sensing* 40:2313–30.

Bajcsy, P., and P. Groves. 2004. "Methodology for Hyperspectral Band Selection." *Photogrammetric Engineering & Remote Sensing* 70:793–802.

Blossey, B., and R. Nötzold. 1995. "Evolution of Increased Competitive Ability in Invasive Non-Indigenous Plants: A Hypothesis." *Journal of Ecology* 83:887–89.

Boyd, D. S., and G. M. Foody. 2010. "An Overview of Recent Remote Sensing and GIS Based Research in Ecological Informatics." *Ecological Informatics* 6(1):26–36.

Bradley, B. A., and J. F. Mustard. 2005. "Identifying Land Cover Variability Distinct from Land Cover Change: Cheatgrass in the Great Basin." *Remote Sensing of Environment* 94:204–13.

Bunting, P., and R. Lucas. 2006. "The Delineation of Tree Crowns in Australian Mixed Species Forests using Hyperspectral Compact Airborne Spectrographic Imager (CASI) Data." *Remote Sensing of Environment* 101:230–48.

Callaway, R. M., and W. M. Ridenour. 2004. "Novel Weapons: Invasive Success and the Evolution of Increased Competitive Ability." *Frontiers in Ecology and the Environment* 2:436–43.

Carlson, K., G. Asner, R. Hughes, R. Ostertag, and R. Martin. 2007. "Hyperspectral Remote Sensing of Canopy Biodiversity in Hawaiian Lowland Rainforests." *Ecosystems* 10(4):536–49.

Carter, G. A., K. L. Lucas, G. A. Blossom, C. L. Lassitter, D. M. Holiday, D. S. Mooneyhan, D. R. Fastring, T. R. Holcombe, and J. A. Griffith. 2009. "Remote Sensing and Mapping of Tamarisk along the Colorado River, USA: A Comparative Use of Summer-Acquired Hyperion, Thematic Mapper and QuickBird Data." *Remote Sensing* 1:318–29.

Catford, J. A., R. Jansson, and C. Nilsson. 2009. "Reducing Redundancy in Invasion Ecology by Integrating Hypotheses into a Single Theoretical Framework." *Diversity and Distributions* 15:22–40.

Clark, M. L., D. A. Roberts, and D. B. Clark. 2005. "Hyperspectral Discrimination of Tropical Rain Forest Tree Species at Leaf to Crown Scales." *Remote Sensing of Environment* 96:375–98.

Colautti, R. I., A. Ricciardi, I. A. Grigorovich, and H. J. MacIsaac. 2004. "Is Invasion Success Explained by the Enemy Release Hypothesis?" *Ecology Letters* 7: 721–33.

Curran, P. J. 1989. "Remote Sensing of Foliar Chemistry." *Remote Sensing of Environment* 30:271–8.

Davis, M. A., J. P. Grime, and K. Thompson. 2000. "Fluctuating Resources in Plant Communities: A General Theory of Invasibility." *Journal of Ecology* 88: 528–34.

DiPietro, D. Y. 2002. "Mapping the Invasive Plant *Arundo Donax* and Associated Riparian Vegetation Using Hyperspectral Remote Sensing." Master's Thesis, University of California, Davis.

Duda, R. O., and P. E. Hart. 1973. *Pattern Classification and Scene Analysis*. New York: John Wiley and Sons.

Elton, C. S. 1958. *The Ecology of Invasions by Animals and Plants*. London: Methuen.

Evangelista, P. H., T. J. Stohlgren, J. T. Morisette, and S. Kumar. 2009. "Mapping Invasive Tamarisk (*Tamarix*): A Comparison of Single-Scene and Time-Series Analyses of Remotely Sensed Data." *Remote Sensing* 1:519–33.

Ge, S., J. Everitt, R. Carruthers, P. Gong, and G. Anderson. 2006. "Hyperspectral Characteristics of Canopy Components and Structure for Phenological Assessment of an Invasive Weed." *Environmental Monitoring and Assessment* 120:109–26.

Gillespie, T. W., G. M. Foody, D. Rocchini, A. P. Giorgi, and S. Saatchi. 2008. "Measuring and Modeling Biodiversity from Space." *Progress in Physical Geography* 32(2):203–21.

Glenn, N. F., J. T. Mundt, K. T. Weber, T. S. Prather, L. W. Lass, and J. Pettingill. 2005. "Hyperspectral Data Processing for Repeat Detection of Small Infestations of Leafy Spurge." *Remote Sensing of Environment* 95:399–412.

Green, A. A., M. Berman, P. Switzer, and M. D. Craig. 1988. "A Transformation for Ordering Multispectral Data in Terms of Image Quality with Implications for Noise Removal." *IEEE Transactions on Geoscience and Remote Sensing* 26:65–74.

Hamada, Y., D. A., Stow, L. L. Coulter, J. C. Jafolla, and L. W. Hendricks. 2007. "Detecting Tamarisk Species (*Tamarix* spp.) in Riparian Habitats of Southern California Using High Spatial Resolution Hyperspectral Imagery." *Remote Sensing of Environment* 109:237–48.

He, K. S., D. Rocchini, M. Neteler, and H. Nagendra. 2011. "Benefits of Hyperspectral Remote Sensing for Tracking Plant Invasions." *Diversity and Distributions* 17:381–92.

Henderson, S., T. P. Dawson, and R. J. Whittaker. 2006. "Progress in Invasive Plants Research." *Progress in Physical Geography* 30:25–46.

Hestir, E. L., S. Khanna, M. E. Andrew, M. J. Santos, J. H. Viers, J. A. Greenberg, S. S. Rajapakse, and S. L. Ustin. 2008. "Identification of Invasive Vegetation Using Hyperspectral Remote Sensing in the California Delta Ecosystem." *Remote Sensing of Environment* 112:4034–47.

Hierro, J. L, J. L. Maron, and R. M. Callaway. 2005. "A Biogeographical Approach to Plant Invasions: The Importance of Studying Exotics in Their Introduced and Native Range." *Journal of Ecology* 93:5–15.

Hughes, F. R., and J. S. Denslow. 2005. "Invasion by a N_2-Fixing Tree Alters Function and Structure in Wet Lowland Forests of Hawaii." *Ecological Applications* 15:1615–28.

Joshi, J., and K. Vrieling. 2005. "The Enemy Release and EICA Hypothesis Revisited: Incorporating the Fundamental Difference between Specialist and Generalist Herbivores." *Ecology Letters* 8:704–14.

Keane, R. M., and M. J. Crawley. 2002. "Exotic Plant Invasion and the Enemy Release Hypothesis." *Trends in Ecology and Evolution* 17(4):164–70.

Kruse, F. A., A. B. Lefkoff, J. W. Boardman, K. B. Heidebrecht, A. T. Shapiro, P. J. Barloon, and A. F. H. Goetz. 1993. "The Spectral Image-Processing System (SIPS)-Interactive Visualization and Analysis of Imaging Spectrometer Data." *Remote Sensing of Environment* 44:145–63.

Khanna, S., M. J. Santos, E. L. Hestir, and S. L. Ustin. 2012. "Plant Community Dynamics Relative to the Changing Distribution of a Highly Invasive Species, *Eichhornia crassipes*: A Remote Sensing Perspective." *Biological Invasions* 14:717–33.

Laba, M., F. Tsai, D. Ogurcak, S. Smith, and M. E. Richmond. 2005. "Field Determination of Optimal Dates for the Discrimination of Invasive Wetland Plant Species Using Derivative Spectral Analysis." *Photogrammetric Engineering and Remote Sensing* 71(5):603–11.

Laba, M., R. Downs, S. Smith, S. Welsh, C. Neider, S. White, M. Richmond, W. Philpot, and P. Baveye. 2008. "Mapping Invasive Wetland Plants in the Hudson River National Estuarine Research Reserve Using QuickBird Satellite Imagery." *Remote Sensing of Environment* 112:286–300.

Lass, W. L., T. S. Prather, N. F. Glenn, K. T. Weber, J. T. Mundt, and J. Pettingill. 2005. "A Review of Remote Sensing of Invasive Weeds and Example of Early Detection of Spotted Knapweed (*Centaurea Maculosa*) and Babysbreath (*Gypsophila paniculata*) with a Hyperspectral Sensor." *Weed Science* 53:242–51.

Lawrence, R. L., S. D. Wood, and R. L. Sheley. 2006. "Mapping Invasive Plants Using Hyperspectral Imagery and Breiman Cutler Classifications (Random Forest)." *Remote Sensing of Environment* 100:356–62.

Levine, J., and C. M. D'Antonio. 1999. "Elton Revisited: A Review of Evidence Linking Diversity and Invasibility." *Oikos* 87:15–26.

Lockwood, J. L., P. Cassey, and T. Blackburn. 2005. "The Role of Propagule Pressure in Explaining Species Invasions." *Trends in Ecology & Evolution* 20:223–8.

Lonsdale, W. M. 1999. "Global Patterns of Plant Invasions and the Concept of Invasibility." *Ecology* 80:1522–36.

Lopez, R. D., C. M. Edmonds, A. C. Neale, T. S. Slonecker, K. B. Jones, and D. T. Heggem. 2004. "Accuracy Assessments of Airborne Hyperspectral Data for Mapping Opportunistic Plant Species in Freshwater Coastal Wetlands." In *Remote Sensing and GIS Accuracy Assessment*, edited by R. S. Lunetta and J. G. Lyon. New York: CRC Press, 318–39.

MacArthur, R. H. 1970. "Species Packing and Competitive Equilibrium for Many Species." *Theoretical Population Biology* 1:1–11.

Mack, R. N., D. Simberloff, W. M. Lonsdale, H. Evans, M. Clout, and F. A. Bazzaz. 2000. "Biotic Invasions: Causes, Epidemiology, Global Consequences, and Control." *Ecological Application* 10:689–710.

Maron, J. L., and M. Vila. 2001. "When do Herbivores Affect Plant Invasions? Evidence for the Natural Enemies and Biotic Resistance Hypotheses." *Oikos* 95:361–73.

Martin, M. E., and J. D. Aber. 1997. "High Spectral Resolution Remote Sensing of Forest Canopy Lignin, Nitrogen, and Ecosystem Processes." *Ecological Applications* 7:431–43.

Melgani, F., and L. Bruzzone. 2004. "Classification of Hyperspectral Remote Sensing Images with Support Vector Machines." *IEEE Transactions on Geoscience and Remote Sensing* 42(8):1778–90.

Miao, X., P. Gong, S. Swope, R. Pu, R. Carruthers, G. L. Anferson, C. R. Heaton, and C. R. Tracy. 2006. "Estimation of Yellow Starthistle Abundance through CASI-2 Hyperspectral Imagery Using Linear Spectral Mixture Models." *Remote Sensing of Environment* 101(3):329–41.

Nagendra, H., and D. Rocchini. 2008. "High Resolution Satellite Imagery for Tropical Biodiversity Studies: The Devil Is in the Detail." *Biodiversity and Conservation* 17:3431–42.

Narumalani, S., D. R. Mishra, R. Wilson, P. Reece, and A. Kohler. 2009. "Detecting and Mapping Four Invasive Species along the Floodplain of North Platte River, Nebraska." *Weed Technology* 23:99–107.

Noujdina, N. V., and S. L. Ustin. 2008. "Mapping Downy Brome (*Bromus Tectorum*) Using Multidate AVIRIS Data." *Weed Science* 56:173–79.

Oldeland, J., D. Wesuls, D. Rocchini, M. Schmidt, and N. Jurgens. 2010. "Does Using Species Abundance Data Improve Estimates of Species Diversity from Remotely Sensed Spectral Heterogeneity?" *Ecological Indicators* 10:390–96.

Pengra, B. W., C. A. Johnston, and T. R. Loveland. 2007. "Mapping an Invasive Plant, *Phragmites australis*, in Coastal Wetlands Using the EO-1 Hyperion Hyperspectral Sensor." *Remote Sensing of Environment* 108:74–81.

Peterson, E. B. 2005. "Estimating Cover of an Invasive Grass (*Bromus tectorum*) Using Tobit Regression and Phenology Derived from Two Dates of Landsat ETM= Data." *International Journal of Remote Sensing* 26:2491–507.

Plaza, A., J. A. Benediktsson, J. W. Boardman, J. Brazile, L. Bruzzone, G. Camps-Valls, et al. 2009. "Recent Advances in Techniques for Hyperspectral Image Processing." *Remote Sensing of Environment* 113:S110–22.

Pu, R., P. Gong, Y. Tian, X. Miao, R. I. Carruthers, and G. L. Anderson. 2008. "Invasive Species Change Detection Using Artificial Neural Networks and CASI Hyperspectral Imagery." *Environmental Monitoring and Assessment* 40:15–32.

Pyšek, P., M. Krivanek, V. Jarosik. 2009. "Planting Intensity, Residence Time, and Species Traits Determine Invasion Success of Alien Woody Species." *Ecology* 90(10):2734–44.

Pyšek, P., and J. Vojtech. 2005. "Residence Time Determines the Distribution of Alien Plants." In *Invasive Plants: Ecological and Agricultural Aspects*, edited by Inderjit. Basel: Birkhauser Verlag, 77–96.

Pyšek, P., and D. M. Richardson. 2006. "The Biogeography of Naturalization in Alien Plants." *Journal of Biogeography* 33:2040–50.

Pyšek, P., and D. M. Richardson. 2010. "Invasive Species, Environmental Change and Management, and Ecosystem Health." *Annual Review of Environment and Resources* 35:25–55.

Rejmánek, M. 2000. "Invasive Plants: Approaches and Predictions." *Austral Ecology* 25:497–506.

Rejmánek, M., and D. M. Richardson. 1996. "What Attributes Make Some Plant Species More Invasive?" *Ecology* 77(6):1655–61.

Richardson, D. M., and P. Pyšek. 2006. "Plant Invasions: Merging the Concepts of Species Invasiveness and Community Invasibility." *Progress in Physical Geography* 30:409–31.

Sakai, A. K., F. W. Allendorf, J. S. Holt, D. M. Lodge, J. Molofsky, A. Kimberly, et al. 2001. "The Population Biology of Invasive Species." *Annual Review of Ecology and Systematics* 32:305–32.

Schaepman, M. E., S. L. Ustin, A. J. Plaza, T. H. Painter, J. Verrelstand, and S. Liang. 2009. "Earth System Science Related Imaging Spectroscopy: An Assessment." *Remote Sensing of Environment* 113:S123–37.

Shanmugam, P., Y. Ahn, and S. Sanjeevi. 2006. "A Comparison of the Classification of Wetland Characteristics by Linear Spectral Mixture Modelling and Traditional Hard Classifiers on Multispectral Remotely Sensed Imagery in Southern India." *Ecological Modelling* 194(4):394–97.

Singh, N., and N. F. Glenn. 2009. "Multitemporal Spectral Analysis for Cheatgrass (*Bromus tectorum*) Classification." *International Journal of Remote Sensing* 30(13):3441–62.

Somers, B., and G. P. Anser. 2012. "Hyperspectral Time Series Analysis of Native and Invasive Species in Hawaiian Rainforests." *Remote Sensing* 4:2510–29.

Townsend, A. R., G. P. Asner, and C. C. Cleveland. 2008. "The Biogeochemical Heterogeneity of Tropical Forests." *Trends in Ecology and Evolution* 23:424–31.

Townshend, J., C. Justice, W. Li, C. Gurney, and J. McManus. 1993. "Global Land Cover Classification by Remote Sensing: Present Capabilities and Future Possibilities." *Remote Sensing of Environment* 35(2–3):243–55.

Turner, W., S. Spector, N. Gardiner, M. Fladeland, E. Sterling, and M. Steininger. 2003. "Remote Sensing for Biodiversity Science and Conservation." *Trends in Ecology and Evolution* 18:306–14.

Underwood, E. C., and S. L. Ustin. 2007. "A Comparison of Spatial and Spectral Image Resolution for Mapping Invasive Plants in Coastal California." *Environmental Management* 39:63–83.

Underwood, E., S. Ustin, and D. DiPietro. 2003. "Mapping Nonnative Plants Using Hyperspectral Imagery." *Remote Sensing of Environment* 86:150–61.

Ustin, S. L., D. DiPietro, K. Olmstead, E. Underwood, and G. J. Scheer. 2002. "Hyperspectral Remote Sensing for Invasive Species Detection and Mapping." In *2002 IEEE International Geoscience and Remote Sensing Symposium, June 24–28. Vol 3*, 1658–60. Toronto, Canada: IEEE International.

Ustin, S. L., and J. A. Gamon. 2010. "Remote Sensing of Plant Functional Types." *New Phytologist* 186(4):795–816.

Vitousek, P. M., C. M. D'Antonio, and G. P. Asner. 2011. "Invasions and Ecosystems: Vulnerabilities and the Contribution of New Technologies." In *Fifty Years of Invasion Ecology: The Legacy of Charles Elton*, edited by D. M. Richardson, 277–88. Oxford: Blackwell Publishing.

Walsh, S. J., A. L. McCleary, C. F. Mena, S. Yang, J. P. Tuttle, A. Gonzalez, and R. Atkinson. 2008. "QuickBird and Hyperion Data Analysis of an Invasive Plant Species in the Galapagos Islands of Ecuador: Implications for Control and Land Use Management." *Remote Sensing of Environment* 112:1927–41.

Wilfong, B. N., D. L. Gorchov, and M. C. Henry. 2009. "Detecting an Invasive Shrub in Deciduous Forest Understories Using Remote Sensing." *Weed Science* 57:512–20.

Williams, A. P. and E. R. Hunt. 2004. "Accuracy Assessment for Detection of Leafy Spurge with Hyperspectral Imagery." *Journal of Range Management* 57:106–12.

Williamson, M., K. Dehnen-Schmutz, I. Kühn, M., Hill, S. Klotz, A. Milbau, J. Stout, and P. Pyšek. 2009. "The Distribution of Range Sizes of Native and Alien Plants in Four European Countries and the Effects of Residence Time." *Diversity and Distributions* 15:158–66.

Wilson, J. R. U., E. E. D. Dormontt, P. J. Prentis, A. J. Lowe, and D. M. Richardson. 2009. "Something in the Way You Move: Dispersal Pathways Affect Invasion Success." *Trends in Ecology and Evolution* 24(3):136–44.

Wilson, J. R., D. M. Richardson, M. Rouget, S. Proches, A. A. Amis, L. Henderson, and W. Thuiller. 2007. "Residence Time and Potential Range: Crucial Considerations in Modeling Plant Invasions." *Diversity and Distributions* 13(1):11–22.

12

Assessing Military Training–Induced Landscape Fragmentation and Dynamics of Fort Riley Installation Using Spatial Metrics and Remotely Sensed Data

Steve Singer, Guangxing Wang, Heidi R. Howard, and Alan B. Anderson

CONTENTS

12.1 Introduction 207
12.2 Study Area and Datasets 208
12.3 Methods 211
12.4 Results 213
12.5 Conclusions and Discussion 219
Acknowledgments 220
References 220

12.1 Introduction

The U.S. Army has the responsibility of managing over 5,500 training sites that cover more than 12 million ha of land for various training programs ranging from small fire to vehicular combat maneuvers (DEPARC 2007). These training exercises inevitably cause degradation of landscape quality on the bases by reducing vegetation cover, increasing soil erosion, and fragmenting landscape. These degraded landscapes restrict not only the environment, but also the Army's land carrying capacity and readiness. Thus, assessing landscape quality at military installations and monitoring its dynamics is becoming very important. This study is aimed at demonstrating a method to assess and monitor landscape fragmentation for U.S. installations.

Spatial or landscape metrics are attributes and algorithms that are used to characterize environmental changes of landscapes and indicate whether the changes are positive or negative in terms of environmental health (McGarigal et al. 2002). So far hundreds of spatial metrics have been developed and applied, which has led to difficulties in selecting the most appropriate spatial metrics for a specific study.

First of all, scale is an interesting issue when discussing spatial metrics. One prefers the metrics that are sensitive to landscape changes but insensitive to different scales. When comparing maps at different scales, it would be convenient if the metrics were insensitive to scale changes, but studies have shown that the spatial metrics are very sensitive to scale (O'Neil et al. 1999). Over extreme scales, the metrics are obviously very sensitive, but at a more common scale such as pixel sizes of 10–100 m, they seem to be relatively insensitive, although the results vary depending on the patterns and specific metrics used (O'Neil et al. 1999). Further solidifying the importance of scale has three key points: (1) explicitly defining the scale of the study area, (2) describing patterns relative to the scale of the study area, and (3) being very cautious of comparisons of landscapes at different scales (McGarigal et al. 2002).

Second, redundancy among spatial metrics is another issue. Although redundancy should be expected with the number of spatial metrics, it has been found to happen too often (Hargis et al. 1998). For example, edge density and contagion are both highly correlated. Contagion quantifies tendencies of patches to occur in large aggregated distributions, whereas edge density is the length of patch edge per unit area within each landscape (McGarigal et al. 2002). Although the results of the two spatial metrics are inversely correlated, the same conclusions can be seen from both measurements (Hargis et al. 1998).

Moreover, spatial metrics should be selected to capture different, independent aspects in spatial patterns of landscapes (O'Neil et al. 1999). For instance, Riiters et al. (1995) investigated 55 metrics but concluded that those metrics could be generalized into six categories: average perimeter–area ratio, contagion, standardized patch shape, patch perimeter–area scaling, number of attribute classes, and large-patch density–area scaling.

In addition, the features of a landscape can be generally quantified at three levels including patch, class, and landscape (Narumalani et al. 2004). Thus, selection of spatial metrics is definitely dependent on the analysis level at which landscape quality is assessed. This study intends to select, compare, and integrate spatial metrics that can be used to quantify landscape fragmentation owing to military training activities. The spatial metrics chosen for this study focuses on fragmentation at all three levels. Patch perimeter–area ratio and contagion are important for all temperate zone ecosystems. Moreover, fractal dimension and cover types must also be included (O'Neil et al. 1999). Fractal dimension measures complexities within landscapes. Fractal geometry proposed by Mandelbrot (1977) can be used to describe complex forms of geometry such as coastlines and landscapes (Sun et al. 2006). The main advantage of fractal analysis is the ability to use continuous dimensions unlike the traditional Euclidean. Euclidean dimensions are integers (lines 1, areas 2, and volumes 3), whereas fractal dimensions have a range. For example, a line will have a dimension ranging from 1 to 2, and it becomes more of a surface as its dimension (D) gets closer to 2 (Sun et al. 2006).

The objective of this study is to assess and monitor landscape fragmentation owing to military training activities using spatial metrics and remotely sensed data and further analyze their potential as indicators of landscape quality measures for a large U.S. Army installation at Fort Riley, Kansas. This study expectedly answers the following questions: (1) Are higher spatial resolution India remote sensing (IRS) images better than Landsat Thematic Mapper™ images for landscape segmentation, land use and land cover (LULC) classification, and further assessment of landscape quality? (2) What spatial metrics can be used to quantify the landscape fragmentation of the Fort Riley installation caused by military training?

12.2 Study Area and Datasets

This study was conducted in Fort Riley, Kansas. Fort Riley is located in the northern Flint Hills of Kansas and has a tall grass prairie landscape. The base is 41,154 ha and dominated by prairie grasses, with shrubs and woodlands existing along the streams (Figure 12.1a). The landscape composition varies over the study area. The eastern portion of the base has a limestone and shale substrate dominated by perennial forbs and grasses, and the western portion has less relief and deeper soils resulting in more woodlands (Quist et al. 2002).

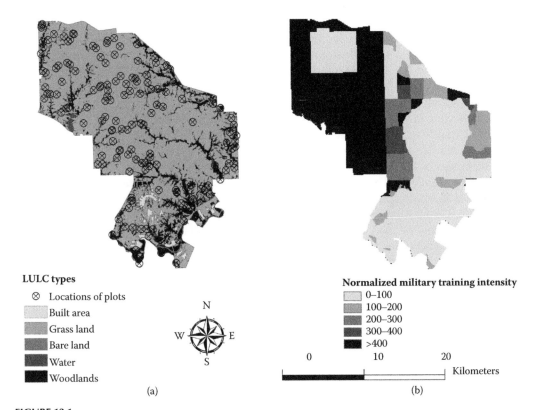

LULC types

⊗ Locations of plots

Built area

Grass land

Bare land

Water

Woodlands

N
W ⊕ E
S

(a)

Normalized military training intensity

0–100
100–200
200–300
300–400
>400

0   10   20
                              Kilometers

(b)

FIGURE 12.1
(a) Land use and land cover types and Range and Training Land Assessment plots used to measure various environmental characteristics such as ground cover, canopy cover, disturbance, and so on and (b) normalized military training intensity (total training days per year per unit) for Fort Riley.

A succession of native plants has been restored since the 1960s as a result of the Army's restoration of the study area (Quist et al. 2002).

A variety of military training activities have taken place at Fort Riley since the 1950s, and plot data from the Range and Training Land Assessment (RTLA) program has been documented (Figure 12.1a). As an example, Figure 12.1b shows the spatial distribution of 1998's military training intensity measured as total training days per year and normalized using training area. There was one impact area located slightly at the east portion of the study area, where no training activities took place. A total of 154 permanent plots were placed at Fort Riley in the form of stratified random sampling and remeasured annually between 1989 and 2001 (Figure 12.1a). The sizes of the plots were 100 m × 6 m with a 100 m transect line along the central line. Within each plot, measurements of vegetation height, ground cover, and canopy cover were collected on an annual basis using a point-intercept method with 1 m intervals (Diersing et al. 1992). Moreover, the type of LULC for each of the plots was recorded and used in landscape segmentation and classification and in result assessment of LULC categories.

There have been some reports on the impact of military training on land condition and assessment of individual environmental functions at Fort Riley (Anderson et al. 2005; Gertner et al. 2002, 2004; Singer 2010; Singer et al. 2012; Wang et al. 2004, 2007). Thus, various datasets were available, including military training intensity, vegetation maps, data for burned areas, and soil erosion maps. In addition, Landsat TM images, IRS satellite images, and historical aerial photographs were also available.

In this study, Landsat TM images acquired for each of the years from 1989 to 2001 and IRS images for the year 1999 were used (Singer 2010; Singer et al. 2012; Wang et al. 2007, 2008). The TM images have a spatial resolution of 30 m × 30 m and consist of channels band 1, 0.45–0.53 µm; band 2, 0.52–0.60 µm; band 3, 0.63–0.69 µm; band 4, 0.76–0.90 µm; band 5, 1.55–1.75 µm; and band 7, 2.08–2.35 µm. Moreover, a set of color infrared orthophotos with a ground resolution of 0.5 m was acquired on July 20, 2004. Using ground control points, the orthophotos were georeferenced to Universal Transverse Mercator (UTM) and put together as a mosaic that covered the study area. The mosaic was used as a base image, and the 1999 cloud-free TM images were georeferenced to this base image (Singer 2010; Singer et al. 2012). The TM images of all other years were then co-registered to the 1999 TM images (Singer 2010; Singer et al. 2012). The root-mean-square error (RMSE) for geometric rectification of the photos was smaller than 3.33 m. The RMSE for georeferencing the TM images varied from 4.61 to 9.33 m. Moreover, radiometric rectification of the TM images for atmospheric effects was done using the Hall et al. (1991) algorithm. In this method, deep water bodies—reservoirs as black bodies and concrete areas like the airport as white bodies—were identified and used to calculate linear transforms relating digital count values among the images.

Moreover, the IRS imagery consisted of multispectral images, including a green channel (0.52–0.59 µm), a red channel (0.62–0.68 µm), a near-infrared channel (0.77–0.86 µm), and panchromatic image (0.5–0.75 µm). The multispectral channels had a spatial resolution of 23 m × 23 m, and the panchromatic image had a pixel size of 5 m × 5 m. In this study, the IRS image was sharpened to a spatial resolution of 5 m × 5 m. The geometric and radiometric correction of the IRS image was also conducted. As examples, a TM color composite image, IRS color composite image, and the aerial orthophoto are shown in Figure 12.2a

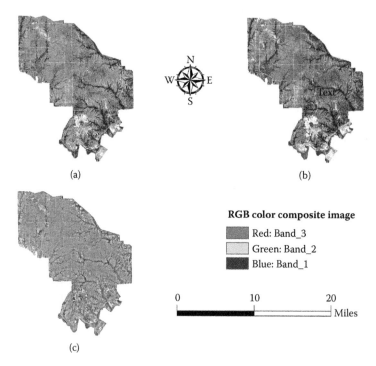

FIGURE 12.2
(**See color insert.**) (a) India remote sensing image for year 1999, (b) Landsat TM image for year 2000, and (c) aerial orthophoto for year 2004 for Fort Riley.

through c, respectively. In addition, the data for burned areas were collected and used to assess LULC classification. The burned areas could be seen in the north and east portions of the Landsat TM image for the installation (Figure 12.2b).

12.3 Methods

This study was conducted by integrating remote sensing and landscape modeling methods. The used methods included landscape segmentation, LULC classification, and selection and calculation of spatial metrics (Singer 2010; Singer et al. 2012). Landscape segmentation consisted of image processing and segmentation. For image processing, the key is to develop image transformations and determine the most appropriate spatial resolution (i.e., pixel size).

Extracting the spatial patterns of landscape elements from images is greatly dependent on spatial resolution of the used images, and some spatial patterns that can be captured at a pixel size may disappear at another (Wang et al. 2008). In this study, the IRS multispectral image (G, green band; R, red band; and NIR, near infrared band) for the year 1999 was sharpened using ERDAS Imagine's Brovey transformation (BT) (ERDAS 2012; Jensen 2005; Singer 2010; Singer et al. 2012).

$$I = NIR + R + G \qquad (12.1)$$

$$\begin{bmatrix} NIR_{BT} \\ R_{BT} \\ G_{BT} \end{bmatrix} = \frac{Pan}{I} \begin{bmatrix} NIR \\ R \\ G \end{bmatrix} \qquad (12.2)$$

where Pan means the IRS panchromatic image at the pixel size of 5 m × 5 m. The sharpened IRS multispectral image had a resolution of 5 m × 5 m and was then scaled up to pixel sizes of 10 m × 10 m, 15 m × 15 m, 20 m × 20 m, 25 m × 25 m, and 30 m × 30 m using the degrading function of ERDAS Imagine (ERDAS 2012). The degrading function calculates and assigns average values from smaller pixels to larger pixels. The principal component analysis (PCA) of the obtained IRS images at each of the spatial resolutions was further carried out. The objective of conducting PCA was to reduce the dimensionality of image feature space and duplication of information (Jensen 2005). For each of the years from 1989 to 2001, BT of the Landsat TM image was also performed on the basis of Equations 12.1 and 12.2 and at the same time using red, green, and blue bands to replace near infrared, red, and green. The results were images at a spatial resolution of 5 m × 5 m. The PCA of both the original and transformed TM image for each year was then conducted.

Landscape segmentation was conducted using IDRISI Taiga Edition (IDRISI 2012). IDRISI Taiga has superior ability for segmenting and classifying images through three functions: SEGMENT, SEGTRAIN, and SEGCLASS. IDRISI's image segmentation is a process by which pixels are classified into homogeneous polygons based on spectral similarity quantified using variance of pixel values within a moving window—a user-defined filter. Segments are determined according to a user-defined threshold. The segmentation module first creates a variance image using the moving window. The more homogeneous the pixels are within the window, the lower the variance values. The pixels at the boundaries of homogeneous regions will naturally have higher variance values than those within homogeneous polygons. Based on the variance values, pixels are grouped into one watershed, image segment, and assigned an ID. The image segments are then merged to form new segments if their differences are smaller than the given threshold value. A smaller threshold produces smaller polygons within which the pixels are more homogenous, whereas a larger threshold leads to larger polygons within which the pixels are more heterogeneous.

In this study, a window of 3 × 3 pixels was used, and different segment thresholds were researched for each spatial resolution of the used images. The smaller the threshold value, the more time required to run the segmentation. Moreover, the finer the spatial resolution, the more storage needed. Thus, both threshold value and spatial resolution became important factors. As the resolutions of images became coarser, the required storage for segmentation decreased dramatically. A sufficient segmentation threshold was found for each resolution.

After segmentation, spectral training profiles were created for each class. With IDRISI, this process is relatively simple and easy as each of the segmented images is simply overlaid on a RGB composite image. Using the composite image as the bottom layer and the segmented image as the top layer, the segments were identified for profiles to be used for classification later. Nearly a total of 100 signatures were created for each image to acquire a sufficient range for each of the LULC categories that were identified based on the composite image, old LULC map, and higher resolution IRS panchromatic image. The profile signature files were then used for classification. IDRISI has many classification algorithms (IDRISI 2012). In this research, both maximum likelihood and SEGCLASS were used. Maximum likelihood creates a more pixilated map showing transitions between polygons, whereas SEGCLASS creates more smoothed maps with immediate changes between polygons. The spectral classes of the obtained classification images were then assigned LULC categories.

The LULC categories used include water, grassland, trees, bare ground, and built-up areas. Accuracy assessments of classification maps were done in ERDAS Imagine using the LULC information of RTLA plots, 1989's LULC map (Figure 12.1), and high-resolution IRS sharpened images by visual interpretation. Error matrices that summarize the correct classification and corresponding omission and commission errors were obtained. In addition, the Kappa values that measure the agreement between an image-derived classification map and classification obtained by chance matching were calculated (Jensen 2005).

The classified images were then exported in an ASCII file format and input into Fragstats (McGarigal et al. 2002; Singer 2010; Singer et al. 2012), in which spatial metrics were calculated to quantify landscape fragmentation (Riiters et al. 1995). The spatial metrics should be selected considering the following criteria: (1) they can characterize the whole landscape structure and play important roles in assessment of comprehensive environmental and landscape quality; (2) they are able to account for most of the landscape variability and have low correlation among them so that the duplication of information can be minimized; (3) they are easily calculated and have the capability to be integrated with other environmental factors because environment is a comprehensive concept.

Based on the aforementioned criteria, two metrics at the landscape level were chosen, including interspersion and juxtaposition index (IJI) and landscape shape index (LSI). IJI is a measure that explicitly takes the spatial configuration of patch types into account. This index accounts for the neighborhood relations between patches. Each patch is analyzed for adjacency with all other patch types and measures the extent to which patch types are interspersed, that is, equally bordering other patch types (Griffith et al. 2000; McGarigal et al. 2002). IJI can be calculated using Equation 12.3:

$$IJI = \frac{-\sum_{i=1}^{m} \sum_{k=i+1}^{m} \left[LE_{ik} \times \ln(LE_{ik}) \right]}{\ln\left(\frac{m(m-1)}{2} \right)}$$

(12.3)

where LE_{ik} is the length of edge between patch types i and k and m is the number of patch types. The value of IJI ranges from 0 to 100 and increases as patches tend to be more evenly interspersed in a "salt and pepper" mixture. That is, the landscapes in which patch types are equally adjacent to each other have higher values of IJI, while the landscapes that have disproportionate distribution of patch type adjacencies possess lower values of IJI. Therefore, IJI quantifies the degree of landscape fragmentation (McGarigal et al. 2002).

LSI measures and standardizes total edge as the perimeter-to-area ratio for the landscape as a whole (McGarigal et al. 2002; Patton 1975). LSI is similar to the habitat diversity index proposed by Patton (1975). LSI quantifies the amount of edge present in a landscape in comparison with a landscape that has the same size but a simple geometric shape such as circle or square and no internal edge. As LSI increases, patches become more fragmented. There is no limitation on scale. As an experiment of examining similarity of two spatial metrics, an aggregation index (AI) that may be highly correlated with LSI was also selected. Theoretically, AI measures the percentage of like adjacencies between cells of same patch type. The lower index values indicate there is less like adjacencies, that is, the class is greatly disaggregated, and the higher values imply the class is greatly aggregated (McGarigal et al. 2002). The difference between AI and LSI also exists in this way: LSI considers landscape boundary as edge (or perimeter), while AI does not take the landscape boundary into account. The objective of comparing LSI and AI was to see if both spatial metrics duplicate the measure of landscape structure information.

The Pearson product–moment correlation coefficients among the time series of the landscape spatial metrics and between each of them and the military training intensity were calculated, and their significance of differences from zero was statistically tested using the equation $r_\alpha = \sqrt{t_\alpha^2 / (n - 2 + t_\alpha^2)}$ based on the student's distribution at the level $\alpha = 5\%$, where n is the number of years for the time series used.

At patch level, patch perimeter and fractal shape were used. Patch perimeter measures length of patch edge. The patch fractal shape measures shape complexity of patches with a range of values from 1 to 2. As the index approaches 2, the shape complexity increases. This fractal shape index is based on perimeter–area and characterizes patch shapes via fractal dimension (McGarigal et al. 2002). Each of the spatial metrics captures a different characteristic of fragmentation, and at the patch level, both fractal shape and perimeter index can be thus combined. If a patch has a large perimeter and simple shape, then it can be concluded that the patch is not as fragmented as a patch with lots of perimeter and a complex shape.

12.4 Results

The landscape segmentations at three spatial resolutions (5 m × 5 m, 20 m × 20 m, and 30 m × 30 m) of IRS images using maximum likelihood classification for the year 1999 were obtained as shown in Figure 12.3. The results for other spatial resolutions including

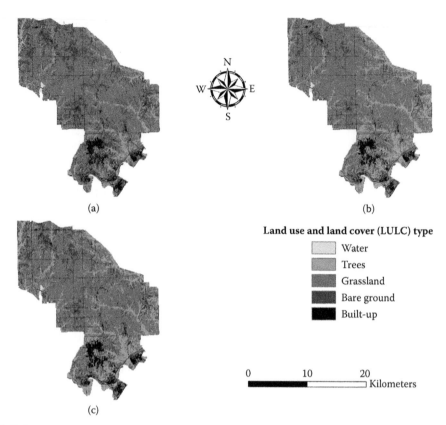

Land use and land cover (LULC) type

☐ Water
☐ Trees
☐ Grassland
☐ Bare ground
■ Built-up

0   10   20
                    Kilometers

(c)

FIGURE 12.3
Landscape segmentations at different spatial resolutions using IDRISI image segmentation and maximum likelihood classification with 1999's IRS images scaled up from (a) 5 m × 5 m to (b) 20 m × 20 m and (c) 30 m × 30 m.

10 m × 10 m, 15 m × 15 m, and 25 m × 25 m were omitted because of limited space. Scaling up the IRS images from a finer spatial resolution to a coarser one improved landscape segmentation, and the improvement was observable as the pixel size increased to 20 m × 20 m and disappeared after this spatial resolution. When the TM and IRS multispectral images were combined with the IRS panchromatic image using BT and the resulting images were scaled up, the obtained classification maps were similar to those above. However, when the resolutions were under 20 m × 20 m, the classification maps looked very pixilated and were not able to effectively differentiate the categories of LULC. That is, the finer spatial resolutions led to too many small patches.

Threshold value was another factor that affected the results of landscape segmentation. A reasonable threshold value meant that when using it, the obvious LULC such as roads and small water bodies could be segmented. As expected, for both IRS images and their BT, the thresholds consistently decreased as resolutions became coarser. For the IRS images, the segmentation threshold varied from 115 at 5 m × 5 m to 45 at 30 m × 30 m. A good segmentation threshold for the TM images was found to be 75. The thresholds higher than that lacked the ability to segment the roads and small water bodies, while thresholds lower than that commonly led to pixilated maps. In addition, when a too small threshold was used, the requirements of storage and running time became a factor.

The BT of the IRS images (Figure 12.4a) were able to identify the main lakes, but incorrectly classified the creeks at the south part of the study area and other water bodies in

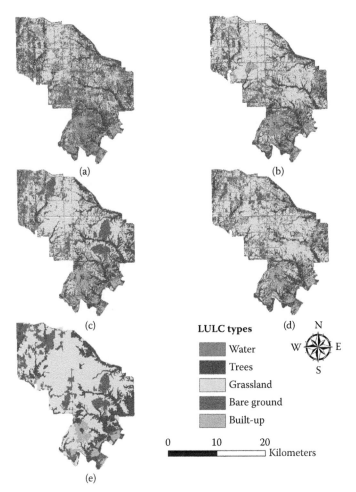

FIGURE 12.4
(See color insert.) (a) Misclassification of water bodies in the west, east, and south parts using 1999's IRS Brovey transformations and maximum likelihood compared to (b) land use and land cover classification using 1999's TM images, (c) burned areas in the north and east parts misclassified into bare grounds using 2000's TM images and ML, (d) 1996's TM images led to misclassification of built-up areas in the east parts by ML, and (e) landscape segmentation and LULC classification using TM images and SEGCLASS.

the northeast corner (out of the study area). Especially, the BT led to the misclassification of water bodies throughout the installation and also had a tendency to mix trees and water bodies in the west parts. All the TM images resulted in correct classification of the water bodies (Figure 12.4b–e). Correct classification of water bodies is important in the U.S. installations because they might impede military training activities such as off-road vehicle traffic. This implied that in this study, the LULC maps obtained using the IRS images would be problematic for their use of planning military training activities by the manager of Fort Riley installation.

When the Landsat TM images acquired in 2000 were used, on the other hand, several areas in the north and central east parts of Fort Riley were classified into bare grounds (Figure 12.4c). In fact, these were burned areas in 2000 based on the TM images in Figure 12.2b and the records of burned areas. That is, misclassification took place. This also happened in some other years. Moreover, the use of 1999's TM image also resulted in misclassification of a large patch of burned area into built-up areas. When the TM images were used, in addition, the built-up areas were many times intermingled with bare lands. Because of cloud cover, the 1995's and 1996's TM images also resulted in misclassification of the grasslands into built-up areas at the east parts of Fort Riley (Figure 12.4d). The misclassifications caused by the TM images were also noticed in the LULC maps obtained using the IRS images.

Regardless of IRS and TM images, the PCA did not improve the performance of landscape segmentation and LULC classification compared to their original images. Given the same spatial resolution and threshold value, maximum likelihood classification produced the maps that showed the transition between LULC types and appeared to be more natural (Figure 12.4c), whereas SEGCLASS produced larger and more smoothed polygons (Figure 12.4e).

As examples, the accuracy assessment of landscape segmentation and LULC maps produced using 1989's TM and 1999's IRS and TM images were accounted for here, but the error matrices were omitted because of space limitations. Based on 128 field plots, 1989's TM images led to the overall accuracy of 84.9% correct with a kappa value of 0.6185. Bare ground had a producer accuracy of 100%, grassland 83.0%, and tree 81.8%. But grassland had the highest user accuracy of 95.4%, tree 66.7%, and bare ground 40.0%. The reason for low user accuracies of tree and bare ground was the misclassification of 9 and 6 grassland plots into tree and bare ground.

Based on 146 field plots, 1999's TM image resulted in a percentage correct of 81.8% with kappa value of 0.7114, which were more accurate compared to the percentage correct of 75.3% and kappa value of 0.6178 obtained using the 1999's IRS image. Misclassifications mainly took place among grassland, bare ground, and built-up areas. The reason was mainly because when the vegetation cover was low, the LULC classes were hard to separate from each other. Another reason was because of transitions, for example, on the sides of roads within the installation; it was difficult to discern where the road began and bare ground existed.

When spatial metrics were calculated at patch level, the LULC classification maps from SEGCLASS led to more reasonable results than those from the maximum likelihood. The reason was because overall the LULC maps from SEGCLASS produced nearly 250 patches every year, whereas the corresponding maps from maxiumum likelihood resulted in more than 20,000 patches, especially many single pixels were regarded as patches. This large number of patches impeded the calculation of spatial metrics.

The selected spatial metrics were calculated using the maps generated using TM images for all the years at both landscape and patch levels. The dynamics of landscape level metrics in Figure 12.5 emphasize the trend of landscape fragmentation that was accounted for by the spatial metrics including IJI, AI, and LSI. Compared to IJI and AI, LSI had a greater variation with larger spikes between years and the largest range. LSI is a standardized measure of total edge and quantifies landscape fragmentation. The higher the LSI, the more fragmented the landscape. Overall, the years 2001, 1999, and 1992 had larger LSI values, while the years 1990, 1991, and 2000 had smaller LSI values.

The aforementioned results of LSI seemed not fully consistent with military training intensity. Overall, the military training intensity had higher values before 1995, and after that the training intensity decreased by about 30%. The highest training intensity took place in 1991, while 1996 had the lowest intensity values. The conflict can be explained by the following: (1) the LSI value associated with how small the patches are and (2) a higher military training intensity leading to more disturbance of ground and vegetation cover, but might not definitely result in more smaller patches. In fact, 2001's landscape segmentation

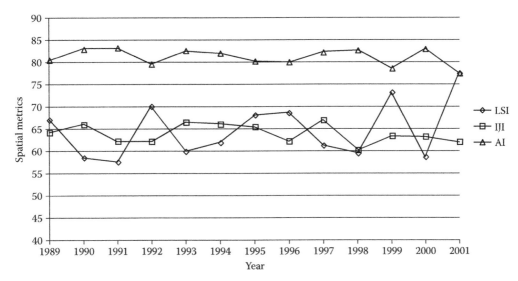

FIGURE 12.5
Dynamics of spatial metrics including landscape shape index (LSI), aggregation index (AI), and interspersion and juxtaposition index (IJI) at landscape level for all years from 1989 to 2001 using TM images for Fort Riley, Kansas.

map showed that the disturbance areas were smaller and spatially distributed almost everywhere compared to 1993's segmentation map in which vegetation cover was heavily degraded, but the disturbed areas mainly concentrated in a central belt from the north to the south. Thus, the patch edge density of 2001's landscape greatly increased the LSI value.

As IJI increases, landscape becomes more disaggregated. Compared to LSI, IJI was more constant. IJI had a mean of 63.54 across the years, showing a slight tendency to have common borders with different LULC classes. This was because of the small amount of water bodies on the landscape as well as the built-up areas located in the southern portion of the installation. The year 1997 had the highest IJI value of 67.06, but the training intensity in 1997 was relatively low. The reasons may be partly because the overall differences of IJI among the years were not obviously great and partly because of the small misclassified areas within the high impact area. In 1997, many burned or bare ground areas were misclassified as built-up areas. These small misclassified areas along with the small bare ground areas throughout the center of the installation could have possibly raised the IJI value.

AI measures the percentage of cell adjacencies for the same patch type instead of patch adjacencies. That is, AI focuses on class aggregation. AI stayed relatively constant over time. The lowest value took place in 2001 with a value of 77.5 and the highest value occurred in 1991 with a value of 83.12. AI also had the smallest range of the values compared to other landscape metrics. Because AI concentrates on class aggregations and the study area was dominated by grass, AI had high landscape value.

On the other hand, the year 2001 had the smallest values for both AI and IJI, which was attributed to the many small patches of bare ground. This characteristic of the landscape in 2001 was thus consistent with the earlier finding from LSI, in which LSI had the highest value in 2001 because of the highly fragmented landscape in that year. Moreover, 1990 had higher values of AI and IJI but lower value of LSI because of larger homogeneous patches.

Pearson product–moment correlation coefficients between the spatial metrics were calculated in Figure 12.6a. The correlations of IJI with LSI and AI, respectively, were not

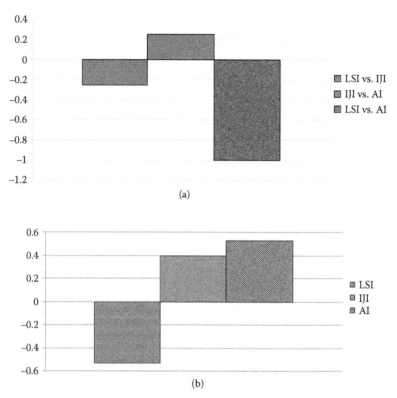

FIGURE 12.6
(a) Correlation among the spatial metrics including landscape shape index (LSI), aggregation index (AI), and interspersion and juxtaposition index (IJI) and (b) correlation between each of the spatial metrics and military training intensity in terms of Pearson product–moment correlation coefficient.

significant based on the value of 0.514 obtained using 13 degrees of freedom at a significance level of 0.05. But there was a significant and almost perfectly negative correlation between LSI and AI. This implied that the information obtained from LSI was highly duplicated by that from AI in a reverse way. Moreover, the Pearson product–moment correlation coefficients between military training intensity and each of the spatial metrics in Figure 12.6b showed that both AI and LSI had significant correlation with the military training intensity at the significance level of 0.05.

At patch level, the segmentation maps obtained using SEGCLASS were used to calculate the spatial metrics because for some of the years, maximum likelihood classification led to too many small patches. In Figure 12.7, the fractal shape index at patch level emphasized the spatial patterns and variability of landscape fragmentation. Compared to others, the shapes of patches were more complex in the central parts of the study area because military training activities took place. In the south parts, the patch shapes varied over space and time because of uneven frequencies and distributions of the activities. A typical example for the fractal shape index quantifying the complexity of the patches is shown in Figure 12.7e, in which there were two large burned areas in the east part and they had relative simple shapes that were captured by this index. Moreover, the fractal shape index could be combined with patch perimeter to create a normalized fragmentation layer at patch level that was used to account for landscape fragmentation at patch level. Because of space limition, these results were omitted here.

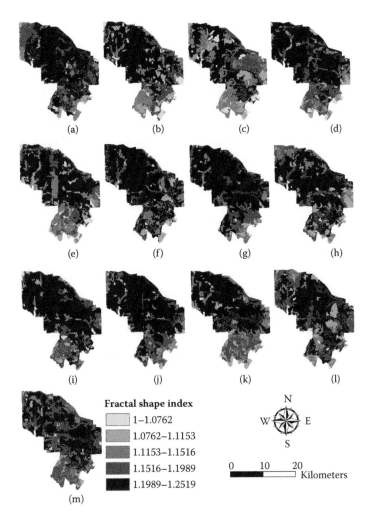

FIGURE 12.7
Time series of fractal shape index maps of Fort Riley landscape at patch level for years (a) 1989, (b) 1990, (c) 1991, (d) 1992, (e) 1993, (f) 1994, (g) 1995, (h) 1996, (i) 1997, (j) 1998, (k) 1999, (l) 2000, and (m) 2001.

12.5 Conclusions and Discussion

For Fort Riley, landscape segmentation for LULC classification could be better conducted using TM images instead of higher spatial resolution IRS images. The IRS image and its BT and PCA lacked the ability to identify water bodies. The main reasons may include the following: (1) when the IRS image was acquired, the water bodies were turbid; (2) the Landsat TM had more channels than the IRS; and (3) although the IRS image had higher spatial resolutions than the TM images, the finer spatial resolutions led to a great number of small patches and degraded the accuracy of landscape segmentation and LULC classification and further resulted in difficulty in calculating the landscape metrics at patch level. In addition, maximum likelihood had greater potential to segment the landscape into homogeneous polygons than the SEGCLASS algorithm. But, sometimes the maximum likelihood produced too many small patches, which usually were not a good thing for calculation of spatial metrics.

In this study, it was found that both LSI and AI had significant correlation with military training intensity and better quantified the landscape fragmentation than IJI. The finding has not been reported in any previous studies. On the other hand, there was almost perfect negative correlation between LSI and AI. This means it is not necessary to use both LSI and AI at the same time to measure landscape fragmentation because of high duplication of the captured landscape structure.

In this study, the landscape level metrics including LSI or AI and IJI well quantified the overall landscape fragmentation of Fort Riley, while the patch level metrics such as fractal shape measured spatial patterns and variability of the landscape quality in detail. The time series of Landsat TM images provided the potential to assess and monitor the spatial variability and temporal dynamics of landscape quality for large areas.

Acknowledgments

We are grateful to the U.S. Army Corps of Engineers, Construction Engineering Research Laboratory (USA-CERL) for providing support (CERL W9132T-08-2-0019) and data sets for this study and also to the editors and reviewers.

References

Anderson, A. B., G. Wang, S. Fang, G. Z. Gertner, B. Güneralp, and D. Jones. 2005. "Assessing and Predicting Changes in Vegetation Cover Associated with Military Land Use Activities Using Field Monitoring Data at Fort Hood, Texas." *Journal of Terramechanics* 42:207–29.

DEPARC (Defense Environmental Programs Annual Report to Congress). 2007. http://www.acq .osd.mil/ie/ (accessed on October 1, 2012).

Diersing, V., R. Shaw, and D. Tazik. 1992. "US Army Land Condition-Trend Analysis (LCTA) Program." *Environmental Management* 16:405–14.

ERDAS. 2012. http://www.erdas.com/ (accessed on October 1, 2012).

Gertner, G. Z., G. Wang, S. Fang, and A. B. Anderson. 2002. "Mapping and Uncertainty of Predictions Based on Multiple Primary Sources from Joint Co-Simulation with Landsat TM Image and Polynomial Regression." *Remote Sensing of Environment* 83:498–510.

Gertner, G. Z., G. Wang, S. Fang, and A. B. Anderson. 2004. "Partitioning Spatial Model Uncertainty When Inputs Are from Joint Simulations of Correlated Multiple Attributes." *Transactions in GIS* 8:441–58.

Griffith, J. A., E. A. Marinko, and K. P. Price. 2000. "Landscape Structure Analyses of Kansas in Three Scales." *Landscape and Urban Planning* 52:45–61.

Hall, F. G., D. E. Strebel, J. E. Nickeson, and S. J. Goetz. 1991. "Radiometric Rectification: Toward a Common Radiometric Response among Multidate, Multisensor Images." *Remote Sensing of Environment* 35:11–27.

Hargis, C., J. Bissonette, and J. David. 1998. "The Behavior of Landscape Metrics Commonly Used in the Study of Habitat Fragmentation." *Landscape Ecology* 13: 167–86.

IDRISI. 2012. http://www.clarklabs.org/ (accessed on October 1, 2012).

Jensen, J. R. 2005. *Introductory Digital Image Processing: A Remote Sensing Perspective*. 3rd ed. Upper Saddle River, NJ: Pearson Prentice Hall.

Mandelbrot, B. B. 1977. *Fractals: Form, Chance and Dimension*. San Francisco, CA: W.H. Freeman and Company.

McGarigal, K., S. A. Cushman, M. C. Neel, and E. Ene. 2002. "FRAGSTATS: Spatial Pattern Analysis Program for Categorical Maps." Computer software program produced by the authors at the University of Massachusetts, Amherst. www.umass.edu/landeco/research/fragstats/fragstats.html.

Narumalani, S., D. Mishra, and R. Rothwell. 2004. "Change Detection and Landscape Metrics for Inferring Anthropogenic Processes in the Greater EFMO Area." *Remote Sensing of Environment* 91:478–89.

O'Neil, R., K. Vitters, J. Wickham, and K. Jones. 1999. "Landscape Pattern Metrics and Regional Assessment." *Ecosystem Health* 5:225–33.

Patton, D. R. 1975. "A Diversity Index for Quantifying Habitat 'Edge'." *Wildlife Society Bulletin* 3:171–73.

Quist, M., P. Fay, C. Guy, A. Knapp, and B. Rubenstein. 2002. "Military Training Effects on Terestrial and Aquatic Communities on a Grassland Military Installation." *Ecological Applications* 13:432–42.

Riiters, K., R. O'Neill, C. Hunsaker, J. Wickham, D. Yankee, S. Timmins, K. Jones, and B. Jackson. 1995. "A Factor Analysis of Landscape Pattern and Structure Metrics." *Landscape Ecology* 10:23–29.

Singer S. 2010. *Assessment of Cumulative Training Impacts for Sustainable Military Land Carrying Capacity and Environment: Quantifying Quality of Environment and Landscape*. Master's thesis, Southern Illinois University at Carbondale.

Singer, S., G. Wang, H. Howard, and A. B. Anderson. 2012. "Comprehensive Assessment Indicator of Environmental Quality for Military Land Management." *Environmental Management* 50:529–40.

Sun, W., G. Xu, P. Gong, and S. Liang. 2006. "Fractal Analysis of Remotely Sensed Images: A Review of Methods and Applications." *International Journal of Remote Sensing* 27:4963–90.

Wang, G., G. Z. Gertner, A. B. Anderson, H. R. Howard, D. Gebhard, D. Althoff, T. Davis, and P. Woodford. 2007. "Spatial Variability and Temporal Dynamics Analysis of Soil Erosion due to Military Land Use Activities: Uncertainty and Implications for Land Management." *Land Degradation and Development* 18:519–42.

Wang, G., G. Z. Gertner, S. Fang, and A. B. Anderson. 2004. "Mapping Vegetation Cover Change Using Geostatistical Methods and Bitemporal Landsat TM Images." *IEEE Transactions on Geoscience and Remote Sensing* 42:632–43.

Wang, G., G. Z. Gertner, H. R. Howard, and A. B. Anderson. 2008. "Optimal Spatial Resolution for Collection of Ground Data and Multi-Sensor Image Mapping of a Soil Erosion Cover Factor." *Journal of Environmental Management* 88:1088–98.

13

Automated Individual Tree-Crown Delineation and Treetop Detection with Very-High-Resolution Aerial Imagery

Le Wang and Chunyuan Diao

CONTENTS

13.1 Introduction..223
13.2 Study Sites and Data Preparation..225
    13.2.1 Study Sites..225
    13.2.2 Data Preparation 225
13.3 Methods..226
    13.3.1 Enhance Tree-Crown Boundaries 226
        13.3.1.1 Dyadic Wavelet Decomposition.......................................226
        13.3.1.2 Edge Probability with the Magnitude Information 227
        13.3.1.3 Scale and Geometric Consistency Constraints...............228
    13.3.2 Edge Detection 228
    13.3.3 Treetop Identification 229
        13.3.3.1 Treetop Detection Based on Radiometric Characteristics................230
        13.3.3.2 Treetop Detection Based on Spatial Characteristics 230
        13.3.3.3 Marker Image Generation...231
    13.3.4 Marker-Controlled Watershed Segmentation.................................231
13.4 Results 232
13.5 Discussion..235
13.6 Conclusion 235
References...236

13.1 Introduction

Forest stands, as the basic units in forest management, play a pivotal role in understanding the function and service of the forest system. A stand is a contiguous area that contains a number of trees that are relatively homogeneous or similar in species composition or age and different from adjacent areas (Lindenmayer and Franklin 2002). Several parameters of the stand are of particular interest to foresters, including tree density, stand basal area, stand diameter, stand height, crown closure, stand volume, stand table, and site index. Traditionally, to acquire those parameters, field plots with a random, stratified, or systematical sampling scheme have to be designed and measured, which is usually expensive and labor-intensive. Nevertheless, timely and accurately obtaining the stand information is critically important for updating the forest inventory (Spurr 1948) and for conducting ecological studies with those parameters as the input (Palace et al. 2007). As remote sensing imagery is more readily accessible, information gathering becomes more frequent and cost-effective.

In the 1940s, visual interpretation of medium- and large-scale aerial imagery for forestry emerged (Brandtberg 1999). However, the manual interpretation method is usually time-consuming, labor-intensive, and biased by the interpreter's experience, which to a great extent triggered the development of automated or semiautomated methods for individual tree recognition. With the increasing availability of very-high-resolution (VHR: meter or submeter level) imagery, the development of automated computer-based photo interpretation has been spurred (Gong et al. 1999), and various algorithms have been developed for automatically delineating individual trees, which basically falls into four major types: local maximum (LM)-based methods (Blazquez 1989; Dralle and Rudemo 1996), contour-based (CB) methods (Gougeon 1995; Pinz et al.1993), three-dimensional (3D) model-based methods (Gong et al. 2002; Sheng et al. 2001), and template matching (TM)-based methods (Pollock 1996; Tarp-Johansen 2002).

The LM method attempts to detect the treetops by finding the local maximum of the image with the assumption that the peak of the tree-crown reflectance is located at or very close to the treetop (Brandtberg and Walter 1998). Despite being fast and simple, it performs poorly as image illumination conditions vary. The TM method characterizes the tree morphology at different locations of the image by considering the trees' geometric and radiometric properties with a series of models. With the information gained, the TM procedure (Pollock 1996) is implemented to search for the locus of best matching trees. In comparison, 3D-based methods have been utilized by fewer researchers. Sheng et al. (2001) applied parametric tree-crown surface model-based image matching to obtain an improved tree-crown surface reconstruction. Gong et al. (2002) developed an interactive tree interpreter for the semiautomatic tree-crown segmentation as an improvement. This method, however, has to determine the treetop locations separately on the left and right epipolar images to be fully automated. The CB method searches for the delimiter between tree crowns and their background by following the intensity valleys underlying the image (Gougeon 1995) or by detecting the crown boundary with edge-detection methods (Brandtberg and Walter 1998). Edge-detection methods are seldom applied in tree-crown delineation mainly because of two difficulties. On the one hand, intensity changes vary with the scale. At the finer scale, tree branches are the major components accounting for the changes in intensity, whereas at the increasing coarser scale, tree internal structures are gradually suppressed and a tree crown tends to merge with its neighbor (Brandtberg and Walter 1998). In that sense, the clusters of trees are where the changes occur. Hence, finding the appropriate scale for delimiting the individual tree-crown boundary is difficult. Considering that tree crowns tend to have different sizes within the forest stand, it is impossible to choose a scale applicable for all the individual trees. Accordingly, information on multiple scales should be investigated and integrated in order to more accurately delineate the individual tree crown. On the other hand, the edge-detection method can only create raw primal sketches as a low-level image processing method. Incorporating the expert knowledge or biological knowledge about the tree-crown shapes into high-level individual tree-crown delineation is inevitably crucial for producing a more accurate result.

To bridge the gap between edge detection and tree-crown delineation, three objectives are set in this study: (1) to incorporate the multiscale scheme for enhancing tree-crown boundaries while suppressing excessive texture inside; (2) to locate treetops with consideration of both geometry and radiometry information; (3) to develop a more advanced two-stage tree-crown detection method, namely edge detection followed by marker controlled watershed segmentation. Hopefully, with the multiscale preprocessing method and the two-stage approach, treetops can be more accurately located and tree-crown boundaries can be more efficiently delineated.

13.2 Study Sites and Data Preparation

13.2.1 Study Sites

The study area is a young ponderosa pine forest stand located at 38°53′42.9″N, 120°37′57.9″W, adjacent to Blodgett Forest Research Station, a research forest of the University of California, Berkeley. The stand has the following characteristics: an average diameter of 9.81 cm at breast height (DBH), a density of 420 stems/hectare, and an average height of 4.05 m. In 2000, a precommercial thinning took place and most of the shrubs and grass were cut down. The dominant species thereafter in the stand was almost the 10- to 11-year-old ponderosa pine (*Pinus ponderosa*).

13.2.2 Data Preparation

A 1:8000 aerial photograph was acquired by an aerial camera with a focal length of 152.9 mm in May 2000, under the uniform cloud cover condition. Then it was scanned at 1000 dpi, with a 20.3 cm spatial resolution digital image produced. A subset image with the size of 500 × 500 pixels was chosen in this study, approximately a ground area of 10,404 m^2 (Figure 13.1). A total of 58 trees, identified both on the ground and on the aerial image, were selected in the subset image. For each tree, the crown diameter was

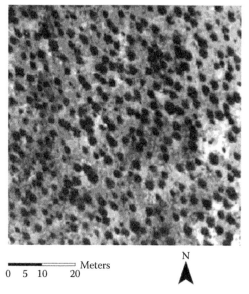

FIGURE 13.1
The scanned aerial photograph in our study area.

measured on the ground in two directions: one along the maximum axis and the other along the perpendicular direction. As reference data, the radius of the tree crown can then be calculated by averaging the two measurements. More details of the data can be found in Wang (2010).

13.3 Methods

The tree-crown delineation algorithm can be divided into three stages. The first stage applies the scale-space theory to preprocess the image in the multiscale scheme for enhancing the true tree-crown boundaries (Wang 2010). The second step utilizes an edge-detection method to obtain primal tree-crown boundaries (Wang et al. 2004). The third stage can be separated into two main parts: treetop marker selection and marker-controlled watershed segmentation (Wang et al. 2004). The following sections describe each step in detail. Further information can be found in Wang et al. (2004) and Wang (2010).

13.3.1 Enhance Tree-Crown Boundaries

To produce accurate crown delineation, the tree-crown information should be investigated at multiple scales. Wavelet-based methods (Mallat 1989) based on scale-space theory can be applied to decompose the original information with multiresolution approximation. By exploring the evolution of the edges, the true tree-crown boundaries can be distinguished from edges that correspond to the tree branches and twigs, and thus, textures of the true tree can be strengthened, while those within a single tree will be suppressed.

13.3.1.1 Dyadic Wavelet Decomposition

Scale-space theory aims at making significant structures and scales explicit (Lindeberg 1993). The idea is to link low-level features detected at different scales in scale space, thus facilitating the identification of high-level objects (Lu and Jain 1992). By convolving the original image with the transformed and dilated wavelet, the original image can be decomposed as one approximation image together with various difference images at different scales. The advantages of the wavelet decomposition include no correlation among images of different scales and the acquisition of local edges' information with coefficients of the wavelet orthonormal basis expansion.

In this study, the cubic spline function mentioned in Mallat (1989) was adopted as the scaling function or as a low-pass smoothing filter. The wavelet function can be considered as the derivative of the smoothing function with orthonormal characteristics. Since calculating the wavelet coefficients at every possible scale is impractical, the dyadic wavelet decomposition with only scales of the power of 2 was used. At each decomposed level, four wavelet coefficients, namely approximate, horizontal, vertical, and diagonal coefficients, can be obtained, and then the gradient magnitude (Equation 13.1) and orientation (Equation 13.2) can be calculated with the purpose of quantifying the evolution of edge pixels from different sources, that is, tree crown or branches, across the scale space.

$$M_{edge} = \sqrt{\left(W_{2^j}^h\right)^2 + \left(W_{2^j}^v\right)^2} \tag{13.1}$$

$$A_{edge} = \arctan\left(\frac{W_{2^j}^v}{W_{2^j}^h}\right) \quad (13.2)$$

where M_{edge} denotes the gradient magnitude and A_{edge} is the gradient orientation. $W_{2^j}^h$ and $W_{2^j}^v$ stand for the horizontal and vertical coefficients, respectively, in the wavelet decomposition at the scale 2^j.

13.3.1.2 Edge Probability with the Magnitude Information

With the wavelet decomposition coefficients, the gradient magnitude can be obtained at each scale and the threshold can be set for separating the edge pixels from the background. However, using a single threshold to judge the edge pixel is apparently inappropriate or inefficient on a complex forest image. Edge probability, from another perspective, provides more information than the single threshold method and can distinguish the edge pixel more accurately.

Scharcanski et al. (2002) modeled the gradient magnitude of background-related pixels with a Rayleigh probability density function (Equation 13.3). Similarly, the magnitude of edge-related pixels can also be modeled by the Rayleigh probability density function with the variance of edge pixels (Equation 13.4). With those two probabilities, the overall probability of a pixel being the gradient magnitude of r is shown in Equation 13.5.

$$P_j\left(r/\text{background}\right) = \frac{r}{\left[\sigma_{background}^j\right]^2}\exp^{-r^2}/2\left[\sigma_{background}^j\right]^2 \quad (13.3)$$

$$P_j\left(r/\text{edge}\right) = \frac{r}{\left[\sigma_{edge}^j\right]^2}\exp^{-r^2}/2\left[\sigma_{edge}^j\right]^2 \quad (13.4)$$

$$P_j\left(r\right) = w_{background}^j P_j\left(r/\text{background}\right) + (1 - w_{background}^j)P_j\left(r/\text{edge}\right) \quad (13.5)$$

where $P_j(r/\text{background})$ is the probability of a pixel having gradient magnitude equal to r, given that it belongs to a background pixel at the scale of 2^j, and $\sigma_{background}^j$ is the standard deviation of a background pixel's gradient magnitude at the scale of 2^j. It is also the same for $P_j(r/\text{edge})$ and σ_{edge}^j. $w_{background}^j$ is the prior probability of a pixel being background or noise.

To calculate $P_j(r)$, three parameters, $\sigma_{background}^j$, σ_{edge}^j, and $w_{background}^j$, have to be known first. A typical method for solving the unknown parameters is using the maximum likelihood function (Equation 13.6). Then from the Bayes theorem, the probability of a pixel belonging to the edge, given the gradient magnitude being r, can be calculated using Equation 13.7. As a result, the gradient magnitude will be replaced with $p(\text{edge}/r)$ in the following process of enhancing the tree-crown boundaries.

$$[w_{background}^j, \sigma_{edge}^j, \sigma_{background}^j] = \arg\max\left(\prod P_j\left(r\right)\right) \quad (13.6)$$

$$p(\text{edge}/r) = \frac{\left(1 - w_{background}\right)p(r/\text{edge})}{p(r)} \quad (13.7)$$

13.3.1.3 Scale and Geometric Consistency Constraints

Since texture information within the tree crown, that is, branches or twigs, is hard to distinguish from that of true tree crown with the edge probability at a single scale, the evolution of pixels along various scales is checked with the assumption that true-crown boundaries will consistently exhibit a large edge probability, while pixels with undesired texture will only show a large edge probability in a small range of scales. Harmonic mean (Equation 13.8) is chosen in this case to evaluate the scale consistency with the aim of further enhancing the true tree-crown texture and suppressing the undesired ones.

$$P_j\left(\text{edge}/r\right) = \frac{M+1}{\dfrac{1}{P_j\left(\text{edge}/r\right)} + \dfrac{1}{P_{j+1}\left(\text{edge}/r\right)} + \cdots + \dfrac{1}{P_{j+m}\left(\text{edge}/r\right)}} \tag{13.8}$$

where $p_j(\text{edge}/r)$ is the edge probability at the scale 2^j, $p_{j+m}(\text{edge}/r)$ is the edge probability at the scale 2^{j+m}, and $M + 1$ is the total number of scales that are included in this analysis.

Besides the scale consistency constraint, tree crowns should also satisfy the condition of geometric consistency, since tree crowns are usually curved in shape and the orientation along the tree-crown boundary should not change significantly. Using Equation 13.2, the gradient direction of each pixel can be assigned and evaluated. A typical example of edge pixels is shown in Figure 13.2 with the gradient direction denoted by an arrow. Thus, the edge probability $p_j(\text{edge}/r)$ can be updated by assigning the weight by the Gaussian function along the gradient direction. The geometric consistency tends to augment the true tree-crown boundary by strengthening the edge pixels with a continuous smooth curve and suppressing the isolated ones. Finally, an inverse wavelet transformation can be used to reconstruct the image with enhanced tree-crown boundary for the subsequent edge detection and tree-crown delineation.

13.3.2 Edge Detection

A forest image is usually composed of tree crown, understory vegetation, and bare soil. Masking out nontree areas and retaining tree-crown objects are inevitably the first step

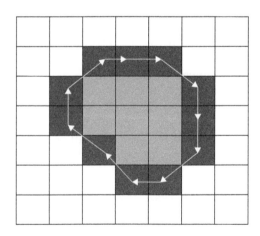

FIGURE 13.2
An example of tree crown's gradient direction of the boundary pixels.

for individual tree-crown recognition. Edge-detection methods, by deriving the initial boundary of the tree crown, can achieve this purpose to a large extent. However, current edge-detection methods can only be applied to a single band. Hence, we applied the intensity–hue–saturation (IHS) transformation to the original aerial-colored image and utilized only the intensity image in the subsequent edge detection.

The Laplacian of the Gaussian (LOG) operator was selected in this study for detecting the edge of the tree crown. The LOG method can be partitioned into two parts. The first part involves a Gaussian smoothing for removing the noise and intensity variation by virtue of a tree's internal structure. For the second part, the pixels corresponding to the zero of the second derivative of the smoothed image were marked as edge pixels, since in an image, an edge indicates the intensity discontinuity and can be captured by the derivative function. The LOG detector can be written as (Marr and Hildreth 1980)

$$LOG(x,y) = -\frac{1}{\pi\sigma^4}\left[1 - \frac{x^2 + y^2}{2\sigma^2}\right]\exp\left(-\frac{x^2 + y^2}{2\sigma^2}\right) \qquad (13.9)$$

The smoothing scale σ in the LOG method implies the minimum width of the edge that can be captured. In our study, the smoothing scale is one pixel, which represents the smallest tree-crown diameter in the image by visual inspection. The LOG operator can also bring about artifacts or phantom edges. To distinguish those phantom edges from the true edges, a method proposed by Clark (1989) was used to remove the phantom edges for the subsequent tree-crown delineation.

With the edge pixels detected by LOG operator, a series of closed contours, indicating the tree-crown boundaries, can be formed with an eight–connectivity scheme. However, those contours obtained may not represent the individual tree crown and typically three scenarios may be included, namely isolated trees, slightly touching trees, or tree clumps (Brandtberg 1999) (Figure 13.3). Isolated single trees tend to form a single circular-shaped contour, while slightly touching trees or tree clumps are more inclined to have irregular or oblong shapes. Hence, further segmentation of those contours is necessary for obtaining the tree-crown boundary on an individual basis. In our study, we identified the treetops within each contour first and then utilized the treetop information as a guide for acquiring the final individual tree-crown boundaries.

13.3.3 Treetop Identification

We treat each closed contour as an object, and for each object, we determine the number of trees it contains by locating the treetops. Treetops can be identified by their unique radiometric and spatial characteristics. With regard to the radiometric intensity, it usually varies in different parts of the tree and reaches the highest on the uppermost

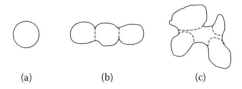

(a)   (b)   (c)

FIGURE 13.3
Three typical cases of objects after edge detection: (a) isolated trees, (b) slightly touching trees, and (c) tree clumps.

sunlit portion of the tree crown. Hence, in our study, a local nonmaximum suppression method was adopted to obtain the treetops in each contour. As regards the spatial characteristic, a treetop is located at or near the center of the tree crown when it is viewed from angles near nadir. Correspondingly, an LM distance method was applied to obtain another set of treetops. The intersection of the two sets of treetops by both methods is identified as the authentic treetop and will be denoted as the marker for the following segmentation.

13.3.3.1 Treetop Detection Based on Radiometric Characteristics

A local nonmaximum suppression method was utilized to detect the pixel with highest radiometric intensity for each crown object as the treetop, based on the gray values of the intensity image from IHS transformation. Specifically, it will use a sliding window to assign a value of one to the center pixel only if all the surrounding pixels' gray values within the window are less than that of the center pixel for locating the treetops (Dralle and Rudemo 1996) and finally create a binary image in which pixels representing the treetops were labeled one and all others were labeled with a value of zero. With respect to the sliding window, the size of the window is vitally important for accurately and efficiently locating the treetop. If the window size is too small, the tree crown with large radius may be assigned more than one treetop. Contrarily, if the window size is too large, trees having smaller crowns may not be detected and assigned a treetop. In our study, a relatively small size of sliding window (3×3 window of pixels) was selected with the purpose of not missing the small treetop. For those false treetops assigned by this method, they can be identified and filtered out by the subsequent algorithm in view of treetops' spatial properties.

13.3.3.2 Treetop Detection Based on Spatial Characteristics

Apart from the treetops selected by virtue of the radiometric features, treetops can also be identified from a spatial perspective. An LM transformed distance method can be applied for locating the treetops spatially, with the assumption that treetops are located in the vicinity of the center of the tree crown from the near nadir view. The method can be divided into two parts. First, the geodesic distance between the pixels within each object of closed-contour and the set of exterior pixels was calculated. Then, the regional maximum of the distance image was extracted and labeled as a treetop.

The geodesic distance is a concept borrowed from mathematical morphology by defining the distance between two pixels p and q within the set A as the length of the shortest path connecting p and q in A and defining the distance from any pixel in set A to its complementary set as the path joining that pixel in A with the nearest pixel in the complement of A. To calculate the distance between each interior pixel and the set of exterior pixels, an elementary disk structure element (SE; 3×3 window of pixels whose values are equal to 1) was defined by considering the eight-connected neighborhoods of the center location, as opposed to an elementary cross SE with only four-connected neighborhoods. As a consequence, the distance can be measured only along connected paths defined by the SE, and the length of each step is determined by the value of each pixel in the SE. Normally, an SE can represent the geometry of the object to be measured and is usually built as a small window of pixels with values either 0 or 1 (Soille 2003).

With morphologically transformed distance calculated for each interior pixel in the object, a resultant distance image, signifying the distance from that interior pixel to the nearest exterior pixel, can be formed. Accordingly, the regional maximum of the distance image for each object can be extracted. As for the regional maximum, it is defined as a connected group of pixels with a single distance value, and for each pixel in the group, the distance value is greater than or equal to that of the surrounding eight-connectivity neighborhood. As a consequence, the regional maximum is usually located near the center of the object, and thus, those pixels corresponding to the regional maximum are marked as treetops spatially.

13.3.3.3 Marker Image Generation

Two sets of treetops can be formed in light of the radiometric and spatial characteristics. To satisfy two conditions simultaneously and set the final treetop marker for the subsequent segmentation, we intersect two sets of treetops for each object by testing the proximity of each treetop detected by the nonmaximum suppression method to that of the maximum-distance method. If a treetop identified by the gray-level method is also located within a 3 × 3 window of the surrounding distance-based treetop, it will be labeled as the final treetop. For example, in Figure 13.4, there are five treetops identified by the gray-level nonmaximum suppression method, four of which coincidentally fall into the window of the surrounding distance-based treetops. Thus, those four treetops were recognized as the final treetops and markers, while the fifth one was filtered out as a pseudo treetop.

13.3.4 Marker-Controlled Watershed Segmentation

With the treetop markers generated by satisfying both radiometric and spatial characteristics, we utilized the marker-controlled watershed segmentation method with the goal of obtaining the individual-based tree-crown boundaries in each object. Watershed

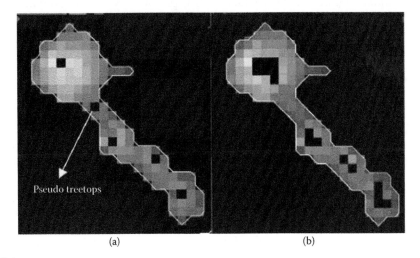

(a)   (b)

FIGURE 13.4
Treetops detection with both (a) radiometric—Markers from gray level—and (b) distance—Markers from geodesic distance—methods.

segmentation, as a nonlinear image processing method in mathematical morphology, was first introduced by Beucher and Lantuejoul and later defined mathematically by both Meyer and others (Pesaresi and Benediktsson 2001). The advantage of this method is to selectively preserve geometric or structural information while fulfilling the required tasks on the image. A further variant of this method is the so-called marker-controlled watershed segmentation, and it can be understood as follows. If we treat the gray-scale image as a topographic model, each pixel in the image will stand for the elevation at that point (Vincent and Soille 1991) and treetop markers will be the local maxima. By inverting the gray tone of the image, those markers will become the local minima lying in the valley. If water is introduced into this topographic model, each valley will collect water with the marker as the starting point, until the water runs over the watershed and infiltrates into the adjacent valley. Each watershed of the valley corresponds to one closed contour containing unique marker, and it can partition the whole area into different catchment basins. Consequently, those contours generated by the segmentation become the desired boundaries of individual tree crowns within the object.

One crucial factor for generating the desired contour is its correlation with the gray-level image patterns. In our study, we determine the tree-crown boundary by creating the geodesic skeleton with the use of the influence zones (Soille 2003). To begin with, the geodesic distance from each interior pixel to each treetop marker is calculated with the elementary disk SE. Next, the influence zone for a specified marker K_i is determined by counting the pixels whose geodesic distance to K_i is smaller than that to any other markers. Subsequently, the individual-based tree-crown boundary can be formed by delimiting the boundaries of those influence zones.

Compared to the traditional marker-controlled watershed segmentation method, our study has two apparent advantages. On the one hand, our segmentation task performs on the individual objects, as opposed to the entire image of the traditional segmentation method. In this way, we can eliminate the error that would otherwise be introduced by the background. On the other hand, the traditional marker-controlled watershed segmentation often leads to severe over-segmentation because of the presence of spurious local minima and maxima (Soille 2003); we tackled this problem by defining the marker first in view of the treetops' characteristics.

13.4 Results

To enhance the tree-crown boundaries, dyadic wavelet decomposition was implemented with the first three levels (2^1, 2^2, 2^3) considered. Accordingly, the gradient magnitude and direction were calculated in virtue of the wavelet coefficients obtained at each level. Edge probability could then be derived at the scale 2^1, 2^2, and 2^3 according to Equation 13.7, and the results are shown in Figure 13.5. By analyzing the evolution of image gradients over the scale space, we found that tree-crown boundaries were effectively strengthened, while the excessive textures that resulted from tree branches and twigs were largely suppressed. Moreover, the scale consistency was checked in terms of the harmonic mean, and the geometric consistency was conducted by using the gradient directions. With the updated edge probability achieved, the magnitude of the horizontal and vertical wavelet coefficients was adjusted and then an inverse wavelet transform was carried out.

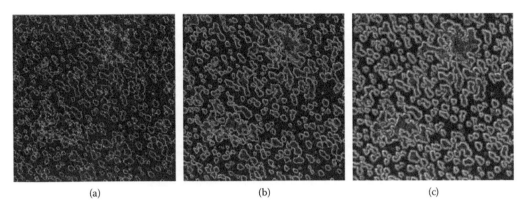

(a)   (b)   (c)

FIGURE 13.5
Edge probabilities at the scale 2^1 (a), 2^2 (b), and 2^3 (c).

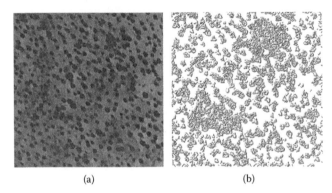

(a)   (b)

FIGURE 13.6
Enhanced tree-crown boundaries from the multiscale wavelet decomposition (a) and the shaded relief of difference between the enhanced and original image (b).

After the edge-enhanced image was acquired (Figure 13.6), it was subtracted from the original image for illustrating the effect of the wavelet decomposition method. A shaded relief image denoting the difference between two images is presented in Figure 13.6. With a significant number of small textures within the tree crowns standing out in the difference image, our method can be proved efficient in strengthening the tree-crown boundaries, as well as in suppressing the textures inside. The enhanced image greatly alleviates the difficulty in the subsequent treetop detection and tree-crown delineation.

With the enhanced version of the image, we applied a two-stage approach, an edge-detection method followed by the marker-controlled watershed segmentation, and the final delineated tree crowns are shown in Figure 13.7. To evaluate our result, a total of 58 trees were selected and measured from the ground; 56 trees were correctly identified from the automated tree-crown delineation method, while the other two were undetected. For those 56 correctly identified trees, we first extracted the crown area from the image and then converted it to the radius based on a circular crown shape assumption. Subsequently, the radius derived from the image and that measured in the field were compared and regressed (Figure 13.8). The slope value in the regression model is 0.875, signifying that the crown size was underestimated from the automated method. The reason can be attributed to the fact that pixels on the tree-crown boundaries were not

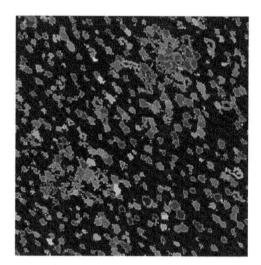

FIGURE 13.7
Final delineated tree crowns from the marker-controlled watershed segmentation method.

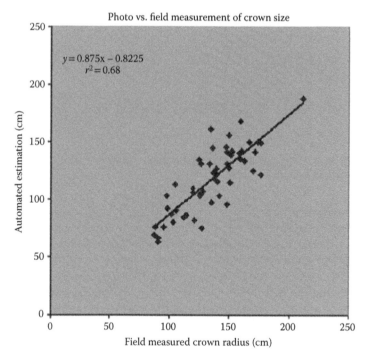

FIGURE 13.8
Regression of the tree-crown radius measured on the ground and that from the image based on 56 identified trees.

obviously detectable and may not be well recorded in the image. The R^2 in this case is 0.68, and it indicates that 68% variance of the automatically delineated tree-crown size can be explained by the regression model. The unexplained variance can actually be caused by the noise introduced when a circular shape was used to convert the irregular crown area to the radius value.

13.5 Discussion

With the developed multiscale scheme, promising tree-crown delineation was achieved. Out of 58 field-surveyed trees, 56 were identified in the image with the automated method. A comparison of crown size for 56 trees with R^2 value of 0.68 indicates the feasibility of delineating the tree crown from remote sensing imagery. Hence, it is necessary to perform an effective enhancement before the tree-crown boundary was delineated.

Traditionally, treetop extraction and tree-crown delineation are treated as separate procedures by most researchers. However, the solution of one part can usually assist in deriving the solution of the other. Owing to their close relationship, the integration of two parts tends to produce a more accurate result. In our algorithm, we filtered out spurious local minima and maxima by locating the treetops and then utilized those treetops as markers for delineating the tree-crown boundaries. The combination of two parts helps avoid the problem of over-segmentation to a great extent. Therefore, the derivation of treetop is indispensable and essential for determining the success of the individual tree-crown delineation.

With respect to the treetop detection, we utilize two methods by considering both radiometric and geometric characteristics. Our results have shown that the morphological information plays a crucial role in detecting the pseudo treetops generated by the radiometric nonmaximum suppression methods. However, existing algorithms have not effectively exploited such shape information. By superimposing a geometric restriction on the traditional gray-level method, we actually reduced the errors to a large degree.

By using treetops as the markers for delineating the individual tree-crown boundaries, we have effectively overcome the shortage of the traditional watershed segmentation method and generated a more accurate boundary image. Although our algorithm takes advantage of both spectral and spatial information for locating treetops and separating individual trees, there are still some potential problems deserving further research. One problem is the limitation of the assumptions. The spatially morphological algorithm for locating the treetops can be implemented only with the assumption that treetops are located around the vicinity of the center of a crown. However, it can only be satisfied within 15° of the nadir, and it may not be applicable to the trees outside of this range. To build a more robust algorithm, different treetop models based on the location of trees can be incorporated. Another problem is the inflexibility of the algorithm. In our methods, the σ value in the LOG edge detection, the window size of the nonmaximum suppression, the window size of the treetop intersection method, and the SE were all assigned a fixed value by virtue of the minimum tree-crown size. However, those parameters could be varied. Different scenarios should be evaluated to produce a more accurate result. Finally, tree-crown boundaries are sometimes inconsistent with gray-scale boundaries. The problem does not affect the trees viewed from the near-nadir direction, but may haunt those from outside the near-nadir range. For those regions, the silhouettes detected from edge-detection methods are sometimes inconsistent with the real tree-crown boundaries, which may be solved by using a 3D-based model.

13.6 Conclusion

In summary, tree-crown boundaries can be effectively enhanced and the internal texture can be largely suppressed by applying the multiscale wavelet decomposition method. The scale and geometric consistency check are critically important in the process of enhancement by accounting for the tree's radiometric and morphological characteristics. Treetops can be more accurately detected by exploring the radiometric and spatial characteristics simultaneously. However, the spatial information is rarely utilized for locating the treetops because of the difficulty of expressing shape information in discrete image space. With the aid of mathematical morphology, we developed an approach to integrate the spatial information with the traditional radiometric method, which filtered out the pseudo treetops to a great extent. Tree crowns and treetops are closely related parameters, and the integration of two parts tends to produce a more accurate delineation result. By locating treetops with their special characteristics and using them as the markers for generating tree-crown contour, the marker-controlled watershed segmentation method sets a good example for combining treetop detection and tree-crown delineation under a unified framework. The result of marker-controlled watershed segmentation achieves a promising agreement with that from the field survey, indicating the feasibility of obtaining the tree-crown area from remote sensing imagery. In future work, a more robust tree-crown algorithm should be developed and tested in a range of forests, especially the undisturbed forest that has not been thinned.

References

Blazquez, C. 1989. "Computer-Based Image Analysis and Tree Counting with Aerial Color Infrared Photography." *Journal of Imaging Technology* 15:163–68.

Brandtberg, T. 1999. "Remote Sensing for Forestry Applications—A Historical Retrospect." CVonline: On-Line Compendium of Computer Vision [Online]. Accessed on August 6, 2012: http://www.dai.ed.ac.uk/CVonline/LOCAL_COPIES/BRANDTBERG/UK.html

Brandtberg, T., and F. Walter. 1998. "Automated Delineation of Individual Tree Crowns in High Spatial Resolution Aerial Images by Multiple-Scale Analysis." *Machine Vision and Applications* 11:64–73.

Clark, J. J. 1989. "Authenticating Edges Produced by Zero-Crossing Algorithms." *Pattern Analysis and Machine Intelligence, IEEE Transactions on* 11:43–57.

Dralle, K., and M. Rudemo. 1996. "Stem Number Estimation by Kernel Smoothing of Aerial Photos." *Canadian Journal of Forest Research* 26:1228–36.

Gong, P., G. S. Biging, S. Lee, X. Mei, Y. Sheng, R. Pu, B. Xu, K. P. Schwarzr, and M. Mostafa. 1999. "Photo Ecometrics for Forest Inventory." *Geographic Information Sciences* 5:9–14.

Gong, P., Y. Sheng, and G. Biging. 2002. "3D Model-Based Tree Measurement from High-Resolution Aerial Imagery." *Photogrammetric Engineering and Remote Sensing* 68:1203–12.

Gougeon, F. A. 1995. "A Crown-Following Approach to the Automatic Delineation of Individual Tree Crowns in High Spatial Resolution Aerial Images." *Canadian Journal of Remote Sensing* 21:274–84.

Lindeberg, T. 1993. "Detecting Salient Blob-Like Image Structures and Their Scales with a Scale-Space Primal Sketch: A Method for Focus-of-Attention." *International Journal of Computer Vision* 11:283–318.

Lindenmayer, D. B., and J. F. Franklin. 2002. *Conserving Forest Biodiversity: A Comprehensive Multi-Scaled Approach*. Washington, DC: Island Press.

Lu, Y., and R. C. Jain. 1992. "Reasoning About Edges in Scale Space." *Pattern Analysis and Machine Intelligence, IEEE Transactions on* 14:450–68.

Mallat, S. G. 1989. "A Theory for Multiresolution Signal Decomposition: The Wavelet Representation." *Pattern Analysis and Machine Intelligence, IEEE Transactions on* 11:674–93.

Marr, D., and E. Hildreth. 1980. "Theory of Edge Detection." *Proceedings of the Royal Society of London. Series B. Biological Sciences* 207:187–217.

Palace, M., M. Keller, G. P. Asner, S. Hagen, and B. Braswell. 2007. "Amazon Forest Structure from IKONOS Satellite Data and the Automated Characterization of Forest Canopy Properties." *Biotropica* 40:141–50.

Pesaresi, M., and J. A. Benediktsson. 2001. "A New Approach for the Morphological Segmentation of High-Resolution Satellite Imagery." *Geoscience and Remote Sensing, IEEE Transactions on* 39:309–20.

Pinz, A., M. B. Zaremba, H. Bischof, F. A. Gougeon, and M. Locas. 1993. "Neuromorphic Methods for Recognition of Compact Image Objects." *Machine Graphics and Vision* 2:209–29.

Pollock, R. 1996. *The Automatic Recognition of Individual Trees in Aerial Images of Forests Based on a Synthetic Tree Crown Image Model.* Department of Computer Science, Vancouver, Canada: University of British Columbia.

Scharcanski, J., C. R. Jung, and R. T. Clarke. 2002. "Adaptive Image Denoising Using Scale and Space Consistency." *Image Processing, IEEE Transactions on* 11:1092–101.

Sheng, Y., P. Gong, and G. Biging. 2001. "Model-Based Conifer-Crown Surface Reconstruction from High-Resolution Aerial Images." *PE & RS Photogrammetric Engineering and Remote Sensing* 67:957–65.

Soille, P. 2003. *Morphological Image Analysis: Principles and Applications.* New York: Springer-Verlag.

Spurr, S. H. 1948. *Aerial Photographs in Forestry.* New York: Ronald Press Company.

Tarp-Johansen, M. J. 2002. "Automatic Stem Mapping in Three Dimensions by Template Matching from Aerial Photographs." *Scandinavian Journal of Forest Research* 17:359–68.

Vincent, L., and P. Soille. 1991. "Watersheds in Digital Spaces: An Efficient Algorithm Based on Immersion Simulations." *IEEE Transactions on Pattern Analysis and Machine Intelligence* 13:583–98.

Wang, L. 2010. "A Multi-Scale Approach for Delineating Individual Tree Crowns with Very High Resolution Imagery." *Photogrammetric Engineering and Remote Sensing* 76:371–78.

Wang, L., P. Gong, and G. S. Biging. 2004. "Individual Tree-Crown Delineation and Treetop Detection in High-Spatial-Resolution Aerial Imagery." *Photogrammetric Engineering and Remote Sensing* 70:351–58.

14

Tree Species Classification

Ruiliang Pu

CONTENTS

14.1 Introduction 239
14.2 Advanced Remote Sensing Sensors/Systems 240
14.3 Techniques and Methods 242
    14.3.1 Spectral Mixture Analysis 242
    14.3.2 Object-Based Image Analysis Method 244
    14.3.3 Hierarchical Mapping System 245
    14.3.4 Hyperspectral Transformation and Feature Extraction 246
    14.3.5 Advanced Classifiers 248
14.4 Considerations and Future Directions 251
    14.4.1 Considerations 251
    14.4.2 Future Directions 253
References 254

14.1 Introduction

Timely and accurate acquisition of information on the status and structural change of forest composition is crucial to develop strategies for sustainable management of natural resources and ecosystem modeling (Pu and Landry 2012). In practice, field surveys and aerial photograph interpretation are two traditional ways to obtain the information about tree canopy and species composition. Both methods are time-consuming, expensive, and usually cannot provide complete coverage of large areas. Satellite remote sensing has an advantage of being able to obtain data for large areas simultaneously. However, previous studies have demonstrated that accurately mapping individual tree species and tree canopy using moderate-resolution satellite imagery is difficult or impossible (Katoh 1988; Congalton et al. 1991; Brockhaus and Khorram 1992; Franklin 1994; Carreiras et al. 2006).

During the past couple of decades, remote sensing systems/sensors have advanced in increasing spectral resolution (e.g., Hyperion hyperspectral sensor at 10 nm spectral resolution) and spatial resolution (e.g., WorldView-2 multispectral images at 2 m resolution and panchromatic band at 0.5 m resolution). This provides opportunities to differentiate species and map individual trees. High spatial resolution commercial satellite imagery has shown to be a cost-effective alternative to aerial photography for generating digital image base maps (Davis and Wang 2003) and extracting individual tree crowns (Ke and Quackenbush 2007). Meanwhile, researchers have also tried to use hyperspectral data (e.g., Airborne Visible/Infrared Imaging Spectrometer [AVIRIS] and Airborne

Hyperspectral Scanners [HyMap]) to identify and map forest species and have achieved a certain degree of success (e.g., Townsend and Walsh 2001; Xiao et al. 2004; Buddenbaum et al. 2005).

Based on the existing literature review, this section provides an overview on remote sensing applications to tree species classification, especially use of high spatial resolution satellite data and airborne and satellite hyperspectral images. The main objectives of this chapter are as follows:

- To review optical high spatial/spectral resolution remote sensing sensors/systems frequently used for tree species classification
- To review suitable image processing and information extraction techniques/methods and their applications in mapping tree species
- To address relevant considerations and challenges for tree species classification and to point out future directions for identifying and mapping tree species using advanced remote sensing techniques

14.2 Advanced Remote Sensing Sensors/Systems

Currently, optical remote sensing sensors/systems suitable for tree species classification mainly include high spatial resolution airborne and satellite sensors/systems (note that airborne digital images will not be reviewed in this subsection) and various airborne and satellite hyperspectral sensors/systems. Table 14.1 summarizes frequently used optical high spatial/spectral resolution sensors/systems using sensors' names and characteristics including band setting and platforms, and so on. High spatial resolution sensors are mostly commercial satellite sensors, including the commonly used IKONOS, QuickBird, GeoEye-1, WorldView-2, and so on, whereas hyperspectral images include both airborne and satellite sensors/systems. Hyperspectral sensors/systems include airborne sensors— AISA, AVIRIS, CASI, HyMap, HYDICE—and the first satellite hyperspectral sensor, Hyperion. All airborne hyperspectral sensors' data are limited in some areas in the world and are high cost to use. Satellite Hyperion hyperspectral data are free to access but the data are rarely available in the world. All commercial satellite sensors' data are accessible worldwide but need a relatively high cost compared to other satellite images. However, the commercial satellite images are still much cheaper than airborne digital images for mapping tree species.

During the past two decades, the advanced optical remote sensing sensors/systems listed in Table 14.1 have been utilized by many researchers for identifying and mapping forest tree species and species compositions. Compared to using moderate resolution satellite data to map forest type and species composition, high spatial resolution satellite images have demonstrated their potential to improving tree species classification. For example, Carleer and Wolff (2004) conducted a study to identify seven tree species/groups in a relatively homogenous forest area using two scenes of IKONOS image data acquired from two different seasons (summer and fall) and were able to achieve the modest overall accuracy (OA) of 82%. Using IKONOS image data for the Iwamizawa region in the center of Hokkaido, northern Japan, an area dominated by conifer plantations and typical mixed forests consisting of broad-leaved trees with large crowns, Katoh (2004) classified 14 tree

TABLE 14.1

A List of Optical High Spatial/Spectral Remote Sensing Sensors/Systems Frequently Used for Tree Species Classification

Sensor/System	Platform	Revisit Period (Days)	No. of Bands	Spectral Range/Band	Spatial Resolution (m)		Spectral Resolution (nm)
					MS	Pan	
High spatial resolution							
GeoEye-1	Satellite	2–3	5	B, G, R, NIR, VNIR	1.65	0.41	
IKONOS	Satellite	3	5	B, G, R, NIR, VNIR	4	1	
QuickBird	Satellite	3–4	5	B, G, R, NIR, VNIR	2.4	0.6	
WorldView-2	Satellite	1.1–3.7	9	2B, G, Y, R, RE, 2NIR, VNIR	2	0.5	
High spectral resolution							
AISA (Eagle, Hawk, Dual)	Airborne		244, 254	400–970 nm, 1000–2500 nm, 400–2500 nm	Variable		1.6–9.0
AVIRIS	Airborne		224	400–2500 nm	Variable		10
CASI (CASI-2)	Airborne		288	405–950 nm	Variable		2.2
HyMap	Airborne		126	400–2500 nm	Variable		10–20
HYDICE	Airborne		206	400–2500 nm	Variable		7.6–14.9
Hyperion, Hyperspectral imager	Satellite, EO-1	>15	220	400–2500 nm	30		10

Notes: EO-1, Earth Observing-1; B, blue; G, green; R, red; RE, red edge; NIR, near infrared; SWIR, shortwave infrared; V, visible; VNIR, visible and near infrared; Pan, pancromatic; MS, multispectral.

species/groups and obtained an average classification accuracy of 52%. By using QuickBird data to identify four leading tree species, Mora et al. (2010) have achieved an OA of 73% for tree species classification. In addition, Kim et al. (2011) obtained a 77% OA using an optimal segmentation of IKONOS image data to identify seven tree species/stands. Similar work by using IKONOS and QuickBird data for tree species classification has been done by Sugumaran et al. (2003), Hájek (2006), and Ke and Quackenbush (2007). Recently, even higher spatial/spectral resolution satellite image data, WorldView-2, have shown a greater potential for mapping tree species than other high-resolution satellite sensors (e.g., IKONOS and QuickBird). For instance, in a comparative analysis of high spatial resolution IKONOS and WorldView-2 imagery for mapping urban tree species, Pu and Landry (2012) demonstrated that the WorldView-2 sensor has a greater capability (OA increased 16%–18%) to map seven urban tree species/groups than that of IKONOS likely due to improved spatial resolution (4 to 2 m) and additional bands (coastal, yellow, red-edge, and NIR2). Sridharan (2011)'s work also proved the potential of WorldView-2 data for tree species classification.

As reviewed earlier, the preliminary results of evaluating the capabilities of those high spatial resolution sensors in identifying tree species and mapping tree canopy indicate that the accuracy generally is not sufficient. During the last couple of decades, the airborne and satellite hyperspectral remote sensing techniques (Table 14.1) have provided new tools for mapping tree species with an emphasis on utilizing subtle spectral information. Researchers have tried to use hyperspectral sensors/systems (such as AVIRIS, HyMap, and Hyperion) to identify and map forest species and have achieved a certain degree of success. For instance, Buddenbaum et al. (2005) classified coniferous tree species with HyMap using geostatistical methods and obtained a classification accuracy of OA = 78%, a result comparable to that obtained with stem density information derived from high spatial resolution imagery. In mapping urban forest species with hyperspectral AVIRIS image data, Xiao et al. (2004) successfully discriminated between three forest types with an OA of 94% although they also reported a relatively low OA (70%) for identifying 16 tree species with the data. With the first satellite hyperspectral sensor, Hyperion data for mapping tropical emergent trees in the Amazon Basin, Papeş et al. (2010) concluded that when using 25 selected narrow bands and pixels that represented >40% of tree crowns, classification was 100% successful for the five taxa. Similar work by using airborne hyperspectral sensors, including AVIRIS, CASI, HyMap, HYDICE, and satellite hyperspectral sensor Hyperion data for tree species classification, has been done by Goodwin et al. (2005), Leckie et al. (2005), Zhang et al. (2006), Tsai et al. (2007), Pu et al. (2008b), Walsh et al. (2008), Jones et al. (2010), and Banskota et al. (2011).

As summarized in Table 14.1, currently, the list of optical high spatial/spectral resolution remote sensing sensors/systems is frequently used and effective for mapping tree species. Particularly, various airborne/satellite hyperspectral sensors/systems and WorldView-2 data are ideal for mapping forest species, especially for mapping invasive tree species (because invasive tree species often express biochemical and physiological properties unique from those of native trees) (Pu 2012).

14.3 Techniques and Methods

14.3.1 Spectral Mixture Analysis

Compared to individual tree crown size, many advanced remote sensing sensors' images have a relatively low spatial resolution, especially most airborne/satellite hyperspectral sensors, such as AVIRIS (usually >10 m for most images acquired before 2000) and Hyperion (30 m).

For this case, under most circumstances, several individual tree crowns are mixed in single pixels. Therefore, to effectively map tree species and estimate species abundance, it is necessary to conduct a spectral mixture analysis (SMA) to measure individual tree species abundance within the mixed pixels. Note that the technique is more suitable for those relatively low spatial resolution multi/hyperspectral sensors. There are two types of spectral mixing: linear spectral mixing and nonlinear spectral mixing. In a linear SMA, the spectral signature of the mixed pixel is assumed to be a linear combination of spectral signatures of surface materials with their areal proportions as weighting factors (Gong and Zhang 1999). The basic physical assumption underlying the linear SMA is that there is no significant amount of multiple scattering between the different constituents (end-members); each photon that reaches the sensor has interacted with just one constituent (Settle and Drake 1993). However, a nonlinear SMA considers that there is significant amount of multiple scattering between the different end-members (Zhang et al. 1998). A nonlinear spectral mixture model can be found in Sasaki et al. (1984) and Zhang et al. (1998). A linear spectral mixing modeling and its inversion have been widely used since the late 1980s, including applications for extracting the abundance of individual tree species within mixed pixels. In a linear spectral mixture model analysis, there are two solutions: a linear least squares solution and a nonlinear solution (e.g., a neural network algorithm). At present, since the SMA method is easy to use, it has been widely and successfully applied for mapping abundance of individual tree species with hyperspectral image data.

This simple mixing model analysis addressed previously has an advantage in which it is relatively simple and provides a physically meaningful measure of tree species abundance within mixed pixels. However, there are a number of limitations to the simple mixing concept: (1) the end-members (tree species) used in SMA are the same for each pixel; (2) the SMA cannot account for subtle spectral differences between tree species efficiently; and (3) the maximum number of end-members (species) that the SMA can map is limited by the number of bands in the image data. Therefore, Roberts et al. (1998) introduced multiple end-member spectral mixture analysis (MESMA), a technique for identifying materials in a hyperspectral image using end-members from a spectral library. MESMA overcomes these limitations of the simple SMA. Using MESMA, the number of end-members (species) and types are allowed to vary for each pixel in the image. The general procedure of the MESMA approach is to start with a series of candidate two-end-member models, evaluate each model based on selection criteria, and then, if required, construct candidate models that incorporate more end-members (Roberts et al. 1998). To select end-members (tree species) during the processing of MESMA, three selection criteria are fraction, root-mean-square error (RMSE), and the residuals of contiguous bands (Roberts et al. 1998). The minimum RMSE model is assigned to each pixel and can be used to map tree species and fractions within the image (Painter et al. 1998) with the MESMA approach.

A number of researchers have applied SMA to hyperspectral data to estimate the abundance of general vegetation cover type or tree species (Lass and Prather 2004; Ramsey et al. 2005a,b; Walsh et al. 2008). A neural network–based, nonlinear solution was also applied to hyperspectral data to estimate the abundance of specific tree species (e.g., Walsh et al. 2008). A few researchers have applied the MESMA approach in various environments for vegetation mapping including tree species mapping. For example, Roberts et al. (1998, 2003) used MESMA and hyperspectral image data AVIRIS to map vegetation species and land cover type in southern California chaparral.

As a result of spectral unmixing, the fractions or species abundance represent the areal proportions of end-members (species), but we do not know where the proportioned areas of the end-members locate within a mixed pixel. A key factor to unmix mixed spectra is to identify suitable/pure end-members/species and extract their corresponding individual spectra for training and test purposes.

14.3.2 Object-Based Image Analysis Method

The use of high spatial resolution imagery also poses a new challenge because the spectral response of an individual tree is influenced by variation in crown illumination and background effects, and thus accuracy is reduced by using conventional pixel-based classifications (Quackenbush et al. 2000). To overcome this problem, region-based or object-based classification can be used. Object-based image analysis (OBIA) techniques first use image segmentation to produce discrete regions or image objects (IOs) that are more homogeneous in themselves than with nearby regions (e.g., a tree crown), and then these IOs rather than pixels are used as the classification unit (Carleer and Wolff 2006; Blaschke 2010). An IO-based strategy can potentially improve tree classification accuracy compared to pixel-based approaches (Pu et al. 2011). It is worthy to note that the OBIA method is suitable for high spatial resolution data such as IKONOS, QuickBird, and WorldView-2 but not for moderate or coarse resolution sensors.

There are many image segmentation techniques currently reported in literature and used for creating IOs. Generally, we can categorize the techniques into four groups: pixel-based (clustering), thresholding-based, edge-based, and region-based methods. Clustering or pixel-based methods are a process whereby a dataset (pixels) is replaced by cluster; pixels may belong together because of the same gray level, color, texture, and so on. Thresholding image segmentation is an operation of converting a multilevel image into a binary image or more values through comparison of pixel values with the predefined threshold value individually. Edge-based segmentation is a location of pixels in the image that corresponds to the boundaries of the objects seen in the image. The region-based segmentation is partitioning of an image into similar/homogenous areas of connected pixels through the application of homogeneity/similarity criteria among candidate sets of pixels. See Spirkovska (1993), Varshney et al. (2009), and Zuva et al. (2011) for a detailed review of those different groups of image segmentation techniques. An ideal image segmentation method adopted to produce appropriate IOs is vital because the OBIA technique can improve tree species classification. Several criteria have been developed for quantitative evaluation of image segmentation results (e.g., Möller et al. 2007; Ke et al. 2010). Pu and Landry (2012) assessed image segmentation quality by measuring both topological and geometric similarities between segmented objects and reference objects. The metrics included (1) the relative area of the ith overlapped (at least partially) segmented object to the ith reference object (R_{Asri}, %) and (2) the position discrepancy of the ith segmented object to the ith reference object (D_{sri}, m). The average R_{Asr} and average D_{sr} are calculated as follows:

$$R_{Asr} = \frac{1}{n}\sum_{i=1}^{n}\frac{A_{si}}{A_{ri}} \times 100\% \qquad (14.1)$$

$$D_{sr} = \frac{1}{n}\sum_{i=1}^{n}\sqrt{(X_{si} - X_{ri})^2 + (Y_{si} - Y_{ri})^2} \qquad (14.2)$$

where A_{si} and A_{ri} are areas (m^2) of segmented object i and reference object i, respectively; X_{si} and Y_{si} are the coordinates (m) of the centroid of segmented object i and X_{ri} and Y_{ri} are the coordinates (m) of the centroid of reference object i; and n is number of reference objects used for the evaluation. Here, an overlapped segmented object to a reference object means that the centroid of the segmented object is within the reference object (polygon). Usually, a well-segmented image with ideal IO should have an R_{Asr} close to 100% and D_{sr} close to 0. However, unlike the research by Möller et al. (2007), Ke et al. (2010), and Wang et al. (2010), R_{Asr} of IO matching tree crowns can be >100% because there are many tree crowns connected together or intersecting each other (Pu and Landry 2012).

Previous studies mapping tree species composition have demonstrated the advantages of OBIA (Leckie et al. 2005; Hájek 2006; Ke and Quackenbush 2007; Voss and Sugumaran 2008; Sridharan 2011; Kim et al. 2011; Pu et al. 2011; Pu and Landry 2012). For example, Ke and Quackenbush (2007) used OBIA with QuickBird multispectral imagery and obtained a 66% average accuracy when they classified five tree species/groups (spruce, pine, hemlock, larch, and deciduous) in a forested area of New York, United States. In an analysis of object-based urban detailed land cover classification with high spatial resolution IKONOS imagery, Pu et al. (2011) demonstrated a significantly higher classification accuracy of five vegetated types (including broad-leaved, needle-leaved, and palm trees) using OBIA compared to pixel-based approaches. In addition, Kim et al. (2011) obtained a 13% higher OA with OBIA by using an optimal segmentation of IKONOS image data to identify seven tree species/stands.

14.3.3 Hierarchical Mapping System

To improve identifying and mapping accuracy of forest tree species, a hierarchical or stepwise masking system is used for two main reasons: (1) the system matches the logical structure of most land cover classification schemes and (2) it is able to increase the relative spectral separability at higher masking levels. Such multilevel classification systems demonstrated the potential of increasing the accuracy of mapping tree species and species composition (Townsend and Walsh 2001; Xiao et al. 2004; Pu et al. 2008c; Sridharan 2011; Pu 2011; Pu and Landry 2012). For example, in mapping urban tree species/groups with WorldView-2 data, Pu and Landry (2012) developed a stepwise masking system that improved classification accuracy for sunlit versus shadow/shaded IOs. Their results suggested that it was effective to improve the accuracy of identifying tree species/groups at a high level. The system consisted of three masking levels and five steps (Pu and Landry 2012): (1) using the threshold of a normalized difference vegetation index to separate the study area into vegetated and nonvegetated areas (mask level 1); (2) using textural information along with hue or intensity information to separate the vegetated area into grass/lawn/shrub and tree canopies (mask level 2); (3) using the brightness threshold of the near-infrared (NIR) band to separate tree canopies into sunlit and shadow/shaded areas (mask level 3); (4) extracting training samples (IOs) separately from the sunlit and shadow areas with references to ground survey data and 0.3 m resolution digital aerial photographs; and (5) using classifiers linear discriminant analysis (LDA) and classification and regression trees (CART) to identify and map urban tree species/groups. Based on spectral and structural differences between vegetated and nonvegetated areas and between grass/turf and tree canopies, it may be expected that all tree species/groups would be classified in one class at level 1 and level 2. In addition, because the spectrum of the shadow/shaded IOs is very different from that of sunlit IOs of different tree species/groups, it is also expected that shadow/shaded tree canopy IOs might be classified as a single class if tree canopies are not first separated into sunlit and shadow/shaded canopies. Their study confirms the effectiveness of the stepwise masking system.

14.3.4 Hyperspectral Transformation and Feature Extraction

When hyperspectral data are used for tree species classification, it is necessary to conduct hyperspectral transformation and extract features from the hyperspectral data. Usually, effective data transformation and feature extraction techniques include principal component analysis (PCA), minimum noise fraction (MNF), canonical discriminant analysis (CDA), partial least squares regression (PLSR), wavelet transform (WT), and so on. Table 14.2 presents a summary of the five data transformation techniques with their characteristics, major advantages and limitation, major factors to be considered, and application examples.

The PCA technique has been applied to reduce the data dimension and feature extraction from hyperspectral data for assessing leaf or canopy biophysical and biochemical parameters (e.g., Pu and Gong 2004; Tsai et al. 2007; Pu et al. 2008a, 2008b). With a covariance (or correlation) matrix calculated from vegetated pixels only, it is commonly believed that the eigenvalues and the corresponding eigenvectors computed from the covariance (or correlation) matrix are expected to be able to enhance vegetation variation information in the first several principal components (PCs). The MNFs transform to maximize the signal-to-noise ratio when choosing PCs with increasing component number. Then several MNFs to maximize the signal-to-noise ratio are selected for further analysis of hyperspectral data, such as for determining end-member spectra for SMA (e.g., Walsh et al. 2008), tree species identification (e.g., Narumalani et al. 2009), and so on.

CDA also is a dimension-reduction technique equivalent to canonical correlation analysis that can be used to determine the relationship between the quantitative variables and a set of dummy variables coded from class variables in a low-dimensional discriminant space (Khattree and Naik 2000; Zhao and Maclean 2000). CDA derives canonical variables, linear combinations of the quantitative variables that summarize between-class variation in much the same way that PCA summarizes most variation in the first several PCs. However, CDA involves human effort and knowledge derived from training samples, and derived canonical variables are directly linked to properties of classes while PCA performs a relatively automatic data transformation and tries to concentrate the majority of data variance in the first several PCs. Compared to using PCs by PCA, use of CDA-derived canonical variables can improve tree species identification accuracy (van Aardt and Wynne 2001, 2007; Pu and Liu 2011).

PLSR is a technique that reduces the large number of measured collinear spectral variables to a few noncorrelated latent variables or PCs. The PCs represent the relevant structural information present in the measured predictor variables (e.g., quantitative spectral variables extracted from image data) and are used to predict the dependent variables (e.g., qualitative tree species coded as "0" or "1") (Apan et al. 2009). PLSR produces the weight (coefficient) matrix that reflects the covariance structure between the predictor and the response variables. The optimal number of factors (or "latent" variables) in the PLSR analysis is usually determined by minimizing the prediction residual error sum of squares (PRESS) statistic (SAS 1991; Chen et al. 2004). The PRESS statistic is calculated through a cross-validation prediction for each model (factor). Because of the capability of PLSR to perform dimension reduction and effective feature extraction from multi/hyperspectral remote sensing data, it is not surprising that many studies have explored the potential of PLSR to map forest species composition and abundance (e.g., Apan et al. 2009; Wolter and Townsend 2011) using multi/hyperspectral data.

WT is a relatively new signal-processing tool that provides a systematic means for analyzing signals at various scales or resolutions and shifts. WT can decompose a spectral signal into a series of shifted and scaled versions of the mother wavelet function, and the

TABLE 14.2

Summary of Hyperspectral Transform and Features Extraction Techniques and Methods Frequently Used for Tree Species Mapping

Approach	Characteristics and Description	Advantages and Limitations	Major Factors	Examples
Principal component analysis (PCA)	A linear combination of raw data to reduce dimensionality and preserve variance contained in raw data as much as possible in the first several component images	Dimension reduction and usefully informative feature extraction; not easy to identify which are more signal components	Identify informative features/components	Pu et al. (2008a) and Pu and Liu (2011)
Minimum noise fraction (MNF)	A linear combination of raw data to reduce dimensionality and preserve minimal noise or maximal signal-to-noise ratio in the first several component images	Dimension reduction and usefully informative feature extraction; not easy to identify which are more informative components	Identify informative features/components	Walsh et al. (2008) and Narumalani et al. (2009)
Canonical discriminant analysis (CDA)	CDA searches for a linear combination of independent variables to achieve maximum separation of classes (populations)	Reduction of dimensionality of the raw data while keeping maximum separability of different classes; need knowledge derived from training samples	Large training sample size needed	van Aardt and Wynne (2001, 2007) and Pu and Liu (2011)
Partial least squares regression (PLSR)	Reduce the large number of measured collinear spectral variables to a few noncorrelated latent variables or principal components (PCs)	PLSR produces the weight (coefficient) matrix that reflects the covariance structure between the predictor and the response variables; need knowledge derived from training samples	Large training sample size needed	Apan et al. (2009) and Wolter and Townsend (2011)
Wavelet transform (WT)	WT can decompose signals over scaled and shifted wavelets. The energy feature of the wavelet decomposition coefficients is computed at each scale for both approximation and details and is used to form an energy feature vector that can serve as a feature extraction through a dimension reduction	Dimension reduction and usefully informative feature extraction; the transform procedure is relatively complicated	There exists a close relationship between biophysical parameters related to tree species and spectral features	Zhang et al. (2006) and Banskota et al. (2011)

local energy variation (indicated with "peaks and valleys") of a spectral signal in different bands at each scale can be detected automatically and provides some useful information for further analysis of hyperspectral data (Pu and Gong 2004). With discrete wavelet transforms (DWT), the energy feature of the wavelet decomposition coefficients is computed at each scale for both approximation and details and is used to form an energy feature vector (Pittner and Kamarthi 1999; Bruce et al. 2001; Li et al. 2001; Pu and Gong 2004). This can become a feature extraction through a dimension reduction. With hyperspectral data of vegetation and the WT technique, several studies have already demonstrated the benefits of wavelet analysis. For example, Banskota et al. (2011) discriminated tree species (within *Pinus* genus) by using the DWT applied to airborne hyperspectral data, AVIRIS, and concluded that the WT can improve species discrimination within the *Pinus* genus. Zhang et al. (2006) also used the WT technique and airborne hyperspectral data, HYDICE, to identify tropical tree species at La Selva, Costa Rica. They thought that the WT spectra were useful for the identification of tree species, and the wavelet coefficients at coarse spectral scales and the wavelet energy feature were more capable of reducing variation within crowns/species and capturing spectral differences between species, thus improving tree species identification.

14.3.5 Advanced Classifiers

At present, while traditional multispectral classifiers (e.g., maximum likelihood classifier [MLC] and LDA) are still used in mapping tree species with high spatial resolution satellite data or low-dimensional data transformed from hyperspectral data (e.g., Goodwin et al. 2005; Leckie et al. 2005; Tsai et al. 2007; Banskota et al. 2011; Pu 2011), more advanced algorithms/methods have received attention as a possible improvement in methodology for mapping tree species. They include spectral matching (SM), support vector machines (SVMs), CART, and artificial neural networks (ANN). The characteristics of these methods/algorithms are summarized in Table 14.3.

While a cross-correlogram spectral matching technique (van der Meer and Bakker 1997) and spectral information divergence (Chang 2000) are important algorithms of the SM method and are extensively utilized in extracting and mapping biophysical and biochemical parameters from multi/hyperspectral data, in mapping tree species, the most commonly used algorithm of the SM method is spectral angle mapper (SAM). For example, Hasmadi et al. (2010) used airborne hyperspectral sensor data, AISA, and the SAM algorithm to discriminate eight species of emergent tree crown in Gunung Stong Forest Reserve, Kelantan, Peninsular Malaysia. Their study concluded that high spectral and spatial resolution imagery acquired over tree crown canopy of tropical forest has a substantial potential for individual tree species mapping. In the conventional SAM classification, each species or vegetation class is assumed to have a unique spectral identity or signature. The assumption for each species means that intraspecies variability is not accounted for in the classification (Cho et al. 2010). Consequently, Cho et al. (2010) developed and evaluated the classification performance of a multiple-end-member spectral angle mapper (MESAM) classification approach in discriminating 10 common African savanna tree species with airborne hyperspectral imagery over Kruger National Park, South Africa, and compared the results with the traditional SAM classifier based on a single end-member per species. The application of the MESAM approach increased the producer's and user's accuracies when compared with the conventional SAM, based on the mean spectra of the training data. Similar work using the SAM algorithm with hyperspectral data for tree species classification has been done by Buddenbaum et al. (2005), Zhang et al. (2006), Narumalani et al. (2009),

TABLE 14.3

Summary of Advanced Classification Algorithms and Methods Frequently Used for Tree Spectral Mapping

Algorithm	Characteristics and Description	Advantages and Limitations	Major Factors	Typical Examples
Spectral matching (SM)	With n-dimensional angles (distances, correlations) to match pixels to reference spectra with smaller angles (shorter distances, higher correlations) representing closer matches to the reference spectrum, otherwise the pixel spectrum does not match the reference spectrum	It is a physically based spectral classification and is less sensitive to differences in curve magnitude caused by variation in lighting across a scene, but it is sensitive to noise in any particular band	Determination of threshold of angle (distance, correlation)	Zhang et al. (2006), Narumalani et al. (2009), Cho et al. (2010), and Hasmadi et al. (2010).
Support vector machines (SVMs)	Nonparameteric and supervised classifier that maps data from spectral space into feature space, wherein continuous predictors are partitioned into binary categories by an optimal n-dimensional hyperplane	(1) Handle data efficiently in high dimensionality, (2) deal with noisy samples in a robust way, (3) make use of only those most characteristic samples as the support vectors in construction of the classification models. The mapping data procedure is relatively complicated	Mapping data from the original input feature space to a kernel feature space	Jones et al. (2010)
Classification and regression tree (CART)	Nonparameteric classifier, used for both feature selection and target classification; achieving satisfactory results depends on the determination of the "best" tree structure and the decision boundaries	Computationally fast and makes no statistical assumptions regarding the distribution of data. The "best" tree structure and the decision boundaries are not easy to "find"	Find a "best" tree structure	Santos et al. (2010) and Pu and Landry (2012)
Artificial neural networks (ANN)	A back-propagation algorithm is often used to train the multilayer perception neural network model with input of spectral variables and output of tree species	Nonparametric supervised classifier and has a ability to estimate the properties of data based on limited training samples; the nature of hidden layers is poorly known and it takes time to find a set of ideal structure parameters	Find an ideal architecture: number of hidden layers, learning rate/moment coefficient, and no. of iterations	Pu et al. (2008), Walsh et al. (2008), and Wang et al. (2008)

and Sánchez-Azofeifa et al. (2011). In SM, it is worth noting that the accuracy of these SM techniques is directly affected by the geometry of sensors' observations and target size. Such effect can be minimized by spectral normalization before conducting SM processing (Pieters 1983). In general, such matching techniques are more favorable to change detection of scene components than identifying unknown scene components (Yasuoka et al. 1990).

In machine learning, SVMs are supervised learning models with associated learning algorithms that analyze data and recognize patterns, used for classification and regression analysis (Wikipedia 2012a). SVMs as a new type of classifiers have been successfully applied to the classification of hyperspectral remote sensing data. Traditionally, classifiers first model the density of the various classes and then find a separating surface for classification. However, estimation of density for various classes with hyperspectral data suffers from the Hughes phenomenon (Hughes 1968), which is that for a limited number of training samples, the classification rate decreases as the dimension increases. The SVM approach directly seeks a separating surface through an optimization procedure that finds so-called support vectors that form the boundaries of the classes. This is an interesting property for hyperspectral image processing because usually there is only a set of limited training samples available to define the separating surface for classification. SVMs are considered kernel-based classifiers that are based on mapping data from the original input feature space to a kernel feature space of higher dimensionality and then solving a linear problem in that space (Burges 1998). The SVM approach can effectively map tree species with multi/hyperspectral data. For example, Heikkinen et al. (2010, 2011) focused on the use of multispectral (or simulated hyperspectral) measurements to classify remotely sensed radiance and reflectance information into three tree species, using an SVM algorithm. They demonstrated that the SVM algorithm was effective to identify the tree species, especially mapping through a Mahalanobis kernel that led to a 5%–10% point improvement in classification performance when compared with other kernels. With airborne hyperspectral sensor AISA data, Dalponte et al. (2011) mapped 11 tree species in the southern Alps by using the SVM algorithm. Their experimental results made it clear that airborne hyperspectral data are effective for tree species classification in complex mountain areas ($\kappa = 0.78$). With an object-based classification of mangroves using a hybrid decision tree, SVM approach, Heumann (2011) also achieved a κ value of 0.86 for classifying true mangroves species and other dense coastal vegetation at the object level using a WorldView-2 image.

CART is a nonparametric decision tree learning technique that produces either classification or regression trees, depending on whether the dependent variable is categorical or numeric, respectively (Wikipedia 2012b). Decision trees are formed by a collection of rules based on variables in the modeling dataset: (1) rules based on variables' values are selected to get the best split to differentiate observations based on the dependent variable; (2) once a rule is selected and splits a node into two, the same process is applied to each "child" node; and (3) splitting stops when CART detects no further gain can be made, or some preset stopping rules are met. Alternatively, the data are split as much as possible and then the tree is later pruned. CART can handle a large set of records of training data and can also incorporate ancillary datasets. Based on the training datasets, several decision trees are generated for your choosing. The binary CART for classification has been utilized for land use/land cover classification and change assessment (e.g., Lawrence and Wright 2001; Bittencourt and Clarke 2003; Otukei and Blaschke 2010) and also been used for tree species classification. For example, when testing the feasibility of high-resolution satellite and airborne imagery for the identification of urban forests and to assist planners in preserving them, Sugumaran et al. (2003) identified urban climax forests using a traditional MLC and a rule-based classifier CART. Results showed that both classifiers were effective to map the urban climax forest species. In a study classifying forest species and delineating tree crown using QuickBird imagery, Ke and Quackenbush (2007) examined the capability of the multispectral imagery for species-level forest classification using a rule-based classifier CART with the assistance of ancillary topographic data. Their preliminary results showed that the tree identification and tree crown delineation algorithms were most applicable for coniferous trees in the image. Mora et al. (2010) also explored the use of tree crown metrics derived from high-resolution satellite images for identifying leading species at four study sites in the Yukon Territory, Canada, with a nonparametric classifier CART. The results indicated that the classification tree accurately identified leading species in 72.5% of the stands. Although the CART approach offers advantages not provided by other approaches (e.g., computationally fast and makes no statistical assumptions regarding the distribution of data), achieving satisfactory results depends on the determination of the "best" tree structure and the decision boundaries (Otukei and Blaschke 2010), which may require substantial trial and error to find the "best" tree structure and optimal input features (predictors).

A multilayered feed-forward ANN algorithm is frequently used for classifying vegetation type and mapping other biophysical and biochemical parameters with multi/hyperspectral data. The network training mechanism is an error-propagation algorithm (Rumelhart et al. 1986; Pao 1989). In a layered structure for mapping tree species, as extensively used, the input to each node is the sum of the weighted outputs of the nodes in the prior layer, except for the nodes in the input layer, which are connected to the input features, that is, a set of spectral variables exacted from multi/hyperspectral sensors' data, and each node in output layer corresponds each tree species or group. As a nonparametric and nonlinear classifier, ANN has been utilized by many researchers in tree species mapping in past decades. For example, Pu et al. (2008a) mapped an invasive species (salt cedar) using airborne hyperspectral sensor CASI data and two classifiers, ANN and LDA. Their experimental results indicate that the ANN outperforms the LDA due to the ANN's nonlinear property and ability of handling data without a prerequisite of a certain distribution of the analysis data. By using QuickBird and Hyperion multi/hyperspectral data, Walsh et al. (2008) complement a spatially and spectrally explicit satellite assessment of an important invasive plant species, guava, on Isabela Island, Ecuador, that integrates diverse remote sensing systems, data types, spatial and spectral resolutions, and analytical and image processing approaches: OBIA, linear mixture model, and ANN. They found a similar characterization of the percentage of guava at the pixel level (from QuickBird) when plotted against the area metric (unmixed from Hyperion), whereas ANN produced a higher spectral unmixed result for guava than that using the linear mixture model. Wang et al. (2008) also achieved a success with an ANN algorithm to map mangrove species from multiseasonal IKONOS imagery.

14.4 Considerations and Future Directions

14.4.1 Considerations

To improve tree species mapping results with multi/hyperspectral data, there are several factors that should be considered when classifying tree species and mapping forest species composition. The factors may include spatial/spectral resolution effect, temporal resolution or seasonal effect, forest environmental setting, shadow and shaded effect, spectral variation in IOs or tree crowns when using OBIA, and different tree species overlapping effect.

(1) Given the relatively small individual tree crown size and spectral similarity among tree species relative to existing remote sensing sensors' spatial/spectral resolution, we should consider appropriate remote sensing data to be chosen for mapping tree species. In a tree species classification project, people should consider minimal individual tree crown size in the study area to determine image spatial resolution. Generally accepted guidelines for minimum image spatial resolution suggest that an object must be covered by at least four pixels to accurately identify the object (Jensen 2007). That means that an ideal image spatial resolution should be higher than the radius of the minimal tree crown in the study area. Choosing hyperspectral data, if available, for mapping tree species is a smart way to improve tree species classification or species abundance estimation. (2) Seasonal effect is an important factor. If using single-sensor data to map tree species/ groups, an effort should be made to select an optimal image acquisition season when there exists a great spectral variation among tree species/groups under investigation. Therefore, selecting an optimal season for acquiring the satellite data should be considered as a research topic to improve tree species identification accuracy. Unfortunately, as was our experience, competition for use of commercial sensors and atmospheric conditions often prevents image acquisition on the optimal target dates. (3) Given the complexity and heterogeneity of artificial materials in the urban environment, the impact of urban tree background (i.e., reflection of below-canopy surfaces) on crown spectra is very different from that of forest areas in wildland. The complex and heterogeneous urban environment, and the relatively low spatial resolution of imagery relative to crown size of some tree species, makes tree crown spectra mix with varying background spectra. For example, the 2 m spatial resolution of WorldView-2 multispectral bands is insufficient to capture pure pixels of the small crown diameter of most small magnolia and palm (i.e., <2 m) in the City of Tampa area, FL, United States (Pu and Landry 2012). (4) Spectral contrast among tree species/groups is often reduced in the urban environment as a result of the shadow/shaded impact on tree crown spectra caused by adjacent buildings, other architectures, and large tree crowns. In a forest wildland, there also exists such a shadow/ shaded effect on tree crown spectrum. Although many vegetation indices and other ratio features extracted from images can reduce or compress the effect of shadow/shaded on tree canopy spectrum, the remaining effect still exists. Tree species classification that separates sunlit and shadow/shaded areas may be one way to reduce the effect on tree species classification. (5) Although OBIA outperforms pixel-based approaches in vegetation classification (e.g., Yu et al. 2006; Pu et al. 2011) and tree species identification (e.g., Kim et al. 2011), the image segmentation process may also cause a greater spectral variation among some individual tree species (IOs) compared to that among pixel-based individual tree species. This is easy to understand when we consider that the spectral variation within individual tree canopy IOs always exists whereas pure pixel-based individual tree canopy spectra do not have the spectral variation. Therefore, it is important to execute the image segmentation process with a set of ideal parameters (scale, color, smoothness, etc.) to create optimal IOs for mapping tree species/groups. (6). In dense urban forests and mixed species forest areas, two different tree species can be found very close to each other with their canopies overlapping. Even with high spatial resolution imagery, the overlapping region will still produce mixed pixels. For this issue, an ideal solution key may be utilizing an MESMA technique to obtain species abundance from the mixed pixels with hyperspectral imagery.

14.4.2 Future Directions

In spite of advances in recent years in remote sensing in improving both spatial and spectral resolutions, relative to minimal individual tree crown size in the study area (e.g., the diameter of most magnolia and palm trees in the City of Tampa, FL, United States, is smaller than 2 m) (Pu and Landry 2012), the spatial resolution of current high spatial resolution sensors (e.g., WorldView-2, multispectral data with 2 m resolution) is still relatively low (although we can use WorldView-2 pan-sharpening data at a nominal spatial resolution of 0.5 m); given the spectral similarity among tree species, the spectral resolution of current multispectral (e.g., WorldView-2 and IKONOS) and some hyperspectral (e.g., AVIRIS and Hyperion) sensors is also relatively low. Therefore, it is necessary to continuously improve image quality from spatial/spectral resolutions of sensors. Pu (2009) used in situ hyperspectral measurements to demonstrate such spectral similarity among 11 urban tree species, especially the five oak tree species that had a very similar reflectance pattern across wavelengths 400–2400 nm. However, because there were a few narrow bands that presented large variation of spectra among the 11 tree species (e.g., at 620 and 900 nm), further increasing spectral resolution is a direction to improve the accuracy of mapping tree species.

The use of multisensor data, which can synergize the high spatial resolution (e.g., commercial satellite sensors: WorldView-2 and IKONOS) and high temporal resolution (e.g., satellite hyperspectral sensors: Hyperion and HyspIRI [NRC 2007]), provides the potential to more accurately map tree species through integration of different features of sensor data. Different forest tree species and types have various phenologies. High temporal resolution satellite hyperspectral image data (with relatively low spatial resolution) acquired with suitable phenologic states to different tree species may provide sufficient information to classify them, allowing the applications of more complex spectral analyses and spectral unmixing techniques to map forest species and species composition. The disadvantage of using multisensor data for mapping tree species is the difficulty in various image acquisition/ processing and use of appropriate identifying and mapping techniques. In the future, after satellite hyperspectral sensor HyspIRI is in operation, the high frequent hyperspectral data will be available globally. Application of multisensor data will become increasingly important in the future study of identifying and mapping tree species (abundance), and thus more advanced image processing, tree species classification, and mapping techniques are needed. Accurately identifying and mapping tree species with multi/hyperspectral data remains an active research topic and new techniques continue to be developed. For an advanced tree species mapping technique, it is required to be easy to use and it should provide accurate classification and mapping results.

In the future, the richness of spectral information available in the continuous spectral coverage afforded by both airborne and spaceborne hyperspectral sensors/systems makes it possible to map tree species or estimate tree species abundance more correctly and more accurately. While some airborne hyperspectral sensors (e.g., AISA sensor) can collect hyperspectral data with a spatial resolution up to <1 m, which should have potential for urban forest mapping in the future, future satellite hyperspectral systems/sensors (e.g., HyspIRI sensor at 60 m) will collect hyperspectral data at a relatively low spatial resolution. To efficiently use the richer and more delicate spectral information of the satellite hyperspectral data (mostly with a high temporal resolution) for mapping tree species, SMA, or multi-end-member SMA to automatically estimate individual tree species abundance within pixels at a relatively low spatial resolution remains to be an important information extraction task for hyperspectral data analysis. Therefore, developing various spectral unmixing algorithms, especially nonlinear spectral unmixing techniques, continues to be an important research topic.

References

Apan, A., S. Phinn, and T. Maraseni. 2009. "Discrimination of Remnant Tree Species and Regeneration Stages in Queensland, Australia Using Hyperspectralimagery." *The First IEEE GRSS Workshop on Hyperpsectral Image and Signal Processing—Evolution in Remote Sensing (WHISPERS'09)*, August 26–28, 2009, Grenoble, France.

Banskota, A., R. H. Wynne, and N. Kayastha. 2011. "Improving Within-Genus Tree Species Discrimination Using the Discrete Wavelet Transform Applied to Airborne Hyperspectral Data." *International Journal of Remote Sensing* 32(13):3551–63.

Bittencourt, H. R., and R. T. Clarke. 2003. "Use of Classification and Regression Trees (CART) to Classify Remotely-Sensed Digital Images." *IGARSS '03 Proceedings*. Accessed February 27, 2013, http://ieeexplore.ieee.org/stamp/stamp.jsp?arnumber=01295258.

Blaschke, T. 2010. "Object Based Image Analysis for Remote Sensing." *ISPRS Journal of Photogrammetry and Remote Sensing* 65:2–16.

Brockhaus, J. A., and S. Khorram. 1992. "A Comparison of SPOT and Landsat TM Data for Use in Conducting Inventories of Forest Resources." *International Journal of Remote Sensing* 13:3035–43.

Bruce, L. M., C. Morgan, and S. Larsen. 2001. "Automated Detection of Subpixel Hyperspectral Targets with Continuous and Discrete Wavelet." *IEEE Transactions on Geoscience and Remote Sensing* 39:2217–26.

Buddenbaum, H., M. Schlerf, and J. Hill. 2005. "Classification of Coniferous Tree Species and Age Classes Using Hyperspectral Data and Geostatistical Methods." *International Journal of Remote Sensing* 26(24):5453–65.

Burges, C. J. C. 1998. "A Tutorial on Support Vector Machines for Pattern Recognition." *Data Mining and Knowledge* Discover 2(2):121–67.

Carleer, A., and E. Wolff. 2004. "Exploitation of very high resolution satellite data for tree species identification." *Photogrammetric Engineering and Remote Sensing* 70(1):135–40.

Carleer, A. P., and E. Wolff. 2006. "Region-Based Classification Potential for Land-Cover Classification with Very High Spatial Resolution Satellite Data. *Proceedings of 1st International Conference on Object-Based Image Analysis (OBIA 2006)*, July 4–5, 2006, Vol. XXXVI, ISSN 1682-1777. Salzburg University, Austria.

Carreiras, J. M. B., J. M. Pereira, and J. S. Pereira. 2006. "Estimation of Tree Canopy Cover in Evergreen Oak Woodlands Using Remote Sensing." *Forest Ecology and Management* 223(1–3):45–53.

Chang, C.-T. 2000. "An Information-Theoretic Approach to Spectral Variability, Similarity, and Discrimination for Hyperspectral Image Analysis." *IEEE Transactions on Geoscience and Remote Sensing* 46:1927–32.

Chen, S., X. Hong, C. J. Harris, and P. M. Sharkey. 2004. "Spare Modeling Using Orthogonal Forest Regression with PRESS Statistic and Regularization." *IEEE Transaction on Systems, Man and Cybernetics* 34:898–911.

Cho, M. A., P. Debba, R. Mathieu, L. Naidoo, J. van Aardt, and G. P. Asner. 2010. "Improving Discrimination of Savanna Tree Species through a Multiple-Endmember Spectral Angle Mapper Approach: Canopy-Level Analysis." *IEEE Transactions on Geoscience and Remote Sensing* 48(11):4133–42.

Congalton, R., J. Miguel-Ayanz, and B. Gallup. 1991. *Remote Sensing Techniques for Hardwood Mapping*. Contract Report to California Department of Forestry and Fire Protection, Sacramento, CA.

Dalponte, M., L. Bruzzone, and D. Gianelle. 2011. "Tree Species Classification in the Southern Alps with Very High Geometrical Resolution Multispectral and Hyperspectral Data." *The Third IEEE GRSS Workshop on Hyperpsectral Image and Signal Processing—Evolution in Remote Sensing (WHISPERS'11)*, June 6–9, 2011, Lisbon, Portugal.

Davis, C. H., and X. Wang. 2003. "Planimetric Accuracy of Ikonos 1 m Panchromatic Orthoimage Products and Their Utility for Local Government GIS Basemap Applications." *International Journal of Remote Sensing* 24(22):4267–88.

Franklin, S. E. 1994. "Discrimination of Subalpine Forest Species and Canopy Density Using Digital CASI, SPOT, and Landsat TM Data." *Photogrammetric Engineering and Remote Sensing* 60:1233–41.

Gong, P., and A. Zhang. 1999. "Noise Effect on Linear Spectral Unmixing." *Geographic Information Sciences* 5:52–7.

Goodwin, N., R. Turner, and R. Merton. 2005. "Classifying Eucalyptus Forests with High Spatial and Spectral Resolution Imagery: An Investigation of Individual Species and Vegetation Communities." *Australian Journal of Botany* 53:337–45.

Hájek, F. 2006. "Object-Oriented Classification of Ikonos Satellite Data for the Identification of Tree Species Composition." *Journal of Forest Science* 52(4):181–87.

Hasmadi, I. M., J. Kamaruzaman, and M. A. N. Hidayah. 2010. "Analysis of Crown Spectral Characteristic and Tree Species Mapping of Tropical Forest Using Hyperspectral Imaging." *Journal of Tropical Forest Science* 22(1):67–73.

Heikkinen, V., I. Korpela, T. Tokola, E. Honkavaara, and J. Parkkinen. 2011. "An SVM Classification of Tree Species Radiometric Signatures Based on the Leica ADS40 Sensor." *IEEE Transactions on Geoscience and Remote Sensing* 49(11):4539–51.

Heikkinen, V., T. Tokola, J. Parkkinen, I. Korpela, and T. Jääskeläinen. 2010. "Simulated Multispectral Imagery for Tree Species Classification Using Support Vector Machines." *IEEE Transactions on Geoscience Remote Sensing* 48(3):1355–64.

Heumann, B. W. 2011. "An Object-Based Classification of Mangroves Using a Hybrid Decision Tree—Support Vector Machine Approach." *Remote Sensing* 3:2440–60.

Hughes, G. F. 1968. "On the Mean Accuracy of Statistical Pattern Recognizers." *IEEE Transactions on Information Theory* 14(1):55–63.

Jensen, J. R. 2007. *Remote Sensing of the Environment: An Earth Resource Perspective*. 2nd ed., 444–450. Upper Saddle River, NJ: Prentice Hall.

Jones, T. G., N. C. Coops, and T. Sharma. 2010. "Assessing the Utility of Airborne Hyperspectral and LiDAR Data for Species Distribution Mapping in the Coastal Pacific Northwest, Canada." *Remote Sensing of Environment* 114:2841–52.

Katoh., M. 1988. "Estimation of rates of the crown area in yezo spruce plantations with Landsat Thematic Mapper data." In: *Proceedings of the 16th ISPRS/IUFRO Kyoto*, 23–29.

Katoh, M. 2004. "Classifying Tree Species in a Northern Mixed Forest Using High-Resolution IKONOS Data." *Journal of Forest Research* 9:7–14.

Ke, Y., and L. J. Quackenbush. 2007. "Forest Species Classification and Tree Crown Delineation Using QuickBird Imagery." *ASPRS 2007, Annual Conference*. May 7–11, 2007, Tampa, FL.

Ke, Y., L. J. Quackenbush, and J. Im. 2010. "Synergistic Use of QuickBird Multispectral Imagery and LIDAR Data for Object-Based Forest Species Classification." *Remote Sensing of Environment* 114(6):1141–54.

Khattree, R., and D. N. Naik. 2000. *Multivariate Data Reduction and Discrimination with SAS Software*, 558pp. Cary, NC: SAS Institute Inc.

Kim, S.-R., W.-K. Lee, D.-A. Kwak, G. S. Biging, P. Gong, J.-H. Lee, and H.-K. Cho. 2011. "Forest Cover Classification by Optimal Segmentation of High Resolution Satellite Imagery." *Sensors* 11:1943–58.

Lass, L. W., and T. S. Prather. 2004. "Detecting the Locations of Brazilian Pepper Trees in the Everglades with a Hyperspectral Sensor." *Weed Technology* 18:437–42.

Lawrence, R. L., and A. Wright. 2001. "Rule-Based Classification Systems Using Classification and Regression Tree (CART) Analysis." *Photogrammetric Engineering & Remote Sensing* 67(10):1137–42.

Leckie, D. G., F. A. Gougeon, S. Tinis, T. Nelson, C. N. Burnett, and D. Paradine. 2005. "Automated Tree Recognition in Old Growth Conifer Stands with High Resolution Digital Imagery." *Remote Sensing of Environment* 94:311–26.

Li, J., L. M. Bruce, J. Byrd, and J. Barnett. 2001. "Automated Detection of *Pueraria montana* (Kudzu) through Haar Analysis of Hyperspectral Reflectance Data." *IEEE International Geoscience and Remote Sensing Symposium*, July 9–13, 2001, Sydney, Australia.

Möller, M., L. Lymburner, and M. Volk. 2007. "The Comparison Index: A Tool for Assessing the Accuracy of Image Segmentation." *International Journal of Applied Earth Observation and Geoinformation* 9(3):311–21.

Mora, B., M. A. Wulder, and J. C. White. 2010. "Identifying Leading Species Using Tree Crown Metrics Derived from Very High Spatial Resolution Imagery in a Boreal Forest Environment." *Canadian Journal of Remote Sensing* 36(4):332–44.

Narumalani, S., D. R. Mishra, R. Wilson, P. Reece, and A. Kohler. 2009. "Detecting and Mapping Four Invasive Species along the Floodplain of North Platte River, Nebraska." *Weed Technology* 23:99–107.

NRC's Decadal Survey report. 2007. *Earth Science and Applications from Space: National Imperatives for the Next Decade and Beyond*. ISBN-13: 978-0-309-14090-4. The National Academic Press. http://www.nap.edu/catalog/11820.html.

Otukei, J. R., and T. Blaschke. 2010. "Land Cover Change Assessment Using Decision Trees, Support Vector Machines and Maximum Likelihood Classification Algorithms." *International Journal of Applied Earth Observation and Geoinformation* 12S:S27–31.

Painter, T. H., D. A. Roberts, R. O. Green, and J. Dozier. 1998. "The Effect of Grain Size on Spectral Mixture Analysis of Snow-Covered Area from AVIRIS Data." *Remote Sensing of Environment* 65:320–32.

Pao, Y. 1989. *Adaptive Pattern Recognition and Neural Networks*. New York: Addison and Wesley.

Papeş, M., R. Tupayachi, P. Martínez, A. T. Peterson, and G. V. N. Powell. 2010. "Using Hyperspectral Satellite Imagery for Regional Inventories: A Test with Tropical Emergent Trees in the Amazon Basin." *Journal of Vegetation Science* 21:342–54.

Pieters, C. M. 1983. "Strength of Mineral Absorption Features in the Transmitted Component of Near-Infrared Reflected Light: First Results from RELAB." *Journal of Geophysical Research* 88:9534–44.

Pittner, S., and S. V. Kamarthi. 1999. "Feature Extraction from Wavelet Coefficients for Pattern Recognition Tasks." *IEEE transactions on Pattern Analysis and Machine Intelligence* 21:83–8.

Pu, P., S. Landry, and Q. Yu. 2011. "Object-Based Urban Detailed Land Cover Classification with High Spatial Resolution IKONOS Imagery." *International Journal of Remote Sensing* 32(12):3285–308.

Pu, R. 2009. "Broadleaf Species Recognition with In Situ Hyperspectral Data." *International Journal of Remote Sensing* 30(11):2759–79.

Pu, R. 2011. "Mapping Urban Forest Tree Species Using IKONOS Imagery: Preliminary Results." *Environmental Monitoring and Assessment* 172:199–214.

Pu, R. 2012. "Detecting and Mapping Invasive Plant Species by Using Hyperspectral Data." Chap. 19 in *Hyperspectral Remote Sensing of Vegetation*, edited by P. S. Thenkabail, J. G. Lyon, and A. Huete, 447–65 CRC Press Taylor & Francis Group.

Pu, R., and D. Liu. 2011. "Segmented Canonical Discriminant Analysis of In Situ Hyperspectral Data for Identifying Thirteen Urban Tree Species." *International Journal of Remote Sensing* 32(8): 2207–26.

Pu, R., and P. Gong. 2004. "Wavelet Transform Applied to EO-1 Hyperspectral Data for Forest LAI and Crown Closure Mapping." *Remote Sensing of Environment* 91:212–24.

Pu, R., P. Gong, Y. Tian, X. Miao, and R. Carruthers. 2008a. "Invasive Species Change Detection Using Artificial Neural Networks and CASI Hyperspectral Imagery." *Environmental Monitoring and Assessment* 140:15–32.

Pu, R., P. Gong, Y. Tian, X. Miao, R. Carruthers, and G. L. Anderson. 2008b. "Using Classification and NDVI Differencing Methods for Monitoring Sparse Vegetation Coverage: A Case Study of Saltcedar in Nevada, USA." *International Journal of Remote Sensing* 29(14):1987–4011.

Pu, R., M. Kelly, G. L. Anderson, and P. Gong. 2008c. "Using CASI Hyperspectral Imagery to Detect Mortality and Vegetation Stress Associated with a New Hardwood Forest Disease." *Photogrammetric Engineering and Remote Sensing* 74(1):65–75.

Pu, R., and S. Landry. 2012. "A Comparative Analysis of High Resolution Ikonos and Worldview-2 Imagery for Mapping Urban Tree Species." *Remote Sensing of Environment* 124:516–33.

Quackenbush, L. J., P. F. Hopkins, and G. J. Kinn. 2000. "Developing Forestry Products from High Resolution Digital Aerial Imagery." *Photogrammetric Engineering and Remote Sensing* 66(11):1337–46.

Ramsey III, E., A. Rangoonwala, G. Nelson, R. Ehrlich, and K. Martella. 2005a. "Generation and Validation of Characteristic Spectra from EO1 Hyperion Image Data for Detecting the Occurrence of the Invasive Species, Chinese Tallow." *International Journal of Remote Sensing* 26(8):1611–36.

Ramsey III, E., A. Rangoonwala, G. Nelson, and R. Ehrlich. 2005b. "Mapping the Invasive Species, Chinese Tallow, with EO1 Satellite Hyperion Hyperspectral Image Data and Relating Tallow Occurrences to a Classified Landsat Thematic Mapper Land Cover Map." *International Journal of Remote Sensing* 26(8):1637–57.

Roberts, D. A., P. E. Dennison, M. Gardner, Y. Hetzel, S. L. Ustin, and C. Lee. 2003. "Evaluation of the Potential of Hyperion for Fire Danger Assessment by Comparison to the Airborne Visible/Infrared Imaging Spectrometer." *IEEE Transactions on Geoscience and Remote Sensing* 41(6):1297–310.

Roberts, D. A., M. Gardner, R. Church, S. Ustin, G. Scheer, and R. O. Green. 1998. "Mapping Chaparral in the Santa Monica Mountains Using Multiple Endmember Spectral Mixture Models." *Remote Sensing of Environment* 65:267–79.

Rumelhart, D. E., G. E. Hinton, and R. J. Williams. 1986. "Learning Internal Representations by Error Propagation." In *Parallel Distributed Processing—Explorations in the Microstructure of Cognition*, edited by D. E. Rumelhart, J. L. McClelland, and the PDP Research Group, Vol. 1, 318–62.

Sánchez-Azofeifa, A., B. Rivard, J. Wright, J.-L. Feng, P. Li, M. M. Chong, and S. A. Bohlman. 2011. "Estimation of the Distribution of *Tabebuia guayacan* (Bignoniaceae) Using High-Resolution Remote Sensing Imagery." *Sensors* 11:3831–51.

Santos, M. J., J. A. Greenberg, and S. L. Ustin. 2010. "Using hyperspectral remote sensing to detect and quantify southeastern pine senescence effects in red-cockaded woodpecker (Picoides borealis) habitat." *Remote Sensing of Environment* 114:1242–50.

Sasaki, K., S. Kawata, and S. Minami. 1984. "Estimation of Component Spectral Curves from Unknown Mixture Spectra." *Applied Optics* 23:1955–59.

SAS Institute Inc. 1991. *SAS/STA User's Guide*, Release 6.03 ed., 1028pp. Gary, NC: SAS Institute Inc.

Settle, J. J., and N. A. Drake. 1993. "Linear Mixing and the Estimation of Ground Cover Proportions." *International Journal of Remote Sensing* 14:1159–77.

Spirkovska, L. 1993. *A Summary of Image Segmentation Techniques, NASA Technical Memorandum 104022*, 11pp. Ames Research Center, Moffett Field, California. Accessed October 11, 2012, http://ntrs.larc.nasa.gov/search.jsp?R=19940006802&qs=Ns%3DNASA-Center%7C0%26N%3D4294706898.

Sridharan, H. 2011. "Urban Forest Mapping Using Worldview-2 8 Band Imagery." Submitted to Student Honors Paper Competition of the Remote Sensing Specialty Group, Association of American Geographers, 2012 Annual Meeting. New York. Accessed February 27, 2013. http://meridian.aag.org/callforpapers/program/AbstractDetail.cfm?.

Sugumaran, R., M. K. Pavuluri, and D. Zerr. 2003. "The Use of High-Resolution Imagery for Identification of Urban Climax Forest Species Using Traditional and Rule-Based Classification Approach." *IEEE Transactions on Geoscience and Remote Sensing* 41(9):1933–39.

Townsend, P. A., and S. J. Walsh. 2001. "Remote Sensing of Forested Wetlands: Application of Multitemporal and Multispectral Satellite Imagery to Determine Plant Community Composition and Structure in Southeastern USA." *Plant Ecology* 157:129–49.

Tsai, F., E. E. Lin, and K. Yoshino. 2007. "Spectrally Segmented Principal Component Analysis of Hyperspectral Imagery for Mapping Invasive Plant Species." *International Journal of Remote Sensing* 28(5):1023–39.

van Aardt, J. A. N., and R. H. Wynne. 2001. "Spectral Separability among Six Southern Tree Species." *Photogrammetric Engineering and Remote Sensing* 67(12):1367–75.

van Aardt, J. A. N., and R. H. Wynne. 2007. "Examining Pine Spectral Separability Using Hyperspectal Data from an Airborne Sensor: An Extension of Field-Based Results." *International Journal of Remote Sensing* 28(2):431–36.

van der Meer, F., and W. Bakker. 1997. "CCSM: Cross Correlogram Spectral Matching." *International Journal of Remote Sensing* 18(5):1197–201.

Varshney, S. S., N. Rajpa, and R. Purwar. 2009. "Comparative Study of Image Segmentation Techniques and Object Matching Using Segmentation." *International Conference on Methods and Models in Computer Science*, 6pp. Proceedings of a meeting held December 14-15, 2009, New Delhi, India. Accessed February 27, 2013, http://ieeexplore.ieee.org/xpls/abs_all .jsp?arnumber=5397985.

Voss, M., and R. Sugumaran. 2008. "Seasonal Effect on Tree Species Classification in an Urban Environment Using Hyperspectral Data, LiDAR, and an Object-Oriented Approach." *Sensors* 8:3020–36.

Walsh, S. J., A. L. McCleary, C. F. Mena, Y. Shao, J. P. Tuttle, A. González, and R. Atkinson. 2008. "QuickBird and Hyperion Data Analysis of an Invasive Plant Species in the Galapagos Islands of Ecuador: Implications for Control and Land Use Management." *Remote Sensing of Environment* 112:1927–41.

Wang, Z., J. R. Jensen, and J. Im. 2010. "An Automatic Region-Based Image Segmentation Algorithm for Remote Sensing Applications." *Environmental Modelling and Software* 25:1149–65.

Wang, L., J. L. Silván-Cárdenas, and W. P. Sousa. 2008. "Neural Network Classification of Mangrove Species from Multi-Seasonal IKONOS Imagery." *Photogrammetric Engineering & Remote Sensing* 74(7):921–27.

Wikipedia, the free encyclopedia. 2012a. *Support Vector Machine*. Accessed October 11, 2012, http:// en.wikipedia.org/wiki/Support_vector_machine.

Wikipedia, the free encyclopedia. 2012b. *Predictive Analytics*. Accessed October 11, 2012, http:// en.wikipedia.org/wiki/Predictive_analytics#Classification_and_regression_trees.

Wolter, P. T., and P. A. Townsend. 2011. "Multi-Sensor Data Fusion for Estimating Forest Species Composition and Abundance in Northern Minnesota." *Remote Sensing of Environment* 115:671–91.

Xiao, Q., S. L. Ustin, and E. G. McPherson. 2004. "Using AVIRIS Data and Multiple-Masking Techniques to Map Urban Forest Tree Species." *International Journal of Remote Sensing* 25(24):5637–54.

Yasuoka, Y., T. Yokota, T. Miyazaki, and Y. Iikura. 1990. "Detection of Vegetation Change from Remotely Sensed Images Using Spectral Signature Similarity." *Proceedings of the International Geoscience and Remote Sensing Symposium (IGRSS'90)*, May 20–24, 1990, 1609–12. College Park, MD, Piscataway, NJ: IEEE.

Yu, Q., P. Gong, N. Clinton, G. Biging, M. Kelly, and D. Schirokauer. 2006. "Object-Based Detailed Vegetation Classification with Airborne High Spatial Resolution Remote Sensing Imagery." *Photogrammetric Engineering and Remote Sensing* 72(7):799–811.

Zhang, J., B. Rivard, A. Sánchez-Azofeifa, and K. Castro-Esau. 2006. "Intra- and Inter-Class Spectral Variability of Tropical Tree Species at La Selva, Costa Rica: Implications for Species Identification Using HYDICE Imagery." *Remote Sensing of Environment* 105:129–41.

Zhang, L., D. Li, Q. Tong, and L. Zheng. 1998. "Study of the Spectral Mixture Model of Soil and Vegetation in Poyang Lake Area, China." *International Journal of Remote Sensing* 19:2077–84.

Zhao, G., and A. L. Maclean. 2000. "A Comparison of Canonical Discriminant Analysis and Principal Component Analysis for Spectral Transformation." *Photogrammetric Engineering & Remote Sensing* 66(7):841–47.

Zuva, T., O. O. Olugbara, S. O. Ojo, and S. M. Ngwira. 2011. "Image Segmentation, Available Techniques, Developments and Open Issues." *Canadian Journal on Image Processing and Computer Vision* 2(3):20–9.

15

Estimation of Forest Stock and Yield Using LiDAR Data

Markus Holopainen, Mikko Vastaranta, Xinlian Liang,
Juha Hyyppä, Anttoni Jaakkola, and Ville Kankare

CONTENTS

15.1 State of the Art in Forest Inventory 260
15.2 Laser Scanning in Measuring Forests 260
15.3 Comparison of Airborne, Mobile, and Terrestrial Laser Scanning 262
15.4 Point Cloud Metrics and Surface Models 265
    15.4.1 Creation of Terrain, Surface, and Canopy Height Models 265
    15.4.2 Geometric Features 266
    15.4.3 Vertical Point Height Distributions 266
15.5 Area-Based Approach 266
15.6 Individual Tree Detection 267
15.7 Applications of Airborne Laser Scanning 268
    15.7.1 Estimation of Tree and Stand Variables 268
        15.7.1.1 An Area-Based Approach 268
        15.7.1.2 Individual Tree Detection 270
        15.7.1.3 Tree Cluster Approach 271
    15.7.2 Predicting Forest Growth and Site Type 272
    15.7.3 Mapping of Forest Management Operations 273
    15.7.4 Forest Biomass and Disturbance Monitoring 273
    15.7.5 Predicting Forest Stock and Yield Value 274
    15.7.6 Sampling-Based Forest Inventories Using Airborne Laser Scanning 275
    15.7.7 Value of ALS Inventory Information in Forest Management Planning 276
15.8 Terrestrial and Mobile Laser Scanning Applications 276
    15.8.1 TLS Applications 276
        15.8.1.1 Multiscan or Single-Scan TLS Measurements? 277
        15.8.1.2 Stem Location Mapping with TLS 278
        15.8.1.3 Determination of Diameter Breast Height and Stem Curve
                with TLS 278
        15.8.1.4 Acquisition of Other Tree Attribute Data from TLS 279
    15.8.2 MLS Applications 279
15.9 Conclusions 280
References 281

15.1 State of the Art in Forest Inventory

The starting point of this chapter is Finnish forest inventory (FI) and methods developed mainly in Finland and in other Nordic countries, that is, we mainly focus on studies performed here. However, for comparison, methodologies developed in Central Europe, the United States, and Canada are also discussed.

Forest resource information is needed for large-scale strategic planning, operative forest management, and preharvest planning. National forest inventories (NFIs) are examples of inventories undertaken for large-scale strategic planning for gathering information about nationwide forest resources, such as growing stock volume, forest cover, growth and yield, biomass, carbon balance, and large-scale wood procurement potential. In NFIs, it is important to have unbiased estimates and to obtain information also from small strata. The making of inventories of forest resources has a long tradition in Finnish forest sciences, making Finland among the first countries in the world to take such measures: A sampling-based FI covering the whole country was introduced more than 90 years ago (NFI1 1920–1924). Finnish foresters were also pioneers in developing new inventory methodologies when multisource forest inventories were introduced in the early 1990s (Kilkki and Päivinen 1987; Tokola 1988; Muinonen and Tokola 1990; Tomppo 1990). However, operational forest management planning has been based on stand-wise field inventory (SWFI) for more than 60 years in Finland. The potential of remote sensing (RS), such as the utilization of satellite—radar—and aerial images in the estimation of forest variables, has been studied intensively, but the methodologies have not become generally used in practice. The reason is simple: the accuracy obtained in forest variable estimation at the stand level using RS data has not been adequate for forest management or preharvest planning.

During the past decade, RS has taken a significant technological leap forward, as it became possible to acquire three-dimensional (3D), spatially accurate information from forest resources using active RS methods. In practical applications, mainly airborne laser scanning (ALS) has opened up groundbreaking potential in natural resource mapping and monitoring (see the principle of ALS in Figure 15.1). ALS collects 3D information from forest resources, which enables a highly accurate estimation of tree or stand variables.

Forest mapping and monitoring is carried out to support decision making by the forest owner. In SWFI, wood procurement potential, the amount of roundwood removal, and forest management proposals are mapped and determined. In addition to stand variables, site types are classified into map forest growth potential, thinning regime, and biodiversity. Forest growth and yield are also highly correlated with forest estate value. The wood procurement chain from forest to users starts with knowledge of the stands available for harvesting.

Currently, the retrieval of stand variables, which is needed in forest management planning, is being replaced by ALS-based inventory methodologies in the Nordic countries. Relatively new ALS-based inventory methodologies were adopted quickly after the first promising studies (e.g., Nilsson 1996; Næsset 1997a, 1997b, 2002; Hyyppä and Hyyppä 1999; Hyyppä and Inkinen 1999).

15.2 Laser Scanning in Measuring Forests

Laser scanning (LS or light detection and ranging [LiDAR]) is an active RS technique that uses the time-of-flight measurement principle to measure the distance to an object. With

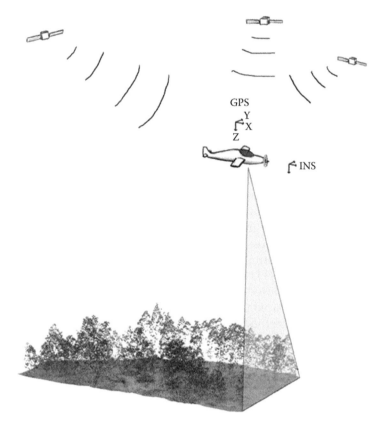

FIGURE 15.1
Principle of airborne laser scanning. (Copyright Ville Kankare.)

the known position of the sensor and precise orientation of these range measurements between the sensor and a reflecting object, the position (x, y, or z) of an object is defined. The principle of LS measurements is the same regardless of the placement of the scanner.

A laser pulse that strikes the forest canopy can produce one or more returns. In the simplest case, a laser pulse scatters directly from the top of the dense forest canopy or from the ground, resulting in a single return. Since the forest canopy is not a solid surface and gaps are present in the canopy cover, the situation becomes more complex when a laser pulse that hits the forest canopy passes through the top of the canopy and intercepts different parts of the canopy, such as the trunk, branches, and leaves, before reaching the ground. This series of events may result in several returns being recorded for a single laser pulse, which is referred to as multiple returns. In most cases, these multiple returns are recorded. Some systems record the full waveform of the reflected laser pulse as well. The first returns are mainly assumed to come from the top of the canopy and the last returns mainly from the ground, which is important for extracting the terrain surface. Multiple returns produce useful information regarding the forest structure (Hyyppä et al. 2009b).

The trunks, branches, and leaves in dense vegetation tend to cause multiple scattering or absorption of the emitted laser energy so that fewer backscattered returns are reflected directly from the ground (Harding et al. 2001; Hofton et al. 2002). This effect increases when the canopy closure, canopy depth, and structure complexity increase because the laser pulse is greatly obscured by the canopy. In practice, the laser system specification and configurations also play an important role in how the laser pulse interacts with the forest. For example, it has been found that a small footprint laser tends to penetrate to the tree crown before reflecting a signal (Gaveu and Hill 2003); ground returns decrease as the scanning angle increases (TopoSys 1996); the penetration rate is affected by the laser beam divergence (Aldred and Bonnor 1985; Næsset 2004); a higher flight altitude alters the distribution of laser returns from the top and within the tree canopies (Næsset 2004); and the distribution of laser returns through the canopy varies with the change in laser pulse repetition frequency (Chasmer et al. 2006). Furthermore, the sensitivity of the laser receiver, wavelength, laser power, and total backscattering energy from the treetops are also factors that may influence the ability of laser pulses to penetrate and distribute laser returns from the forest canopy (Baltavias 1999).

Two main approaches to derive forest information from ALS data have been used: area-based approach (ABA) (Næsset 2002) and individual tree detection (ITD) (Hyyppä and Inkinen 1999). In the former method, statistics calculated from the laser point cloud are used as predictors, and the retrieval of forest variables is typically based on nearest-neighbor (NN) or regression estimation using the laser-derived metrics and tree-by-tree measured field plots. With the ITD method, individual trees are recognized or segmented from the laser point cloud, and tree-level variables are determined either straight from the point cloud or are estimated based on various other ALS features that are extracted for the tree segments using similar methodologies as in ABA. Beyond these two approaches, it is worth mentioning the tree cluster approach (TCA), which can be seen as a combination of these two.

15.3 Comparison of Airborne, Mobile, and Terrestrial Laser Scanning

Today, the terms "ALS," "mobile laser scanning" (MLS), and "terrestrial laser scanning" (TLS) are applied depending on whether the platform is an aircraft, moving vehicle (refer to MLS, Figure 15.2b), or tripod (refer to TLS, Figure 15.2a), respectively. In MLS, the data collection can be performed either in so-called stop-and-go mode or in continuous mode. The stop-and-go mode corresponds to conventional TLS measurements, and therefore, the term "MLS" refers to the continuous model, that is, the use of continuous scanning measurements along the drive track. Also, the use of an unmanned aircraft vehicle (UAV, Figure 15.3) as a carrying platform can be better characterized by the term "MLS" rather than ALS. Both ALS and MLS are multisensor systems that integrate various types of navigation and data acquisition. The navigation sensors typically include global navigation satellite system receivers and an inertial measurement unit, while the data acquisition sensors typically include terrestrial laser scanners and digital cameras. With conventional TLS measurements, TLS is mounted on a tripod, and no navigation sensors are included in the system. The tripod is placed in the desired location, and the scanner measures the 3D locations of the targets within reach of the scanner. Scanners measuring phase shifts are mainly used in measuring individual trees or field plots, while "pulse scanners" can be used in mapping larger areas with a maximum distance of around 1 km to the target. TLS produces a dense point cloud from the surrounding trees. For example, with current phase-shift scanners, it takes 2–4 minutes to measure the surrounding area with a radius of 70–120 m, as the applied pulse density at a 10 m distance is still 6.3 mm. This corresponds to 25,000 points/m^2. In practice, the trees shadow each other, and the maximum usable range is a few tens of meters when collecting accurate plot-wise field reference with TLS.

(a)

(b)

FIGURE 15.2
(See color insert.) Principle of (a) terrestrial laser scanning (TLS) and (b) mobile laser scanning (MLS): MLS measurement unit, Roamer. (Copyright Ville Kankare; Harri Kaartinen, FGI.)

MLS and TLS have so far been used mainly for research purposes. From the forest mapping point of view, MLS could be linked to a logging machine to collect tree quality data, while TLS (or MLS) could be used in acquiring a plot-level reference for ALS-based inventories.

The instruments used for LS FI purposes typically emit very short (3–10 ns), narrow-beamwidth (0.15–2.0 mrad), and infrared (0.80–1.55 μm) laser pulses at near-nadir incidence angles (<30° with ALS) with high pulse repetition frequencies (50–400 kHz). In general, when operated at flying altitudes of around 500–3000 m, ALS sensors generate

(a)

(b)

FIGURE 15.3
(**See color insert.**) (a) ALS in UAV and (b) collected data. (Copyright Anttoni Jaakkola, FGI.)

a dense sample pattern (0.5–20 pulses/m²) with a small footprint (<1 m) on the ground. Elevation accuracy (Z) is typically between 5 and 30 cm, and planimetric accuracy (XY) is typically between 20 and 80 cm with ALS. ALS is feasible for large-area mapping and cost-effective in areas larger than 50 km², but due to transfer costs, multitemporal laser surveys are seldom studied and developed. Current ALS acquisition costs for large area collections are less than €0.5/ha.

Data provided by TLS systems can be characterized by the following technical parameters: (1) point density in the range of 10,000 pulses/m² at the 10 m distance, (2) distance accuracy of a few millimeters to 1–2 cm, and (3) operational scanning range from 1 to 300 m. TLS is feasible for detailed small-area surveys typically having a radius of a maximum of few tens of meters. TLS is also feasible for detailed measurements of areas less than several hectares, which still require merging of dozens or hundreds of scans of forest conditions. TLS is currently actively studied so that it can provide a plot reference with lower costs than conventional plot inventories.

Data provided by MLS systems can be characterized by the following technical parameters: (1) point density in the range of 100–1000 pulses/m² at a 10 m distance, (2) distance measurement accuracy of 2–5 cm, and (3) operational scanning range from 1 to 100 m. The operational range is better than that of TLS but less than that of ALS. MLS potential for forestry includes plot reference collection, collection of very detailed data of small areas, creation of virtual reality of forests, and multitemporal studies of forest conditions.

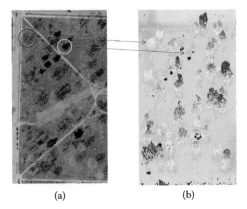

(a)   (b)

FIGURE 1.1
LiDAR data used to identify tree shape, height, and life form. The red circle indicates a deciduous tree (sycamore) and the yellow circle a conifer tree. Courtesy of Jim Stout, IMGIS, IN. (a) Aerial photo and (b) LiDAR point cloud.

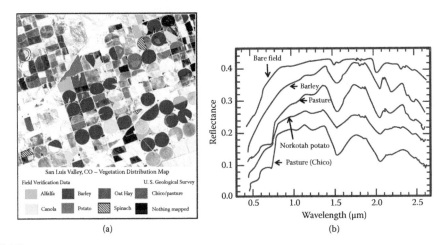

(a)   (b)

FIGURE 1.3
Vegetation species identification using hyperspectral imaging technique. (a) AVIRIS image showing circular fields in San Luis Valley, CO, and (b) reference spectra used in the mapping of vegetation species. (From Clark, R.N. et al., *Proceedings: Summitville Forum '95*, edited by H.H. Posey, J.A. Pendelton, and D. Van Zyl, 64–9. Colorado Geological Survey Special Publication 38, 1995. With permission.) Available at http://speclab.cr.usgs.gov/PAPERS.veg1/vegispc2.html.

FIGURE 4.1
One pixel in which we can observe more than one class. The red color represents the mountainous arid class and the tan-orange color represents the sand class.

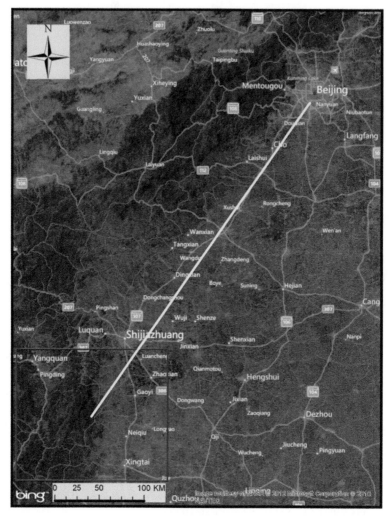

FIGURE 4.2
The location of our study area, southwest of Beijing in the North China Plain. Source: Bing Maps.

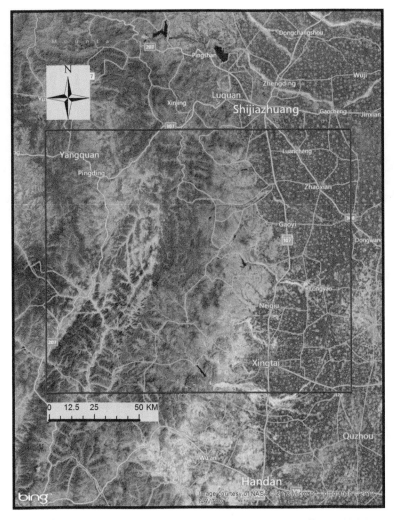

FIGURE 4.3
Close-up of our study area near Shijiazhuang, China. (Courtesy of NASA and Bing Maps.)

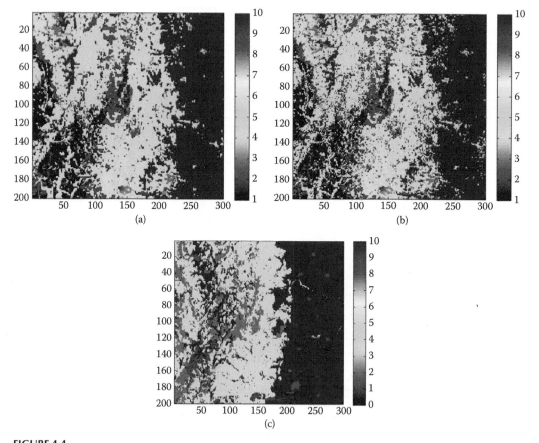

FIGURE 4.4

(a, b, c) Classification maps made by the classifiers C_1, C_2, and C_3, respectively. The color numbers (right-hand vertical scale) represent the class of the crisp classifier. Color 1 corresponds to class 21, color 2 corresponds to class 22, and so on (see Table 4.1 for definition of classes).

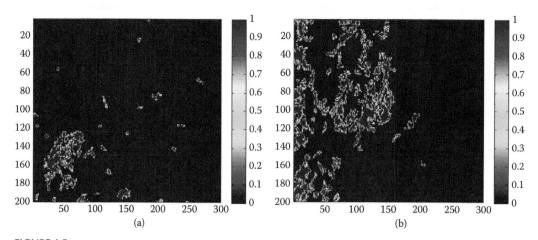

FIGURE 4.5

Visualization of the soft reference data. In each image, we show the degree of membership (on a scale of 0–1) of each pixel to each of the nine classes (a) forest, (b) forest with shrub, (c) forest with low cover, (d) grassland with medium cover, (e) grassland with low cover, (f) water, (g) impervious land, (h) unused land, and (i) arid land.

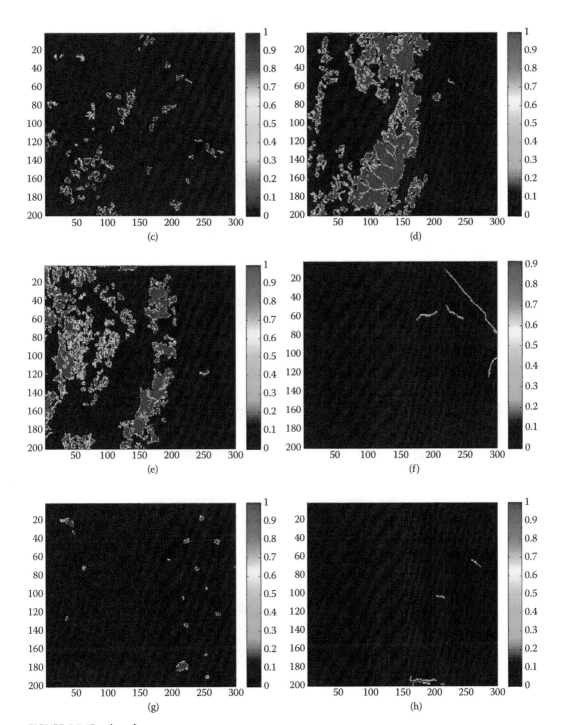

FIGURE 4.5 (*Continued*)

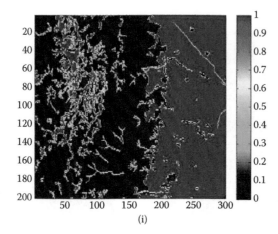

FIGURE 4.5 (*Continued*)

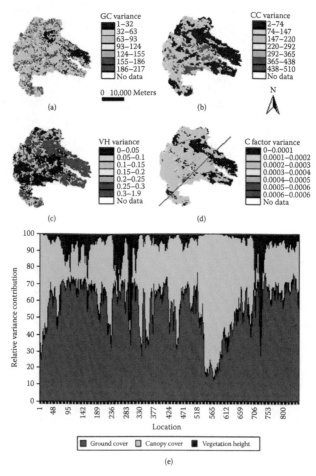

FIGURE 5.2
Variance maps of predicted ground cover (a), canopy cover (b), vegetation height (c), and vegetation cover factor (d) related to soil erosion at Fort Hood, Texas, and relative variance contributions (e) of the input variables to uncertainty of the predicted cover factor for the pixels at a transect line marked at the vegetation cover factor variance (d). (From Gertner, G. et al., *Remote Sens. Environ.*, 83, 498–510, 2002a.)

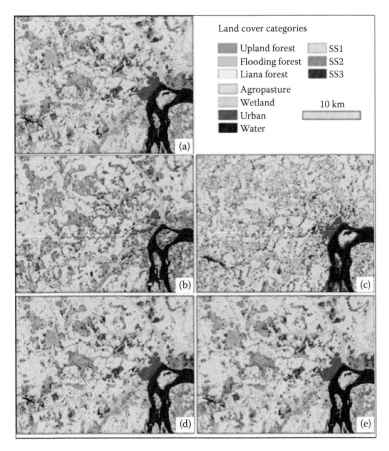

FIGURE 6.3
Comparison of classification results based on (a) Landsat Thematic Mapper (TM) multispectral image, (b, c) HH, HV, and their corresponding textural images from ALOS PALSAR L-band and RADARSAT-2 C-band, respectively, (d) multisensor wavelet-based fusion images from TM multispectral bands and PALSAR L-band, and (e) multisensor wavelet-based fusion images from TM multispectral bands and RADARSAT C-band.

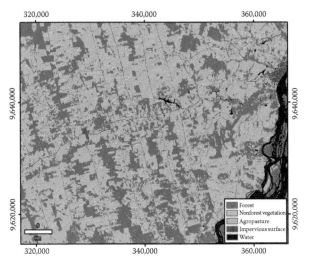

FIGURE 7.7
Land use/cover distribution from the 2008 Landsat Thematic Mapper image.

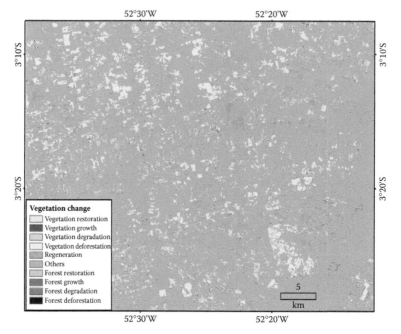

FIGURE 7.8
Detailed vegetation change trajectories developed from the combination of the 2008 land use/cover classification and vegetation gain/loss images.

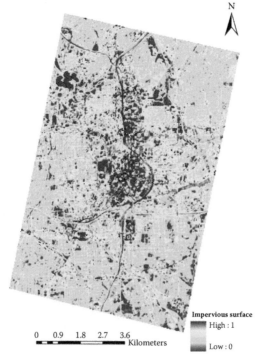

FIGURE 8.3
Fraction image of impervious surface derived from linear spectral mixture analysis.

FIGURE 8.4
Testing samples in downtown Atlanta.

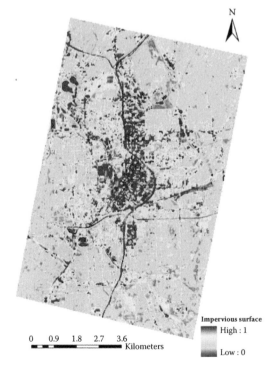

Impervious surface

High : 1

Low : 0

0   0.9   1.8   2.7   3.6
Kilometers

FIGURE 8.6
Fraction image of impervious surface derived from multilayer perceptron.

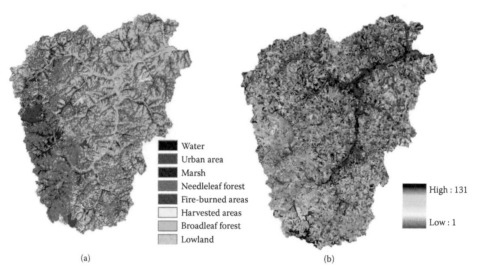

Water	
Urban area	
Marsh	
Needleleaf forest	
Fire-burned areas	High : 131
Harvested areas	
Broadleaf forest	Low : 1
Lowland	

(a)   (b)

FIGURE 10.3
The use of remote sensing images to initialize the LANDIS model: (a) shows the landscape types classified from Landsat imagery and (b) shows 131 unique forest stands that have different species compositions and age structures, by merging the remote sensing products (a) with an elevation map and forest inventory data at different landscape locations. See Xu et al. (2004) for details.

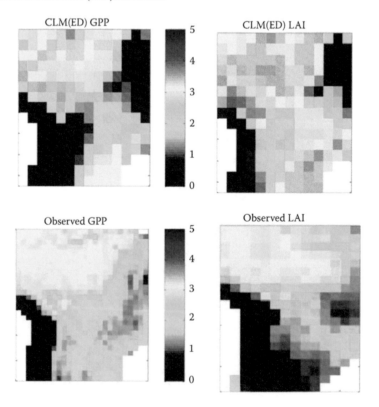

FIGURE 10.6
Benchmarking of CLM(ED) model simulation against observed gross primary production (GPP) (in kg C/m^2/year) and leaf area indices (LAIs) (m^2 leaf/m^2 ground) in the Amazon derived from MODIS products. Courtesy of Rosie Fisher from the National Center for Atmospheric Research, Colorado.

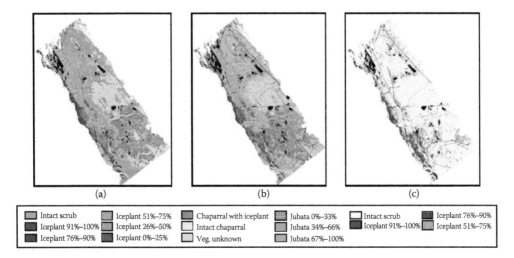

☐ Intact scrub	☐ Iceplant 51%–75%	☐ Chaparral with iceplant	☐ Jubata 0%–33%	☐ Intact scrub	☐ Iceplant 76%–90%
☐ Iceplant 91%–100%	☐ Iceplant 26%–50%	☐ Intact chaparral	☐ Jubata 34%–66%	☐ Iceplant 91%–100%	☐ Iceplant 51%–75%
☐ Iceplant 76%–90%	☐ Iceplant 0%–25%	☐ Veg. unknown	☐ Jubata 67%–100%		

FIGURE 11.2
Mapping of iceplant (*Carpobrotus edulis*) and jubata grass (*Cortaderia jubata*) using (a) minimum noise fraction, (b) band ratio indexes, and (c) continuum removal processing algorithms at Vandenberg Air Force Base in California. (Reproduced from Ustin et al., *2002 IEEE International Geoscience and Remote Sensing Symposium, June 24–28, Vol 3*, 1658–60. Toronto, Canada: IEEE International. With permission.)

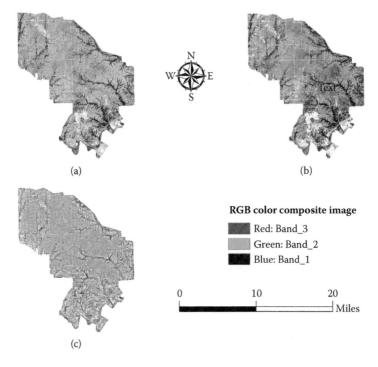

FIGURE 12.2
(a) India remote sensing image for year 1999, (b) Landsat TM image for year 2000, and (c) aerial orthophoto for year 2004 for Fort Riley.

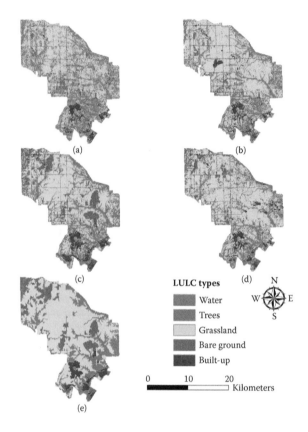

LULC types

- Water
- Trees
- Grassland
- Bare ground
- Built-up

0   10   20
⌐_____⌐ Kilometers

FIGURE 12.4
(a) Misclassification of water bodies in the west, east, and south parts using 1999's IRS Brovey transformations and maximum likelihood compared to (b) land use and land cover classification using 1999's TM images, (c) burned areas in the north and east parts misclassified into bare grounds using 2000's TM images and ML, (d) 1996's TM images led to misclassification of built-up areas in the east parts by ML, and (e) landscape segmentation and LULC classification using TM images and SEGCLASS.

FIGURE 15.2
Principle of (a) terrestrial laser scanning (TLS) and (b) mobile laser scanning (MLS): MLS measurement unit, Roamer. (Copyright Ville Kankare; Harri Kaartinen, FGI.)

(a)

(b)

FIGURE 15.3
(a) ALS in UAV and (b) collected data. (Copyright Anttoni Jaakkola, FGI.)

FIGURE 15.4
An example of ALS CHM. (Copyright Hannu Hyyppä.)

FIGURE 15.9
Mobile laser scanning in Evo study area. (Copyright Harri Kaartinen & Antero Kukko, FGI.)

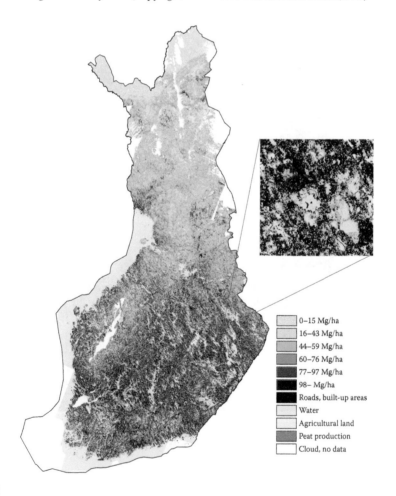

	0–15 Mg/ha
	16–43 Mg/ha
	44–59 Mg/ha
	60–76 Mg/ha
	77–97 Mg/ha
	98– Mg/ha
	Roads, built-up areas
	Water
	Agricultural land
	Peat production
	Cloud, no data

FIGURE 16.1
A map showing predictions of the total biomass of trees, above-and belowground per hectare (Mg/ha), produced using the multisource national forest inventory of Finland (excluding Åland). The estimates correspond to the year 2009. Some areas against the east and west border in North Finland are indicated by the cloud color (white), in addition to some areas inside the borders, owing to lack of satellite images caused by clouds. Map data: National Land Survey of Finland MML/VIR/MYY/328/08.

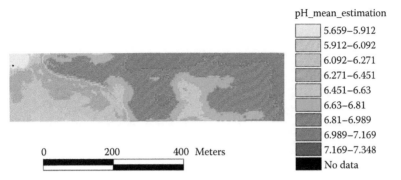

pH_mean_estimation
	5.659–5.912
	5.912–6.092
	6.092–6.271
	6.271–6.451
	6.451–6.63
	6.63–6.81
	6.81–6.989
	6.989–7.169
	7.169–7.348
	No data

0   200   400 Meters

FIGURE 18.2
In-field variability of soil pH as indicated by the sequential Gaussian cosimulation method using aerial hyper-spectral data (From Yao, H., *Hyperspectral Imagery for Precision Agriculture*, Department of Agricultural and Biological Engineering, University of Illinois at Urbana-Champaign, Urbana, IL, 2004. With permission.)

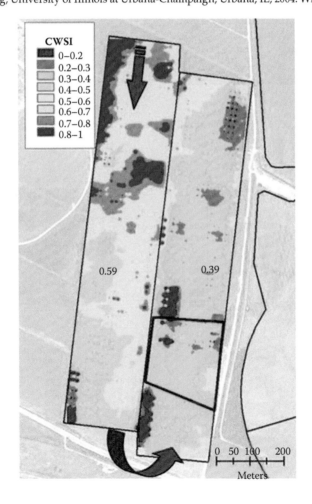

FIGURE 18.3
Water stress map of a cotton field before last irrigation on August 20, 2007. Arrows indicate lateral move position and pivoting directions of the irrigation rig. Numbers are the mean Crop Water Stress Index (CWSI) levels for the east and west parts of the field. The bold polygon marks the ground monitored part of the field. (From Meron, M., et al., *Precis. Agric.*, 11, 148, 2010.)

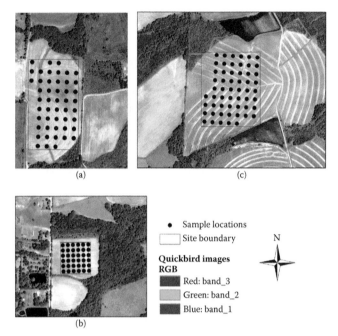

(a)

(c)

- Sample locations
  Site boundary   N

**Quickbird images
RGB**
  Red: band_3
  Green: band_2
  Blue: band_1

(b)

FIGURE 19.1
(a) Site 2, (b) site 5, and (c) site 6 with sample locations shown on the QuickBird images.

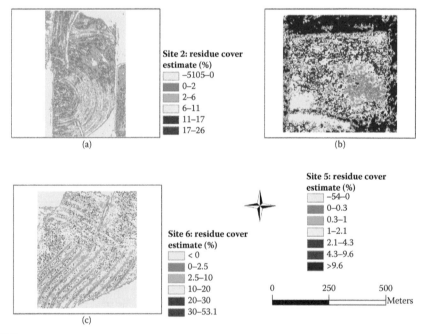

**Site 2: residue cover
estimate (%)**
  −5105–0
  0–2
  2–6
  6–11
  11–17
  17–26

(a)

(b)

**Site 5: residue cover
estimate (%)**
  −54–0
  0–0.3
  0.3–1
  1–2.1
  2.1–4.3
  4.3–9.6
  >9.6

**Site 6: residue cover
estimate (%)**
  < 0
  0–2.5
  2.5–10
  10–20
  20–30
  30–53.1

0   250   500
                              Meters

(c)

FIGURE 19.9
The results of sites 2, 5, and 6 using regression modeling and QuickBird images that have the highest correlation
with residue cover at a spatial resolution of 1.2 m × 1.2 m: (a) site 2: residue cover estimate; (b) site 5: residue cover
estimate; and (c) site 6: residue cover estimate.

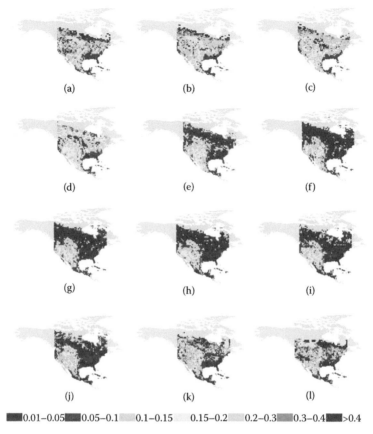

0.01–0.05 ▮ 0.05–0.1 ▮ 0.1–0.15 ▮ 0.15–0.2 ▮ 0.2–0.3 ▮ 0.3–0.4 ▮ >0.4

FIGURE 20.5
Seasonal variations of vegetation background red and near-infrared reflectance over the North America continent derived from Multiangle Imaging SpectroRadiometer data. (a) January, (b) February, (c) March, (d) April, (e) May, (f) June, (g) July, (h) August, (i) September, (j) October, (k) November, and (l) December. (From Pisek, J., and J.M. Chen, *Remote Sens. Environ.*, 113, 2412, 2009.)

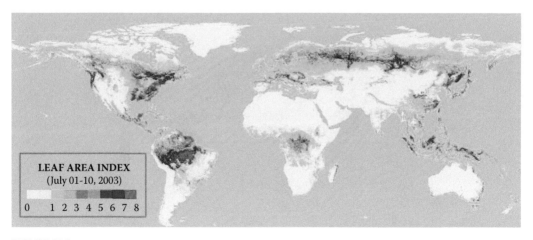

FIGURE 20.8
Global leaf area index map at 1 km resolution generated by the GLOBCARBON algorithm using cloud-free VEGETATION data on July 1–10, 2003. (From Deng, F., et al., *IEEE Trans. Geosci. Remote Sens.* 44, 2219, 2006.)

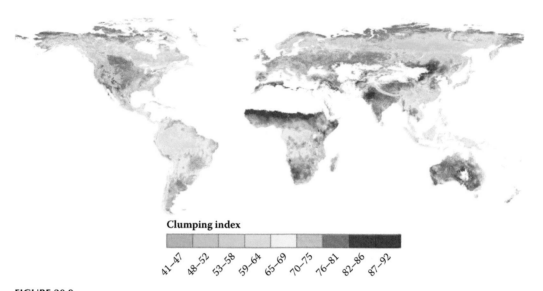

FIGURE 20.9
Global clumping index map at 0.5 km resolution derived using the MODIS BRDF product for 2006. The numerical value should be divided by 100. (From He, L., et al., *Remote Sens. Environ.*, 119, 118, 2012.)

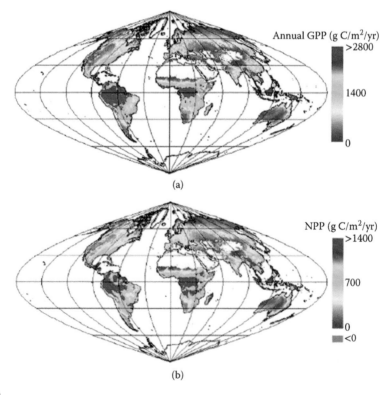

FIGURE 22.3
Three-year (2001–2003) mean global 1 km Moderate Resolution Imaging Spectroradiometer (MODIS) annual gross photosynthetic production (GPP) and net primary production (NPP) images: (a) GPP and (b) NPP. (Modified from Zhao, et al., *Remote Sens. Environ.*, 95, 164–76, 2005.)

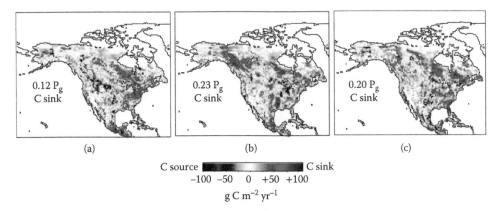

C source ░░░░░░░░░░ C sink
-100 -50 0 $+50$ $+100$
$g\ C\ m^{-2}\ yr^{-1}$

FIGURE 22.5
Predicted North American interannual variation in annual NEP flux from 1996 to 1998: (a) 1996, (b) 1997, and (c) 1998. (From Potter, C. et al., *Global Planetary Change*, 39, 201–213, 2003.)

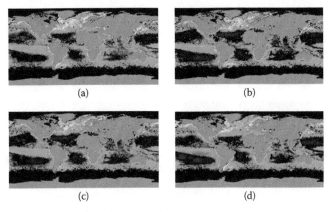

FIGURE 23.2
Global monthly composite from SeaWiFS, May 2002, with Chl-a calculated using the (a) OC4v4 Chl-a algorithm and the GIOP outputs for (b) Chl-a concentration, (c) absorption by phytoplankton at 443 nm, and (d) absorption by CDOM at 443 nm. The products are displayed using the SeaWiFS rainbow palette. (Courtesy of the NASA OBPG group.)

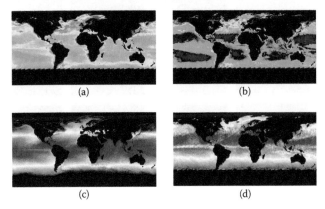

FIGURE 23.3
Global monthly composite from MODIS, May 2010, with NPP calculated using the (a) VGPM model, (b) MODIS Chl-a, (c) SST, and (d) PAR products. The NPP and Chl-a products are displayed using the SeaWiFS rainbow palette, Chl-a as a log value, while SST and PAR are displayed using a blue to white to red color palette. (Courtesy of the NASA OBPG group and ocean productivity site, http://orca.science.oregonstate.edu/.)

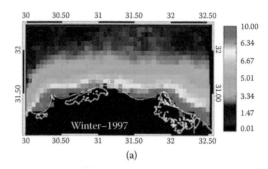

(a)

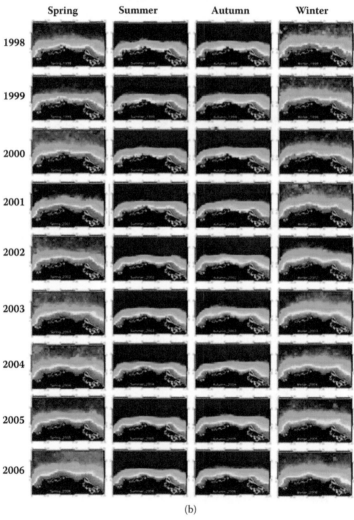

(b)

FIGURE 23.5
(a) Enlarged view of the Winter 1997 mean seasonal Chl-a concentrations. (b) Mean seasonal Chl-a concentrations off the Nile delta during 1997–2006, as calculated using the MedOC4 algorithm of Volpe et al. (2007) applied to the GlobColour dataset. (From Moufaddal, W.M., and S. Lavender, Assessment of the Possible Key Factors for Fall and Rise of the Coastal Fisheries off the Nile Delta: A Remote Sensing Approach, *Proceedings of the Second Gulf Conference & Exhibition on Environment and Sustainability*, 620–635, Kuwait, 2009a.)

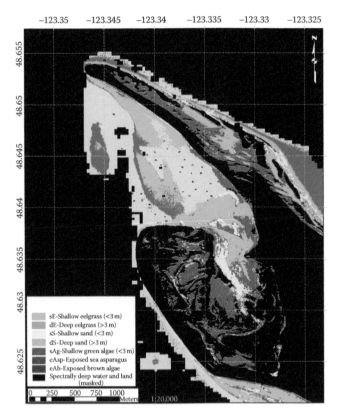

FIGURE 24.1
Airborne imaging spectroradiometer for applications (AISA) classification. (From O'Neill et al. 2011.)

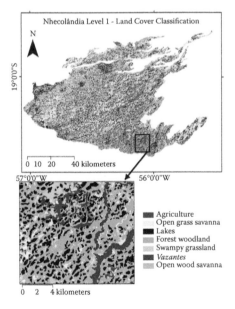

FIGURE 24.2
Classification Level 1 of wetland habitats: Pantanal. (Adapted from Evans T. L. and M. Costa, *Remote Sensing of Environment*, 128, 118–37, 2013.)

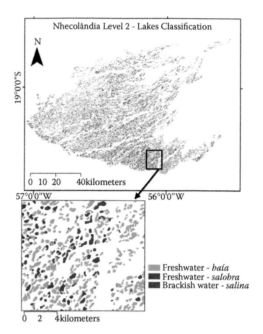

FIGURE 24.3
Classification Level 2 of wetland lakes: Pantanal. (Adapted from Evans T. L. and M. Costa, *Remote Sensing of Environment*, 128, 118–37, 2013).

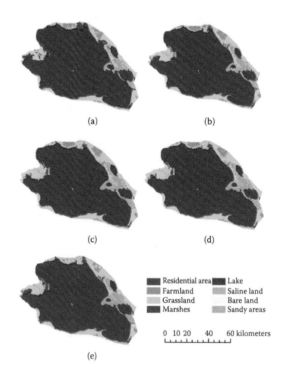

FIGURE 25.2
LULC classification maps of the wetland landscape of Qinghai Lake for the years (a) 1977, (b) 1987, (c) 2000, (d) 2005, and (e) 2011.

FIGURE 26.1
Range of study area: red points are the sampling sites. (Modified from Lau et al. (2008).)

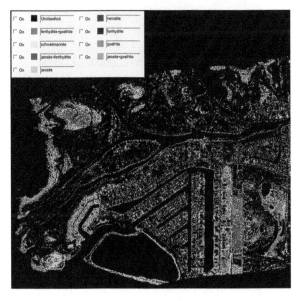

FIGURE 26.6
Image classification of iron-bearing minerals.

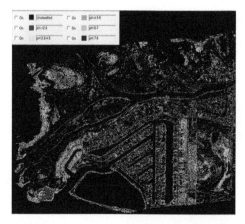

FIGURE 26.7
Map of soil acidity deduced from the distribution of iron-bearing minerals.

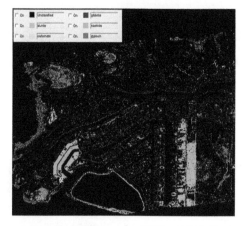

FIGURE 26.8
Image classification of non-iron-bearing minerals.

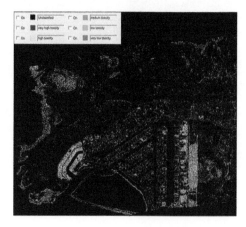

FIGURE 26.9
Map of soil toxicity deduced from the distribution of aluminum-bearing minerals and pH.

The amount of data produced by such systems is huge (at the rate of 0.25–1 M pts/s), and manual processing of the data is very time-consuming. This prompts a need for automated methods that decrease the amount of manual work required to produce accurate 3D models. Due to different geometry with ALS and MLS/TLS as well as different point density, the processing methods of MLS/TLS are quite different from the methods of ALS.

15.4 Point Cloud Metrics and Surface Models

15.4.1 Creation of Terrain, Surface, and Canopy Height Models

LS data is 3D point data in its discrete form with additional characteristics recorded for every return, such as echo type and intensity. The most frequently used method for the creation of a digital surface model (DSM) is to take the highest first echo within a given neighborhood and interpolate the missing heights. Following the creation of the digital terrain (or elevation) model (DTM or DEM), a canopy height model (CHM) (Figure 15.4) can be calculated by subtracting the height of the ground from the DSM. The DSM is calculated from the highest echoes as the height of the ground, while the DTM is calculated from the lowest. The accuracy of the DTM varies in forest conditions by around 10–50 cm (Kraus and Pfeifer 1998; Hyyppä et al. 2000; Ahokas et al. 2002; Reutebuch et al. 2003; Takeda 2004). The structure of the forest, variations in the terrain, and scanning parameters such as opening angle and pulse density affect the accuracy of the DTM (Ahokas et al. 2005).

Airborne laser measurements tend to underestimate tree height (Nelson et al. 1988; Hyyppä and Inkinen 1999; Lefsky et al. 2002; Rönnholm et al. 2004). The first echo return reflects more often from the shoulder of the tree instead of the top. Although a laser pulse hits the top, the treetop may not be dense enough to reflect a recordable return signal. On the other hand, dense undervegetation causes overestimation in the DTM. Mainly for these reasons, the CHM is underestimated. Other factors affecting tree height measurement accuracy are flying height, pulse density, pulse footprint, modeling algorithms, scanner properties (e.g., sensitivity to record return signals, field of view, zenith scan angle, and beam divergence), and structure and density of the tree crown (Holmgren et al. 2003; Hopkinson et al. 2006).

FIGURE 15.4
(**See color insert.**) An example of ALS CHM. (Copyright Hannu Hyyppä.)

15.4.2 Geometric Features

Tree variables can be measured directly from a laser point cloud. Traditionally, the characteristic geometric features have been the tree height and crown dimensions. The crown dimensions are determined by segmentation, as the tree height is the highest CHM value, or a laser return inside the segment. With dense ALS data (>5 hits/m^2), it is also feasible to calculate various other geometric features to be used in estimating tree variables. These features describe the crown volume, shape, and structure (Holmgren and Persson 2004; Vauhkonen et al. 2010).

15.4.3 Vertical Point Height Distributions

The prediction of stand variables in ABA is based mainly on point height metrics calculated from ALS data. Features such as percentiles calculated from a normalized point height distribution, mean point height, densities of the relative heights or percentiles, standard deviation, and coefficient of variation are generally used (Næsset 2002).

The percentiles are down to the top heights calculated from the vertical distribution of the point heights, that is, the percentile describes the height at which a certain number of cumulative point heights occur. The proportion of vegetation hits compared with all hits is also used as a predictor feature describing the crown density. A hit is seen as a vegetation hit from trees or bushes if it has been reflected from over some threshold limit above ground level. All the features are calculated separately for every echo type. The reason for this is that the sampling between echo types is somewhat different (Korpela et al. 2012). With dense ALS data, the predictors used in ABA and ITD have become similar (i.e., Villikka et al. 2007). The point height metrics that have generally been used in ABA are also used in the ITD approach. In ITD, these features are calculated at the tree crown level and used in the estimation of tree-level variables analogously to the estimation of stand variables in ABA.

15.5 Area-Based Approach

The area-based prediction of forest variables is based on a statistical dependency between the variables measured in the field and predictor features derived from RS data. In the case of ALS, the method in which this kind of two-stage procedure is used to produce stand-level information from wall-to-wall grid-level predictions is called an ABA (Næsset 2002). In a more general context, two-stage predictions using field and RS data have a long history in FI (i.e., Poso et al. 1984), and this process could also be called ABA.

When ABA is applied, accurate training data must be on hand (Poso et al. 1984; Næsset 2002). Training plots should represent the whole population and cover the variations in it as much as possible. The efficient selection of the training plot locations requires preknowledge of the inventory area (i.e., Maltamo et al. 2011). A sample unit in ABA is most often a grid cell, the size of which refers to the size of the field-measured training plot. Then, the laser-derived features are extracted from the grid cell areas and used as possible predictors. The statistical relation between the predictors and response variables is modeled using training data when both of them are on hand. The response variables are predicted for grid cells without training data using regression or NN methods (Figure 15.5). If stand-level variables are needed, they are calculated by weighting the grid-level predictions inside the stand.

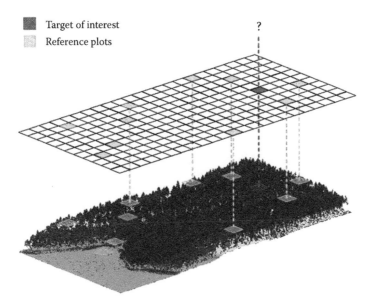

Target of interest
Reference plots

FIGURE 15.5
NN method. (Copyright Ville Kankare.)

15.6 Individual Tree Detection

Individual trees can be detected from the laser point clouds. Two main approaches exist for detecting single trees from ALS data: point-based and surface model-based approaches (i.e., Hyyppä and Inkinen 1999; Persson et al. 2002; Wang et al. 2008; Gupta et al. 2010). The main tree detection methods developed in boreal forest conditions belong to the latter and are usually based on finding local maxima on a smoothed CHM. After the local maxima are found, the boundaries of the crown are extracted, that is, using a watershed-based region detector (Figure 15.6). The accuracy of the CHM and the corresponding accuracy of the ITD depend on the pulse density applied. A pulse density of around 5–6 hits/m² is seen as a prerequisite for the ITD, although even with pulse densities of around 2 hits/m², it is possible to detect individual dominant trees or tree groups (i.e., Breidenbach et al. 2010; Vastaranta et al. 2011c). Tree variables such as height and location can be measured directly from the laser point cloud. The estimate for tree height can be the highest echo or the CHM value within the tree crown. The XY location of the tree is generally the location of the treetop, that is, the XY location of the highest echo or CHM value.

Laser-based tree height is underestimated (e.g., Rönnholm et al. 2004) and usually calibrated using field data. Tree *dbh* cannot be measured directly from the point cloud and must be predicted. In *dbh* prediction, general allometric models, local models, or NN methods can be used (Kalliovirta and Tokola 2005; Peuhkurinen et al. 2007; Yu et al. 2010, 2011).

Besides tree height and crown dimensions, the use of other geometric features and point height distributions has become more common in ITD predictions (Holmgren and Persson 2004; Villikka et al. 2007; Vauhkonen et al. 2010; Yu et al. 2011). In tree species classification, spectral features from aerial images or laser pulse return intensity are applied (i.e., Korpela et al. 2010). When NN methods are applied for the prediction of tree variables, training data is required. With respect to the plot level training data used in ABA, ITD training data should cover the whole population variation at the tree level. However, ITD

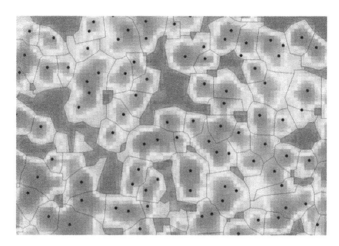

FIGURE 15.6
Trees crowns are delineated from the CHM using watershed segmentation.

can be carried out without any field measurements if desired. In this case, missing variables are measured straight from the point cloud and/or modeled with existing models. Major error sources in ITD include the detection of the trees and the modeling of the missing variables, especially *dbh* and tree species classification (Holopainen et al. 2010a; Vastaranta et al. 2011b).

15.7 Applications of Airborne Laser Scanning

15.7.1 Estimation of Tree and Stand Variables

15.7.1.1 An Area-Based Approach

In the first ABA studies, single-forest variables were predicted. Næsset (1997a) predicted stand mean height using the highest laser returns in grid cells within a stand. The use of all returns resulted in the underestimation of the mean height. Stand mean volume was predicted in Næsset (1997b) with regression. In the model, the predictors used were the mean height of the laser returns, laser-derived canopy cover, and mean height. Magnussen and Boudewyn (1998) calculated quantiles from the laser point height distribution and used those as predictors of mean height. Later, these types of features were used in many ABA and ITD studies to predict variables of particular interest.

Hyyppä and Hyyppä (1999) predicted forest variables using area-based features as predictors. For the first time, ground elevation was subtracted from laser point heights, which enabled the use of point heights as predictors that were directly comparable to the tree heights. In the study, ALS inventories were compared to various other optical RS methodologies, and it was concluded that ALS inventories had superior accuracy compared to others.

Næsset (2002) formulated data-specific regression models to predict forest stand variables using plot-wise tree-by-tree field-measured modeling data and laser point height distribution metrics. With the developed models, stand variables were predicted for grid cells, and from them, stand-level variables were calculated. The standard deviation of the predicted stand variables varied between stand development classes and site types. The variations were 0.61–1.17 m in mean height (Hg), 1.37–1.61 cm in mean diameter (Dg), 8.6%–11.7% in basal area (BA), and 11.4%–14.2% in stem volume (VOL).

In Finland, the ABA was tested by Suvanto et al. (2005). Regression models were developed using laser height metrics for Dg, Hg, stem number, BA, and VOL of 472 reference plots. The predicted accuracies for 67 stands were 9.5%, 5.3%, 18.1%, 8.3%, and 9.8%, respectively. The predictions outperformed the accuracy of conventional SWFI (Poso 1983; Haara and Korhonen 2004; Saari and Kangas 2005; Vastaranta et al. 2010a). In forest management planning inventories in Scandinavia, species-specific information is needed for growth projections and simulated bucking. Tree species composition also has a major effect on forest value. The formulation of data-specific models for every stratum is thus laborious, and NN methodologies are more suitable for that estimation task. Maltamo et al. (2006) added features from aerial photographs and variables from existing stand registers as predictors, in addition to ALS height metrics and the NN imputation of VOL. The *k*-most-similar-neighbor (*k*-MSN) imputation method was used, and the plot-level VOL accuracy varied from 13% to 16% depending on the predictors used. Packalén and Maltamo (2007) used the *k*-MSN method to impute species-specific stand variables using ALS metrics and aerial photographs. Basically, they used the same dataset as in Suvanto et al. (2005), and the accuracies for species-specific VOLs at the stand level were 62.3%, 28.1%, and 32.6% for deciduous, Scots pine (*Pinus sylvestris*, L.), and Norway spruce [*Picea abies* (L.) H. Karst], respectively. Holopainen et al. (2010b) predicted timber assortment volumes with corresponding data and methodologies. At the stand level, the saw wood prediction accuracies (RMSE) were 79.2% (7.0 m³/ha), 33.6% (35.5 m³/ha), and 78.6% (6.2 m³/ha) for Scots pine, Norway spruce, and birch, respectively. Vastaranta et al. (2012a, 2012b) combined LS inventory methods. ITD was used to measure training data for the ABA. The RMSE in the imputed VOL was 24.8%, 25.9%, and 27.2% for the ABA trained with field measurements, ITD$_{auto}$, and ITD$_{visual}$, respectively. The developed method could be applied in areas with sparse road networks or when the costs of fieldwork must be minimized. The method is especially suitable for large-scale biomass or tree volume mapping.

ABA has been intensively studied in the Nordic countries because of the practical need to replace SWFI. However, the methodology is applicable and has also been studied outside boreal forest regions. ABA has proven to be suitable for forest variable estimation in an alpine environment. Hollaus et al. (2007) obtained a cross-validated accuracy (RMSE) of 21.4% for VOL prediction, which is in line with Nordic studies. Hudak et al. (2007) tested several NN-imputation methodologies in ABA. They concluded that Random Forest (RF) was the most robust and flexible among the imputation methods tested. Latifi et al. (2010) tested ABA in a temperate forest for timber volume prediction. Their results strengthen the findings by Hudak et al. (2007). Falkowski et al. (2010) imputed tree-level inventory data to parameterize a forest growth simulator. The results were validated with independent inventory data, and the root-mean-square differences in BA and VOL were 5 m²/ha and 16 m³/ha, respectively. They concluded that ABA was effective in generating tree-level FI data from ALS metrics. Only a few studies have tested ABA in tropical forest conditions. Hou et al. (2011) compared ALS, airborne color infrared (CIR), and Advanced Land Observation Satellite (ALOS) Advanced Visible and Near-Infrared Radiometer (AVNIR)-2 datasets to estimate VOL and BA in Laos. The prediction procedure followed Nordic experiences (i.e., Næsset 2002). In the study, ALS data proved to be superior, with an RMSE of 36.9% for VOL and 47.4% for BA. Integrating ALS metrics with other predictors from airborne CIR or ALOS AVNIR-2 did not improve the prediction accuracies significantly.

15.7.1.2 Individual Tree Detection

ITD is based on detecting trees from a 3D point cloud (see Figures 5 and 8), and tree variables are either directly measured or predicted using derived ALS features. Hyyppä and Inkinen (1999) showed that by segmenting tree crowns from the CHM, 40%–50% of the trees in coniferous forests could be correctly segmented. Persson et al. (2002) improved the crown delineation and were able to link 71% of the tree heights to the reference trees. The linked trees represented 91% of the total volume. When trees are detected by segmenting the CHM, only trees that contribute to the CHM can be detected (Kaartinen and Hyyppä 2008). Therefore, forest structure has a major influence on tree detection accuracy (i.e., Falkowski et al. 2008; Vauhkonen et al. 2012). Tree detection accuracy results from heterogeneous boreal forests are presented in Pitkänen et al. (2004), where the overall detection accuracy was only 40% (70% for dominant trees). Yu et al. (2011) presented an accuracy of 69% for tree detection in various managed forest conditions. These results are on a completely different scale from those in Peuhkurinen et al. (2007), where ITD was carried out for two mature conifer stands (density ~465 stems/ha) and the number of harvestable trees was underestimated by only <3%—a result that may, however, include some commission errors (segmentation of a single tree into several segments). Koch et al. (2006) detected individual trees using a local maximum filter and delineated crowns using watershed analyses. The obtained results were encouraging in coniferous stands, but dense stands of deciduous trees were more problematic. Heinzel et al. (2011) used crown size as prior information for tree detection and improved the tree delineation accuracy by about 30% for deciduous and mixed stands compared with the result of using a non-crown-size-dependent algorithm. In general, CHM-based tree detection approaches are at their best in single-layered, mature stands (i.e., Peuhkurinen et al. 2007). Point-based approaches are needed to discriminate nearby or subdominant trees. However, this has proven to be a rather challenging task (e.g., Wang et al. 2008; Gupta et al. 2010; Vauhkonen et al. 2012). Tree detection errors were studied with 12 different ITD algorithms by Kaartinen and Hyyppä (2008) and with six algorithms by Vauhkonen et al. (2012). Kaartinen and Hyyppä (2008) concluded that the most important factor in tree detection is the algorithm used, while the effect of pulse density (2–8 returns/m^2) was observed to be marginal. In that study, all of the algorithms were tested within two nearby study areas consisting of a few stands. In addition to several ITD algorithms, Vauhkonen et al. (2012) used test sites varying from tropical pulpwood plantations to managed boreal forests. Their main finding was that forest structure, such as tree density and clustering, strongly affects the performance of the tree detection algorithm used. The difference between algorithms was not seen to be as significant as in Kaartinen and Hyyppä (2008).

In ITD, tree species classification has proven to be a challenging task, especially using only ALS data. Holmgren and Persson (2004) classified Scots pines and Norway spruces by their structural differences with >90% accuracy. In recent years, even more promising tree species classification results have been reported when high point density data have been used in combination with aerial images or ALS intensity. Liang et al. (2007) classified deciduous-coniferous trees in leaf-off conditions with an accuracy of 89.8%, taking advantage of differences in first-last pulse data. Holmgren et al. (2008) combined high-density laser data with multispectral images. Canopy-related metrics such as height distribution and canopy shape were calculated along with spectral features. A classification accuracy of 96% was achieved with 1711 trees. Vauhkonen et al. (2009) used solely high-intensity ALS data (~40 returns/m^2) and calculated so-called alpha shape metrics describing the canopy structure for the identification of tree species. The overall classification accuracy was 95%. When a method similar to that was tested with a larger dataset (1249 vs. 92 trees) and a more practical point density (6–8 returns/m^2), an identification accuracy of 78% was obtained for three tree species (Vauhkonen et al. 2010). Korpela et al. (2010) obtained an 88%–90% classification accuracy for Scots pine, Norway spruce, and birch using ALS intensity statistics. Puttonen et al. (2010) used illuminated-shaded area separation from aerial photographs combined with ALS data in tree species classification and achieved an overall accuracy of 70.8% with three species. Thus, taking the latest results into consideration, a solution for practical tree species determination can be said to be within reach, at least in the Nordic countries, where the number of commercially important tree species is rather low.

At the individual tree level, the most important variable is the diameter at breast height (*dbh*), from which the stem form, volume, and timber assortments are estimated. ITD yields direct information about tree height and crown dimensions, on which *dbh* predictions have traditionally been based (i.e., Kalliovirta and Tokola 2005). The allometric relation between height and *dbh* is not as strong as the relation between *dbh* and height. Thus, *dbh* predictions based on tree height involve uncertainty.

More dense laser data have enabled the calculation of several laser height metrics for individual trees that can be used in the NN imputation of tree variables (Villikka et al. 2007; Maltamo et al. 2009; Vauhkonen et al. 2010; Yu et al. 2011). These features are also used in tree species classification, as mentioned earlier. Maltamo et al. (2009) predicted tree variables, including tree quality variables, of Scots pines using *k*-MSN estimation combined with plot- and tree-level height metrics calculated from ALS data. The RMSEs for *dbh*, height, and volume were 5.2%, 2.0%, and 11%, respectively, when 133 accurately matched trees were used in the validation. The respective accuracies were 13%, 3%, and 31% in Vauhkonen et al. (2010) and 21%, 10%, and 46% in Yu et al. (2011). Vauhkonen et al. (2010) used 1249 trees, and Yu et al. (2011) used 1476 trees for validation. In Yu et al. (2011) in particular, the mismatching of reference and laser tree candidates may have affected the results. Furthermore, tree height determination from CHM is highly accurate but is prone to underestimation (i.e., Rönnholm et al. 2004). If the ground elevation and the uppermost proportion of a crown are not detected, then the tree height is automatically underestimated. Laser tree height is usually calibrated against field trees to reduce the bias caused by several scanning parameters and data processing steps, such as the filtering used in producing surface models (see, i.e., Hyyppä et al. 2009a). However, as shown by the aforementioned studies, tree height is the most accurately determined variable in ITD.

15.7.1.3 Tree Cluster Approach

In the TCA, the CHM is first segmented, as in ITD. In the second phase, accurately located field trees are linked to the segments (Hyyppä et al. 2005, 2006; Breidenbach et al. 2010; Lindberg et al. 2010). In contrast to ITD, it is not assumed that a single segment represents a single tree (see Figure 15.4). In the TCA, all the field trees are linked to the nearest segment. Thus, segments may include no, one, two, or even more trees. All of the other methodologies are adapted from ITD or ABA. The TCA requires accurate tree-by-tree measured reference data. Field trees used in the modeling have to be positioned with an accuracy that enables reliable linking to the corresponding CHM segments. The TCA could be described as an ABA that operates at the segment level instead of the grid level.

Tree detection is the main error source in ITD (Vastaranta et al. 2011b). This method practically solves the tree detection problems, resulting in unbiased estimates for certain area levels. The TCA does not provide information as detailed as ITD could, in theory, but it is still capable of capturing the spatial variation in stand variables better than ABA. Lindberg et al. (2010) used the TCA to predict consistent tree height and stem diameter distributions. Breidenbach et al. (2010) obtained a plot-level RMSE of 17.1% for VOL compared with 20.6% with ABA.

15.7.2 Predicting Forest Growth and Site Type

ALS has a high geometric accuracy, which makes it suitable for monitoring forest growth (Yu et al. 2004). The growth of an individual tree can be monitored in several ways with two-time-point laser data: as differences in laser-measured tree heights (Figure 15.7), as differences in CHMs or DSMs, as differences in laser height metrics, or as differences between tree volume estimates (Yu et al. 2004, 2006; Næsset and Gobakken 2005; Yu et al. 2008; Vastaranta et al. 2011a).

Yu et al. (2006) demonstrated that the growth of an individual tree can be measured with a standard error of only 0.14 m using multitemporal high-density ALS data (10 hits/m²). The time period between the data acquisitions affects the accuracy of the measurements. In boreal forests, where the growth of stands is relatively slow, 1-year growth is not measurable with a high degree of accuracy using either ALS or the traditional forester's field measurement equipment. Næsset and Gobakken (2005) observed statistically significant changes in bitemporal ALS height metrics. However, the volume growth estimates had poor accuracy due to the short 2-year time interval between the ALS acquisitions. Yu et al. (2008) and Hopkinson et al. (2008) concluded that the longer the growth period was, the more accurate the growth detection would be. In temperate forests, Hopkinson et al. (2008) used multitemporal ALS data and showed that even annual forest *h*-growth was detectable. The relative standard error of the stand-level annual growth estimates was still high (ca. 100%) but decreased rapidly when the time interval was extended (~10% after 3 years).

Site type classification is needed to describe the production potential of forest stands, select optimal harvesting strategies, and determine nature protection and recreational values. Site type can be predicted from laser data using height-over-age curves or differences between site types in laser point height distributions (Gatziaolis 2007; Vehmas et al. 2008, 2009; Korpela et al. 2009; Holopainen et al. 2010c).

Site type estimation through site indexes provides a useful method for the determination of stand productivity. ALS-based forest mapping will open new opportunities for the implementation of site indexing in practice: in operative forest management planning, estimating the value of forest estates, and mapping ecologically important habitats. In the future, site indexing could be based on multitemporal ALS.

FIGURE 15.7

A 150 m long and 6 m wide cross section produced from an ALS point cloud taken at two times: white in summer 2003 and gray in summer 1998. The change of point clouds indicates removal of trees and growth of small trees. (Copyright Xiaowei Yu.)

15.7.3 Mapping of Forest Management Operations

In SWFI, forest management proposals for the next 10 years are determined for every stand. Proposals cover the whole rotation from renewal to the final cutting, and the timing varies from "immediate" to "rest" within the next 10-year period. When ABA is applied, only the forest variables are inventoried by RS, and forest management proposals are determined through additional field work. If laser data could be applied in the determination and timing of forest management proposals as well, this would enhance the efficiency of the ABA (Räsänen 2010; Vastaranta et al. 2010b). The mapping of harvesting sites is also one of the key decision points for large-scale forest owners (Laamanen and Kangas 2012). Räsänen (2010) used low-density laser data in determining micro-stand first-thinning maturity with a classification accuracy rate of more than 97% without using separate test and training sets. Vastaranta et al. (2011d) predicted the thinning maturity of stands using ABA. Logistic regression models based on ALS point height metrics predicted the thinning maturity with a classification accuracy rate of 79% (1) and 83% (2). A study by Vastaranta et al. (2011d) concluded the ALS-based prediction of forest management proposals could provide a practical future means of locating stands with operational needs.

15.7.4 Forest Biomass and Disturbance Monitoring

One of the biggest challenges in programs that aim to reduce global emissions from deforestation and forest degradation (i.e., Reducing Emissions from Deforestation and Forest Degradation, or REDD) is how to measure and monitor forest biomass and its changes effectively and accurately. Stand biomass is highly correlated with tree heights, which can be determined accurately by ALS (Kellndorfer et al. 2010). ALS-based RS capabilities, such as the direct measurement of vegetation structure or tree and stand variables, should enhance the accuracy of the current biomass estimation means at all levels from single-tree to nationwide inventory applications (i.e., Koch 2010; Holopainen et al. 2011a).

The inventory of stands' above-ground biomass (AGB) can be based on single-time-point ALS acquisition. Multitemporal ALS can be used when monitoring biomass changes. Lefsky et al. (1999) showed that a single profiling LiDAR-derived feature such as the quadratic mean of the canopy height could explain 80% of the variance in AGB. The structure of the forest canopy and the leaf area index (LAI) affects the penetration of the laser pulse in the crowns (Solberg et al. 2009). Changes in AGB have also been estimated using changes in LAI. The ground truth of LAI can be determined using a special measuring device or estimated from the ALS data (i.e., Solberg et al. 2006, 2009; Solberg 2008; Korhonen et al. 2011).

Popescu et al. (2004) combined small-footprint ALS and multispectral data to estimate plot-level volume and AGB in deciduous and pine forests using ITD. The maximum R^2 values were 0.32 for deciduous trees and 0.82 for pines. The respective RMSEs were 44 t/ha and 29 t/ha. Bortolot and Wynne (2005) also used ITD in AGB estimation, and the correlation (r) varied from 0.59 to 0.82 and the RMSEs from 13.6 t/ha to 140.4 t/ha. Jochem et al. (2011) used a semiempirical model that was originally developed for VOL estimation to estimate AGB in spruce-dominated alpine forests. The model was extended with three canopy transparency parameters (CTPs) extracted from ALS. The R^2 values for the fitted AGB models were 0.70 without any CTP and varied from 0.64 to 0.71 with different CTPs. The standard deviations varied from 87.4 t/ha (35.8%) to 101.9 t/ha (41.7%). Latifi et al. (2010) tested ABA in southwestern Germany in timber volume and biomass mapping. They obtained accuracies of 23.3%–31.4% in plot-level timber volume and 22.4%–33.2% in AGB prediction, depending on the feature sets and feature selection used.

Räty et al. (2011) made one of the pilot studies modeling single-tree AGB using dense ALS data. In the study, 38 trees consisting of 19 Scots pines and Norway spruces were analyzed in the laboratory after dense ALS data was acquired. Trees were segmented from the CHM, and features used as biomass predictors were calculated at tree segment level. In linear regression, AGB estimation accuracy was 21% and 40% for Scots pines and Norway spruces, respectively.

The risk of forest hazards is growing partly because of climate change, which affects the natural forest dynamics. Damage caused by drought, snow, wind, and insects is more common. Forest damage can be monitored, for example, by measuring changes in the LAI. This kind of approach is suitable for damage that causes defoliation. Solberg (2008) studied the use of multitemporal ALS in monitoring insect-related defoliation in Norway. Solberg had ALS data from three different time points and used changes in LAI as an indicator of defoliation. Solberg observed that the LAI values were high before the damage in July, as natural growth during the summer was also detected and affected the high values of LAI. Multitemporal data is expensive to use in practice. Thus, Solberg (2008) proposed an indicator calculated for single-pass ALS data to be used in forest health monitoring. The proposed indicator was the relation between ALS-derived LAI and forest stand density.

Kantola et al. (2010) tested the use of ITD in the classification of defoliated and healthy trees using dense ALS data (10 hits/m^2) in conjunction with aerial images. Predictions were made using logistic LASSO regression, RF, and k-MSN. The classification accuracy ranged between 83.7% and 88.1% (kappa value 0.67–0.76).

Vastaranta et al. (2012b) developed a ΔCHM method for the detection of snow-damaged crowns. In it, bitemporal ALS CHMs were contrasted, and the resulting difference image was analyzed using binary image operations to extract the damaged crowns. This kind of method has been used before in detecting harvested trees (Yu et al. 2004). Performance of the method was evaluated by errors of omission and commission as well as the error in the estimated damaged crown projection area (DCPA). The method makes use of two threshold parameters, the required height difference (Δh) in the contrasted CHMs and the minimum plausible area of damage (mCC). The best-case performance was evaluated for these parameters, and the optimal values were ~1.0 m for Δh and ~5 m^2 for mCC. The plot-level omission error rates were 19%–75%, while the commission error rates were 0%–21%. The relative estimation accuracy rate of the DCPA was −16.4%–5.4%. Vastaranta et al. (2011a) tested the area-based classification of snow damage with multitemporal ALS data and concluded that area-based estimation is also suitable for snow-induced change detection. Area-based estimation could also detect changes in trees that are not contributing to CHM, which is not possible with methodologies that only use changes in CHMs, as in Vastaranta et al. (2012b).

15.7.5 Predicting Forest Stock and Yield Value

The economical value of forests is crucial information for landowners and various forestry organizations. Estimates of the value of forest estates are needed for many purposes, for example, in real estate business, in land divisions and exchanges, and for considering forestry investment. The need for determining the value of forests has become more important since forests are increasingly considered to be one possible investment outlet among other real or financial assets. International Financial Reporting Standards (IFRS) also require that forest enterprises present systematically computed estimates of the value of their forested land annually. The methodologies used for assessing forest value vary among organizations. Probably, the most common method currently used for assessing forest estate value is computing the net present value (NPV) of forests based on predicted future cash transactions. The sales comparison approach—in which the value of a forest estate is determined using historical data covering market prices or realized forest estate sales in the same region—is also used to some extent.

Holopainen et al. (2010e, 2010f), Mäkinen et al. (2010), and Holopainen (2011) assessed the magnitude of uncertainty of ALS-based FI data in forest NPV computations. A starting point was the current state of change in operative forest planning in which traditional SWFIs are being replaced by area-based ALS ABA inventories. It was shown that at the stand level, the growth models used in forest planning simulation computations were the greatest source of uncertainty with respect to NPVs computed throughout the rotation period. Uncertainty almost as great was caused by ABA and SWFI data uncertainty, while the uncertainty caused by fluctuation in timber prices was considerably lower in magnitude (Holopainen et al. 2010e). Regarding forest property level deals had a considerably lesser degree of NPV deviation than did stand-level: ABA inventory errors were the most prominent source of uncertainty, leading to a 5.1%–7.5% relative deviation in property-level NPV when an interest rate of 3% was applied (Holopainen et al. 2010f). ABA inventory error-related uncertainty resulted in significant bias in property-level NPV estimates. The Holopainen (2011) study forms a basis for developing practical methodologies for taking uncertainty into account in forest property valuation. The study also furthered the development of a forest property valuation methodology totally based on ALS inventorying.

15.7.6 Sampling-Based Forest Inventories Using Airborne Laser Scanning

LS data are a far more expensive type of auxiliary data than, for example, satellite images. Thus, strategic large-area forest inventories are still based either solely on field measurements (national level) or on a fusion of field data and satellite images (county level). At the county level, ALS inventory has been studied. Næsset (2004) tested ABA in a 65 km^2 area in Norway. The relative error of the predicted plot level volume was 17.5%–22.5%, and the respective stand-level accuracy was 9.3%–12.2%. Holmgren and Jonson (2004) conducted a similar study in a 50 km^2 area in Sweden with a stand-level volume RMSE of 14.1%. In the aforementioned studies, ALS data covered the whole study area. In recent years, far larger areas have been inventoried operationally using ABA. At the national inventory level, it is not feasible to acquire wall-to-wall ALS data for FI purposes. Holopainen and Hyyppä (2003) and Næsset et al. (2006) suggested the use of ALS data in strip-based sampling. There have been many studies in which profiling LiDAR has been used to acquire sampled FI data (i.e., Nelson et al. 2003a, 2003b, 2004). Nelson et al. (2004) inventoried forest resources in the state of Delaware in the United States using 14 flight lines with a 4 km sampling distance. Their timber volume estimate at the county and state levels differed from the U.S. Forest Service estimate by 21% and 1%, respectively. The corresponding differences in the AGB estimates were 22% and 16%. However, only a few studies have used a sampling procedure with ALS data. Gautam et al. (2011) used a two-phase sampling procedure to estimate the forest AGB in Laos. The procedure integrates sample plots with ALS transects (10% coverage tested) and satellite images, and it attains a relative RMSE of 25%–35% in AGB in an area of 0.5 ha. The first sampling phase is based on full coverage by satellite imagery, and the second phase is based on ALS data and field measurements. A broad stratification is made based on satellite images. Then, a sample of ALS transects is collected and the field plots are positioned based on ALS characteristics. Field plots are used to calibrate statistical models based on ALS. Finally, variables predicted using ALS models are used as references when estimation is carried out for a complete wall-to-wall area using satellite images. A somewhat similar approach was used in Gregoire et al. (2011) and in Ståhl et al. (2011) for a large-area FI in Norway. They used NFI field plots, ALS transects, and profiling LiDAR data. The two-phase laser sampling estimates for AGB were close to the estimates predicted using only field plots. However, the corresponding standard errors were larger. Ståhl et al. (2011) obtained standard errors close to those of systematic field sampling with laser sampling. In this case, the predictions were overestimations. In both studies, profiles of LiDAR and ALS were also compared, and the ALS was found to be more useful. Review of the LiDAR sampling for large-area forest characterization can be found in Wulder et al. (2012).

15.7.7 Value of ALS Inventory Information in Forest Management Planning

Reliable inventory data are essential for forest planning simulation and optimization calculations. In assessing the state of a stand, the estimates may differ significantly from the real situation due to the inventory method and simulation models (i.e., growth models) used. This aspect can be studied using cost-plus-loss analyses, in which the expected losses due to nonoptimal decision making are added to the total FI costs. The cost-plus-loss approach was widely utilized in recent FI- and planning-related research (e.g., Holmström et al. 2003; Eid et al. 2004; Holopainen and Talvitie 2006; Mäkinen et al. 2012).

In addition to using the cost-plus-loss approach, the value of inventory data can be analyzed by means of focusing on the effect of inventory data accuracy on the timing of loggings and the NPV of thinning and clear-cutting revenues. Holopainen et al. (2010a) showed that input data accuracy significantly affects both logging timing and logging revenue NPV. With respect to logging timing, inaccurate inventory data caused an error in thinning or clear-cutting timing, ranging from 6.5 to 10.3 years, depending on input data source and simulation methodology. With respect to simulated logging revenue NPV, inaccurate inventory data caused a relative error ranging from 28.2% to 57%. Holopainen et al. (2010a) also showed that ALS ITD leads to a more accurate simulated timing of loggings than the traditional SWFI. However, it was concluded that current simulation models are not able to fully utilize the accurate tree height information produced by the ALS ITD inventory methodology.

15.8 Terrestrial and Mobile Laser Scanning Applications

15.8.1 TLS Applications

The studies of using TLS in forest inventories have been concentrated on the collection of main tree attributes such as the dbh, tree height, and stem location (e.g., Watt and Donoghue 2005; Tansey et. al. 2009; Huang et al. 2011; Liang et al. 2011a; Lovell et al. 2011). The applicability of TLS data for collecting additional tree attributes has been reported for, for example, leaf orientation, LAI, volume, subcanopy architecture, and biomass (e.g., Hosoi and Omasa 2006; Moorthy et al. 2008; Lefsky and McHale 2008; Strahler et al. 2008; Hilker et al. 2010; Holopainen et al. 2011a; Zheng and Moskal 2012). Detailed models of the stems, foliage, and branches may be created from TLS data at tree level (Pfeifer and Winterhalder 2004; Cote et al. 2009) or at plot level (Hopkinson et al. 2004; Henning and Radtke 2006; Van der Zande et al. 2011; Holopainen et al. 2011a; Liang et al. 2012a). Canopy structure has also been studied using a combination of TLS and ALS data (Lovell et al., 2003; Hilker et al., 2010; Hosoi et al., 2010; Jung et al., 2011). Only the trees visible to the scanner can be measured; thus, tree density, visibility, and measuring geometry strongly affect how accurately tree variables can be measured (Liang et al. 2012a).

15.8.1.1 Multiscan or Single-Scan TLS Measurements?

Forest plots are measured with one scan from the center of the plot (Holopainen et al. 2011a; Liang et al. 2011a) or with several scans around the plot (Hopkinson et al. 2004; Henning and Radtke 2006) (Figure 15.8). In the single-scan mode, the amount of 3D data obtained is smaller and the field measurements are faster compared with multiscan mode. The major drawback is that trees located in blind spots, that is, shadowed by other trees, cannot be measured. In multiscan measurement, several scans are made from different positions inside and outside of the plot. Scans are accurately coregistered typically using artificial reference targets to obtain the complete dataset. In the multiscan mode, blind spots are reduced, and distribution of points is more even. However, additional work needs to be done in the field and in postprocessing, especially in the integration of several scans into a single point cloud. Fully automated processing techniques are currently being developed for TLS applications.

The tree stems estimated from TLS data can be used as training data to estimate stem attributes from ALS data, as pointed out by Lindberg (2012). For this kind of purpose, it might not be necessary to find all trees in a given area from the TLS data as long as the selection is representative of all trees. In this context (ITD technique), the question whether to use multiscan or single-scan TLS measurements is not so vital. Multiscan measurements become necessary if all trees on the plot need to be covered, especially on the dense forest plot.

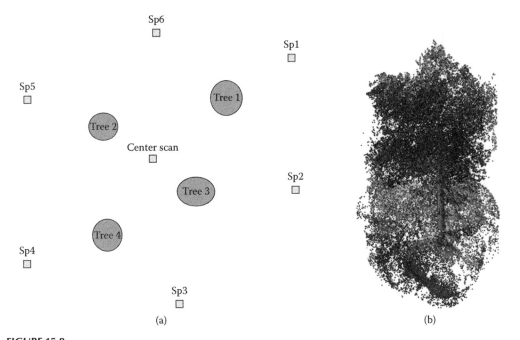

FIGURE 15.8
(a) Multiscan principle. Squares represent different scan points (Sp) in one-tree group, and green circles represent sample trees. (b) Example of the acquired data using three individual scans. (Copyright Ville Kankare.)

15.8.1.2 Stem Location Mapping with TLS

TLS data can be used to determine the relative stem location within a plot. Absolute mapping requires matching of the result to known coordinates. The stem location mapping can be performed in two-dimensional (2D) or 3D space. In the 2D space, the stem location is mapped in a layer sliced from the 3D point cloud, or in the image built from the row/column information based on the scanning geometry (Liang et al. 2011a). The advantage of this idea lies in its simplicity. The amount of computation is usually modest.

In the 2D-layer searching technique, a sliced layer with a certain thickness is cut from the original point cloud. Stems are identified by means of point clustering, circle finding, or waveform analysis from this layer. Multilayer estimation has also been proposed to improve the detection performance. This method assumes that the trees on the plots have a clear stem. The detection becomes a problem if branches are present or nearby branches are overlapped in the layer. Additionally, knowledge of the terrain is necessary to enable constructing the layer at a defined height above the elevation model.

When applying the range-image clustering method, points—or pixels—in the range image are grouped according to local properties, that is, the distance or surface curvature. This technique benefits from the image structure, where the searching of neighboring points in the 3D space can be performed in an approximate manner in the 2D space. Another potential benefit is that detecting trunks from far away is possible.

In the 3D space, the attributes of an individual point are estimated in a local neighborhood, and the stem is identified by means of semantic interpretations. In general, this technique needs more computation than the methods performed in 2D space, but it does not require particular plot knowledge or data structure. The method processes the point cloud itself and can be used in different types of forest.

As an example of stem mapping accuracies, the proportion of located stems from single- and multiscan data has been rather limited, that is, 22% and 52% with density of 556 stems/ha, 97% with density of 321 stems/ha, and 100% with density of 310 stems/ha (Thies and Spiecker 2004; Maas et al. 2008). In Litkey et al. (2008), 85% of the trunks, which could be manually identified from the TLS data, were automatically detected in single-scan data. Thus, automated processing in more dense forests is still under development.

In Liang et al. (2012a), the average stem count was 1022 stems/ha, and the obtained overall detection accuracy was 73%. The depicted method in Liang et al. (2012) was successful in finding 85% of the trunks within the range of 5 m. This shows that in a dense managed forest stand, the majority of trunks can be located using single-scan TLS data only for limited range, but the combination of several single-scans or the usage of multiscan is needed to provide a practical system that is feasible in various forest environments if most of the trees in the plots need to be mapped. Liang et al. (2012b) presented a fully automated method for detecting changes in forest structure over time using bitemporal TLS data. They demonstrated that 90% of tree stem changes could be automatically located from single-scan TLS data.

15.8.1.3 Determination of Diameter Breast Height and Stem Curve with TLS

Conventionally, diameter breast height (at the height of 1.3 m) is needed in detailed forest measurements and calculations. Therefore, Pfeifer and Winterhalder (2004) modeled, in addition to stem diameters, branch diameters with an accuracy of better than 1 cm.

Henning and Radtke (2006) measured tree stem diameters up to the crown base height with accuracies <1 cm and <2 cm under 13 m of stem height. Vastaranta et al. (2009) used TLS to measure tree dbhs with a standard error of 4.5% (8.3 mm). Since point clouds include data beyond dbh, the tree stem curve also can be obtained from the TLS data. The tree stem curve is the tapering of the stem as a function of the height. It is the key input needed in the harvesting to maximize the profit at the individual tree level. In modern cut-to-length harvesters, the bucking of stem is controlled by value and demand matrices (Uusitalo 2010). The stem optimization calculation requests the knowledge of the shape of the stem as accurately as possible. The stem curve information is also directly turned to many FI and ecology parameters. Measuring the stem curve from a point cloud was first shown by Thies and Spiecker (2004) and followed by Henning and Radtke (2006). In Maas et al. (2008), a spruce was scanned using single-scan mode. The RMSE was reported as 4.7 cm. In Liang et al. (2011b), the RMSE of stem curve estimation is 1.8 cm for a pine and 0.6 cm for a spruce using single-scan measurement. When using multiscan data, the RMSE is 1.3 cm for the pine and 0.6 cm for the spruce. The automated stem curve estimation is still under study.

15.8.1.4 Acquisition of Other Tree Attribute Data from TLS

Hopkinson et al. (2004) showed that TLS is capable of measuring forest stand variables. Moorthy et al. (2008) determined in laboratory conditions the canopy gap fraction and LAI from TLS data with R^2 of 0.95 and 0.98, respectively. Hyyppä et al. (2009b) and Kaasalainen et al. (2010) conducted defoliation and biomass change measurements using TLS. The diminished number of point returns estimated the level of defoliation: The change in point returns correlated with an R^2 of 0.99 with a change in biomass. Holopainen et al. (2011a) modeled tree AGBs for Scots pines and Norway spruces. The stem and crown dimensions measured from the TLS point clouds correlated strongly ($r = 0.98 – 0.99$) with laboratory biomass measurements carried out after scanning.

15.8.2 MLS Applications

MLS can be seen as a method falling between ALS and TLS. MLS is laser scanning that is done from a moving vehicle such as a car or a logging machine. The application of MLS in forestry is being actively studied (Lin et al. 2010; Lin and Hyyppä 2011; Holopainen et al. 2011b) (Figure 15.9). In the near future, MLS can be seen as a practical means to produce tree maps or inventories in urban forest environments. Holopainen et al. (2011b) obtained promising results with TLS and MLS compared with the accuracy of LS-based results and method efficiencies in Helsinki city street and park tree mapping. MLS and a logging machine could enable the automatic selection of harvestable trees and enhancements in stem bucking. However, MLS is still far from a widely used practical application in forestry, but the situation may change due to the rapid development of automatic MLS and TLS data processing.

As an example of multitemporal applications using the mini-UAV-based LS, a single coniferous tree was measured and it was manually defoliated (Jaakkola et al. 2010), and it was shown that the biomass change can be relatively easily mapped using consecutive laser surveys from the same tree ($R^2 = 0.92$). Mini-UAV-based LS allows, for example, multitemporal surveys and especially the collection of ALS point clouds from needed sample plots simultaneously with field reference collection.

FIGURE 15.9
(See color insert.) Mobile laser scanning in Evo study area. (Copyright Harri Kaartinen & Antero Kukko, FGI.)

The main challenge for MLS usage in forestry is the need to acquire high quality positioning inside the forest. This is needed in the continuous mode of the MLS. That would allow the quick collection of plot-wise point clouds representing field reference. The MLS could be mounted on a snowmobile or on a small forest tractor, as shown in Figure 15.9. The processing of MLS data would be similar to that represented in the examples of the TLS.

15.9 Conclusions

During the past decade in forest mapping and monitoring applications, the possibility of acquiring spatially accurate active 3D RS information instead of 2D has been a major turning point. When the aim is to produce forest resource information that is as accurate as possible for forest managers, this change has opened up totally new possibilities. ALS is an efficient tool for 3D probing of the forest from above, and it is very promising concerning forest mapping and monitoring needs. Operational ALS-based forest inventories apply the ABA method. In it, the scale is at the stand-sub-stand levels. The total timber volume is obtained at high accuracy, while the information about the size distribution, species, timber assortments, or the number of trees has limited reliability (i.e., Holopainen et al. 2010b).

To increase wood value and productivity in industry, information on wood raw material quantity and quality—combined with logistic concepts that integrate transport systems and management models throughout the wood raw material supply chain—is needed. ALS could pave the way for "precision forestry," in which forest resource monitoring would be carried out even at the single-tree level. Many possible additional benefits such as operational planning for optimal timing of operations, better scheduling and logistics, and accurate information on stem dimensions as well as quality and optimal cutting of stems could be achieved (Holopainen et al. 2010d).

Estimation of timber assortments with ABA is slightly more accurate than it is when based on SWFI data. In special cases, as in Peuhkurinen et al. (2007), promising results are achieved by using ITD. Information about raw wood quality is difficult to add when inventory is carried out by using ABA. In principle, there are means in ITD to measure exact distributions of timber assortment. However, problems in tree detection and species identification have to be worked out first. Tree detection is the most important error component that must receive attention (Vastaranta et al. 2011b; Kaartinen et al. 2012).

In the near future, ALS inventories should be developed further from two standpoints. First, the information content should meet the forest management requirements better. Second, if the current accuracy level is accepted, the cost efficiency of the inventory can be developed in many ways. These methodologies could involve the use ITD with low-pulse densities or the use of ITD in acquiring training data for ABA (Vastaranta et al. 2011c, 2012a).

Multitemporal ALS and change detection based on it enable monitoring systems when long time-series data start to be more widely available. Bitemporal or multitemporal ALS datasets could even be applied in operational forest management systems for the determination of forest growth potential as well as monitoring forest operations and health. On the other hand, these kinds of datasets provide a base for studying forest growth dynamics in a 3D canopy environment in a way that has been, more or less, impossible. Forest growth that can be determined with bitemporal data correlates with site type.

TLS and MLS are efficient and objective options for acquiring accurate field data, and they could be utilized, for example, in collecting ground truth data for ALS estimations. The applications of TLS and MLS for forestry, that is, in forest biomass estimation and monitoring, have not been widely studied, although their potential for forest-related measurements has been better understood in recent years. TLS and MLS are capable of measuring all important tree characteristics such as diameter, height, and location, and they can also provide information on canopy-related characteristics and stem form—information that has not been achievable before. However, previous studies have shown that ALS, TLS, and MLS have similar problems with tree detection caused by lack of visibility (i.e., Holopainen et al. 2011a; Liang et al. 2012a). Ground-based measurements are highly affected by understory shrubs, and airborne methods are affected by canopy closure.

Due to the shorter operating distances of TLS and MLS, they can easily have a higher pulse rate than ALS. At present, it is possible to use software and methods developed for ALS, but due to the different scanning geometry, changing point density as a function of range, and the fast processing needed, algorithms for TLS and MLS data processing need to be developed separately. It can be concluded that automatic processing methods should still be improved before use of the TLS and MLS data in forestry is fully operational.

Acceleration of FI and forest management planning is just a beginning step in the technology leap that LS provides. Logistics in wood procurement and timber harvesting, besides monitoring forest biomass changes, constitute one area where LS data will be applied in the near future. Data acquisition costs are dropping all of the time, and data processing is becoming more automated, thus enabling use of ALS for larger areas, even in global applications.

References

Ahokas, E., H. Kaartinen, L. Matikainen, J. Hyyppä, and H. Hyyppä. 2002. "Accuracy of High-Pulse-Rate Laser Scanners for Digital Target Models." *Observing our Environment from Space. New Solutions for New Millennium, Proceedings of the 21st EARSeL Symposium*, May 14–16, 2001, 175–78. Paris: Balkema.

Ahokas, E., X. Yu, J. Oksanen, J. Hyyppä, H. Kaartinen, and H. Hyyppä. 2005. "Optimization of the Scanning Angle for Countrywide Laser Scanning." *Proceedings of ISPRS Workshop Laser Scanning*, September 12–14, 2005, Enschede, The Netherlands, *International Archives of Photogrammetry, Remote Sensing and Spatial Information Sciences* XXXVI (Part 3/W19):115–19.

Aldred, A., and G. Bonnor 1985. *Application of Airborne Lasers to Forest Surveys*. Information Report PI-X-51, Canadian Forestry Service, Petawawa National Forestry Institute, Canada, 62pp.

Baker, J., P. Mitchell, R. Cordey, G. Groom, J. Settle, and M. Stileman. 1994. "Relationships between Physical Characteristics and Polarimetric Radar Backscatter for Corsican Pine Stands in Thetford Forest, UK." *International Journal of Remote Sensing* 15:2827–49.

Baltavias, E. P. 1999. "Airborne Laser Scanning: Basic Relations and Formulas." *ISPRS Journal of Photogrammetry and Remote Sensing* 54(2–3):199–214.

Bortolot, Z., and R. Wynne. 2005. "Estimating Forest Biomass Using Small Footprint LiDAR Data: An Individual Tree-Based Approach That Incorporates Training Data." *ISPRS Journal of Photogrammetry and Remote Sensing* 59:342–60.

Breidenbach, J., E. Næsset, V. Lien, T. Gobakken, and S. Solberg. 2010. "Prediction of Species Specific Forest Inventory Attributes Using a Nonparametric Semi-Individual Tree Crown Approach Based on Fused Airborne Laser Scanning and Multispectral Data." *Remote Sensing of Environment* 114:911–24.

Chasmer, L., C. Hopkinson, and P. Treizt. 2006. "Investigating Laser Pulse Penetration through a Conifer Canopy by Integrating Airborne and Terrestrial LiDAR." *Canadian Journal of Remote Sensing* 32(2):116–25.

Cote, J. F., J. L. Widlowski, R. A. Fournier, and M. M. Verstraete. 2009. "The Structural and Radiative Consistency of Three-Dimensional Tree Reconstructions from Terrestrial LiDAR." *Remote Sensing of Environment* 113(5):1067–81.

Eid, T., T. Gobakken, and E. Næsset. 2004. "Comparing Stand Inventories for Large Areas Based on Photo-Interpretation and Laser Scanning by Means of Cost-Plus-Loss Analyses." *Scandinavian Journal of Forest Research* 19:512–23.

Falkowski, M., A. Hudak, N. Crookston, P. Gessler, and A. Smith. 2010. "Landscape-Scale Parameterization of a Tree-Level Forest Growth Model: A k-NN Imputation Approach Incorporating LiDAR Data." *Canadian Journal of Forest Research* 40:184–99.

Falkowski, M., A. Smith, P. Gessler, A. Hudak, L. Vierling, and J. Evans. 2008. "The Influence of the Conifer Forest Canopy Cover on the Accuracy of Two Individual Tree Detection Algorithms Using LiDAR Data." *Canadian Journal of Remote Sensing* 34(2):1–13.

Gatziolis, D. 2007. "LiDAR-Derived Site Index in the U.S. Pacific Northwest—Challenges and Opportunities." *The International Archives of the Photogrammetry, Remote Sensing and Spatial Information Sciences* XXXVI (Part 3/W52):136–43. Espoo, Finland.

Gautam, B., T. Tokola, J. Hämäläinen, M. Gunia, J. Peuhkurinen, H. Parviainen, V. Leppanen, et al. 2011. "Integration of Airborne LiDAR, Satellite Imagery, and Field Measurements Using a Two-Phase Sampling Method for Forest Biomass Estimation in Tropical Forests." *International Symposium on "Benefiting from Earth Observation"*, October 4–6, 2010, Kathmandu, Nepal.

Gaveu, D., and R. Hill. 2003. "Quantifying Canopy Height Underestimation by Laser Pulse Penetration in Small-Footprint Airborne Laser Scanning Data." *Canadian Journal of Remote Sensing* 29:650–57.

Gregoire, T., G. Ståhl, E. Næsset, T. Gobakken, R. Nelson, and S. Holm. 2011. "Model-Assisted Estimation of Biomass in a LiDAR Sample Survey in Hedmark County, Norway." *Canadian Journal of Forest Research* 41:83–95.

Gupta, S., B. Koch, and H. Weinacker. 2010. "Tree Species Detection Using Full Waveform LiDAR Data in a Complex Forest." In *Technical Commission VII Symposium—100 Years of ISPRS*, edited by W. Wagner and B. Székely, July 5–7, 2010, 249–54. Vienna, Austria: Vienna University of Technology.

Haara, A., and K. Korhonen, 2004. "Kuvioittaisen arvioinnin luotettavuus." *Metsätieteen aikakauskirja* 4:489–508.

Harding, D., M. Lefsky, G. Parke, and J. Blair. 2001. "Laser Altimeter Canopy Height Profiles. Methods and Validation for Closed Canopy, Broadleaved Forests." *Remote Sensing of Environment* 76(3):283–97.

Heinzel, J., H. Weinacker, and B. Koch. 2011. "Prior Knowledge Based Single Tree Extraction." *International Journal of Remote Sensing* 32:4999–5020.

Henning, J. G., and P. J. Radtke. 2006. "Detailed Stem Measurements of Standing Trees from Ground-Based Scanning LiDAR." *Forest Science* 52(1):67–80.

Hilker, T., M. van Leeuwen, N. C. Coops, M. A. Wulder, G. J. Newnham, D. L. B. Jupp, and D. S. Culvenor. 2010. "Comparing Canopy Metrics Derived from Terrestrial and Airborne Laser Scanning in a Douglas-Fir Dominated Forest Stand." *Trees-Structure and Function* 24(5):819–32.

Hofton, M., L. Rocchio, J. Blair, and R. Dubayah. 2002. "Validation of Vegetation Canopy LiDAR Sub-Canopy Topography Measurements for a Dense Tropical Forest." *Journal of Geodynamics* 34(3–4):491–502.

Hollaus, M., W. Wagner, B. Maier, and K. Schadauer. 2007. "Airborne Laser Scanning of Forest Stem Volume in a Mountainous Environment." *Sensors* 7:1559–77.

Holmgren, J. 2003. "Estimation of Forest Variables Using Airborne Laser Scanning." PhD thesis, Acta Universitatis Agriculturae Sueciae, Silvestria 278, Swedish University of Agricultural Sciences, Umeå, Sweden.

Holmgren, J., and T. Jonsson. 2004. "Large Scale Airborne Laser Scanning of Forest Resources in Sweden." *International Archives of Photogrammetry, Remote Sensing and Spatial Information Sciences* XXXVI (8/W2):157–60.

Holmgren, J., and Å. Persson. 2004. "Identifying Species of Individual Trees Using Airborne Laser Scanner." *Remote Sensing of Environment* 90:415–23.

Holmgren, J., Å. Persson, and U. Söderman. 2008. "Species Identification of Individual Trees by Combining High Resolution LiDAR Data with Multi-Spectral Images." *International Journal of Remote Sensing* 29:1537–52.

Holmström, H., H. Kallur, and G. Ståhl. 2003. "Cost-Plus-Loss Analyses of Forest Inventory Strategies Based on kNN-Assigned Reference Sample Plot Data." *Silva Fennica* 37(3):381–98.

Holopainen, M. 2011. "Effect of Airborne Laser Scanning Accuracy on Forest Stock and Yield Estimates." Doctoral thesis, Aalto University, School of Engineering, Department of Surveying, Finland, Aalto University Doctoral Dissertations 6/2011, 160pp.

Holopainen, M., and J. Hyyppä. 2003. "Possibilities with Laser Scanning in Practical Forestry." In *Proceedings of the ScandLaser Scientific Workshop on Airborne Laser Scanning of Forests*, edited by J. Hyyppä, E. Næsset, H. Olsson, T. Granqvist Pahlen, and H. Reese, 264–73.

Holopainen, M., J. Hyyppä, L.-M. Vaario, and K. Yrjälä. 2010d. "Implications of Technological Development to Forestry." In *Forest and Society—Responding to Global Drivers of Change*, edited by G. Mery, P. Katila, G. Galloway, R. I. Alfaro, M. Kanninen, M. Lobovikov, and J. Varjo, Invited book chapter, IUFRO-World Series Vol. 25, 157–82. (Convening lead authors of the chapter: Hetemäki, L. and G. Mery).

Holopainen, M., A. Mäkinen, J. Rasinmäki, J. Hyyppä, H. Hyyppä, H. Kaartinen, R. Viitala M. Vastaranta, and A. Kangas. 2010a. "Effect of Tree Level Airborne Laser Scanning Accuracy on the Timing and Expected Value of Harvest Decisions." *European Journal of Forest Research* 129:899–910.

Holopainen, M., A. Mäkinen, J. Rasinmäki, K. Hyytiäinen, S. Bayazidi, and I. Pietilä. 2010e. "Comparison of Various Sources of Uncertainty in Stand-Level Net Present Value Estimates." *Forest Policy and Economics* 12(2010):377–86. doi 10.1016/j.forpol.2010.02.009.

Holopainen, M., A. Mäkinen, J. Rasinmäki, K. Hyytiäinen, S. Bayazidi, M. Vastaranta, and I. Pietilä. 2010f. "Uncertainty in Forest Net Present Value Estimations." *Forests* 2010(1):177–93. doi:10.3390/f1030177.

Holopainen, M., and M. Talvitie. 2006. "Effects of Data Acquisition Accuracy on Timing of Stand Harvests and Expected Net Present Value." *Silva Fennica* 40(3):531–43.

Holopainen, M., M. Vastaranta, R. Haapanen, X. Yu, J. Hyyppä, H. Kaartinen, R. Viitala, and H. Hyyppä. 2010c. "Site-Type Estimation Using Airborne Laser Scanning and Stand Register Data." *The Photogrammetric Journal of Finland* 22(1):16–32.

Holopainen, M., M. Vastaranta, V. Kankare, J. Hyyppä, X. Liang, P. Litkey, X. Yu, et al. 2011b. "The Use of ALS, TLS and VLS Measurements in Mapping and Monitoring Urban Trees." In *JURSE 2011—Joint Urban Remote Sensing Event*, edited by U. Stilla, P. Gamba, C. Juergens, and D. Maktav, April 11–13, 2011. Munich, Germany.

Holopainen, M., M. Vastaranta, V. Kankare, M. Räty, M. Vaaja, X. Liang, X. Yu, et al. 2011a. "Biomass Estimation of Individual Trees Using TLS Stem and Crown Diameter Measurements." In *Laser Scanning 2011 Proceedings*, edited by D. Lichti and A. Habib.

Holopainen, M., M. Vastaranta, J. Rasinmäki, J. Kalliovirta, A. Mäkinen, R. Haapanen, T. Melkas, X. Yu, and J. Hyyppä. 2010b. "Uncertainty in Timber Assortment Estimates Predicted from Forest Inventory Data." *European Journal of Forest Research* 129:1131–42.

Hopkinson, C., L. Chasmer, and R. Hall. 2008. "The Uncertainty in Conifer Plantation Growth Prediction from Multi-Temporal LiDAR Datasets." *Remote Sensing of Environment* 112:1168–80.

Hopkinson, C., L. Chasmer, K. Lim, P. Treitz, and I. Creed. 2006. "Towards a Universal LiDAR Canopy Height Indicator." *Canadian Journal of Remote Sensing* 32(2):139–52.

Hopkinson, C., L. Chasmer, C. Young-Pow, and P. Treitz. 2004. "Assessing Forest Metrics with a Ground-Based Scanning LiDAR." *Canadian Journal of Forest Research* 34:573–83.

Hosoi, F., Y. Nakai, and K. Omasa. 2010. "Estimation and Error Analysis of Woody Canopy Leaf Area Density Profiles Using 3-D Airborne and Ground-Based Scanning LiDAR Remote-Sensing Techniques." *IEEE Transactions on Geoscience and Remote Sensing* 48(5):2215–23.

Hosoi, F. K., and K. Omasa. 2006. "Voxel-Based 3-D Modeling of Individual Trees for Estimating Leaf Area Density Using High-Resolution Portable Scanning LiDAR." *IEEE Transactions on Geoscience and Remote Sensing* 44(12):3610–18.

Hou, Z., Q. Xu, and T. Tokola. 2011. "Use of ALS, Airborne CIR and ALOS AVNIR-2 Data for Estimating Tropical Forest Attributes in Lao PDR." *ISPRS Journal of Photogrammetry and Remote Sensing* 66(6):776–86.

Huang, H., Z. Li, P. Gong, X. Cheng, N. Clinton, C. Cao, W. Ni, and L. Wang. 2011. "Automated Methods for Measuring DBH and Tree Heights with a Commercial Scanning LiDAR." *Photogrammetric Engineering and Remote Sensing* 77(3):219–27.

Hudak, A., N. Crookston, J. Evans, D. Hall, and M. Falkowski. 2008. "Nearest Neighbor Imputation of Species-Level, Plot-Scale Forest Structure Attributes from LiDAR Data." *Remote Sensing of Environment* 112:2232–45.

Hyyppä, H., and J. Hyyppä. 1999. "Comparing the Accuracy of Laser Scanner with Other Optical Remote Sensing Data Sources for Stand Attribute Retrieval." *The Photogrammetric Journal of Finland* 16(2):5–15.

Hyyppä, J., H. Hyyppä, X. Yu, H. Kaartinen, A. Kukko, and M. Holopainen. 2009b. "Forest Inventory Using Small-Footprint Airborne LiDAR." In *Topographic Laser Ranging and Scanning: Principles and Processing*, edited by J. Shan and C. Toth, Invited Book Chapter, 335–70.

Hyyppä, J., and M. Inkinen. 1999. "Detecting and Estimating Attributes for Single Trees Using Laser Scanner." *The Photogrammetric Journal of Finland* 16:27–42.

Hyyppä, J., A. Jaakkola, H. Hyyppä, H. Kaartinen, A. Kukko, M. Holopainen, L. Zhu, et al. 2009a. "Map Updating and Change Detection Using Vehicle-Based Laser Scanning." *2009 Urban Remote Sensing Joint Event*, May 20–22. Shanghai.

Hyyppä, J., T. Mielonen, H. Hyyppä, M. Maltamo, X. Yu, E. Honkavaara, and H. Kaartinen. 2005. "Using Individual Tree Crown Approach for Forest Volume Extraction with Aerial Images and Laser Point Clouds." *Proceedings of ISPRS Workshop Laser Scanning*, September 12–14, 2005, Enschede, The Netherlands. *International Archives of Photogrammetry, Remote Sensing and Spatial Information Sciences* XXXVI (Part 3/W19):144–49.

Hyyppä, J., U. Pyysalo, H. Hyyppä, H. Haggrén, and G. Ruppert. 2000. "Accuracy of Laser Scanning for DTM Generation in Forested Areas." *Proceeding of SPIE*, Vol. 4035, 119–30.

Hyyppä, J., X. Yu, H. Hyyppä, and M. Maltamo. 2006. "Methods of Airborne Laser Scanning for Forest Information Extraction." *Workshop on 3D Remote Sensing in Forestry*, February 14–15, 2006, 16. Vienna: CD-ROM.

Jaakkola, A., J. Hyyppä, A. Kukko, X. Yu, H. Kaartinen, M. Lehtomäki, and Y. Lin. 2010. "A Low-Cost Multi-Sensoral Mobile Mapping System and its Feasibility for Tree Measurements." *ISPRS Journal of Photogrammetry and Remote Sensing* 65(6):514–22.

Jochem, A., M. Hollaus, M. Rutzinger, and B. Höfle. 2011. "Estimation of Aboveground Biomass in Alpine Forests: A Semi-Empirical Approach Considering Canopy Transparency Derived from Airborne LiDAR Data." *Sensors* 11:278–95.

Jung, S., D. Kwak, T. Park, W. Lee, S. Yoo, S. E. Jung, D. A. Kwak, T. J. Park, W. K. Lee, and S. J. Yoo. 2011. "Estimating Crown Variables of Individual Trees Using Airborne and Terrestrial Laser Scanners." *Remote Sensing* 3(11):2346–63.

Kaartinen, H., and J. Hyyppä. 2008. EuroSDR/ISPRS Project, Commission II "Tree Extraction." Final Report, EuroSDR. European Spatial Data Research, Official Publication No 53.

Kaartinen, H., J. Hyyppä, X. Yu, M. Vastaranta, H. Hyyppä, A. Kukko, M. Holopainen, et al. 2012. "An International Comparison of Individual Tree Detection and Extraction Using Airborne Laser Scanning." *Remote Sensing* 4(4):950–74.

Kaasalainen, S., J. Hyyppä, A. Krooks, M. Karjalainen, P. Lyytikäinen-Saarenmaa, M. Holopainen, and A. Jaakkola. 2010. "Comparison of Terrestrial Laser Scanner and Synthetic Aperture Radar Data in the Study of Forest Defoliation." *ISPRS Commission VII Symposium*, July 5–7, 2010. Vienna, Austria.

Kalliovirta, J., and T. Tokola. 2005. "Functions for Estimating Stem Diameter and Tree Age Using Tree Height, Crown Width and Existing Stand Database Information." *Silva Fennica* 39(2):227–48.

Kantola, T., M. Vastaranta, X. Yu, P. Lyytikäinen-Saarenmaa, M. Holopainen, M. Talvitie, S. Kaasalainen, S. Solberg, and J. Hyyppä. 2010. "Classification of Defoliated Trees Using Tree-Level Airborne Laser Scanning Data Combined with Aerial Images." *Remote Sensing* 2:2665–79.

Kellndorfer, J., W. Walker, E. LaPoint, K. Kirsch, J. Bishop, and G. Fiske. 2010. "Statistical Fusion of LiDAR, INSAR, and Optical Remote Sensing Data for Forest Stand Height Characterization: A Regional-Scale Method Based on LVIS, SRTM, Landsat ETM+, and Ancillary Data Sets." *Journal of Geophysical Research* 115:G00E08. doi:10.1029/2009JG000997.

Kilkki, P., and R. Päivinen. 1987. "Reference Sample Plots to Combine Field Measurements and Satellite Data in Forest Inventory." Department of Forest Mensuration and Management, University of Helsinki. *Research Notes* 19:210–15.

Koch, B. 2010. "Status and Future of Laser Scanning, Synthetic Aperture Radar and Hyperspectral Remote Sensing Data for Forest Biomass Assessment." *ISPRS Journal of Photogrammetry and Remote Sensing* 65(6):581–90.

Koch, B., U. Heyder, and H. Weinacker. 2006. "Detection of Individual Tree Crowns in Airborne LiDAR Data." *Photogrammetric Engineering and Remote Sensing* 72:357–63.

Korhonen, L., I. Korpela, J. Heiskanen, and M. Maltamo. 2011. "Airborne Discrete-Return LiDAR Data in the Estimation of Vertical Canopy Cover, Angular Canopy Closure and Leaf Area Index." *Remote Sensing of Environment* 115(4):1065–80.

Korpela, I., A. Hovi, and F. Morsdorf. 2012. "Understory Trees in Airborne LiDAR Data—Selective Mapping due to Transmission Losses and Echo-Triggering Mechanisms." *Remote Sensing of Environment* 119:92–104.

Korpela, I., M. Koskinen, H. Vasander, M. Holopainen, and K. Minkkinen. 2009. "Airborne Small-Footprint Discrete-Return LiDAR Data in the Assessment of Boreal Mire Surface Patterns, Vegetation And Habitats." *Forest Ecology and Management* 258:1549–66.

Korpela, I., H. O. Ørka, M. Maltamo, T. Tokola, and J. Hyyppä. 2010. "Tree Species Classification Using Airborne LiDAR—Effects of Stand and Tree Parameters, Downsizing of Training Set, Intensity Normalization and Sensor Type." *Silva Fennica* 44:319–39.

Kraus, K., and N. Pfeifer. 1998. "Determination of Terrain Models in Wooded Areas with Airborne Laser Scanner Data." *ISPRS Journal of Photogrammetry and Remote Sensing* 53:193–203.

Laamanen, R., and A. Kangas. 2012. "Large-Scale Forest Owner's Information Needs in Operational Planning of Timber Harvesting—Some Practical Views in Metsähallitus, Finnish State-Owned Enterprise." *Silva Fennica* 45(4):711–27.

Latifi, H., A. Nothdurft, and B. Koch. 2010. "Non-Parametric Prediction and Mapping of Standing Timber Volume and Biomass in a Temperate Forest: Application of Multiple Optical/LiDAR–Derived Predictors." *Forestry* 83:395–407.

Lefsky, M., W. Cohen, G. Parker, and D. Harding. 2002. "LiDAR Remote Sensing for Ecosystem Studies." *Bioscience* 52:19–30.

Lefsky, M., D. Harding, W. Cohen, G. Parker, and H. Shugart. 1999. "Surface LiDAR Remote Sensing of Basal Area and Biomass in Deciduous Forests of Eastern Maryland, USA." *Remote Sensing of Environment* 67:83–98.

Lefsky, M., and M. R. McHale. 2008. "Volume Estimates of Trees with Complex Architecture from Terrestrial Laser Scanning." *Journal of Applied Remote Sensing* 2(1):023521.

Liang, X., J. Hyyppä, H. Kaartinen, M. Holopainen, and T. Melkas. 2012b. "Detecting Changes in Forest Structure over Time with Bi-Temporal Terrestrial Laser Scanning Data." *ISPRS International Journal of Geo-Information* 2012(1):242–55.

Liang, X., J. Hyyppä, V. Kankare, and M. Holopainen. 2011b. "Stem Curve Measurement Using Terrestrial Laser Scanning." *Proceedings Of SilviLaser Conference*, 16–20 October, 6pp. Hobart, Tasmania.

Liang, X., J. Hyyppä, and L. Matikainen. 2007. "Deciduous-Coniferous Tree Classification Using Difference between First and Last Pulse Laser Signatures." In *Proceedings of the ISPRS Workshop on Laser Scanning 2007 and SilviLaser 2007*, edited by P. Rönnholm, H. Hyyppä, and J. Hyyppä, September 12–14, 2007, Espoo, Finland. *IAPRS*, Vol. XXXVI (Part 3/W52):253–57.

Liang, X., P. Litkey, J. Hyyppä, H. Kaartinen, A. Kukko, and M. Holopainen. 2011a, "Automatic Plot-Wise Tree Location Mapping Using Single-Scan Terrestrial Laser Scanning." *Photogrammetric Journal of Finland* 22:37–48.

Liang, X., P. Litkey, J. Hyyppä, H. Kaartinen, M. Vastaranta, and M. Holopainen. 2012a. "Automatic Stem Mapping Using Single-Scan Terrestrial Laser Scanning." *IEEE Transactions on Geoscience and Remote Sensing* 50(2):661–70.

Lin, Y., and J. Hyyppä. 2011. "k-Segments-Based Geometric Modeling of VLS Scan Lines." *IEEE Geoscience and Remote Sensing Letter* 8(1):93–97.

Lin, Y., A. Jaakkola, J. Hyyppä, and H. Kaartinen, 2010. "From TLS to VLS: Biomass Estimation at Individual Tree Level." *Remote Sensing* 2(8):1864–79. http://www.mdpi.com/2072-4292/2/8/1864/.

Lindberg, E. 2012. *Estimation of Canopy Structure and Individual Trees from Laser Scanning Data.* Acta Universitatis Agriculturae Sueciae, 2012:33, ISSN 1652-6880, ISBN 978-91-576-7669-6, Umeå, Print: Arkitektkopia, Umeå 2012.

Lindberg, E., J. Holmgren, K. Olofsson, J. Wallerman, and H. Olsson. 2010. "Estimation of Tree Lists from Airborne Laser Scanning by Combining Single-Tree and Area-Based Methods." *International Journal of Remote Sensing* 31(5):1175–92.

Litkey, P., X. Liang, J. Hyyppä, A. Kukko, H. Kaartinen, and M. Holopainen. 2008. "Single-Scan TLS Methods for Forest Parameter Retrieval. *Proceedings of SilviLaser 2008*, 295–304. Edinburgh.

Lovell, J. L., D. L. B. Jupp, D. S. Culvenor, and N. C. Coops. 2003. "Using Airborne and Ground-Based Ranging LiDAR to Measure Canopy Structure in Australian Forests." *Canadian Journal of Remote Sensing* 29(5):607–22.

Lovell, J. L., D. L. P. Jupp, G. J. Newnham, and D. S. Culvenor. 2011. "Measuring Tree Stem Diameters Using Intensity Profiles from Ground-Based Scanning LiDAR from a Fixed Viewpoint. *ISPRS Journal of Photogrammetry and Remote Sensing* 66(1):46–55.

Maas, H. G., A. Bienert, S. Scheller, and E. Keane. 2008. "Automatic Forest Inventory Parameter Determination from Terrestrial Laser Scanner Data." *International Journal of Remote Sensing* 29(5):1579–93.

Maltamo, M., O. M. Bollandsas, E. Næsset, T. Gobakken, and P. Packalén. 2011. "Different Plot Selection Strategies for Field Training Data in ALS-Assisted Forest Inventory." *Forestry* 84(1):23–31.

Magnussen, S., and P. Boudewyn. 1998. Derivations of Stand Heights from Airborne Laser Scanner Data with Canopy-Based Quantile Estimators." *Canadian Journal of Forest Research* 28:1016–31.

Mäkinen, A., M. Holopainen, J. Rasinmäki, and A. Kangas. 2010. "Propagating the Errors of Initial Forest Variables through Stand- and Tree-Level Growth Simulators." *European Journal of Forest Research* 129:887–97. doi 10.1007/s10342-009-0288-0.

Mäkinen, A., A. Kangas, and M. Nurmi. 2012. "Using Cost-Plus-Loss Analysis to Define Optimal Forest Inventory Interval and Forest Inventory Accuracy." *Silva Fennica* 46(2):211–26.

Maltamo, M., J. Malinen, P. Packalén, A. Suvanto, and J. Kangas. 2006. "Non-Parametric Estimation of Stem Volume Using Laser Scanning, Aerial Photography and Stand Register Data." *Canadian Journal of Forest Research* 36:426–36.

Maltamo, M., J. Peuhkurinen, J. Malinen, J. Vauhkonen, P. Packalén, and T. Tokola. 2009. "Predicting Tree Attributes and Quality Characteristics of Scots Pine Using Airborne Laser Scanning Data." *Silva Fennica* 43(3):507–21.

Moorthy, I., J. R. Miller, B. Hu, J. Chen, and Q. Li. 2008. "Retrieving Crown Leaf Area Index from an Individual Tree Using Ground-Based LiDAR Data." *Canadian Journal of Remote Sensing* 34(3):320–32.

Muinonen, E., and T. Tokola. 1990. "An Application of Remote Sensing for Communal Forest Inventory." *Proceedings from SNS/IUFRO Workshop: The Usability of Remote Sensing for Forest Inventory and Planning*, February 26–28, 1990. Umeå, Sweden, Remote Sensing Laboratory, Swedish University of Agricultural Sciences, Report 4, 35–42.

Nelson, R., W. Krabill, and J. Tonelli. 1988. "Estimating Forest Biomass and Volume Using Airborne Laser Data." *Remote Sensing of Environment* 24:247–67.

Nelson, R., G. Parker, and M. Hom. 2003a. "A Portable Airborne Laser System for Forest Inventory." *Photogrammetric Engineering and Remote Sensing* 69:267–73.

Nelson, R., A. Short, and M. Valenti. 2004. "Measuring Biomass and Carbon in Delaware Using an Airborne Profiling LiDAR." *Scandinavian Journal of Forest Research* 19:500–11.

Nelson, R., M. A. Valenti, A. Short, and C. Keller. 2003b. "A Multiple Resource Inventory of Delaware Using Airborne Laser Data." *BioScience* 53:981–92.

Nilsson, M. 1996. "Estimation of Tree Heights and Stand Volume Using Airborne LiDAR System." *Remote Sensing of Environment* 56(1):1–7.

Næsset, E. 1997a. "Determination of Mean Tree Height of Forest Stands Using Airborne Laser Scanner Data." *ISPRS Journal of Photogrammetry and Remote Sensing* 52:49–56.

Næsset, E. 1997b. "Estimating Timber Volume of Forest Stands Using Airborne Laser Scanner Data." *Remote Sensing of Environment* 61(2):246–53.

Næsset, E. 2002. "Predicting Forest Stand Characteristics with Airborne Scanning Laser Using a Practical Two-Stage Procedure and Field Data." *Remote Sensing of Environment* 80:88–99.

Næsset, E. 2004. "Practical Large-Scale Forest Stand Inventory Using a Small-Footprint Airborne Scanning Laser." *Scandinavian Journal of Forest Research* 19:164–79.

Næsset, E., and T. Gobakken. 2005. "Estimating Forest Growth Using Canopy Metrics Derived from Airborne Laser Scanner Data." *Remote Sensing of Environment* 96:453–65.

Næsset, E., T. Gobakken, J. Holmgren, H. Hyyppä, J. Hyyppä, M. Maltamo, M. Nilsson, H. Olsson, Å. Persson, and U. Söderman. 2004. "Laser Scanning of Forest Resources: The Nordic Experience." *Scandinavian Journal of Forest Research* 19 (6): 482–99.

Næsset, E., T. Gobakken, and R. Nelson. 2006. "Sampling and Mapping Forest Volume and Biomass Using Airborne LiDARs." *Proceedings of the Eighth Annual Forest Inventory and Analysis Symposium* 2006.

Nyström, M., J. Holmgren, and H. Olsson. 2011. "Change Detection of Mountain Vegetation Using Multi-Temporal ALS Point Clouds." *SilviLaser 2011 Proceedings*, October 16–20, 2011. Hobart, Australia.

Packalén, P., and M. Maltamo. 2007. "The k-MSN Method in the Prediction of Species Specific Stand Attributes Using Airborne Laser Scanning and Aerial Photographs." *Remote Sensing of Environment* 109:328–41.

Persson, Å., J. Holmgren, and U. Söderman. 2002. "Detecting and Measuring Individual Trees Using an Airborne Laser Scanner." *Photogrammetric Engineering and Remote Sensing* 68:925–32.

Peuhkurinen, J., M. Maltamo, J. Malinen, J. Pitkänen, and P. Packalén. 2007. "Preharvest Measurement of Marked Stands Using Airborne Laser Scanning." *Forest Science* 53(6):653–61.

Pfeifer, N., and D. Winterhalder. 2004. "Modelling of Tree Cross Sections from Terrestrial Laser-Scanning Data with Free-Form Curves." *International Archives of Photogrammetry, Remote Sensing and Spatial Information Sciences* 36(8/W2):76–81.

Pitkänen, J., M. Maltamo, J. Hyyppä, and X. Yu. 2004. "Adaptive Methods for Individual Tree Detection on Airborne Laser Based Canopy Height Model." In *Proceedings of ISPRS Working Group VIII/2: Laser-Scanners for Forest and Landscape Assessment*, edited by M. Theis, B. Koch, H. Spiecker, and H. Weinacker, 187–91. Germany: University of Freiburg.

Popescu, S., R. Wynne, and J. Scrivani. 2004. "Fusion of Small-Footprint LiDAR and Multispectral Data to Estimate Plot-Level Volume and Biomass in Deciduous and Pine Forests in Virginia, USA." *Forest Science* 50:551–65.

Poso, S. 1983. "Kuvioittaisen Arvioimismenetelmän Perusteita." *Silva Fennica* 17:313–43.

Poso, S., T. Häme, and R. Paananen. 1984. "A Method of Estimating the Stand Characteristics of a Forest Compartment Using Satellite Imagery." *Silva Fennica* 18:261–92.

Puttonen, E., P. Litkey, and J. Hyyppä. 2010. "Individual Tree Species Classification by Illuminated-Shaded Area Separation." *Remote Sensing* 2(1):19–35.

Räsänen, I. 2010. "Determining the Need for First Thinning and the Spatial Pattern of Trees Based on Laser Scanning." Master's thesis in Forest Sciences, University of Eastern Finland, Faculty of Science and Forestry, Finland, 59pp.

Räty, M., V. Kankare, X. Yu, M. Holopainen, M. Vastaranta, T. Kantola, J. Hyyppä, and R. Viitala. 2011. "Tree Biomass Estimation Using ALS Features." *SilviLaser 2011 Proceedings*, October 16–20, 2011. Hobart, Australia.

Reutebuch, S., R. McGaughey, H. Andersen, and W. Carson. 2003. Accuracy of High Resolution LiDAR Terrain Model Under a Conifer Forest Canopy. *Canadian Journal of Remote Sensing* 29:527–35.

Rönnholm, P., J. Hyyppä, H. Hyyppä, H. Haggrén, X. Yu, and H. Kaartinen. 2004. "Calibration of Laser-Derived Tree Height Estimates by Means of Photogrammetric Techniques." *Scandinavian Journal of Forest Research* 19:524–28.

Saari, A., and A. Kangas. 2005. "Kuvioittaisen arvioinnin harhan muodostuminen." *Metsätieteen aikakauskirja* 1/2005:5–18.

Solberg, S. 2008. "Mapping Gap Fraction, LAI and Defoliation Using Various ALS Penetration Variables." *International Journal of Remote Sensing* 31(5):1227–44.

Solberg, S., A. Brunner, K. H. Hanssen, H. Lange, E. Næsset, M. Rautiainen, and P. Stenberg. 2009. "Mapping LAI in a Norway Spruce Forest Using Airborne Laser Scanning." *Remote Sensing of Environment* 113:2317–27.

Solberg, S., E. Næsset, K. Hanssen, and E. Christiansen. 2006. "Mapping Defoliation during a Severe Insect Attack on Scots Pine Using Airborne Laser Scanning." *Remote Sensing of Environment* 102:364–76.

Ståhl, G., S. Holm, T. G. Gregoire, T. Gobakken, E. Næsset, and R. Nelson. 2011. "Model-Based Inference for Biomass Estimation in a LiDAR Sample Survey in Hedmark County, Norway." *Canadian Journal of Forest Research* 41:96–107.

Strahler, A. H., D. L. B. Jupp, C. E. Woodcock, C. B. Schaaf, T. Yao, F. Zhao, X. Yang, et al. 2008. "Retrieval of Forest Structural Parameters Using a Ground-Based LiDAR Instrument (Echidna®)." *Canadian Journal of Remote Sensing* 34(S2):426–40.

Suvanto, A., M. Maltamo, P. Packalén, and J. Kangas. 2005. "Kuviokohtaisten puustotunnusten ennustaminen laserkeilauksella." *Metsätieteen aikakauskirja* 4/2005:413–28.

Takeda, H. 2004. "Ground Surface Estimation in Dense Forest." *The International Archives of Photogrammetry, Remote Sensing and Spatial Information Sciences* 35(Part B3):1016–23.

Tansey, K., N. Selmes, A. Anstee, N. J. Tate, and A. Denniss. 2009. "Estimating Tree and Stand Variables in a Corsican Pine Woodland from Terrestrial Laser Scanner Data." *International Journal of Remote Sensing* 30(19):5195–209.

Thies, M., and H. Spiecker. 2004. "Evaluation and Future Prospects of Terrestrial Laser Scanning for Standardized Forest Inventories." *International Archives of the Photogrammetry, Remote Sensing and Spatial Information Sciences* 36:192–97.

Tokola, T. 1988. "Satelliittikuvien käyttö koealaotantaan perustuvassa suuralueiden inventoinnissa." Master's thesis, Faculty of Forestry, University of Joensuu, Finland, 72pp.

Tomppo, E. 1990. "Satellite Image-Based National Forest Inventory of Finland." *The Photogrammetric Journal of Finland* 12(1):115–20.

TopoSys. 1996. *Digital Elevation Models, Services and Products.* Toposys GMbH: Germany.

Uusitalo, J. 2010. *Introduction to Forest Operations and Technology.* Hämeenlinna: JVP Forest Systems Oy.

Van der Zande, D., J. Stuckens, W. W. Verstraeten, S. Mereu, B. Muys, and P. Coppin. 2011. "3D Modeling of Light Interception in Heterogeneous Forest Canopies Using Ground-Based LiDAR Data." *International Journal of Applied Earth Observation and Geoinformation* 13(5):792–800.

Vastaranta, M., M. Holopainen, X. Yu, R. Haapanen, T. Melkas, J. Hyyppä, and H. Hyyppä. 2011c. "Individual Tree Detection and Area-Based Approach in Retrieval of Forest Inventory Characteristics from Low-Pulse Airborne Laser Scanning Data." *The Photogrammetric Journal of Finland* 22(2):1–13.

Vastaranta, M., M. Holopainen, X. Yu, J. Hyyppä, H. Hyyppä, and R. Viitala. 2010b. "Determination of Stands First Thinning Maturity Using Airborne Laser Scanning." *Silvilaser 2010 Conference Proceedings.*

Vastaranta, M., M. Holopainen, X. Yu, J. Hyyppä, H. Hyyppä, and R. Viitala. 2011d. "Predicting Stand-Thinning Maturity from Airborne Laser Scanning Data." *Scandinavian Journal of Forest Research* 26:187–96. doi:10.1080/02827581.2010.547870.

Vastaranta, M., M. Holopainen, X. Yu, J. Hyyppä, A. Mäkinen, J. Rasinmäki, T. Melkas, H. Kaartinen, and H. Hyyppä. 2011b. "Effects of Individual Tree Detection Error Sources on Forest Management Planning Calculations." *Remote Sensing* 3(8):1614–26.

Vastaranta, M., V. Kankare, M. Holopainen, X. Yu, J. Hyyppä, and H. Hyyppä. 2012a. "Combination of Individual Tree Detection and Area-Based Approach in Imputation of Forest Variables Using Airborne Laser Data." *ISPRS Journal of Photogrammetry and Remote Sensing* 67:73–79.

Vastaranta, M., I. Korpela, A. Uotila, A. Hovi, and M. Holopainen. 2011a. "Area-Based Snow Damage Classification of Forest Canopies Using Bi-Temporal LiDAR Data." In *Laser Scanning 2011 Proceedings,* edited by D. Lichti and A. Habib.

Vastaranta, M., I. Korpela, A. Uotila, A. Hovi, and M. Holopainen. 2012b. "Mapping of Snow-Damaged Trees in Bi-Temporal Airborne LiDAR Data." *European Journal of Forest Research* 131:1217–28. doi: 10.1007/s10342-011-0593-2.

Vastaranta, M., T. Melkas, M. Holopainen, H. Kaartinen, J. Hyyppä, and H. Hyyppä. 2009. "Laser-Based Field Measurements in Tree-Level Forest Data Acquisition." *The Photogrammetric Journal of Finland* 21(2):51–61.

Vastaranta, M., R. Ojansuu, and M. Holopainen. 2010a. "Puustotietojen ajantasaistuksen luotettavuus." *Metsätieteen aikakauskirja* 4/2010:367–81.

Vauhkonen, J., L. Ene, S. Gupta, J. Heinzl, J. Holmgren, J. Pitkänen, S. Solberg, Y., et al. 2012. "Comparative Testing of Single-Tree Detection Algorithms Under Different Types of Forest." *Forestry* 85(1):27–40.

Vauhkonen, J., I. Korpela, M. Maltamo, and T. Tokola. 2010. "Imputation of Single-Tree Attributes Using Airborne Laser Scanning-Based Height, Intensity, and Alpha Shape Metrics." *Remote Sensing of Environment* 114(6):1263–76.

Vauhkonen, J., T. Tokola, P. Packalén, and M. Maltamo. 2009. "Identification of Scandinavian Commercial Species of Individual Trees from Airborne Laser Scanning Data Using Alpha Shape Metrics." *Forest Science* 55(1):37–47.

Vehmas, M., K. Eerikäinen, J. Peuhkurinen, P. Packalén, and M. Maltamo. 2008. "Airborne Laser Scanning for the Identification of Boreal Forest Site Types." In *Silvilaser 2008 Proceedings,* edited by R. Hill, J. Rossette, and J. Suárez, 58–65.

Vehmas, M., K. Eerikäinen, J. Peuhkurinen, P. Packalén, and M. Maltamo. 2009. "Identification of Boreal Forest Stands with High Herbaceous Plant Diversity Using Airborne Laser Scanning." *Forest Ecology and Management* 257:46–53.

Villikka, M., M. Maltamo, P. Packalén, M. Vehmas, and J. Hyyppä. 2007. "Alternatives for Predicting Tree-Stem Volume of Norway Spruce Using Airborne Laser Scanning." *Photogrammetric Journal of Finland* 20(2):33–42.

Wang, Y., H. Weinacker, B. Koch, and K. Sterenczak. 2008. "LiDAR Point Cloud Based Fully Automatic 3D Single Tree Modelling in Forest and Evaluations of the Procedure." *International Archives of the Photogrammetry, Remote Sensing and Spatial Information Sciences* XXXVII:45–51.

Watt, P., and D. Donoghue. 2005. "Measuring Forest Structure with Terrestrial Laser Scanning." *International Journal of Remote Sensing* 26(7):1437–46.

Wulder, M. A., J. C. White, R. F. Nelson, E. Næsset, H. O. Ørka, N. C. Coops, T. Hilker, C. W. Bater, and T. Gobakken. 2012. "LiDAR Sampling for Large-Area Forest Characterization: A Review." *Remote Sensing of Environment* 121:196–209.

Yu, X., J. Hyyppä, M. Holopainen, and M. Vastaranta. 2010. "Comparison of Area Based and Individual Tree Based Methods for Predicting Plot Level Attributes." *Remote Sensing* 2:1481–95. doi:10.3390/rs2061481.

Yu, X., J. Hyyppä, H. Kaartinen, and M. Maltamo. 2004 "Automatic Detection of Harvested Trees and Determination of Forest Growth Using Airborne Laser Scanning." *Remote Sensing of Environment* 90:451–62.

Yu, X., J. Hyyppä, H. Kaartinen, M. Maltamo, and H. Hyyppä. 2008. "Obtaining Plot-Wise Mean Height and Volume Growth in Boreal Forests Using Multi-Temporal Laser Surveys and Various Change Detection Techniques." *International Journal of Remote Sensing* 29:1367–86.

Yu, X., J. Hyyppä, A. Kukko, M. Maltamo, and H. Kaartinen. 2006. "Change Detection Techniques for Canopy Height Growth Measurements Using Airborne Laser Scanning Data." *Photogrammetric Engineering and Remote Sensing* 72(12):1339–48.

Yu, X., J. Hyyppä, M. Vastaranta, and M. Holopainen. 2011. "Predicting Individual Tree Attributes from Airborne Laser Point Clouds Based on Random Forest Technique." *ISPRS Journal of Photogrammetry and Remote Sensing* 66:28–37.

Zheng, G., and M. Moskal. 2012. "Leaf Orientation Retrieval from Terrestrial Laser Scanning (TLS) Data." *IEEE Transactions on Geoscience and Remote Sensing* 99:1–10.

16

National Forest Resource Inventory and Monitoring System

Erkki Tomppo, Matti Katila, and Kai Mäkisara

CONTENTS

16.1 History of National Forest Inventories 291
16.2 Need for Multisource Methods...293
16.3 Current Multisource Methods 294
    16.3.1 Input Data 294
    16.3.2 Preprocessing 295
    16.3.3 k-NN Estimation and Its Parameters 296
        16.3.3.1 Selecting Estimation Parameters and Their Values for k-NN 298
    16.3.4 Estimates for Areal Units Using k-NN Weights or Predictions.....................299
        16.3.4.1 The Use of Plot Expansion Factors 299
        16.3.4.2 Calibrated MS-NFI Estimators...300
        16.3.4.3 Stratified k-NN 301
        16.3.4.4 Final Plot Weights and Estimates 301
        16.3.4.5 Post-Stratification 301
    16.3.5 Pixel-Level Predictions and Map Production...302
16.4 Assessing the Errors: Current and Potential Methods...302
    16.4.1 Current Methods in Assessing Reliability of the Results 306
    16.4.2 Model-Based Error Estimation 306
        16.4.2.1 Model-Based Error Estimation at a Pixel Level 306
References...308

16.1 History of National Forest Inventories

The history of forest inventories goes back to the end of the Middle Ages when intensive use of forest resources first led to wood shortages, which, in turn, forced users to begin forest planning, particularly near towns and mines (Loetsch and Haller 1973; Tomppo et al. 2010a,b). The first information collected for these purposes was assessments of forest area and crude estimates of growing stock.

The first inventories were often local with the aim of assessing the available timber resources for specific purposes and were often conducted by the timber users, for example, companies (Davis et al. 2001; Loetsch and Haller 1973). It soon became obvious that such inventories could not easily be used to compile national-level forest information for purposes of formulating national forest policy; thus, national forest inventories (NFIs) were initiated.

Since that time, forest information has been collected via user-driven NFIs in many countries. NFIs have different histories in different countries, but some type of forest information has been collected in both European and North American countries since the nineteenth century (Tomppo et al. 2010b).

However, systematic forest assessments based on statistical sampling methods began only in the twentieth century. Forest inventory is a good and far from trivial application for statistical sampling methods. Difficulties arise from the large number of parameters to be estimated and the dependencies between different variables, even at different scales. Nearby trees are more similar than those farther apart from each other. Large-scale trend like changes of forest parameters are common. Some of the classical statistical questions are as follows:

- What kind of sampling design is "optimal"? Simple-minded optimization approaches are not possible in forest inventories, because different variables have different covariance structures and presume different sampling designs. A compromise must often be searched for.
- What type of estimators should be used?
- How to avoid bias?
- How to assess the reliability of the estimates?

A more recent challenge is the utilization of increasingly available supplementary data, such as satellite images and digital maps, to enable accurate small area estimation.

Sample-based NFIs were initiated in the Nordic countries in the late 1910s and early 1920s and were introduced in other European countries in the late 1940s (Tomppo et al. 2010a).

The early national inventories in the Nordic countries included not only information about areas, volume, and increment of growing stock and the amount of timber, but also age, size, and species structure of forests, silvicultural status of forests, accomplished and needed cutting, and silvicultural regimes (Ilvessalo 1927). The purpose was to provide information for forestry authorities, timber users, and planners who developed national forest policies.

In the United States, the earliest large-area, sample-based inventories date to passage of federal legislation in 1928. This enabling legislation is likely due, at least in part, to the influence of Professor Yrjö Ilvessalo, head of the Finnish NFI, on Calvin Coolidge, the President of the United States, made during his visit to President Coolidge in late 1920s (LaBau et al. 2007).

Today, sample-based inventories are conducted in most European and North American countries, although the tradition in Eastern Europe has been to gather national data by aggregating data from stand-level inventories originally designed for management planning purposes. However, many Eastern European countries have recently revised their systems in favor of statistical sample-based NFIs.

The primary purpose of NFIs in the Nordic countries has been to provide accurate information for forest management and forest industry investment planning, although assessing the productivity of forests and possibility of taxing forest income was one of the motivations for starting the Finnish NFI in 1921. Accordingly, future forest development scenarios based on NFI data have been used to propose alternative cutting scenarios that have then formed the basis for forest policy and utilization. For Central European countries, the main purpose for conducting forest inventories has been to monitor sustainable use of forests.

The use of aerial photographs has been developed since 1930s, and the images are globally in wide use in forest inventories in different ways (e.g., Loetsch and Haller 1973;

Tomppo et al. 2010a). Some inventories use aerial photographs to delineate forest land and other land (e.g., Lawrence and Bull 2010). Other early examples are volume estimation for management inventories (Nyyssönen 1955) and the use of the photo plots as additional observations to decrease sampling errors (Poso 1972).

Airborne laser scanner data have rapidly come into use in management inventories (Næsset et al. 2004). None of the NFIs uses the data in estimation so far. Problems are discussed in Section 16.3.1. Gregoire et al. (2011) and Ståhl et al. (2011) have developed sampling-based estimators and error estimators using model-assisted and model-based frameworks, respectively, applicable also in large areas.

16.2 Need for Multisource Methods

Based on the information from the sample plots, estimates for a country, or regions within a country, can be made. The densities of plots are high enough to ensure that the resulting sampling errors are low for core variables such as area of forest land and amount of growing stock. For example, the standard error in the estimation of the volume of the growing stock for a land area of 0.8 to 5.0 million ha is around 2% in Finland, and it is 0.6% at the national level (Tomppo et al. 2011).

The development of the Finnish multisource National Forest Inventory (MS-NFI) began in 1989, and the first operative results were calculated in 1990 (Tomppo 1990, 1991, 1996; Tomppo et al. 2008a,b, 2012). The driving force behind the development was the need to cheaply obtain forest resource information for areas smaller than would be possible using only field data. Furthermore, new natural resource satellite images provided possibilities for increasing the efficiency of inventories at relatively small additional costs. Following the Finnish example, a similar development began in Sweden a few years later. The name multisource forest inventory has been used for forest inventories utilizing satellite image data, in addition to field data, but in some cases, other data sources, such as digital maps, are also used.

In contrast to previous satellite image classification methods, methods were sought that would be able to provide area and volume estimates, possibly broken down into subclasses, such as tree species, timber assortments, and stand-age classes. In the optimal case, a method should be able to provide estimates for small areas equally good as the field data–based method providing estimates at national and regional levels. In the first experiments, regression analysis and discriminant analysis were tested (Tomppo 1988). The variables had to be predicted separately (or in small groups) in these approaches. The main experiences were that (i) it was difficult to obtain a sufficient degree of detail in the information in discriminant or similar analyses and (ii) the dependence structure between the estimates generally was lost when estimates of different parameters were made independently. To avoid these drawbacks, a new method was developed, and the method has become known as the k-nearest-neighbor method (k-NN) (Fix and Hodges 1951; Linton and Härdle 1998; Tomppo 1990). Since the first implementation of the method, it has been modified continuously and new features have been added (Katila and Tomppo 2001, 2002; Katila et al. 2000; Hagner and Olsson 2004; Nilsson 1997). The core of the current Finnish method is presented in Tomppo and Halme (2004). Any digital land use map or land cover data can be used to improve the accuracy of the predictions (Tomppo 1991, 1996). The list of references on k-NN applications and tests in forest inventories are given in Tomppo (2006), Tomppo et al. (2008a,b), and McRoberts et al. (2010a,b).

16.3 Current Multisource Methods

16.3.1 Input Data

The spatial input data for NFIs can include both imagery from different sensors and ancillary data from other sources (maps, digital elevation models, etc.).

The choice of image data for NFI depends on several theoretical and practical things. Theoretically, the input data must provide the desired accuracy of the estimates with the available analysis methods. This limits the acceptable resolution in the spatial, spectral, and radiometric dimensions. The choice is further limited by availability and price of the data. Sampling-based image analysis lowers the price, but the images within the predefined sample must still be available. The amount of data within the national data set may also be constrained by the available analysis capability. One result from these constraints is that many new data types giving accurate results for small areas are not applicable at the national level.

The remote sensing methods for forests are based on electromagnetic radiation reflected from or scattered by the different parts of the trees. The passage of the radiation is also affected by the atmosphere. Good wavelengths for forest analysis are those where the reflectance of the vegetation can be well separated from the other factors affecting the received radiation (e.g., ground, atmosphere, illumination). The applicable wavelength ranges include the optical range from visible to infrared and some microwave ranges.

The optical data is available from several satellite systems. The high-resolution (HR) images (pixel size 5–30 m) are most suitable for large area forest inventory. Each image frame covers a fairly large area (185 × 170 km for Landsat TM). This enables separate analysis of each frame or combination of subsequent frames on same track, simplifying the radiometric processing. This kind of data is available from several sensors (Landsat TM and ETM, SPOT, IRS, ALOS AVNIR-2, RapidEye). Many of these satellites also include channels beyond the near-infrared region. The future Landsat and Sentinel-2 satellites will considerably increase the availability of this kind of data.

The aerial photographs are a popular source for very-high-resolution (VHR) data (pixel size 1 m or smaller). They are available operationally in many countries. The VHR satellite data could provide similar kind of data, but it is not so easily available for large areas.

The clouds severely limit the availability of the optical satellite data for large regions within a 1- or 2-year time frame. For aerial photographs, the same problem exists, but it is not as severe because the timing of image acquisition is not limited by the satellite orbits.

The microwave data are very insensitive to the atmosphere, and the clouds are not a severe problem. The drawback is that the sensitivity to vegetation is not very good. The spatial resolution is usually enhanced using the synthetic aperture radar technique. The microwave bands most suitable for vegetation analysis (L and P bands) are not so good for other applications, and no operational satellites use these bands. Using polarimetric instruments and interferometry have been tested for forest application, but these methods are not yet ready for operational use.

Because of these considerations, the Finnish MS-NFI currently uses HR satellite data. The main data source is the Landsat TM sensor. The data are augmented with IRS and SPOT data when Landsat data are not available for some region at the desired time point.

In addition to remote sensing measurements, other spatial ancillary data can be used as input. It can be used to augment the input vectors or to stratify the data. Examples of this kind of data are digital maps, digital elevation models, and large scale representations of some trends (e.g., temperature sum, trends in forest variables, forest type, and volumes). The spatial accuracy must be considered when using this kind of data.

The ground data are used to build the models used in analysis or to calibrate the model for each image being analyzed. For remote sensing, the field data should characterize areas that are of the same size as the image features (functions of values of pixels or pixel groups) used in the analysis. The distribution of the field data should enable modeling all of the possible forest conditions with good accuracy.

Nationwide field work is expensive. Because of this, the ground data used for remote sensing may be primarily collected for another purpose. It may be field plot data used primarily for computation of very accurate, nationwide estimates. In this case, the field plots probably are smaller than would be ideal for remote sensing. The data may also be stand data collected for forest management. In this case, the within-stand variability, especially near the borders between stands, is not well represented. These shortcomings must then be taken into account in the analysis as far as possible.

The Finnish MS-NFI uses the Finnish NFI field plots as ground data. These are angle count plots that have been truncated to maximum radius of 12.52 or 12.45 m, depending on the region. The number of plots in NFI11 for all of Finland is 60,000 (Tomppo et al. 2011).

16.3.2 Preprocessing

Preprocessing of images can be divided into two parts: geometric and radiometric processing. The geometric processing establishes a mapping from the image pixel coordinate system to the geographic coordinate system. This enables locating the analysis results on the map. The radiometric processing removes from the pixel values effects caused by the instrument and the imaging conditions. The goal is to have pixel values that depend only on the forest variables. Some of the preprocessing steps are already performed by the supplier of the data.

The geometric processing for NFIs does not differ from geometric processing for other purposes. The practical target for geometric accuracy depends on the resolution of the data. A common target is to keep the geometric errors below one-half pixel in the result grid when measured at ground control points or independent checkpoints. This may be a too tight goal if VHR data is used because of the control point accuracy.

The pixel values correspond to some physical measurement, for example, the radiance from the target within the spectral response of the sensor. The results of the physical measurements depend on several things. They depend on the forest variable value that is being estimated, which is desirable. In addition to this, the value depends on the imaging system and the atmospheric conditions. The preprocessing has to separate the target-dependent variations from the other variations as far as the analysis requires.

The optical measurements in the visible and infrared region depend on the atmospheric conditions between the sensor and the target. This is most significant in satellite images, but it is observable also in aerial photographs. The measurements also depend on the imaging geometry, that is, the angles between the surface normal, the light source, and the camera. Within single images, it is enough to remove the contribution of the atmosphere within the image. If several images from different dates (orbits) are mosaicked before analysis, the contribution of the atmosphere must be removed between the images. If the image field of view is wide, the viewing geometry effects must also be removed. In principle, this can be done using atmospheric correction. In practice, however, the atmospheric correction is not enough to equalize the images. One reason for this is that accurate enough information about the exact atmospheric conditions is not available. The differences between images can be reduced by normalizing the images to each other using models. This can be performed instead of, or in addition to, atmospheric correction.

Radar is almost insensitive to atmospheric conditions. No atmospheric correction is necessary, but the effects caused by the viewing geometry must be corrected.

One kind of radiometric processing is also determining whether the quality of a pixel is good enough for analysis or not. One example is locating the clouds and cloud shadows and removing those pixels from further analysis.

Several methods for cloud detection in remote sensing data have been presented (Zhu and Woodcock 2012). However, none of these seems to be good enough for forest analysis. The problem is that the automatic methods cannot detect the haze at the edges of the clouds with sufficient accuracy. The results from automatic methods must be refined manually. In many cases, it turns out that it would have been easier to outline the clouds and shadows manually.

If the spatial resolutions of the images are high, it may not be possible to use the individual pixel data as features in analysis. For instance, if the resolution permits distinguishing of individual trees, the individual pixel values are not useful unless the model is built for individual trees. Features representing larger areas must be computed instead. These can be averages of the spectra over several trees or features characterizing the spatial variation of the intensities (the texture).

The ground data and the images are probably not from the same date. This means that the ground data must be screened so that ground data are not used where the forest has changed radically between the field work and the data of the image, for example, because of clearcuts.

16.3.3 k-NN Estimation and Its Parameters

The nonparametric k-NN estimation method allows calculating field plot weights, also called plot expansion factors. Thus, estimation based on k-NN mimics the common estimation procedures used in forest inventories and, in principle, makes it possible to calculate all kinds of estimates. The plot weights are sums of the weights of the plots calculated for the individual satellite image pixels belonging to a unit of interest and possibly to a land mask defining the pixels for which the estimates are made (e.g., forestry land). The pixel weights $w_{i,p}$ are calculated by the k-NN estimation method, which uses the distance metric d, defined in the feature space of the satellite image data (Nilsson 1997; Tomppo 1991, 2006):

$$w_{i,p} = \frac{1/d^t_{p_i,p}}{\displaystyle\sum_{j\in\{i_1(p),\dots,i_k(p)\}} 1/d^t_{p_j,p}} \quad \text{if and only if } i \in \{i_1(p),\dots,i_k(p)\} \qquad (16.1)$$

$$0 \quad \text{otherwise}$$

Here i is an arbitrary field plot, p is an arbitrary pixel, p_j is the pixel corresponding to field plot j, and $\{i_1(p),\dots,i_k(p)\}$ is the set of the nearest plots in the feature space when using distance metric (Equation 16.1). The power t is a real number, usually $t \in [0,2]$. The value of t and its effect on the pixel level errors of the estimates have been studied by, among others, Franco-Lopez et al. (2001) and Katila and Tomppo (2001).

The distance metric d, used in practical applications, most often is the Euclidean distance or a weighted Euclidean distance or a function of those (Franco-Lopez et al. 2001; McRoberts et al. 2002a; Reese et al. 2003; Tomppo 1996; Tomppo and Halme 2004). Other distance metrics exists as well. These may account for the correlation between spectral bands and varying information content in the spectral bands (e.g., Mouer and Stage 1996; Muinonen et al. 2001; Nilsson 1997; Tokola et al. 1996). Such weighting methods make it possible to use a feature space that is defined by a combination of spectral data and ancillary data.

One problem related to the k-NN technique is a high processing time when there is a large number of plots, a complicated distance metric, and particularly if several test runs are needed. Finley et al. (2006) employed squared mean distance to increase the efficiency in search for the final k-NN estimation using, for example, Euclidean distance.

The k-NN version used in the operative Finnish MS-NFI is described in more detail in the following (see also Tomppo et al. 2008a). Tomppo and Halme (2004) proposed the use of the large-scale variation of forest variables instead of geographical restrictions. The purpose is to direct the selection of the nearest neighbors, on the average, to forests similar to the target pixel. Maximum horizontal geographical distance and a moving window were used in early applications (Katila and Tomppo 2001). The distance metric in this improved k-NN method (called ik-NN) is

$$d^2_{p_j,p} = \sum_{l=1}^{n_f} \omega_{l,f}^2 (f_{l,p_j} - f_{l,p})^2 + \sum_{l=1}^{n_g} \omega_{l,g}^2 (g_{l,p_j} - g_{l,p})^2 \tag{16.2}$$

where f is a vector of image variables, n_f is the number of image variables, g is a vector of large-area forest variables, n_g is the number of large-area forest variables, p is the target pixel and p_j a field plot pixel, and ω is the given weights of the variables.

The second novel feature of ik-NN is that the weights ω are computed using optimization based on a genetic algorithm and a fitness function

$$f(\omega, \gamma, \hat{\sigma}, \hat{\bar{e}}) = \sum_{j=1}^{n_e} \gamma_j \hat{\sigma}_j(\omega) + \sum_{j=1}^{n_e} \gamma_{j+n_e} \hat{\bar{e}}_j(\omega) \tag{16.3}$$

where γ_j is the user-defined coefficient of the estimated pixel-level average standard errors $\hat{\sigma}_j$ and bias $\hat{\bar{e}}_j$ for forest variable j. The variables employed are (1) total volume, (2) volume of pine, (3) volume of spruce, (4) volume of birch, and (5) volume of other broad-leaved tree species. The pixel-level average biases and errors are estimated using field plots and leave-one-out cross-validation, $\hat{\sigma} = \sqrt{\dfrac{\sum_{i \in F} (\hat{m}_i - m_i)^2}{n_F}}$ and the bias $\hat{\bar{e}} = \dfrac{\sum_{i \in F} (\hat{m}_i - m_i)}{n_F}$.

Here m_i is the observed value of the variable to be estimated (e.g., total volume), $\hat{m}_i$ is the estimate on plot i, and n_F is the number of field plots. The values of the elements of weight vector γ were originally determined using an iterative trial and error method when optimizing the values of the elements for vector ω and were constant over the images (Tomppo and Halme 2004). Note that Equation 16.3 presents only the fitness function employed in the optimization in estimating the values of ω applied in the distance metric (Equation 16.2).

Any set of variables m and the reliability statistics of their estimates can be used in Equation 16.3. A prerequisite is that the values of the variables affect the reflectance at the wavelength of the remote sensing instrument. Categorical variables can also be used. Tomppo et al. (2009) used different classification accuracy statistics of categorical variables and a modified version of the method by Tomppo and Halme (2004) to predict forest site fertility and tree species dominance in Finland and forest type in Italy as well as conifer or broadleaved dominance.

16.3.3.1 Selecting Estimation Parameters and Their Values for k-NN

The basic principle of k-NN estimation is quite straightforward. However, practice has shown that the predictions and estimation errors depend largely on the core estimation parameters of the k-NN algorithm and the selections made in estimation (see also McRoberts 2009):

1. The spectral variables (pixel values) used in the distance metric, that is, spectral bands or their transformations
2. Possible correction of the pixel values for variation in illumination angle caused by elevation variation (slope, aspect) (Katila and Tomppo 2001; Tomppo 1996)
3. The distance metric (Tomppo and Halme 2004; Tomppo et al. 2009)
4. The value of k (Katila and Tomppo 2001; Nilsson 1997; McRoberts 2009; McRoberts et al. 2002a,b; Tokola et al. 1996)
5. The weights to be attached to the nearest neighbors, for example, even weights or functions of the employed distance and powers (negative)
6. The variables employed in restricting the area from which the nearest neighbors are sought for a pixel, for example, a geographical area (Katila and Tomppo 2001)
7. The use of additional information, for example, large area variation of forest variables (Tomppo and Halme 2004)
8. The use of ancillary data in the estimation, for example, for stratification (McRoberts 2009)

Examples of the parameters employed and their values in Finland and Sweden are given in Table 16.1. The selection of the parameters for operational applications varies between cases. In Finland, the parameters are selected by image scene and the selection is documented. Two types of criteria are used: (1) the standard error and bias at pixel level and (2) the difference between multisource estimates and estimates based on the field data only for areas large enough with low enough standard errors of the field data–based

TABLE 16.1

The Used k-NN Estimation Parameters in Finland and Sweden

	Finland	Sweden
Variables applied in the distance metric	Illumination-corrected spectral values for satellite image bands (Landsat TM 1–5, 7; Landsat 7 ETM+ 1–8, Pan; Spot HRV XS 1–3; IRS-1 C LISS 1–4) and large area forest variable estimates (NFI9, NFI10)	Haze and illumination-corrected spectral values for Landsat TM or ETM+ bands 3, 4, 5, and 7.
Distance metric	Weighted Euclidean distance	Euclidean distance
Value of k	5–10	15
Weights attached to the nearest neighbors	Weights proportional to the inverse or inverse squared distance ($t = 1$ or 2)	The weights proportional to the inverse squared distance ($t = 2$)
Restrictions for search of nearest neighbors	A maximum vertical (100 m or more in Northern Finland) and horizontal reference area (HRA) (40–120 km), since NFI9 large area forest variable maps are used to direct the NN selection, possibly with a HRA limit	No restrictions were used

Source: Tomppo, E.O., et al. *Remote Sensing of Environment*, 113, 500–17, 2009.

estimates. The field data–based errors are usually 2–5%, and the areas are typically 200,000–400,000 ha. The difference is measured as a function of the standard error (e.g., Katila and Tomppo 2001; Tomppo and Halme 2004; Tomppo et al. 2012). The values of the parameters depend, for example, on imaging conditions, number of available field plots, and variability of forests. In Sweden, the production line is highly automated (Reese et al. 2002). The selections are not independent; a change in one parameter also affects the "optimal value" of the other parameter. More studies are needed to "optimize" the values simultaneously (see also Tomppo et al. 2008b).

16.3.4 Estimates for Areal Units Using k-NN Weights or Predictions

There are two optional main ways to employ k-NN (or ik-NN) in estimating the values of forest parameters for groups of pixels, for example, mean volumes by tree species (m^3/ha) for areal units (e.g., municipalities or groups of municipalities). The two methods are (1) the use of plot expansion factors and (2) the use of post-stratification where the stratification is based on the pixel-level predictions of a variable.

16.3.4.1 The Use of Plot Expansion Factors

To calculate the plot expansion factors, the field plot weights on the pixels, $w_{i,p}$, are summed for the areal units (e.g., municipalities) in an image analysis process extending over the pixels belonging to each unit. The weight of plot i in areal unit u is denoted by

$$c_{i,u} = \sum_{p \in u} w_{i,p} \tag{16.4}$$

Note that the plot expansion factor represents the sums of the areas of the pixels attached to plot i in the analysis and vary by plot and unit u. In a design-based approach, the factor is constant over the areas where the sampling density is constant. Another difference is that plots outside a certain unit (municipality) can get a positive factor, while in a design-based approach, only plots inside the unit in question have a positive factor.

Reduced weight sums $c_{i,u}^r$ are obtained from Equation 16.4 if clouds or their shadows cover part of the areal unit u. The real weight sum for plot i is estimated by means of the equation

$$c_{i,u} = c_{i,u}^r \frac{\hat{A}_{s,u}}{\hat{A}_{s,u}^r} \tag{16.5}$$

where $\hat{A}_{s,u}$ is the estimated area of forestry land in unit u and $\hat{A}_{s,u}^r$ is the estimated area of forestry land in unit u not covered by the cloud mask. The area estimates $\hat{A}_{s,u}$ and $\hat{A}_{s,u}^r$ can be taken from digital maps when forestry land boundaries and cloud mask is available or estimated by means of field plots when total land area, field plot data, and cloud mask are available. Equation 16.5 assumes that the forestry land covered by clouds in areal unit u is on average similar to the rest of the forestry land in that unit with respect to the forest parameters to be estimated (Tomppo 1996).

Expansion factors (Equations 16.4 and 16.5) are calculated separately for the mineral soil stratum and peatland stratum within the forestry land and also for other land use classes such as arable land, built-up land, roads, and water bodies if a stratified estimation is employed (Katila and Tomppo 2002) (see also Sections 16.3.4.2 and 16.3.4.3).

In the MS-NFI, the forestry land is first delineated directly from numerical land use map data. However, the land use map data may be out-of-date or may include location errors. The definitions and classifications used may also differ from those of the NFI. Two optional methods have been developed and used to reduce the effect of map errors on small-area multisource forest resource estimates: a statistical calibration method (Katila et al. 2000) and a k-NN estimation by strata (Katila and Tomppo 2002). Both methods affect the factors in Equations 16.4 and 16.5.

16.3.4.2 Calibrated MS-NFI Estimators

The calibration method is based on the confusion matrix between land use classes of the field sample plots and corresponding map information. The bias in the land use class or other areal cover type estimates obtained from remote sensing or map data can be corrected by means of the error probabilities contained in the confusion matrix (Czaplewski and Catts 1992; Walsh and Burk 1993).

The aim is to define the map strata in such a way that each stratum is reasonably homogeneous with respect to the map errors and the land use class distribution. This enables the use of synthetic small-area estimation, utilizing the proportions that have been estimated from a larger region. An inverse calibration method is used to correct for map errors in the MS-NFI small-area estimates utilizing the field data from large area, for example, forestry centre R in which small area u belongs. First, the proportion of field data–based land use class l within each map stratum is estimated by the corresponding plot count ratio

$$\hat{P}_{h,l} = \frac{n_{h,l}}{n_h} \tag{16.6}$$

computed over the entire large area. Here $n_{h,l}$ is the number of field plots belonging to map strata h and land use class l, and n_h is the number of field plots belonging to map strata h. The calibrated area estimator is then obtained by summing the corresponding proportions of municipality level stratum areas:

$$A_{u,l}^* = \sum_h \hat{P}_{h,l} A_{u,h} \tag{16.7}$$

where $A_{u,h}$ is the area of map stratum h in a particular municipality u. The aggregate of small-area estimates over forestry center R is equal to the unbiased post-stratification estimator (Katila 2006; Katila et al. 2000; Tomppo et al. 2008a).

Errors in the areas of forestry land on map data affect the MS-NFI estimates at the municipality level. The sum of field plot weights over a computation unit is equal to the area of forestry land based on map data. There are two types of error attributable to map errors: (1) by pixels that are falsely classified as forestry land on the basis of the map data, and (2) by pixels in the nonforestry land map strata, which actually belong to forestry land. A heuristically derived method for calibrating the field plot weights $c_{i,u}$ was proposed by Katila et al. (2000), in which the sum of the calibrated weights for computation unit u is equal to the calibrated forestry land F area estimator $A_{u,F}^*$ (Equation 16.7).

Although these weights add up to $A_{u,F}^*$, the nonnegative values of an individual weight are not guaranteed. The calibration typically increases the mean volume estimates and reduces the forestry land area estimates for small areas if forestry land is overrepresented on maps.

16.3.4.3 Stratified k-NN

Another method has been presented to reduce the effect of inaccurate map data on the forest resource estimates (Katila and Tomppo 2001). In this method, called stratified MS-NFI, the k-NN estimation is employed by strata. All the field plots within each map stratum are used for estimating the areas of land use classes and forest variables of the particular stratum, independently of the land use class based on field measurements. In the original and calibrated MS-NFI, only those field plots are used that completely belong to forestry land. The advantage of the stratified MS-NFI method is that all the sample plots within stratum are included in the training data.

The final estimates for the stratified MS-NFI are derived by combining the stratum-wise estimates using Equation 16.8.

The strata employed were formed so as to be as homogeneous as possible with respect to the NFI-based land classes. An example of the use of map stratification is forestry land on mineral soils, forestry land on peatlands, arable land, built-up areas and roads, and water (Tomppo et al. 2008a). The number of strata is restricted by the fact that there should be a sufficient number of field plots for the k-NN estimation. The aim of the method was to obtain simultaneously the forestry land area estimate and accurate forest variable estimates within each stratum. The stratified MS-NFI is essentially a different estimation method compared to the calibrated MS-NFI, in which the MS-NFI estimates are more or less calibrated systematically upwards or downwards (Katila and Tomppo 2001). In the operative MS-NFI, neither of the two methods has proved to be significantly better than the other one.

16.3.4.4 Final Plot Weights and Estimates

The final field plot weights, $c_{i,u}^{f}$, have been calculated using correction arising from map errors, and possible clouds ratio estimation is employed to obtain the estimates (Cochran 1977). In this sense, the estimation procedure is similar to that using field plot data only. Volume estimates, for example, are computed by computation unit u in the following way. Mean volumes are estimated by the equation

$$v = \frac{\sum_{i \in I_s} c_{i,u}^{f} v_{i,t}}{\sum_{i \in I_s} c_{i,u}^{f}} \tag{16.8}$$

where $v_{i,t}$ is the estimated volume per hectare of timber assortment (log product) t for plot i and I_s is the set of field plots belonging to stratum s. The corresponding total volumes are obtained by replacing the denominator in Equation 16.8 by 1.

16.3.4.5 Post-Stratification

An alternative way for making areal estimates to the use of k-NN-based plot expansion factors and calibrating them for k-NN estimates is to use the k-NN maps for post-stratification of the NFI plots. It has been shown that post-stratified estimates, based on satellite data or k-NN predictions, can be used to decrease the sampling errors of the estimates of forest area and other parameters such as volume of growing stock (McRoberts et al. 2002a,b; Nilsson et al. 2003, 2005). Thus, it is possible to calculate post-stratified estimates of variables for smaller areas than what is possible using estimates based on field data alone with a desired accuracy.

In the Swedish NFI, for example, post-stratified estimation of NFI plots with k-NN-based volume estimate maps has reduced the standard errors of estimates (e.g., volume of growing stock, volume by tree species, tree biomass, and forest land area) by 10%–35%, when compared to the estimates calculated for field plots alone at a county level (on approximately 1 million ha forest land) (Nilsson et al. 2005). The improvement is smaller for forest variables that are less correlated with k-NN predictions. Different numbers of strata were tested, and it was found that at least five strata should be used for all selected forest parameters (cf. Tomppo et al. 2008b).

In Finland, post-stratification based on digital map data has been applied to NFI9 field plot data to calculate forestry land area estimates for two forestry centers, each with a forestry land area of 1.4 million hectares (Katila et al. 2000). The sampling errors of the large-area estimates were assessed by generalizing the variance estimator of the field data method (cf. Matérn 1960) to a multistratum case, that is, over map strata (Katila et al. 2000). The post-stratification estimator nearly halved the standard error of forestry land area estimates. Equally, mean volume of growing stock and mean volumes by tree species estimates and their standard errors have been estimated for subregions of forestry centers applying post-stratification and NFI10 field plot data.

16.3.5 Pixel-Level Predictions and Map Production

One important type of product from the multisource inventory is pixel-level predictions of forest variables, which is usually presented in map format. Within forestry land F, the pixel-level prediction, $\hat{m}_p$, of variable M for pixel p is the weighted average of the values of M on the k-nearest field plots when the distance is measured by means of the distance metric and plot weights (Equations 16.1 and 16.2), that is,

$$\hat{m}_p = \sum_{i \in F} w_{i,p}\, m_i \qquad (16.9)$$

where m_i is the value of the variable M on plot i and $w_{i,p}$ is the weight of plot i to pixel p. Examples of the mapped forest variables are stand age, mean stand diameter, mean stand height, and volumes by tree species (pine, spruce, birch, and other broadleaved trees) and by timber assortment class. The mode value is used instead of the weighted average for categorical variables, for example, land use class, main site class, and site fertility class. The total number of maps in the most recent inventory is 43 (Table 16.2 and Figure 16.1). In Sweden, nationwide raster maps of volume by species (pine, spruce, birch, other broadleaved trees, and all species together), stand age, tree height, and biomass have been produced for 2000, 2005, and 2010 (SLU 2012). The errors of pixel-level predictions, error sources, and possibilities to decrease errors are discussed in Section 16.4.

16.4 Assessing the Errors: Current and Potential Methods

This subsection summarizes the current methods employed in assessing the reliability of the pixel-level predictions and the estimates for a group of pixels. Some recent efforts to derive error estimators for an estimate of an arbitrary group of pixels are also described. The error estimators are presented as examples of the approaches for potential operative

TABLE 16.2

Operative Raster Map Products from the Finnish Multisource Inventory 2009, 43 Layers

Site Variables	Biomasses (10 kg/ha)
Land class (1,2,3)	Biomass, pine, stem, and bark
Main site class (1,2,3,4)	Biomass, pine, foliage
Site fertility class (1,…,8)	Biomass, pine, living branches
Stand variables	Biomass, pine, stump
Mean diameter of stand (cm)	Biomass, pine, roots, $d > 1$ cm
Mean height of stand (dm)	Biomass, pine, dead branches
Stand age of growing stock (year)	Biomass, pine, stem residual
Stand basal area (m²/ha)	Biomass, spruce, stem, and bark
Canopy cover (%)	Biomass, spruce, foliage
Canopy cover of broadleaved trees (%)	Biomass, spruce, living branches
Volumes (m³/ha)	Biomass, spruce, stump
Volume of the growing stock	Biomass, spruce, roots, $d > 1$ cm
Pine volume	Biomass, spruce, dead branches
Pine saw timber volume	Biomass, spruce, stem residual
Pine pulpwood volume	Biomass, broadleaved tree, stem, and bark
Spruce volume	Biomass, broadleaved tree, foliage
Spruce saw timber volume	Biomass, broadleaved tree, living branches
Spruce pulpwood volume	Biomass, broadleaved tree, stump
Birch volume	Biomass, broadleaved tree, roots, $d > 1$ cm
Birch saw timber volume	Biomass, broadleaved tree, dead branches
Birch pulpwood volume	Biomass, broadleaved tree, stem residual
Other broadleaved tree volume	
Other broadleaved tree saw timber volume	
Other broadleaved tree pulpwood volume	

error estimators, although they are not yet employed in the operative Finnish MS-NFI. Deriving this type of error estimator has proven to be a challenging task. The problem can be divided into the derivation of (1) an error estimator for a pixel-level prediction and (2) an error estimator for a parameter for an area of interest.

Difficulties arise because of the following reasons:

1. Errors depend on the actual value of the variable to be predicted and so pixel-level errors are spatially dependent.
2. The variables measured or observed on the field plots are also spatially dependent.
3. The spectral values of adjacent pixels of a satellite image are dependent because of the atmospheric properties (scattering) and imaging technique.

Furthermore, several error sources make the error estimation complex (e.g., Tomppo et al. 2008a).

Practical applications have shown that the errors are strongly dependent also upon how the reference set for the potential neighbors is restricted and, thus, on how a satellite image covers the target area in question. This is a particularly important factor in the cases of trend-like changes in forest variables, such as that occurring over most parts of Finland

(Figure 16.1). There are also many error sources that are not easy to handle in an analytic way. Tomppo (2006) and Tomppo et al. (1998, 2008a) and Katila (2004) list the following error sources:

1. Measurement errors in the field data.
2. Model errors in the field data (volume models for predicting volumes for sample trees).
3. Errors in predicting volumes for tallied trees.
4. The fact that information of an image pixel and the corresponding field information often come from different areas; the pixel size or shape is different from field plot size or shape. Sampling may also be employed when measuring trees on a plot so that the field data that may affect the spectral data are not known. Variations in the ground vegetation composition under a similar growing stock can also cause variations in spectral values not represented in the growing stock.
5. Location error in field plots and corresponding image pixels.
6. Temporal difference in field data and image data. This error occurs in cases where image acquisition from the same time point of field measurements is not possible.
7. The radiometric resolutions of the satellite sensors are not able to capture all variations in the field.
8. Scattering of light by the atmosphere so that the information for a pixel is affected by the areas surrounding the corresponding ground element.
9. Limitations in the imaging techniques, for example, some sensors interpolate the spectral values in the boundary regions of an image.
10. Limitations in the reference data. The reference field data do not cover the entire variation of the field variables of the target area.
11. Limitations in any ancillary data employed, for example, numerical map data are out-of-date.
12. In a fragmented landscape, a large number of mixed boundary pixels, pixels including information from several land classes, for example, forest land and arable land, may decrease the accuracy of the estimates. This problem is also related to the spatial resolution of the remote sensing instrument.
13. The topographic correction model or reflectance model is not well defined.

At the pixel level, the prediction errors measured with relative root-mean-square error (RMSE) have been high in several multisource studies, for example, 50%–80% for field plot volume predictions (Katila and Tomppo 2001; Nilsson 1997; Tokola et al. 1996). However, cross-validation probably gives an overestimate of the average error, since the target pixel plot includes the location error of the NFI field plot.

It should also be noted that these are average error estimates, and the errors depend on the actual value of the variable. It should also be noted that RMSE based on leave-one-out cross-validation is not a direct error estimate. Kim and Tomppo (2006) presented a model-based method for assessing the uncertainty of pixel-level predictions. A model-based method was also employed by McRoberts et al. (2007) for assessing errors for an area of an arbitrary size.

On the basis of the comparisons of the MS-NFI9 estimates with the estimates calculated from independent data sets, the coefficient of variation for the mean volume of growing stock for areas of about 10,000 ha is of a magnitude of about 5%, and for areas of 100 ha, it is

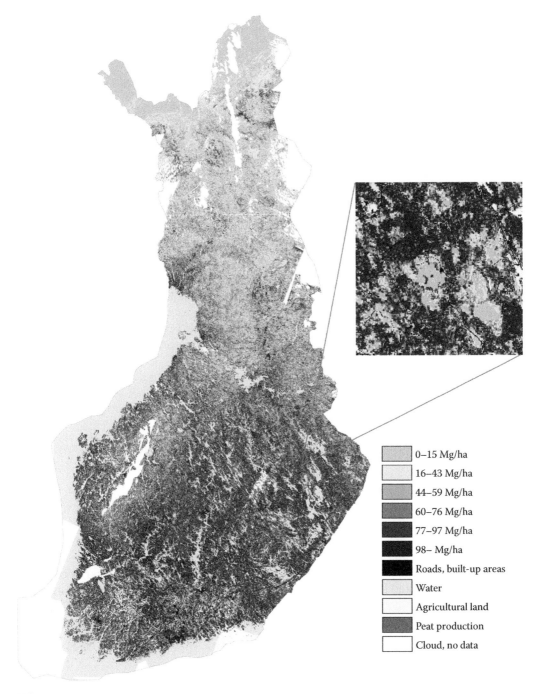

FIGURE 16.1

A map showing predictions of the total biomass of trees, above-and belowground per hectare (Mg/ha), produced using the multisource national forest inventory of Finland (excluding Åland). The estimates correspond to the year 2009. Some areas against the east and west border in North Finland are indicated by the cloud color (white), in addition to some areas inside the borders, owing to lack of satellite images caused by clouds. Map data: National Land Survey of Finland MML/VIR/MYY/328/08.

about 10%–15%. The errors of the area estimates of different development classes for a total forest land area of 10,000 ha varies between 10% and 50% when the areas of the classes varied from 500 to 2800 ha.

16.4.1 Current Methods in Assessing Reliability of the Results

To obtain an idea of the level of error in the Finnish MS-NFI, the pixel-level RMSE and the pixel-level average bias using leave-one-out cross-validation are calculated. For a sufficiently large area consisting of a group of pixels, the MS-NFI estimates are compared to the estimates and error estimates based solely on field data. Some empirical error estimates are also available for reliability assessments (Katila 2006; Tomppo et al. 2008a). Pixel-level error estimation is usually employed in the Finnish MS-NFI when selecting the estimation parameters for k-NN or ik-NN, as described in Section 16.3.3. Standard error estimates for areal units are also used in parameter selection and in evaluating the quality of the estimates.

16.4.2 Model-Based Error Estimation

Although the pixel-level RMSE has been estimated using leave-one-out cross-validation in many k-NN papers, it does not accommodate any of the dependencies listed above and is not a direct measure of the standard error for a prediction as noted by Kim and Tomppo (2006). Furthermore, if the variation of a variable on the field plots does not cover all the variation of the variable in an area of interest (AOI), the leave-one-out RMSE results in an underestimation for the pixel-level error.

The direct statistical squared error estimate for an estimate $\hat{M}$ of the parameter M is the mean square error (MSE)

$$E(\hat{M} - M)^2 \tag{16.10a}$$

and similarly for the prediction $\tilde{y}(\mathbf{x}_i)$ of the variable $y(\mathbf{x}_i)$, the MSE is

$$E(\tilde{y}(\mathbf{x}_i) - y(\mathbf{x}_i))^2 \tag{16.10b}$$

Accommodating the spatial dependencies into the estimators of Equations 16.10a and b requires a model-based approach where the observed value y_i of pixel i is considered to be a realization of a random variable. For example, it can be assumed that y_i is a realization of the distribution of all possible realizations associated with the same covariate data vector, $\mathbf{x}_i$, for example, satellite image data vector. The pixel-level mean and variance of the distribution are denoted by μ_i and σ_i^2, respectively. The realization, y_i, can be expressed as

$$y_i = \mu_i + \varepsilon_i \tag{16.11}$$

where $E(\varepsilon_i) = 0$, $\mathrm{Var}(\varepsilon_i) = \sigma_i^2$ (cf. McRoberts et al. 2007; Tomppo et al. 2008a). Note that the deviations ε_i are spatially correlated.

16.4.2.1 Model-Based Error Estimation at a Pixel Level

When $\hat{M}$ and $\tilde{y}(\mathbf{x}_j)$ in Equation 16.10 are unbiased estimators for M and $y(\mathbf{x}_i)$, respectively, the MSEs presented in Equation 16.10a and b are identical to the error variances; thus, for example, Equation 16.10b is equal to

$$\mathrm{Var}(\tilde{y}(\mathbf{x}_j) - y(\mathbf{x}_j)) \tag{16.12}$$

Kim and Tomppo (2006) proposed a model-based estimator for the error variance of the pixel-level predictions (Equation 16.9) when the weight w is the inverse of the squared Euclidean distance. The error variance was expressed as a function of the distances in the feature space and a variogram. An interesting detail of this approach is that the variance estimator was based on the variogram in the covariate space, for example, satellite image feature space, not in the geographical space as often is the case. The variance (Equation 16.12) can then be presented as (Kim and Tomppo 2006)

$$\mathrm{Var}(\tilde{y}(\mathbf{x}_o) - y(\mathbf{x}_o)) = -\sum_i \sum_j \left(\frac{d_{oi}^{-2}}{\sum_i d_{oi}^{-2}} \right) \left(\frac{d_{oj}^{-2}}{\sum_i d_{oi}^{-2}} \right) \gamma(\mathbf{x}_i - \mathbf{x}_j) + 2\sum_i \left(\frac{d_{oi}^{-2}}{\sum_i d_{oi}^{-2}} \right) \gamma(\mathbf{x}_i - \mathbf{x}_o) \qquad (16.13)$$

where d_{oi} is the distance (Equation 16.2) from the target pixel o to pixel i, γ is the variogram in the covariate space, and x_i is the covariate vector associated to pixel i. A Matérn class model (Matérn 1960) was used to estimate a parametric model for the variogram and was fitted to the prediction residuals.

McRoberts et al. (2007) also used a model-based method, starting from the model in Equation 16.11, and utilized a variogram model to accommodate the spatial dependencies of the residuals. Kim and Tomppo (2006) employed a variogram model in the geographical space.

McRoberts et al. (2007) used an unweighted mean of the values of the variables as the k-NN prediction, that is, the weights in Equation 16.1 were equal.

The following notations were used and the following assumptions were made:

1. An approximate symmetry among the nearest neighbors around x_i in the covariate space.
2. $\mu_j^i \approx \mu_i$, where μ_i is the superpopulation mean corresponding to x_i and μ_j^i is the superpopulation mean of the distribution of which y_j^i is a realization.

These assumptions guarantee unbiasedness of the individual predictors and estimators both at pixel level and at the level of aggregates of pixels. Furthermore, the k-NN prediction $\tilde{y}_i$ was employed as both the estimator of the mean, μ_i, and the realization, y_i; that is, $\hat{\mu}_i = \hat{y}_i = \tilde{y}_i$.

The estimator of the variance of the prediction of a realization, y_i, was expressed as a function of σ_i^2, and the estimator for σ_i^2 was expressed using the variation between the prediction and the values of the k-NN plots taking into account spatial correlation among the realizations y_i, which leads to

$$\hat{\sigma}_i^2 = \frac{\sum_{j=1}^{k} \left(y_j^i - \tilde{y}_i \right)^2}{k - \frac{1}{k}\sum_{j_1=1}^{k}\sum_{j_2=1}^{k} \rho_{j_1 j_2}} \qquad (16.14)$$

This estimator depends on the unknown $\rho_{ij} = \mathrm{Cov}(\varepsilon_i, \varepsilon_j)/\sigma_i\sigma_j$. Its estimator was derived using the connection of variogram and correlation, $\rho_{ij} = 1 - (\gamma(d_{ij})/\gamma_{total})$, and the empirical semivariogram.

Following a model-based approach, McRoberts et al. (2007) also presented a variance estimator for the predictions of a realization, y_i, in addition to the variance of $\hat{\mu}$ and also a variance estimator for the superpopulation means and the mean over predictions of realizations from the superpopulation.

The study of McRoberts et al. (2007) is a promising approach to solve the famous problem. However, some assumptions were made in deriving the variance estimators. For example, the superpopulation means of the neighbors was assumed the same as that of the target pixel which, together with the zero-expectation of the deviations ε of the realizations y_i from the population mean μ_i, symmetry of the distribution of neighbors in covariate space, and adequacy of the range of reference set observations in covariate space, guarantee unbiasedness.

A model-based estimator of the uncertainty of k-NN predictions was also proposed by Magnussen et al. (2009). It has some features common to McRoberts et al. (2007) but differs in many others. One difference is that it also tries to handle a possible bias of the estimators. The starting assumption was same as in McRoberts et al. (2007). For a given unit, that is, pixel i, with covariate data vector $\mathbf{x}_i$ (e.g., satellite image data), the value of the associated variable Y is assumed to be a random realization y_i from a superpopulation with a fixed superpopulation mean μ_i and a variance σ_i. The error estimators were derived in a different way than in McRoberts et al. (2007).

Other approaches to the derivation of the error estimate for an arbitrary group of pixels, not necessarily using k-NN estimation, include the variogram model by Lappi (2001) for a calibration estimator, a subsampling method for nonstationary spatial data by Ekström and Sjöstedt-de Luna (2004), and the design-based approach by Baffetta et al. (2009) for k-NN estimation. Lappi (2001) also used a model-based approach and derived a calibration estimator for the predicted average value (McRoberts et al. 2007), while the error variance of the calibration estimator for the predicted average was derived using a variogram model.

Magnussen et al. (2010) and McRoberts et al. (2011) proposed resampling techniques for variance estimation. A modified balanced resampling replication estimator of variance of a k-NN total was used in the previous one. It is suitable for small area estimation but cannot handle large data sets. McRoberts et al. (2011) investigated the bootstrap and the jackknife estimators and compared to a parametric estimator for estimating uncertainty using the k-NN technique with forest inventory. They evaluated the assumptions underlying a parametric approach to estimating k-NN variances, and assessed the utility of the bootstrap and jackknife methods with respect to the quality of variance estimates, ease of implementation, and computational intensity. Furthermore, adaptation of resampling methods to accommodate cluster sampling was investigated. The general conclusions were that support was provided for the assumptions underlying the parametric approach; the parametric and resampling estimators produced comparable results.

References

Baffetta, F., L. Fattorini, S. Franceschi, and P. Corona. 2009. "Design-Based Approach to k-Nearest Neighbours Technique for Coupling Field and Remotely Sensed Data in Forest Surveys." *Remote Sensing of Environment* 113:463–75.

Cochran, W. G. 1977. *Sampling Techniques*. 3rd ed. New York: Wiley.

Czaplewski, R. L., and G. P. Catts. 1992. "Calibration of Remotely Sensed Proportion or Areal Estimates for Misclassification Error." *Remote Sensing of Environment* 39:29–43.

Davis, L. S., K. N. Johnson, P. Bettinger, and T. E. Howard. 2001. *Forest Management: To Sustain Ecological, Economic, and Social Values*. 4th ed. Long Grove, IL: Waveland Press, Inc.

Ekström, M., and S. Sjöstedt-de Luna. 2004. "Subsampling Methods to Estimate the Variance of Sample Means Based on Nonstationary Spatial Data with Varying Expected Values." *Journal of the American Statistical Association* 99:82–95.

Finley, A., R. E. McRoberts, and A. R. Ek. 2006. "Applying an Efficient Neighbor Search to Forest Attribute Imputation." *Forest Science* 52:130–35.

Fix, E., and J. L. Hodges. 1951. *Discriminatory Analysis—Nonparametric Discrimination: Consistency Properties.* Report no. 4. Random Field, Texas: US Air Force School of Aviation Medicine.

Franco-Lopez, H., A. R. Ek, and M. E. Bauer. 2001. "Estimation and Mapping of Forest Stand Density, Volume, and Cover Type using the k-Nearest Neighbors Method." *Remote Sensing of Environment* 77:251–74.

Gregoire, T. G., G. Ståhl, E. Naesset, T. Gobakken, R. Nelson, and S. Holm. 2011. "Model-Assisted Estimation of Biomass in a LiDAR Sample Survey in Hedmark County, Norway." *Canadian Journal of Forest Research* 41:83–95.

Hagner, O., and H. Olsson. 2004. "Normalisation of Within-Scene Optical Depth Levels in Multispectral Satellite Imagery Using National Forest Inventory Plot Data." In *Proceedings from the 24th EARSeL Symposium, Workshop on "Remote sensing of land use and land cover"*, 28–29 May 2004. Dubrovnik, Croatia.

Ilvessalo, Y. 1927. The Forests of Suomi Finland. Results of the general survey of the forests of the country carried out during the years 1921–1924. *Communicationes ex Instituto Quaestionum Forestalium Finlandie* [Publications of the Forest Research Institute in Finland] 11:1–192 (in Finnish with English summary).

Katila, M. 2004. "Controlling the Estimation Errors in the Finnish Multisource National Forest Inventory." The Finnish Forest Research Institute, Research Papers, 910. http://ethesis.helsinki .fi/julkaisut/maa/mvaro/vk/katila/.

Katila, M. 2006. "Empirical Errors of Small Area Estimates from the Multisource National Forest Inventory in Eastern Finland." *Silva Fennica* 40:729–42.

Katila, M., J. Heikkinen, and E. Tomppo. 2000. "Calibration of Small-Area Estimates for Map Errors in Multisource Forest Inventory." *Canadian Journal of Forest Research* 30:1329–39.

Katila, M., and E. Tomppo. 2001. "Selecting Estimation Parameters for the Finnish Multisource National Forest Inventory." *Remote Sensing of Environment* 76:16–32.

Katila, M., and E. Tomppo. 2002. "Stratification by Ancillary Data in Multisource Forest Inventories Employing k-Nearest Neighbour Estimation." *Canadian Journal of Forest Research* 32:1548–61.

Kim, H. -J., and E. Tomppo. 2006. "Model-Based Prediction Error Uncertainty Estimation for k-NN Method." *Remote Sensing of Environment* 104:257–63.

LaBau, V. J., J. T. Bones, N. P. Kingsley, H. G. Lund, and W. B. Smith. 2007. *A History of Forest Survey in the United States: 1830–2004*. FS–877. Washington, DC: U.S. Department of Agriculture, Forest Service.

Lappi, J. 2001. "Forest Inventory of Small Areas Combining the Calibration Estimator and a Spatial Model." *Canadian Journal of Forest Research* 31:1551–60.

Lawrence, M., and G. Bull. 2010. "National Forest Inventories Reports: Great Britain." In *National Forest Inventories–Pathways for Common Reporting*, edited by E. Tomppo, T. Gschwantner, M. Lawrence, and R. E. McRoberts, 245–58. Dordrecht, The Netherlands: Springer.

Linton, O., and W. Härdle. 1998. "Nonparametric Regression." In *Encyclopedia of Statistical Sciences, update*, edited by S. Kotz, C. B. Read, and D. L. Banks, vol. 2, 470–85. New York: Wiley.

Loetsch, F., and K. E. Haller. 1973. *Forest Inventory*, vol. I. München: BLV Verlagsgesellschaft mbH.

Magnussen, S., R. McRoberts, and E. O. Tomppo. 2009. "Model-Based Mean Square Error Estimators for k-Nearest Neighbour Predictions and Applications Using Remotely Sensed Data for Forest Inventories." *Remote Sensing of Environment* 113:476–88.

Magnussen, S., R. E. McRoberts, and E. O. Tomppo. 2010. "A Resampling Variance Estimator for the k Nearest Neighbours Techniques." *Canadian Journal of Forest Research* 40:648–58.

Matérn, B. 1960. *Spatial Variation. Meddelanden från statens skogsforskningsinstit* [Publications of the Swedish Forest Research Institute] 49:1–144. Also appeared as Lecture Notes in Statistics 36. Springer-Verlag. 1986.

McRoberts, R. 2009. "Diagnostic Tools for Nearest Neighbors Techniques When Used with Satellite Imagery." *Remote Sensing of Environment* 113:489–99.

McRoberts, R., E. O. Tomppo, A. O. Finley, and J. Heikkinen. 2007. "Estimating Areal Means and Variances of Forest Attributes Using the k-Nearest Neighbors Technique and Satellite Imagery." *Remote Sensing of Environment* 111:466–80.

McRoberts, R. E., W. B. Cohen, E. Naesset, S. V. Stehman, and E. O. Tomppo. 2010a. "Using Remotely Sensed Data to Construct and Assess Forest Attribute Maps and Related Special Products." *Scandinavian Journal of Forest Research* 25:340–67.

McRoberts, R. E., S. Magnussen, E. O. Tomppo, and G. Chirici. 2011. "Parametric, Bootstrap, and Jackknife Variance Estimators for the k-Nearest Neighbors Technique with Illustrations Using Forest Inventory and Satellite Image Data." *Remote Sensing of Environment* 115:3165–74.

McRoberts, R. E., M. D. Nelson, and D. G. Wendt. 2002a. "Stratified Estimation of Forest Area Using Satellite Imagery, Inventory Data, and the k-Nearest Neighbors Technique." *Remote Sensing of Environment* 82:457–68.

McRoberts, R. E., E. O. Tomppo, and E. Naesset. 2010b. "Advances and Emerging Issues in National Forest Inventories." *Scandinavian Journal of Forest Research* 25:368–81.

McRoberts, R. E., D. G. Wendt, M. D. Nelson, and M. H. Hansen. 2002b. "Using a Land Cover Classification Based on Satellite Imagery to Improve the Precision of Forest Inventory Area Estimates." *Remote Sensing of Environment* 81:36–44.

Mouer, M., and A. Stage. 1996. "Most Similar Neighbor: An Improved Sampling Inference Procedure for Natural Resource Planning." *Forest Science* 41:337–59.

Muinonen, E., M. Maltamo, H. Hyppänen, and V. Vainikainen. 2001. "Forest Stand Characteristics Estimation Using a Most Similar Neighbor Approach and Image Spatial Structure Information." *Remote Sensing of Environment* 78:223–28.

Næsset, E., T. Gobakken, J. Holmgren, H. Hyyppä, J. Hyyppä, M. Maltamo, M. Nilsson, H. Olsson, Å. Persson, and U. Söderman. 2004. "Review Article, Laser Scanning of Forest Resources: The Nordic Experience." *Scandinavian Journal of Forest Research* 19:482–89.

Nilsson, M. 1997. "Estimation of Forest Variables Using Satellite Image Data and Airborne Lidar." Doctoral thesis. Swedish University of Agricultural Sciences, Umeå.

Nilsson, M., S. Folving, P. Kennedy, J. Puumalainen, G. Chirici, P. Corona, M. Marchetti, et al. 2003. "Combining Remote Sensing and Field Data for Deriving Unbiased Estimates of Forest Parameters Over Large Regions." In *Advances in Forest Inventory for Sustainable Forest Management and Biodiversity Monitoring*, edited by P. Corona, M. Köhl, and M. Marchetti. Dordrecht, The Netherlands: Kluwer Academic Publishers.

Nilsson, M., S. Holm, H. Reese, J. Wallerman, and J. Engberg. 2005. "Improved Forest Statistics from the Swedish National Forest Inventory by Combining Field Data and Optical Satellite Data Using Post-Stratification." In *Proceedings of ForestSAT 2005 in Borås, May 31–June 3, Report 8a*, edited by H. Olsson, 22–26. Sweden: Skogstyrelsen.

Nyyssönen, A. 1955. "On the Estimation of the Growing Stock from Aerial Photographs." *Communicationes Instituti Forestalis Fenniae* 46:1–57.

Poso, S. 1972. "A Method of Combining Photo and Field Samples in Forest Inventory." *Commununicationes Instituti Forestalis Fenniae* 76:1–133.

Reese, H., M. Nilsson, T. G. Pahlén, O. Hagner, S. Joyce, U. Tingelöf, M. Egberth, and H. Olsson. 2003. "Countrywide Estimates of Forest Variables Using Satellite Data and Field Data from the National Forest Inventory." *Ambio* 32:542–48.

Reese, H., M. Nilsson, P. Sandström, and H. Olsson. 2002. "Applications Using Estimates of Forest Parameters Derived from Satellite and Forest Inventory Data." *Computers and Electronics in Agriculture* 37:37–56.

SLU. 2012. kNN-Sweden: Current map data on forest land. Swedish University of Agricultural Sciences. http://skogskarta.slu.se/index.cfm?eng=1. Accessed November 7, 2012.

Ståhl, G., S. Holm, T. G. Gregoire, T. Gobakken, E. Naesset, and R. Nelson. 2011. "Model-based inference for biomass estimation in a LiDAR sample survey in Hedmark County, Norway." *Canadian Journal of Forest Research* 41:96–107.

Tokola, T., J. Pitkänen, S. Partinen, and E. Muinonen. 1996. "Point Accuracy of a Non-Parametric Method in Estimation of Forest Characteristics with Different Satellite Materials." *International Journal of Remote Sensing* 17:2333–51.

Tomppo, E. 1988. "Standwise Forest Variate Estimation by Means of Satellite Images." In *IUFRO S4.02.05 Meeting, August 28–September 2, 1988, Forest Station Hyytiälä, Finland*. University of Helsinki, Department of Forest Mensuration and Management, Research Notes, 21, 103–111.

Tomppo, E. 1990. "Designing a Satellite Image-Aided National Forest Survey in Finland." In *The Usability of Remote Sensing For Forest Inventory and Planning. Proceedings from SNS/IUFRO Workshop in Umeå, 26–28 February 1990*, 43–47. Umeå, Sweden: Swedish University of Agricultural Sciences, Remote Sensing Laboratory, Report 4.

Tomppo, E. 1991. "Satellite Image-Based National Forest Inventory of Finland." In *Proceedings of the Symposium on Global and Environmental Monitoring, Techniques and Impacts*, 17–21 September 1990. Victoria, BC, Canada. *International Archives of Photogrammetry and Remote Sensing*, vol. XXVIII, 419–24.

Tomppo, E. 1996. "Multi-Source National Forest Inventory of Finland." In *New Thrusts in Forest Inventory. Proceedings of the Subject Group S4.02–00 'Forest Resource Inventory and Monitoring' and Subject Group S4.12–00 'Remote Sensing Technology'. vol. 1. IUFRO XX World Congress, 6–12 August 1995,Tampere, Finland*, edited by R. Vanclay, J. Vanclay, and S. Miina, 27–41. EFI Proceedings, 7. Joensuu, Finland: European Forest Institute.

Tomppo, E. 2006. "The Finnish Multi-Source National Forest Inventory— Small Area Estimation and Map Production." In *Forest Inventory: Methodology and Applications. Managing Forest Ecosystems*, edited by A. Kangas, and M. Maltamo, vol. 10, 195–224. Dordrecht, The Netherlands: Springer.

Tomppo, E., T. Gschwantner, M. Lawrence, and R. E. McRoberts, eds. 2010a. *National Forest Inventories—Pathways for Common Reporting*. Dordrecht, The Netherlands: Springer.

Tomppo, E., M. Haakana, M. Katila, and J. Peräsaari. 2008a. *Multi-Source National Forest Inventory–Methods and Applications*. Managing Forest Ecosystems 18. Dordrecht, The Netherlands: Springer.

Tomppo, E., and M. Halme. 2004. "Using Coarse Scale Forest Variables as Ancillary Information and Weighting of Variables in k-NN Estimation: A Genetic Algorithm Approach." *Remote Sensing of Environment* 92:1–20.

Tomppo, E., J. Heikkinen, H. M. Henttonen, A. Ihalainen, M. Katila, H. Mäkelä, T. Tuomainen, and N. Vainikainen. 2011. *Designing and Conducting a Forest Inventory–Case: 9th National Forest Inventory of Finland*. Managing Forest Ecosystems 22. Dordrecht, The Netherlands: Springer.

Tomppo, E., M. Katila, K. Mäkisara, and J. Peräsaari. 2012. The Multi-source National Forest Inventory of Finland—methods and results 2007. *Working Papers of the Finnish Forest Research Institute 227*, Vantaa, Finland. Accessed on February 28, 2013: http://www.metla.fi/julkaisut/workingpapers/2012/mwp227.htm.

Tomppo, E., M. Katila, J. Moilanen, H. Mäkelä, and J Peräsaari. 1998. "Kunnittaiset Metsävaratiedot 1990–94." *Folia Forestalia* 4B:619–839 (in Finnish) [The municipality level forest resource estimates for Finland. Finnish Periodical of Forest Sciences].

Tomppo, E., H. Olsson, G. Ståhl, M. Nilsson, O. Hagner, and M. Katila. 2008b. "Combining National Forest Inventory Field Plots and Remote Sensing Data for Forest Databases." *Remote Sensing of Environment* 112:1982–99.

Tomppo, E., K. Schadauer, R. E. McRoberts, T. Gschwantner, K. Gabler, and G. Ståhl. 2010b. "Introduction." In *National Forest Inventories—Pathways for Common Reporting*, edited by E. Tomppo, T. Gschwantner, M., Lawrence, and R. E. McRoberts, 1–18. Dordrecht, The Netherlands: Springer.

Tomppo, E. O., C. Gagliano, F. De Natale, M. Katila, and R. E. McRoberts. 2009. "Predicting Categorical Forest Variables Using an Improved k-Nearest Neighbour Estimation and Landsat Imagery." *Remote Sensing of Environment* 113:500–17.

Walsh, T. A., and T. E. Burk. 1993. "Calibration of Satellite Classifications of Land Area." *Remote Sensing of Environment* 46:281–90.

Zhu, Z., and C. E. Woodcock. 2012. "Object-Based Cloud and Cloud Shadow Detection in Landsat Imagery." *Remote Sensing of Environment* 118:83–94.

Section V

Agriculture

17

Remote Sensing Applications on Crop Monitoring and Prediction

Bingfang Wu and Jihua Meng

CONTENTS

17.1 Introduction...315
    17.1.1 Early Warning..316
    17.1.2 Crop Production Monitoring...316
    17.1.3 Agricultural Sustainability..316
17.2 Advances in Methodology...317
    17.2.1 Crop Condition Monitoring..317
    17.2.2 Agricultural Drought Monitoring..318
    17.2.3 Crop Acreage Estimation...318
    17.2.4 Crop Yield Estimation..319
    17.2.5 Crop Phenophase Monitoring..319
17.3 Global and National Operational Systems..320
    17.3.1 CropWatch...320
    17.3.2 Global Information and Early Warning System..323
    17.3.3 USDA/Foreign Agricultural Service..323
    17.3.4 Monitoring Agricultural Resources..324
    17.3.5 National Systems...325
    17.3.6 Comments on Crop Monitoring Systems...326
17.4 Next Steps in Crop Monitoring...328
References...328

17.1 Introduction

Agricultural food production is an essential component of societal well-being. The risk of food supply disruptions will continue to grow as our agricultural systems and the land that sustains us continue to respond to the pressures of climate change, energy needs, and population increase. Accurate, objective, reliable, and timely predictions of crop yield over large areas are critical for national food security as they can support policy making on import/export plans and prices (Li et al. 2007b). A number of global trends suggest an urgent need for a comprehensive, systematic, and accurate global agricultural monitoring system (Brown 2005). More frequent extreme climate events such as floods, drought, and frosts are adversely affecting agricultural production worldwide. Changes in precipitation amount, seasonality, intensity, and distribution impact rain-fed agriculture. Further adaptation of agricultural systems to a changing climate can be expected (FAO 2007). The increase in global population, changes in cropland extent, and the draw-down rate of aquifers are also placing more pressure on food security.

Reliable information on agricultural production and production estimates are essential for agricultural markets and for the formulation of effective national and international agricultural policies. More importantly, improving such information would particularly benefit agencies working to increase food security in the developing world. Better information could also benefit those most susceptible to food insecurity, for example, by fostering the development of insurance and microfinance systems for subsistence agricultural producers. The remainder of this section describes the importance of remote sensing (RS)-derived crop production information, while Section 17.2 introduces recent advances in methodology, Section 17.3 presents the main global and national monitoring systems, and Section 17.4 discusses key developments for the future of crop monitoring.

More accurate and timely information on agricultural production is needed for three related domains: early warning of harvest shortfalls, crop production monitoring, and agricultural sustainability.

17.1.1 Early Warning

The relationship between people and food is complex, but in many developing countries, agricultural production and the ability to access food are directly linked. Thus, better information about changes in production can indicate areas where food policies need to be altered or where food aid may be necessary. Development assistance does not always target the countries most in need, in part, because of inaccurate or untimely information and donor policies. More timely information about harvest shortfalls can enable early identification of potential problem areas and, with the necessary international political will, allow for earlier and more widespread support for food programs in affected areas. Persistent shortfalls can help prioritize efforts to develop more sustainable agricultural systems.

17.1.2 Crop Production Monitoring

Large-scale changes are taking place in the distribution of agricultural lands and crop production. International trade, national agricultural policies, commodity prices, and producer decisions are all shaped by information about crop production and demand. Improved crop production monitoring enables more accurate forecasting of commodity prices, reducing risk, and increasing market efficiency. Understanding prices and risks is a key component of effectively addressing food supply problems and is essential for reducing food insecurity. Improved monitoring can help reduce risk and contribute to increasing productivity and efficiency at a range of scales, from the farm unit level to the global level.

17.1.3 Agricultural Sustainability

Crop cultivation is an intensive form of land use that relies on soil resources, climate resources, and farm input. Unless those resources are managed and replaced, arable land may degrade and become unfit for continued agricultural production. The impact of the abuse of arable land can be seen in the form of reduced agricultural production, reduced air and water quality, ecosystem exploitation and degradation, and declines in species diversity (GEO 2007). Over the long term, changes in production can also serve as regional indicators of ecosystem health. In semiarid systems, agricultural irrigation places heavy demands on water resources and requires careful management. Climate variability, extreme weather events, and increased and competing demands on the water supply in the short term can affect productivity and, in the long term, sustainability of agricultural production. In this context, the Nairobi Work Program developed by the Subsidiary Body for Scientific and Technological Advice of the United Nations Framework Convention on Climate Change calls for the development and dissemination of tools and observations that enable assessment of vulnerability to climate variability and change (http://unfccc.int/adaptation/sbsta_agenda_item_adaptation/items/3633.php).

17.2 Advances in Methodology

Advances in RS technology over the last 40 years have improved our ability to monitor crops from space, providing timely data from local to global scales.

Crop-info parameters are indices that can reflect the crop growing process and its output. The key crop-info parameters include crop acreage, crop phenophase, crop condition, drought, and crop yield. Based on extensive research and study, several reliable methods have been developed to monitor crop condition, drought conditions, crop acreage, crop yield, and crop phenophase on the basis of the available RS information. Those five key crop-info parameters are discussed in Sections 17.2.1 through 17.2.5.

17.2.1 Crop Condition Monitoring

Crop condition includes both group and individual crop characteristics. Crop condition information can be acquired using RS technology, which is also very useful for deriving crop yield information (Rao et al. 1982). Macroscale crop condition information can provide decision support for agricultural policy making and grain trade; microscale crop condition information can be used to adjust field management dynamically and realize precision farming.

After more than 30 years of research, quite a few crop condition monitoring models have been developed, including a direct monitoring model (Liu et al. 1997; Shi and Mao 1992), a same period comparing model (Liu et al. 2003; Wu and Yang 2002; Wu et al. 2004), a crop growing process monitoring model (Groten 1993; Jiang et al. 2002; Meng and Wu 2008; Zhang et al. 2004), a crop growing model (Li et al. 2007a), and a diagnosis model (Wang et al. 2001).

Each model has its own advantages and disadvantages. The direct monitoring model uses certain spectral reflectance or vegetation indices to evaluate crop condition. This model has the advantage of being easy to use and requiring little data, but because of its limited theoretic foundation, it is hard to use on a large scale or in a complex agricultural area. The same period comparing model can reflect on the variation in crop conditions between different years, but this model can evaluate crop conditions only over a short time frame, and the monitoring result is easily influenced by changes in crop phenophase and difference in crop types. The crop growing process monitoring model can reflect the crop growing continuance and crop condition throughout the crop growing season, but, as a result of the limitation in spatial resolution of high frequency RS image acquisitions, field-level monitoring cannot be implemented. Next, the crop growing model can actually reflect the crop growing status accurately, but the application of this model needs a lot of agro-parameters, and the model must be calibrated by local field data. The lack of local agro-parameters and field data, together with its complexity on data processing, put a limitation on the application of this model. Finally, the diagnosis model evaluates the crop condition with environmental parameters that influence crop growth, but the mechanisms for how these factors influence crop growth are still unclear.

17.2.2 Agricultural Drought Monitoring

The current drought monitoring models could be classified into nine unique models: the thermal inertia model (Murray and Verhoef 2007; Price 1985), the temperature condition index model (Anup et al. 2006; Kogan 1995), the anomaly vegetation index model (Anyamba et al. 2001; Xiao and Chen 1994), the vegetation condition index model (Kogan 1990; Parinaz 2008), the crop water stress index model (Moran 2000; Vijendra 2009), the vegetation water supply index model (Lambin and Ehrlich 1995), the vegetation temperature condition index model (Wan et al. 2004; Wang et al. 2003), the vertical drought index model (Qin et al. 2007), and the microwave RS model (Rahman et al. 2008; Zhao et al. 2006).

The application, scope, and conditions of the various drought monitoring models are not the same. For example, the thermal inertia model is more suitable for low vegetation coverage areas, while the crop water stress index model is more suitable for high vegetation coverage areas. The anomaly vegetation index model, the vegetation condition index model, and the temperature condition index model are all optimal during healthy vegetation periods, but require a long time series of RS images. Next, the vegetation temperature condition index model requires collection of the RS indices for different soil water statuses, ranging from minimum to maximum in the target region. Finally, the microwave RS model is an effective method to monitor surface soil water content, but further research needs to be carried out on how to evaluate the soil moisture combined with crop growth information (phenophase, density, etc.) and how to estimate deep soil moisture.

17.2.3 Crop Acreage Estimation

Various types of RS data are available to estimate crop acreage at regional scales using digital classification methods, such as optical data with high spatial resolution (Castillejo-González et al. 2009; Oza et al. 2008), medium spatial resolution (Badhwar 1984; Chang et al. 2007; Dutta et al. 1994; Yadav et al. 2002), and radar data (Bouman and Uenk 1992; Chakraborty and Panigrahy 2000; McNairn et al. 2009). However, pixels in RS data do not always correspond to a single crop type or field. Mixed pixels, pixels that include data about multiple land uses or crops, have a serious impact on crop classification accuracy in agricultural regions with small area crop fields. Therefore, crop classification based on RS data needs to be combined with a solid ground survey to accurately estimate crop acreage.

RS-based sampling methods have demonstrated accurate results for large area crop acreage monitoring systems (Cecil and Charles 1984; Gallego and Bamps 2008; Macdonald and Hall 1980; NASS of the USDA 2009b; Sushil 2001; Taylor et al. 1997; Tsiligirides 1998; Wu and Li 2004). To improve crop acreage estimation in China, methods integrating area sampling frames and RS techniques should also be considered. Yet, several obstacles must be overcome. The small field size of cropland requires RS data with spatial resolution finer than 10 m for accurate crop identification, especially in southern China. It is, however, unaffordable and impossible to monitor the entire major grain producing regions with such high resolution data throughout the growing season. Since the introduction of the household contracting responsibility system for farming in 1978, families have been planting at least three different crops and sometimes as many as 10, which greatly complicates the agricultural landscape, especially in the summer and autumn seasons (Tan et al. 2006). In addition to being able to be applied in this complex agricultural landscape, the new method should also be highly efficient and be able to deliver the results before harvest time or at least within 3 months after harvest. The method must also be highly accurate to meet the demands of complex agricultural systems and economically affordable to collect and process the necessary large volume of RS data and ground inventory. Finally, the method should deliver timely, reliable, and useful results that can satisfy and convince grain traders.

Many countries estimate crop acreage using traditional survey and sampling techniques, sometimes in combination with RS. The purpose of those studies, however, was to enhance and not to replace a particular sampling method. A regression estimator and recent cropland mapping based on RS data classification was adopted by the National Agricultural Statistics Service (NASS) of the USDA (2009a,b) (Boryan et al. 2011; Taylor et al. 1997), which leverages intensive field-collected data. A national agricultural-specific land cover classification product called the Cropland Data Layer is disseminated annually via the NASS CropScape (http://nassgeodata.gmu.edu/CropScape) (Han et al. 2012) portal since 2011.

17.2.4 Crop Yield Estimation

Estimating crop yield with RS takes advantage of RS-derived information about crop growth, along with other non-RS information like meteorological and agronomic conditions.

The current RS-based crop yield–predicting models could be classified into the following types: statistical models (Kalubarme et al. 2003; Tennakoon et al. 1992; Xu et al. 2008), agronomic models (Hou and Wang 2002; Xu et al. 1994), potential-stress models (Boken and Shaykewich 2002; Cayci et al. 2009; Ko and Piccinni 2009), and biomass-harvest index models (Bastiaanssen and Ali 2003; Moriondo et al. 2007; Zhang 2000).

The statistical models are simple and easy to use, but have no biophysical basis and cannot reflect the crop growing process, which leads to a scale-matching problem and regional extrapolating problems. Agronomic models are developed on the basis of the two key elements (trend yield and meteorological yield) of crop yield composition. The key of this model is to build a relationship between those three elements and the RS data, which brings forward a new direction for crop yield estimation with RS. Similar to the statistical models, however, if statistical methods are applied to build the relationship, the defects of the statistical methods, which lack biophysical parameters, cannot be avoided. Potential-stress models emphasize the two aspects of yield accumulation: potential and stress or trend and fluctuation. Lots of indices are applied to estimate crop yield, and the important step is to estimate the effect of stress factors (such as temperature or soil moisture) to the final yield. More research is needed to study how these stress factors influence final crop yield. Finally, biomass-harvest index models can express the crop physiological growing process and the mechanism of crop yield accumulation, but estimation of the crop harvest index is still on the leading edge of RS applications and needs further research.

17.2.5 Crop Phenophase Monitoring

Crop phenophase information is crucial for crop condition monitoring, field management, and agricultural decision making. Vegetation indices generated from RS data can accurately reflect seasonal and interannual variations of vegetation greenness on a large scale. As a result, vegetation indices can be used in vegetation monitoring, classification, and phenophase analysis (Meng et al. 2009).

Traditional crop phenophase monitoring is based on field observation, which directly observes the annual and interannual bio-phenological variations at fixed sampling points. This method is difficult to apply in large-scale crop monitoring because of the large investments in time and labor. A number of crop phenophase monitoring methods using time series Normalized Difference Vegetation Index (NDVI) data have been developed, including NDVI threshold method (White et al. 1997), the maximum increasing rate in NDVI curve method (Maignan et al. 2008), the backward average of NDVI time serial method (Reed and Brown 1994), the empirical formula method (Moulin et al. 1997), and other methods that use logarithmic and exponential functions to simulate spectral characteristics of corn and soybean during the growing season and determine the beginning and ending of the crop growing process (Badhwar 1984). Some literatures use the exponential characteristics of the NDVI time-series curves to define the crop phenophase at the inflexion of the NDVI curve (Meng et al. 2009) or monitor the crop growing season based on continuous increasing or decreasing NDVI time-series (Xin et al. 2001). Others carry out research on crop seeding monitoring, crop yield estimation, terrestrial vegetation changes, and shifts of vegetation green wave with NDVI time-series.

Most researches perform pixel-scale or regional-scale monitoring using high temporal and low spatial resolution NDVI time series. Although the spatial diversity of crop phenophase is actually not at pixel or regional scales, but at field-plot scale, restrictions posed by the temporal and spatial resolution of current popular RS data make it difficult to carry out field-scale phenophase monitoring.

17.3 Global and National Operational Systems

The primary international monitoring systems include the CropWatch system at the Institute of Remote Sensing and Digital Earth (RADI) of the Chinese Academy of Sciences, the UNFAO Global Information and Early Warning System (GIEWS), the USDA Foreign Agricultural Service (FAS) Global Agriculture Monitoring (GLAM) System, and the Monitoring Agricultural Resources (MARS) Unit of the European Commission, at the Joint Research Center (JRC) Ispra. Sections 17.3.1 through 17.3.6 describe these four monitoring systems, along with a brief description of four typical national systems in Russia, Brazil, India, and Argentina, followed by a reflection on these various operational systems.

17.3.1 CropWatch

CropWatch is a global crop monitoring system developed by the RADI, Chinese Academy of Sciences (formerly called the Institute of Remote Sensing Applications). CropWatch initiated crop condition monitoring in 1998 and was further developed to include production estimation of wheat, maize, rice, and soybean. Since 2000, CropWatch has expanded to cover most of the prominent grain-producing countries throughout the world (Wu 2000, 2004).

CropWatch consists of seven components, including data preprocessing, crop condition monitoring, drought monitoring, crop acreage monitoring, crop yield prediction, and cropping index monitoring, for the purposes of food supply–demand balance and early warning (Figure 17.1). Operational monitoring is supported by both field and RS data from multiple sources. Model validation and accuracy assessments are used with the annual monitoring processes. Social and economic data are mainly used for grain supply and demand balance analysis.

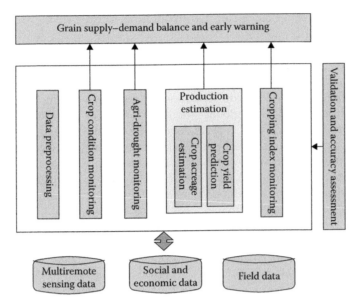

FIGURE 17.1
Structure of CropWatch.

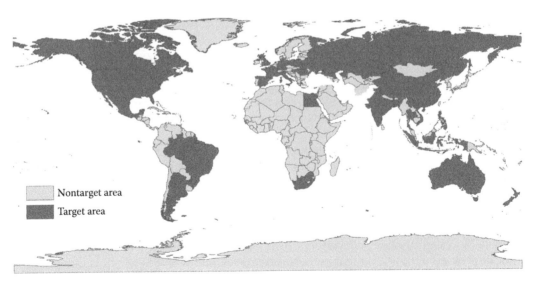

FIGURE 17.2
Monitoring scope of CropWatch.

Wheat, rice, maize, and soybean are the major target crops monitored by CropWatch, which accounts for 25%, 25%, 30%, and 8%, respectively, of overall global grain production according to FAO statistics (http://faostat.fao.org). There were 31 major grain-producing countries in the world selected as the target area (Figure 17.2). According to the results of CropWatch in 2011, the targeted monitored countries contribute 84% of wheat, 82% of rice, 84% of maize, and 91% of soybean for global overall production. Although crop condition monitoring, crop acreage estimation, and crop yield prediction were implemented for the target countries, CropWatch performed drought monitoring and crop index monitoring only inside China.

CropWatch has been evolving since its kickoff in 1998. Huge efforts have been made to improve its products and its operation during the last 13 years. The main improvements have been the standard processing of RS data, updates in the crop condition monitoring and crop acreage estimation methods, crop yield model development using biomass and harvest index estimated from RS data, and global calibration/validation of algorithms and models.

The monitoring algorithms and models developed during the last 13 years have been integrated into graphical user interfaces (GUI) driven software to form an operational system. Systematization of CropWatch mainly includes database development and processing system development. A database for both national and global RS-based crop monitoring was designed and developed with an Oracle database and ArcSDE as a spatial data management engine. Satellite images, meteorological data, basic spatial data, statistical data, and validation data have been stored and managed by the database. The monitoring technologies of CropWatch were developed into GUI software through the Interactive Developing Language (IDL) tools of Exelis Corporation. Specifically, preprocessing modules for Landsat TM, HJ-1 CCD, NOAA AVHRR, MODIS, and FY-3A MERSI (Medium Resolution Spectral Imager on FY-3 satellite) data, a crop condition monitoring module, a crop yield prediction module (Xu et al. 2008), a drought monitoring module, and a cropping index monitoring module were developed using software engineering methods, which were developed in client/server mode with IDL tools.

CropWatch aims to provide timely, independent, and reliable crop information services to support better grain-related decision making. High independence and efficiency are identified as the primary advantages of CropWatch.

Currently, there are quite a number of data sources for regional or national crop production, including statistical departments, market departments, and grain management departments. Independent monitoring of results is a primary requirement for an RS-based operational crop monitoring system. CropWatch maintains the independence of their results by using objective data sources and independent methods to the greatest degree possible.

For crop yield prediction, the "Biomass-Harvest Index" model no longer depends on statistical crop yield data. Crop yields from observation data alone are sufficient for proper calibration of the prediction model. For crop acreage estimation, CropWatch developed a suitable method integrating crop classification and transect sampling. While other surveys adopt manual collection of field information, which inherently are subjective, newly developed crop identification methods have proven to be more accurate than traditional methods and can provide independent estimates. For drought monitoring, all drought indicators or drought assessment models were developed with RS, which is independent from the traditional meteorological methods.

CropWatch implemented new methodologies for crop yield prediction, crop acreage estimation, and drought monitoring. Innovations in methodology were implemented to eliminate the dependence of crop yield prediction and crop acreage estimation on statistical data and intensive field investigation.

CropWatch does not have globally distributed offices that deliver reports on local crop production; all the information is from satellite-based monitoring, combined with meteorological analysis. RS data is the dominant data source for CropWatch. Crop condition monitoring, drought monitoring, and cropping index monitoring are all driven solely by RS data, whereas crop yield prediction and crop acreage estimation are driven mainly by RS data and are integrated with meteorological and field investigation data, respectively.

CropWatch is highly efficient as a result of its systematization, allowing for the timely delivery of its global monitoring results. CropWatch has committed itself to integrating the monitoring technologies into software platforms and making systematization a high priority. CropWatch, therefore, continues to make significant improvements on monitoring efficiency through systematization. Although various tasks are involved with RS-based crop monitoring on a global scale and timeliness is a critical requirement, reliable results can be released on time every month because of the development of a standardized processing system. This system is not only efficient and reliable, but has also reduced uncertainties by reducing the scope of analysis subjectivity from the monitoring staff. As a result, three staff can finish the operation within 2 weeks every month without tremendous repetitive and manual work.

As a result of its high level of systematization, CropWatch has been successfully transplanted in other places. Examples include the drought monitoring system of the Information Centre of the Ministry of Water Resources, the National Disaster Relief Centre of China, and provincial crop monitoring systems of three provinces (Jiangxi, Shaanxi, and Hubei). Furthermore, the CropWatch technology and system on global crop monitoring have been transferred to Chinagrain® web, which provides the largest agricultural product market information service in China.

CropWatch can provide timely monitoring results. At all stages of the crop growing season before harvest, CropWatch releases crop condition and drought monitoring results to support early decisions of crop production. One month before harvest, CropWatch will release crop production estimates. Just after harvest, CropWatch will release the revised production with higher accuracy, which plays an important role in the grain future trading market and assists grain traders' decision making.

17.3.2 Global Information and Early Warning System

GIEWS (http://www.fao.org/giews/english/index.htm) was established in 1975 to monitor food supply and demand at the global scale and to provide early warning of serious regional food shortages. Information from GIEWS is used to identify impending food security crises so that the UN World Food Programme and other international and national agencies can develop country-specific needs assessments. GIEWS integrates satellite-derived information on land cover and land use with in situ data on agricultural statistics, livestock, agricultural markets, and weather. GIEWS monitoring is designed to enable oversight of ground-based sampling to validate crop production estimates and development of quick, early, and partial indemnity for immediate action.

GIEWS does not actually implement satellite-based crop acreage estimation and yield prediction at the global scale. Its primary task involving satellite RS is to provide near real-time meteorological conditions (cold cloud duration, land surface temperature, and precipitation) and crop condition information. The estimates of production for standing crops are collected and revised on the basis of quite a number of factors that might influence planted area and yields. Reports from all over the world, as well as the comprehensive analysis implemented by its experienced analysts, are the key components that decide the accuracy and reliability of its perspective production numbers for major crops.

17.3.3 USDA/Foreign Agricultural Service

The goal of the Office of Global Analysis (OGA) of FAS (http://www.pecad.fas.usda.gov/index.cfm), specifically within the International Production and Assessment Division, is to produce reliable, objective, timely, transparent, and accurate data on global agricultural production. FAS monitors world agricultural production and world supply and demand for agricultural products to provide baseline market information and information for U.S. domestic early warning. FAS analyses rely upon a combination of meteorological data, field reports, and satellite observations at low and moderate spatial resolutions to aid in crop and growth stage identification and yield analysis. These data are used to confirm or deny unsubstantiated information about forecasted crop yields and to identify unreported events likely to impact crop yields. To bring these disparate sources of data together, FAS has developed Crop Explorer at http://www.pecad.fas.usda.gov/cropexplorer/, a GIS-based decision support system. The GLAM Project (Becker-Reshef et al. 2010) jointly funded by USDA and the NASA Applied Sciences Program is updating the FAS decision support system with the newest generation of NASA satellite observations.

USDA FAS utilizes RS in its operational global crop monitoring and provides crop condition information (map/charts/profile) to OGA (Production Estimates and Crop Assessment Division, PECAD) for further analysis. The analysts from OGA derive official production numbers for major crops by analyzing information from different sources, including crop condition information from satellite imagery and production reports from all over the world. In addition to its Washington, D.C., staff, FAS has a global network of 98 offices covering 162 countries. These offices are staffed by locally hired agricultural experts who serve U.S. agricultural monitoring around the world (http://www.fas.usda.gov/aboutfas. asp). While this reporting system provides primary information for the analysis in OGA, satellite-based information is basically used as complementary, not essential, information.

17.3.4 Monitoring Agricultural Resources

The mission of MARS-Food (http://agrifish.jrc.it/marsfood/Default.htm), a program within the European Commission's JRC, is to monitor food security for at-risk regions worldwide. This information contributes to European Union (EU) external aid and development policies, in particular food aid and food security policy. The desired outcome is to avoid food shortages and market disruptions and to better calibrate and direct European food aid. Satellite observations and meteorological data are integrated with baseline data on regional agronomic practices into crop growth models to develop MARS-Food ten-daily and monthly bulletins with yield forecasts by crop. Trends, similarity analysis, regression, and expert assessments are used to produce monthly reports that are intended to be directly used by food security administrators. In addition to quantitative and qualitative crop yield assessments, several indicators, like rainfall, radiation, and temperature, and water satisfaction indices, are published with comparisons to long-term historical average and last year indicators so that food security administrators can have a complete picture of the conditions in food-insecure areas. MARS Stat (http://agrifish.jrc.it/marsstat/default. htm) is a partner program focused on developing early, independent, and objective statistical estimates about the production of the main crops in Europe.

The current operational crop monitoring and yield forecasting program in MARS is AGRI4CAST Action. The Action is centered on the JRC's crop yield forecasting system aiming at providing accurate and timely crop yield forecasts and crop production biomass for the EU territories and other strategic areas of the world (http://mars.jrc.ec.europa.eu/mars/about-us/agri4cast). Crop development and growth is simulated via process-based models to estimate the response of the system from crop-soil to weather and agricultural management. These models are driven mainly by meteorological data, combined with RS-derived vegetation indices. In the first period of the MARS Project (1988–1997), crop area estimation had a central role with two major activities: regional crop inventories and rapid area estimates of crop area change at EU level. The regional crop inventories made a cost-efficiency analysis of the combination of ground observations with classified satellite images in the EU landscape. Further analysis showed that the margin for subjectivity of the method was too much to make it useful. Therefore, the method was abandoned and similar methods were not recommended. The AGRI4CAST Action is now focused on LUCAS (Land Use/Cover Area-frame Survey). The unit supports Eurostat on the design, efficiency analysis, and progressive improvement of the Eurostat LUCAS survey based on an area frame of points. In LUCAS, the crop acreage estimates will be derived from framed field surveys, instead of RS.

17.3.5 National Systems

Besides the aforementioned four global crop monitoring systems, there are many other systems operating at the national or regional level. Selected systems will be briefly introduced.

In Russia, agricultural production is monitored to foster sustainable agricultural development, for environmental assessment and protection and to monitor compliance with international environmental and trade conventions. The national agricultural monitoring system, established within the Ministry of Agriculture in 2003, relies on combined use of information from regional agricultural committees, satellite RS data, and ground agro-meteorological observations. The RS component of the agricultural monitoring system is developed by the Russian Academy of Sciences Space Research Institutes and involves daily MODIS observations as primary sources of the satellite data. The primary user of the information is the Federal Ministry of Agriculture, while the Ministry of Natural Resources, the Federal Statistical Agency and Hydro-Meteorological Service, regional agricultural committees and administrators, and local agricultural producers and enterprises are considered as potential users in the near future. The main foci of the agricultural monitoring system are arable land area, crop land use mapping, crop rotation, and seasonal crop development.

In Brazil, because of regional differences in soils, relief, climate, management practices, diseases, calendar, rotation, and area expansion, crop monitoring and forecasting is a great challenge for the government. The need for more precise, less subjective, and timely information led the Ministry of Agriculture, through its National Supply Agency (CONAB—Companhia Nacional de Abastecimento), which is responsible for the official crop production figures, to create a national agricultural monitoring and forecasting project in 2003, called "GeoSafras." The two main components of the project are area estimates, using statistics and RS, and yield estimates, using agrometerological, spectral, and mixed models. The strategy to put together such a project and gather results in a relatively short period of time was to create a network of 18 partner institutions (universities, research institutes, local and federal government agencies) to develop methodologies for area estimates and yield forecast, based mainly on geospatial technologies.

In India, the National Crop Forecasting Centre of the Department of Agriculture & Cooperation (DAC) of the Government of India was established in 1998, with a mandate to develop a framework for providing crop production forecasts at district, state, and national levels. In addition, to support the high-level decision making and planning, it is responsible for providing information on crop sowing progress, crop condition throughout the growing period, and the effect of episodic events such as floods, drought, hail storms, pests, and disease on crop production. Use of RS has been an important consideration by the DAC, which sponsored the Crop Acreage and Production Estimation (CAPE) project. The Space Applications Centre (SAC) of the Indian Space Research Organization (ISRO) has led the project in developing the following: (1) a RS-based procedure for crop acreage estimation at the district level, (2) spectral and weather models for yield forecasting, (3) a semi-automatic software package called CAPEMAN (later renamed CAPEWORKS) for analysis of RS data, and (4) technology transfer to teams across the country that use these procedures and make in-season crop production forecasts. LISS-III data from the Indian Remote Sensing (IRS) satellites are being regularly used to make crop production forecasts.

In Argentina, the Dirección de Coordinación de Delegaciones (DCD) is the national Argentine government agency responsible for agricultural estimates within the Secretary of Agriculture, Livestock, Fisheries and Food. This mission is accomplished by a network of 34 offices throughout the main agricultural areas of the country, gathering information about different statistics concerning agriculture, including area, yield, seeding and harvesting progress, crop condition, and so on. Each of these offices reports periodically to the DCD in Buenos Aires, where the information is checked, summarized, and released to the user community. Since 1981, RS has been applied to improve the estimates through three approaches: (1) land use/land cover stratification as a basis of area sampling frames; (2) digital analysis of satellite imagery (Landsat) to estimate area planted of extensive crops (i.e., wheat, corn, and soybean) using classification techniques; and (3) support of qualitative estimates of land use through coarse resolution (SAC-C, 175 m pixel size) imagery and e-mailing results in .xls format to local offices. SAC-C is an international cooperative mission between NASA, the Argentine Commission on Space Activities, Centre National d'Etudes Spatiales (the French Space Agency), Instituto Nacional De Pesquisas Espaciais (Brazilian Space Agency), Danish Space Research Institute, and Agenzia Spaziale Italiana (Italian Space Agency).

17.3.6 Comments on Crop Monitoring Systems*

The global crop monitoring systems described previously have different strengths and weaknesses, and also different objectives and approaches. Crop condition, acreage, yield, and agricultural drought are the four primary themes in the above introduced systems.

The methodology in crop acreage estimation and yield estimation in these systems are developed mainly from the achievements of LACIE and AgRISTARS. Field survey–based sampling is still the dominant method. Along with the reduction in RS image prices and the advent of more wide-swath sensors, coverage and temporal capacity of RS have been greatly enhanced. Yet the dependence on field surveys has not been reduced (even strengthened in some systems), which is against the primary goal of using RS technology to reduce field work. There are generally two reasons for this: (1) There is no satisfying crop identification method for operational systems. Both accuracy and efficiency are fundamental in operational systems, yet the current crop identification method cannot satisfy the two requisites at the same time; thus, new crop identification technology (e.g., automated classification methods) has to be developed. (2) In many cases, RS images are acquired only to design sampling frames; the advantage of RS images in dynamic monitoring is not exploited.

In yield estimation, agro-meteorological models are still the primary method, which use meteorological data, statistical data, and sampling survey data as input. Along with the reinforcement of regular statistical agricultural investigations, the result from RS technology should be independent, unbiased, and serve as a supplement to traditional yield investigations. Great progress has been made in RS-based yield estimation, which is substituting traditional meteorological methods after field observation–based calibration.

In crop condition monitoring, the widely used application of AVHRR, VEGETATION, and MODIS promoted the development of RS-based crop condition monitoring, which enables

---

* This section draws heavily from "Review of Overseas Crop Monitoring Systems with Remote Sensing" (Wu et al. 2010).

short-term prediction. Crop production can be estimated with these data, which is also the primary advantage of RS-based crop monitoring and the key difference with traditional agricultural statistics. Another advantage of RS-based crop condition monitoring is comprehensiveness, compared to that in field survey–based crop condition monitoring, which uses limited observation points to represent the whole area.

Drought is a major agricultural disaster that impacts agricultural production, and agricultural drought is a complex process. RS is an effective tool for large-scale agricultural drought monitoring. Although the drought situation could be partly reflected in crop condition anomaly, which can be from disease or nutrition stress, agricultural drought monitoring is still necessary, especially when its influence on crop production is different at different crop growing stages. Necessary attention has not been devoted to agricultural drought monitoring in some of the current operational crop monitoring systems.

The role and the effect of RS technology in these systems should be enhanced by technology updates and methodology improvements. The potential of RS technology should be further leveraged by regularly updating and enhancing agricultural/crop information acquiring and analysis tools. A whole set of crop-relevant information acquiring technology could be constructed using RS.

Countries including the United States and China have more than one operational crop monitoring system. Because of the systems being operated by different departments and because of the independent missions of these different departments, the systems operate independently of each other. Currently, there is little data or information exchange between these unique and different systems, where an appropriate integration would integrate the advantages of different systems, enhance the system operational efficiency, increase the reliability of the monitoring results, and avoid the confusion of end-users when results from different systems are not compatible. The integration can be implemented at the following three levels:

- Data-level: implement data source sharing between different systems and enhance data availability to these systems.
- System-level: integrate different systems by selecting optimal monitoring methods/ models in each monitoring theme; combining different systems into one.
- Information-level: integrate the monitoring results of different systems; this could be achieved by presenting numbers from different systems together to users (leave the judgment to them) or providing only one optimum number to use after further analysis.

The administration structure, functional division, and the sense of ownership make the integration difficult at any level. And a certain degree of overlap is positive if we look at it from another perspective. The competiveness between them will promote the development of new monitoring technology.

All the current monitoring technologies/systems are based on the national boundaries, producing production estimates at regional or national scales. As the world grain transportation and trade markets are more open and accessible, food security issues need to be considered to examine the whole globe as one producer. Instead of producing a global estimate by simply adding up the regional estimates, investments should be made in the research and development of new models and methods that can produce global crop production estimates by considering the whole globe as one producer. To achieve this, indicators for global crop estimation need to be identified so that models/methods can be developed by qualitative and quantitative analysis. After that a "true" global crop production estimation system could be developed.

17.4 Next Steps in Crop Monitoring

Efforts will continue to improve global crop monitoring, focusing on a system update to take advantage of new RS data and model enhancements to address climate change.

Currently, the low-resolution sensors such as MODIS, AVHRR, and VEGETATION, as well as mid-resolution sensors such as IRS, TM, ETM, and DMC are still the major RS data sources for operational crop monitoring systems. RS technology has been and is developing at a fast pace; the last decade has witnessed the advent of quite a number of new satellites and sensors. These sensors bring forward microwave data with higher resolution and multifrequencies (RadarSat-2, TerraSAR, and COSMO-SkyMed), high-resolution optical data with more spectral information (RapidEye and WorldView-2), mid-resolution optical data with shorter revisit cycle (HJ-1 CCD), as well as some continuance of traditional low-resolution data (FY-3 MERSI and VIIRS).

As more and more data can be accessed and acquired free of charge, the data expense for large-area crop monitoring has been greatly reduced. These systems can afford to ingest more and more satellite-based data or even can use RS data from more than one sensor to increase the timeliness of crop information acquisition. Yet the full application of these data sources in crop monitoring needs to be supported by additional research.

Although the amount of farm products, especially grain production, has increased over the past several decades, global climate change brings a new challenge to global agriculture and grain production. Climate change is expected to alter the global food supply, with implications for global and regional agricultural production and food security. Indeed, the effects of climate change on food availability and the stability of the food system are already being felt, especially in rural locations where crops fail or yields decline and in areas where supply chains are disrupted, market prices increase, and livelihoods are lost.

Understanding and predicting regional climate trends and their impact on agriculture and the national food supply is a high priority for governments. Climate change affects commercial farming, which is often an integral part of national economies, and subsistence farming, which is common in developing countries, and determines the livelihood of millions of people. This latter group is by far the most vulnerable to climate variability and change, and it is well recognized that poor, natural resource–dependent, rural households will bear a disproportionate burden of the adverse impacts of climate change.

A satellite-based crop monitoring system can conduct worldwide agriculture monitoring and act as a key tool in studying the influence of global change on agriculture production and food security.

References

Anup, K. P., L. Chai, R. P. Singh, and M. Kafatos. 2006. "Crop Yield Estimation Model for Iowa Using Remote Sensing and Surface Parameters." *International Journal of Applied Earth Observation and Geoinformation* 8:26–33.

Anyamba, A., C. J. Tucker, and J. R. Eastman. 2001. "NDVI Anomaly Patterns over Africa during the 1997/98 ENSO Warm Event." *International Journal of Remote Sensing* 22:1847–59.

Badhwar, G. D. 1984. "Automatic Corn-Soybean Classification Using Landsat MSS Data: Early Season Crop Proportion Estimation." *Remote Sensing of Environment* 14:31–7.

Bastiaanssen, W. G. M., and S. Ali. 2003. "A New Crop Yield Forecasting Model Based on Satellite Measurements Applied across the Indus Basin Pakistan." *Agriculture, Ecosystem and Environment* 94:321–40.

Becker-Reshef, I., C. Justice, M. Sullivan, E. Vermote, C. Tucker, A. Anyamba, J. Small, et al. 2010. "Monitoring Global Croplands with Coarse Resolution Earth Observations:The Global Agriculture Monitoring (GLAM) Project." *Remote Sensing* 2(6):1589–609.

Boken, V. K., and C. F. Shaykewich. 2002. "Improving an Operational Wheat Yield Model Using Phonological Phase-Based Normalized Difference Vegetation Index." *International Journal of Remote Sensing* 23(20):4155–68.

Boryan, C., Z. Yang, R. Mueller, and M. Craig. 2011. "Monitoring US Agriculture:The US Department of Agriculture, National Agricultural Statistics Service, Cropland Data Layer Program." *GeoCarto International* 26(5):341–58.

Bouman, B. A. M., and D. Uenk. 1992. "Crop Classification Possibilities with Radar in ERS-1 and JERS-1 Configuration." *Remote Sensing of Environment* 40:1–13.

Brown, L. 2005. *Outgrowing the Earth*. Washington, DC:Earth Policy Inst.

Castillejo-González, I. L., F. López-Granados, A. García-Ferrer, J. M. Peña-Barragán, M. Jurado-Expósito, M. S. De la Orden, and M. González-Audicana. 2009. "Object- and Pixel-Based Analysis for Mapping Crops and Their Agro-Environmental Associated Measures Using QuickBird Imagery." *Computers and Electronics in Agriculture* 68:207–15.

Cayci, G., L. K. Heng, H. S. Öztürk, D. Sürek, C. Kütük, and M. Sağlam. 2009. "Crop Yield and Water Use Efficiency in Semi-Arid Region of Turkey." *Soil & Tillage Research* 103:65–72.

Cecil, R. H., and R. P. Charles. 1984. "Estimating Optimal Sampling Unit Sizes for Satellite Surveys." *Remote Sensing of Environment* 14:183–96.

Chakraborty, M., and S. Panigrahy. 2000. "A Processing and Software System for Rice Crop Inventory Using Multi-Date RADARSAT ScanSAR Data." *ISPRS Journal of Photogrammetry and Remote Sensing* 55:119–28.

Chang, J., M. C. Hansen, K. Pittman, M. Carroll, and C. DiMicell. 2007. "Corn and Soybean Mapping in the United States Using MODIS Time-Series Data Sets." *Agronomy Journal* 99:1654–64.

Dutta, S., S. A. Sharma, A. P. Khera, M. Ajai Yadav, R. S. Hooda, K. E. Mothikumar, and M. L. Manchanda. 1994. "Accuracy Assessment in Cotton Acreage Estimation Using Indian Remote Sensing Satellite Data." *ISPRS Journal of Photogrammetry and Remote Sensing* 49:21–6.

Food and Agriculture Organization (FAO). 2007. "Adaptation to Climate Change in Agriculture, Forestry and Fisheries:Perspective, Framework and Priorities." *UN FAO, Interdepartmental Working Group on Climate Change*, 24. Rome, Italy.

Gallego, J., and C. Bamps. 2008. "Using CORINE Land Cover and the Point Survey LUCAS for Area Estimation." *International Journal of Applied Earth Observation and Geoinformation* 10:467–75.

GEO. 2007. Developing a Strategy for Global Agricultural Monitoring in the Framework of Group on Earth Observations (GEO) Workshop Report, Rome, Italy.

Groten. S. M. E. 1993. "NDVI-Crop Monitoring and Early Yield Assessment of Burkina Faso." *International Journal of Remote Sensing* 14(8):1495–515.

Han, W., Z. Yang, L. Di, and R. Mueller. 2012. "Cropscape:A Web Service Based Application for Exploring and Disseminating US Conterminous Geospatial Cropland Data Products for Decision Support." *Computers and Electronics in Agriculture* 84:111–23.

Hou, Y., and S. Wang. 2002. "Study on the Model of Crop Yield Estimation Based on NDVI and Temperature." *Geography and Territorial Research* 18(3):105–7.

Jiang, D., N. Wang, X. Yang, and H. Liu. 2002. "Principles of the Interaction between NDVI Profile and the Growing Situation of Crops." *Acta Ecologica Sinica* 22(2):247–52.

Kalubarme, M. H., M. B. Potdar, and K. R. Manjunath. 2003. "Growth Profile Based Crop Yield Models: A Case Study of Large Area Wheat Yield Modeling and Its Extendibility Using Atmospheric Corrected NOAA AVHRR Data." *International Journal of Remote Sensing* 234(10):2037–54.

Ko, J., and G. Piccinni. 2009. "Corn Yield Responses Under Crop Evapotranspiration-Based Irrigation Management." *Agricultural Water Management* 96:799–808.

Kogan, F. N. 1990. "Remote Sensing of Weather Impacts on Vegetation in Non-Homogenous Areas." *International Journal of Remote Sensing* 11:1405–19.

Kogan, F. N. 1995. "Application of Vegetation Index and Brightness Temperature for Drought Detection." *Advances in Space Research* 15:91–100.

Lambin, E. F., and D. Ehrlich. 1995. "Combing Vegetation Indices and Surface Temperature for Land Cover Mapping at Broad Spatial Scales." *International Journal of Remote Sensing* 16(3):573–79.

Li, A., S. L. Liang, A. Wang, and J. Qin. 2007a. "Estimating Crop Yield from Multi-Temporal Satellite Data Using Multivariate Regression and Neural Network Techniques." *Photogrammetric Engineering and Remote Sensing* 73(10):1149–59.

Li, W., C. Zhao, J. Wang, and L. Liu. 2007b. "Research Situation and Prospects of Wheat Condition Monitoring Based on Growth Model and Remote Sensing." *Remote Sensing for Land& Resources* 2:6–9.

Liu, A., C. Wang, and Z. Liu. 2003. "Cotton Information Extraction and Growth Monitoring in Arid Area Based on RS and GIS." *Geography and Geo-Information Science* 19(4):101–4.

Liu, K., X. Zhang, and J. Huang. 1997. "Study on Monitor of Rice Growing and Rice Yield Estimation by Remote Sensing in Jianghan Plain." *Journal of Central China Normal University (Nat. Sci.)* 31(4):482–87.

Macdonald, R. B., and F. G. Hall. 1980. "Global Crop Forecasting." *Science* 208:670–79.

Maignan, F., F. M. Breon, and C. Bacour. 2008. "Interannual Vegetation Phenology Estimates from Global AVHRR Measurements Comparison with In Situ Data and Applications." *Remote Sensing of Environment* 112:496–505.

McNairn, H., C. Champagne, J. Shang, D. Holmstrom, and G. Reichert. 2009. "Integration of Optical and Synthetic Aperture Radar (SAR) Imagery for Delivering Operational Annual Crop Inventories." *ISPRS Journal of Photogrammetry and Remote Sensing* 64:434–49.

Meng, J., and B. Wu. 2008. "Study on the Crop Condition Monitoring Methods with Remote Sensing." In International Society for Photogrammetry and Remote Sensing, 945–50, Beijing, China.

Meng J., B. Wu, Q. Li, X. Du, and K. Jia. 2009. "Monitoring Crop Phenology with MERIS Data—A Case Study of Winter Wheat in North China Plain." *Progress in Electromagnetics Research Symposium.* Beijing, China.

Moran, M. S., D. Hymer, and J. Qi. 2000. "Soil Moisture Evaluation Using Multi-Temporal Synthetic Aperture Radar SAR in Semiarid Rangeland." *Agricultural and Forest Meteorology* 105:69–80.

Moriondo, M., F. Maselli, and M. Bindi. 2007. "A Simple Model of Regional Wheat Yield Based on NDVI Data." *European Journal of Agronomy* 26:267–74.

Moulin, S., L. Kergoat, and N. Viovy. 1997. "Global-Scale Assessment of Vegetation Phenology Using NOAA/AVHRR Satellite Measurements." *Journal of Climate* 10:1154–70.

Murray, T., and A. Verhoef. 2007. "Moving towards a More Mechanistic Approach in the Determination of Soil Heat Flux from Remote Measurements I: A Universal Approach to Calculate Thermal Inertia." *Agricultural and Forest Meteorology* 147:80–7.

NASS of the USDA. 2009a. *History of Remote Sensing for Crop Acreage.* http://www.nass.usda.gov/Surveys/Remotely Sensed Data Crop Acreage/index.asp (accessed February 6, 2011).

NASS of the USDA. 2009b. "The USDA/NASS 2009 Cropland Data Layer: 48 State Continental US Coverage." http://www.nass.usda.gov/research/Cropland/Method/cropland.pdf. (accessed March 20, 2013).

Oza, M. P., M. R. Pandya, and D. R. Rajak. 2008. "Evaluation and Use of Resourcesat-I Data for Agricultural Applications." *International Journal of Applied Earth Observation and Geoinformation* 10:194–205.

Parinaz, R. B., A. A. Darvishsefat, A. Khalili, and M. F. Makhdoum. 2008. "Using AVHRR-Based Vegetation Indices for Drought Monitoring in the Northwest of Iran." *Journal of Arid Environments* 72:1086–96.

Price, J. C. 1985. "On the Analysis of Thermal Infrared Imagery: The Limited Utility of Apparent Thermal Inertia." *Remote Sensing of Environment* 18(1):59.

Qin, Q., A. Ghulam, L. Zhu, L. Wang, J. Li, and P. Nan. 2007. "Evaluation of MODIS Derived Perpendicular Drought Index for Estimation of Surface Dryness Over Northwestern China." *International Journal of Remote Sensing* 2(7):1983–95.

Rahman, M. M., M. S. Moran, D. P. Thoma, R. Bryant, C. D. Holifield Collins, T. Jackson, B. J. Orr, and M. Tischler. 2008. "Mapping Surface Roughness and Soil Moisture Using Multi-Angle Radar Imagery without Ancillary Data." *Remote Sensing of Environment* 112:391–402.

Rao, M. V. K., R. S. Ayyangar, and P. P. N. Rao. 1982. "Role of Multispectral Data in Assessing Crop Management and Crop Yield." In *Machine Processing of Remote Sensed Data Symposium* 1103–12. Minnesota, USA.

Reed, B. C., and J. F. Brown. 1994. "Measuring Phenological Variability from Satellite Imagery." *Journal of Vegetation Science* 5(5):703–14.

Shi, D., and L. Mao. 1992. "A Research on the Method of NOAA/AVHRR Monitor of Remote Sensing the Growing Tendency of Winter Wheat." *Acta Meteorologica Sinica* 50(4):520–23.

Sushil, P. 2001. "Crop Area Estimation Using GIS, Remote Sensing and Area Frame Sampling." *International Journal of Applied Earth Observation and Geoinformation* 3:86–92.

Tan, S. H., N. Heerink, and F. T. Qu. 2006. "Impact of Land Fragmentation on Small Rice Farmers' Technical Efficiency in Southeast China." *Scientia Agricultura Sinica* 39:2467–73 (in Chinese, with English abstract).

Taylor, C., C. Sannier, J. Delincé, and F. J. Gallego. 1997. *Regional Crop Inventories in Europe Assisted by Remote Sensing: 1988–1993.* Synthesis Report of the MARS Project–Action 1. Joint Research Centre, European Commission.

Tennakoon, S. B., V. V. A. Murty, and A. Eiumnoh. 1992. "Estimation of Cropped Area and Grain Yield of Rice Using Remote Sensing Data." *International Journal of Remote Sensing* 13(3):427–39.

Tsiligirides, T. A. 1998. "Remote Sensing as a Tool for Agricultural Statistics: A Case Study of Area Frame Sampling Methodology in Hellas." *Computers and Electronics in Agriculture* 20:45–77.

Vijendra, K. B. 2009. "Improving a Drought Early Warning Model for an Arid Region Using a Soil Moisture Index." *Applied Geography* 29(3):402–8.

Wan, Z., P. Wang, and X. Li. 2004. "Using MODIS Land Surface Temperature and Normalized Difference Vegetation Index Products for Monitoring Drought in the Southern Great Plains, USA." *International Journal of Remote Sensing* 25(1):61–72.

Wang, J., C. Zhao, W. Huang, X. Guo, and H. Li. 2001. "Effect of Soil Water Content on the Wheat Leaf Water Content and the Physiological Function." *Journal of Triticeae Crops* 4:47–52.

Wang, P., Z. Wan, J. Gong, X. Li, and J. Wang. 2003. "Advances in Drought Monitoring by Using Remotely Sensed Normalized Difference Vegetation Index and Land Surface Temperature Products." *Advance in Earth Sciences* 18(4):527–33.

White, M. A., P. E. Thorton, and S. W. Running. 1997. "A Continental Phenology Model for Monitoring Vegetation Responses to Interannual Climatic Variability." *Global Biogeochemical Cycles* 11:217–34.

Wu, B. F. 2000. "Operational Remote Sensing Methods for Agricultural Statistics." *Acta Geographica Sinica* 55(1):23–35.

Wu, B. F. 2004. "China Crop Watch System with Remote Sensing." *Journal of Remote Sensing* 8(6):481–97.

Wu, B. F., J. Meng, and Q. Li. 2010. "Review of Overseas Crop Monitoring Systems with Remote Sensing." *Advances in Earth Science* 25(10):1003–12.

Wu, B. F., and Q. Z. Li. 2004. "Crop Acreage Estimation Using Two Individual Sampling Frameworks with Stratification." *Chinese Journal of Remote Sensing* 8:551–69 (in Chinese, with English abstract).

Wu, B. F., F. Zhang, C. Liu, L. Zhang, and Z. Luo. 2004. "An Integrated Method for Crop Condition Monitoring." *Journal of Remote Sensing* 8(6):498–514.

Wu, J., and Q. Yang. 2002. "Crop Monitoring and Yield Estimation Using Synthetic Methods in Arid Land." *Acta Geographica Sinica* 21(5):593–8.

Xiao, Q., and W. Chen. 1994. "A Study on Soil Moisture Monitoring Using NOAA Satellite." *Quarterly Journal of Applied Meteorology* 5(3):312–17.

Xin, J., Z. Yu, and P. M. Driessen. 2001. "Monitoring Phenological Key Stages of Winter Wheat with NOAA NDVI Data." *Journal of Remote Sensing* 5(6):442–7.

Xu, X., Z. Niu, H. Cao, and X. Fang. 1994. "The Establishment of Remote Sensing Model for Crop Yield Estimation." *Remote Sensing of Environment China* 9(2):100–5.

Xu, X., B. F. Wu, J. Meng, Q. Li, W. Huang, L. Liu, and J. Wang. 2008. "Research Advances in Crop Yield Estimation Models Based on Remote Sensing." *Transactions of the CSAE* 24(2):290–98.

Yadav, I. S., N. K. S. Rao, B. M. C. Reddy, R. D. Rawal, V. R. Srinivasan, N. T. Sujatha, C. Bhattacharya, P. P. N. Rao, K. S. Ramesh, and S. Elango. 2002. "Acreage and Production Estimation of Mango Orchards Using Indian Remote Sensing (IRS) Satellite Data." *Scientia Horticulturae* 93:105–23.

Zhang, F., B. F. Wu, C. Liu, Z. Luo, S. Zhang, and G. Zhang. 2004. "A Method to Extract Regional Crop Growth Profile with Time Series of NDVI Data." *Journal of Remote Sensing* 8(6):515–28.

Zhang, J. 2000. "Imitation Methods of Remote Sensing-Numerical Value for Estimating Yield of Crops." *Journal of Arid Land Resources and Environment* 14(2):82–6.

Zhao, D. M., B. K. Su, and M. Zhao. 2006. "Soil Moisture Retrieval from Satellite Images and Its Application to Heavy Rainfall Simulation in Eastern China." *Advances in Atmospheric Sciences* 23(2):299–316.

18

Remote Sensing Applications to Precision Farming

Haibo Yao and Yanbo Huang

CONTENTS

18.1 Introduction..333
    18.1.1 Precision Farming..333
    18.1.2 Remote Sensing Data...335
18.2 Applications of Remote Sensing Data in Precision Agriculture...............................336
    18.2.1 Precision Farming Management 336
    18.2.2 Spectral, Spatial, and Temporal Considerations..337
    18.2.3 Vegetation Indices...338
    18.2.4 Application 1: Soil Mapping...340
    18.2.5 Application 2: Pest and Disease Detection..341
    18.2.6 Application 3: Plant Water Stress Detection 342
    18.2.7 Application 4: Crop (Nitrogen) Stress Detection..344
    18.2.8 Application 5: Weed Detection and Mapping 345
    18.2.9 Application 6: Herbicide Drift Detection 346
18.3 Conclusions...346
References...347

18.1 Introduction

18.1.1 Precision Farming

Traditional mechanized agriculture treats large fields with uniform agronomic practices, such as field preparation, applying fertilizers and pesticides, water management, as well as planting. Advances in technologies such as modern electronics, variable rate technology (VRT), the global positioning system (GPS), geographic information system (GIS), and remote sensing enable a different approach for farming practices. In this approach, instead of managing the entire field as one uniform unit, a crop field can be handled site specifically based on local field needs. Thus, the concept behind precision agriculture/precision farming is to manage in-field variability. With help from the above technologies, farmers can now decide when, where, and how to apply fertilizers and pesticides. With economic, productivity, and environmental considerations, the goals of precision agriculture can be described as (Yao 2004)

- Greater yield than traditional farming with the same amount of input
- The same yields with reduced input
- Greater yield than traditional farming with reduced input

The precision farming concept has drawn much attention from farmers and researchers since its inception (National Research Council 1997; Zhang et al. 2002; Bongiovanni and Lowenberg-Deboer 2004). There are four indispensable parts for a complete precision farming system: (1) field variability sensing and information extraction, (2) decision making, (3) precision field control, and (4) operation and result assessment. It is important to correctly implement these four parts in order to successfully utilize any precision agriculture system. Among them, the decision-making part is the crucial component (Stafford 2000). In the decision-making process, farmers need to produce the right management decisions based on the variability information derived from data collected in the crop field. An overview of precision agriculture management is shown in Figure 18.1. However, it was acknowledged that decision support remains the least developed area of precision

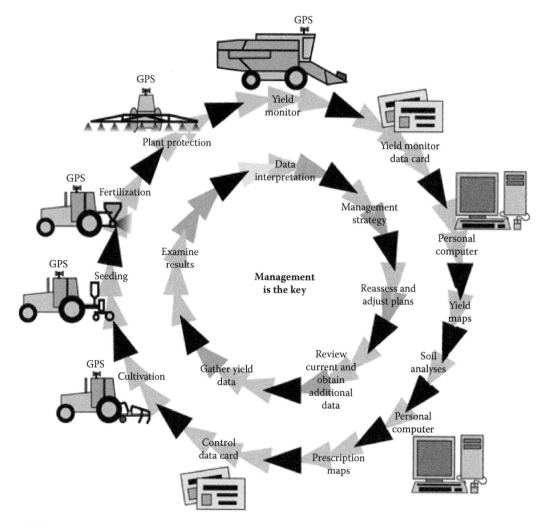

FIGURE 18.1

Decision support combines traditional management skills with precision farming (PF) tools to help precision farmers make the best management choices or "prescriptions" for their crop production system. (From Grisso, R., et al., *Precision Farming: A Comprehensive Approach*, Virginia, Communications and Marketing, College of Agriculture and Life Sciences, Virginia Polytechnic Institute and State University, 2009.)

farming (Grisso et al. 2009). It is still a challenge to build databases based on the relationships between input and potential yields, refining analytical tools, and increasing agronomic knowledge at the local level.

To make proper decisions, it is important to accurately obtain in-field variability information. Sensing and information extraction, which require obtaining field information at the right location at the right time, are crucial parts for a precision farming system. Sensing and information extraction involve using various sensors to capture data associated with field conditions and are generally implemented through on-the-go or remote sensing platforms. For the on-the-go system, the sensors and instruments are normally mounted on a ground vehicle such as a sprayer or a combine. The widely used grain yield monitor is a good example of this application. The remote sensing system can be implemented from either close distance (ground) or remote distance such as from airborne or spaceborne sensors (Scotford and Miller 2005; Omasa et al. 2006; Larson et al. 2008). Once the raw data are obtained, appropriate algorithms can be used to extract the field information for further analysis.

18.1.2 Remote Sensing Data

Agricultural remote sensing typically uses surface reflectance information in the visible and near-infrared (NIR) region of the electromagnetic spectrum. It provides a fast and economical way to acquire detailed field data in a short period. Remote sensing has thus been used in a broad range of applications in the farm industry. Traditionally, agricultural remote sensing used broadband multispectral imagery with a spectral range from visible to NIR. Typical spectral bands are in blue (450 nm), red (650 nm), green (550 nm), and NIR (680–800 nm) regions. The bandwidth of such bands is normally from 40 to 100 nm. Some remote sensing satellites can provide imagery with even longer wavelengths (Omasa et al. 2006). The infrared band with the wavelengths from 8 to 14 µm can be used for thermal remote sensing to assist in detection of crop water stress and spatial irrigation system management (Thomson et al. 2012). An airborne multispectral imager for agricultural remote sensing can provide imagery with high spatial resolution of the target and generally has high georeferencing accuracy. On the other hand, it lacks the detailed spectral information of the target due to the broadband nature of the imaging system.

The introduction of hyperspectral remote sensing imagery to agriculture in the past two decades has provided more opportunities for field level information extraction. A hyperspectral image has more bands (tens to hundreds or even thousands) with a narrow bandwidth (one to several nanometers) in the same spectral range (e.g., 400–2500 nm) as a multispectral image. Each pixel within the hyperspectral image is typically described as a data vector and the entire image as an image cube. With the fine spectral information provided with a hyperspectral image, it is expected that hyperspectral imagery could potentially provide more information for precision agriculture. On the other hand, the increased number of data dimensions in a hyperspectral image also increases complexity in image processing and might impact accuracy. One such example is the Hughes phenomenon (Hughes 1968), which shows that classification accuracy decreases as data dimensions increase, especially when a large number of wavebands are involved. This is exactly the case for hyperspectral imagery when hundreds of bands are used for image classification. Thus, it is desirable to reduce the original image dimensionality in order to reduce image processing complexity and to increase image interpretation accuracy. Feature selection and feature extraction are the two major types of dimension reduction methods (Richard and Jia 1999). The purpose of feature selection is to remove the least effective features (image bands) and select the most effective features. Feature selection is an evaluation of the existing set of features for the hyperspectral image to select the most discriminating features and discard the rest. Feature extraction would transform the pixel vector into a new set of coordinates in which the basis for feature selection is more evident. Common feature extraction techniques used in remote sensing include the linear combination of image bands such as in principal component transformation and canonical analysis, and arithmetic transformation such as vegetation indices.

Section 18.2 first discusses issues with precision farming management, and the spectral, spatial, and temporal aspects of using remote sensing imagery in precision farming. The second part is a discussion of vegetation indices. Finally, applications of remote sensing data in precision farming, including soil mapping, crop water stress detection, insect/pest infestation identification, crop nitrogen stress detection, weed mapping, and herbicide drift detection, are discussed. The use of remote sensing for crop yield estimation has drawn extensive research. It is discussed separately in Chapter 18.

18.2 Applications of Remote Sensing Data in Precision Agriculture

18.2.1 Precision Farming Management

Technology development and implementation plays an important role in the agricultural industry. In a crop production system, crop yield can be regarded as the single most important output. Crop yield can be grain yield, biomass harvest, or other outcomes from the production system. Some aspects related to crop production such as field topography, soil characteristics and fertility, tillage practice, fertilizer application, crop rotation, seeding, weed and pest control, irrigation, and weather can all be regarded as inputs for the crop production system. In this system, remote sensing can provide field variability information of the manageable inputs in a map-driven approach for precision farming practices. One example of this approach is the implementation of VRT. A prescription map based on field variability measured by remote sensing can be generated. Subsequent variable rate application of fertilizer, herbicide, or other agricultural chemicals can be implemented using the prescription map. In this process, the use of GPS and GIS is also necessary (Neményi et al. 2003). Furthermore, the concept of "management zone" is an important topic in precision farming (Khosla et al. 2008). The management zone is defined as a smaller section of a large field where the field properties of interest are regarded as relatively homogeneous.

Remote sensing has proved to be a useful tool for management zone delineation. To define the management zone, both remote sensing imagery and ground samples were used (Song et al. 2009; López-Lozano et al. 2010). In one study (Song et al. 2009), QuickBird imagery, together with soil and wheat yield data, was used for management zone delineation. The satellite imagery was used to generate the Soil Adjusted Vegetation Index for the zone definition. In another study (López-Lozano et al. 2010), the QuickBird imagery was used with soil maps. The imagery was used to derive leaf area index (LAI) maps in a commercial corn field. The soil maps were developed based on ion concentrations (Na, Mg, Ca, K, and P) and texture from soil samples and also from electromagnetic induction readings. Both maps were then linked to create management zones for variable rate applications of certain inputs such as fertilizer and water.

18.2.2 Spectral, Spatial, and Temporal Considerations

Remote sensing imagery can be taken from space, aerial, or terrestrial platforms. There are many sensors that can be used to take gray scale, multispectral, and hyperspectral image data. Generally, there are three issues related to using remote sensing imagery in agricultural applications. They are the spectral, spatial, and temporal characteristics of the image.

Spectral range and resolution of remote sensing image are important factors for agricultural remote sensing. Most image data have spectral range in the visible NIR region from 400 to 1000 nm. This is the region where plant canopy reflectance shows distinct spectral signatures under different conditions. In addition, some applications such as soil property mapping extend the spectral region to longer wavelengths, normally from 1000 to 2500 nm, which is also called the shortwave infrared region. For spectral resolution, agricultural remote sensing has widely used multispectral images with a spectral resolution (or bandwidth) of several hundred nanometers. Multispectral images are sometimes called broadband images, with each broadband covering a specific wavelength range such as blue, green, red, or NIR. More recently, hyperspectral imagery has been used for agricultural remote sensing. A hyperspectral image has a bandwidth of one to several nanometers, which has significantly finer image spectral resolution than a multispectral image. Hyperspectral imagery thus provides the potential for more detailed information extraction in agricultural applications.

Airborne or spaceborne remote sensing images can have large spatial coverage over the target area. One necessary step to process these images is image geometric correction and georeferencing. In this process, on-board sensor attitude information (Yao and Tian 2004) and ground control points (GCPs) (Gómez-Candón et al. 2011) are needed. The GCP can be natural landmarks or artificial targets. The GCPs are normally measured with GPS. Sometimes high accuracy GPS data (centimeter level) measured with real-time kinematic GPS are used depending on the actual application. Since aerial or space remote sensing data can cover a large area in a short period, they provide a fast, accurate, and economical method for precision applications. Spatial resolution is another important factor, which varies considerably (from submeter to several hundred meters) depending on the sensor platform. For proper data interpretation, it is preferred that the spatial resolution of remote sensing data and ground truth be matched. In case of uneven spatial resolution from different data sources, spatial resampling on one dataset is needed to meet this requirement.

The last issue is related to temporal remote sensing data acquisition. It pertains to time of acquisition of each image and the time interval between image acquisitions. The time to take one image is dependent on the actual applications. If the application is to develop a soil map, the image should be taken in the off-growing season. Similarly, mapping crop residue for conservation practices also uses imagery acquired in off-season. On the other hand, if the purpose is for crop monitoring, the remote sensing imagery should be obtained from the growing season. For instance, it is helpful for proper yield estimation and management if the temporal relationship between image and yield could be identified. Naturally, the spatial yield pattern does not appear immediately before harvest. Instead, the yield pattern is built up gradually throughout the growing season. Zwiggelaar (1998) found that the spectral reflectance of plants has both temporal and spatial aspects. Since the variation of crop spectral reflectance during the growing season can be used to relate to yield, it could help growers to estimate yield during the growing season. Furthermore, in-seasonal image data acquisition might need to occur many times to better understand the growth pattern and in-field variability. Bégué et al. (2008) pointed out that single-date images may be insufficient for the diagnosis of crop condition or for prediction. For this reason, research was also carried out to find the optimal imaging window. In one study implemented in Australia (Van Niel and McVicar 2004), multiple images (Landsat Enhance Thematic Mapper) were taken through the growing season to classify four crops, such as rice, maize, sorghum, and soybeans. The classification used a pixel-based maximum likelihood classifier, which is a commonly used supervised classification method for remote sensing image analysis. It was found that late February to middle March was the best window for overall classification of the crops. For individual crop separation, different windows were found.

18.2.3 Vegetation Indices

Vegetation indices have been used widely in remote sensing data analysis. For the simplest format, vegetation indices could be the individual image bands. Filella et al. (1995) used individual image bands located at 430, 550, 680, and 780 nm to build different indices for wheat nitrogen status evaluation. Blackburn (1998) also used single spectral bands for developing different indices for estimating chlorophyll concentrations.

In most cases, vegetation indices are calculated based on band ratios and combinations. The most widely used vegetation index is the Normalized Difference Vegetation Index (NDVI) (Rouse et al. 1973) calculated by using the red and NIR wavelengths as shown in the following equation:

$$\text{NDVI} = \frac{NIR - R}{NIR + R} \tag{18.1}$$

where NIR is the broadband *NIR* reflectance and R is the broadband *Red* reflectance. In case of hyperspectral imagery, narrow-band reflectance from the respective spectral regions can also be used to construct NDVI. Numerous applications have been found for NDVI. Hurcom and Harrison (1998) used the NDVI calculated from 677 to 833 nm to measure vegetation cover in a semiarid area. Serrano et al. (2000) used two image bands at 680 and 900 nm to compute vegetation indices, including the NDVI for estimating the biomass and yield of winter wheat. Meanwhile, other vegetation indices were also developed in the remote sensing community. Some of the common indices have been summarized in Table 18.1. These indices were first developed for broadband imagery but could be used for hyperspectral imagery as well.

The use of hyperspectral images makes it possible to build more refined vegetation indices by using distinct narrow bands. Thus, many hyperspectral vegetation indices have been developed for different applications. For example, Elvidge and Chen (1994) used narrow bands at 674 and 755 nm to calculate several narrow-band indices for the LAI and percent green cover estimation. The results were compared with the corresponding broadband indices. Broge and Leblanc (2000) calculated narrow-band vegetation indices from spectral bands centered at 670 and 800 nm and having a 10 nm bandwidth. Daughtry et al. (2000) used discrete bands at 550, 670, and 801 nm to develop narrow-band indices for nitrogen stress estimation in corn. Haboudane et al. (2002) calculated several vegetation indices using image bands centered at 550, 670, 700, and 800 nm for crop chlorophyll content prediction. The reason for choosing 700 nm is because it is located at the edge between the region where vegetation reflectance is dominated by pigment absorption and the beginning of the red-edge region. The red edge is where reflectance is more influenced by the structural characteristics of the vegetation. Gong et al. (2003) implemented a hyperspectral index to correct the effects of soil background. This study evaluated 12 vegetation indices using 168 bands selected from the Hyperion image after removing the water absorption

TABLE 18.1

Common Vegetation Indices Found in Literature

Acronym	Name	Equation	Reference
DVI	Difference Vegetation Index	$DVI = NIR - R$	Tucker (1979)
OSAVI	Optimized Soil Adjusted Vegetation Index	$OSAVI = \dfrac{NIR - R}{NIR + R + 0.16}$	Rondeaux et al. (1996)
RI	Redness Index	$RI = \dfrac{R - G}{R + G}$	Escadafal and Huete (1991)
RVI	Ratio Vegetation Index	$RVI = \dfrac{NIR}{R}$	Jordan (1969)
SAVI	Soil Adjusted Vegetation Index	$SAVI = \dfrac{NIR - R}{NIR + R + L} \times (1 + L)$, where L is 0.25, 0.5, or 0.75	Huete (1988)
TVI	Transformed Vegetation Index	$TVI = \sqrt{|NDVI + 0.5|}$	Perry and Lautenschlager (1984)

G, green; R, red.

bands and noise bands. These indices were two-band indices and were constructed using all possible two-band combinations.

Another approach to analyze hyperspectral data is to use derivative indices (Thorp et al. 2004). Since the hyperspectral data were discretely sampled for each wavelength based on the predetermined spectral sampling interval, the first derivative spectra can be calculated with the following differential method:

$$R'_\lambda = \frac{\Delta R_\lambda}{\Delta W} = \frac{R_{\lambda+1} - R_{\lambda-1}}{W_{\lambda+1} - W_{\lambda-1}} \tag{18.2}$$

where R is reflectance, R' is first derivative spectra, and ΔR is reflectance change is adjacent bands. W_λ is the wavelength number (nm) at wavelength λ. $\lambda + 1$ indicates the next image band of the band at wavelength λ and vice versa for $\lambda - 1$. Higher order derivatives can also be calculated in the same way.

Derivatives and especially higher order derivatives are relatively insensitive to illumination variations. Another advantage is minimizing soil background effect. These outcomes are also applicable with hyperspectral data, due to its small spectral sampling interval. Derivative analysis is promising for use with remote sensing data (Tsai and Philpot 1998). First- and second-order derivatives are commonly used. Because derivative analysis is quite sensitive to noise, spectral data smoothing is usually applied. Some examples of the filtering techniques are Savitzky–Golay filtering (Savitzky and Golay 1964) and mean low-pass filtering. For applications of derivative analysis, Bajwa and Tian (2005) used first derivatives from aerial hyperspectral data and partial least square regression (PLSR) to model soil fertility factors including pH, organic matter (OM), Ca, Mg, P, K, and soil electrical conductivity. Smith et al. (2004) suggested that derivative analysis in the red-edge range (690–750 nm) could be used for plant stress detection. On the other hand, Estep and Carter (2005) found that when certain derivatives were used for plant nitrogen and water stress detection, there was no advantage of using the derivatives compared to narrowband vegetation indices. In their study, Estep and Carter (2005) used Airborne Visible/

Infrared Imaging Spectrometer (AVIRIS) data over corn plots having different nitrogen fertilization treatments. The predefined first derivatives were at 495, 568, 696, 982, and 1025 nm. Imanishi et al. (2004) applied first- and second-order derivatives for crop drought stress detection with canopy reflectance data measured with a spectrometer.

18.2.4 Application 1: Soil Mapping

In precision farming applications, soil property maps can be used for prescribing variable rate applications. For example, a soil pH map can be used to generate a prescription map for variable rate lime application. Soil property maps are traditionally produced with grid soil sampling and spatial interpolation of the soil data. For rapid data collection, studies have focused on using reflectance in the visible and NIR regions for looking at soil nutrients. It is expected that soil surface spectral reflectance could be used for soil constituent and nutrient content discrimination. Another application of remote sensing data is to use the imagery as an aid for guided target soil sampling (Wetterlind et al. 2008). A recent review on using remote sensing for soil mapping can be found in Ge et al. (2011a). As a summary, to use remotely sensed data for soil property mapping, Moran et al. (1997) recommended "Measurements of soil and crop properties at sample sites combined with multispectral imagery could produce accurate, timely maps of soil and crop characteristics for defining precision management units." Ben-Dor and Banin (1995) used reflectance curves in the infrared region to study six soil properties. The results indicated that a different number of bands ranging from 25 to 3113 are required for different properties. Palacios-Orueta and Ustin (1998) found that the total iron and OM contents were the main factors affecting soil spectral shape. Thomasson et al. (2001) found that the spectral regions from 400 to 800 nm and from 950 to 1500 nm are sensitive to soil nutrient composition.

The above studies pointed out that there are different sensitive regions in the spectrum for different soil nutrient properties. This information is a viable source for remote sensing–based soil nutrient content classification and mapping. When hyperspectral imagery was used, various sensitive ranges from a single hyperspectral image would be applied to classify different soil nutrients. DeTar et al. (2008) found that some soil properties could be detected using aerial hyperspectral data over nearly bare fields. The highest regression R^2 (0.806) was for percentage sand. Other properties such as slit, clay, chlorides, electric conductivity, and P had lower R^2 (0.66–0.76). Ben-Dor (2002) used a 79-band hyperspectral image (0.4–1.4 μm) and multiple regression models to predict four soil properties, OM, soil field moisture, soil saturated moisture, and soil salinity, each with different bands. Ge and Thomasson (2006) combined wavelet analysis with conventional regression methods for soil property determination. It was found that Ca, Mg, P, and Zn could be predicted with reasonable R^2 values (>0.5). Finally, identification of such sensitive spectral regions for soil nutrient mapping remains a major task in hyperspectral remote sensing research.

Another effort in remote sensing–based soil mapping is the incorporation of geostatistical techniques (Yao 2004; López-Granados et al. 2005; Ge et al. 2011b). In one study (Yao 2004), two geostatistical approaches, colocated ordinary cokriging and sequential Gaussian cosimulation, were used to predict soil nutrient variables. The results showed that the cosimulation method yielded the best estimation ($R^2 = 0.71$) for K prediction. Figure 18.2 is the pH map generated from the cosimulation ($R^2 = 0.58$). It can be seen that the pH zones could be divided into two regions along the grass waterway, which is located in the middle-left of the field. To the left of the waterway, the soil is acidic with low pH value estimations. To the right of the waterway, the soil varies from acidic to basic. This analysis could provide information to help decision making on variable rate lime applications. Ge et al. (2007) applied

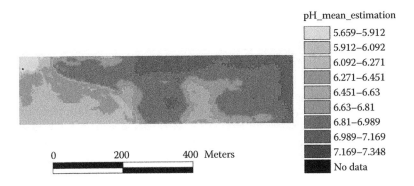

pH_mean_estimation

	5.659–5.912
	5.912–6.092
	6.092–6.271
	6.271–6.451
	6.451–6.63
	6.63–6.81
	6.81–6.989
	6.989–7.169
	7.169–7.348
	No data

0   200   400 Meters

FIGURE 18.2
(See color insert.) In-field variability of soil pH as indicated by the sequential Gaussian cosimulation method using aerial hyperspectral data (From Yao, H., *Hyperspectral Imagery for Precision Agriculture*, Department of Agricultural and Biological Engineering, University of Illinois at Urbana-Champaign, Urbana, IL, 2004. With permission.)

a regression-kriging method to analyze soil sampling and reflectance data. The regression-kriging model led to an R^2 value of 0.65 for Na, greater than that from a principal component regression approach. Bilgili et al. (2011) also found that cokriging and regression-kriging could improve the predictions of soil properties with reflectance data. After reviewing statistical techniques such as simple regression, the "soil line" approach, principal component analysis, and geostatistics for soil OM estimation using remote sensing data, Ladoni et al. (2010) concluded that remote sensing data could help the design of the soil sampling strategy.

18.2.5 Application 2: Pest and Disease Detection

For site-specific pesticide application, it is essential to identify the infestation location, coverage, and severity. When pest invasion or disease infestation happens, plant canopy reflectance could change and the differences in reflectance could be used for pest and disease detection. Some studies using remote sensing techniques for the detection are summarized below.

The Russian wheat aphid, *Diuraphis noxia*, is an important pest of winter wheat and barley. It causes significant loss to the wheat growers (Backoulou et al. 2011). To help develop target pesticide application, it is desired to identify and spatially differentiate the infestations in the wheat field. Backoulou et al. (2011) used airborne multispectral imagery for the detection of crop stress caused by *D. noxia*. The results indicated that *D. noxia*–induced stress varied spatially with regard to size, shape, and spatial arrangement. The cotton plant feeding strawberry spider mite caused leaf puckering and reddish discoloration in early stages of infestation, which eventually led to leaf drop. Fitzgerald et al. (2004) was able to distinguish between adjacent mite-free and mite-infested cotton field areas with a spectral unmixing process on AVIRIS imagery. Bacterial leaf blight is a vascular disease of irrigated rice, which causes up to 50% yield loss with serious infestations.

Fungal pathogen infection is another important disease for many agricultural crops. Identification of fungal infection at early growth stage is essential for site-specific fungicide applications. Franke and Menz (2007) tried to use multispectral remote sensing in a multitemporal analysis for powdery mildew (*Blumeria graminis*) and leaf rust (*Puccinia recondita*) pathogen detection in winter wheat. Two QuickBird (satellite) images and one aerial hyperspectral image were used in the study with a decision tree. Mixture tuned matched filtering and NDVI were used to classify data into the areas showing different levels of disease severity. The results implied that the images were only moderately suitable for early detection of crop infections. In a similar study that only used the above hyperspectral image data, Mewes et al. (2011) applied spectral angle mapper and support vector machines to classify the data. The results showed that a few key spectral features are sufficient to detect powdery mildew infection in wheat.

Moreover, Muhammed (2005) used spectral classification for the estimation of fungal disease severity of winter wheat. In this study, data normalization and a nearest-neighbor classification were applied, which resulted in $R^2 > 0.93$. Zhang et al. (2005) investigated a fungal pathogen causing disease and late blight with vegetation indices. The results showed that the diseased tomatoes could be separated from healthy ones before economic damage happened. Huang et al. (2007) used wavelengths of 531 and 570 nm from airborne data to calculate NDVI for yellow rust detection in wheat. The resulting R^2 was 0.91 between the disease index and the vegetation index. Mahlein et al. (2010) used a vegetation index–based approach to explore the potential of using hyperspectral data to detect and differentiate three fungal leaf diseases in sugar beet. It was concluded that a distinctive differentiation of the three diseases was possible using a combination of several indices.

18.2.6 Application 3: Plant Water Stress Detection

The implementation of site-specific irrigation requires knowledge of the variability of crop water status. Plant canopy temperature has been proven to be a good indicator for the estimation of plant water status. When a plant is stressed because of insufficient water supply, the decreases in transpiration may cause the plant temperature to rise beyond air temperature. On the other hand, well-watered plants have lower canopy temperature than air temperature. This temperature difference can be recorded by infrared thermography and used for plant water stress detection (Alchanatis et al. 2010). Wang et al. (2010) developed an algorithm to automatically estimate canopy temperature using infrared thermography. Once the canopy temperature is obtained, a widely used method to describe water stress is the Crop Water Stress Index (CWSI) (Corp et al. 2006a). The index is defined as a fraction of the canopy temperature between dry and wet reference under ambient conditions. The CWSI can be written as

$$\text{CWSI} = \frac{T_{\text{canopy}} - T_{\text{wet}}}{T_{\text{dry}} - T_{\text{wet}}} \tag{18.3}$$

where T_{dry} and T_{wet} are the reference temperatures of dry (nontranspiring) and wet (fully transpiring) leaf surfaces, respectively. T_{canopy} is the canopy surface temperature measured for water stress estimation. T_{wet} can be collected from the canopy of well-watered crop sections.

A typical way to obtain T_{canopy} is through thermal infrared remote sensing. Cohen et al. (2005) investigated the relationship between CWSI and crop leaf water potential using thermal imagery and spatial analysis. The study found that statistically the relationship between CWSI and crop leaf water potential was more stable and had slightly higher correlation coefficients than the relationship between crop canopy temperature and crop leaf water potential. Moller et al. (2007) further applied thermal and visible imagery for estimating crop water status of irrigated grapevine. Their study indicated that thermal-based CWSI was highly correlated with crop leaf conductance ($R^2 = 0.9$) and moderately correlated with crop stem water potential ($R^2 = 0.8$).

Alchanatis et al. (2010) found that the optimal time of thermal image acquisition for mapping cotton water status was in the midday. In the study, the leaf water potential was estimated with measurements of the temperature of an artificial wet surface and of the ambient air. In another study (Meron et al. 2010) to map cotton water stress, an artificial wet surface was also used to normalize CWSI to ambient condition. It was found that the leaf water potentials of cotton were linearly related to CWSI ($R^2 = 0.816$). Figure 18.3 is a water stress map of a cotton field. In a different approach, DeTar et al. (2006) tried to use airborne multispectral and hyperspectral imagery for full canopy cotton water stress detection. This study used a thermal infrared sensor to provide reference canopy

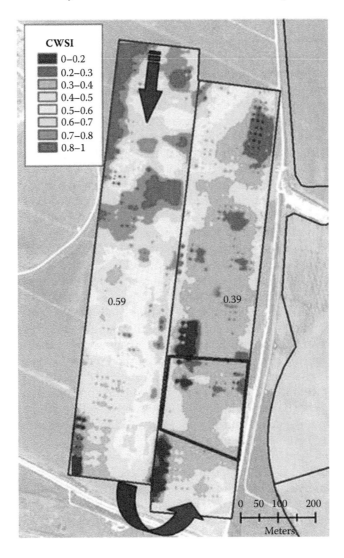

FIGURE 18.3
(See color insert.) Water stress map of a cotton field before last irrigation on August 20, 2007. Arrows indicate lateral move position and pivoting directions of the irrigation rig. Numbers are the mean Crop Water Stress Index (CWSI) levels for the east and west parts of the field. The bold polygon marks the ground monitored part of the field. (From Meron, M., et al., *Precis. Agric.*, 11, 148, 2010.)

temperature information. The results indicated that water stress in full canopy Acala cotton could be detected with remote sensing imagery. The difference in canopy temperature could be related to the average reflectance over a range from 923 to 1010 nm. A recent study (Thomson et al. 2012) showed the use of spatial statistics could enhance the value of thermal imagery. Finally, with knowledge of the crop water stress, site-specific irrigation can be implemented. One example is the automated center pivot irrigation system developed by Peters and Evett (2008). In this system, canopy temperature information was used for irrigation scheduling.

18.2.7 Application 4: Crop (Nitrogen) Stress Detection

Nitrogen is an important fertilizer in crop production. Nitrogen can be applied before planting and/or in the growing season. In either way, a good estimation of field condition is needed to decide the correct amount of nitrogen applied to the field. Historically, farmers tend to overestimate the needed amount, which would bring high fertilizer cost as well as elevated environmental burden due to excessive applications. Since different nitrogen levels in plants affect crop chlorophyll concentration and result in different canopy reflectance spectra (Walburg et al. 1982), many studies have recommended the use of remotely sensed canopy reflectance for crop nitrogen detection. Both high spatial resolution aerial and satellite imagery can be used for this application.

Among all the crops, corn plants attract the most attention in nitrogen sensing and site-specific nitrogen management. Multispectral remote sensing has been used in nitrogen stress detection. Blackmer et al. (1995) used aerial black–and–white imagery filtered at a central wavelength of 550 nm to find the feasibility of nitrogen stress detection in a corn canopy. Williams et al. (2010) used aerial images taken with a 35 mm slide film for corn nitrogen sensing. Color values of corn plants with zero and sufficient (280 kg·N·ha^{-1}) nitrogen applications were used to calculate the relative ratio of unfertilized to fertilized and relative difference color values. This color index approach could be a low-cost approach for N sensing. GopalaPillai et al. (1998) found that canopy reflectance extracted from a color infrared (CIR) image was well correlated to the amount of nitrogen applied to corn. Another study (Kyveryga et al. 2012) with a CIR image showed that late-season images were much better in detecting N deficiency areas than early-season images. Vegetation indices such as NDVI and green NDVI derived from SPOT (Satedlite Pour l'Observation dela Terre) images were found to be highly correlated to corn nitrogen stress (Corp et al. 2006b). Another study (Bausch and Khosla 2010) with satellite images (QuickBird) also indicated that the multispectral imagery could be used to evaluate nitrogen status in irrigated corn at V12 and later growth stages.

Hyperspectral imagery (mostly aerial imagery) has been widely used in nitrogen stress detection in corn. Zara et al. (2000) reported that the slope of the reflectance spectra between 560 and 580 nm generated the best results for corn nitrogen stress detection with AVIRIS hyperspectral images. Cassady et al. (2000) calculated three indices, including NDVI, Photosynthetic Reflectance Index, and Red-edge Vegetation Stress Index (RVSI), using certain bands from AVIRIS images, and pointed out that RVSI had the highest correlation with both applied nitrogen and measured chlorophyll levels in corn. Boegh et al. (2002) found that the CASI (Compact Airborne Spectrographic Imager) image green and NIR bands, which are the maximum reflectance bands of chlorophyll, were the most important predictors. Haboudane et al. (2002) developed a combined modeling- and indices-based approach to predict corn chlorophyll content using CASI imagery. This method used the ratio of an index sensitive to low chlorophyll values and a Soil Adjusted Index to build the prediction model. Another study (Miao et al. 2009) with aerial hyperspectral imagery pointed out that multiple regression analysis was advantageous to single band or vegetation index in explaining and estimating chlorophyll meter reading variability.

The use of remote sensing for nitrogen detection has been also implemented for other crops other than corn. Christensen et al. (2004) indicated that N content in barley could be predicted with 81% of accuracy. Min and Lee (2005) found the R^2 for N content in citrus tree leaves was 0.839. Zhao et al. (2005) attained an accuracy of 62.4% in discriminating N in cotton plants. El-Shikha et al. (2008) used the Canopy Chlorophyll Content Index (CCCI) for cotton content nitrogen estimation. The CCCI could account for seasonal changes in canopy density and changes in canopy chlorophyll. The results showed that CCCI was promising to estimate cotton nitrogen status for well-irrigated plants after the canopy reached about 30% cover. Jain et al. (2007) used a regression model for N estimation in potato plants and obtained R^2 of 0.551. Li et al. (2010) estimated N content of winter wheat with the R^2 value of 0.58 in an experimental field and 0.51 in a farmer's field. For rice plant N estimation, Nguyen et al. (2006) had the results of R^2 from 0.76 to 0.87 based on the validation data. In a 3-year study, a regression model was reported to have an $R^2 = 0.938$ for N estimation in rice (Ryu et al. 2009). Bajwa (2006) concluded that the PLSR models could explain 47% to 71% of the variability in rice plant N. A further study (Bajwa et al. 2010) also showed that vegetation indices could be used for mid-season nitrogen estimation in rice plant.

18.2.8 Application 5: Weed Detection and Mapping

Effective weed management is very important for the agricultural industry. Traditionally, weed control has relied heavily on herbicide application. One main consequence of herbicide application is increased environmental contamination. This situation calls for more effective use of herbicide, that is, using variable rate and site-specific technology to apply appropriate dosage of herbicide to the weeds. Less herbicide usage could also bring economic benefit to the farmers. For instance, it was pointed out that herbicide savings could be from 71.7% to 95.4% for the no-treatment areas and from 4.3% to 12% for the low-dose herbicide area (de Castro et al. 2011). Implementing site-specific herbicide application requires accurate detection and mapping of weeds in the agricultural field. Many sensing technologies have been developed for weed detection, among which optical remote sensing (Thorp and Tian 2004; Shaw 2005) and machine vision systems (Tang et al. 2003) dominate. In the remote sensing approach, soil background is a main factor that affects the detection outcomes. Similar to other applications, vegetation indices and spectral mixture analysis could be used for weed sensing.

When using hyperspectral images for weed sensing, the application can be grouped into three typical areas: mapping invasive weed species (Glenn et al. 2005; Hestir et al. 2008), weed stress characterization (Goel et al. 2003; Karimi et al. 2006), and weed species identification (Koger et al. 2003; Piron et al. 2008). For mapping invasive species, Hestir et al. (2008) used remotely sensed hyperspectral images in wetland systems. The images were analyzed with spectral angle mapping and spectral mixture analysis. A moderate mapping accuracy was obtained, which was primarily due to significant spectral variation of the mapped weed species. Goel et al. (2003) used artificial neural network with hyperspectral data acquired by a CASI. The task was to classify four different weed management strategies in corn fields where three different nitrogen application rates were also used. The study found that when only one factor (weed or nitrogen) was considered at a time, the classification accuracy was higher than the situations with both factors. For weed species identification, Vrindts et al. (2002) used canopy reflectance to classify sugar beet, maize, and seven weed species under controlled laboratory conditions. Crop and weeds were separated with more than 97% accuracy using a selected number of wavelength band ratios. While under field conditions, over 90% of crop and weed spectra can be classified correctly with the prevailing light conditions.

18.2.9 Application 6: Herbicide Drift Detection

In modern farm industry, herbicide application is the major means for weed control. Since the mid-1990s, one type of herbicide, glyphosate, has seen a significant increase because of the increased adoption of genetically modified glyphosate-resistant (GR) crops. Glyphosate is a nonselective herbicide used for weed control in genetically modified GR crops. Glyphosate application for weed control in GR crops can drift onto an off-target area, causing unwanted damage to non-GR plants. Consequently, early detection of crop injury from off-target drift of herbicide is critical in crop production. Remote sensing has been used in an attempt to detect crop injury because of herbicide drift. In some studies, multispectral imagery with several broadband reflectance measurements including at least one red and one NIR band was used (Thelen et al. 2004; Huang et al. 2012). The multispectral images generally had good spatial resolution in the data. However, the broadband imagery had lower spectral resolution in the reflectance measurement, and the standard bands customarily assigned to multispectral systems might not be appropriate for optimal damage detection. Thus, the use of multispectral imagery lacked the ability to explore the fine spectral reflectance features of a plant at the red edge (~700 nm) of the electromagnetic spectrum range. A recent greenhouse-based study (Yao et al. 2012) used hyperspectral imagery to monitor postapplication injury of soybean plants at different time frames. Both vegetation indices and derivatives were used in the study. The results showed that hyperspectral imaging of plant canopy reflectance could be a useful tool for early detection of soybean crop injury from glyphosate. The technique could potentially differentiate crop injury at 4 hours after herbicide application.

18.3 Conclusions

The development of precision farming technologies has enabled farmers to manage in-field variability and to treat farms site-specifically. With this practice, farmers are able to increase farming productivity, improve profit, and at the same time ease the tension on the evergrowing environmental burden caused by the modern agricultural industry. In summary, remotely sensed data are important data sources for field variability sensing and information extraction. This is an essential step for the implementation of precision farming technology, which also includes management decision making, precision field operation control, and result assessment. However, little research has been found on the actual adoption of using remote sensing by farmers in precision farming practices. It was suggested (Larson et al. 2008) that value-added services offered by remote sensing service providers and crop consultants might be an important factor to explain the use of remote sensing imagery by cotton growers. McBratney et al. (2005) pointed out some future directions of precision agriculture, listed as following in the order of their importance:

1. Appropriate criteria for the economic assessment of precision agriculture
2. Insufficient recognition of temporal variation
3. Lack of whole-farm focus
4. Crop quality assessment methods
5. Product tracking and traceability
6. Environmental auditing

Among them, insufficient recognition of temporal variation, lack of whole-farm focus, crop quality assessment methods, and environmental auditing are all related to field and crop information sensing and extraction. In this aspect, remote sensing technology would provide important help in dealing with the above concerns. In this chapter, the value of agricultural remote sensing for precision farming was investigated. Specific information of the following six agricultural remote sensing applications were reviewed: (1) soil mapping, (2) insect/pest infestation identification, (3) crop water stress detection, (4) in-field nitrogen stress detection, (5) weed sensing and mapping, and (6) herbicide drift detection. In addition, this chapter has summarized many vegetation indices used for different applications in agricultural remote sensing. The spectral, spatial, and temporal issues related to remote sensing data are also important considerations to precision farming applications.

References

Alchanatis, V., Y. Cohen, S. Cohen, M. Moller, M. Spristin, M. Meron, J. Tsipris, Y. Saranga, and E. Sela. 2010. "Evaluation of Different Approaches for Estimating and Mapping Crop Water Status in Cotton with Thermal Imaging." *Precision Agriculture* 11:27–41.

Backoulou, G. F., N. C. Elliott, K. Giles, M. Phoofolo, and V. Catana. 2011. "Development of a Method Using Multispectral Imagery and Spatial Pattern Metrics to Quantify Stress to Wheat Fields Caused by *Diuraphis Noxia*." *Computers and Electronics in Agriculture* 75(1):64–70.

Bajwa, S. G. 2006. "Modeling Rice Plant Nitrogen Effect on Canopy Reflectance with Partial Least Square Regression (PLSR)." *Transactions of the American Society of Agricultural Engineers* 49(1):229–37.

Bajwa, S. G., A. R. Mishra, and R. J. Norman. 2010. "Canopy Reflectance Response to Plant Nitrogen Accumulation in Rice." *Precision Agriculture* 11:488–506.

Bajwa, S. G., and L. F. Tian. 2005. "Soil Fertility Characterization in Agricultural Fields Using Hyperspectral Remote Sensing." *Transactions of the American Society of Agricultural Engineers* 48 (6):2399–406.

Bausch, W. C., and R. Khosla. 2010. "QuickBird Satellite versus Ground-Based Multi-Spectral Data for Estimating Nitrogen Status of Irrigated Maize." *Precision Agriculture* 11:274–90.

Bégué, A., P. Todoroff, and J. Pater. 2008. "Multi-Time Scale Analysis of Sugarcane Within-Field Variability: Improved Crop Diagnosis Using Satellite Time Series?" *Precision Agriculture* 9:161–71.

Ben-Dor, E. 2002. "Quantitative Remote Sensing of Soil Properties." *Advances in Agronomy* 75:173–243.

Ben-Dor, E., and A. Banin. 1995. "Near-Infrared Analysis as a Rapid Method to Simultaneously Evaluate Several Soil Properties." *Soil Science Society of America Journal* 59(2):364–72.

Bilgili, A. V., F. Akbas, and H. M. van Es. 2011. "Combined Use of Hyperspectral VNIR Reflectance Spectroscopy and Kriging to Predict Soil Variables Spatially." *Precision Agriculture* 12:395–420.

Blackburn, G. A. 1998. "Quantifying Chlorophylls and Carotenoids at Leaf and Canopy Scales: An Evaluation of Some Hyperspectral Approaches." *Remote Sensing of Environment* 66(3):273–85.

Blackmer, T. M., J. S. Schepers, and G. E. Meyer. 1995. "Remote Sensing to Detect Nitrogen Deficiency in Corn." In *Proceedings of the Second International Conference on Site-Specific Management for Agricultural Systems*. Minneapolis, MN: ASA-CSSA-SSSA.

Boegh, E., H. Soegaard, N. Broge, C. B. Hasager, N. O. Jensen, K. Schelde, and A. Thomsen. 2002. "Airborne Multispectral Data for Quantifying Leaf Area Index, Nitrogen Concentration, and Photosynthetic Efficiency in Agriculture." *Remote Sensing of Environment* 81(2–3):179–93.

Bongiovanni, R., and J. Lowenberg-Deboer. 2004. "Precision Agriculture and Sustainability." *Precision Agriculture* 5:359–87.

Broge, N. H., and E. Leblanc. 2000. "Comparing Prediction Power and Stability of Broad-Band and Hyperspectral Vegetation Indices for Estimation of Green Leaf Area Index and Canopy Chlorophyll Density." *Remote Sensing of Environment* 76:156–72.

Cassady, P. E., E. M. Perry, M. E. Gardner, and D. A. Roberts. 2000. "Airborne Hyperspectral Imagery for the Detection of Agricultural Crop Stress." In *Proceedings of SPIE* 4151, Hyperspectral Remote Sensing of the Land and Atmosphere, 197 (February 8, 2001). doi: 10.1117/12.417008.

Christensen, L. K., B. S. Bennedsen, R. N. Jørgensen, and H. Nielsen. 2004. "Modelling Nitrogen and Phosphorus Content at Early Growth Stages in Spring Barley Using Hyperspectral Line Scanning." *Biosystems Engineering* 88(1):19–24.

Cohen, Y., V. Alchanatis, M. Meron, Y. Saranga, and J. Tsipris. 2005. "Estimation of Leaf Water Potential by Thermal Imagery and Spatial Analysis." *Journal of Experimental Botany* 56(417):1843–52.

Corp, L. A., E. M. Middleton, C. S. T. Daughtry, and P. K. E. Campbell. 2006a. "Solar Induced Fluorescence and Reflectance Sensing Techniques for Monitoring Nitrogen Utilization in Corn." In *IEEE International Conference on Geoscience and Remote Sensing*, July 31–August 4, 2006, Denver, CO, 2267–70.

Corp, L. A., E. M. Middleton, J. E. McMurtrey, P. K. E. Campbell, and L. M. Butcher. 2006b. "Fluorescence Sensing Techniques for Vegetation Assessment." *Applied Optics* 45(5):1023–33.

Daughtry, C. S. T., C. L. Walthall, M. S. Kim, E. B. D. Colstoun, and J. E. McMurtrey. 2000. "Estimating Corn Leaf Chlorophyll Concentration from Leaf and Canopy Reflectance." *Remote Sensing of Environment* 74(2):229–39.

de Castro, A. I., M. Jurado-Expósito, J. M. Peña-Barragán, and F. López-Granados. 2011. "Airborne Multi-Spectral Imagery for Mapping Cruciferous Weeds in Cereal and Legume Crops." *Precision Agriculture* 13:302–21.

DeTar, W. R., J. H. Chesson, J. V. Penner, and J. C. Ojala. 2008. "Detection of Soil Properties with Airborne Hyperspectral Measurements of Bare Fields." *Transactions of the American Society of Agricultural Engineers* 51(2):463–70.

DeTar, W. R., J. V. Penner, and H. A. Funk. 2006. "Airborne Remote Sensing to Detect Plant Water Stress in Full Canopy Cotton." *Transactions of the American Society of Agricultural Engineers* 49(3):655–65.

El-Shikha, D. M., E. M. Barnes, T. R. Clarke, D. J. Hunsaker, J. A. Haberland, P. J. Pinter Jr., P. M. Waller, and T. L. Thompson. 2008. "Remote Sensing of Cotton Nitrogen Status Using the Canopy Chlorophyll Content Index (CCCI)." *Transactions of the American Society of Agricultural and Biological Engineers* 51(1):73–82.

Elvidge, C. D., and Z. Chen. 1994. "Comparison of Broad-Band and Narrow-Band Red and Near-Infrared Vegetation Indices." *Remote Sensing of Environment* 54:38–48.

Escadafal, R., A. R. Huete. 1991. "Étude des propriétés spectrales des sols arides appliquée à l'amélioration des indices de vegetation obtenus par télédection." *Comptes rendus de l'Académie des Sciences* 312:1385–91.

Estep, L., and G. A. Carter. 2005. "Derivative Analysis of AVIRIS Data for Crop Stress Detection." *Photogrammetric Engineering and Remote Sensing* 71:1417–21.

Filella, I., L. Serrano, J. Serra, and J. Penuelas. 1995. "Evaluating Wheat Nitrogen Status with Canopy Reflectance Indices and Discriminant Analysis." *Crop Science* 35:1400–05.

Fitzgerald, G. J., S. J. Maas, and W. R. Deter. 2004. "Spider Mite Detection and Canopy Component Mapping in Cotton Using Hyperspectral Imagery and Spectral Mixture Analysis." *Precision Agriculture* 5:275–89.

Franke, J., and G. Menz. 2007. "Multi-Temporal Wheat Disease Detection by Multi-Spectral Remote Sensing." *Precision Agriculture* 8:161–72.

Ge, Y., C. L. S. Morgan, J. A. Thomasson, and T. Waiser. 2007. "A New Perspective to Near-Infrared Reflectance Spectroscopy: A Wavelet Approach." *Transactions of the American Society of Agricultural Engineers* 50(1):303–11.

Ge, Y., and J. A. Thomasson. 2006. "Wavelet Incorporated Spectral Analysis for Soil Property Determination." *Transactions of the American Society of Agricultural Engineers* 49(4):1193–201.

Ge, Y., J. A. Thomasson, and R. Sui. 2011a. "Remote Sensing of Soil Properties in Precision Agriculture: A Review." *Frontiers of Earth Science* 5(3):229–38.

Ge, Y., J. A. Thomasson, R. Sui, and J. Wooten. 2011b. "Regression-Kriging for Characterizing Soils with Remote Sensing Data." *Frontiers of Earth Science* 5(3):239–44.

Glenn, N. F., J. T. Mundt, K. T. Weber, T. S. Prather, L. W. Lass, and J. Pettingill. 2005. "Hyperspectral Data Processing for Repeat Detection of Small Infestations of Leafy Spurge." *Remote Sensing of Environment* 95(3):399–412.

Goel, P. K., S. O. Prasher, R. M. Patel, J. A. Landry, R. B. Bonnell, A. A. Viau, and J. R. Miller. 2003. "Potential of Airborne Hyperspectral Remote Sensing to Detect Nitrogen Deficiency and Weed Infestation in Corn." *Computers and Electronics in Agriculture* 38(2):99–124.

Gómez-Candón, D., F. López-Granados, J. J. Caballero-Novella, M. Gomez-Casero, M. Jurado-Exposito, and L. Garcia-Torres. 2011. "Geo-Referencing Remote Images for Precision Agriculture Using Artificial Terrestrial Targets." *Precision Agriculture* 12:876–91.

Gong, P., R. L. Ru, and G. S. Biging. 2003. "Estimation of Forest Leaf Area Index Using Vegetation Indices Derived from Hyperion Hyperspectral Data." *IEEE Transactions on Geoscience and Remote Sensing* 41:1355–62.

GopalaPillai, S., L. Tian, and J. Beal. 1998. *Detection of Nitrogen Stress in Corn using Digital Aerial Imaging*. ASAE Meeting Paper No. 983030. St. Joseph, MI.

Grisso, R., M. Alley, P. McClellan, D. Brann, and S. Donohue. 2009. "*Precision Farming: A Comprehensive Approach*." Virginia: Communications and Marketing, College of Agriculture and Life Sciences, Virginia Polytechnic Institute and State University.

Haboudane, D., J. R. Miller, N. Tremblay, P. J. Zarco-Tejada, and L. Dextraze. 2002. "Integrated Narrow-Band Vegetation Indices for Prediction of Crop Chlorophyll Content for Application to Precision Agriculture." *Remote Sensing of Environment* 18:416–26.

Hestir, E. L., S. Khanna, M. E. Andrew, M. J. Santos, J. H. Viers, J. A. Greenberg, S. S. Rajapakse, and S. L. Ustin. 2008. "Identification of Invasive Vegetation Using Hyperspectral Remote Sensing in the California Delta Ecosystem." *Remote Sensing of Environment* 112(11):4034–47.

Huang, W., D. W. Lamb, Z. Niu, Y. Zhang, L. Liu, and J. Wang. 2007. "Identification of Yellow Rust in Wheat Using In-Situ Spectral Reflectance Measurements and Airborne Hyperspectral Imaging." *Precision Agriculture* 8:187–97.

Huang, Y., S. J. Thomson, W. T. Molin, K. N. Reddy, and H. Yao. 2012. "Early Detection of Soybean Plant Injury from Glyphosate by Measuring Chlorophyll Reflectance and Fluorescence." *Journal of Agricultural Science* 4:117–24.

Huete, A.R. 1988. "A Soil-Adjusted Vegetation Index (SAVI)." *Remote Sensing of Environment* 25:295–309.

Hughes, G. F. 1968. "On the Mean Accuracy of Statistical Pattern Recognizers." *IEEE Transactions on Information Theory* 14(1):55–63.

Hurcom, S. J., and A. R. Harrison. 1998. "The NDVI and Spectral Decomposition for Semi-Arid Vegetation Abundance Estimation." *International Journal of Remote Sensing* 19(16): 3109–125.

Imanishi, J., K. Sugimoto, and Y. Morimoto. 2004. "Detecting Drought Status and Lai of Two Quercus Species Canopies Using Derivative Spectra." *Computers and Electronics in Agriculture* 43:109–29.

Jain, N., S. S. Ray, J. P. Singh, and S. Panigrahy. 2007. "Use of Hyperspectral Data to Assess the Effects of Different Nitrogen Applications on a Potato Crop." *Precision Agriculture* 8:225–39.

Jordan, C.F. 1969. "Derivation of Leaf Area Index from Quality of Light on the Forest Floor." *Ecology* 50:663–6.

Karimi, Y., S. O. Prasher, R. M. Patel, and S. H. Kim. 2006. "Application of Support Vector Machine Technology for Weed and Nitrogen Stress Detection in Corn." *Computers and Electronics in Agriculture* 51(1–2):99–109.

Khosla, R., D. Inman, D. G. Westfall, R. M. Reich, M. Frasier, M. Mzuku, B. Koch, and A. Hornung. 2008. "A Synthesis of Multi-Disciplinary Research in Precision Agriculture: Site-Specific Management Zones in the Semi-Arid Western Great Plains of the USA." *Precision Agriculture* 9:85–100.

Koger, C., L. M. Bruce, D. R. Shaw, and K. N. Reddy. 2003. "Wavelet Analysis of Hyperspectral Reflectance Data for Detecting Pitted Morningglory (Ipomoea Lacunosa) in Soybean (Glycine Max)." *Remote Sensing of Environment* 86(1):108–19.

Kyveryga, P. M., T. M. Blackmer, and R. Pearson. 2012. "Normalization of Uncalibrated Late-Season Digital Aerial Imagery for Evaluating Corn Nitrogen Status." *Precision Agriculture* 13:2–16.

Ladoni, M., H. A. Bahrami, S. K. Alavipanah, and A. A. Norouzi. 2010. "Estimating Soil Organic Carbon from Soil Reflectance: A Review." *Precision Agriculture* 11:82–99.

Larson, J. A., R. K. Roberts, B. C. English, S. L. Larkin, M. C. Marra, S. W. Martin, K. W. Paxton, and J. M. Reeves. 2008. "Factors Affecting Farmer Adoption of Remotely Sensed Imagery for Precision Management in Cotton Production." *Precision Agriculture* 9:195–208.

Li, F., Y. Miao, S. D. Hennig, M. L. Gnyp, X. Chen, L. Jia, and G. Bareth. 2010. "Evaluating Hyperspectral Vegetation Indices for Estimating Nitrogen Concentration of Winter Wheat at Different Growth Stages." *Precision Agriculture* 11:335–57.

López-Granados, F., M. Jurado-Expósito, J. M. Peña-Barragán, and L. García-Torres. 2005. "Using Geostatistical and Remote Sensing Approaches for Mapping Soil Properties." *European Journal of Agronomy* 23(3):279–89.

López-Lozano, R., M. A. Casterad, and J. Herrero. 2010. "Site-Specific Management Units in a Commercial Maize Plot Delineated Using Very High Resolution Remote Sensing and Soil Properties Mapping." *Computers and Electronics in Agriculture* 73(2):219–29.

Mahlein, A.-K., U. Steiner, H.-W. Dehne, and E.-C. Oerke. 2010. "Spectral Signatures of Sugar Beet Leaves for the Detection and Differentiation of Diseases." *Precision Agriculture* 11:413–31.

McBratney, A., B. Whelan, and T. Ancev. 2005. "Future Directions of Precision Agriculture." *Precision Agriculture* 6:7–23.

Meron, M., J. Tsipris, V. Orlov, V. Alchanatis, and Y. Cohen. 2010. "Crop Water Stress Mapping for Site-Specific Irrigation by Thermal Imagery and Artificial Reference Surfaces." *Precision Agriculture* 11:148–62.

Mewes, T., J. Franke, and G. Menz. 2011. "Spectral Requirements on Airborne Hyperspectral Remote Sensing Data for Wheat Disease Detection." *Precision Agriculture* 12:795–812.

Miao, Y., D. J. Mulla, G. W. Randall, J. A. Vetsch, and R. Vintila. 2009. "Combining Chlorophyll Meter Readings and High Spatial Resolution Remote Sensing Images for In-Season Site-Specific Nitrogen Management of Corn." *Precision Agriculture* 10:45–62.

Min, M., and W. S. Lee. 2005. "Determination of Significant Wavelengths and Prediction of Nitrogen Content for Citrus." *Transactions of the American Society of Agricultural Engineers* 48(2):455–61.

Moller, M., V. Alchanatis, Y. Cohen, M. Meron, J. Tsipris, A. Naor, V. Ostrovsky, M. Sprintsin, and S. Cohen. 2007. "Use of Thermal and Visible Imagery for Estimating Crop Water Status of Irrigated Grapevine." *Journal of Experimental Botany* 58(4):827–38.

Moran, M., Y. Inoue, and E. M. Barnes. 1997. "Opportunities and Limitations for Image-Based Remote Sensing in Precision Crop Management." *Remote Sensing of Environment* 61(3):319–46.

Muhammed, H. H. 2005. "Hyperspectral Crop Reflectance Data for Characterising and Estimating Fungal Disease Severity in Wheat." *Biosystems Engineering* 91(1):9–20.

National Research Council. 1997. *Precision Agriculture in the 21st Century: Geospatial and Information Technologies in Crop Management*. Washington, DC: National Academy Press.

Neményi, M., P. Á. Mesterházi, Z. Pecze, and Z. Stépán. 2003. "The Role of GIS and GPS in Precision Farming." *Computers and Electronics in Agriculture* 40(1–3):45–55.

Nguyen, H. T., J. H. Kim, A. T. Nguyen, L. T. Nguyen, J. C. Shin, and B.-W. Lee. 2006. "Using Canopy Reflectance and Partial Least Squares Regression to Calculate Within-Field Statistical Variation in Crop Growth and Nitrogen Status of Rice." *Precision Agriculture* 7:249–64.

Omasa, K., K. Oki, and T. Suhama. 2006. "Section 5.2 Remote Sensing from Satellites and Aircraft." In *CIGR Handbook of Agricultural Engineering*, edited by Axel Munack. St. Joseph, MI: American Society of Agricultural and Biological Engineers.

Palacios-Orueta, A., and S. L. Ustin. 1998. "Remote Sensing of Soil Properties in the Santa Monica Mountains I. Spectral Analysis." *Remote Sensing of Environment* 65(2):170–83.

Perry, C.R., L. F. Lautenschlager. 1984. "Functional Equivalence of Spectral Vegetation Indices." *Remote Sensing of Environment* 14:169–82.

Peters, R. T., and S. R. Evett. 2008. "Automation of a Center Pivot Using the Temperature-Time-Threshold Method of Irrigation Scheduling." *Journal of Irrigation and Drainage Engineering* 134 (3):286–91.

Piron, A., V. Leemans, O. Kleynen, F. Lebeau, and M.-F. Destain. 2008. "Selection of the Most Efficient Wavelength Bands for Discriminating Weeds from Crop." *Computers and Electronics in Agriculture* 62:141–48.

Richard, J. A., and X. Jia. 1999. *Remote Sensing Digital Image Analysis: An Introduction*. Berlin Heidelberg, Germany: Springer-Verlag.

Rondeaux, G., M. Steven, and F. Baret. 1996. "Optimization of Soil-Adjusted Vegetation Indices." *Remote Sensing of Environment* 55(2):95–107.

Rouse, J. W., R. H. Haas, J. A. Schell, and D. W. Deering. 1973. "Monitoring Vegetation Systems in the Great Plains with ERTS." In *Third ERTS Symposium, NASA SP-351*. Washington, DC: NASA.

Ryu, C., M. Suguri, and M. Umeda. 2009. "Model for Predicting the Nitrogen Content of Rice at Panicle Initiation Stage Using Data from Airborne Hyperspectral Remote Sensing." *Biosystems Engineering* 104:465–75.

Savitzky, A., and M. J. E. Golay. 1964. "Smoothing and Differentiation of Data by Simplified Least Squares Procedures." *Analytical Chemistry* 36(8):1627–39.

Scotford, I. M., and P. C. H. Miller. 2005. "Applications of Spectral Reflectance Techniques in Northern European Cereal Production: A Review." *Biosystems Engineering* 90(3):235–50.

Serrano, L., I. Filella, and J. Peuelas. 2000. "Remote Sensing of Biomass and Yield of Winter Wheat under Different Nitrogen Supplies." *Crop Science* 40(3):723–31.

Shaw, D. R. 2005. "Translation of Remote Sensing Data into Weed Management Decisions." *Weed Science* 53(2):264–73.

Smith, K. L., M. D. Steven, and J. J. Colls. 2004. "Use of Hyperspectral Derivative Ratios in the Red-Edge Region to Identify Plant Stress Responses to Gas Leaks." *Remote Sensing of Environment* 92:207–17.

Song, X., J. Wang, W. Huang, L. Liu, G. Yan, and R. Pu. 2009. "The Delineation of Agricultural Management Zones with High Resolution Remotely Sensed Data." *Precision Agriculture* 10:471–87.

Stafford, J. V. 2000. "Implementing Precision Agriculture in the 21st Century." *Journal of Agricultural Engineering Research* 76(3):267–75.

Tang, L., L. F. Tian, and B. L. Steward. 2003. "Texture-Based Real-Time Broadleaf and Grass Classification for Selective Weed Control." *Transactions of the ASAE* 46(4):1247–54.

Thelen, K. D., A. N. Kravchenko, and C. Lee. 2004. "Use of Optical Remote Sensing for Detecting Herbicide Injury in Soybean." *Weed Technology* 18:292–97.

Thomasson, J. A., R. Sui, M. S. Cox, and A. Al-Rajehy. 2001. "Soil Reflectance Sensing for Determining Soil Properties in Precision Agriculture." *Transactions of the American Society of Agricultural Engineers* 44(6):1445–53.

Thomson, S. J., C. M. Ouellet-Plamondon, S. L. DeFauw, Y. Huang, D. K. Fisher, and P. J. English. 2012. "Potential and Challenges in Use of Thermal Imaging for Humid Region Irrigation System Management." *Journal of Agricultural Science* 4(4):103–16.

Thorp, K. R., L. Tian, H. Yao, and L. Tang. 2004. "Narrow-Band and Derivative-Based Vegetation Indices for Hyperspectral Data." *Transactions of the American Society of Agricultural Engineers* 47 (1):291–99.

Thorp, K. R., and L. F. Tian. 2004. "A Review on Remote Sensing of Weeds in Agriculture." *Precision Agriculture* 5:477–508.

Tsai, F., and W. Philpot. 1998. "Derivative Analysis of Hyperspectral Data." *Remote Sensing of Environment* 66(1):41–51.

Tucker, C. J. 1979. "Red and Photographic Infrared Linear Combinations for Monitoring Vegetation." *Remote Sensing of Environment* 8:127–50.

Van Niel, T. G., and T. R. McVicar. 2004. "Determining Temporal Windows for Crop Discrimination with Remote Sensing: A Case Study in South-Eastern Australia." *Computers and Electronics in Agriculture* 45(1–3):91–108.

Vrindts, E., J. De Baerdemaeker, and H. Ramon. 2002. "Weed Detection Using Canopy Reflection." *Precision Agriculture* 3:63–80.

Walburg, G., M. E. Bauer, C. S. T. Daughtry, and T. L. Housley. 1982. "Effects of Nitrogen Nutrition on the Growth, Yield, and Reflectance Characteristics of Corn Canopies." *Agronomy Journal* 74:677–83.

Wang, X., W. Yang, A. Wheaton, N. Cooley, and B. Moran. 2010. "Automated Canopy Temperature Estimation via Infrared Thermography: A First Step Towards Automated Plant Water Stress Monitoring." *Computers and Electronics in Agriculture* 73(1):74–83.

Wetterlind, J., B. Stenberg, and M. Söderström. 2008. "The Use of Near Infrared (NIR) Spectroscopy to Improve Soil Mapping at the Farm Scale." *Precision Agriculture* 9(1–2):57–69.

Williams, J. D., N. R. Kitchen, P. C. Scharf, and W. E. Stevens. 2010. "Within-Field Nitrogen Response in Corn Related to Aerial Photograph Color." *Precision Agriculture* 11:291–305.

Yao, H. 2004. *Hyperspectral Imagery for Precision Agriculture.* Urbana, IL: Department of Agricultural and Biological Engineering, University of Illinois at Urbana-Champaign.

Yao, H., Y. Huang, Z. Hruska, S. J. Thomson, and K. N. Reddy. 2012. "Using Vegetation Index and Modified Derivative for Early Detection of Soybean Plant Injury from Glyphosate." *Computers and Electronics in Agriculture* 89:145–57.

Yao, H., and L. Tian. 2004. "Practical Methods for Geometric Distortion Correction of Aerial Hyperspectral Imagery." *Applied Engineering in Agriculture* 20(3):367–75.

Zara, P. M., P. C. Doraiswamy, and J. McMutrey. 2000. "Assessing Variability of Nitrogen Status in Corn Plants with Hyperspectral Remote Sensing." In *Proceedings of ASPRS Annual Meeting.*

Zhang, M., Z. Qin, and X. Liu. 2005. "Remote Sensed Spectral Imagery to Detect Late Blight in Field Tomatoes." *Precision Agriculture* 6:489–508.

Zhang, N., M. Wang, and N. Wang. 2002. "Precision Agriculture: A Worldwide Overview." *Computers and Electronics in Agriculture* 36(2–3):113–32.

Zhao, D. H., J. L. Li, and J. G. Qi. 2005. "Identification of Red and NIR Spectral Regions and Vegetative Indices for Discrimination of Cotton Nitrogen Stress and Growth Stage." *Computers and Electronics in Agriculture* 48:155–69.

Zwiggelaar, R. 1998. "A Review of Spectral Properties of Plants and Their Potential Use for Crop/Weed Discrimination in Row-Crops." *Crop Protection* 17:189–206.

19

Mapping and Uncertainty Analysis of Crop Residue Cover Using Sequential Gaussian Cosimulation with QuickBird Images

Cha-Chi Fan, Guangxing Wang, George Z. Gertner, Haibo Yao, Dana G. Sullivan, and Mark Masters

CONTENTS

19.1 Introduction..353
19.2 Materials and Methods..355
    19.2.1 Study Area and Datasets 355
    19.2.2 Image Transformation...356
    19.2.3 Sequential Gaussian Cosimulation for Mapping............................358
19.3 Results 360
19.4 Conclusions and Discussion 369
19.5 Summary..370
Acknowledgments 371
References..371

19.1 Introduction

In the United States, about 30% of agricultural lands have been classified as highly erodible, contributing to continuous degradation of soil productivity. Conservation tillage is a well-known best management practice to conserve soil water resources. Crop residue reduces soil losses from water and wind erosion and increases sequestration of carbon in the soil. Crop residue retention is considered a soil conservation practice, and accurately estimating and mapping crop residue percentage cover is critical (Pacheco and McNairn 2010). Traditionally, ground survey–based methods are often used to estimate crop residue percentage cover (Bannari et al. 2006; Pacheco et al. 2006). However, this method is very time-consuming. Thus, developing a satellite-derived mapping algorithm to rapidly quantify crop residue percentage cover at field and landscape scales has become very important.

Various sensor images have been used for classifying and mapping crop residue. Landsat Thematic Mapper (TM) images are widely used in this area. Gowda et al. (2001) and Van Deventer et al. (1997) used TM image–based logistic regression models to distinguish conventional and conservation tillage and concluded that the best estimates were obtained using ratio and difference images containing TM band 5. Thoma et al. (2004) applied Enhanced TM Plus (ETM+) data to estimate crop residue cover. For differentiating no-till and traditional tillage cropping, South et al. (2004) used ETM+ imagery and found that spectral angle mapping and cosine of the angle concept were better than minimum distance, Mahalanobis distance, and maximum likelihood.

Chevrier et al. (2002) found that hyperspectral PROBE-1 data well discriminated crop residue and bare soil at the combination of bands 36 (943 nm) and 115 (2303 nm). Yang et al. (2003) used hyperspectral imagery with the classification and regression tree approach to classify tillage practice and residue levels. Liu et al. (2004) examined the relationships between optical indices and dry biomass and crop height of corn, soybean, and wheat using Compact Airborne Spectrographic Imager (CASI) hyperspectral data. In particular, Nagler et al. (2003), Daughtry et al. (2004, 2005), and McMurtrey et al. (2005) proposed a cellulose absorption index (CAI) from hyperspectral image bands from 2.00 to 2.24 µm and found a significant linear relationship between CAI and crop residue cover. Furthermore, Daughtry et al. (2005) evaluated several spectral indices for measuring crop residue cover based on ground-based and airborne hyperspectral data.

On the other hand, the methods used for mapping crop residue can be divided into two groups: point-in-polygon or point-in-stratum and regression modeling. For the first group, homogeneous polygons or strata are first derived by supervised or unsupervised classification of all pixels. Within each polygon or stratum, the cells are assumed to be homogeneous and an average is calculated and assigned to each cell. Regression modeling first develops the relationship of a primary variable with remotely sensed images based on the sample locations and then uses the relationship to estimate the unobserved locations. Moreover, classification and regression modeling can be combined (Wang et al. 2002), that is, image classification is first made and a regression model is then constructed within each polygon or stratum. The aforementioned classification and mapping methods are unbiased for populations. However, the estimates obtained are not reliable for subareas and specific locations. Moreover, classification leads to smooth maps, whereas regression modeling may result in illogical estimates such as negative values.

In addition, spectral unmixing that estimates the fractional abundances of surface targets at a subpixel level has also been applied to map crop residue percentage cover (Bannari et al. 2006, 2007; Pacheco et al. 2006; Pacheco and McNairn 2010). This technique helps to map and monitor residue cover, and its success depends on accurately extracting end-members. But low spectral contrast between soil and crop residue will lead to large errors of estimates (Pacheco et al. 2006).

Moreover, all the methods neglect regionalized variable theory—spatial autocorrelation or spatial variability of a variable—that is, the closer the locations, the more similar the values of a variable. Thus, they usually overestimate uncertainties of population estimates or neglect uncertainties of local estimates. The aforementioned shortcomings for the methods can be overcome by geostatistical methods such as cokriging and sequential Gaussian cosimulation (Gomez-Hernandez and Journel 1992; Xu et al. 1992; Barata et al. 1996; Hunner et al. 2000; Dungan 2002; Wang et al. 2007).

Geostatistical methods are based on regionalized variable theory, provide estimates with minimum error variances, and are unbiased for both populations and local areas (Goovaerts 1997). When a geostatistical method is used to map crop residue, the residue cover is considered a primary variable and the remote sensing image is treated as a secondary variable (Wang et al. 2007). Remote sensing data are reflectance records of objects on the ground and represent the spatial characteristics and variability of the surface. The spatial autocorrelation of variables and spatial cross-correlation between them are expected. The variograms and cross-variograms that quantify the spatial correlation can be taken into account in the mapping of crop residue.

In contrast to sequential Gaussian cosimulation, cokriging methods tend to "smooth out" local details of spatial variability. This means that cokriging methods preserve fewer details of ground variation as the distances between estimated points and ground data locations increase. The estimates are smooth and uniform in the map, but the ground may actually contain much more variability. This is because the error variances calculated in cokriging depend only on local data configuration and not on the data values themselves (Goovaerts 1997). On the other hand, sequential Gaussian cosimulation with remote sensing data overcomes this shortcoming and more accurately reproduces the spatial variability of the primary variable. For example, Wang et al. (2002, 2003) compared several methods to map soil erosion–relevant vegetation cover factor and suggested that the cosimulation algorithm was better than cokriging, a linear and nonlinear regression in terms of map accuracy and uncertainty.

The objectives of this study are to demonstrate a sequential Gaussian cosimulation algorithm and compare its results with traditional regression modeling in terms of map accuracy and uncertainty analysis of crop residue percentage cover.

19.2 Materials and Methods

19.2.1 Study Area and Datasets

Study sites were located at three farms (sites 2, 5, and 6) near Tifton, Georgia (31°26′ N, 83°35′ W) (Figure 19.1). Tillage regimes comprised one conventional tillage site (site 2) and two strip tillage sites (sites 5 and 6). All sites consisted of a cotton–peanut rotation with 1 to 2 years duration for each crop. In 2006, all the sites were planted with cotton in late April. For strip tillage sites, winter cover crops were planted in December 2005 (site 5—wheat [*Triticum aestivum* L.], site 6—rye [*Secale cereale*]) and were killed approximately 3 weeks prior to spring planting.

Ground data for crop residue cover were collected on May 9–11, 2006, proximate to remotely sensed image acquisition. Each site was grid sampled (0.20 ha) for digital images and crop residue estimation (line transect). Digital photographs were acquired at nadir from the center of each grid point, using a 5 megapixel Olympus C-505 Zoom (London, UK). The images were taken from a height of 1 m with a spatial resolution of 1.4 m². Percentages cover (residue, vegetation, and soil) were obtained via a supervised classification having 5–15 classes, using ERDAS Imagine 8.4 (Leica Geosystems, Heerbrugg, Switzerland). Residue cover percentage was calculated by dividing the pixels that were classified as residue by the total pixel count within each image (Sullivan et al. 2006). A 30 m line transect marked in 10 cm intervals was also used to estimate crop residue cover. A tick mark was counted each time a piece of residue touched the outside, left edge of the tape (Shelton et al. 1993; Eck et al. 2001).

QuickBird images were acquired on May 10, 2006, from DigitalGlobe, Inc. The images consisted of four multispectral bands (band 1—blue: 0.45–0.52 μm; band 2—green: 0.52–0.6 μm; band 3—red: 0.63–0.69 μm; and band 4—near-infrared: 0.76–0.90 μm) and one panchromatic band (0.45–0.90 μm). The spatial resolutions varied from 0.61 to 0.72 m for the panchromatic image and 2.44 to 2.88 m for multispectral images. The panchromatic image was used to sharpen the multispectral images to a spatial resolution of 0.6 m. Moreover, geometric corrections and relative radiometric corrections based on $q_c = (q_r - a_{(\text{dark-offset})}) / b_{\text{relative-gain}}$ were made, where q_c is the corrected digital number (DN) of

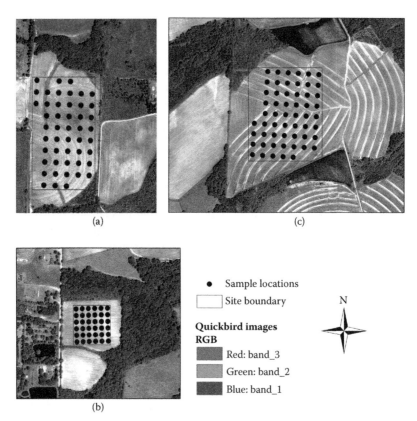

FIGURE 19.1
(**See color insert.**) (a) Site 2, (b) site 5, and (c) site 6 with sample locations shown on the QuickBird images.

a pixel, q_r the raw DN, $a_{dark-offset}$ the dark offset for a given image acquisition, and $b_{relative-gain}$ the relative gain. In addition, the QuickBird images were aggregated to spatial resolutions of 1.2 m × 1.2 m and 30 m × 30 m for approximating two types of plot support sizes mentioned earlier and used to collect data of crop residue cover. Figure 19.1 shows composites of QuickBird images (red, green, and blue) for the three sites and the corresponding sampled locations. The QuickBird images present the details of spatial variability of the site surfaces, for example, the rugged fields ploughed were obviously noted.

The sample locations and values of crop residue cover at spatial resolutions of 1.4 m² squares and 30 m transect lines for sites 2, 5, and 6 are shown in Figure 19.2. At site 2, crop residue cover had higher values to the southeastern parts and lower values to the west and northeast. At site 5, crop residue cover was higher to the west and center than anywhere else at this site. At site 6, crop residue cover had higher values along the eastern and northeastern boundaries compared to the southern region of the field. For all three sites, the spatial distribution of crop residue cover values at a plot size of 1.4 m² squares is similar to that at a plot size of 30 m transect lines.

19.2.2 Image Transformation

Some transformations of QuickBird images, including normalized difference vegetation index (NDVI), 10 band ratios, and principal component analysis, were carried out in

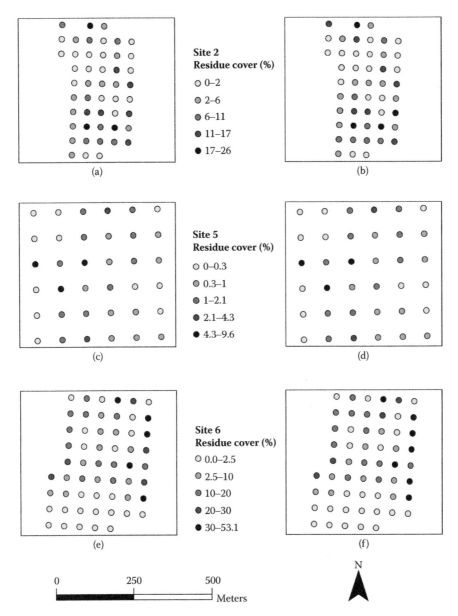

FIGURE 19.2
Sample data of crop residue cover for three sites: (a) site 2 for a plot size of 1.4 m², (b) site 2 for a plot size of 30 m transect line, (c) site 5 for a plot size of 1.4 m², (d) site 5 for a plot size of 30 m transect line, (e) site 6 for a plot size of 1.4 m², and (f) site 6 for a plot size of 30 m transect line.

this study. NDVI was defined using near-infrared and red channels. The 10 band ratios include QB1/QB2, QB1/QB3, QB1/QB4, QB2/QB3, QB2/QB4, QB3/QB4, (QB1 + QB2)/QB3, (QB1 + QB2)/QB4, (QB1 + QB3)/QB4, and (QB2 + QB3)/QB4. The coefficients of correlation between the crop residue cover and the original and transformed images were then calculated. The best images that had the highest correlation with crop residue percentage cover were used for mapping.

19.2.3 Sequential Gaussian Cosimulation for Mapping

An image-based sequential Gaussian cosimulation algorithm was used to map crop residue percentage cover (Goovaerts 1997; Wang et al. 2007). This algorithm is based on a random process model. A random process is considered as a collection of random variables that are distributed in a two-dimensional space and spatially autocorrelated. This autocorrelation is also called the spatial variability of a random process and can be quantified using a variogram (Dungan 2002). In this study, crop residue cover Z was regarded as a random process; $Z(u)$ and $Z(u+h)$ were random variables defined at locations u and $u+h$, respectively, in two-dimensional space; and h was a separation vector called lag. Let $z(u_\alpha)$ and $z(u_\alpha + h)$ be the sample data values of the crop residue cover at locations u_α and $u_\alpha + h$, respectively; the variogram $\gamma_{zz}(h)$ was estimated using sample data:

$$\hat{\gamma}_{zz}(h) = \frac{1}{2N(h)} \sum_{\alpha=1}^{N(h)} \left(z(u_\alpha) - z(u_\alpha + h)\right)^2 \tag{19.1}$$

where $N(h)$ is the number of pairs for data locations. The obtained variogram is called the sample variogram. Furthermore, a covariance function $C_{zz}(h)$ was estimated:

$$\hat{C}_{zz}(h) = \frac{1}{N(h)} \sum_{\alpha=1}^{N(h)} z(u_\alpha)z(u_\alpha + h) - m_{-h}m_{+h} \tag{19.2}$$

$$\text{with } m_{-h} = \frac{1}{N(h)} \sum_{\alpha=1}^{N(h)} z(u_\alpha), \qquad m_{+h} = \frac{1}{N(h)} \sum_{\alpha=1}^{N(h)} z(u_\alpha + h)$$

If the crop residue cover is not only intrinsically stationary but also second-order stationary, that is, it has both a constant mean and a constant variance, then its covariance function exists and has the following relationship with the sample variogram (Goovaerts 1997):

$$\hat{C}_{zz}(h) = \hat{C}_{zz}(0) - \hat{\gamma}_{zz}(h) \tag{19.3}$$

where $\hat{C}_{zz}(0)$ is the traditional variance. In addition, theoretically, a spatial cross-correlation between the crop residue Z and a spectral variable Y can be estimated using collocated sample and image data:

$$\hat{\gamma}_{zy}(h) = \frac{1}{2N(h)} \sum_{\alpha=1}^{N(h)} \left(z(u_\alpha) - z(u_\alpha + h)\right)\left(y(u_\alpha) - y(u_\alpha + h)\right) \tag{19.4}$$

where $y(u_\alpha)$ and $y(u_\alpha + h)$ are the values of the spectral variable at spatial locations u_α and $u_\alpha + h$, respectively. In this study, however, the cross-variogram was approximated using the following Markov model (Goovaerts 1997):

$$\hat{\gamma}_{zy}(h) = \frac{C_{zy}(0)}{C_{zz}(0)} \hat{\gamma}_{zz}(h) \tag{19.5}$$

where $C_{yz}(0)$ is the covariance between the crop residue cover and the spectral variable. Moreover, the sample variogram was fit using a permissible spherical function. The aforementioned variogram and cross-variogram were used to derive the statistical parameters of the conditional distributions in the sequential Gaussian cosimulation algorithm.

In this study, the value of a random process $Z(u)$ at a location u was considered as a realization of the random process. The realization was obtained by randomly drawing a value from its conditional cumulative distribution function (CCDF). The CCDF for each location was determined using a simple collocated cokriging estimator (Goovaerts 1997):

$$z^{\text{sck}}(u) = \sum_{\alpha=1}^{n(u)} \lambda_\alpha^{\text{sck}}(u)\left[z(u_\alpha) - m_z\right] + \lambda_y^{\text{sck}}(u)\left[y(u) - m_y\right] + m_z \tag{19.6}$$

$$\sigma^{2(\text{sck})}(u) = C_{zz}(0) - \sum_{\alpha=1}^{n(u)} \lambda_\alpha^{\text{sck}}(u)C_{zz}(u_\alpha - u) - \lambda_y^{\text{sck}}(u)C_{zy}(0) \tag{19.7}$$

where $n(u)$ is the number of the used neighboring sampled data and previously simulated values, if any, for the estimation of the conditional distribution. This number varied from location to location. It is noted that m_z and m_y are the means of sample and image data. $\lambda_\alpha^{\text{sck}}(u)$ are the weights for the sample data and previously simulated values, and $\lambda_y^{\text{sck}}(u)$ is the weight for the image datum.

In this study, Equations 19.6 and 19.7, respectively, provided a conditional expectation and a conditional variance used to estimate the mean and variance parameters of the CCDF at an unknown location. The data used in cokriging were searched for within a given neighborhood and included the sample data and previously simulated values, if any, and collocated image data. From the CCDF, a value was randomly drawn and assigned to the unknown location. The value was then used as a realization of the random process at this location and, at the same time, it became a conditional datum for the next estimation. We continued the simulation for the next location, and the order of visiting each location was determined by setting up a random path. We visited each of the locations until all the locations or cells had obtained estimates, which led to a realization of the expected map for the study area. By repeating the simulation 400 times using different random paths, 400 realizations of the expected map were obtained. This meant that 400 values were generated for each location and from them a sample mean and a sample variance were thus calculated. The sample mean was regarded as the predicted value of this random process at this location and the sample variance meant its uncertainty.

Moreover, theoretically, the cosimulation assumes that variables are multivariate Gaussian distributed. In this study, however, the variables showed asymmetric distributions. Thus, a normal score transformation of both the sample data and the images was conducted. To determinate an appropriate number of replications, the relationship of variance of the predicted values across the whole study area with the number of simulations was explored and it was found that the variance decreased rapidly at the beginning and then slowly and eventually got stable as the number of runs increased. When the number of runs reached 400, the variance stabilized.

In this study, the neighborhood within which the sample data of crop residue percentage cover, remotely sensed data, and previously simulated values if any were searched for was defined based on the range parameter of the fitted variograms. It was found that the neighborhood varied depending on different sites and spatial resolutions of crop residue cover sample data and images. When the spatial resolution was 1.4 m, the radii of the used neighborhood were 131.6, 40.3, and 165.6 m for sites 2, 5, and 6, respectively. When the spatial resolution was 30 m, the radii of the neighborhood were 134.4, 21.9, and 149.2 m for sites 2, 5, and 6, respectively. Moreover, to avoid the case in which there were no data or too many data to be used, a minimum number and a maximum number of the data, 2 and 7, respectively, were employed for both sample data and previously simulated values.

In addition to the aforementioned cosimulation, in this study regression modeling was applied to map crop residue percentage cover. The same images were used in both cosimulation and regression modeling. Thus, both methods were compared to verify the advantages of the cosimulation in terms of map accuracy and uncertainty analysis. The comparison was conducted by using cross validation and calculating root-mean-square errors (RMSEs) between the estimated and observed values of crop residue cover. The cross validation meant that one sample plot was drawn and used as a reference and the rest of the sample plots were applied to develop the cosimulation and regression model that predicted the value of the reference location. This process was repeated until all the sample plots were estimated.

19.3 Results

Experimental variograms of crop residue cover were calculated using sample data and fit using the spherical model for each site at two spatial resolutions. At a spatial resolution of 1.4 m², the obtained variograms included the following:

Site 2:

$$\hat{\gamma}_{Res.Cov.}(|h|) = 0.63 + 0.37\left[1.5\left(\frac{|h|}{131.6}\right) - 0.5\left(\frac{|h|}{131.6}\right)^3\right] \quad 0<|h|\leq 131.6 \text{ m} \qquad (19.8)$$

Site 5:

$$\hat{\gamma}_{Res.Cov.}(|h|) = 0.47 + 0.53\left[1.5\left(\frac{|h|}{40.3}\right) - 0.5\left(\frac{|h|}{40.3}\right)^3\right] \quad 0<|h|\leq 40.3 \text{ m} \qquad (19.9)$$

Site 6:

$$\hat{\gamma}_{Res.Cov.}(|h|) = 0.61 + 0.39\left[1.5\left(\frac{|h|}{165.6}\right) - 0.5\left(\frac{|h|}{165.6}\right)^3\right] \quad 0<|h|\leq 165.6 \text{ m} \qquad (19.10)$$

At a spatial resolution of 30 transect lines, the variograms included the following:

Site 2:

$$\hat{\gamma}_{Res.Cov.}(|h|) = 0.65 + 0.35\left[1.5\left(\frac{|h|}{134.4}\right) - 0.5\left(\frac{|h|}{134.4}\right)^3\right] \quad 0<|h|\leq 134.4 \text{ m} \qquad (19.11)$$

Site 5:

$$\hat{\gamma}_{\text{Res.Cov.}}(|h|) = 0.4 + 0.6\left[1.5\left(\frac{|h|}{21.944}\right) - 0.5\left(\frac{|h|}{21.944}\right)^3\right] \quad 0 < |h| \le 21.944 \text{ m} \quad (19.12)$$

Site 6:

$$\hat{\gamma}_{\text{Res.Cov.}}(|h|) = 0.5 + 0.5\left[1.5\left(\frac{|h|}{149.2}\right) - 0.5\left(\frac{|h|}{149.2}\right)^3\right] \quad 0 < |h| \le 149.2 \text{ m} \quad (19.13)$$

The variograms were standardized by forcing the sill parameters to be equal to 1 unit for their use in the sequential Gaussian cosimulation. Moreover, the original coordinates were transformed to smaller values just for simplification. A common feature was that all the variograms had relatively large nugget parameters, and the reason for this might be measurement errors and the lack of sample data at short sampling distances.

The obtained coefficients of correlation between the crop residue cover and the original and transformed images are listed in Tables 19.1 through 19.3 for the three sites. For site 2, the correlation varied from −0.0029 to 0.2962 at a spatial resolution of 1.2 m and from −0.4166 to 0.4076 at a spatial resolution of 30 m. The images that had the highest correlation with crop residue were principal component 1 (PC1) for a spatial resolution of 1.2 m and original band 4 for a spatial resolution of 30 m. For site 5, the correlation varied from −0.1229 to 0.2087 at a spatial resolution of 1.2 m and from −0.0759 to 0.2432 at a spatial resolution of 30 m. The images that had the highest correlation with crop residue were ratio image

TABLE 19.1

Correlation of QuickBird Images with the Crop Residue Cover for Site 2

Spatial Resolution: 1.2 m × 1.2 m

Image	QB1	QB2	QB3	QB4	NDVI
Correlation	0.2922	0.2888	0.2873	0.1740	−0.2293
Image	QB1/QB2	QB1/QB3	QB1/QB4	QB2/QB3	QB2/QB4
Correlation	0.0461	0.0276	0.2480	−0.0029	0.2683
Image	QB3/QB4	(QB1 + QB2)/QB3	(QB1 + QB2)/QB4	(QB1 + QB3)/QB4	(QB2 + QB3)/QB4
Correlation	0.2436	0.0152	0.2625	0.2612	0.2608
Image	PC1	PC2	PC3	PC4	
Correlation	**0.2962**	0.1677	−0.0561	0.0923	

Spatial Resolution: 30 m × 30 m

Image	QB1	QB2	QB3	QB4	NDVI
Correlation	0.0186	−0.0246	−0.0567	**−0.4166**	−0.0350
Image	QB1/QB2	QB1/QB3	QB1/QB4	QB2/QB3	QB2/QB4
Correlation	0.2226	0.2381	0.0968	0.2084	0.0670
Image	QB3/QB4	(QB1 + QB2)/QB3	(QB1 + QB2)/QB4	(QB1 + QB3)/QB4	(QB2 + QB3)/QB4
Correlation	0.0388	0.2345	0.0820	0.0667	0.0524
Image	PC1	PC2	PC3	PC4	
Correlation	−0.0269	0.3835	0.4076	−0.0053	

Note: Significant value at the significant level of 0.05 was 0.273.

TABLE 19.2

Correlation of QuickBird Images with the Crop Residue Cover for Site 5

Spatial Resolution: 1.2 m × 1.2 m

Image	QB1	QB2	QB3	QB4	NDVI
Correlation	−0.0842	−0.1123	−0.1063	−0.1229	0.0236
Image	QB1/QB2	QB1/QB3	QB1/QB4	QB2/QB3	QB2/QB4
Correlation	**0.2087**	0.1001	0.0645	−0.0755	−0.0605
Image	QB3/QB4	(QB1 + QB2)/QB3	(QB1 + QB2)/QB4	(QB1 + QB3)/QB4	(QB2 + QB3)/QB4
Correlation	−0.0272	0.0287	−0.0081	0.0095	−0.0440
Image	PC1	PC2	PC3	PC4	
Correlation	−0.1029	−0.1224	−0.1150	0.2013	

Spatial Resolution: 30 m × 30 m

Image	QB1	QB2	QB3	QB4	NDVI
Correlation	0.0870	0.0923	0.0894	0.2403	−0.0759
Image	QB1/QB2	QB1/QB3	QB1/QB4	QB2/QB3	QB2/QB4
Correlation	0.0614	0.0538	0.0746	−0.0432	0.0773
Image	QB3/QB4	(QB1 + QB2)/QB3	(QB1 + QB2)/QB4	(QB1 + QB3)/QB4	(QB2 + QB3)/QB4
Correlation	0.0756	0.0159	0.0759	0.0752	0.0764
Image	PC1	PC2	PC3	PC4	
Correlation	0.0887	**0.2432**	0.1071	0.1993	

Note: Significant value at the significant level of 0.05 was 0.325.

TABLE 19.3

Correlation of QuickBird Images with the Crop Residue Cover for Site 6

Spatial Resolution: 1.2 m × 1.2 m

Image	QB1	QB2	QB3	QB4	NDVI
Correlation	−0.3584	−0.3739	−0.3387	−0.3370	0.3105
Image	QB1/QB2	QB1/QB3	QB1/QB4	QB2/QB3	QB2/QB4
Correlation	0.2243	−0.1339	−0.3188	**−0.4475**	−0.3852
Image	QB3/QB4	(QB1 + QB2)/QB3	(QB1 + QB2)/QB4	(QB1 + QB3)/QB4	(QB2 + QB3)/QB4
Correlation	−0.3157	−0.2929	−0.3607	−0.3289	−0.3558
Image	PC1	PC2	PC3	PC4	
Correlation	−0.3555	0.2828	0.2724	0.4352	

Spatial Resolution: 30 m × 30 m

Image	QB1	QB2	QB3	QB4	NDVI
Correlation	−0.2340	−0.2575	−0.2049	−0.3877	−0.0469
Image	QB1/QB2	QB1/QB3	QB1/QB4	QB2/QB3	QB2/QB4
Correlation	0.3881	**−0.4374**	−0.0115	−0.4185	−0.0628
Image	QB3/QB4	(QB1 + QB2)/QB3	(QB1 + QB2)/QB4	(QB1 + QB3)/QB4	(QB2 + QB3)/QB4
Correlation	0.0465	−0.4256	−0.0386	0.0208	−0.0050
Image	PC1	PC2	PC3	PC4	
Correlation	−0.2755	−0.3468	0.3705	0.0309	

Note: Significant value at the significant level of 0.05 was 0.261.

band 1/band 2 for a spatial resolution of 1.2 m and principal component 2 (PC2) for a spatial resolution of 30 m. For site 6, the correlation varied from −0.4475 to 0.4352 at a spatial resolution of 1.2 m and from −0.4374 to 0.3881 at a spatial resolution of 30 m. The images that had the highest correlation with crop residue were ratio image band 2/band 3 for a spatial resolution of 1.2 m and ratio image band 1/band 3 for a spatial resolution of 30 m.

Overall, the correlation of images with crop residue cover was low but better in sites 2 and 6 than in site 5. Compared to the original images, most of the transformed images did not significantly improve the correlation with the crop residue cover. However, some band ratios and principal components did increase the coefficients of the correlation. In addition, aggregating the image data from a spatial resolution of 1.2 m to one of 30 m increased the correlation in site 2 but not in sites 5 and 6. It was also found that the highest correlation for each of two spatial resolutions and each of the sites 2 and 6 was statistically significant based on the equation $r_\alpha = \sqrt{t_\alpha^2 / (n - 2 + t_\alpha^2)}$ from the Student's distribution at the level $\alpha = 5\%$, where n is the number of sample plots used. The images that had the highest correlations were thus used in the sequential Gaussian cosimulation to map the crop residue cover.

Using the sequential Gaussian cosimulation and the images that had the highest correlation with the crop residue cover, that is, PC1 and original band 4 for spatial resolutions of 1.2 and 30 m, respectively, we mapped the crop residue cover for site 2. Figure 19.3 shows

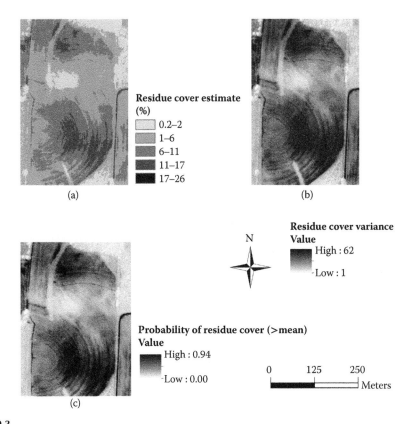

FIGURE 19.3
The results of site 2 using the sequential Gaussian cosimulation and the first principal component obtained from sharpened QuickBird multispectral images at a spatial resolution of 1.2 m × 1.2 m: (a) residue cover estimate, (b) residue cover variance, and (c) probability for residue cover larger than the sample mean.

the maps of crop residue cover estimate, variance, and probability for estimates larger than sample mean at a spatial resolution of 1.2 m (i.e., 1.4 m²) for site 2. The maps had similar spatial distributions of values. Thus, the estimates of crop residue cover, their variances, and probabilities for the estimates larger than the sample mean were consistently greater at the southeastern and northern parts of site 2 and smaller at the northeastern, western, and up-central parts. The spatial patterns were also similar to those of crop residue cover values from the sample plots in Figure 19.2a.

The variance map (Figure 19.3b) indicated the measure of spatial uncertainties of the crop residue cover estimates. Comparison of Figure 19.2a with Figure 19.3a and b showed that the greater the variation of sample plot data, the greater the uncertainty of the estimate and that the variances varied depending on both the sample data values and their spatial configurations. Figure 19.3c presents the probability of crop residue cover estimates larger than the sample mean. In the areas where the estimates were larger, the probabilities were also greater. This meant in those areas the soils were more likely protected than the other areas if the sample mean was a critical value for soil and water conservation. Both variance and probability maps can be used to help managers in decision making.

In Figure 19.4, the crop residue cover of site 2 was mapped at a spatial resolution of 30 m. When the spatial resolution decreased from 1.2 m × 1.2 m to 30 m × 30 m, the estimate, variance, and probability maps showed similar spatial distributions to those at a spatial

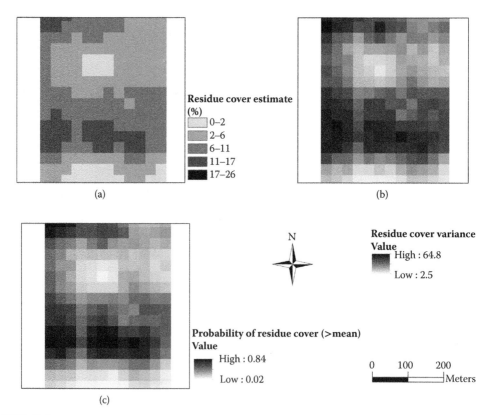

FIGURE 19.4
The results of site 2 using the sequential Gaussian cosimulation and QuickBird band 4 image at a spatial resolution of 30 m × 30 m: (a) residue cover estimate, (b) residue cover variance, and (c) probability for residue cover larger than the sample mean.

resolution of 1.2 m. The map of crop residue cover estimates at the spatial resolution of 30 m had a smaller RMSE between the estimated and measured values of crop residue cover than that at the spatial resolution of 1.2 m, but the details of spatial variability disappeared there.

The descriptions of the aforementioned maps for site 2 can be applied to the maps obtained using the cosimulation algorithm and the best images for both sites 5 (Figures 19.5 and 19.6) and 6 (Figures 19.7 and 19.8). For example, at site 6 the estimates of crop residue cover, their variances, and probabilities for the estimates larger than the sample mean had higher values along the eastern and northeastern boundaries compared to the southern and southwest regions of the field (Figure 19.7). The spatial patterns of estimates were consistent with those from the sample plots in Figure 19.2c. When the crop residue cover was mapped at a spatial resolution of 30 m (Figure 19.8), similar spatial distributions of estimates to those at the spatial resolution of 1.2 m were obtained. Overall, the map of crop residue cover estimates at the spatial resolution of 30 m for site 6 had a smaller RMSE than that at the spatial resolution of 1.2 m. But the original larger values along the northeastern boundaries became smaller and the reason was that the coarse spatial resolution smoothed the estimates of the crop residue cover. In addition, the spatial patterns lost the details of spatial variability.

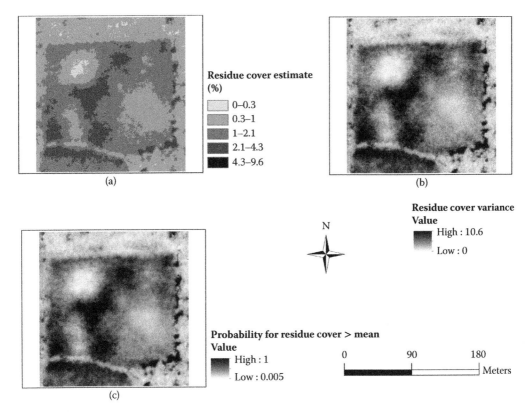

FIGURE 19.5
The results of site 5 using the sequential Gaussian cosimulation and the QuickBird ratio image band 1/band 2 at a spatial resolution of 1.2 m × 1.2 m: (a) residue cover estimate, (b) residue cover variance, and (c) probability for residue cover larger than the sample mean.

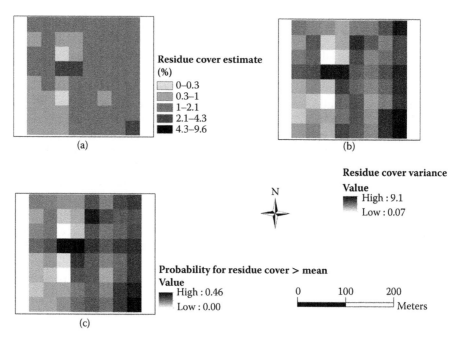

FIGURE 19.6
The results of site 5 using the sequential Gaussian cosimulation and the QuickBird ratio image (band 1 + band 2)/band 3 at a spatial resolution of 30 m × 30 m: (a) residue cover estimate, (b) residue cover variance, and (c) probability for residue cover larger than the sample mean.

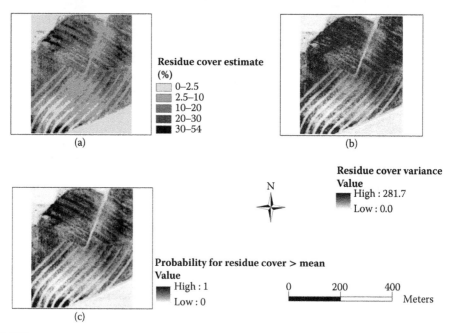

FIGURE 19.7
The results of site 6 using the sequential Gaussian cosimulation and the QuickBird ratio image band 2/band 3 at a spatial resolution of 1.2 m × 1.2 m: (a) residue cover estimate, (b) residue cover variance, and (c) probability for residue cover larger than the sample mean.

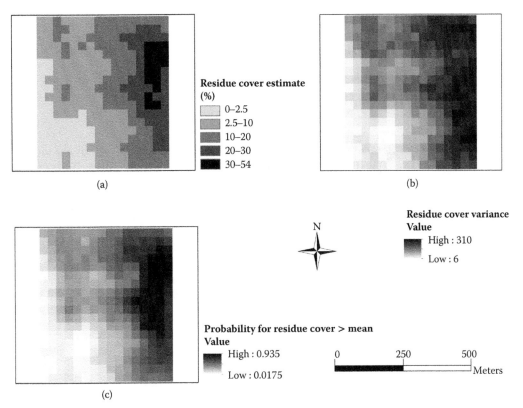

FIGURE 19.8
The results of site 6 using the sequential Gaussian cosimulation and the QuickBird ratio image band 1/band 3 at a spatial resolution of 30 m × 30 m: (a) residue cover estimate, (b) residue cover variance, and (c) probability for residue cover larger than the sample mean.

Moreover, we obtained regression models of the crop residue cover (denoted as RC in equations) with the same images used in the aforementioned cosimulation for three sites. The models included the following:

Site 2 at the spatial resolution of 1.2 m is as follows:

$$RC = -5412.981 + 115.7129 PC_1 - 0.9185 PC_1^2 + 0.003209 PC_1^3 - 0.000004163 PC_1^4 \quad (19.14)$$

Site 2 at the spatial resolution of 30 m is as follows:

$$RC = 102442400 - 2572206 B_4 + 24219.3 B_4^2 - 101.352 B_4^3 + 0.15905 B_4^4 \quad (19.15)$$

Site 5 at the spatial resolution of 1.2 m is as follows:

$$RC = 410782.4 - 1801051 RB_{12} + 2954925 RB_{12}^2 - 2149769 RB_{12}^3 + 585059.9 RB_{12}^4 \quad (19.16)$$

Site 5 at the spatial resolution of 30 m is as follows:

$$RC = -2 \times 10^{13} + 5 \times 10^{11} PC_2 - 4 \times 10^9 PC_2^2 + 2 \times 10^7 PC_2^3 - 22939 PC_2^4 \quad (19.17)$$

Site 6 at the spatial resolution of 1.2 m is as follows:

$$RC = -2479287 + 11280110B_4 - 19235830B_4^2 + 14571980B_4^3 - 4137724B_4^4 \qquad (19.18)$$

Site 6 at the spatial resolution of 30 m is as follows:

$$RC = 6416.637 - 8032.022RB_{13} \qquad (19.19)$$

The coefficients of determination for these regression models were relatively low and varied from 0.165 to 0.3056. Figure 19.9 presents the estimate maps of crop residue cover at the spatial resolution of 1.2 m for sites 2, 5, and 6. For sites 2 and 6, the spatial distributions of the estimates from the regression modeling (Figure 19.9a and c) were similar to those derived using the sequential Gaussian cosimulation (Figures 19.3a and 19.7a). However, the estimates had smaller ranges of variation. For site 5, compared with the estimate map from the cosimulation (Figure 19.5a), the estimate map from the regression method (Figure 19.9b) had greater values at the northern parts. Overall, the cosimulation better captured the spatial feature of crop residue cover than the regression modeling. Thus, if the corresponding site estimate maps in Figures 19.3a and 19.9a are compared with that in

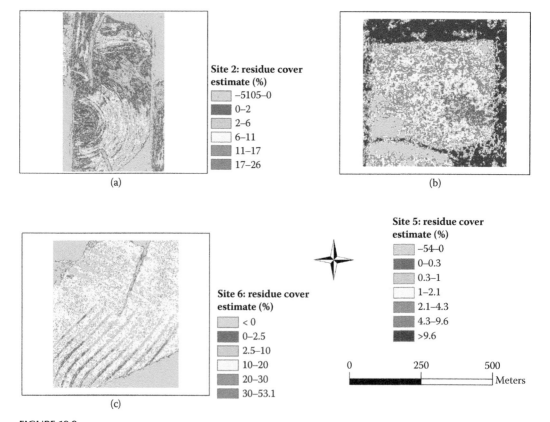

(a)

(b)

(c)

Site 2: residue cover estimate (%)
- −5105–0
- 0–2
- 2–6
- 6–11
- 11–17
- 17–26

Site 6: residue cover estimate (%)
- < 0
- 0–2.5
- 2.5–10
- 10–20
- 20–30
- 30–53.1

Site 5: residue cover estimate (%)
- −54–0
- 0–0.3
- 0.3–1
- 1–2.1
- 2.1–4.3
- 4.3–9.6
- >9.6

0   250   500
Meters

FIGURE 19.9
(See color insert.) The results of sites 2, 5, and 6 using regression modeling and QuickBird images that have the highest correlation with residue cover at a spatial resolution of 1.2 m × 1.2 m: (a) site 2: residue cover estimate; (b) site 5: residue cover estimate; and (c) site 6: residue cover estimate.

TABLE 19.4

Comparison of Estimate Maps between Sequential Gaussian Cosimulation and Regression Modeling for the Same Site and Same Spatial Resolution in Terms of RMSEs between the Estimated and Observed Values of Crop Residue Cover

	Spatial Resolution of 1.2 m × 1.2 m	Spatial Resolution of 30 m × 30 m
Site 2		
Cosimulation	6.8971	5.2305
Regression	5.7233	5.9697
Site 5		
Cosimulation	2.1749	2.3479
Regression	2.1088	2.4078
Site 6		
Cosimulation	12.4939	10.7274
Regression	10.6153	12.3696

Figure 19.2a, the spatial patterns of the estimates from the cosimulation better match the spatial patterns of the sample data than that of the regression modeling.

In Table 19.4, the RMSE values from the sequential Gaussian cosimulation and regression modeling were compared for three sites. For both sites 2 and 6, the regression modeling resulted in smaller RMSEs than the cosimulation at the spatial resolution of 1.2 m, whereas at the spatial resolution of 30 m the RMSEs from the regression modeling were greater than those from the cosimulation. For site 5, both methods had similar RMSEs. However, a significant disadvantage for the regression modeling was that this method led to some negative and illogical values of crop residue cover for all three sites 2, 5, and 6. Extremely large values took place along the boundaries of site 5, and this shortcoming was especially obvious outside the sampled area. More important is that the regression modeling lacks the ability to provide the measure of uncertainty of the estimate at each location, including variances of local estimates and their probabilities for the estimates larger or lesser than a given value, which is useful for soil and water conservation.

19.4 Conclusions and Discussion

In this study, we used high spatial resolution QuickBird images and two mapping methods including regression modeling and sequential Gaussian cosimulation to map the crop residue cover for three farm sites. The correlation of remotely sensed images with crop residue cover was found to be very low mainly because of too low crop residue cover at these sites. Compared with the original images, the principal components and band ratios slightly improved the correlation with crop residue cover.

When the QuickBird images were aggregated from a spatial resolution of 0.6 m to 1.2 m and 30 m, respectively, corresponding to the spatial resolutions of sample plot data, the estimate maps obtained using the cosimulation were slightly more accurate at the spatial resolution of 30 m than those at the spatial resolution of 1.2 m for both sites 2 and 6. However, the improvement of map accuracy was not obvious. Moreover, aggregating the images to the spatial resolution of 30 m led to the disappearance of the details of spatial variability for the crop residue cover. On the other hand, the estimate maps of the crop residue cover at the finer spatial resolution presented the details of spatial variability of estimates and their uncertainties.

When the same QuickBird images were used for mapping the crop residue cover, the regression modeling resulted in smaller RMSEs than the cosimulation at the spatial resolution of 1.2 m, whereas the cosimulation led to more accurate estimates than the regression modeling at the spatial resolution of 30 m. Overall, the spatial patterns of crop residue cover estimates from the cosimulation looked more reasonable than those from the regression modeling compared with the spatial distributions of the sample plot data. Moreover, the regression modeling generated illogical estimates of crop residue cover. This shortcoming did not exist for the cosimulation. In addition, the cosimulation was able to reproduce the spatial uncertainty of crop residue cover by providing variance and probability maps that could be used to perform uncertainty analysis and risk assessment of land management. These are the advantages that the regression modeling cannot provide.

Overall, the combination of high-resolution QuickBird images and sequential Gaussian cosimulation provides an alternative to improve the maps of crop residue cover. The maps show detailed spatial distribution of estimates. If these map estimates are input into the revised universal soil loss equation (RUSLE) (Renard et al. 1997), they will provide the potential to obtain the spatial distribution of soil erosion and thus make it possible for decision makers to work out more detailed plans of land management and conservation than traditional methods. Moreover, the cosimulation method generates the variance and probability maps of the estimates for crop residue cover, and if the uncertainty maps are inputted into the RUSLE model, the risk assessment of decision making for land management plans can be carried out. Thus, it provides the potential to improve RUSLE model–based decision making.

19.5 Summary

In the United States, about 30% of agricultural lands have been classified as highly erodible, contributing to the continuous degradation of soil productivity. Conservation tillage is a well-known best management practice to conserve soil water resources. Accurate assessments of conservation tillage adoption are critical. Taking advantage of the intensive data that satellite imageries can provide, developing a satellite-derived mapping algorithm poses a great benefit to rapidly quantify crop residue cover at field and landscape scales. For this purpose, we combined ground sample data of crop residue cover with high-resolution QuickBird images and compared an image-based sequential Gaussian cosimulation with a traditional regression approach to map crop residue cover for three sites in the Little River Experimental Watershed in Georgia. The results showed that although the aggregation of the QuickBird images to a spatial resolution of 30 m slightly increased the map accuracy of crop residue cover, the details of spatial variability disappeared. Moreover, there were no significant differences between the cosimulation and the regression modeling in terms of RMSE from the crop residue cover maps created by both methods. However, the regression method produced illogical estimates outside the range of ground sample data and the cosimulation was able to quantify spatial prediction uncertainty for uncertainty analysis and risk assessment of land management. These are the advantages of the sequential cosimulation over the traditional regression method.

Acknowledgments

We are grateful to the National Aeronautics and Space Administration for providing funding support for this study; the U.S. Department of Agriculture–Agricultural Research Service Southeast Watershed Research Laboratory and Flint River Water Planning and Policy Center, Albany, Georgia, for coordination and assistance in ground data sampling and collection; and the Institute for Technology Development for other support and assistance.

References

Bannari, A., A. Pacheco, and K. S. Staenz. 2006. "Estimating and Mapping Crop Residues Cover on Agricultural Lands Using Hyperspectral and IKONOS Data." *Remote Sensing of Environment* 104:447–59.

Bannari, A., K. Staenz, and K. S. Khurshid. 2007. "Remote Sensing of Crop Residue Using Hyperion (EO-1) Data." *IEEE International Symposium on Geoscience and Remote Sensing (IGARSS)* 2007:2795–99.

Barata, M. T., M. C. Nunes, A. J. Sousa, F. H. Muge, and M. T. Albuquerque. 1996. "Geostatistical Estimation of Forest Cover Areas Using Remote Sensing Data." In *Geostatistics Wollongong'96*, 2, edited by E. Y. Baafi and N. A. Schofield, 1244–57. Dordrecht, The Netherlands: Kluwer Academic Publishers.

Chevrier, M., A. Bannari, J. C. Deguise, H. McNairn, and K. Staenz. 2002. "Hyperspectral Narrow-Wavebands for Discriminating Crop Residue from Bare Soil." *IEEE International Geoscience and Remote Sensing Symposium (IGARSS 2002)* 4:2202–04.

Daughtry, C. S. T., E. R. Hunt Jr., and J. E. McMurtrey III. 2004. "Assessing Crop Residue Cover Using Shortwave Infrared Reflectance." *Remote Sensing of Environment* 90:126–34.

Daughtry, C. S. T., E. R. Hunt Jr., P. C. Doraiswamy, and J. E. McMurtrey III. 2005. "Remote Sensing the Spatial Distribution of Crop Residues." *American Society of Agronomy* 97:864–71.

Dungan, J. L. 2002. "Conditional Simulation: An Alternative to Estimation for Achieving Mapping Objective." In *Spatial Statistics for Remote Sensing*, edited by A. Stein, F. V. D. Meer, and B. Gorte, 135–52. Dordrecht, The Netherlands: Kluwer Academic Publishers.

Eck, K. J., D. E. Brown, and A. B. Brown. 2001. *Estimating Corn and Soybean Residue Cover*. Ext. Circ. AT–269. West Lafayette, IN: Purdue University.

Gomez-Hernandez, J. J., and A. G. Journel. 1992. "Joint Sequential Simulation of MultiGaussian Fields." In *Geostatistics Tróia 1992*, edited by A. Soars, Vol. 1, 85–94. Dordrecht, The Netherlands: Kluwer Academic Publishers.

Goovaerts, P. 1997. *Geostatistics for Natural Resources Evaluation*. New York: Oxford University Press.

Gowda, P. H., B. J. Dalzell, D. J. Mulla, and F. Kollman. 2001. "Mapping Tillage Practices with Landsat Thematic Mapper Based Logistic Regression Models." *Journal of Soil and Water Conservation* 56:91–97.

Hunner, G., H. T. Mowrer, and R. M. Reich. 2000. "An Accuracy Comparison of Six Spatial Interpolation Methods for Modeling Forest Stand Structure on the Fraser Experimental Forest." In *Proceedings of the 4th International Symposium on Spatial Accuracy Assessment in Natural*

Resources and Environmental Sciences, edited by G. B. M. Heuvelink, and M. J. P. M. Lemmens. Amsterdam, The Netherlands: Delft University Press.

Liu, J. G., J. R. Miller, E. Pattey, D. Haboudane, I. B. Strachan, and M. Hinther. 2004. "Monitoring Crop Biomass Accumulation Using Multi-Temporal Hyperspectral Remote Sensing Data." *IEEE International Symposium on Geoscience and Remote Sensing (IGARSS)* 2004:1637–40.

McMurtrey, J. E., C. S. T. Daughtry, and T. E. Devine. 2005. "Spectral Detection of Crop Residues for Soil Conservation from Conventional and Large Biomass Soybean." *Agronomy for Sustainable Development* 25:25–33.

Nagler, P. L., Y. Inoue, E. P. Glenn, A. L. Russ, and C. S. T. Daughtry. 2003. "Cellulose Absorption Index (CAI) to Quantify Mixed Soil–Plant Litter Scenes." *Remote Sensing of Environment* 87:310–25.

Pacheco, A., and H. McNairn. 2010. "Evaluating Multi Spectral Remote Sensing and Spectral Unmixing Analysis for Crop Residue Mapping." *Remote Sensing of Environment* 114:2219–28.

Pacheco, A., H. McNairn, and M. S. Anne. 2006. "Multispectral Indices and Advanced Classification Techniques to Detect Percent Residue Cover over Agricultural Crops Using Landsat Data." In *Proceeding of the Society of Photo-Optical Instrumentation Engineers (SPIE) - Conference on Remote Sensing and Modeling of Ecosystems for Sustainability III*, edited by W. Gao and S. L. Ustin, Vol. 6298, Article Number: 62981C, doi:10.1117/12.694675, Spie-International Society for Optical Engineering, Bellingham, WA.

Renard, K. G., C. R. Foster, G. A. Weesies, D. K. McCool, and D. C. Yoder. 1997. *Predicting Soil Erosion by Water: A Guide to Conservation Planning with the Revised Universal Soil Loss Equation (RUSLE)*. Agriculture Handbook Number 703, U.S. Government Printing Office, SSOP Washington DC: U.S. Department of Agriculture.

Shelton, D. P., R. Kanable, and P. L. Jasa. 1993. *Estimating Percent Residue Cover Using the Line-Transect Method*. Ext. Circ. G93–1133. Lincoln, NE: University of Nebraska.

South, S., J. G. Qi, and D. P. Lusch. 2004. "Optimal Classification Methods for Mapping Agricultural Tillage Practices." *Remote Sensing of Environment* 91:90–97.

Sullivan, D. G., C. C. Truman, T. C. Strickland, H. H. Schomberg, D. M. Endale, and D. H. Franklin. 2006. "Evaluating Techniques for Determining Tillage Regime in the Southeastern Coastal Plain and Piedmont." *Agronomy Journal* 98(5):1236–46.

Thoma, D. P., S. C. Gupta, and M. E. Bauer. 2004. "Evaluation of Optical Remote Sensing Models for Crop Residue Cover Assessment." *Journal of Soil and Water Conservation* 59:224–33.

Van Deventer, A. P., A. D. Ward, P. H. Gowda, and J. G. Lyonl. 1997. "Using Thematic Mapper Data to Identify Contrasting Soil Plains and Tillage Practices." *Photogrammetric Engineering and Remote Sensing* 63:87–93.

Wang, G., G. Z. Gertner, A. B. Anderson, H. Howard, D. Gebhart, D. Althoff, T. Davis, and P. Woodford. 2007. "Spatial Variability and Temporal Dynamics Analysis of Soil Erosion Due to Military Land Use Activities: Uncertainty and Implications for Land Management." *Land Degradation and Development* 18:519–42.

Wang, G., G. Z. Gertner, S. Fang, and A. B. Anderson. 2003. "Mapping Multiple Variables for Predicting Soil Loss by Joint Sequential Co-Simulation with TM Images and Slope Map." *Photogrammetric Engineering & Remote Sensing* 69:889–98.

Wang, G., S. Wente, G. Z. Gertner, and A. B. Anderson. 2002. "Improvement in Mapping Vegetation Cover Factor for Universal Soil Loss Equation by Geo-Statistical Methods with Landsat TM Images." *International Journal of Remote Sensing* 23:3649–67.

Xu, W., T. T. Tran, R. M. Srivastava, and A. G. Journel. 1992. "Integrating Seismic Data in Reservoir Modeling: The Collocated Cokriging Alternative." In *The 67th Annual Technical Conference and Exhibition of the Society of Petroleum Engineers*, October 4–7, 833–42. Washington, DC, Richardson, TX: Society of Petroleum Engineers Inc.

Yang, C. C., S. O. Prasher, P. Enright, C. Madramootoo, M. Burgess, P. K. Goel, and I. Callum. 2003. "Application of Decision Tree Technology for Image Classification Using Remote Sensing Data." *Agricultural Systems* 76:1101–17.

Section VI

Biomass and Carbon Cycle Modeling

20

Remote Sensing of Leaf Area Index of Vegetation Covers

Jing M. Chen

CONTENTS

20.1 Introduction...375
20.2 Ground-Based LAI Measurement Theory and Techniques 376
20.3 Principles of LAI Retrieval Using Remote Sensing Data.......................................379
    20.3.1 Leaf Optical Properties...380
    20.3.2 Influence of Canopy Architecture on Reflectance 380
    20.3.3 Influence of Vegetation Background on Reflectance 383
    20.3.4 Influence of Illumination and Observation Angles on Reflectance 384
20.4 Algorithms for Retrieving LAI Using Optical Remote Sensing Data.......................385
    20.4.1 LAI Algorithms Based on Vegetation Indices 385
    20.4.2 LAI Algorithms Based on Radiative Transfer Models 388
    20.4.3 Examples of Global LAI and Clumping Index Maps 390
20.5 Remaining Issues in LAI Retrieval 391
    20.5.1 Issue 1: Large Differences among Existing Global LAI Products..................392
    20.5.2 Commonly Distorted Seasonal Variation Patterns of LAI.......................392
20.6 Concluding Remarks...393
Acknowledgments 393
References..393

20.1 Introduction

Since leaf surface is a substrate on which major physical and biological processes of plants occur, leaf area index (LAI) is arguably the most important vegetation structural parameter and indispensible for all process-based models for estimating terrestrial fluxes of energy, water, carbon, and other masses. It is therefore of interest not only to the remote sensing community that produces LAI maps but also to ecological, meteorological, and hydrological communities that use LAI products for various modeling purposes (Sellers et al. 1997; Dai et al. 2003; Chen et al. 2005).

LAI is defined as one-half of the total (all-sided) leaf area per unit ground surface area (Chen and Black 1992; see also the review by Jonckheere et al. 2004). This definition is suitable for all convex forms of leaves, and for flat leaves, it is the same as that proposed by Ross (1981). From the radiation interception point of view, a concave leaf surface behaves the same as the flat surface covering the concave surface, and the concave area should be replaced by the flat area called "the intercepting area" in the LAI definition (Chen and Black 1992). By this argument, the definition of a curled broadleaf area for the radiation interception purpose is neither the flattened area nor the largest projected area in its *in situ* form, and it should be half the sum of all intercepting and convex areas on both sides. Depending on the concaveness of a leaf, this refined definition can make up to 50% difference in the "directly" measured leaf area in the laboratory.

Plant leaves intercept solar radiation and selectively absorb part of it for conversion of stored chemical energy by photosynthesis. The unabsorbed radiation is reflected by the leaf surface and transmitted through the leaves. Healthy plant leaves, therefore, have distinct reflectance and transmittance spectra relative to soil and other nonliving materials (Lillesand et al. 2008). Optical remote sensing makes use of the contrast between leaf and soil spectral characteristics for retrieving LAI of vegetation covers. However, vegetation stands have complex three-dimensional (3D) canopy architecture, such as tree crowns, branches, and shoots in forests, plantation rows in crops, and foliage clumps in shrubs. Remote sensing signals acquired from a vegetation cover are influenced not only by the amount of leaf area in the canopy but also by the canopy architecture. Seasonal variations of the vegetation background, such as moss/grass cover and snow cover on the forest floor, also greatly influence the total reflectance from a vegetated surface. It has, therefore, been a challenge for the remote sensing community to produce consistent and accurate LAI products using satellite measurements.

Since the launches of advanced satellite sensors such as MODerate resolution Imaging Spectroradiometer (MODIS) and VEGETATION in late 1990s, various remote sensing algorithms have been developed to retrieve LAI using optical remote sensing data (see Garrigues et al. 2008). Much progress has also been made in improving the techniques of indirect measurements of LAI on the ground and in developing LAI algorithms with consideration of the effects of canopy architecture and its background. The purposes of this chapter are, therefore, threefold: (1) to review recent progress in ground LAI measurement techniques, (2) to introduce the basic principles of LAI remote sensing, and (3) to discuss remaining issues in existing LAI algorithms.

20.2 Ground-Based LAI Measurement Theory and Techniques

There are direct and indirect techniques for measuring LAI on the ground. Direct techniques include destructive sampling and allometry (Smith et al. 1993; Chen 1996; Gower et al. 1999), litterfall collection (Cutini et al. 1998; Breda 2003), and point contact (Warren Wilson 1960) methods. These techniques are generally accurate but labor intensive. Indirect techniques utilize optical instruments, which can acquire rapid measurements of LAI based on radiative transfer principles. These measurements can also be accurate, if the instruments are well calibrated and are based on sound theories. In developing LAI algorithms for remote sensing applications, a large amount of ground-based LAI data is required, and therefore indirect techniques have been widely used for this purpose (Jonckheere et al. 2004). The principles of indirect LAI measurement techniques are briefly introduced below.

Optical instruments measure transmitted radiation through a plant canopy, from which the canopy gap fraction is derived. The canopy gap fraction, $P(\theta)$, at zenith angle θ, is related to the plant area index, denoted as L_t, which includes both green leaves and nongreen materials such as stems and branches that intercept radiation. This relation is given by the following equation:

$$P(\theta) = e^{-G(\theta)\Omega L_t/\cos\theta} \qquad (20.1)$$

where $G(\theta)$ is the projection coefficient, which is determined by the leaf angular distribution (Monsi and Saeki 1953; Campbell 1990), and Ω is the clumping index, which is related to the leaf spatial distribution pattern (Nilson 1971). If the leaf angle distribution is spherical (i.e., all leaves in the canopy can form a sphere by adjusting their spatial positions while preserving their angular positions), $G(\theta) = 0.5$. The leaf angle distribution of a canopy is planophile, if it has more horizontal leaves than the spherical distribution, and is erectophile, if it has more vertical leaves than the spherical distribution (Ross 1981). Planophile and erectophile distributions make the transmission of radiation through the canopy vary with solar incidence angle in different ways, that is, different $G(\theta)$ functions. If $P(\theta)$ is measured at one angle and $G(\theta)$ and Ω are known, L_t can be inversely calculated using Equation 20.1. However, both $G(\theta)$ and Ω are generally unknown, and, therefore, different optical instruments have been developed to measure these unknown quantities.

The LI-COR LAI-2000 Plant Canopy Analyzer is an optical instrument developed to address the issue of unknown $G(\theta)$ because of nonspherical leaf angle distribution. It measures the diffuse radiation transmission simultaneously in five concentric rings covering the zenith angle range from 0° to 75°, that is, $P(\theta)$ at five angles. These measurements are used to calculate the LAI based on Miller's theorem (Miller 1967):

$$L_e = 2\int_0^{\pi/2} \ln\frac{1}{P(\theta)}\cos\theta\sin\theta d\theta \qquad (20.2)$$

The original Miller's equation was developed for canopies with random leaf spatial distributions, that is, $\Omega = 1$, and allows the calculation of LAI without the knowledge of $G(\theta)$ when $P(\theta)$ is measured over the full zenith angle range and its azimuthal variation is ignored. There is also a method to derive LAI and $G(\theta)$ simultaneously using multiple angle measurements (Norman and Campbell 1989). For spatially nonrandom canopies, Miller's theorem actually calculates the effective LAI (Chen et al. 1991), expressed as

$$L_e = \Omega L \qquad (20.3)$$

Equation 20.2 can be discretized to calculate L_e using the $P(\theta)$ measurements at five zenith angles by LAI-2000. L_e calculated this way includes all green and nongreen materials above the instrument. With measured L_e, the following equation is proposed to calculate LAI (Chen 1996):

$$L = \frac{(1-\alpha)L_e}{\Omega} \qquad (20.4)$$

where α is the woody-to-total area ratio. The total area includes both green leaves and nongreen materials such as stems, branches, and attachments (e.g., moss) on branches. The α value is generally in the range of 0.05–0.3 depending mostly on forest age (Chen et al. 2006). It can be estimated through destructive sampling or optical measurements (Kucharik et al. 1998). For deciduous canopies, it can be measured using optical instruments such as LAI-2000 during the leafless period in winter (Barr et al. 2004). However, caution should be taken in using α values measured in this way as the underlying assumption of using the leafless α value in Equation 20.4 is that leaves and nonleaf materials are distributed randomly and independently in space while in reality much of the branches are directly under leaves. This assumption may cause overcorrection for these nongreen materials using the leafless α value. Accurate determination of the α value remains a challenge in indirect LAI measurements (Chen et al. 2006), although there have been some attempts to develop multispectral optical instruments to differentiate between green and nongreen materials from the ground perspective (Kucharik et al. 1998; Zou et al. 2009).

There are also optical techniques for indirect measurement of the clumping index (Chen and Cihlar 1995). These techniques are based on the canopy gap size distribution theory of Miller and Norman (1971). An optical instrument named Tracing Radiation and Architecture of Canopies (TRAC) (Chen and Cihlar 1995) was developed to measure the canopy gap size distribution using the solar beam as the probe. In conifer canopies, gaps between needles within a shoot (a basic collection of needles around the smallest twig) are obscured due to the penumbra effect, and the clumping index derived from TRAC measurements represents the clumping effects at scales larger than the shoot (treated as the foliage element), denoted as Ω_E. According to a random gap size distribution curve based on Miller and Norman's theory, large gaps caused by the nonrandom foliage element distribution, that is, those caused by tree crowns and branches, are identified and removed to reconstruct a random gap size distribution. With this gap removal technique, Ω_E is calculated from the following equation (Chen and Cihlar 1995; Leblanc 2002):

$$\Omega_E(\theta) = \frac{\ln\left[F_m(0,\theta)\right]\left[1 - F_{mr}(0,\theta)\right]}{\ln\left[F_{mr}(0,\theta)\right]\left[1 - F_m(0,\theta)\right]} \tag{20.5}$$

where $F_m(0,\theta)$ is the total canopy gap fraction at zenith angle θ, that is, the accumulated gap fraction from the largest to smallest gaps, and $F_{mr}(0,\theta)$ is the total canopy gap fraction after removing large gaps resulting from the nonrandom foliage element distribution due to canopy structures such as tree crowns and branches.

Clumping within individual shoots depends on the density of needles on a shoot. This level of foliage clumping was recognized and estimated in various ways by Oker-Blom (1986), Gower and Norman (1990), Stenberg et al. (1994), Fassnacht et al. (1994), and others. Based on a theoretical development by Chen (1996), this clumping is quantified using the needle-to-shoot area ratio (γ_E) as follows:

$$\gamma_E = \frac{A_n}{A_s} \tag{20.6}$$

where A_n is half the total needle area (including all sides) in a shoot and A_s is half the shoot area (for a shoot that can be approximated by a sphere, the total shoot area is the spherical surface area, not the projected disk area). To obtain γ_E, shoots need to be sampled from trees of different sizes at different heights, and A_n and A_s need to be measured using laboratory equipment (Chen et al. 1997). For broadleaf forests, the individual leaves are the foliage elements, and therefore $\gamma_E = 1$.

The total clumping of a stand can, therefore, be written as

$$\Omega = \frac{\Omega_E}{\gamma_E} \tag{20.7}$$

and the final equation for deriving LAI from indirect measurements is

$$L = \frac{(1-\alpha)L_e\gamma_E}{\Omega_E} \tag{20.8}$$

The following protocol has been suggested by Chen et al. (2002) for indirect measurement of LAI based on Equation 20.8:

1. To measure the effective LAI (L_e) using LAI-2000.
2. To measure the element clumping index (ΩE) using TRAC.
3. To measure the needle-to-shoot area ratio (γ_E), where possible. Otherwise suggested default values for various forest types and ages can be used (Chen 1996a; Chen et al. 2006).
4. To measure the woody-to-total area ratio (α) where possible. Otherwise, they can be estimated based on forest type and age according to previous experimental results (Chen 1996a; Chen et al. 2006).

To follow this protocol, it is required to visit a site twice. Once is to use LAI-2000 under diffuse light conditions near dawn and dusk, and the other time is to use TRAC under direct sunlight conditions near noon. Since LAI-2000 covers the zenith angle range from $0°$ to $75°$, it is preferably operated within the narrow durations shortly after sunrise or shortly before sunset, when the solar zenith angle is larger than $75°$ to minimize the effect of direct sunlight on its measurements. An overcast sky with uniform diffuse radiation would be another preferred condition for using LAI-2000, but rainless overcast conditions are rare. TRAC requires steady direct sunlight under clear sky conditions. It can also be operated between a large cloud gap allowing for an uninterrupted run along a transect. To measure the canopy gap size distribution, TRAC uses a walking and high-frequency sampling technique and therefore requires some preparation of the transect, such as clearing falling tree trunks and major dead brunches along the transect to allow walking at reasonably constant speed. Some precautions should also be taken in selecting the transect direction to make the angle between the transect and the solar azimuth larger than $30°$.

Hemispherical photography techniques have been developed for measuring LAI over the last several decades (Anderson 1964; Olsson et al. 1982; Chen et al. 1991; Baret et al. 1993; Whitford et al. 1995; Englund et al. 2000; Frazer et al. 2001; Weiss et al. 2004; Ryu et al. 2012). These techniques can be used to measure both L_e and Ω_E (Walter et al 2003; Leblanc et al. 2005), but the measurements are less reliable because of issues with accurate determination of photographic exposure, especially the apparent variable degrees of exposure across the zenith angle range (Chen et al. 2006). However, digital cameras cost much less than LAI-2000 and TRAC and can provide reasonably accurate measurements. If photographic exposure (Zhang et al. 2005) and processing (Leblanc et al. 2005) are performed correctly, digital cameras can provide LAI measurements that are comparable to those measured through the combined use of LAI-2000 and TRAC (Chen et al. 2006). One practical advantage of choosing a digital camera over LAI-2000 and TRAC may be that we only need to visit one site once to acquire all information needed to calculate LAI.

20.3 Principles of LAI Retrieval Using Remote Sensing Data

Optical remote sensing signals acquired from vegetation are influenced by many factors including vegetation structure, leaf optical property, and background optical property. Vegetation structure refers to both the density of the reflective materials (i.e., LAI) and their spatial organization (canopy architecture). Remote sensing signals are therefore not uniquely related to LAI but also dependent on many other factors influencing the reflection of solar radiation. This complication presents challenges in mapping LAI from remote sensing imagery.

20.3.1 Leaf Optical Properties

Healthy plant leaves absorb most visible solar radiation and also scatter (reflect and transmit) most nonvisible solar radiation (Figure 20.1). In the visible spectrum, chlorophyll pigments in leaves absorb strongly at violet and blue (380–480 nm) and red (620–680 nm) wavelengths, resulting in low reflectance and transmittance at these wavelengths. Radiation in green wavelengths (495–570 nm) is much less absorbed by leaf pigments than that in blue and red wavelengths, and therefore leaves have relative high reflectance and transmittance in green wavelengths, making leaves appear green. Liquid water contained in leaves also absorbs solar radiation, and the absorptivity increases with wavelength and is the largest in middle infrared (MIR) wavelengths (1300–2500 nm). The magnitude of reflectance in MIR decreases with increasing leaf water content (Figure 20.1). In the MIR spectral range, there are two peak reflectance and transmittance wavelengths at about 1700 nm and 2200 nm, which correspond to atmospheric windows and many sensors acquire measurements in these windows. These measurements are useful for assessing the vegetation liquid water content and thus LAI.

The leaf reflectance and transmittance spectral patterns are similar between broadleaf and needleleaf species (Figure 20.2). The magnitudes of the reflectance are also similar, but the transmittance of needleleaves is usually smaller than that of broadleaves because needleleaves are generally thicker.

20.3.2 Influence of Canopy Architecture on Reflectance

Remote sensing imagery consists of picture elements named pixels, and when a pixel has a sufficient size, such as those of Landsat (30 m × 30 m), ASTER (20 m × 20 m), and MODIS (250 m × 250 m), the reflectance of a pixel can be conceptually regarded as that of

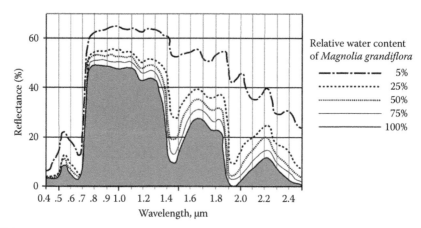

FIGURE 20.1
Typical leaf reflectance and transmittance spectra from 0.4 to 2.5 μm and the influence of leaf water content on its reflectance spectrum. (Reprint from Jensen, J. R., *Remote Sensing of the Environment: An Earth Resource Perspective*, 2nd ed., Prentice Hall, 2007.)

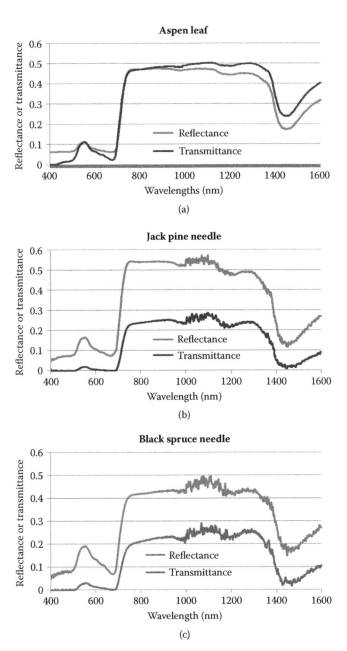

FIGURE 20.2
Reflectance and transmittance spectra of (a) an aspen leaf, (b) jack pine needleleaves, and (c) black spruce needle leaves.

a plant canopy consisting of an extensive slab of foliage (e.g., many tree crowns). Although broadleaf and needleleaf reflectances are similar in magnitude (Figure 20.2), their forest canopy reflectances are quite different, as shown in Figure 20.3 with measurements from the satellite sensor CHRIS-PROBA over boreal forests in Canada at five angles. The positive and negative angles denote the forward and backward scattering directions relative to the incoming solar radiation direction. The reflectance from the black spruce forest is

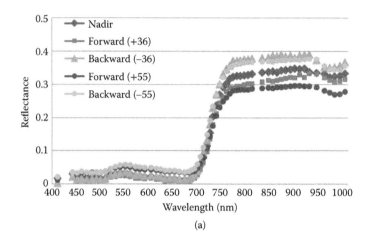

(a)

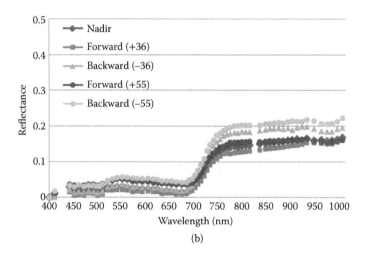

(b)

FIGURE 20.3
Comparison of canopy level reflectance spectra between (a) broadleaf (aspen) and (b) conifer (black spruce) forests measured by the CHIRS-PROBA sensor near Sudbury, Ontario, Canada, on August 10, 2007. These reflectance spectra were acquired with 34 m spatial resolution at five zenith angles (nadir, ±36°, ±55°), when the solar zenith angle was 30°, and the difference between the sun and sensor azimuth angles is 6.5° in the backward scattering direction and 33.7° in the forward scattering direction.

almost only half of that of the aspen forest at all five angles. The LAI and the forest backgrounds are similar in these two canopies (3.2 for black spruce and 2.9 for aspen), and therefore the difference is mainly caused by the canopy architecture and to some extent by the difference in leaf transmittance. For the same LAI, broadleaf forests intercept and reflect more solar radiation than conifer forests because broadleaf forests are less clumped than conifer forests (Chen et al. 1997; Chen et al. 2006). As shown in Equation 20.1, foliage clumping, that is, grouping of leaves in crowns, whorls, branches, and shoots, increases the canopy gap fraction and decreases solar radiation interception by the canopy even if the LAI is the same. In addition to clumping within tree crowns, conifer needles are also clumped within shoots, and the overall clumping index of conifer forests is smaller than that of broadleaf forests (by definition, the smaller the clumping index, the more clumped the canopy is). The canopy level reflectance is also influenced by leaf transmittance as

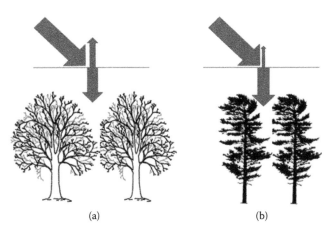

(a)   (b)

FIGURE 20.4
Schematic illustration of the difference in reflectance between (a) broadleaf and (b) conifer canopies. The higher reflectance of broadleaf canopies than conifer canopies is mostly due to their more horizontal and less clumped canopy architecture and larger leaf transmittance.

radiation transmitted through leaves can also be reflected back to space through multiple scattering within the canopy. Since the transmittance of broadleaves is considerably higher than that of conifer needles (Figure 20.2), the multiple scattering contribution to the total canopy level reflectance in broadleaf forests is much larger than that in conifer forests, and this contribution is especially pronounced in the near-infrared (NIR) wavelengths. Figure 20.4 shows a schematic of the influence of the canopy architecture on the reflectance from broadleaf and conifer forests, in which the same incoming radiation is reflected more by broadleaf than by conifer forests because of the reasons mentioned above.

20.3.3 Influence of Vegetation Background on Reflectance

The reflectance of a vegetated pixel comprises contributions from the vegetation cover and its background, which is the underlying soil in cropland and grassland but often includes other covers of the forest floor such as litter, moss, and understory vegetation. Depending on the height of the understory (grass, shrubs, and tree saplings), it may or may not be included in forest LAI measurements because optical instruments are often operated conveniently at the waist height in forests and most of the understory is excluded in the LAI measurements. For carbon cycle modeling, there is a need to differentiate between the overstory (trees) and understory because of their different carbon residence times (Liu et al. 2002). Trees accumulate carbon in stems and coarse roots and have much longer carbon residence times than grass and shrubs.

Because the greenness of moss and understory varies greatly during the growing season, the influence of this variation on forest LAI retrieval is considerable and cannot be ignored. Figure 20.5 shows the seasonal variations of the background red and NIR reflectance over the North America continent derived from Multiangle Imaging SpectroRadiometer (MISR) data (Pisek and Chen 2009). Using MISR observations at nadir and 45° zenith angle, the vegetation background reflectances were derived based on the difference in the probability of observing the background between these two angles. The seasonal variations of both red and NIR reflectances of the background are quite pronounced (Figure 20.5), and these variations are found to reduce the retrieved LAI in the summer in the range from 0.0 to 1.5, with larger differences at higher latitudes where the forests are open.

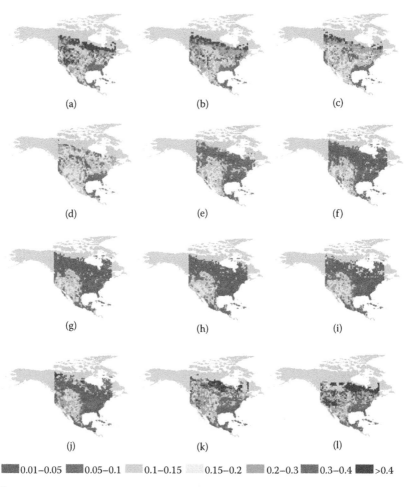

0.01–0.05 ▮ 0.05–0.1 ▮ 0.1–0.15 ▮ 0.15–0.2 ▮ 0.2–0.3 ▮ 0.3–0.4 ▮ >0.4

FIGURE 20.5
(**See color insert.**) Seasonal variations of vegetation background red and near-infrared reflectance over the North America continent derived from Multiangle Imaging SpectroRadiometer data. (a) January, (b) February, (c) March, (d) April, (e) May, (f) June, (g) July, (h) August, (i) September, (j) October, (k) November, and (l) December. (From Pisek, J., and J.M. Chen, *Remote Sens. Environ.*, 113, 2412, 2009.)

20.3.4 Influence of Illumination and Observation Angles on Reflectance

For the same vegetation, measured reflectance at a wavelength can vary greatly with a sensor's observation angle and solar illumination angle. As the solar zenith angle increases, the path length of a solar beam through the canopy increases, causing more solar radiation to be intercepted by the canopy. The hemispherically integrated reflectance (e.g., albedo) from vegetation, therefore, increases with solar zenith angle (Yang et al. 2008). At different view angles relative to the sun, observed reflectance can vary dramatically depending on the proportion of the sunlit canopy appearing in the view direction, with the maximum reflectance in the backward scattering direction when the view direction coincides with the sun's direction and the minimum reflectance in the forward scattering direction when canopy shadows are maximally observed. Figure 20.3 shows reflectance spectra measured by the CHRIS-PROBA sensor at five zenith angles. These variations of observed reflectance with view angle at different solar zenith angles are often described using the bidirectional reflectance distribution function (BRDF), and many semiempirical kernel-based models (Walthall et al. 1985; Roujean et al. 1992; Chen and Cihlar 1997) and physics-based models (Li and Strahler 1985; Chen and Leblanc 1997, 2001; Kuusk and Nilson 2000; Pinty et al. 2004) have been developed to simulate forest and nonforest BRDF. These models can be utilized for LAI algorithm development.

20.4 Algorithms for Retrieving LAI Using Optical Remote Sensing Data

Many algorithms have been developed to retrieve LAI from remote sensing imagery with full or partial consideration of the aforementioned factors influencing remote sensing measurements from vegetation. These algorithms are developed based on different methods as described in the following sections.

20.4.1 LAI Algorithms Based on Vegetation Indices

Reflectance spectra of healthy leaves show distinct low values in the red (620–750 nm) wavelengths and high values in NIR (800–1300 nm) wavelengths (Figure 20.1), and therefore many vegetation indices (VIs) have been developed using remote sensing measurements in red and NIR bands for estimating LAI and other vegetation parameters (Table 20.1). Liquid water in aboveground living biomass absorbs MIR (1300–2500 nm) radiation, lowering the reflectance in the MIR band. Since foliage biomass interacts most with solar radiation, the MIR reflectance is expected to correlate well with LAI. Some two-band and three-band VIs utilizing the additional information from MIR have also been developed for LAI retrieval (Table 20.1).

Not all two-band and three-band VIs are well correlated to LAI (Figure 20.6). The significance level of the correlation of two-band VIs with LAI varies greatly even though they are constructed using the same two-band reflectance data because these two data are combined in different ways under different assumptions. An ideal VI for LAI retrieval should preferably have the following properties: (1) it is more or less linearly related to LAI, and (2) it can minimize the impacts of both random and bias remote sensing errors on its value. A linear relationship between a VI and LAI is preferred because it is insensitive to the surface heterogeneity within a pixel and induces less error in spatial scaling (Chen 1999). No VIs have so far been found to be linearly related to LAI for all plant functional types. However, some are more linearly related to LAI than others. Simple ratio (SR), for example, is more linearly related to LAI than Normalized Difference Vegetation Index (NDVI) and Soil-Adjusted Vegetation Index (SAVI), and therefore SR is preferred in our studies (Chen and Cihlar 1996; Chen et al. 2002). Ideally, VIs would vary with LAI only, or the effects of surface variations other than LAI can be considered by adjusting coefficients or constants in the algorithm. Measured reflectances in different spectral bands are affected by environmental noise, such as subpixel clouds and their shadows that are not identified in image processing, mixtures of nonvegetative surface features (small water bodies, rock, etc.), fogs, smokes, and so on. This unwanted noise frequently exists in remote sensing imagery and can dramatically alter the values of VIs. However, the impacts of these types of noise on the reflectances in different spectral bands are often

TABLE 20.1

VIs Useful for LAI Retrieval

Vegetation Index	Definition	Reference
NDVI	$\dfrac{(\rho_n - \rho_r)}{(\rho_n + \rho_r)}$	Rouse et al. (1974)
SR	$\dfrac{\rho_n}{\rho_r}$	Jordan (1969)
MSR	$\dfrac{\dfrac{\rho_n}{\rho_r} - 1}{\sqrt{\dfrac{\rho_n}{\rho_r} + 1}}$	Chen (1996)
RDVI	$\dfrac{\rho_n - \rho_r}{\sqrt{\rho_n + \rho_r}}$	Roujean and Bren (1995)
WDVI	$a = \dfrac{\rho_{n,\,soil}}{\rho_{r,\,soil}}$ $\rho_n - a \cdot \rho_r$	Clevers (1989)
SAVI	$\dfrac{(\rho_n - \rho_r)(1 + L)}{(\rho_n + \rho_r + L)}$ $L = 0.5$	Huete (1988)
SAVI1	$\dfrac{(\rho_n - \rho_r)(1 + L)}{(\rho_n + \rho_r + L)}$ $L = 1 - 2.12 \cdot NDVI \cdot WDVI$	Qi et al. (1994)
GEMI	$\dfrac{\eta(1 - 0.25 \cdot \eta) - (\rho_r - 0.125)}{(1 - \rho_r)}$ $\eta = \dfrac{2(\rho_n{}^2 - \rho_r{}^2) + 1.5\,\rho_n + 0.5\rho_r}{\rho_n + \rho_r + 0.5}$	Pinty and Verstraete (1992)
NLI	$\dfrac{(\rho_n{}^2 - \rho_r)}{(\rho_n{}^2 + \rho_r)}$	Goel and Qin (1994)
ARVI	$\dfrac{(\rho_n - \rho_{rb})}{(\rho_n + \rho_{rb})}$ $\rho_{rb} = \rho_r - \gamma(\rho_b - \rho_r)$	Kaufman and Tanre (1992)
SARVI	$\dfrac{(\rho_n - \rho_{rb})(1 + L)}{(\rho_n + \rho_{rb} + L)}$ $L = 0.5$	Huete and Liu (1994)
SARVI2	$\dfrac{2.5(\rho_n - \rho_r)}{(1 + \rho_n + 6\rho_r - 7.5\rho_b)}$	
MNDVI	$\dfrac{(\rho_n - \rho_r)}{(\rho_n + \rho_r)}(1 - \dfrac{\rho_s - \rho_{smin}}{\rho_{smax} - \rho_{smin}})$	Nemani et al. (1993)
RSR	$\dfrac{\rho_n}{\rho_r}(1 - \dfrac{\rho_s - \rho_{smin}}{\rho_{smax} - \rho_{smin}})$	Brown et al. (2000)

Source: Chen, J.M., *Can. J. Remote Sens.*, 22, 229, 1996b.

Note: ρ_r is reflectance in the red band, ρ_n is reflectance in the NIR band, ρ_b is reflectance in the blue band, and ρ_s is reflectance in the shortwave infrared band. ρ_{smin} and ρ_{smax} are the minimum and maximum reflectances, respectively, in a shortwave infrared image (usually taken as the 5% and 95% cutoff points of the histogram of the image). WDVI, Weighted Difference Vegetation Index.

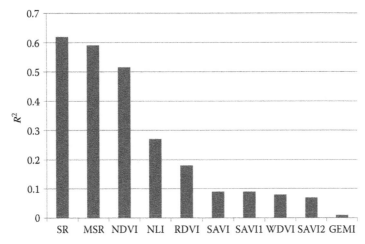

FIGURE 20.6
Correlation coefficient (R^2) between vegetation indices from Landsat Thematic Mapper and leaf area index (LAI) measured in conifer forests in Saskatchewan and Manitoba, Canada. (From Chen, J.M., *Can. J. Remote Sens.*, 22, 229, 1996b.)

correlated. For example, subpixel clouds would cause red and NIR reflectances to increase simultaneously and cloud shadows would decrease them simultaneously. The same is true for all other types of noise mentioned above. Variations in the solar and view angles also cause variations of reflectances in the various spectral bands in the same directions and in about the same proportions. VIs that are based on ratios of these two bands, such as NDVI and SR, can greatly reduce the impacts of various sources of noise. However, some VIs with sophisticated manipulations of the two-band data, such as Global Environmental Monitoring Index (GEMI), would amplify the noise. VIs that cannot be expressed as a function of the ratio of these two-band reflectances, such as SAVI, Nonlinear Index (NLI) and Renormalized Difference Vegetation Index (RDVI) will retain the noise. Modified simple ratio (MSR), for example, is developed with the same purpose as RDVI to increase its linearity with LAI, but it is better correlated to LAI than RDVI because it can be expressed as a function of the ratio of NIR and red reflectances while RDVI cannot. The ability of a VI to minimize the unwanted measurement noise is of paramount importance in LAI retrieval because noise in the reflectance measurements can come from many sources and is not avoidable.

Three-band VIs have been developed for various purposes. Atmospherically Resistant Vegetation Index (ARVI) and Soil and Atmospheric Resistant Vegetation Index (SARVI) modify NDVI and SAVI, respectively, with the reflectance in the blue band to reduce the atmospheric effect. They are useful when there are insufficient simultaneous atmospheric data to conduct atmospheric correction. Modified NDVI (MNDVI) and reduced simple ratio (RSR) introduce a multiplier to NDVI and SR, respectively, based on the reflectance in an MIR band (1600–1800 nm or 2100–2300 nm). Figure 20.7 shows how SR and RSR are correlated to LAI for conifer, broadleaf, and mixed forests. In these cases, RSR has several advantages over SR for LAI retrieval (Brown et al. 2000): (1) it is more significantly correlated with LAI for these forest types individually because it is more sensitive to LAI variation; (2) the differences in the LAI–RSR relationship among these forest types are greatly reduced from those in the LAI–SR relationship, and therefore RSR is particularly useful

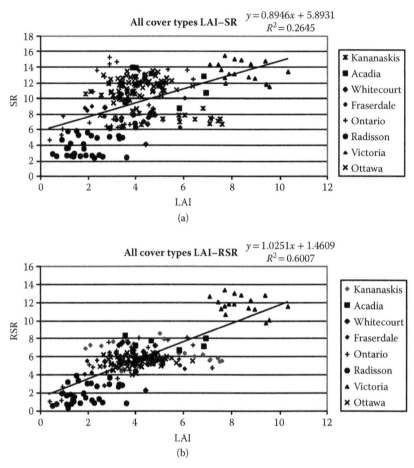

FIGURE 20.7
Relationships between (a) leaf area index (LAI) and simple ratio (SR) and between (b) LAI and reduced simple ratio (RSR) for all cover types in various locations in Canada, with deciduous forests and crops in Ottawa, deciduous forests in Ontario (several locations), and conifer forests in other locations. (From Chen, J.M., et al., *Remote Sens. Environ.* 80, 165, 2002.)

for mixed cover types; and (3) the influence of the variable background optical properties is much smaller on RSR than on SR because MIR reflectance is highly sensitive to the greenness of the background due to the strong absorption of MIR radiation by grass, moss, and understory. These advantages of RSR over SR for forest LAI retrieval are confirmed by several independent studies (Eklundh et al. 2003; Stenberg et al. 2004, 2008; Wang et al. 2004; Chen et al. 2005; Tian et al. 2007; Heiskanen et al. 2011). However, RSR is sensitive to soil and vegetation wetness and can increase greatly immediately after rainfall or irrigation, and therefore in some LAI algorithms, it is used only for forests (Deng et al. 2006).

20.4.2 LAI Algorithms Based on Radiative Transfer Models

The relationships between LAI and reflectances in individual spectral bands can be simulated using plant canopy radiative transfer models, and LAI algorithms can be developed based on these modeled relationships. Models are useful alternatives to empirical relationships established through correlating VIs or reflectances with LAI measurements because the empirical data are often limited in spatial and temporal coverage and are often location specific. These empirical relationships are also dependent on the quality of ground LAI data, the spectral response functions of remote sensing sensors, the angle of measurements, atmospheric effects, and so on. The quality of LAI data can be influenced by the method of LAI measurements, the definition of LAI, and the measurement protocol. Some reported LAI values are actually the effective LAI without considering the clumping effect, and some optical measurements do not include the correction for nongreen materials (Equation 20.8). Radiative transfer models can theoretically avoid these shortcomings of empirical data, but they need to be calibrated with ground data. In this calibration process, misconceptions and errors in empirical data can also bias the model outcome. For example, some destructive LAI values used for model validation are incorrectly based on the projected area rather than half the total leaf area.

There have been many LAI algorithms developed using radiative transfer models (Table 20.2) for regional and global LAI retrieval. These algorithms are characterized by the radiative transfer modeling method, the ways to consider foliage clumping and background optical properties, and the ways to combine the individual bands. A radiative transfer model, however sophisticated it may be, is an abstract representation of the complex reality, and therefore the modeled relationship between LAI and remote sensing data depends not only on the aforementioned factors but also on how radiative transfer is simulated, such as the ways to consider multiple scattering in the canopy, the assumed leaf angle distributions, the treatments of diffuse sky radiation, and so on. As radiative transfer methods are diverse, it is expected that the simulated relationships between remote sensing data and LAI are quite different among the existing model-based global LAI algorithms. There is a need to calibrate radiative transfer models and LAI algorithms against an accurate ground and remote sensing dataset covering the diverse plant structural types

TABLE 20.2

Global LAI Products and Their Main Characteristics

	CYCLOPES	**ECOCLIMAP**	**GLOBCARBON**	**MODIS**
Algorithm development	1D turbid media radiative transfer model	Empirical LAI-NDVI relationships	Geometric-Optical Model	Lookup tables produced using a 3D radiative transfer model
Clumping consideration	No clumping consideration except consideration of the differences among cover types at the landscape level	Clumping within shoot and canopy is considered, but clumping at the landscape level is not considered	Clumping is fully considered based on TRAC-measured cover type–specific values	Clumping is considered through a parameter related to the 3D canopy structure
Background optical property	Assigned constant values	Assigned constant values	Assigned constant values	Assigned constant values
Seasonal smoothing	No	No	Yes	No
Reference	Baret et al. (2007)	Masson et al. (2003)	Deng et al. (2006)	Knyazikhin et al. (1998) and Yang et al. (2006)

Source: Modified from Garrigues, S., et al., *J. Geophys. Res.* 113, G02028, 2008.

Note: Global LAI products include CYCLOPES (Carbon Cycle and Change in Land Observational Products from an Ensemble of Satellites), ECOCLIMAP, GLOBCARBON, and MODIS.

around the globe. The radiation transfer model intercomparison efforts (Pinty et al. 2004; Widlowski et al. 2007) have laid a foundation for further activities to satisfy this need.

20.4.3 Examples of Global LAI and Clumping Index Maps

All algorithms shown in Table 20.2 have been employed to generate global LAI products using data from different satellite sensors and are available on the Internet for different periods of time. Figure 20.8 shows an example of a global LAI map generated using the GLOBCARBON (GLOBal Biophysical Products for Terrestrial CARBON Studies) algorithm (Deng et al. 2006). This product is unique in the following aspects: (1) the effective LAI is first mapped from RSR and SR, and cover type–specific clumping index values (or a clumping index map) are used to convert the effective LAI into LAI (Equation 20.3), and (2) the angular variations of reflectances and RSR and SR are considered using lookup tables developed with a geometrical optical model (Chen and Leblanc 1997, 2001). The general spatial pattern of LAI shown in Figure 20.8 is similar to other LAI products, but there are many significant differences from them (Garrigues et al. 2008). The GLOBCARBON LAI algorithm has recently been used to generate a 30-year LAI product from 1981 to 2010 using the combination of Advanced Very High Resolution Radiometer and MODIS data (Liu et al. 2012). There are also other initiatives to produce global LAI products from multisensors using data fusion techniques (Xiao et al. 2013).

To assist the production of improved global LAI maps, global clumping index maps have also been produced from POLDER data at 6 km resolution (Chen et al. 2005) and MODIS data at 0.5 km resolution (He et al. 2012). Figure 20.9 shows an example of a global clumping index map retrieved using the MODIS BRDF product. This product provides the accumulated angular variation information over 16-day periods, from which the angular index of the normalized difference between hotspot and darkspot (NDHD) is derived from each 0.5 km pixel. Based on a geometric optical model (Chen and Leblanc 1997, 2001), the clumping index is related to NDHD, and this relationship is used for mapping the clumping index from measured NDHD. This example clumping index map shows that conifer forests at high latitudes are most clumped with values about 0.5 and grassland and cropland

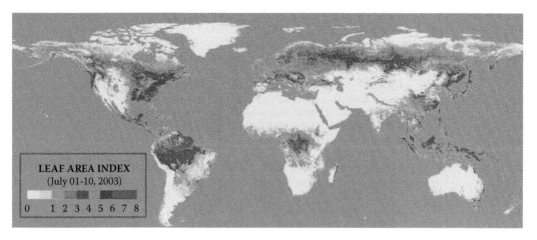

FIGURE 20.8
(**See color insert.**) Global leaf area index map at 1 km resolution generated by the GLOBCARBON algorithm using cloud-free VEGETATION data on July 1–10, 2003. (From Deng, F., et al., *IEEE Trans. Geosci. Remote Sens.* 44, 2219, 2006.)

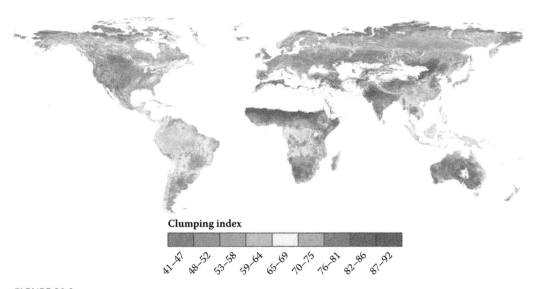

FIGURE 20.9
(**See color insert.**) Global clumping index map at 0.5 km resolution derived using the MODIS BRDF product for 2006. The numerical value should be divided by 100. (From He, L., et al., *Remote Sens. Environ.*, 119, 118, 2012.)

are least clumped with values close to 1, while broadleaf forests and other vegetation types are the intermediate cases.

For many applications, the clumping index map may have equal importance to LAI. The effective LAI determines the amount of radiation intercepted and absorbed by the canopy, while the true LAI represents the total leaf area in the canopy responsible for exchanges of water, carbon, and other masses. For a typical conifer forest with a clumping index of 0.5, for example, clumping effectively reduces the LAI by half in terms of the radiation interception ability of the canopy. If the LAI product is accurate but clumping is ignored, the radiation interception would be greatly overestimated. In this case, it has been found that ignoring this clumping effect, global gross primary productivity (GPP) would be overestimated by 12% (Chen et al. 2012). However, if gap fraction retrieved from remote sensing is directly converted to LAI without considering the clumping effect, LAI in this case would be underestimated (for the typical conifer forest, it would be underestimated by 50%). In this case, the radiation interception and sunlit leaf area can still be accurately estimated, but shaded leaf area is greatly underestimated. This underestimation of shaded leaf area can cause an underestimation of the global GPP by 9% (Chen et al. 2012). Since transpiration of plants is a passive result of photosynthesis, it can be inferred that clumping would affect transpiration by similar magnitudes. Consideration of foliage clumping is therefore important in both production and application of LAI products.

20.5 Remaining Issues in LAI Retrieval

Since the launches of the VEGETATION sensor in 1998 and the MODIS sensor in 1999, there have been several global LAI products that are routinely produced and freely distributed online. These products have greatly enhanced applications of satellite remote sensing data to ecological, hydrological, meteorological, and climatological studies at regional and global scales. However, these LAI products have not been used as widely as they should have been because of several outstanding issues in these products. Some of the issues are analyzed below to stimulate further development of LAI algorithms.

20.5.1 Issue 1: Large Differences among Existing Global LAI Products

Although differences among global products for a surface parameter are common, the differences among existing global LAI products are alarmingly large (Garrigues et al. 2008). LAI may be the most complex surface parameter that can be retrieved from remote sensing data. Many factors in the algorithms can cause the differences among LAI products, including the relationships between LAI and reflectance in individual bands simulated using models of different complexities and assumptions, the ways in which the complex 3D vegetation architecture is represented, the assignments of the background optical properties, the quality of ground-based LAI data that are used for algorithm validation, and so on. These factors are related in certain ways. The modeled relationships between LAI and reflectance, for example, are necessarily calibrated against ground LAI and remote sensing reflectance data, and the qualities of ground LAI and remote sensing data as well as a sensor's spectral response characteristics would influence the finalization of these relationships.

20.5.2 Commonly Distorted Seasonal Variation Patterns of LAI

Conifer forests keep leaves all year round and generally have small seasonal variations in LAI. The amplitude of seasonal LAI variation in conifer stands depends on the average needle life span, ranging from 2 years for some pine species to over 10 years for fir species (White et al. 2000), and therefore we expect that LAI of conifer forests in the winter would be about 10% to 33% lower than the maximum value in the peak growing season. However, many remote sensing LAI products show large seasonal variations for conifer forests, and many LAI algorithms produce near zero conifer LAI in the winter (Garrigues et al. 2008). These fictitiously large variations are caused by several factors that influence remote sensing measurements: (1) variations in the background optical properties, especially the large variations due to the occurrences of the snow cover in the winter and understory in the summer and (2) variations in leaf optical properties, such as pigments. Through the use of multiple angle remote sensing, the seasonal variations in the background reflectances can be estimated at regional scale (Figure 20.5) (Pisek and Chen 2009) and much of the fictitious seasonal variations in LAI can be remedied. However, the seasonal variations in leaf optical properties are very difficult to separate from the seasonal variations in LAI itself as the reflectances in the individual bands as well as VIs respond to LAI and leaf greenness variations in the same direction. Leaf greenness is determined by the contents of chlorophyll and other pigments in the leaf. Many empirical and modeled VIs or reflectances, in fact, respond to the surrogate of LAI and leaf greenness. For photosynthesis modeling, this surrogate can be treated as LAI and the error caused by not separating the leaf area and greenness may be tolerated (in winter, plants are not photosynthesizing anyway). However, for energy balance estimation, this treatment would cause considerable errors that cannot be tolerated. For example, if LAI in the winter is too small, the albedo of the conifer forest will be too large, resulting in unrealistic energy budget of the surface. For meteorological and hydrological applications, it is therefore critical to produce LAI products that can represent both the seasonal maximum as well as the seasonal variation pattern.

20.6 Concluding Remarks

LAI may be arguably the most complex land surface parameter and also most difficult to retrieve using remote sensing data. However, it is also arguably the most useful parameter for many earth science applications. Since the launches of VEGETATION and MODIS sensors in late 1990s, much progress has been made in producing global and regional LAI products that are available for a wide array of users. Much progress has also been made in many aspects associated with LAI mapping using remote sensing data including the unified LAI definition, indirect methods for measuring LAI that include the clumping effect, the LAI measurement protocol that makes combined use of several optical instruments, LAI algorithms based on empirical data and physical models, algorithms for mapping the clumping index, and methods for mapping the seasonal variation of vegetation background. Nevertheless, existing global LAI products still have several serious shortcomings and are not yet reliable for operational use in meteorological, hydrological, and ecological models. These shortcomings, including the distorted seasonal variations for conifer forests and incomplete consideration of vegetation clumping, should be overcome before LAI products can be widely used in earth sciences.

Acknowledgments

Anita Simic provided some of the leaf reflectance data and CHRIS data that are used in Figures 20.2 and 20.3, and Yongqin Zhang and Holly Croft provided leaf reflectance and transmittance data shown in Figure 20.2. Holly Croft also provided useful comments on the manuscript.

References

Anderson, M. C. 1964. "Studies of Woodland Light Climate. 1. The Photographic Computation of Light Conditions." *Journal of Ecology* 52:27–41.

Baret, F., B. Andrieu, J. C. Folmer, J. F. Hanocq, and C. Sarrouy. 1993. "Gap Fraction Measurement Using Hemispherical Infrared Photographies and Its Use to Evaluate PAR Interception Efficiency." In *Crop Structure and Light Microclimate: Characterization and Applications*, edited by C. Varlet-Grancher, R. Bonhomme, and H. Sinoquet, 359–72. Paris: INRA.

Baret, F., O. Hagolle, B. Geiger, P. Bicheron, B. Miras, M. Huc, et al. 2007. "LAI, fAPAR and fCover CYCLOPES Global Products Derived from VEGETATION: Part 1: Principles of the Algorithm." *Remote Sensing of Environment* 110:275–86.

Barr, A. G., T. A. Black, E. H. Hogg, N. Kljun, K. Morgenstern, and Z. Nesic. 2004. "Inter-Annual Variability of Leaf Area Index of Boreal Aspen-Hazelnut Forest in Relation to Net Ecosystem Production." *Agricultural Forest Meteorology* 126:237–55.

Breda, N. J. J. 2003. "Ground-Based Measurements of Leaf Area Index: A Review of Methods, Instruments and Current Controversies." *Journal of Experimental Botany* 54(392):2403–17.

Brown, L. J., J. M. Chen, and S. G. Leblanc. 2000. "Short Wave Infrared Correction to the Simple Ratio: An Image and Model Analysis." *Remote Sensing of Environment* 71:16–25.

Campbell, G. S. 1990. "Derivation of an Angle Density Function for Canopies with Ellipsoidal Leaf Angle Distributions." *Agricultural Forest Meteorology* 49:173–76.

Chen, J. M. 1996a. "Optically-Based Methods for Measuring Seasonal Variation in Leaf Area Index of Boreal Conifer Forests." *Agricultural Forest Meteorology* 80:135–63.

Chen, J. M. 1996b. "Evaluation of Vegetation Indices and a Modified Simple Ratio for Boreal Applications." *Canadian Journal of Remote Sensing* 22:229–42.

Chen, J. M. 1999. "Spatial Scaling of a Remotely Sensed Surface Parameter by Contexture." *Remote Sensing of Environment* 69:30–42.

Chen, J. M., and T. A. Black. 1992. "Defining Leaf Area Index for Non-Flat Leaves." *Plant, Cell and Environment* 15:421–29.

Chen, J. M., T. A. Black, and R. S. Adams. 1991. "Evaluation of Hemispherical Photography for Determining Plant Area Index and Geometry of a Forest Stand." *Agricultural and Forest Meteorology* 56:129–43.

Chen, J. M., X. Chen, W. Ju, and X. Geng. 2005. "Distributed Hydrological Model for Mapping Evapotranspiration Using Remote Sensing Inputs." *Journal of Hydrology* 305:15–39.

Chen, J. M., and J. Cihlar. 1995. "Plant Canopy Gap-Size Analysis Theory for Improving Optical Measurements of Leaf Area Index." *Applied Optics* 34(27):6211–22.

Chen, J. M., and J. Cihlar 1996. "Retrieving Leaf Area Index for Boreal Conifer Forests Using Landsat TM images." *Remote Sensing of Environment* 55:153–62.

Chen, J. M., and J. Cihlar. 1997. "A Hotspot Function in a Simple Bidirectional Reflectance Model for Satellite Applications." *Journal of Geophysical Research* 102:25907–13.

Chen, J. M., A. Govind, O. Sonnentag, Y. Zhang, A. Barr, and B. Amiro. 2006. "Leaf Area Index Measurements at Fluxnet Canada Forest Sites." *Agricultural and Forest Meteorology* 140:257–68.

Chen, J. M., and S. G. Leblanc. 1997. "A 4-Scale Bidirectional Reflection Model Based on Canopy Architecture." *IEEE Transactions on Geoscience and Remote Sensing* 35:1316–37.

Chen, J. M., and S. G. Leblanc. 2001. "Multiple-Scattering Scheme Useful for Hyperspectral Geometrical Optical Modelling." *IEEE Transactions on Geoscience and Remote Sensing* 39(5):1061–71.

Chen, J. M., L. Liu, S. G. Leblanc, R. Lacaze, and J. L. Roujean. 2003. "Multi-Angular Optical Remote Sensing for Assessing Vegetation Structure and Carbon Absorption." *Remote Sensing of Environment* 84:516–25.

Chen, J. M., G. Mo, J. Pisek, F. Deng, M. Ishozawa, and D. Chan. 2012. "Effects of Foliage Clumping on Global Terrestrial Gross Primary Productivity." *Global Biogeochemical Cycles* 26: GB1019, 18.

Chen, J. M., G. Pavlic, L. Brown, J. Cihlar, S. G. Leblanc, H. P. White, et al. 2002. "Derivation and Validation of Canada-Wide Coarse Resolution Leaf Area Index Maps Using High-Resolution Satellite Imagery and Ground Measurements." *Remote Sensing of Environment* 80:165–84.

Chen, J. M., P. M. Rich, S. T. Gower, J. M. Norman, and S. Plummer. 1997. "Leaf Area Index of Boreal Forests: Theory, Techniques, and Measurements." *Journal of Geophysical Research* 102(D24):29429–43.

Chen, X., L. Vierling, D. Deering, and A. Conley. 2005. "Monitoring Boreal Forest Leaf Area Index across a Siberian Burn Chronosequence: A MODIS Validation Study." *International Journal of Remote Sensing* 26(24):5433–51.

Cutini, A., G. Matteucci, and G. S. Mugnozza. 1998. "Estimation of Leaf Area Index with the LI-COR LAI-2000 in Deciduous Forests." *Forest Ecology and Management* 105:55.

Clevers, J. G. P. W. 1989. "The Applications of a Weighted Infrared-Red Vegetation Index for Estimating Leaf Area Index by Correcting for Soil Moisture." *Remote Sensing of Environment* 29:25–37.

Dai, Y., X. Zeng, R. E. Dickinson, I. Baker, G. B. Bonan, M. G. Bosilovich, et al. 2003. "The Common Land Model." *Bulletin of the American Meteorological Society* 84(8):1013–23.

Deng, F., J. M. Chen, S. Plummer, and M. Chen. 2006. "Global LAI Algorithm Integrating the Bidirectional Information." *IEEE Transactions on Geoscience and Remote Sensing* 44:2219–29.

Eklundh, L., K. Hall, H. Eriksson, J. Ardö, and P. Pilesjö. 2003. "Investigating the Use of Landsat Thematic Mapper Data for Estimation of Forest Leaf Area Index in Southern Sweden." *Canadian Journal of Remote Sensing* 29(3):349–62.

Englund, S. R., J. J. O'Brien, and D. B. Clark. 2000. "Evaluation of Digital and Film Hemispherical Photography and Spherical Densiometry for Measuring Forest Light Environments." *Canadian Journal for Forest Research* 30:1999–2005.

Fassnacht, K., S. T. Gower, J. M. Norman, and R. E. McMurtrie. 1994. "A Comparison of Optical and Direct Methods for Estimating Foliage Surface Area Index in Forests." *Agricultural Forest Meteorology* 71:183–207.

Frazer, G. W., R. A. Fournier, J. A. Trofymow, and J. R. Hall. 2001. "A Comparison of Digital and Film Fisheye Photography for Analysis of Forest Canopy Structure and Gap Light Transmission." *Agricultural Forest Meteorology* 109:249–63.

Garrigues, S., R. Lacaze, F. Baret, J. T. Morisette, M. Weiss, J. E. Nickeson, et al. 2008. "Validation of Intercomparison of Global Leaf Area Index Products Derived from Remote Sensing Data." *Journal of Geophysical Research* 113:G02028.

Goel, N. S., and W. Qin. 1994. "Influences of Canopy Architecture on Relationships between Various Vegetation Indices and LAI and FPAR: A Computer Simulation." *Remote Sensing Reviews* 10:309–47.

Gower, S. T., C. J. Kucharik, and J. M. Norman. 1999. "Direct and Indirect Estimation of Leaf Area Index, fAPAR, and Net Primary Production of Terrestrial Ecosystem." *Remote Sensing of Environment* 70:29–51.

Gower, S. T., and J. M. Norman. 1990. "Rapid Estimation of Leaf Area Index in Forests Using the LI-COR LAI-2000." *Ecology* 72:1896–900.

He, L., J. M. Chen, J. Pisek, C. B. Schaaf, and A. H. Strahler. 2012. "Global Clumping Index Map Derived from the MODIS BRDF Product." *Remote Sensing of Environment* 119:118–30.

Heiskanen, J., M. Rautiainen, L. Korhonen, M. Mõttus, and P. Stenberg 2011. "Retrieval of Boreal Forest LAI Using a Forest Reflectance Model and Empirical Regressions." *International Journal of Applied Earth Observation and Geoinformation* 13(4):595–606.

Huete, A. R. 1988. "A Soil Adjusted Vegetation Index (SAVI)." *Remote Sensing of Environment* 25:295–309.

Huete, A. R., and H. Q. Liu 1994. "An Error and Sensitivity Analysis of the Atmospheric- and Soil-Correcting Variants of the NDVI for the MODIS-EOS." *IEEE Transaction on Geoscience and Remote Sensing* 32:897–905.

Jensen, J. R. 2007. *Remote Sensing of the Environment: An Earth Resource Perspective*. 2nd ed. Prentice Hall.

Jonckheere, I., S. Fleck, K. Nackaerts, B. Muys, P. Coppin, M. Weiss, and F. Baret. 2004. "Methods for Leaf Area Index Determination. Part I:Theories, Techniques, and Instruments." *Agricultural and Forest Meteorology* 121:19–35.

Jordan, C. F. 1969. "Derivation of Leaf Area Index from Quality of Light on the Forest Floor." *Ecology* 50:663–66.

Kaufman, Y. J., and D. Tanre. 1992. "Atmospherically Resistant Vegetation Index (ARVI) for EOS-MODIS." *IEEE Transactions on Geoscience and Remote Sensing* 30:261–70.

Knyazikhin, Y., J. V. Martonchik, R. B. Myneni, D. J. Diner, and S. W. Running. 1998. "Synergistic Algorithm for Estimating Vegetation Canopy Leaf Area Index and Fraction of Absorbed Photosynthetically Active Radiation from MODIS and MISR Data." *Journal of Geophysical Research* 103(D24):32, 257–75.

Kucharik, C. J., J. M. Norman, and S. T. Gower. 1998. "Measurements of Branch Area and Adjusting Leaf Area Index Indirect Measurements." *Agricultural and Forest Meteorology* 91:69–88.

Kuusk, A., and T. Nilson. 2000. "A Directional Multispectral Forest Reflectance Model." *Remote Sensing of Environment* 72:244–62.

Leblanc, S. G. 2002. "Correction to the Plant Canopy Gap Size Analysis Theory Used by the Tracing Radiation and Architecture of Canopies (TRAC) Instrument." *Applied Optics* 31(36):7667–70.

Leblanc, S. G., J. M. Chen, R. Fernandes, D. W. Deering, and A. Conley. 2005. "Methodology Comparison for Canopy Structure Parameters Extraction from Digital Hemispherical Photography in Boreal Forests." *Agricultural and Forest Meteorology* 129:187–207.

Li, X., and A. H. Strahler. 1985. "Geometric-Optical Modeling of a Conifer Forest Canopy." *IEEE Transactions on Geoscience and Remote Sensing* 23:705–21.

Lillesand, T. M., R. W. Kiefer, and J. W. Chipman. 2008. *Remote Sensing and Image Interpretation*. 6th ed. John Wiley & Sons.

Liu, J., J. M. Chen, J. Cihlar, and W. Chen. 2002. "Net Primary Productivity Mapped for Canada at 1-km Resolution." *Global Ecology and Biogeography* 11:115–29.

Liu, Y., R. Liu, and J. M. Chen. 2012. "Retrospective Retrieval of Long-Term Consistent Global Leaf Area Index (1991–2010) Maps from Combined AVHRR and MODIS Data." *Journal of Geophysical Research-Biogeosciences* 117:G04003.

Masson, V., J. L. Champeaux, F. Chauvin, C. Meriguer, and R. Lacaze. 2003. "A Global Database of Land Surface Parameters at 1 km Resolution in Meteorological and Climate Models." *Journal of Climate* 16:1261–82.

Miller, J. B. 1967. "A Formula for Average Foliage Density." *Australian Journal of Botany* 15:141–44.

Miller, E. E., and J. M. Norman, 1971. "Sunfleck Theory for Plant Canopies 1. Lengths of Sunlit Segments along a Transect." *Agronomy Journal* 63(5):735–38.

Monsi, and Saeki. 1953. "Uber Den Lichifktor in Den Pflanzengesellschaften und Scine Bedeutung Fur Die Stoffprodcktion." *Japanese Journal of Botany* 14:22–52.

Nemani, R., L. Pierce, S. Running, and L. Band. 1993. "Forest Ecosystem Processes at the Watershed Scale: Sensitivity to Remotely Sensed Leaf Area Index Estimates." *International Journal of Remote Sensing* 14:2519–34.

Nilson, T. 1971. "A Theoretical Analysis of the Frequency of Gaps in Plant Stands." *Agricultural Meteorology* 8:25–38.

Norman, J. M., and G. S. Campbell. 1989. "Canopy Structure." In *Plant Physiological Ecology: Field Methods and Instrumentation*, edited by R. W. Pearcy, J. Ehlerlnger, H. A. Mooney, and P. W. Rundel, 301–25. New York: Chapman and Hall.

Oker-Blom, P. 1986. "Photosynthetic Radiation Regime and Canopy Structure in Modeled Forest Stands." *Acta Forest Fennica* 197:1–44.

Olsson, L., K. Carlsson, H. Grip, and K. Perttu. 1982. "Evaluation of Forest-Canopy Photographs with Diode-Array Scanner OSIRIS." *Canadian Journal for Forest Research* 12:822–28.

Pinty, B., N. Gobron, J. -L. Widlowski, T. Lavergne, and M. M. Verstraete. 2004. "Synergy between 1-D and 3-D Radiation Transfer Models to Retrieve Vegetation Canopy Properties from Remote Sensing Data." *Journal of Geophysical Research* 109:D21205.

Pinty, B., J. L. Widlowski, M. Taberner, N. Gobron, M. M. Verstraete, M. Disney, et al. 2004. "Radiation Transfer Model Intercomparison (RAMI) Exercise: Results from the Second Phase." *Journal of Geophysical Research* 109:D06210.

Pinty, B., and M. M. Verstrate. 1992. "GEMI: A Non-Linear Index to Monitor Global Vegetation from Satellites." *Vegetation* 101:15–20.

Pisek, J., and J. M. Chen. 2009. "Mapping of Forest Background Reflectivity over North America with the NASA Multiangle Imaging SpectroRadiometer (MISR)." *Remote Sensing of Environment* 113:2412–23.

Qi, J., A. Chehbouni, A. R. Huete, Y. H. Kerr, and S. Sorooshian. 1994. "A Modified Soil Adjusted Vegetation Index." *Remote Sensing of Environment* 48:119–126.

Rich, P. 1990. "Characterising Plant Canopies with Hemispherical Photographs." *Remote Sensing Reviews* 5(1):13–29.

Ross, J. 1981. *The Radiation Regime and Architecture in Plant Stands*. New York: Springer.

Roujean, J. L., and F. M. Breon. 1995. "Estimating PAR Absorbed by Vegetation from Bidirectional Reflectance Measurements." *Remote Sensing of Environment* 51:375–84.

Roujean, J. L., M. J. Leroy, and P. Y. Deschamps. 1992. "A Birectional Reflectance Model of the Earth's Surface for the Correction of Remote Sensing Data." *Journal of Geophysical Research* 97:20455–68.

Rouse, J. W., R. H. Hass, J. A. Shell, and D. W. Deering. 1974. "Monitoring Vegetation Systems in the Great Plains with ERTS-1." *Third Earth Resources Technology Satellite Symposium* Vol. 1, 309–17.

Ryu, K., J. Verfaillie, C. Macfarlane, H. Kobayashi, O. Sonnentag, R. Vargas, S. Ma, and D. Baldocchi. 2012. "Continuous Observations of Tree Leaf Area Index at Ecosystem Scale Using Upward-Pointing Digital Cameras." *Remote Sensing of Environment* 126:116–25.

Sellers, P., P. J. Sellers, R. E. Dickinson, D. A. Randall, A. K. Betts, F. G. Hall, et al. 1997. "Modeling the Exchange of Energy, Water, and Carbon Between Continents and the Atmosphere." *Science* 275:502–09.

Smith, N. J., J. M. Chen, and T. A. Black. 1993. "Effects of Foliage Clumping on Estimates of Stand Leaf Area Density Using the LICOR LAI-2000." *Canadian Journal of Forest Research* 23:1940–43.

Stenberg, P., S. Linder, H. Smolander, and J. Flower-Ellis. 1994. "Performance of the LAI-2000 Plant Canopy Analyzer in Estimating Leaf Area Index of Some Scots Pine Stands." *Tree Physiology* 14:981–95.

Stenberg, P., M. Rautiainen, T. Manninen, and P. Voipio. 2008. "Boreal Forest Leaf Area Index from Optical Satellite Images: Model Simulations and Empirical Analyses Using Data from Central Finland." *Boreal Environment Research* 13:433–43.

Stenberg, P., M. Rautiainen, T. Manninen, P. Voipio, and H. Smolander. 2004. "Reduced Simple Ratio Better than NDVI for Estimating LAI in Finnish Pine and Spruce Stands." *Silva Fennica* 38(1):3–14.

Tian, Q., Z. Luo, J. M. Chen, M. Chen, and F. Hui. 2007. "Retrieving Leaf Area Index for Coniferous Forest in Xingguo County, China with Landsat ETM+ Images." *Journal of Environmental Management* 85(3):624–27.

Wagner, S. 2001. "Relative Radiance Measurements and Zenith Angle Dependent Segmentation in Hemispherical Photography." *Agricultural and Forest Meteorology* 107:103–15.

Walter, J. M., R. A. Fournier, K. Soudani, and E. Meyer. 2003. "Integrating Clumping Effects in Forest Canopy Structure: An Assessment through Hemispherical Photographs." *Canadian Journal of Remote Sensing* 29(3):388–410.

Walthall, C. L., J. M. Norman, J. M. Welles, G. Campbell, and B. L. Blad. 1985. "Simple Equation to Approximate the Bidirectional Reflectance from Vegetation Canopies and Bare Soil Surfaces." *Applied Optics* 24:383–87.

Wang, Y., C. E. Woodcock, W. Buermann, P. Stenberg, P. Voipio, H. Smolander, et al. 2004. "Evaluation of the MODIS LAI Algorithm at a Coniferous Forest Site in Finland." *Remote Sensing of Environment* 91(1):114–27.

Warren Wilson, J. 1960. "Inclined Point Quadrats." *New Phytologist* 59:1–8.

Weiss, M., F. Baret, G. J. Smith, I. Jonckheere, and P. Coppin. 2004. "Review of Methods for In Situ Leaf Area Index (LAI) Determination Part II: Estimation of LAI, Errors and Sampling." *Agricultural and Forest Meteorology* 121:37–53.

White, M. A., P. E. Thornton, S. W. Running, and R. R. Nemani. 2000. "Parameterization and Sensitivity Analysis of the BIOME–BGC Terrestrial Ecosystem Model: Net Primary Production Controls." *Earth Interactions* 4(3):1–85.

Whitford, K. R., I. J. Colquhoun, A. R. G. Lang, and B. M. Harper. 1995. "Measuring Leaf Area Index in a Sparse Eucalypt Forest: A Comparison of Estimate from Direct Measurement, Hemispherical Photography, Sunlight Transmittance, and Allometric Regression." *Agricultural and Forest Meteorology* 74:237–49.

Widlowski, J. L., M. Taberner, B. Pinty, V. Bruniquel-Pinel, M. Disney, R. Fernandes, et al. 2007. "Third Radiation Transfer Model Intercomparison (RAMI) Exercise: Documenting Progress in Canopy Reflectance Models." *Journal of Geophysical Research* 112:D09111.

Xiao, Z., S. Liang, J. Wang, P. Chen, and X. Yin 2013. "Leaf Area Index Retrieval from Multi-Sensor Remote Sensing Data Using General Regression Neural Networks." *IEEE Transactions on Geoscience and Remote Sensing*, doi: 10.1109/TGRS.2013.2237780.

Yang, W., D. Huang, B. Tan, J. C. Stroeve, N. V. Shabanov, Y. Knyazikhin, R. R. Nemani, and R. B. Myneni. 2006. "Analysis of Leaf Area Index and Fraction Vegetation Absorbed PAR Products from the Terra MODIS Sensor:2000–2005." *IEEE Transactions on Geoscience and Remote Sensing* 44(7):1829–42.

Yang, F., K. Mitchell, Y. T. Hou, Y. Dai, X. Zeng, Z. Wang, and X. Z. Liang. 2008. "Dependence of Land Surface Albedo on Solar Zenith Angle: Observations and Model Parameterization." *Journal of Applied Meteorology and Climatology* 47:2963–82.

Zhang, Y., J. M. Chen, and J. R. Miller. 2005. "Determining Exposure of Digital Hemispherical Photographs for Leaf Area Index Estimation." *Agricultural and Forest Meteorology* 133:166–81.

Zou, J., G. Yan, L. Zhu, and W. Zhang. 2009. "Woody-to-Total Area Ratio Determination with a Multispectral Canopy Imager." *Tree Physiology* 29(8):1069.

21

LiDAR Remote Sensing of Vegetation Biomass

Qi Chen

CONTENTS

21.1 Introduction...399
21.2 Remote Sensing of Vegetation Biomass Using Different LiDAR Systems.................401
    21.2.1 Airborne Small-Footprint Discrete-Return Scanning LiDAR.......................401
        21.2.1.1 Area-Based Methods 402
        21.2.1.2 Individual Tree Analysis Methods..405
    21.2.2 Airborne Small-Footprint Discrete-Return Profiling LiDAR.......................405
        21.2.2.1 Useful Profiling LiDAR Metrics for Biomass Estimation.................406
        21.2.2.2 Challenges of Profiling LiDAR for Biomass Mapping......................406
    21.2.3 Airborne Medium- and Small-Footprint Waveform LiDARs 407
        21.2.3.1 Useful Waveform Metrics for Biomass Estimation 407
        21.2.3.2 Importance of Correcting Attenuated Waveforms 407
    21.2.4 Satellite Large-Footprint Waveform LiDAR 408
        21.2.4.1 Challenges of Large-Footprint Satellite LiDAR for Biomass
                Mapping 408
    21.2.5 Ground-Based Discrete-Return Scanning LiDAR 409
        21.2.5.1 Challenges of Ground-Based LiDAR for Biomass Estimation 409
21.3 General Challenges and Future Directions 409
    21.3.1 Fusion of LiDAR with Other Remotely Sensed Data 409
        21.3.1.1 Fusion with Optical Imagery 409
        21.3.1.2 Fusion with Radar...410
    21.3.2 Generalizability of Biomass Models 410
    21.3.3 Biological and Ecological Interpretation of Biomass Models 412
21.4 Conclusions...412
References..413

21.1 Introduction

Accurate estimates of vegetation biomass are critical for calibrating and validating biogeochemical models (Hurtt et al. 2010), quantifying carbon fluxes from land use and land cover change (Shukla et al. 1990; Houghton et al. 2001), and supporting the United Nations Framework Convention on Climate Change (UNFCCC) program to reduce deforestation and forest degradation (Reducing Emissions from Deforestation and Forest Degradation) (Asner 2009). For instance, it was argued that at least half of the uncertainty in the estimates of emissions of carbon from land use change results from uncertain estimates of biomass density (Houghton 2005; Houghton et al. 2009).

Vegetation biomass is commonly defined as the amount of organic matter in living and dead plant materials expressed as dry weight per unit area (Brown 1997). Estimates of biomass over large areas have been determined by a wide range of approaches, including interpolating measurements from a relatively small number of plots typically for ecological studies (Houghton et al. 2001; Malhi et al. 2006); averaging measurements from a large number of forest inventory plots determined by rigorous statistical sampling (Brown and Schroeder 1999; Jenkins et al. 2001); stratifying and multiplying field measurements with categorical land cover or biome maps (Fearnside 1997; Gibbs et al. 2007); geographic information system modeling with environmental variables such as soil, elevation, and precipitation (Olson et al. 1983; Brown et al. 1993; Iverson et al. 1994; Brown and Gaston 1995; Gaston et al. 1998); process-based biogeochemical modeling (Potter 1999; Potter et al. 2001); and direct mapping of the continuous biomass field based on remotely sensed imagery and relevant vegetation products (Houghton et al. 2001; Blackard et al. 2008). Among them, the remote sensing–based approach is appealing because it can provide spatially explicit estimates of the actual biomass at each pixel location, instead of only the average/total biomass within a given inventory unit (such as county, state, and country) (Jenkins et al. 2001) or the potential/equilibrium biomass (Iverson et al. 1994; Potter et al. 2001). The fine-grained biomass maps generated from remotely sensed data will enhance our understanding of contemporary environmental changes since natural and human disturbances can occur at scales as small as 10–1000 m (Houghton 2005; Houghton et al. 2009).

Remote sensing of biomass has been based on reflectance (e.g., Blackard et al. 2008), vegetation indices (e.g., Myneni et al. 2001; Tan et al. 2007), leaf area index (e.g., Feng et al. 2007), vegetation cover (e.g., Houghton et al. 2001), or their combinations (e.g., Blackard et al. 2008). Conventionally, the remotely sensed data used are optical imagery and radio detecting and ranging (radar), such as data from Landsat (e.g., Foody et al. 2003), Moderate Resolution Imaging Spectroradiometer (e.g., Blackard et al. 2008), advanced very high radiation radiometer (e.g., Myneni et al. 2001), hyperspectral imagery such as Hyperion (e.g., Thenkabail et al. 2004), high-spatial-resolution imagery such as IKONOS and QuickBird (e.g., Clark et al. 2004), and synthetic aperture radar (SAR) (e.g., Le Toan et al. 1992; Kasischke et al. 1997). The main challenge for optical and radar data is the insensitivity or saturation of remotely sensed signals at moderate to high biomass levels (Waring et al. 1995). For example, Le Toan et al. (2004) summarized that the saturation of AirSAR and E-SAR occurs at around 30, 50, and 150–200 Mg/ha at C, L, and P bands, respectively, over different types of temperate, boreal, and tropical forests. Relatively new radar technologies such as InSAR (Balzter et al. 2007; Treuhaft et al. 2009; Brown et al. 2010) and PolInSAR (Cloude and Papathanassiou 1998; Mette et al. 2003; Garestier et al. 2008; Neumann et al. 2010) can retrieve canopy height (CH), which can be used to improve biomass mapping. However, their performance over complex forests of high biomass is to be rigorously tested.

A promising new approach for estimating biomass is light detection and ranging (LiDAR), which does not saturate even at very high biomass levels (>1000 Mg/ha) (Means et al. 1999). LiDAR provides unprecedented accuracy in measuring vegetation structures such as height (Nelson et al. 2000; Lefsky et al. 2005a; Chen et al. 2006; Dolan et al. 2009; Chen 2010b), crown size (Riano et al. 2003; Chen et al. 2006), basal area (Nelson et al. 1997; Gobakken and Naesset 2005; Chen et al. 2007a), stem volume (Holmgren et al. 2003; Hollaus et al. 2007; Chen et al. 2007a;), and vertical profile (Harding et al. 2001; Zhao et al. 2009), all of which can be related to biomass. The potential of LiDAR for vegetation biomass retrieval was demonstrated more than two decades ago using airborne profiling LiDAR systems (Nelson et al. 1984, 1988a). During the last decade, the body of literature on LiDAR remote sensing of biomass grew rapidly, mainly due to the advances made in sensors, software, and data availability (Chen 2007). Recent reviews on LiDAR remote sensing of biomass can be found in the works of Koch (2010), Gleason and Im (2011), Lu et al. (2012), and Rosette et al. (2012). This chapter summarizes past research advances on biomass estimation using airborne, spaceborne, and ground-based LiDAR systems and discusses key research questions and challenges and future directions.

21.2 Remote Sensing of Vegetation Biomass Using Different LiDAR Systems

LiDAR measures distances between a sensor and objects based on the time lags between transmitting light amplification by stimulated emission of radiation (laser) beams from the sensor and receiving signals reflected from the illuminated objects. The distances derived from LiDAR, combined with the position of the sensor and the direction of the laser beam, uniquely determine the three-dimensional (3D) coordinates of the objects illuminated. The errors of 3D coordinates vary with a myriad of factors such as laser range sampling interval, global positioning system (GPS) positioning, inertial measurement unit, flying altitude, and surface reflectivity. But, in general, the vertical precision of position is on the order of decimeters for airborne and satellite LiDAR (Zwally et al. 2002; Chen 2010a).

LiDAR remote sensing systems can be distinguished based on the way in which returned signals are recorded (discrete return or waveform), scanning patterns (profiling or scanning), platforms (airborne, spaceborne, or ground based), and footprint sizes (small footprint: ~1 m or smaller, medium footprint: ~10–30 m, or large footprint: ~50 m or larger) (Lu et al. 2012). The most common configurations of LiDAR systems are airborne small-footprint discrete-return scanning LiDAR, airborne small-footprint discrete-return profiling LiDAR, airborne medium-footprint waveform LiDAR, and satellite large-footprint waveform systems. Ground-based systems and airborne small-footprint waveform systems are also emerging. The key findings and issues on the use of these LiDAR systems for biomass retrieval are summarized and discussed in Sections 21.2.1 through 21.2.5.

21.2.1 Airborne Small-Footprint Discrete-Return Scanning LiDAR

An airborne small-footprint discrete-return scanning LiDAR system typically includes three key components: (1) a laser range finder to measure distance, (2) a differential GPS system to measure the 3D position of the airplane/sensor, and (3) an inertial navigation system to measure the attitude of the airplane. Such systems can record the information of one or multiple returns of a transmitted laser pulse. The information of a laser return includes, at a minimum, 3D coordinates and usually intensity, laser pulse angle, the number of the return (first, last, or other), and so on, among which the 3D coordinates are the key information. So, the data acquired by a discrete-return LiDAR system is also called a point cloud. Over the last decade, airborne small-footprint discrete-return scanning LiDAR has experienced thrilling developments in hardware and software so that today data can be acquired with a much lower cost and be processed more effectively (Chen 2007, 2009; Chen et al. 2007b). For example, the laser pulse repetition rate has reached 500 kHz, whereas it was only 10–20 kHz a decade ago. Some earlier systems only record either the first or the last return (Omasa et al. 2003), whereas most current systems can record both first and last returns and even up to four returns (Table 21.1). As a mainstream LiDAR system for terrain mapping and various environmental applications, airborne discrete-return scanning LiDAR has been used in a large number of studies for mapping biomass, mainly using two kinds of approaches: (1) area-based and individual tree–based methods.

21.2.1.1 Area-Based Methods

Area-based methods develop statistical models to relate biomass with metrics derived from a LiDAR point cloud at the plot or stand level and apply the models over the whole study area. Such methods have been used to estimate the biomass of boreal forests (Lim et al. 2003; Naesset and Gobakken 2008; St-Onge et al. 2008; Thomas et al. 2006), temperate forests (Hall et al. 2005; Bortolot 2006; van Aardt et al. 2006; Zhao et al. 2009; Chen et al. 2012; Gleason and Im 2012; Lu et al. 2012), savanna woodlands (Lucas et al. 2006), and tropical forests (Asner 2009; Asner et al. 2009; Kronseder et al. 2012). The development of statistical models requires field data for calibration and validation. Almost all of these studies were conducted at the landscape level; so the numbers of field plots or stands used were relatively small (~20–60) except in a few studies such as the one by Naesset and Gobakken (2008), which used more than 1000 forest inventory plots to estimate the biomass for 10 different forest areas in south Norway. Most of the field plots had a fixed radius of ~10–50 m, whereas some used plots of variable radii (e.g., van Aardt et al. 2006). Besides estimating total aboveground biomass (AGBM), some also modeled the biomass

TABLE 21.1

Summary of Different LiDAR Systems

Type	Characteristics	Examples of Developers (Systems)
Airborne small-footprint discrete-return scanning LiDAR	High horizontal and vertical position accuracy (on the order of 10 cm); footprint size less than 1 m (at the flying altitude of 1000 m); PRF up to a few hundred kHz; usually records both first and last returns and up to four returns	Optech (ALTM); Lecia (ALS); Riegl (LMS)
Airborne small-footprint waveform scanning LiDAR	The same as above, except that waveform can be recorded with the addition of a waveform digitizer	Optech (ALTM); Leica (ALS); Riegl (LMS, VQ)
Airborne small-footprint discrete-return profiling LiDAR	Horizontal positional accuracy is low (on the order of 10 m); footprint size less than 1 m (at the flying altitude of 1000 m); PRF of a few kHz; usually records only first return and/or last return	NASA (PALS)
Airborne medium-footprint waveform LiDAR	High horizontal and vertical position accuracy (on the order of decimeters); footprint size of 10–30 m with a flying altitude of ~10 km; PRF of tens to hundreds of Hz; records the received waveform	NASA (SLICER and LVIS)
Satellite large-footprint waveform LiDAR	High vertical position accuracy (on the order of 10 cm); nominal footprint size of 60 m with a flying altitude of ~600 km; PRF of 40 Hz; records the received waveform	NASA (GLAS)
Ground-based small-footprint discrete-return scanning LiDAR	Very high horizontal and vertical position accuracy (on the order of millimeters depending on the distance); PRF of a few to tens of kHz; records the first or last return; range varies from a few hundred to a few thousand meters	Leica (ScanStation); Optech (ILRIS); Riegl (LMS, VZ)

Note: PRF, pulse repetition frequency, kHZ, thousand Hertz.

of individual components such as foliage (Lim and Treitz 2004; Hall et al. 2005; Thomas et al. 2006), branches (Lim and Treitz 2004; Thomas et al. 2006), stems (Lim and Treitz 2004; Thomas et al. 2006), and roots (Thomas et al. 2006). All studies modeled live vegetation biomass. Kim et al. (2009) reported the results for modeling the biomass of dead trees as well.

21.2.1.1.1 Different Ways of Deriving Area-Based LiDAR Metrics

The most widely used area-based LiDAR metrics for biomass prediction are various height metrics (Lim et al. 2003; Lim and Treitz 2004; Patenaude et al. 2004; Hall et al. 2005). Height metrics are usually calculated from vegetation returns, which are typically defined as returns with a certain height (such as 0.5 m [Lucas et al. 2006], 2 m [Naesset and Gobakken 2008], or 3 m [Hall et al. 2005]; Kim et al. 2009) above the ground surface. More complicated methods than using thresholds were also developed. For example, in a mature hardwood forest in Ontario, Canada, Lim and Trietz (2004) used the expectation–maximization algorithm to identify overstory canopy returns and obtained an R^2 of 0.90 using the 25th percentile height. However, this algorithm assumes that overstory and understory returns are relatively well separated. Therefore, its application over dense and continuous canopy could be unsuccessful, especially when LiDAR point density is low (Thomas et al. 2006).

LiDAR metrics can be calculated based on first, last, or all returns. There is no one-to-one correspondence between first returns and vegetation or last returns and the ground. However, first returns are usually more related to canopy surface structure than last returns (Thomas et al. 2006). A few studies have used LiDAR metrics derived from first returns for predicting biomass. For example, Hall et al. (2005) found that a canopy cover metric derived from first returns was able to predict foliage biomass and total AGBM in a Ponderosa pine forest in Colorado with R^2 values of 0.79 and 0.74, respectively. Kim et al. (2009) found that the model using first returns improved R^2 by 0.1 for predicting the total AGBM in the mixed coniferous forest in Arizona compared with the one using all returns.

Instead of directly using the point cloud, height metrics can also be calculated from grids of the CH model (CHM) (Asner et al. 2009; Patenaude et al. 2004; St-Onge et al. 2008; Zhao et al. 2009). CHM usually records the maximum CH within each grid cell. Therefore, the height metrics derived from CHM are more related to upper canopy instead of the complete vertical profile of vegetation structure. Lu et al. (2012) found that the LiDAR metrics generated from CHM, that is, first returns, last returns, and all returns, have a similar performance (with an R^2 difference of up to 0.02) in predicting biomass over a mixed coniferous forest in Sierra Nevada of California.

Different cell sizes have been used in generating CHM, ranging from 0.5 m (Zhao et al. 2009) and 1 m (Patenaude et al. 2004; St-Onge et al. 2008) to 5 m (Asner et al. 2009). There was no discussion in these studies on how cell sizes were determined. However, it is expected that a larger cell size will produce a smoother CHM that can weigh taller canopy elements more since a CHM usually records the height of the highest canopy returns within a grid cell.

The intensity of a laser return has also been used in a few studies to generate metrics for predicting biomass. For example, Lim et al. (2003) used the mean height of points with intensity >200 to predict the biomass of a mature hardwood forest in Ontario using a simple power model. It was found that using such a subset of laser points increases the coefficient of determination, R^2, from 0.72 to 0.85. Kim et al. (2009) used a stepwise regression model to predict the biomass in a mixed coniferous forest in Grand Canyon National Park, Arizona, with 17 LiDAR metrics extracted from the height and intensity information of LiDAR returns. It was found that the canopy volume of the LiDAR returns of high intensity and the low intensity peak count of the intensity distribution were among the best for predicting live and dead tree biomass, respectively.

21.2.1.1.2 Useful Height Metrics for Biomass Estimation

It would be ideal if a LiDAR height metric that can predict biomass well over different sites is identified. Unfortunately, the best height metric reported in the literature differs a lot in previous studies, including the 80th (Patenaude et al. 2004), 75th (St-Onge et al. 2008), 50th (Thomas et al. 2006), 30th (Stephens et al. 2012), and 25th (Lim and Treitz 2004) percentile heights. Such variability is likely the result of differences in vegetation structure, model form, and data processing procedure. But the dominant cause is unclear. Data processing procedure and/or model form might be as important as, if not more important than, vegetation structure. For example, the three studies conducted by Lim and Treitz (2004), St-Onge et al. (2008), and Thomas et al. (2006) all focused on boreal forests, but their best LiDAR metrics for biomass estimation differ significantly. These three studies differ in model form and data processing procedure: Lim and Trietz (2004) and St-Onge et al. (2008) used a simple power model, whereas Thomas et al. (2006) used a simple linear model with square root—transformed biomass; Lim and Treitz (2004) and Thomas et al. (2006) derived height metrics from a raw point cloud, whereas St-Onge et al. (2008) calculated height metrics based on CHM. In particular, the percentile height in the study by Lim and Treitz (2004) was calculated based on overstory canopy returns, which might explain why a much lower percentile was chosen. On the other hand, the best metrics in the studies of Patenaude et al. (2004) and St-Onge et al. (2008) are similar (80th and 75th percentile heights, respectively) even though they focused on temperate deciduous forest and boreal forest, respectively. This is likely because the data processing procedure and the model form in the two studies are the same in that percentile heights were extracted from a $1 \times 1\ m^2$ CHM and biomass was modeled with a simple power model.

21.2.1.1.3 Other Useful Metrics for Biomass Estimation

Besides height metrics, other metrics such as the ones related to canopy cover (Hall et al. 2005), canopy density (Naesset and Gobakken 2008), and canopy volume (Kim et al. 2009) have been proved to be useful for predicting biomass as well. Canopy cover is useful especially over areas with low biomass density (Hall et al. 2005). Canopy cover and density can complement height metrics for a better characterization of 3D canopy structure because they are related to horizontal and vertical vegetation structures, respectively (Hall et al. 2005; Bortolot 2006; Naesset and Gobakken 2008). Among these metrics, canopy volume is unique given that it is a composite index for characterizing both horizontal and vertical canopy structures. This metric has been successfully used for predicting biomass (Hall et al. 2005), stem volume, and basal area (Chen et al. 2007a).

21.2.1.1.4 Common Area-Based Statistical Models

Power models have been used in a large number of studies for biomass estimation, probably because most allometric equations for calculating biomass in the field are power models. At the log–log scale, simple power models (Lim et al. 2003; Lim and Treitz 2004; Patenaude et al. 2004; St-Onge et al. 2008) correspond to simple linear regression models, whereas multiplicative power models (Naesset and Gobakken 2008) correspond to multiple linear regression models. Many studies have performed nonlinear transformation on LiDAR metrics or biomass. For example, the quadratic term of LiDAR metrics might be used (Asner 2009). Biomass might be log transformed (Hall et al. 2005) or square root transformed (Thomas et al. 2006) for reducing the heterogeneity of regression residual variance. Although nonlinear models have been widely used, there are a few studies that have used linear models for predicting biomass (Bortolot 2006; Lucas et al. 2006; van Aardt et al. 2006; Asner et al. 2009; Zhao et al. 2009). The area-based models are mostly statistical and thus have the risk of overfitting. If stepwise regression or similar statistical (e.g., CART [classification and regression tree]) methods are used to select the best subset of LiDAR metrics, it is not so intuitive to interpret the physical meaning of the model. More research will be needed to develop models with good generalizability in the estimation of biomass and canopy structure (Chen et al. 2007a; Chen 2010b).

21.2.1.2 Individual Tree Analysis Methods

Individual tree analysis (ITA) methods are to automatically identify individual tree crowns and extract individual tree information such as tree height and crown size, which can be related to biomass and other canopy structure variables through allometric equations (Chen et al. 2006; Chen et al. 2007a). If the individual tree boundaries can be delineated, LiDAR statistical metrics such as those used in area-based approaches can also be extracted by analyzing the LiDAR point cloud within individual tree polygons. Compared to area-based approaches, theoretically no fieldwork is needed for model development if tree crowns can be delineated with no errors and allometric equations exist to estimate biomass based on LiDAR metrics such as tree height and/or crown size. If there are no such allometric equations, the amount of fieldwork required is still much smaller than that for area-based approaches because field data are needed only for a sample of trees instead of a sample of plots or stands. In terms of biomass mapping, this is probably the most appealing factor of ITA methods compared with area-based approaches.

21.2.1.2.1 Challenges of ITA Methods for Biomass Estimation

Despite the benefits discussed in Section 21.2.1.2, ITA methods face several challenges that limit their broader applications. First of all, LiDAR data of high point density are usually needed to measure canopy surface morphology in sufficient detail so that trees can be automatically identified with intelligent algorithms. Although a minimal point density is not established in the literature, it is recommended to have LiDAR data with a spot spacing of 0.5 m or smaller for effective tree isolation. The LiDAR data density might not be a big issue nowadays due to the significant increase in laser repetition rates of recent airborne LiDAR sensors. The major limitation of ITA might be that tree recognition algorithms are prone to errors given the complexity of canopy surface, especially for broadleaf trees that do not have regular geometry and sharp treetops as coniferous trees do. Tree recognition is challenging if trees of different heights are clustered and there are no distinct elevation "valleys" among trees. It is also more difficult to detect understory and smaller trees than dominant/codominant trees in a forest. These problems can be alleviated if we can develop more accurate and computationally efficient algorithms for tree isolation (Chen et al. 2006) and develop LiDAR metrics and allometric equations that are less sensitive to errors in tree recognition (Chen et al. 2007a), which obviously are not trivial tasks. The advent of commercial airborne small-footprint waveform LiDAR provides better opportunities for detecting small trees under complex canopy, but the data analysis methods usually are complicated (Reitberger et al. 2009). Given the aforementioned challenges, only a few studies have used ITA methods to estimate biomass (e.g., Omasa et al. 2003; Bortolot and Wynne 2005; Popescu 2007).

21.2.2 Airborne Small-Footprint Discrete-Return Profiling LiDAR

Unlike scanning LiDAR, airborne discrete-return profiling LiDAR does not include an inertial navigation system. So the positional accuracy of laser returns is much lower (on the order of 10 m) (Nelson et al. 2004). However, profiling LiDAR systems are cost-effective for large-area vegetation inventory. For example, the cost of the portable airborne LiDAR system (PALS), a profiling LiDAR developed by the National Aeronautics and Space Agency (NASA), was around $30,000 (Nelson et al. 2003), whereas commercial airborne discrete-return scanning systems typically cost 1–2 million dollars. Nelson and colleagues have been pioneering the use of airborne profiling LiDAR for vegetation studies and have estimated the biomass of pine plantations in Georgia (Nelson et al. 1988a, b); tropical wet forests in Costa Rica (Nelson et al. 1997, 2000); statewide vegetation in Delaware (Nelson et al. 2004); the pinelands of New Jersey (Skowronski et al. 2007); and the forest south of the tree line in Québec, Canada (Boudreau et al. 2008; Nelson et al. 2009).

21.2.2.1 Useful Profiling LiDAR Metrics for Biomass Estimation

Similar to the studies that use scanning LiDAR systems, various metrics can be derived from LiDAR returns to estimate biomass. A variety of metrics have been demonstrated to be useful for predicting biomass, including canopy profile area (Nelson et al. 1988a, b), average CH of canopy returns (Nelson et al. 1997, 2000), coefficients of variation of the average height of canopy returns and all returns (Nelson et al. 1997, 2000), quadratic mean height (QMH) of canopy returns (Nelson et al. 2004), QMH of all returns (Boudreau et al. 2008; Nelson et al. 2009), and 80th percentile height (Skowronski et al. 2007). Note that the two-dimensional (2D) canopy profile area proposed by Nelson et al. (1988a, b) is analogous to the 3D canopy geometric volume generated from an airborne scanning LiDAR (Hall et al. 2005; Chen et al. 2007a).

21.2.2.2 Challenges of Profiling LiDAR for Biomass Mapping

Airborne profiling LiDARs have a low pulse repetition rate and sparse laser pulse density. Therefore, one of the main challenges of profiling LiDAR data processing is to automatically derive the ground elevation so that CH can be accurately estimated for each laser pulse. Ground returns were manually determined and connected in early studies (Nelson et al. 1988a). Later, Nelson et al. (2004) and Boudreau et al. (2008) proposed a semiautomatic approach to fit the lowest returns within a moving window with splines, followed by inspection and, if necessary, adjustment of the splines by an operator.

Another challenge of using profiling LiDAR is related to the relatively coarse positional accuracy of its LiDAR pulses, which makes it difficult to accurately register laser returns with field measurements of biomass, especially over areas such as the tropics where accessibility in the field is limited. To handle this problem, Nelson et al. (1997) proposed a simulator of laser returns based on the field data of tree characteristics when estimating biomass of three sites in the tropical wet forests of Costa Rica. Statistical models were developed to relate the biomass measured in the field with the simulated laser returns. Then, the developed statistical models were used to estimate biomass for a larger area. Such an approach was also used to estimate biomass for the state of Delaware (Nelson et al. 2004).

With its low cost and portability, PALS has been used to estimate biomass for relatively large areas such as the state of Delaware of 5205 km² (Nelson et al. 2004). Note that profiling LiDAR systems essentially can only sample vegetation along transects. Unlike scanning LiDAR systems, it is difficult to use them to directly generate wall-to-wall maps of biomass. To calculate the total biomass, a study area might need to be stratified with land cover maps and biomass is then estimated for each stratum (Nelson et al. 2004).

21.2.3 Airborne Medium- and Small-Footprint Waveform LiDARs

Compared to discrete-return LiDAR, waveform LiDAR can better characterize vegetation structure by recording the whole vertical profile of returned energy reflected from vegetation. However, its advantages might be marginal in terms of digital elevation model mapping, which is the major conventional application of airborne LiDAR. This might explain why there were no commercial airborne waveform LiDAR systems until 2004 (Mallet and Bretar 2009), almost 10 years behind the advent of the first commercial discrete-return LiDAR systems.

A number of studies have used experimental airborne medium-footprint waveform systems, specifically Scanning LiDAR Imager of Canopies by Echo Recovery (SLICER) (Blair et al. 1994) and Laser Vegetation Imaging Sensor (LVIS) (Blair et al. 1999), both developed by NASA. These two airborne waveform systems act as simulators for satellite LiDAR missions, so they were flown at relatively high altitudes (~10 km) and have a medium-sized (~10–30 m, equivalent to one or two typical crown width) footprint. SLICER has been used to estimate biomass for temperate deciduous forests in Maryland (Lefsky et al. 1999b); boreal coniferous forests in Manitoba, Canada (Lefsky et al. 2002); and temperate coniferous forests in the Pacific Northwest (Means et al. 1999; Lefsky et al. 1999a, 2001, 2005b). Means et al. (1999) found that a SLICER of footprint size 10 m can predict total biomass and leaf biomass with R^2 values of 0.96 and 0.84, respectively, for coniferous forests in the Pacific coast with maximum biomass density up to 1300 Mg/ha, which provides strong evidence that LiDAR can map forests with very high biomass where conventional optical and radar remote sensing systems have saturation problems.

There are only a few studies that have used small-footprint waveform LiDAR for biomass mapping (e.g., Sarrazin et al. 2011; Kronseder et al. 2012). The data volume of small-footprint waveform LiDAR is at least one magnitude higher than discrete-return LiDAR data of comparable pulse density, and there is a lack of methods specifically designed for processing small-footprint waveform LiDAR data. For example, Kronseder et al. (2012) simply used the point clouds generated from waveform LiDAR for biomass estimation, making the methods not essentially different from the ones for discrete-return LiDAR. There is an urgent need for more sophisticated methods for fully exploiting the rich information in waveforms.

21.2.3.1 Useful Waveform Metrics for Biomass Estimation

A large number of metrics can be generated from waveforms. In the studies that use SLICER for biomass estimation, the best LiDAR metrics include mean CH (MCH) (Lefsky et al. 2001, 2002; Means et al. 1999), quadratic MCH (QMCH) (Means et al. 1999, Lefsky et al. 1999b), canopy volume (Lefsky et al. 1999a), and the number of waveforms taller than 55 m (Lefsky et al. 1999a). LVIS has been used to estimate biomass for tropical wet forests in Costa Rica (Drake et al. 2002a, b; Dubayah et al. 2010), coniferous forests in the Sierra Nevada mountain of California (Hyde et al. 2005), and temperate mixed forests in New Hampshire (Anderson et al. 2006). The best metrics for biomass estimation in LVIS-based studies include height of medium waveform elevation (HOME) relative to the ground peak elevation (Anderson et al. 2006; Drake et al. 2002a; Hyde et al. 2005), quantile heights (Drake et al. 2002b), mean, and minimum height (Hyde et al. 2005).

21.2.3.2 Importance of Correcting Attenuated Waveforms

The vertical profile of waveform energy does not necessarily resemble the vertical profile of vegetation structure because laser energy attenuates as it penetrates through a canopy.

Lefsky et al. (1999a) applied Beer's law to estimate the canopy vertical profile (CHP) from raw waveform signals before calculating LiDAR metrics. However, when Drake et al. (2002b) used the quantile metrics from raw LiDAR waveforms as well as from the CHP estimated by Beer's law to estimate total AGBM in the tropical wet forest in Costa Rica, it was found that the stepwise regression does not include the quantile metrics from CHP. In other words, this study indicated that the waveform transformation did not lead to better results for estimating total AGBM. Since the numbers of studies addressing this issue are few, more research is needed along this direction to reach a more general conclusion.

21.2.4 Satellite Large-Footprint Waveform LiDAR

The Geoscience Laser Altimeter System (GLAS) on board the Ice, Cloud, and Land Elevation Satellite (ICESat) was the only operating satellite LiDAR system from 2003 to 2009; today, a new-generation photon-counting satellite LiDAR is being developed (Abdalati et al. 2010). We focus on GLAS in this section. The lasers on GLAS transmit pulses from an altitude of ~600 km, producing 60 m nominal footprint size and ~170 m shot spacing along track on the ground. The receiver on GLAS records the vertical canopy structure up to 81.6 m for early laser campaigns (laser 1a and 2a periods) or 150 m for later campaigns to avoid the truncation of signals by tall objects or steep slopes (Chen 2010a).

Since its launch in 2003, GLAS has produced unprecedented datasets at the global scale. However, only a few studies have explored the use of GLAS in biomass mapping. This might be because of the high cost of collecting field measurements of biomass at the scale of GLAS footprints due to their large sizes. Also, a relatively large area has to be visited to get a representative sample for statistical modeling due to the sampling nature of GLAS waveforms, especially when the accessibility to specific shots is limited. One common solution to this problem is to use estimates from airborne LiDAR data as a proxy of ground truth (Chen 2010a, b) given the high accuracy and precision of airborne LiDAR–derived biomass estimates. Such a strategy has been used for GLAS-based biomass modeling in the boreal forest in Canada (Boudreau et al. 2008; Duncanson et al. 2010) and the peat swamp forest in Kalimantan, Indonesia (Ballhorn et al. 2011). Since GLAS is essentially sampling the vegetation along the orbit, it is usually combined with optical imagery to produce continuous maps of biomass (e.g., Duncanson et al. 2010; Saatchi et al. 2011; Mitchard et al. 2012).

21.2.4.1 Challenges of Large-Footprint Satellite LiDAR for Biomass Mapping

One of the main challenges for GLAS data processing is that signals from terrains and vegetation are convoluted and waveforms are broadened over mountainous areas due to its large footprint size. Usually, Gaussian decomposition is used to disentangle the terrain and vegetation signals to estimate terrain elevation (Chen 2010a; Liu and Chen 2012) and vegetation height (Chen 2010b), based on which biomass can be estimated. The large footprint size of GLAS was optimally set for detecting ice sheets and glaciers, not necessarily for retrieving vegetation information. Chen (2010b) simulated the different footprint sizes from 10 to 40 m for retrieving maximum vegetation height. It was found that a smaller footprint size can increase the performance in vegetation information retrieval and the footprint size of 10 m has a bias and a root-mean-square error of 1 m and 1–2 m, respectively, for estimating maximum CH even when models are applied across coniferous and woodland sites. Since the next satellite LiDAR mission, ICESat-2, carries a sensor of 10 m footprint (Abdalati et al. 2010), we expect that the terrain convolution problem will be greatly alleviated for vegetation information retrieval in the future.

21.2.5 Ground-Based Discrete-Return Scanning LiDAR

Compared with other LiDAR systems, ground-based LiDAR is unique in many aspects. Ground-based LiDAR systems can measure vegetation with very high precision without destructive measurement/sampling. It is also possible to distinguish different vegetation components (stem, branches, and foliage) from a point cloud. Besides, the upward scanning geometry of most ground-based LiDAR systems can help to better characterize understory and crown base height, which are important parameters for fire behavior prediction modeling. A few studies have used ground-based LiDAR for biomass estimation (e.g., Loudermilk et al. 2009; Seidel et al. 2011; Yao et al. 2011; Ku et al. 2012). Some studies (e.g., Loudermilk et al. 2009) have found that terrestrial LiDAR can measure canopy attributes with higher accuracy than the conventional field-based approach.

21.2.5.1 Challenges of Ground-Based LiDAR for Biomass Estimation

When ground-based LiDAR systems are used to measure trees in a plot or stand, the trees usually have to be scanned from multiple positions due to the shadowing effects caused by objects closer to the scanners. The 3D coordinates of laser returns collected at individual scanning positions are local and relative to the scanners. The individual datasets have to be georeferenced to a common coordinate system based on features visible to multiple positions, which is not a trivial task. To make it more difficult, ground-based LiDAR systems can acquire data with point densities 100–1000 times higher than the average point density of airborne small-footprint LiDAR systems. Such massive volumes of data pose a significant challenge in developing fast, automatic, and memory-efficient software for data processing and information extraction.

21.3 General Challenges and Future Directions

21.3.1 Fusion of LiDAR with Other Remotely Sensed Data

21.3.1.1 Fusion with Optical Imagery

LiDAR has limited spectral information usually with only one wavelength in the near-infrared region. It would be interesting to investigate whether the integration of multispectral or hyperspectral imagery with LiDAR can improve biomass estimation. Some previous studies have reported that the addition of multispectral or hyperspectral features to LiDAR resulted in only slight or no improvements in biomass estimation over mixed coniferous forests in California (Hyde et al. 2006), tropical forests in Costa Rica (Clark et al. 2011), and temperate coniferous forest in Germany (Latifi et al. 2012). The study by Anderson et al. (2008) is among the few studies that have reported significant improvements when integrating LiDAR with spectral features from optical data. They integrated LVIS and Airborne Visible/Infrared Imaging Spectrometer (AVIRIS) data to estimate AGBM in a northern temperate mixed coniferous and deciduous forest in Bartlett Experimental Forest (BEF) in central New Hampshire. It was found that 8%–9% more of the variation in AGBM was explained by the use of integrated sensor data in comparison with either AVIRIS or LVIS metrics applied singly.

21.3.1.2 Fusion with Radar

Radar is another technology that can potentially be fused with LiDAR data for biomass estimation. Interestingly enough, mixed results also have been reported in the literature. Hyde et al. (2006, 2007) have found that simply adding radar backscatters InSAR height at the locations of LiDAR data does not necessarily improve the performance of their biomass estimation for coniferous forests in California and Arizona. Nelson et al. (2007) made a similar conclusion when they used a low frequency (80–120 MHz) VHF radar (BioSAR) and a profiling LiDAR (PALS) to estimate biomass over loblolly pine plantations in North Carolina. In contrast, Banskota et al. (2011) obtained very difficult results when they used BioSAR and airborne profiling and scanning LiDAR for estimating forest biomass in Virginia: the combination of BioSAR with LiDAR can significantly increase model performance. They attributed the discrepancy between their findings and the earlier ones to the fact that the previous studies focused on more or less even-aged pine forests, whereas their studies involve not only conifers but also hardwood and mixed forests. Tsui et al. (2012) also found that the integration of C-band HH backscatter to a LiDAR-only biomass model can explain an additional 8.9% and 6.5% of the variability in total aboveground and stem biomass, respectively, whereas C-band polarimetric entropy explained an additional 17.9% of the variability in crown biomass.

In general, the dominant finding from previous studies is that the addition of radar and optical spectral features into LiDAR do not, or only slightly, improve biomass estimation. This is reasonable given that biomass is highly related to tree height, which can be accurately measured by LiDAR. The predictors derived from radar and image spectral features usually become insignificant once the LiDAR-derived height metrics are included in the model. One promising method for integrating LiDAR with other remote sensing data is to use the other data for vegetation classification and develop vegetation type–dependent biomass modeling approaches (e.g., Swatantran et al. 2011), especially with novel statistical skills such as mixed-effects modeling (e.g., Chen et al. 2012). Note that the discussion here only involves the integration of airborne LiDAR data and other remotely sensed imagery. Satellite LiDAR could have a very different strategy for data fusion because of its nature of sampling instead of wall-to-wall mapping. The data fusion issue related to satellite LiDAR is not included here due to space limitation.

21.3.2 Generalizability of Biomass Models

A key issue in biomass estimation with LiDAR data is model generalizability, the extent to which a model developed at one study area can be applied to other areas (Chen 2010b). This is critical for biomass estimation at the regional and global scales or for areas where field measurements of biomass are difficult or impossible to collect. There are two aspects from which model generalizability can be improved: first, to choose LiDAR metrics that are mostly related to biomass over a wide range of vegetation conditions and, second, to choose a model form that is applicable over different areas based on these metrics.

Several studies have attempted to identify LiDAR metrics with which a model can be applied across different sites. Lim et al. (2003) found that when maximum height was used to estimate biomass, different models should be fitted for different groups of plots in their mature hardwood forest in Ontario. However, when the mean height of the pulses with an intensity >200 is considered, there is no need to develop different models. Lefsky et al. (1999b) tested four height indices for estimating biomass. Of the four indices predicting AGBM, they found that QMCH had the best performance when the model based on it was applied over two different datasets.

A few studies have found that MCH is the best LiDAR metric that can be applied over different study areas (e.g., Lefsky et al. 2002; Asner et al. 2012). Lefsky et al. (2002) found that a simple model based on MCH (AGBM = 0.378MCH^2, where AGBM is in megagrams per hectare) can explain 84% of the biomass variability for 149 plots across three types of forests (temperate conifer, temperate deciduous, and boreal conifer) in North America. Asner et al. (2009, 2012) confirmed the usefulness of MCH for biomass mapping, but they (Asner et al. 2012) found that biomass is affected not only by MCH but also by basal area and wood density when they developed carbon (biomass) models across four different tropical study areas (Hawaii, Panama, Peru, and Madagascar). They found that the regional variations of MCH-based biomass models are dominated by the basal area–MCH relationship. This might partially explain why Anderson et al. (2006) found that the model by Lefsky et al. (2002) performed worse than HOME for estimating the AGBM of 20 circular plots in BEF using LVIS.

There is converging evidence showing that the relationships between biomass and LiDAR height metrics are vegetation type dependent (Maclean and Krabill 1986; Drake et al. 2003; Naesset and Gobakken 2008; Ni-Meister et al. 2010; Asner et al. 2012; Chen et al. 2012). This is reasonable given that vegetation biomass is determined not only by height but also by factors such as stem tapering and wood density (Niklas 1995; Chave et al. 2006), which could be vegetation specific and difficult to directly measure using LiDAR. There are at least two key research issues about incorporating vegetation type information into biomass modeling: First, how do we develop vegetation type–dependent biomass models? The most straightforward method is to fit a model per vegetation type, but the problem is that it will reduce the sample size per model if the total number of field plots is small and the number of vegetation types is large. Second, what level of detail (e.g., species, alliance, or association) should vegetation be stratified at? A very detailed stratification requires vegetation maps at the fine taxonomic level and a significant amount of effort for collecting field measurements for model calibration. On the other hand, a very coarse stratification can lead to a poor performance in biomass estimation. Chen et al. (2012) addressed the first issue using a mixed-effects model, a novel statistical technique, and the second issue by comparing two alliance-level vegetation classification systems (Society of American Foresters [SAF] and National Vegetation Classification [NVC]). They found that the mixed-effects model can deal with the issue of small sample size per vegetation type and improve biomass estimation, and the coarser SAF vegetation type has almost the same performance as the finer NVC types, indicating that the height–biomass relationships do not vary much at the fine taxonomic level. This is consistent with the finding of Nelson et al. (1988a), who showed that stratifying a pine plantation in Georgia into four southern pine species (loblolly pine, shortleaf pine, slash pine, and longleaf pine) did little to increase the accuracy of biomass estimation using airborne profiling LiDAR.

A more thorough study of the aforementioned issue depends on a good understanding of plant allometry between vegetation structure variables such as biomass, height, diameter-at-breast height, and wood density. Unfortunately, allometric equations are available only for a limited number of species, most of which are from developed countries (TerMikaelian and Korzukhin 1997; Jenkins et al. 2004; Lambert et al. 2005; Zhao et al. 2012). Over areas such as the tropics where species diversity is high, generalized biomass equations (Brown 1997; Chave et al. 2005) have been used, which may lead to a bias in estimating biomass for a particular species (Litton and Kauffman 2008). Moreover, developing or validating allometric equations is strenuous because it usually involves harvesting different tree components if a direct assessment is required. It is expected that accumulation of field measurements (Clark and Kellner 2012) and progress in ecological theory in plant allometry (Brown et al. 2005; Kozlowski and Konarzewski 2004) can shed more light on this research direction (Chen et al. 2007a).

It is very common for stepwise regression to be used to select predictors from a large number of LiDAR metrics for biomass estimation. However, stepwise regression usually suffers from the overfitting problem. Instead, models based on a priori knowledge and ecological theories might have better model generalizability when LiDAR metrics and model form are carefully chosen (Chen et al. 2007a). Magnussen et al. (2012) proposed a generic model to predict tree-size-related canopy attributes from the CH of LiDAR first returns. The model they proposed includes two predictors: mean and variance of CH. Using three separate datasets from Norway, they obtained compelling results when they estimated a number of canopy attributes (such as volume and basal area), not including biomass. It would be interesting to test such models for biomass estimation in the future.

21.3.3 Biological and Ecological Interpretation of Biomass Models

An issue relevant to the generalizability of LiDAR-based biomass models is the biological and ecological interpretation of the models. Most previous studies ran stepwise regression models with a large number of LiDAR metrics and reported the best model without much interpretation. As LiDAR-based biomass studies are rapidly accumulating, we should put more emphasis than before on interpreting the LiDAR metrics selected by a model so that other researchers have better ideas on what LiDAR metrics they should try for their specific studies and are able to make better sense of their own models.

For example, QMH has been found to be the best predictor for biomass in a couple of studies (e.g., Lefsky et al. 1999b; Chen et al. 2012; Lu et al. 2012). QMH can be calculated as

$\sqrt{\left(\frac{1}{n}\right) \times \sum_{i=1}^{n} h_i^2}$, where n is the number of laser points and h_i is the height above the ground

for point i. As indicated in the formula, QMH gives a higher weight to higher points. This is analogous to the fact that the biomass of a plot or stand is dominated by tall trees. QMH is also an index integrating the height of all points, similar to the fact that the biomass at a given area is an integration of the biomass of all trees. This makes it theoretically appealing for biomass estimation. As introduced earlier, MCH was also the best predictor of biomass in many studies (e.g., Lefsky et al. 2002; Asner et al. 2012). In most studies, the best models include either QMH or MCH, but not both. And there usually is a positive relationship between biomass and either metric. However, in coniferous forests on the Pacific coast, Means et al. (1999) found that their best model included both QMH and MCH. Moreover, they found that QMH is negatively related to biomass, whereas MCH has a positive effect on biomass. This was explained by the fact that mature stands have relatively lower biomass compared with old-growth stands at a given height, whereas the foliage of mature stands mostly is in the upper half of the canopy (note that QMH weighs higher canopy elements more than lower ones). This indicates that in the future both QMH and MCH should be attempted for predicting biomass if a study area includes old-growth forests.

21.4 Conclusions

Information on biomass is useful for numerous environmental applications and is critical for understanding and predicting global environmental changes. LiDAR is a state-of-the-art remote sensing technology of tremendous potential in biomass mapping without the limitation of saturation at high biomass levels. This chapter gives a thorough and in-depth review of the current status, future directions, and main challenges of biomass estimation using various satellite, airborne, and ground-based LiDAR systems.

Each type of LiDAR systems has its strengths and weaknesses, and it is possible to combine them for better biomass mapping (Nelson et al. 2009; Chen 2010a, b). Airborne and satellite LiDAR systems measure vegetation from above the canopy, missing much information on stem and understory. Instead, ground-based LiDAR systems can measure stem, branches, and understory in much more detail and thus can be used to calibrate and validate airborne and satellite LiDAR data and products. It is unlikely to generate a wall-to-wall map of biomass directly from satellite LiDAR due to its sparse spatial distribution of measurements. However, airborne LiDAR can be used to assess the sampling errors of satellite LiDAR and help optimize the design of next-generation satellite systems such as ICESat-2. It is expected that this comprehensive review will facilitate the use of different LiDAR systems in tandem for biomass mapping. Overall, research on using many LiDAR systems for biomass mapping is still in its infancy; much remains to be explored in the future to address issues such as data fusion and model generalizability.

References

Abdalati, W., H. Zwally, R. Bindschadler, B. Csatho, S. Farrell, H. Fricker, D. Harding, et al. 2010. "The ICESat-2 Laser Altimetry Mission." *Proceedings of the IEEE* 98(5):735–51.

Anderson, J., M. E. Martin, M. L. Smith, R. O. Dubayah, M. A. Hofton, P. Hyde, B. E. Blair, J. B. Peterson, and R. G. Knox. 2006. "The Use of Waveform Lidar to Measure Northern Temperate Mixed Conifer and Deciduous Forest Structure in New Hampshire." *Remote Sensing of Environment* 105:248–61.

Anderson, J. E., L. C. Plourde, M. E. Martin, B. H. Braswell, M. L. Smith, R. O. Dubayah, M. A. Hofton, and J. B. Blair. 2008. "Integrating Waveform Lidar with Hyperspectral Imagery for Inventory of a Northern Temperate Forest." *Remote Sensing of Environment* 112:1856–70.

Asner, G. P. 2009. "Tropical Forest Carbon Assessment:Integrating Satellite and Airborne Mapping Approaches." *Environmental Research Letters* 4:11. doi:10.1088/1748-9326/1084/1083/034009.

Asner, G. P., R. F. Hughes, T. A. Varga, D. E. Knapp, and T. Kennedy-Bowdoin. 2009. "Environmental and Biotic Controls over Aboveground Biomass throughout a Tropical Rain Forest." *Ecosystems* 12:261–78.

Asner, G. P., J. Mascaro, H. C. Muller-Landau, G. Vieilledent, R. Vaudry, M. Rasamoelina, J. S. Hall, and M. van Breugel. 2012. "A Universal Airborne LiDAR Approach for Tropical Forest Carbon Mapping." *Oecologia* 168:1147–60.

Ballhorn, U., J. Jubanski, and F. Siegert. 2011. "ICESat/GLAS Data as a Measurement Tool for Peatland Topography and Peat Swamp Forest Biomass in Kalimantan, Indonesia." *Remote Sensing* 3:1957–82.

Balzter, H., C. S. Rowland, and P. Saich. 2007. "Forest Canopy Height and Carbon Estimation at Monks Wood National Nature Reserve, UK, Using Dual-Wavelength SAR Interferometry." *Remote Sensing of Environment* 108:224–39.

Banskota, A., R. H. Wynne, P. Johnson, B. Emessiene. 2011. "Synergistic Use of Very High-Frequency Radar and Discrete-Return Lidar for Estimating Biomass in Temperate Hardwood and Mixed Forests." *Annals of Forest Science* 68:347–56.

Blackard, J. A., M. V. Finco, E. H. Helmer, G. R. Holden, M. L. Hoppus, D. M. Jacobs, A. J. Lister, et al. 2008. "Mapping US Forest Biomass Using Nationwide Forest Inventory Data and Moderate Resolution Information." *Remote Sensing of Environment* 112:1658–77.

Blair, J. B., D. B. Coyle, J. L. Bufton, and D. J. Harding. 1994. "Optimization of an Airborne Laser Altimeter for Remote Sensing of Vegetation and Tree Canopies." In *International Geoscience and Remote Sensing Symposium*, 939–41. Pasadena, CA.

Blair, J. B., D. L. Rabine, and M. A. Hofton. 1999. "The Laser Vegetation Imaging Sensor: A Medium-Altitude, Digitisation-Only, Airborne Laser Altimeter for Mapping Vegetation and Topography." *ISPRS Journal of Photogrammetry and Remote Sensing* 54:115–22.

Bortolot, Z. J. 2006. "Using Tree Clusters to Derive Forest Properties from Small Footprint Lidar Data." *Photogrammetric Engineering and Remote Sensing* 72:1389–97.

Bortolot, Z. J., and R. H. Wynne. 2005. "Estimating Forest Biomass Using Small Footprint LiDAR Data: An Individual Tree-Based Approach that Incorporates Training Data." *ISPRS Journal of Photogrammetry and Remote Sensing* 59:342–60.

Boudreau, J., R. F. Nelson, H. A. Margolis, A. Beaudoin, L. Guindon, and D. S. Kimes. 2008. "Regional Aboveground Forest Biomass Using Airborne and Spaceborne LiDAR in Quebec." *Remote Sensing of Environment* 112:3876–90.

Brown, S. 1997. "Estimating Biomass and Biomass Change of Tropical Forests: A Primer." *Food and Agriculture Organization*. http://www.fao.org/docrep/w4095e/w4095e00.HTM. Accessed on March 13, 2013.

Brown, S., and G. Gaston. 1995. "Use of Forest Inventories and Geographic Information Systems to Estimate Biomass Density of Tropical Forests: Application to Tropical Africa." *Environmental Monitoring and Assessment* 38:157–68.

Brown, S., L. R. Iverson, A. Prasad, and D. Liu. 1993. "Geographical Distributions of Carbon in Biomass and Soils of Tropical Asian Forests." *Geocarto International* 8:45–59.

Brown, C. G., K. Sarabandi, and L. E. Pierce. 2010. "Model-Based Estimation of Forest Canopy Height in Red and Austrian Pine Stands Using Shuttle Radar Topography Mission and Ancillary Data: A Proof-of-Concept Study." *IEEE Transactions on Geoscience and Remote Sensing* 48:1105–18.

Brown, S. L., and P. E. Schroeder. 1999. "Spatial Patterns of Aboveground Production and Mortality of Woody Biomass for Eastern US Forests." *Ecological Applications* 9:968–80.

Brown, J. H., G. B. West, and B. J. Enquist. 2005. "Yes, West, Brown and Enquist's Model of Allometric Scaling Is Both Mathematically Correct and Biologically Relevant." *Functional Ecology* 19: 735–38.

Chave, J., C. Andalo, S. Brown, M. A. Cairns, J. Q. Chambers, D. Eamus, H. Folster, et al. 2005. "Tree Allometry and Improved Estimation of Carbon Stocks and Balance in Tropical Forests." *Oecologia* 145:87–99.

Chave, J., H. C. Muller-Landau, T. R. Baker, T. A. Easdale, H. Ter Steege, and C. O. Webb. 2006. "Regional and Phylogenetic Variation of Wood Density Across 2456 Neotropical Tree Species." *Ecological Applications* 16:2356–67.

Chen, Q. 2007. "Airborne Lidar Data Processing and Information Extraction." *Photogrammetric Engineering and Remote Sensing* 73:109–12.

Chen, Q. 2009. "Improvement of the Edge-Based Morphological (EM) Method for Lidar Data Filtering." *International Journal of Remote Sensing* 30:1069–74.

Chen, Q. 2010a. "Assessment of Terrain Elevation Derived from Satellite Laser Altimetry over Mountainous Forest Areas Using Airborne Lidar Data." *ISPRS Journal of Photogrammetry and Remote Sensing* 65:111–22.

Chen, Q. 2010b. "Retrieving Canopy Height of Forests and Woodlands over Mountainous Areas in the Pacific Coast Region Using Satellite Laser Altimetry." *Remote Sensing of Environment* 114:1610–17.

Chen, Q., D. Baldocchi, P. Gong, and M. Kelly. 2006. "Isolating Individual Trees in a Savanna Woodland Using Small Footprint Lidar Data." *Photogrammetric Engineering and Remote Sensing* 72:923–32.

Chen, Q., P. Gong, D. Baldocchi, and Y. Q. Tian. 2007a. "Estimating Basal Area and Stem Volume for Individual Trees from Lidar Data." *Photogrammetric Engineering and Remote Sensing* 73:1355–65.

Chen, Q., P. Gong, D. Baldocchi, and G. Xie. 2007b. "Filtering Airborne Laser Scanning Data with Morphological Methods." *Photogrammetric Engineering and Remote Sensing* 73:175–85.

Chen, Q., G. Vaglio Laurin, J. Battles, and D. Saah. 2012. "Integration of Airborne Lidar and Vegetation Types Derived from Aerial Photography for Mapping Aboveground Live Biomass." *Remote Sensing of Environment* 121:108–17.

Clark, D. B., and J. R. Kellner. 2012. "Tropical Forest Biomass Estimation and the Fallacy of Misplaced Concreteness." *Journal of Vegetation Science* 23:1191-6.

Clark, D. B., J. M. Read, M. L. Clark, A. M. Cruz, M. F. Dotti, and D. A. Clark. 2004. "Application of 1-M and 4-M Resolution Satellite Data to Ecological Studies of Tropical Rain Forests." *Ecological Applications* 14:61–74.

Clark, M. L., D. R. Roberts, J. J. Ewel, and D. B. David. 2011. "Estimation of Tropical Rain Forest Aboveground Biomass with Small-Footprint Lidar and Hyperspectral Sensors." *Remote Sensing of Environment* 115:2931–42.

Cloude, S. R., and K. P. Papathanassiou. 1998. "Polarimetric SAR Interferometry." *IEEE Transactions on Geoscience and Remote Sensing* 36:1551–65.

Dolan, K., J. G. Masek, C. Q. Huang, and G. Q. Sun. 2009. "Regional Forest Growth Rates Measured by Combining ICESat GLAS and Landsat Data." *Journal of Geophysical Research-Biogeosciences* 114:G00E05. doi:10.1029/2008JG000893.

Drake, J. B., R. O. Dubayah, D. B. Clark, R. G. Knox, J. B. Blair, M. A. Hofton, R. L. Chazdon, J. F. Weishampel, and S. D. Prince. 2002a. "Estimation of Tropical Forest Structural Characteristics Using Large-Footprint Lidar." *Remote Sensing of Environment* 79:305–19.

Drake, J. B., R. O. Dubayah, R. G. Knox, D. B. Clark, and J. B. Blair. 2002b. "Sensitivity of Large-Footprint Lidar to Canopy Structure and Biomass in a Neotropical Rainforest." *Remote Sensing of Environment* 81:378–92.

Drake, J. B., R. G. Knox, R. O. Dubayah, D. B. Clark, R. Condit, J. B. Blair, and M. Hofton. 2003. "Above-Ground Biomass Estimation in Closed Canopy Neotropical Forests using Lidar Remote Sensing: Factors Affecting the Generality of Relationships." *Global Ecology and Biogeography* 12:147–59.

Dubayah, R. O., S. L. Sheldon, D. B. Clark, M. A. Hofton, J. B. Blair, G. C. Hurtt, and R. L. Chazdon. 2010. "Estimation of Tropical Forest Height and Biomass Dynamics Using Lidar Remote Sensing at La Selva, Costa Rica." *Journal of Geophysical Research* 115:G00E09. doi:10.1029/2009JG000933.

Duncanson, L. I., K. O. Niemann, and M. A. Wulder. 2010. "Integration of GLAS and Landsat TM Data for Aboveground Biomass Estimation." *Canadian Journal of Remote Sensing* 36(2):129–41.

Fearnside, P. M. 1997. "Greenhouse Gases from Deforestation in Brazilian Amazonia: Net Committed Emissions." *Climatic Change* 35:321–60.

Feng, X., G. Liu, J. M. Chen, M. Chen, J. Liu, W. M. Ju, R. Sun, and W. Zhou. 2007. "Net Primary Productivity of China's Terrestrial Ecosystems from a Process Model Driven by Remote Sensing." *Journal of Environmental Management* 85:563–73.

Foody, G. M., D. S. Boyd, and M. E. J. Cutler. 2003. "Predictive Relations of Tropical Forest Biomass from Landsat TM Data and Their Transferability between Regions." *Remote Sensing of Environment* 85:463–74.

Garestier, F., P. C. Dubois-Fernandez, and I. Champion. 2008. "Forest Height Inversion Using High-Resolution P-Band Pol-InSAR Data." *IEEE Transactions on Geoscience and Remote Sensing* 46:3544–59.

Gaston, G., S. Brown, M. Lorenzini, and K. D. Singh. 1998. "State and Change in Carbon Pools in the Forests of Tropical Africa." *Global Change Biology* 4:97–114.

Gibbs, H. K., S. Brown, J. O. Niles, and J. A. Foley. 2007. "Monitoring and Estimating Tropical Forest Carbon Stocks: Making REDD a Reality." *Environmental Research Letters* 2:045023. doi:10.1088/1748-9326/2/4/045023.

Gleason, C., and J. Im. 2011. "A Review of Remote Sensing of Forest Biomass and Biofuel: Options for Small-Area Applications." *GIScience and Remote Sensing* 48(2):141–70.

Gleason, C., and J. Im. 2012. "Forest Biomass Estimation from Airborne LiDAR Data Using Machine Learning Approaches." *Remote Sensing of Environment* 125:80–91.

Gobakken, T., and E. Naesset. 2005. "Weibull and Percentile Models for Lidar-Based Estimation of Basal Area Distribution." *Scandinavian Journal of Forest Research* 20:490–502.

Hall, S. A., I. C. Burke, D. O. Box, M. R. Kaufmann, and J. M. Stoker. 2005. "Estimating Stand Structure Using Discrete-Return Lidar: An Example from Low Density, Fire Prone Ponderosa Pine Forests." *Forest Ecology and Management* 208:189–209.

Harding, D. J., M. A. Lefsky, G. G. Parker, and J. B. Blair. 2001. "Laser Altimeter Canopy Height Profiles: Methods and Validation for Closed-Canopy, Broadleaf Forests." *Remote Sensing of Environment* 76:283–97.

Hollaus, M., W. Wanger, B. Nmaier, and K. Schadauer. 2007. "Airborne Laser Scanning of Forest Stem Volume in a Mountainous Environment." *Sensors* 7:1559–77.

Holmgren, J., M. Nilsson, and H. Olsson. 2003. "Estimation of Tree Height and Stem Volume on Plots-Using Airborne Laser Scanning." *Forest Science* 49:419–28.

Houghton, R. A. 2005. "Aboveground Forest Biomass and the Global Carbon Balance." *Global Change Biology* 11:945–58.

Houghton, R. A., F. Hall, and S. J. Goetz. 2009. "Importance of Biomass in the Global Carbon Cycle." *Journal of Geophysical Research-Biogeosciences* 114:G00E03. doi:10.1029/2009JG000935.

Houghton, R. A., K. T. Lawrence, J. L. Hackler, and S. Brown. 2001. "The Spatial Distribution of Forest Biomass in the Brazilian Amazon: A Comparison of Estimates." *Global Change Biology* 7:731–46.

Hurtt, G. C., J. Fisk, R. Q. Thomas, R. O. Dubayah, P. R. Moorcroft, and H. H. Shugart. 2010. "Linking Models and Data on Vegetation Structure." *Journal of Geophysical Research-Biogeosciences* 115:G00E10. doi:10.1029/2009JG000937.

Hyde, P., R. Dubayah, B. Peterson, J. B. Blair, M. Hofton, C. Hunsaker, R. Knox, and W. Walker. 2005. "Mapping Forest Structure for Wildlife Habitat Analysis Using Waveform Lidar: Validation of Montane Ecosystems." *Remote Sensing of Environment* 96:427–37.

Hyde, P., R. Dubayah, W. Walker, J. B. Blair, M. Hofton, and C. Hunsaker. 2006. "Mapping Forest Structure for Wildlife Habitat Analysis using Multi-sensor (LiDAR, SAR/InSAR, ETM plus, Quickbird) Synergy." *Remote Sensing of Environment* 102:63–73.

Hyde, P., R. Nelson, D. Kimes, and E. Levine. 2007. "Exploring LIDAR-RaDAR Synergy: Predicting Aboveground Biomass in a Southwestern Ponderosa Pine Forest Using LiDAR, SAR and InSAR." *Remote Sensing of Environment* 106:28–38.

Iverson, L. R., S. Brown, A. Prasad, H. Mitasova, A. J. R. Gillespie, and A. E. Lugo. 1994. "Use of GIS for Estimating Potential and Actual Forest Biomass for Continental South and Southeast Asia." *Ecological Studies* 101:67–116.

Jenkins, J. C., R. A. Birdsey, and Y. Pan. 2001. "Biomass and NPP Estimation for the Mid-Atlantic Region (USA) Using Plot-Level Forest Inventory Data." *Ecological Applications* 11:1174–93.

Jenkins, J. C., D. C. Chojnacky, L. S. Heath, and R. A. Birdsey. 2004. *Comprehensive Database of Diameter-Based Biomass Regressions for North American Tree Species*. General Technical Report NE-319. Newtown Square, PA: US Department of Agriculture, Forest Service, Northeastern Research Station.

Kasischke, E. S., J. M. Melack, and M. C. Dobson. 1997. "The Use of Imaging Radars for Ecological Applications: A Review." *Remote Sensing of Environment* 59:141–56.

Kim, Y., Z. Q. Yang, W. B. Cohen, D. Pflugmacher, C. L. Lauver, and J. L. Vankat. 2009. "Distinguishing between Live and Dead Standing Tree Biomass on the North Rim of Grand Canyon National Park, USA Using Small-Footprint Lidar Data." *Remote Sensing of Environment* 113:2499–510.

Koch, B. 2010. "Status and Future of Laser Scanning, Synthetic Aperture Radar and Hyperspectral Remote Sensing Data for Forest Biomass Assessment." *ISPRS Journal of Photogrammetry and Remote Sensing* 65(6):581–90.

Kozlowski, J., and M. Konarzewski. 2004. "Is West, Brown and Enquist's Model of Allometric Scaling Mathematically Correct and Biologically Relevant?" *Functional Ecology* 18:283–89.

Kronseder, K., U. Ballhorn, V. Bohm, and F. Siegert. 2012. "Above Ground Biomass Estimation across Forest Types at Different Degradation Levels in Central Kalimantan Using LiDAR Data." *International Journal of Applied Earth Observation and Geoinformation* 18:37–48.

Ku, N., S. Popescu, R. Ansley, H. Perotto-Baldivieso, and A. Filippi. 2012. "Assessment of Available Rangeland Woody Plant Biomass with a Terrestrial Lidar System." *Photogrammetric Engineering & Remote Sensing* 78(4):349–61.

Lambert, M. C., C. H. Ung, and F. Raulier. 2005. "Canadian National Tree Aboveground Biomass Equations." *Canadian Journal of Forest Research* 35:1996–2018.

Latifi, H., F. Fassnacht, and B. Koch. 2012. "Forest Structure Modeling with Combined Airborne Hyperspectral and LiDAR Data." *Remote Sensing of Environment* 121:10–25.

Le Toan, T., A. Beaudoin, J. Riom, and D. Guyon. 1992. "Relating Forest Biomass to SAR Data." *IEEE Transactions on Geoscience and Remote Sensing* 30:403–11.

Le Toan, T., Quegan, S., Woodward, I., Lomas, M., Delbart, N. and Picard, G. 2004. "Relating Radar Remote Sensing of Biomass to Modelling of Forest Carbon Budgets." *Climatic Change* 67:379–402.

Lefsky, M. A., W. B. Cohen, S. A. Acker, G. G. Parker, T. A. Spies, and D. Harding. 1999a. "Lidar Remote Sensing of the Canopy Structure and Biophysical Properties of Douglas-Fir Western Hemlock Forests." *Remote Sensing of Environment* 70:339–61.

Lefsky, M. A., W. B. Cohen, D. J. Harding, G. G. Parker, S. A. Acker, and S. T. Gower. 2002. "Lidar Remote Sensing of Above-Ground Biomass in Three Biomes." *Global Ecology and Biogeography* 11:393–99.

Lefsky, M. A., W. B. Cohen, and T. A. Spies. 2001. "An Evaluation of Alternate Remote Sensing Products for Forest Inventory, Monitoring, and Mapping of Douglas-Fir Forests in Western Oregon." *Canadian Journal of Forest Research* 31:78–87.

Lefsky, M. A., D. Harding, W. B. Cohen, G. Parker, and H. H. Shugart. 1999b. "Surface Lidar Remote Sensing of Basal Area and Biomass in Deciduous Forests of Eastern Maryland, USA." *Remote Sensing of Environment* 67:83–98.

Lefsky, M. A., D. J. Harding, M. Keller, W. B. Cohen, C. C. Carabajal, F. D. Espirito-Santo, M. O. Hunter, and R. de Oliveira. 2005a. "Estimates of Forest Canopy Height and Aboveground Biomass Using ICESat." *Geophysical Research Letters* 32:L22S02. doi:10.1029/2005gl023971.

Lefsky, M. A., A. T. Hudak, W. B. Cohen, and S. A. Acker. 2005b. "Geographic Variability in Lidar Predictions of Forest Stand Structure in the Pacific Northwest." *Remote Sensing of Environment* 95:532–48.

Lim, K. S., and P. M. Treitz. 2004. "Estimation of Above Ground Forest Biomass from Airborne Discrete Return Laser Scanner Data Using Canopy-Based Quantile Estimators." *Scandinavian Journal of Forest Research* 19:558–70.

Lim, K., P. Treitz, K. Baldwin, I. Morrison, and J. Green. 2003. "Lidar Remote Sensing of Biophysical Properties of Tolerant Northern Hardwood Forests." *Canadian Journal of Remote Sensing* 29:658–78.

Litton, C. M., and J. B. Kauffman. 2008. "Allometric Models for Predicting Aboveground Biomass in Two Widespread Woody Plants in Hawaii." *Biotropica* 40:313–20.

Liu, W. and Q. Chen. 2013. "Synergistic Use of Satellite Laser Altimetry and Shuttle Radar Topography Mission DEM for Estimating Ground Elevation over Mountainous Vegetated Areas." *IEEE Geoscience and Remote Sensing Letters.* 10:481-5.

Loudermilk, E. L., J. K. Hiers, J. J. O'Brien, R. J. Mitchell, A. Singhania, J. C. Fernandez, W. P. Cropper, and K. C. Slatton. 2009. "Ground-Based LIDAR: A Novel Approach to Quantify Fine-Scale Fuelbed Characteristics." *International Journal of Wildland Fire* 18:676–85.

Lu, D., Q. Chen, G. Wang, E. Moran, M. Batistella, M. Zhang, G. Vaglio Laurin, and D. Saah. 2012. "Aboveground Forest Biomass Estimation with Landsat and LiDAR Data and Uncertainty Analysis of the Estimates." *International Journal of Forestry Research.* 2012:16. doi:10.1155/2012/436537.

Lucas, R. M., N. Cronin, A. Lee, M. Moghaddam, C. Witte, and P. Tickle. 2006. "Empirical Relationships between AIRSAR Backscatter and LiDAR-Derived Forest Biomass, Queensland, Australia." *Remote Sensing of Environment* 100:407–25.

Maclean G. A. and W. B. Krabill. 1986. "Gross-Merchantable Timber Volume Estimation using an Airborne LIDAR System." *Canadian Journal of Remote Sensing* 12(1):7–18.

Magnussen, S., E. Nasset, T. Gobakken, and G. Frazer. 2012. "A Fine-Scale Model for Area-Based Predictions of Tree-Size-Related Attributes Derived from LiDAR Canopy Heights." *Scandinavian Journal of Forest Research* 27:312–22.

Malhi, Y., D. Wood, T. R. Baker, J. Wright, O. L. Phillips, T. Cochrane, P. Meir, et al. 2006. "The Regional Variation of Aboveground Live Biomass in Old-Growth Amazonian Forests." *Global Change Biology* 12:1107–38.

Mallet, C., and F. Bretar. 2009. "Full-Waveform Topographic Lidar: State-of-the-Art." *ISPRS Journal of Photogrammetry and Remote Sensing* 64:1–16.

Means, J. E., S. A. Acker, D. J. Harding, J. B. Blair, M. A. Lefsky, W. B. Cohen, M. E. Harmon, and W. A. McKee. 1999. "Use of Large-Footprint Scanning Airborne Lidar to Estimate Forest Stand Characteristics in the Western Cascades of Oregon." *Remote Sensing of Environment* 67:298–308.

Mette, T., I. Hajnsek, K. Papathanassiou, and D. Center. 2003. "Height-Biomass Allometry in Temperate Forests Performance Accuracy of Height-Biomass Allometry." In *IEEE International Geoscience and Remote Sensing Symposium*, 1942–44. Toulouse, France.

Mitchard E. T. A., S. S. Saatchi, L. J. T. White, K. A. Abernethy, K. J. Jeffery, S. L. Lewis, M. Collins, M. A. Lefsky, M. E. Leal, I. H. Woodhouse, and P. Meir. 2012. "Mapping Tropical Forest Biomass with Radar and Spaceborne LiDAR in Lopé National Park, Gabon: Overcoming Problems of High Biomass and Persistent Cloud." *Biogeosciences* 9:179–91.

Myneni, R. B., J. Dong, C. J. Tucker, R. K. Kaufmann, P. E. Kauppi, J. Liski, L. Zhou, V. Alexeyev, and M. K. Hughes. 2001. "A Large Carbon Sink in the Woody Biomass of Northern Forests." *Proceedings of the National Academy of Sciences of the United States of America* 98:14784–89.

Naesset, E., and T. Gobakken. 2008. "Estimation of Above- and Below-Ground Biomass across Regions of the Boreal Forest Zone Using Airborne Laser." *Remote Sensing of Environment* 112:3079–90.

Nelson, R., J. Boudreau, T. G. Gregoire, H. Margolis, E. Naesset, T. Gobakken, and G. Stahl. 2009. "Estimating Quebec Provincial Forest Resources Using ICESat/GLAS." *Canadian Journal of Forest Research* 39:862–81.

Nelson, R., P. Hyde, P. Johnson, B. Emessiene, M. Imhoff, R. Campbell, and W. Edwards. 2007. "Investigating RaDAR-LiDAR Synergy in a North Carolina Pine Forest." *Remote Sensing of Environment* 110:98–108.

Nelson, R., J. Jimenez, C. E. Schnell, G. S. Hartshorn, T. G. Gregoire, and R. Oderwald. 2000. "Canopy Height Models and Airborne Lasers to Estimate Forest Biomass: Two Problems." *International Journal of Remote Sensing* 21:2153–62.

Nelson, R., W. Krabill, and G. Maclean. 1984. "Determining Forest Canopy Characteristics using Airborne Laser Data." *Remote Sensing of Environment* 15:201–12.

Nelson, R., W. Krabill, and J. Tonelli. 1988a. "Estimating Forest Biomass and Volume Using Airborne Laser Data." *Remote Sensing of Environment* 24:247–67.

Nelson, R., R. Oderwald, and T. G. Gregoire. 1997. "Separating the Ground and Airborne Laser Sampling Phases to Estimate Tropical Forest Basal Area, Volume, and Biomass." *Remote Sensing of Environment* 60:311–26.

Nelson, R., G. Parker, and M. Hom. 2003. "A Portable Airborne Laser System for Forest Inventory." *Photogrammetric Engineering and Remote Sensing* 69:267–73.

Nelson, R., A. Short, and M. Valenti. 2004. "Measuring Biomass and Carbon in Delaware Using an Airborne Profiling LIDAR." *Scandinavian Journal of Forest Research* 19:500–11.

Nelson, R., R. Swift, and W. Krabill. 1988b. "Using Airborne Lasers to Estimate Forest Canopy and Stand Characteristics." *Journal of Forestry* 86:31–38.

Nelson, R., W. Krabill, and G. Maclean. 1984. "Determining forest canopy characteristics using airborne laser data." Remote Sensing of Environment 15:201–212.

Neumann, M., L. Ferro-Famil, and A. Reigber. 2010. "Estimation of Forest Structure, Ground, and Canopy Layer Characteristics from Multibaseline Polarimetric Interferometric SAR Data." *IEEE Transactions on Geoscience and Remote Sensing* 48:1086–104.

Niklas, K. J. 1995. "Size-Dependent Allometry of Tree Height, Diameter and Trunk-Taper." *Annals of Botany* 75:217–27.

Ni-Meister, W., S. Lee, A. H. Strahler, C. E. Woodcock, C. Schaaf, T. Yao, J. Ranson, G. Sun, and J. B. Blair. 2010. "Assessing General Relationships between Aboveground Biomass and Vegetation Structure Parameters for Improved Carbon Estimate from Lidar Remote Sensing." *Journal of Geophysical Research* 115:G00E11. doi:10.1029/2009JG000936.

Olson, J. S., J. A. Watts, and L. J. Allison. 1983. *Carbon in Live Vegetation of Major World Ecosystems*. Report ORNL-5862. Oak Ridge, TN: Oak Ridge National Laboratory.

Omasa, K., G. Y. Qiu, K. Watanuki, K. Yoshimi, and Y. Akiyama. 2003. "Accurate Estimation of Forest Carbon Stocks by 3-D Remote Sensing of Individual Trees." *Environmental Science & Technology* 37:1198–201.

Patenaude, G., R. A. Hill, R. Milne, D. L. A. Gaveau, B. B. J. Briggs, and T. P. Dawson. 2004. "Quantifying Forest Above Ground Carbon Content Using LiDAR Remote Sensing." *Remote Sensing of Environment* 93:368–80.

Popescu, S. C. 2007. "Estimating Biomass of Individual Pine Trees Using Airborne Lidar." *Biomass & Bioenergy* 31:646–55.

Potter, C. S. 1999. "Terrestrial Biomass and the Effects of Deforestation on the Global Carbon Cycle." *Bioscience* 49:769–78.

Potter, C., V. Brooks Genovese, S. Klooster, M. Bobo, and A. Torregrosa. 2001. "Biomass Burning Losses of Carbon Estimated from Ecosystem Modeling and Satellite Data Analysis for the Brazilian Amazon Region." *Atmospheric Environment* 35:1773–81.

Reitberger, J., C. Schnorr, P. Krzystek, and U. Stilla. 2009. "3D Segmentation of Single Trees Exploiting Full Waveform LIDAR Data." *ISPRS Journal of Photogrammetry and Remote Sensing* 64:561–74.

Riano, D., E. Meier, B. Allgower, E. Chuvieco, and S. L. Ustin. 2003. "Modeling Airborne Laser Scanning Data for the Spatial Generation of Critical Forest Parameters in Fire Behavior Modeling." *Remote Sensing of Environment* 86:177–86.

Rosette, J., J. Suarez, R. Nelson, S. Los, B. Cook, and P. North. 2012. "Lidar Remote Sensing for Biomass Assessment." In *Remote Sensing of Biomass:Principles and Applications*, edited by T. Fatoyinbo, 3–21. InTech. ISBN 978-953-51-0313-4.

Saatchi, S., N. Harris, S. Brown, M. Lefsky, E. Mitchard, W. Salas, B. Zutta, et al. 2011. "Benchmark Map of Forest Carbon Stocks in Tropical Regions across Three Continents." *Proceedings of the National Academy of Sciences of the United States of America* 108(24):9899–04.

Sarrazin, M., J. van Aardt, G. Asner, J. McGlinchy, D. Messinger, and J. Wu. 2011. "Fusing Small-Footprint Waveform LiDAR and Hyperspectral Data for Canopy-Level Species Classification and Herbaceous Biomass Modeling in Savanna Ecosystems." *Canadian Journal of Remote Sensing* 37(6):653–65.

Seidel, D., F. Beyer, D. Hertel, S. Fleck, and C. Leuschner. 2011. "3D-Laser Scanning: A Non-Destructive Method for Studying Above-Ground Biomass and Growth of Juvenile Trees." *Agricultural and Forest Meteorology* 151:1305–11.

Shukla, J., C. Nobre, and P. Sellers. 1990. "Amazon Deforestation and Climate Change." *Science* 247:1322–25.

Skowronski, N., K. Clark, R. Nelson, J. Hom, and M. Patterson. 2007. "Remotely Sensed Measurements of Forest Structure and Fuel Loads in the Pinelands of New Jersey." *Remote Sensing of Environment* 108:123–29.

Stephens, P. R., M. O. Kimberley, P. N. Beets, T. S. Paul, N. Searles, A. Bell, C. Brack, and J. Broadley. 2012. "Airborne Scanning LiDAR in a Double Sampling Forest Carbon Inventory." *Remote Sensing of Environment* 117:348–57.

St-Onge, B., Y. Hu, and C. Vega. 2008. "Mapping the Height and Above-Ground Biomass of a Mixed Forest Using Lidar and Stereo Ikonos Images." *International Journal of Remote Sensing* 29:1277–94.

Swatantran, A., R. Dubayah, D. Roberts, M. Hofton, and J. B. Blair. 2011. "Mapping Biomass and Stress in the Sierra Nevada Using Lidar and Hyperspectral Data Fusion." *Remote Sensing of Environment* 115:2917–30.

Tan, K., S. L. Piao, C. H. Peng, and J. Y. Fang. 2007. "Satellite-Based Estimation of Biomass Carbon Stocks for Northeast China's Forests between 1982 and 1999." *Forest Ecology and Management* 240:114–21.

TerMikaelian, M. T., and M. D. Korzukhin. 1997. "Biomass Equations for Sixty-Five North American Tree Species." *Forest Ecology and Management* 97:1–24.

Thenkabail, P. S., E. A. Enclona, M. S. Ashton, C. Legg, and M. J. De Dieu. 2004. "Hyperion, IKONOS, ALI, and ETM Plus Sensors in the Study of African Rainforests." *Remote Sensing of Environment* 90:23–43.

Thomas, V., P. Treitz, J. H. McCaughey, and I. Morrison. 2006. "Mapping Stand-Level Forest Biophysical Variables for a Mixedwood Boreal Forest Using Lidar: An Examination of Scanning Density." *Canadian Journal of Forest Research* 36:34–47.

Treuhaft, R. N., B. D. Chapman, J. R. dos Santos, F. G. Goncalves, L. V. Dutra, P. Graca, and J. B. Drake. 2009. "Vegetation Profiles in Tropical Forests from Multibaseline Interferometric Synthetic Aperture Radar, Field, and Lidar Measurements." *Journal of Geophysical Research-Atmospheres* 114(D23110). doi:10.1029/2008JD011674.

Tsui, O. W., N. C. Coops, M. A. Wulder, P. L. Marshall, and A. McCardle. 2012. "Using Multi-frequency Radar and Discrete-Return LiDAR Measurements to Estimate Above-Ground Biomass and Biomass Components in a Coastal Temperate Forest." *ISPRS Journal of Photogrammetry and Remote Sensing* 69:121–33.

van Aardt, J. A. N., R. H. Wynne, and R. G. Oderwald. 2006. "Forest Volume and Biomass Estimation Using Small-Footprint Lidar-Distributional Parameters on a Per-Segment Basis." *Forest Science* 52:636–49.

Waring, R. H., J. B. Way, E. R. Hunt, L. Morrissey, K. J. Ranson, J. F. Weishampel, R. Oren, and S. E. Franklin. 1995. "Imaging Radar for Ecosystem Studies." *Bioscience* 45:715–23.

Yao, T., X. Yang, F. Zhao, Z. Wang, Q. Zhang, D. Jupp, J. Lovell, et al. 2011. "Measuring Forest Structure and Biomass in New England Forest Stands Using Echidna Ground-Based Lidar." *Remote Sensing of Environment* 115:2965–74.

Zhao, F., Q. Guo, and M. Kelly. 2012. "Allometric Equation Choice Impacts Lidar-Based Forest Biomass Estimates: A Case Study from the Sierra National Forest, CA." *Agricultural and Forest Meteorology* 165:64–72.

Zhao, K. G., S. Popescu, and R. Nelson. 2009. "Lidar Remote Sensing of Forest Biomass: A Scale-Invariant Estimation Approach Using Airborne Lasers." *Remote Sensing of Environment* 113:182–96.

Zwally, H. J., B. Schutz, W. Abdalati, J. Abshire, C. Bentley, A. Brenner, J. Bufton, et al. 2002. "ICESat's Laser Measurements of Polar Ice, Atmosphere, Ocean, and Land." *Journal of Geodynamics* 34:405–45.

22

Carbon Cycle Modeling for Terrestrial Ecosystems

Tinglong Zhang and Changhui Peng

CONTENTS

22.1 Introduction...421
22.2 RS Data and C Cycle Models...422
    22.2.1 RS-Based Vegetation Properties 422
    22.2.2 Integrating RS and C Cycle Modeling..424
22.3 RS-Based C Cycle Modeling for Terrestrial Ecosystems............................425
    22.3.1 CASA Model..425
    22.3.2 Global Production Efficiency Model 426
    22.3.3 VPM Model..427
    22.3.4 Biome-BGC Model 428
    22.3.5 BEPS Model...428
    22.3.6 MODIS Primary Production Product (MOD17) Algorithm............431
22.4 Case Studies Integrating RS with Models...432
22.5 Summary..436
References...437

22.1 Introduction

The terrestrial biosphere plays a vital role in global carbon (C) cycling (Prentice et al. 2000; IPCC 2000, 2007). Due to the complexity and heterogeneity of terrestrial ecosystems, however, accurately quantifying terrestrial C cycling and predicting the impacts of global environmental change is immensely challenging. Much effort has been devoted to understanding and quantifying global C cycling through a variety of methods. These include field observation, model simulation, and remote sensing (RS) (Ito and Oikawa 2002).

The modeling approach is essential for investigations related to C cycling. It can be applied to all aspects of varying types of research whether local or global or research that integrates empirical information, bridges different spatial scales, or makes predictions. A wide variety of models have been developed, ranging from purely statistical ones to models that simulate basic ecophysiological and demographic processes (Peng 2000).

A growing body of research has demonstrated the complementary nature of combining RS and ecosystem modeling for terrestrial C cycling studies (David et al. 2004). Ecosystem process models are important tools used in employing information provided by RS products to quantify C flux and other such key elements. Physiologically based process models applied in a spatially distributed manner can assimilate and effectively integrate a diverse assemblage of environmental data, such as information related to soil, climate, and vegetation. To date, RS has acquired and made available a great deal of vegetation data. At the same time, the integration of RS and process models is a rapidly evolving field (Cohen and Goward 2004).

22.2 RS Data and C Cycle Models

22.2.1 RS-Based Vegetation Properties

Vegetation is an important component of life-supporting natural systems and ecosystems because it provides the basic foundation for all living beings. The satellite-based RS approach has begun to serve an essential role in obtaining global data, improving how land-surface vegetation is represented in C cycle modeling. Accordingly, various attempts have been made via RS to obtain vegetation properties required by many C cycle models:

Land cover: Land cover is a fundamental variable that impacts and links various components of human and physical environments. Specifically, determining land cover type is an important first step in the implementation of a spatially distributed C cycle model (Reich et al. 1997). RS has been widely applied as a basis for mapping regional or global land cover. DeFries and Townshend (1994) compiled the first global land cover map using the maximum likelihood classification approach of monthly composites of data from the advanced very high resolution radiometer (AVHRR) Normalized Difference Vegetation Index (NDVI). Subsequently, DeFries et al. (1998) used a decision tree classification technique to generate a global land cover map with an 8 km spatial resolution as determined by AVHRR data. Today, timeliness and quality of land cover maps generated from NOAA-AVHRR (GLCC) (Hansen et al. 2000; Loveland et al. 2000), SPOT-VEGETATION (GLC2000) (JRC 2003), the Moderate Resolution Imaging Spectroradiometer (MODIS) (MOD12) (Friedl et al. 2002), and even a new joint 1 km global land cover product (SYNMAP) (Jung et al. 2006) are useful for a wide array of scientific applications that require land cover information for both regional and global scales.

Leaf Area Index (LAI): LAI is the basic canopy structural parameter required for process-based canopy photosynthesis models. Specifying LAI using RS data is particularly informative when models include canopy light extinction and other processes that vary with canopy depth (Chen et al. 2000; Peng et al. 2002). There is usually a significant correlation between LAI and spectral vegetation indexes (such as NDVI). However, relationships tend to be asymptotic with saturation at LAI levels that increase on the order of three to five (Turner et al. 1999). Forest LAI has also been estimated using airborne light detection and ranging (LiDAR) sensors. LiDAR, comparable to the functionality of hyperspectral imaging and interferometric synthetic aperture radar (InSAR) combined, has the added advantage of being able to characterize the distribution of canopy foliage alongside canopy height (Lefsky et al. 2002; Treuhaft et al. 2002, 2004). Besides maximum seasonal LAI, some C cycle models also assimilate spatial information on seasonal changes in LAI or dates that mark the beginning of leaf out and the beginning and end of leaf drop. For effective use in global-scale modeling, this variable must be collected over a long period of time and should represent every region of the terrestrial surface under investigation. LAI is operationally produced from MODIS data. The Terra platform, incorporating

MODIS and other such instruments, was launched in December 1999, and data collection began in March 2000. LAI products are now available at 8-day intervals of 1 km resolution. The product was released in August 2000, made available through the Earth Resources Observation and Science (EROS) Land Processes Distributed Active Archive Center. Presently, MODIS emphasis is centered on algorithm validation and its functionality with the product (Myneni et al. 2002; Tian et al. 2002).

Biomass: Accurate estimation of biomass is essential to better understand terrestrial ecosystem C cycles by which forests serve as the primary reservoir of terrestrial C. Living material (biomass) is often used in initializing vegetation C pools for regional model simulations (Kimball et al. 2000) and is particularly important in estimating autotrophic respiration (Ra). Stem mass is also indicative of detrital residue that supports increased heterotrophic respiration after a disturbance takes place. Optical and microwave wavelengths vary in sensitivity to aboveground vegetation biomass. Optical RS methods such as empirical and statistical regression models that employ NDVI have been used to estimate the amount of and temporal variability in aboveground biomass (Dong et al. 2003). Synthetic aperture radar (SAR) and other RS systems based on active microwave imaging are also sensitive to vegetation structure as well as the amount of biomass present, including both photosynthetic and nonphotosynthetic vegetation components. Microwave wavelengths penetrate to greater depths in plant canopies than do optical sensors and generally show more promise in assessing standing woody biomass (Kasischke et al. 1997). Radio detection and ranging (radar) sensitivity to vegetation biomass is strongly dependent on wavelength. Longer wavelengths (L-band) generally are able to detect vegetation volume and biomass levels better than shorter wavelengths (C-band). Like optical RS, radar shows an asymptotic relationship to vegetation biomass, although the saturation levels for longer microwave wavelengths are much higher than those for optical sensors. Recent research employing LiDAR and InSAR sensors has shown promise in increasing the maximum biomass potential that RS is capable of detecting (Lefsky et al. 1999; Treuhaft et al. 2004; Zhao et al. 2009).

Canopy and tree height: C exchange between forests and the atmosphere is a vital component of the global C cycle. Satellite light amplification by simulated emission of radiation (laser) altimetry has a unique capability in estimating forest canopy height, which has a direct and increasingly well-understood relationship with aboveground C storage. Canopy height and density are also important factors that regulate wind velocity and surface roughness as well as canopy resistance to evapotranspiration and C exchange. These mechanisms are beginning to be incorporated into ecosystem process models (Williams et al. 2001). In addition, LiDAR sensors are very effective in determining both canopy and tree heights (Lefsky et al. 2002). Although InSAR is less accurate, it has the advantage of being able to cover larger areas (Treuhaft et al. 2004). Thus, as more extensive tree height mapping becomes possible, the resultant data can be readily applied as model inputs. Moreover, since canopy height is a function of stand age as well as a standard indicator of site quality, the combination of stand height obtained from RS and stand age obtained from change detection analysis (also based on RS) can be used to validate modeled bole wood production. Recently, Vauhkonen et al. (2011) used airborne laser scanning (ALS) and planting distance data in combination to detect trees and extract their heights. This suggests that ALS data have a greater potential in characterizing canopy structure than other RS technologies (Coops et al. 2004; Magnusson 2006).

Forest stand age: Forest stand age is in itself a powerful variable of considerable value by which one can model the health of forest-dominant ecosystems. Forest net primary production (NPP) changes with stand age (He et al. 2012). Temporal patterns in net ecosystem production (NEP) for cover types over relatively long successional stages are also generally well understood because of the strong C source associated with woody residue during early successional stages and the accumulation of stem wood and woody debris during mid-to-late successional stages (Janisch and Harmon 2002). If stand age is specified by RS, C cycle process models can be run over the course of succession, thereby improving their ability in accurately simulating NPP and NEP (Thornton et al. 2002). Classification based on RS is the most common approach in assessing regional patterns of forest stand age because it requires only one image and is relatively efficient compared to field inventory approaches. Changes in reflectance subject to stand age (those changes that allow for classification) are associated with differences in the proportion of ground surface, represented by spectral characteristics of the foliage and structural properties of the canopy, which influence shading patterns. In moist tropical forests, successional stages can be partitioned using the Landsat Enhanced Thematic Mapper Plus (ETM+) sensor. This is accomplished by the differences exhibited in red and near-infrared reflectance (Moran et al. 1994). An alternative and more precise approach in estimating stand age is to employ multiple images to change detection analysis (Hall et al. 1991; Cohen et al. 2002).

22.2.2 Integrating RS and C Cycle Modeling

There are a variety of ways to integrate RS and C cycle models. First, RS can be used as input data for various models, such as the Carnegie-Ames-Stanford approach (CASA) (Potter et al. 1993; Field et al. 1995), the boreal ecosystem productivity simulator (BEPS) (Chen et al. 1999; Liu et al. 2002, 2003; Ju et al. 2006), and the biome-biogeochemical cycles (Biome-BGC) model (Running and Coughlan 1988; Kimball et al. 1997; Thornton 1998). Each model uses different vegetation variables obtained from RS inversion. For example, LAI is the most important input variable for BEPS as the fraction of absorbed photosynthetically active radiation (fAPAR) is for CASA (as it is for other light use efficiency [LUE] models). Second, RS data can be used to optimize model parameters through data assimilation methods and adjust and calibrate model simulation results (see Stöckli et al. 2008). Third, RS and C cycle models can be used to validate each other. A typical example is the BigFoot project (2005).

A global terrestrial observation system is required to assist in the validation of global products, such as land cover and NPP from MODIS as well as other sensors and modeling programs. A key component of such a system is the Eddy Flux Tower Network (FLUXNET). However, flux sensors measure NEP and not NPP. BigFoot is learning how NEP and NPP are related and how to integrate a wide range of C cycle observations through modeling. Another key component of observational systems is the use of RS and models to scale up tower fluxes and site-level field measurements to larger landscape or regional scales (Xiao et al. 2010, 2011). Although this practice may be relatively common at a given site, no other project is endeavoring to do so using standardized methods across so many biomes. As such, BigFoot is a path-finding pursuit that will ultimately contribute to the development of useful scaling principles. The overall goal of BigFoot is to provide validation for MODIS Land Science Team (MODLAND) science products, including land cover, LAI, fAPAR, and NPP (Campbell et al. 1999). To do so, ground measurements, RS data, and ecosystem process models were used at sites representative of different biomes. The project can also serve as a nucleus for the global terrestrial observing system that is necessary to validate global and generalized products used to monitor terrestrial biosphere health.

22.3 RS-Based C Cycle Modeling for Terrestrial Ecosystems

Based on a profusion of observational data, Lieth (1975) developed the first global terrestrial ecosystem model in the early 1970s. The model (named the Miami model) provided a good approximation of the global distribution of potential vegetation productivity. Esser (1987) extended the productivity model to simulate the global C cycle, including biomass growth and soil C dynamics. Meanwhile, Uchijima and Seino (1985) developed a micrometeorological model (named the Chikugo model) of vegetation productivity by which net radiation is converted into dry matter production. These empirical models were useful in retrieving present state conditions but were unsuitable for prediction under changing environments. For example, they could not provide a plausible estimation when atmospheric carbon dioxide (CO_2) increased appreciably compared to present levels.

Accordingly, a mechanistic (i.e., process-based) model based on the physiological regulation of processes was required to extrapolate different environmental conditions. Melillo et al. (1993) published the first simulation results derived from a mechanistic model (Terrestrial Ecosystem Model [TEM]). They estimated the response of terrestrial NPP to elevated CO_2 and climate change. Subsequently, various mechanistic models have been developed, including Biome-BGC by Running and Hunt (1993), CENTURY by Parton et al. (1993), CASA by Potter et al. (1993), CARAIB by Warnant et al. (1994), SLAVE by Friedlingstein et al. (1995), FBM by Lüdeke et al. (1994), GTEC by Post et al. (1997), CEVSA by Cao and Woodward (1998), TsuBiMo by Alexandrov et al. (2002), and Sim-CYCLE by Ito and Oikawa (2002). In the literature, more than a hundred C cycle models are presently available. Only those models related to RS approaches are discussed in this chapter.

22.3.1 CASA Model

The CASA model (Potter et al. 1993; Field et al. 1995) estimates NPP for each time step using Equation 22.1:

$$NPP = \sum PAR \times fAPAR \times \varepsilon \times \Delta t \qquad (22.1)$$

where PAR is the total photosynthetically active radiation (megajoules), corresponding to incoming incident radiation here; fAPAR is the fraction of PAR absorbed by photosynthetic tissues (unitless); ε is LUE (gram per megajoule PAR); and summation is carried out over the growing season. fAPAR can be estimated from the simple ratio (SR) or NDVI through linear functions, depending on the model and the NDVI dataset.

fAPAR was first calculated as a linear function of SR, following the work of Sellers et al. (1996):

$$SR = \frac{NIR}{red} \qquad (22.2)$$

where NIR and red are the near-infrared and red reflectance, respectively.

$$fAPAR = (SR - SR_{min})(fAPAR_{max} - fAPAR_{min})/(SR_{max} - SR_{min}) + fAPAR_{min} \qquad (22.3)$$

where SR is the value of the simple ratio at a given pixel; SR_{min} and SR_{max} correspond to second and ninety-eighth percentiles of SR, respectively, for the entire upland cropland region; and $fAPAR_{min}$ and $fAPAR_{max}$ are defined as 0.01 and 0.95, respectively. SR_{min} and SR_{max} are computed every 10 days. LUE is calculated as a product of optimal LUE (ε^*) and its temperature and water stressors:

$$\varepsilon = \varepsilon^* T_1 T_2 W_s \qquad (22.4)$$

where ε^* is the global maximum LUE for aboveground biomass when environmental conditions are optimal, and T_1, T_2, and W_s are scalars representing environmental stressors that reduce LUE (Field et al. 1995). T_1 represents a physiological reduction of LUE when temperatures are higher or lower than optimal value (T_{opt}), defined as the mean temperature in the month of maximum NDVI. T_2 reduces LUE as temperatures deviate from 20°C, representing constraints beyond physiological compensation at extreme temperatures. W is a soil moisture downregulator. T_1, T_2, and W_s are calculated by Equations 22.5, 22.6, and 22.7, respectively:

$$T_1 = 0.8 + 0.02T_{opt} - 0.0005T_{opt}^2 \qquad (22.5)$$

$$T_2 = \frac{1}{1 + \exp\{0.2(T_{opt} - 10 - T)\}} \times \frac{1}{1 + \exp\{0.3(-T_{opt} - 10 + T)\}} \qquad (22.6)$$

where T is the monthly mean temperature.

$$W_s = 0.5 + 0.5\frac{ET}{PET} \qquad (22.7)$$

where ET and PET are monthly actual evapotranspiration and potential evapotranspiration, respectively.

22.3.2 Global Production Efficiency Model

The Global Production Efficiency Model (GLO-PEM) consists of linked components that describe canopy radiation absorption, utilization, and R_a processes and the regulation of these processes through environmental factors (Prince and Goward 1995; Goetz and Prince 2000):

$$NPP = \sum_t \left[(S_t \times N_t)\varepsilon_g - R \right] \qquad (22.8)$$

where S_t is incident PAR in time t; N_t is the fraction of incident PAR absorbed by the vegetation canopy (fAPAR), calculated as a function of NDVI (Prince and Goward 1995; Goetz and Prince 1999); ε_g is radiation utilization efficiency (RUE) as it relates to gross primary production (GPP); and R is autotrophic respiration calculated as a function of vegetation, biomass, air temperature, and photosynthetic rate. Algorithms for calculating these variables are described by Prince and Goward (1995) and Goetz et al. (1999). For this model, two factors are affected by environmental factors such as air temperature, water vapor pressure deficit (VPD), soil moisture, and atmospheric CO_2 concentration. The following relationship can be used to determine ε_g:

$$\varepsilon_g = \varepsilon_g^* \times \sigma \qquad (22.9)$$

where ε_g^* is the maximum possible LUE of PAR absorbed by vegetation, determined by photosynthetic enzyme kinetics (a function of photosynthetic pathway, temperature, and the CO_2/oxygen [O_2] ratio); and σ is the reduction of ε_g^* caused by environmental factors that control stomatal conductance, determined by the following relationship:

$$\sigma = f(T)f(\delta_q)f(\delta_\theta) \tag{22.10}$$

where $f(T)$, $f(\delta_q)$, and $f(\delta_\theta)$ represent air temperature, VPD, and soil moisture stress, respectively

22.3.3 VPM Model

For the VPM model, leaves and canopy are composed of photosynthetically active vegetation (PAV) (chloroplasts) and nonphotosynthetic vegetation (NPV) (e.g., stem, branch, cell wall, and vein). Based on the conceptual partitioning of PAV and NPV, the VPM model was recently developed to estimate forest GPP (Xiao et al. 2004a, 2004b). A brief description of the VPM model is as follows:

$$GPP = \varepsilon_g \times fAPAR_{PAV} \times PAR \tag{22.11}$$

$$\varepsilon_g = \varepsilon_0 \times T_{scalar} \times W_{scalar} \times P_{scalar} \tag{22.12}$$

where PAR is photosynthetically active radiation ($\mu mol/m^2/s$ photosynthetic photon flux density [PPFD]); $fAPAR_{PAV}$ is the fraction of PAR absorbed by PAV (chloroplasts); and ε_g is LUE ($\mu mol\ CO_2/\mu mol$ PPFD). The parameter ε_0 is apparent quantum yield or maximum LUE ($\mu mol\ CO_2/\mu mol$ PPFD), and T_{scalar}, W_{scalar}, and P_{scalar} are the downregulation scalars for temperature effects, water, and leaf phenology, respectively, on vegetation LUE. For the current version of the VPM model, $fAPAR_{PAV}$ is assumed to be a linear function of the Enhanced Vegetation Index (EVI). The coefficient a in Equation 22.13 is simply set to 1.0 (Xiao et al. 2004a, 2004b):

$$FAPAR_{PAV} = a \times EVI \tag{22.13}$$

T_{scalar} is estimated at each time step using the equation developed for TEM (Raich et al. 1991):

$$T_{scalar} = \frac{(T - T_{min})(T - T_{max})}{\left[(T - T_{min})(T - T_{max})\right] - (T - T_{opt})^2} \tag{22.14}$$

where T_{min}, T_{max}, and T_{opt} are the minimum, maximum, and optimal temperatures, respectively, for photosynthetic activity. T_{scalar} is set to zero if air temperature falls below T_{min}. The effect of water on plant photosynthesis (W_{scalar}) has been estimated as a function of soil moisture and/or VPD in a number of production efficiency models (Field et al. 1995; Prince and Goward 1995; Running et al. 2000). An alternative and simple approach that uses a satellite-derived water index was proposed to estimate the approximate seasonal dynamics of W_{scalar} (Xiao et al. 2004a, 2004b).

$$W_{scalar} = \frac{1 + LSWI}{1 + LSWI_{max}} \tag{22.15}$$

where $LSWI_{max}$ is the maximum Land Surface Water Index (LSWI) during the plant growing season for individual pixels. P_{scalar} is included to account for the effect of canopy level leaf phenology (leaf age) on photosynthesis. In this version of the VPM model, the calculation of P_{scalar} is dependent on the longevity of leaves (deciduous vs. evergreen). For a canopy that is dominated by leaves with a life expectancy of 1 year (i.e., one season deciduous tree growth), P_{scalar} is calculated as a linear function at two different phases (Xiao et al. 2004b):

$$P_{scalar} = \frac{1+LSWI}{2} \qquad (22.16)$$

From bud burst to full expansion of leaves,

$$P_{scalar} = 1 \qquad (22.17)$$

After the full expansion of leaves, LSWI values range from −1 to +1 (a value range of 2). The simplest formulation of P_{scalar} (Equation 22.16) is a linear scalar with a value range from 0 to 1.

22.3.4 Biome-BGC Model

Biome-BGC is a process-based model that simulates the storage and flux exchange between ecosystem water, C, and nitrogen (N) and the atmosphere (Kimball et al. 1997; Thornton 1998); it is the successor of the Forest-BGC model (Running and Coughlan 1988). It uses daily time steps for input and output variables and, as a rule, the procedure can by itself satisfactorily model day-to-day variations in ecosystem circulation (Kimball et al. 1997). Biome-BGC is a biogeochemical and ecophysiological model that applies general stand information and daily meteorological data (such as daily maximum temperature, minimum temperature, and daytime temperature as well as precipitation, daytime VPD, daytime radiation, and day length) to simulate energy, C, water, and N cycling (Figure 22.1). Biome-BGC emphasizes LAI as a key structural attribute having considerable control over ecosystem processes. Numerous studies have shown that GPP and NPP are positively correlated to LAI. This is because LAI controls the amount of incident radiation absorbed and converted into photosynthate. Therefore, there exists a sound physiological basis for LAI being an important ecosystem attribute. Moreover, LAI can be remotely sensed and used as a spatial input in Biome-BGC by which to carry out regional to global C simulations. For this model, canopy is treated as one layer (BIG-leaf), but leaf area is divided into shaded and sunlit sections. Photosynthesis is based on the Farquhar model (Farquhar et al. 1980). CO_2 and water vapor derived from canopy conductance are regulated by air temperature, VPD, radiation, and modeled soil water potential. Litter and soil are each represented by one layer; but layers are divided into four pools, each having different decomposition parameters. Initial stocks of C are either input as initial values or estimated by spin-up simulations. The N quantity within different pools is estimated on the basis of the C quantity and the corresponding C:N ratio of the respective pool. Turnover rates and allocation parameters decide how production is distributed between compartments.

22.3.5 BEPS Model

BEPS was primarily developed to simulate forest ecosystem C budgets and water balances (Chen et al. 1999; Liu et al. 2002, 2003; Ju et al. 2006). It is a process-based ecosystem model that incorporates energy partitioning, photosynthesis, R_a, soil organic matter (SOM)

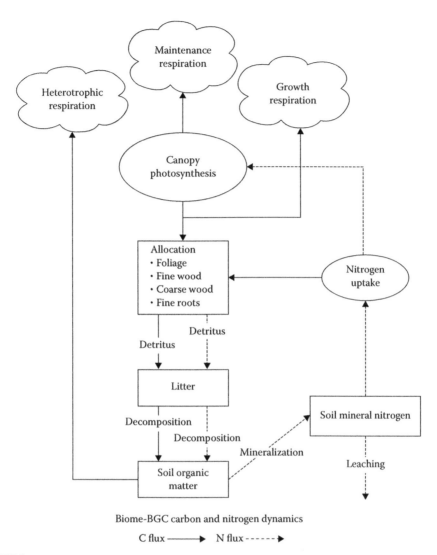

Biome-BGC carbon and nitrogen dynamics

C flux ———▶   N flux ------▶

FIGURE 22.1
Flowchart diagram describing modeled carbon (C) and nitrogen (N) dynamics of the biome-biogeochemical cycles (Biome-BGC) model. (From http://www.ntsg.umt.edu/project/biome-bgc).

decomposition, hydrological processes, and soil thermal transfer modules. Within the framework of the model, the canopy is stratified into overstory and understory layers, each separated into sunlit and shaded leaf groups. Snow packing and melting and rainfall infiltration and runoff as well as soil vertical percolation are simulated within the hydrological processes module. To estimate the vertical distribution of soil moisture and temperature, the soil profile is divided into five layers of differing depths. Ecosystem respiration (RE) includes both plant autotrophic respiration and soil heterotrophic respiration. Plant R_a is separated into two components: growth respiration (GR) and maintenance respiration (MR). Soil heterotrophic respiration results from the decomposition of the nine SOM pools, similar to the CENTURY model (Parton et al. 1993; Ju and Chen 2005). The following sentences briefly describe processes regulated by parameters: first, plant leaf photosynthetic rates are simulated by the photosynthesis–transpiration coupling method where leaf photosynthetic rates are estimated using Farquhar's biochemical model (Farquhar et al. 1980; Baldocchi 1994). The C assimilation process is coupled with leaf stomatal conductance through the empirical relationship of the Ball–Woodrow–Berry model (Ball et al. 1987). Photosynthetic rates and stomatal conductance are calculated as follows:

$$A = \min(A_c, A_j) - R_d \tag{22.18}$$

$$A_c = V_{cmax} f_V(T) \frac{C_i - \Gamma}{C_i + K_c(1 + O_i / K_o)} \tag{22.19}$$

$$A_j = J_{max} f_J(T) \frac{C_i - \Gamma}{4(C_i - 2\Gamma)} \tag{22.20}$$

and

$$g_s = m \frac{A_{hr}}{C_s} + g_o \tag{22.21}$$

where A, A_c, and A_j are the net photosynthetic, Rubicon-limited, and light-limited gross photosynthetic rates (micromole per square meter per second), respectively; R_d is daytime leaf dark respiration; V_{cmax} is the maximum carboxylation rate; J_{max} is the electron transport rate; C_i and O_i are intercellular CO_2 and O_2 concentrations, respectively; Γ is the CO_2 compensation point devoid of dark respiration; K_c and K_o are the Michaelis–Menten constants for CO_2 and O_2, respectively; $f_V(T)$ and $f_J(T)$ are the air temperature (T) response functions for V_{cmax} and J_{max}, respectively; g_s is bulk stomatal conductance (micromole per square meter per second); g_o is residual conductance; hr and C_s are leaf surface relative humidity and CO_2 concentrations, respectively; and m is an empirical coefficient. Plant GR is set at 20% of A. MR, which is dependent on temperature, is expressed as follows:

$$R_m = \sum_{i=1}^{4} r_i M_i f_m(T) \tag{22.22}$$

where r is the reference respiration rate at base temperature, M is biomass (kilogram per square meter) (subscript $i = 1, \ldots, 4$ corresponds to leaf, sapwood, coarse root, and fine root separately), and $f_m(T)$ is the air temperature response function of r. Heterotrophic respiration (R_h) is derived from the five litter pools and the four soil C pools:

$$R_h = \sum_{j=1}^{9} \tau_j k_j C_j \tag{22.23}$$

where τ is the respiration coefficient, which is equal to the percentage of decomposed C released into the atmosphere; k is the decomposition rate of a soil C pool affected by several environmental and biochemical factors, such as temperature, moisture, lignin fraction, and soil texture; and C is the C pool size. The simulation time step is set to 30 minutes. Forcing datasets include atmospheric variables (temperature, relative humidity, wind speed, precipitation, solar irradiance, and sky long-wave irradiance), vegetation type and stand age, canopy clumping index, soil texture and physical properties, and initial C pool values. Model outputs include GPP, R_a, heterotrophic respiration, net radiation, latent and sensible heat fluxes, soil moisture and temperature profiles, and so on. RS data are used in BEPS to obtain spatial distribution of LAI, land cover type, and biomass.

It has been widely used at various geographical locations and at different timescales (Liu et al. 1997, 1999, 2002).

22.3.6 MODIS Primary Production Product (MOD17) Algorithm

MOD17 is the first regular, near-real-time dataset for the continuous monitoring of primary production of vegetated landmasses at 8-day intervals of 1 km resolution. The core science behind the MOD17 algorithm is its application of radiation conversion efficiency logic that predicts daily GPP using the satellite-derived fraction of photosynthetically active radiation (fPAR) (from MOD15) and independent PAR estimates as well as other surface meteorological fields (from NASA Data Assimilation Office [DAO]) and the ensuing estimation of MR and GR terms that are subtracted from GPP to arrive at annual NPP. MR and GR components are based on allometric relationships linking daily biomass and annual growth of plant tissues to MOD15 satellite-derived estimates of LAI. These allometric relationships were derived from an extensive literature review, and they incorporate the same parameters used in Biome-BGC (Running and Hunt 1993). Parameters relating active phased array radar (APAR) to GPP and parameters relating LAI to MR and GR are estimated separately for each unique MOD12 vegetation type. GPP parameters are derived empirically from the output of Biome-BGC simulations carried out over a gridded global domain, using multiyear gridded global daily meteorological observations. MR and GR parameters are directly taken from the Biome-BGC ecophysiological parameter lists organized by plant functional type (White et al. 1998).

The final estimation of daily GPP is illustrated in the top panel of Figure 22.2.

A comparison between the broad features of C cycle models based on RS technologies (cited in Sections 22.3.1 through 22.3.6) was done (see Table 22.1).

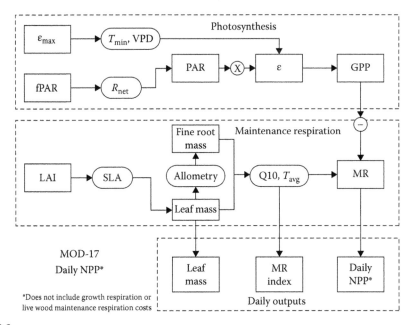

FIGURE 22.2
Flowchart illustrating the data flow of the daily component of the MOD17 algorithm (*Note*: NPP* indicates that not all autotrophic respiration [R_a] terms have been subtracted). (Modified from Heinsch, F. A., et al., *User's Guide GPP and NPP (MOD17A2/A3): Products NASA MODIS Land Algorithm*; Version 2. Missoula, MT: The University of Montana, 2003.)

TABLE 22.1

Comparison between the Broad Features of C Cycle Models Based on RS Technologies

| | GPP/NPP | | | | |
	Temporal Resolution	Calculated as	Influenced by	Strategy	Key References
CASA	1 month	NPP	$NPP = f(R_s, fPAR, T, AET/PET)$ $R_a = f(VegC, GPP)$	PEM, LUE derived empirically, applied to NPP	Potter et al. (1993); Field et al. (1995)
GLO-PEM	10 days	$GPP-R_a$	$GPP = f(R_s, fPAR, SW, VDP)$	PEM, LUE derived from a mechanistic model, applied to GPP	Prince and Goward (1995)
VPM	1 day	GPP	$GPP = f(R_s, fPAR_{PAV}, T, W,$ etc.$)$	PEM, LUE derived empirically, applied to GPP	Xiao et al. (2004a, 2004b)
Biome-BGC	1 day	$GPP-R_a$	$GPP = f(R_s, LAI, T, SW, VPD, CO_2, Leaf-N)$ $R_a = f(VegC, T)$	Estimates LAI from water balance, no phenology	Running and Hunt (1993)
BEPS	1 day	$GPP-R_a$	$GPP = f(R_s, LAI, T, SW, VPD, CO_2, Leaf-N)$ $R_a = f(VegC, T)$	Utilizes LAI derived from R_s, to calculate GPP and R_a	Chen et al. (1999); Liu et al. (2002)
MOD17	8 days	$GPP-R_a$	$GPP = f(R_s, LAI, fPAR, T, VPD)$ $R_a = f(VegC, LAI, T, GPP)$	PEM, LUE, and other vegetation parameters derived from the Biome-BGC model	Running et al. (1999)

Source: Cramer, W., et al., *Glob. Change Biol.*, 5(S1), 1–15, 1999.
Note: R_s—solar radiation; T—(air) temperature; SW—soil water; AET—actual evapotranspiration; VegC—vegetation C (i.e., C in leaves, sapwood, heartwood, roots, etc.); leaf-N—N content of leaves.

22.4 Case Studies Integrating RS with Models

Recent scaling studies have integrated RS with ecosystem process models to evaluate C cycle components in different regions of the world and on a global scale:

Global mapping of terrestrial primary productivity from MOD17 data: MOD17 has two subproducts: MOD17A2, which stores 8-day composite GPP, net photosynthesis (PSN net), and corresponding quality control (QC); and MOD17A3, which contains annual NPP and QC. These data are freely available to the public from the Numerical Terradynamic Simulation Group (http://www.ntsg.umt.edu) or the EROS Data Center Distributed Active Archive Center.

In this case study, results from Zhao's global mapping on terrestrial primary productivity was used. It is noted that 3-year (2001–2003) average annual global 1 km GPP and NPP images are provided in Figure 22.3. Results show high values for MODIS GPP and NPP in areas covered by forest and woody savanna, as expected, especially in tropical regions. Low NPP occurred in areas dominated by adverse environments, such as high latitudes with short growing seasons as

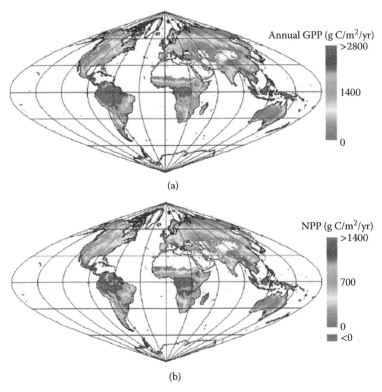

FIGURE 22.3
(See color insert.) Three-year (2001–2003) mean global 1 km Moderate Resolution Imaging Spectroradiometer (MODIS) annual gross photosynthetic production (GPP) and net primary production (NPP) images: (a) GPP and (b) NPP. (Modified from Zhao, et al., *Remote Sens. Environ.*, 95, 164–76, 2005.)

a result of low temperatures as well as dry areas with limited water availability. Global mean total GPP was 109.29 P_g C/year and NPP was 56.02 P_g C/year (not taking into account barren land cover) (Zhao et al. 2005).

Continental-scale estimation of terrestrial C sinks obtained from satellite data and ecosystem modeling (using the CASA model): In this case study, the NASA-CASA model was used to estimate the monthly C flux of terrestrial ecosystems from 1982 to 1998 based on satellite observations of monthly vegetation cover. The model was driven by vegetation properties derived from AVHRR as well as radiative transfer algorithms developed for MODIS. The terrestrial NEP sink for atmospheric CO_2 on the North American continent remained fairly consistent (between +0.2 and +0.3 P_g C/year), except during relatively cool annual periods when continental NEP flux was predicted to amount to nearly zero. The predicted NEP sink for atmospheric CO_2 over Eurasia increased notably in the late 1980s and has remained fairly consistent (between +0.3 and +0.55 P_g C/year) since 1988. In contrast, southern hemisphere tropical zones (latitudes between 0° and 30° south) have a major influence on predicted global trends in NEP interannual variability. For the terrestrial biosphere, atmospheric CO_2–predicted NEP flux has varied widely between an annual source of 0.9 P_g C/year and a sink of +2.1 P_g C/year. Selected results are provided in Figures 22.4 and 22.5 (data from North America).

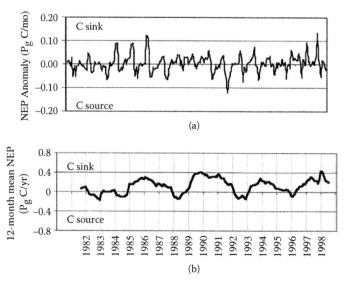

FIGURE 22.4
National Aeronautics and Space Administration–Carnegie-Ames-Stanford approach net ecosystem production (NEP) interannual simulation results for North America: (a) monthly predictions and (b) annual running mean. (From Potter, C. et al., *Global Planetary Change*, 39, 201–213, 2003.)

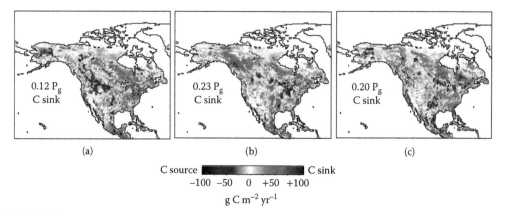

FIGURE 22.5
(**See color insert.**) Predicted North American interannual variation in annual NEP flux from 1996 to 1998: (a) 1996, (b) 1997, and (c) 1998. (From Potter, C. et al., *Global Planetary Change*, 39, 201–213, 2003.)

Mapping China's terrestrial ecosystem NPP using BEPS: In this case study, terrestrial NPP in China was simulated using BEPS, a carbon–water-coupled process model based on RS inputs. Required BEPS input data included land cover, LAI, available soil water-holding capacity (AWC), soil water content, digital elevation model (DEM), and daily meteorological data. Data were processed using the same coordinate system (Albers equal-area conic projection). At 1 km resolution, the image size was 5300 × 4300 pixels. Using these databases in addition to BEPS, daily NPP maps for the entire landmass of China for 2001 were produced and GPP and R_a estimated (Feng et al. 2007). The NPP map of China for 2001 is provided in Figure 22.6 (pixel size is 1 km × 1 km), and the corresponding GPP and R_a maps are provided in Figures 22.7 and 22.8.

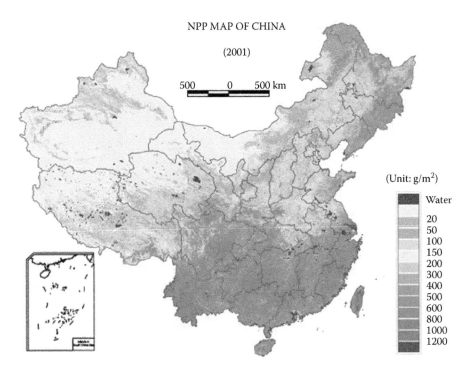

FIGURE 22.6
NPP map of China for 2001. (From Feng, X., et al., *J. Environ. Manage.*, 85(3), 563–73, 2007.)

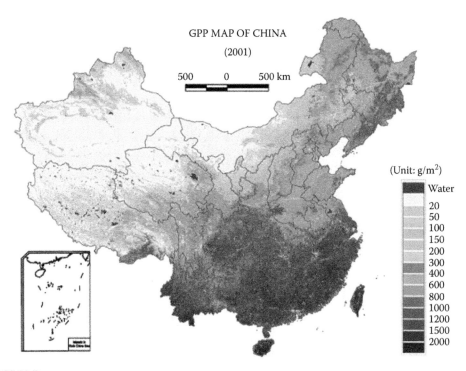

FIGURE 22.7
GPP map of China for 2001. (From Feng, X., et al., *J. Environ. Manage.*, 85(3), 563–73, 2007.)

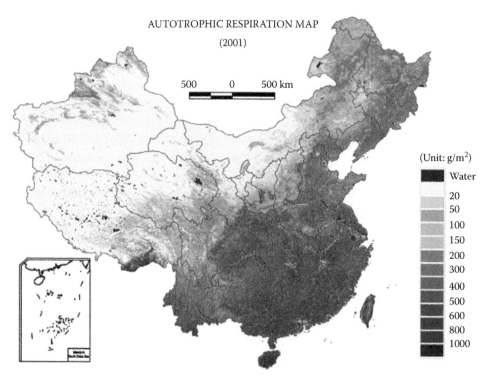

FIGURE 22.8
R_a map of China for 2001. (From Feng, X., et al., *J. Environ. Manage.*, 85(3), 563–73, 2007.)

22.5 Summary

Ecosystem process models have become an important tool to investigate terrestrial eco-system C cycling, especially when scaling GPP, NPP, and even NEP over landmasses that range from landscapes to regional domains. The use of models as a platform for data synthesis and integration can distill a wide array of diverse data into useful infor-mation and force consistency among numerous discrete observational datasets (David et al. 2004). Satellite-based RS is providing increasing spatial data variety, reflecting environmental and vegetation characteristics that are potentially usable as model inputs or for the validation of model outputs. The integration of process models and RS is particularly effective for monitoring vegetation dynamics and C cycles that range from landscapes to regional scales. Research challenges in this field include optimizing spatial and temporal resolutions for specific applications, differentiating relative influ-ences of structural and chemical variables on ecosystem C fluxes, and systematically validating model-based site flux and RS observations. For the future, the development of C cycle models must provide greater emphasis on using remotely sensed data as input variables, thus enhancing model spatial scale expansion capability. Furthermore, RS must develop more technology for monitoring vegetation and ecosystems, addi-tional algorithms to carry out inversions of various types of vegetation parameters, and more generalized data production that models require for operation. Although C cycle models that use RS data to optimize parameters are fundamental (Peng et al. 2011), integrating these models with RS data through data assimilation methods is also an important undertaking.

_____

References

Alexandrov, G. A., T. Oikawa, and Y. Yamagata. 2002. "The Scheme for Globalization of a Process Based Model Explaining Gradations in Terrestrial NPP and Its Applications." *Ecological Modelling* 148:293–306.

Baldocchi, D. 1994. "An Analytical Solution for Coupled Photosynthesis and Stomatal Conductance Model." *Tree Physiology* 14:1069–79.

Ball, J. T., I. E. Woodrow, and J. A. Berry. 1987. "A Model Predicting Stomatal Conductance and Its Contribution to the Control of Photosynthesis under Different Environmental Conditions." In *Progress in Photosynthesis Research*, edited by Biggins J. Martinus, 221–24. Dordrecht: Nijhoff Publishers.

BigFoot. 2005. http://www.fsl.orst.edu/larse/bigfoot/index.html.

Campbell, J. L., S. Burrows, S. T. Gower, and W. B. Cohen. 1999. *BigFoot: Characterizing Land Cover, LAI, and NPP at the Landscape Scale for EOS/MODIS Validation*. Field Manual 2.1. Oak Ridge, Tennessee: Oak Ridge National Laboratory, Environmental Science Division.

Cao, M., and F. I. Woodward. 1998. "Dynamic Responses of Terrestrial Ecosystem Carbon Cycling to Global Climate Change." *Nature* 393:249–52.

Chen, J. M., W. Chen, J. Liu, and J. Cihlar. 2000. "Annual Carbon Balance of Canada's Forests during 1895–1996." *Global Biogeochemical Cycles* 14:839–49.

Chen, J. M., J. Liu, J. Cihlar, and M. L. Goulden. 1999. "Daily Canopy Photosynthesis Model through Temporal and Spatial Scaling for Remote Sensing Applications." *Ecological Modelling* 124:99–119.

Cohen, W. B., and S. N. Goward. 2004. "Landsat's Role in Ecological Applications of Remote Sensing." *BioScience* 54:535–45.

Cohen, W. B., T. A. Spies, R. J. Alig, D. R. Oetter, T. K. Maiersperger, and M. Fiorella. 2002. "Characterizing 23 Years (1972–1995) of Stand Replacement Disturbance in Western Oregon Forests with Landsat Imagery." *Ecosystems* 5:122–37.

Coops, N. C., M. A. Wulder, D. S. Culvenor, and B. St-Onge. 2004. "Comparison of Forest Attributes Extracted from Fine Spatial Resolution Multispectral and Lidar Data." *Canadian Journal of Remote Sensing* 30(6):855–66.

Cramer, W., D. W. Kicklighter, A. Bondeau, B. Moore III, G. Churkina, B. Nemry, A. Ruimy, A. L. Schloss, and the Participants of the Potsdam Intercomparison. 1999. "Comparing Global Models of Terrestrial Net Primary Productivity (NPP): Overview and Key Results." *Global Change Biology* 5(S1):1–15.

DeFries, R. S., M. Hansen, J. R. G. Townshend, and R. Sohlberg. 1998. "Global Land Cover Classification at 8km Spatial Resolution: Use of Training Data Derived from Landsat Imagery in Decision Tree Classifiers." *International Journal of Remote Sensing* 19:3141–68.

DeFries, R. S. and J. G. R. Townshend. 1994. "NDVI Derived Land Cover Classifications at a Global Scale." *International Journal of Remote Sensing* 5:3567–86.

Dong, J., R. K. Kaufmann, R. B. Myneni, C. J. Tucker, P. E. Kauppi, J. Liski, W. Buermann, V. Alexeyev, and M. K. Hughes. 2003. "Remote Sensing Estimates of Boreal and Temperate Forest Woody Biomass: Carbon Pools, Sources, and Sinks." *Remote Sensing of Environment* 84:393–410.

Esser, G. 1987. "Sensitivity of Global Carbon Pools and Fluxes to Human and Potential Climatic Impacts." *Tellus* 39B:245–60.

Farquhar, G. D., S. V. Caemmerer, and J. A. Berry. 1980. "A Biochemical Model of Photosynthetic CO_2 Assimilation in Leaves of C-3 Species." *Planta* 149:78–90.

Feng, X., G. Liu, J. M. Chen, M. Chen, J. Liu, W. M. Ju, R. Sun, and W. Zhou. 2007. "Net Primary Productivity of China's Terrestrial Ecosystems from a Process Model Driven by Remote Sensing: Carbon Sequestration in China's Forest Ecosystems." *Journal of Environmental Management* 85(3):563–73.

Field, C. B., J. T. Randerson, and C. M. Malmstrom. 1995. "Global Net Primary Production:Combining Ecology and Remote Sensing." *Remote Sensing of Environment* 51:74–88.

Friedl, M. A., D. K. McIver, J. C. F. Hodges, X. Y. Zhang, D. Muchoney, A. H. Strahler, C. E. Woodcock, et al. 2002. "Global Land Cover Mapping from MODIS: Algorithms and Early Results." *Remote Sensing of Environment* 83:287–302.

Friedlingstein, P., I. Fung, J. John, G. Grasseur, D. Erickson, and D. Schimel. 1995. "On the Contribution of CO_2 Fertilization to the Missing Biospheric Sink." *Global Biogeochemical Cycles* 9:541–56.

Goetz, S. J., and S. D. Prince. 1999. "Satellite Remote Sensing of Primary Production: An Improved Production Efficiency Modeling Approach." *Ecological Modelling* 122:239–55.

Goetz, S. J., and S. D. Prince. 2000. "Interannual Variability of Global Terrestrial Primary Production: Results of a Model Driven with Satellite Observations." *Journal of Geophysical Research* 105:20077–91.

Goetz, S. J., S. D. Prince, S. N. Goward, M. M. Thawley, and J. Small. 1999. "Satellite Remote Sensing of Primary Production: An Improved Production Efficiency Modeling Approach." *Ecological Modelling* 122:239–55.

Hall, F. G., D. B. Botkin, D. E. Strebel, K. D. Woods, and S. J. Goetz. 1991. "Large-Scale Patterns of Forest Succession as Determined by Remote Sensing." *Ecology* 72:628–40.

Hansen, M. C., R. S. DeFries, J. R. G. Townshend, and R. Sohlberg. 2000. "Global Land Cover Classification at 1 km Spatial Resolution Using a Classification Tree Approach." *International Journal of Remote Sensing* 21:1331–64.

He, L., J. M. Chen, Y. Pan, R. Birdsey, and J. Kattge. 2012. "Relationships between Net Primary Productivity and Forest Stand Age in U.S. Forests." *Global Biogeochemical Cycles* 26:GB3009.

Heinsch, F. A., M. Reeves, P. Votava, S. Kang, C. Milesi, M. Zhao, J. Glassy, et al. 2003. *User's Guide GPP and NPP (MOD17A2/A3):Products NASA MODIS Land Algorithm, Version 2*. Missoula, MT: The University of Montana.

IPCC. 2000. *Land Use, Land-Use Change and Forestry*. Intergovernmental Panel on Climate Change. Cambridge: Cambridge University Press (ISBN:92-9169-114-3).

IPCC. 2007. *Climate Change 2007: The Physical Science Basis*. Summary for Policymakers. Contribution of Working Group I to the Fourth Assessment Report of the Intergovernmental Panel on Climate Change. Cambridge: Cambridge University Press. http://www.ipcc.ch/publications_and_data/publications_ipcc_fourth_assessment_report_wgl_report_the_physical_science_basis.htm.

Ito, A., and T. Oikawa. 2002. "A Simulation Model of the Carbon Cycle in Land Ecosystems (Sim-CYCLE): A Description Based on Dry-Matter Production Theory and Plot-Scale Validation." *Ecological Modelling* 151:147–79.

Janisch, J. E., and M. E. Harmon. 2002. "Successional Changes in Live and Dead Wood Carbon Stores: Implications for Net Ecosystem Productivity." *Tree Physiology* 22:77–89.

JRC. 2003. *Global Land Cover 2000 Database*. Brussels, Belgium: European Commission, Joint Research Centre.

Ju, W., and J. M. Chen. 2005. "Distribution of Soil Carbon Stocks in Canada's Forest and Wetlands Simulated Based on Drainage Class, Topography and Remotely Sensed Vegetation Parameters." *Hydrological Processes* 19:77–94.

Ju, W., J. M. Chen, T. A. Black, A. G. Barr, J. Liu, and B. Chen. 2006. "Modelling Multi-Year Coupled Carbon and Water Fluxes in a Boreal Aspen Forest." *Agricultural and Forest Meteorology* 140:136–51.

Jung, M., K. Henkel, M. Herold, and G. Churkina. 2006. "Exploiting Synergies of Global Land Cover Products for Carbon Cycle Modeling." *Remote Sensing of Environment* 101(4):534–53.

Kasischke, E. S., J. M. Melack, and M. C. Dobson. 1997. "The Use of Imaging Radars for Ecological Applications: A Review." *Remote Sensing of Environment* 59:141–56.

Kimball, J. S., A. R. Keyser, S. W. Running, and S. S. Saatchi. 2000. "Regional Assessment of Boreal Forest Productivity Using an Ecological Process Model and Remote Sensing Parameter Maps." *Tree Physiology* 20:761–75.

Kimball, J. S., S. W. Running, and R. Nemani. 1997. "An Improved Method for Estimating Surface Humidity from Daily Minimum Temperatures." *Agricultural and Forest Meteorology* 85:87–98.

Lefsky, M. A., W. B. Cohen, S. A. Acker, G. G. Parker, T. A. Spies, and D. Harding. 1999. "Lidar Remote Sensing of the Canopy Structure and Biophysical Properties of Douglas-Fir Western Hemlock Forests." *Remote Sensing of Environment* 70:339–61.

Lefsky, M. A., W. B. Cohen, G. G. Parker, and D. J. Harding. 2002. "Lidar Remote Sensing for Ecosystem Studies." *BioScience* 52:19–30.

Lieth, H. 1975. "Modeling the Primary Productivity of the World." In *Primary Productivity of the Biosphere*, edited by H. Lieth and R. H. Whittaker, 237–63. Berlin: Springer-Verlag.

Liu, J., J. M. Chen, and J. Cihlar. 2003. "Mapping Evapotranspiration Based on Remote Sensing: An Application to Canada's Landmass." *Water Resources Research* 39:1189.

Liu, J., J. M. Chen, J. Cihlar, and W. Chen. 1999. "Net Primary Productivity Distribution in the BOREAS Region from a Process Model Using Satellite and Surface Data." *Journal of Geophysical Research* 104(D22):27735–54.

Liu, J., J. M. Chen, J. Cihlar, and W. Chen. 2002. "Net Primary Productivity Mapped for Canada at 1-km Resolution." *Global Ecology and Biogeography* 11:115–29.

Liu, J., J. M. Chen, J. Cihlar, and W. M. Park. 1997. "A Process-Based Boreal Ecosystem Productivity Simulator Using Remote Sensing Inputs." *Remote Sensing of Environment* 62:158–75.

Loveland, T. R., B. C. Reed, J. F. Brown, D. O. Ohlen, Z. Zhu, L. Yang, and J. W. Merchant. 2000. "Development of a Global Land Cover Characteristics Database and IGBP DISCover from 1km AVHRR Data." *International Journal of Remote Sensing* 21:1303–30.

Lüdeke, M. K. B., F. W. Badeck, and R. D. Otto. 1994. "The Frankfurt Biosphere Model: A Global Process-Oriented Model of Seasonal and Long-Term CO_2 Exchange between Terrestrial Ecosystems and the Atmosphere. I. Model Description and Illustrative Results for Cold Deciduous and Boreal Forests." *Climate Research* 4:143–66.

Magnusson, M. 2006. "Evaluation of Remote Sensing Techniques for Estimation of Forest Variables at Stand Level." Doctoral thesis. Umeå, Sweden: Swedish University of Agricultural Sciences, 85. http://pub.epsilon.slu.se/1211/

Melillo, J. M., A. D. McGuire, D. W. Kicklighter, B. Moore III, C. J. Vörösmarty, and A. L. Schloss. 1993. "Global Climate Change and Terrestrial Net Primary Production." *Nature* 363:234–40.

Moran, E. F., E. Brondizio, P. Mausel, and Y. Wu. 1994. "Integrating Amazonian Vegetation, Land-Use, and Satellite Data." *BioScience* 44:329–38.

Myneni, R. B., S. Hoffman, Y. Knyazikhin, J. L. Privette, J. Glassy, Y. Tian, Y. Wang, et al. 2002. "Global Products of Vegetation Leaf Area and Fraction Absorbed Par from Year One of MODIS Data." *Remote Sensing of Environment* 83:214–31.

Parton, W. J., J. M. O. Scurlock, D. S. Ojima, T. G. Gilmanov, R. J. Scholes, D. S. Schime, T. Kirchner, et al. 1993. "Observations and Modeling of Biomass and Soil Organic Matter Dynamics for the Grassland Biome Worldwide." *Global Biogeochemical Cycles* 7:785–809.

Peng, C. 2000. "From Static Biogeographical Model to Dynamic Global Vegetation Model: A Global Perspective on Modelling Vegetation Dynamics." *Ecological Modelling* 135:33–54.

Peng, C., J. Guiot, H. B. Wu, H. Jiang, and Y. Q. Luo. 2011. "Integrating Models with Data in Ecology and Paleoecology: Advances toward a Model-Data Fusion Approach." *Ecology Letters* 14:522–36.

Peng, C., J. X. Liu, Q. L. Dang, A. J. Michael, and H. Jiang. 2002. "TRIPLEX: A Generic Hybrid Model for Predicting Forest Growth, and Carbon and Nitrogen Dynamics." *Ecological Modelling* 153:109–30.

Post, W. M., A. W. King, and S. D. Wullschleger. 1997. "Historical Variations in Terrestrial Biospheric Carbon Storage." *Global Biogeochemical Cycles* 11:99–109.

Potter, C., S. Klooster, R. Myneni, V. Genovese, P. N. Tan, and V. Kumar. 2003. "Continental-Scale Comparisons of Terrestrial Carbon Sinks Estimated from Satellite Data and Ecosystem Modeling 1982–1998." *Global Planetary Change* 39:201–13.

Potter, C. S., J. T. Randerson, C. B. Field, P. A. Matson, P. M. Vitousek, H. A. Mooney, and S. A. Klooster. 1993. "Terrestrial Ecosystem Production: A Process Model Based on Global Satellite and Surface Data." *Global Biogeochemical Cycles* 7:811–41.

Prentice, I. C., M. Heimann, and S. Sitch. 2000. "The Carbon Balance of the Terrestrial Biosphere: Ecosystem Models and Atmospheric Observations." *Ecological Applications* 10(6):1553–73.

Prince, S. D., and S. N. Goward. 1995. "Global Primary Production: A Remote Sensing Approach." *Journal of Biogeography* 22:815–35.

Raich, J. W., E. B. Rastetter, J. M. Melillo, D. W. Kicklighter, P. A. Steudler, B. J. Peterson, A. L. Grace, B. Moore III, and C. J. Vorosmarty. 1991. "Potential Net Primary Productivity in South America-Application of a Global Model." *Ecological Applications* 1:399–429.

Reich, P. B., M. B. Walters, and D. S. Ellsworth. 1997. "From Tropics to Tundra: Global Convergence in Plant Functioning." *Proceedings of the National Academy of Sciences* 94:13730–34.

Running, S. W., and J. C. Coughlan. 1988. "A General Model of Forest Ecosystem Processes for Regional Applications. I. Hydrological Balance, Canopy Gas Exchange and Primary Production Processes." *Ecological Modelling* 42:125–54.

Running, S. W., and E. R. J. Hunt. 1993. "Generalization of a Forest Ecosystem Process Model for Other Biomes, BIOME-BGC, and an Application for Global-Scale Models." In *Scaling Physiological Processes*, edited by J. R. Ehleringer and C. B. Field, 141–58. San Diego, CA: Academic Press.

Running, S. W., R. Nemani, J. M. Glassy, and P. E. Thornton. 1999. "MODIS Daily Photosynthesis (PSN) and Annual Net Primary Production (NPP) Product (MOD17)." Algorithm Theoretical Basis Document Version 3.0.

Running, S. W., P. E. Thornton, R. Nemani, and J. M. Glassy. 2000. "Global Terrestrial Gross and Net Primary Productivity from the Earth Observing System." In *Methods in Ecosystem Science*, edited by O. E. Sala, R. B. Jackson, H. A. Mooney, and R. W. Howarth, 44–57. New York: Springer-Verlag.

Sellers, P. J., D. A. Randall, G. J. Collatz, D. A. Dazlich, and C. Zhang. 1996. "A Revised Land Surface Parameterization (SiB2) for Atmospheric GCMs. Part 1. Model Formulation." *Journal of Climate* 9:676–705.

Stöckli, R., T. Rutishauser, D. Dragoni, J. O'Keefe, P. E. Thornton, M. Jolly, L. Lu, and A. S. Denning. 2008. "Remote Sensing Data Assimilation for a Prognostic Phenology Model." *Journal of Geophysical Research* 113:G04021.

Thornton, P. E. 1998. "Description of a Numerical Simulation Model for Predicting the Dynamics of Energy, Water, Carbon, and Nitrogen in a Terrestrial Ecosystem." PhD Dissertation. Missoula, MT: University of Montana, 280.

Thornton P. E., B. E. Lawb, H. L. Gholzc, K. L. Clark, E. Falge, D. S. Ellsworth, A. H. Goldstein, et al. 2002. "Modeling and Measuring the Effects of Disturbance History and Climate on Carbon and Water Budgets in Evergreen Needle Leaf Forests." *Agricultural and Forest Meteorology* 113:185–222.

Tian, Y., Y. Wang, Y. Zhang, Y. Knyazikhin, J. Bogaert, and R. B. Myneni. 2002. "Radiative Transfer Based Scaling of LAI/FPAR Retrievals from Reflectance Data of Different Resolutions." *Remote Sensing of Environment* 84:143–59.

Treuhaft, R. N., G. P. Asner, B. E. Law, and S. VanTuyl. 2002. "Forest Leaf Area Density Profiles from the Quantitative Fusion of Radar and Hyperspectral Data." *Journal of Geophysical Research* 107:4568–80.

Treuhaft, R. N., B. E. Law, and G. P. Asner. 2004. "Forest Attributes from Radar Interferometric Structure and Its Fusion with Optical Remote Sensing." *BioScience* 54:561–71.

Turner, D. P., W. B. Cohen, R. E. Kennedy, K. S. Fassnacht, and J. M. Briggs. 1999. "Relationships between Leaf Area Index and TM Spectral Vegetation Indices across Three Temperate Zone Sites." *Remote Sensing of Environment* 70:52–68.

Turner, D. P., S. V. Ollinger, and J. S. Kimball. 2004. "Integrating Remote Sensing and Ecosystem Process Models for Landscape- to Regional-Scale Analysis of the Carbon Cycle." *BioScience* 54(6):573–84.

Uchijima, Z., and H. Seino. 1985. "Agroclimatic Evaluation of Net Primary Productivity of Natural Vegetations (1) Chikugo Model for Evaluating Net Primary Productivity." *Journal of Agricultural Meteorology* 40:353–52.

Vauhkonen, J., L. Mehtätalo, and P. Packalén. 2011. "Combining Tree Height Samples Produced by Airborne Laser Scanning and Stand Management Records to Estimate Plot Volume in Eucalyptus Plantations." *Canadian Journal of Forest Research* 41(8):1649–58.

Warnant, P., L. François, D. Strivay, and J. C. Gérard. 1994. "CARAIB: A Global Model of Terrestrial Biological Productivity." *Global Biogeochemical Cycles* 8:255–70.

Williams, M., B. J. Bond, and M. G. Ryan. 2001. "Evaluating Different Soil and Plant Hydraulic Constraints on Tree Function Using a Model and Sap Flow Data from Ponderosa Pine." *Plant, Cell and Environment* 24:679–90.

White, J. D., S. W. Running, P. E. Thornton, R. E. Keane, K. C. Ryan, D. B. Fagre, C. H. Key. 1998. "Assessing Simulated Ecosystem Processes for Climate Variability Research at Glacier National Park, USA." *Ecological Applications* 8:805–23.

Xiao, X., D. Hollinger, J. D. Aber, M. Goltz, E. A. Davidson, and Q. Y. Zhang. 2004a. "Satellite-Based Modeling of Gross Primary Production in an Evergreen Needleleaf Forest." *Remote Sensing of Environment* 89:519–34.

Xiao, J., Q. Zhuang, B. E. Law, J. Chen, D. D. Baldocchi, D. R. Cook, R. Oren, et al. 2010. "A Continuous Measure of Gross Primary Production for the Conterminous U.S. Derived from MODIS and AmeriFlux Data." *Remote Sensing of Environment* 114:576–91.

Xiao, J. F., Q. L. Zhuang, B. E. Law, D. D. Baldocchi, J. Chen, A. D. Richardson, J. M. Melillo, et al. 2011. "Assessing Net Ecosystem Carbon Exchange of U.S. Terrestrial Ecosystems by Integrating Eddy Covariance Flux Measurements and Satellite Observations." *Agricultural and Forest Meteorology* 151:60–69.

Xiao, X., Q. Zhang, B. Braswell, S. Urbanski, S. Boles, S. Wofsy, B. Moore, and D. Ojima. 2004b. "Modeling Gross Primary Production of a Deciduous Broadleaf Forest Using Satellite Images and Climate Data." *Remote Sensing of Environment* 91:256–70.

Zhao, M. S., F. A. Heinsch, R. R. Nemani, and S. W. Running. 2005. "Improvements of the MODIS Terrestrial Gross and Net Primary Production Global Data Set." *Remote Sensing of Environment* 95:164–76.

Zhao, K., S. Popescu, and R. Nelson. 2009. "Lidar Remote Sensing of Forest Biomass: A Scale-Invariant Estimation Approach Using Airborne Lasers." *Remote Sensing of Environment* 113:182–96.

23

Remote Sensing Applications to Modeling Biomass and Carbon of Oceanic Ecosystems

Samantha Lavender and Wahid Moufaddal

CONTENTS

23.1 Introduction...443
23.2 Overview of Algorithms and Methods 444
23.3 Case Study—Fisheries...448
23.4 Summary and Comments on the Future 454
References...454

23.1 Introduction

It is understood by the majority of the world's population that we live in a rapidly changing environment, but for many, the marine environment and hence changes occurring within it remain hidden. A significant challenge for scientists is to not only monitor these changes but also understand and predict the consequences so that we can both change our behavior and mitigate the impacts.

Although interactions between the ocean, atmospheric, and terrestrial systems are primarily driven by physical or chemical processes, biological processes are an important component. Almost three-quarters of the Earth's surface is ocean, covered with oceanic waters containing most of the volatile gases and hence playing a significant role in the global biogeochemical cycles (Allen et al. 2010).

Climate is affected by processes such as the carbon biological pump (Volk and Hoffert 1985; Longhurst and Harrison 1989), primarily by processes such as the phytoplankton photosynthesis and carbon export that remove an estimated 11–16 PgC (petagrams of carbon) from oceanic surface waters each year (Falkowski et al. 2000). The oceanic ecosystem also plays a key role in cycling nitrogen, phosphorus, and oxygen with the distribution of phytoplankton biomass defined by the availability of growth-limiting factors such as light and nutrients. These are in turn regulated by physical processes including ocean circulation, mixed-layer dynamics, upwelling, atmospheric dust deposition, and the solar cycle.

Understanding the oceanic ecosystem requires measurements. For example, marine primary production can be assessed directly using flux measurements in the field such as 14C fixation experiments. However, ecosystems are dynamic, with the annual mean value of marine photosynthesis being 45–50 Gt C coming from a biomass of 1 Gt (Carr et al. 2006). Therefore, in situ ship-based measurements cannot resolve low-frequency spatial and temporal variability; much less make direct observations of mesoscale variability beyond isolated snapshots. However, this has improved through the use of floats and glider observations, for example, there are now over 3500 (milestone reached as of February 2012 with the real-time situation available from http://www.argo.ucsd.edu/) Argo floats with an International Ocean Colour Coordinating Group (IOCCG) Working Group looking at bio-optical sensor deployment.

To aid the understanding of the three-dimensional (3D) environment and predict changes, numerical ocean modeling is used alongside in situ measurements. Biological modeling has progressed from simple three-component or NPZ-type (nutrients, phytoplankton, and zooplankton) models to those with increasing complexity in terms of model structure, parameterization of biochemical laws, and a link with hydrodynamical models (coupled ocean-ecosystem models) to improve dynamics and feedback. Increasing computer power has supported this development through the handling of complex mathematical solutions and physical downscaling (finer resolutions and/or nested models); see Fasham (1993) and Allen et al. (2010) for detailed reviews of the modeling of oceanic ecosystems.

Also, from the start of the Joint Global Ocean Flux Study (JGOFS), remote measurements of near-surface chlorophyll a (Chl-a) concentrations were envisaged as the major tool for extrapolating upper-ocean chemical and biological measurements in time and space and linking calculations of new and primary production with the flux of particulate material through the water column (Yoder et al. 2001). Satellite ocean color radiometry (OCR) adds another dimension to in situ/modelling marine biology and ecosystem studies as it provides global measurements on timescales that are directly related to biogeochemical distributions and processes. Based on NASA's Coastal Zone Color Scanner (CZCS), launched in late 1978, it was possible to identify ecological provinces and describe the seasonal cycle of phytoplankton distribution (Longhurst 1998).

23.2 Overview of Algorithms and Methods

There is no direct in situ measurement of phytoplankton carbon biomass; for in situ measurements, it is usually estimated from cell biology through microscopic measurements and/or flow cytometry, which can be time consuming (Wang et al. 2009), although modern techniques such as cameras and image processing are being applied to speed up the process, for example, the FlowCAM (Flow Cytometer And Microscope) instrument from Fluid Imaging. Therefore, phytoplankton biomass is often inferred from Chl-a, a pigment that is common to all phytoplankonic autotrophs and can easily be measured in situ. Techniques include the following:

- High-performance liquid chromatography (HPLC): It is often used within the Earth observation (EO) community for looking at multiple pigments, but requires expensive equipment and standards plus an experienced technician.

- Fluorometry: It is based on fluorescence and the most common approach as it is much cheaper than HPLC and reasonably accurate, but often restricted in terms of pigments detected—primarily Chl-a and phaeopigments. In situ depth profiling by fluorometers allows the vertical structure of phytoplankton populations to be sampled, which can include interpolating between HPLC samples, but the measurements will be affected by photoacclimation (i.e., the modification of the in vivo Chl response to the surrounding average light level) and fluorescence from other substances. For underway sampling, using a pumped system, in vivo fluorometry in conjunction with HPLC provides temporal interpolation so that fine-scale resolutions features can be observed.

- Spectrophotometry: It is a less common approach based on absorption, but can be useful for looking at pigments other than Chl-a, for example, Chl-b and Chl-c. It is not recommended where there are significant amounts of phaeophytin, a natural degradation product of chlorophyll, although acidification can be used to determine the contribution from phaeophytin.

While each of these methods has advantages and disadvantages (Mantoura et al. 1997), the precision and accuracy of the final pigment concentration is ultimately determined by the efficiency of the technique used to extract the pigments from the cell. It is important that internationally agreed protocols are followed and that uncertainties are calculated.

Remote sensing needs to be able to detect a change in the electromagnetic radiation—backscatter, reflectance, fluorescence, or emission depending on the frequency/wavelength. By going through a series of images for 1 month (May 2010), this section aims to show the derivation of products from OCR. As this chapter is primarily application focused, it will not review the processing that leads up to the point where we have normalized water leaving radiances/reflectances, so further detail can be found in papers and books focused on OCR techniques, for example, IOCCG Reports No. 5 and 10 (2006, 2010), Robinson (2010), and Lavender (2012). The reality is that there are many different algorithms and models available in the peer-reviewed literature and so one challenge for the end user is to decide what is most appropriate for their own application. In this case, this chapter shows globally available (for free) products from NASA's Ocean Biology Processing Group (OBPG) (http://oceancolor.gsfc.nasa.gov/) and a European Space Agency (ESA) project called GlobColour (http://www.globcolour.info).

ESA initiated the Data User Element program to build a user community for EO data by running projects that develop and demonstrate user-driven applications; marine-focused projects have included GlobColour (for OCR), Medspiration (for sea surface temperature [SST]), and GlobWave (for satellite wave data). GlobColour (Pinnock et al. 2007) kicked off in November 2005 with the first year being a feasibility demonstration phase for multisensor data merging, including the quantification of error statistics and intersensor biases, and the second/third years dedicated to the production of the 10-year (September 1997 to December 2006) time series from MERIS, MODIS-Aqua, and SeaWiFS (Maritorena et al. 2010). This followed on from the recommendations in the IOCCG Report 6 (2007), and since then, further merged datasets have been published (e.g., Mélin et al. 2009).

Satellite-based observations of Chl-a are detected through the change in the reflectance as Chl-a naturally absorbs blue light and emits, or fluoresces, red light while having no strong absorption in the green. This provides an estimate of the near-surface layer biomass whose depth is dependent on the turbidity of the water, that is, the depth to which photons penetrate to and subsequently return to the surface from.

Figure 23.1 shows the variation in normalized water leaving radiance (which is equivalent to a reflectance as it is in the absence of an atmosphere and with the sun directly overhead, so the term reflectance will be used from this point onwards) with wavelength, with Figure 23.1a through d displaying bands 443, 490, 510, and 560 nm, respectively. In the open ocean, such as the middle of the oceanic gyres, the highest reflectance values are at the shortest wavelengths reflecting the shape of an expected open ocean reflectance spectra; the shape is dominated by the absorption of water itself. If, for example the North Atlantic, there are lower blue reflectances reflecting the presence of Chl-a and hence absorption at these wavelengths.

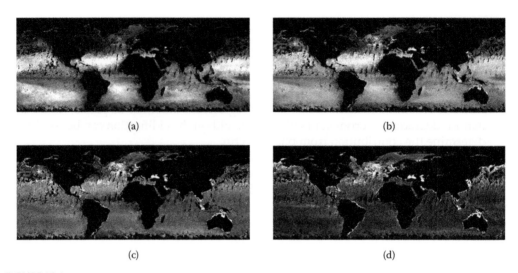

(a) (b)

(c) (d)

FIGURE 23.1
Global monthly composite from MERIS, May 2010, shown as the individual bands of normalized water leaving radiance at (a) 443 nm, (b) 490 nm, (c) 510 nm, and (d) 560 nm. (Courtesy of ESA/GlobColour project.)

The green wavelength (560 nm) has high values around the coast and for some pixels in the North Atlantic, which shows the presence of scattering by particles.

Space agencies have developed standard Chl-a absorption algorithms, which can be applied to global EO datasets, for example, as follows:

- OC3M and OC4v6 from NASA for MODIS and SeaWiFS, respectively (O'Reilly et al. 2000): based on two reflectance ratios for MODIS, 443/555 and 490/555, and on three reflectance ratios for SeaWiFS, 443/555, 490/555, and 510/555.

- OC4Me from ESA for MERIS (Morel et al. 2007): based on three reflectance ratios 443/560, 490/560, and 510/560.

In the equation $\log_{10}[\text{Chl-a}] = a_0 + \sum_{n=1}^{N} a_n \left[\log_{10} \left({}_j R_{rs} \right) \right]^n$, N is the number of band ratios, and the ratio with the highest value is used for each pixel.

An alternative approach is semianalytical algorithms (SAAs), which involves a simplified radiative transfer model based on multispectral absorption and backscattering. NASA's OBPG initiated the GIOP (Generalized Inherent Optical Property) algorithm process by offering to provide the datasets, processing framework, and an international forum within which a new generation of global inherent optical property (IOP) products could be developed and evaluated; an extension of the activity reported in IOCCG (2006). By deconstructing the SAAs, NASA was able to identify similarities and uniqueness with the main aim of achieving community-wide consensus on a unified SAA with which to generate global satellite IOP products and also to determine an approach to implement uncertainties (Franz and Werdell 2010); example products are available on the web, http://oceancolor.gsfc.nasa.gov/WIKI/GIOP.html.

In addition to Chl-a algorithms, one should take into account chromophoric dissolved organic matter (CDOM), which is the light-absorbing fraction of the total oceanic dissolved organic matter pool characterized by high absorption across ultraviolet (UV) wavelengths decreasing monotonically toward longer wavelengths (Bricaud et al. 1981). It is quantifiable from OCR (Siegel et al. 2002), with satellite estimates revealing that CDOM accounts for nearly half of total nonwater light absorption in the global surface ocean at 440 nm, near the maximum of spectral light absorption by chlorophyll in phytoplankton (Swan et al. 2009). Regional variation in CDOM is not accounted for in standard Chl-a algorithms and hence estimations of global phytoplankton biomass can be overestimated, which in turn would impact global biogeochemical predictions (Siegel et al. 2005). Furthermore, CDOM absorbs over 90% of total UV radiation entering the euphotic layer, thereby serving a photoprotective role in biological processes (Zepp 2002), and the UV photochemistry contributes to the sea surface flux of climate-relevant trace gases such as CO_2 (Johannessen and Miller 2001).

Figure 23.2a and b shows a comparison of the OC4v4 Chl-a algorithm and GIOP Chl-a concentration for the May 2010 monthly composite from SeaWiFS. Compared to the standard Chl-a absorption algorithm, the GIOP SAA has lower values due to an improved separation of Chl-a and CDOM absorption. Also, in waters with nonbiological particulate scattering, there will be a separation of Chl-a from suspended particulate matter.

Once we have an estimate of biomass, then the next step would be to obtain a value for the net primary production (NPP). Integrated biomass can be obtained by assuming a vertical profile and carbon to Chl-a ratio. Then, to go from biomass to a photosynthesis rate, a time-dependent variable is needed with solar radiation being the normal choice. Therefore, simple mechanistic models can compute productivity from biomass, photosynthetically available radiation (PAR), and a transfer or yield function that incorporates the physiological response of the measured Chl-a to light, nutrients, temperature, and other environmental variables. In addition, SST is often used to parameterize the photosynthetic potential.

Figure 23.3 shows NPP estimates from MODIS based on the Vertically Generalized Production Model (VGPM) by Behrenfeld and Falkowski (1997); euphotic depths are calculated from MODIS Chl-a following Morel and Berthon (1989). The NPP estimates (Figure 23.3a) have a similar pattern to that of Chl-a (Figure 23.3b), reflecting the strong dependence on this input. The modification comes from PAR (Figure 23.3c) and SST (Figure 23.3d), which have higher values at lower altitudes compared to higher latitudes although PAR is strongly

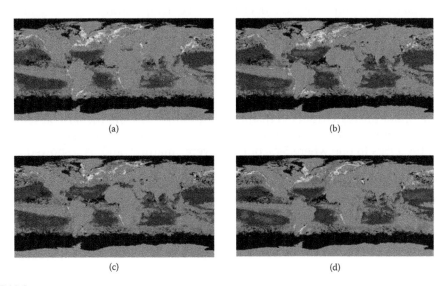

(a)   (b)

(c)   (d)

FIGURE 23.2
(**See color insert.**) Global monthly composite from SeaWiFS, May 2002, with Chl-a calculated using the (a) OC4v4 Chl-a algorithm and the GIOP outputs for (b) Chl-a concentration, (c) absorption by phytoplankton at 443 nm, and (d) absorption by CDOM at 443 nm. The products are displayed using the SeaWiFS rainbow palette. (Courtesy of the NASA OBPG group.)

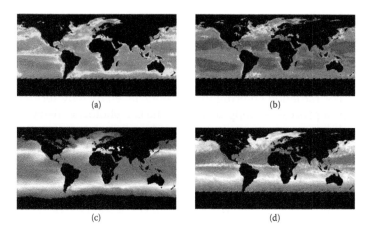

FIGURE 23.3
(See color insert.) Global monthly composite from MODIS, May 2010, with NPP calculated using the (a) VGPM model, (b) MODIS Chl-a, (c) SST, and (d) PAR products. The NPP and Chl-a products are displayed using the SeaWiFS rainbow palette, Chl-a as a log value, while SST and PAR are displayed using a blue to white to red color palette. (Courtesy of the NASA OBPG group and ocean productivity site, http://orca.science.oregonstate.edu/.)

influenced by cloudiness and so there are low values at the equator and also SST shows the westward-propagating tongue of cold water in the equatorial Pacific, from South America, that is associated with the El Niño–Southern Oscillation (ENSO); strongest during La Niña conditions and may disappear when strong El Niño conditions are present.

The third primary production algorithm round robin (PPARR3) compared outputs from 24 ocean color–based models and model variants (Carr et al. 2006), concluding that parameterization of the maximum or optimal photosynthetic rate has a large impact on the variability, plus there is a large divergence in response to SST perturbations illustrating the need to improve our understanding of, and ability to model, the effect of temperature on photosynthesis.

23.3 Case Study—Fisheries

Marine and freshwater fish stocks are an important high-protein human food source, but they come under pressure from overfishing; for example, during the past 50 years, global fisheries have moved from a situation where 90% of them were classified as "undeveloped" in the 1950s to one where 34% of them were "mature" and 34% showed declining yields in the 1990s (Garibaldi and Grainger 2002). As the demographic pressure and the international demand for fish and fish products continues to increase, an amplification of the fishing pressure on the living marine resources is expected.

Sustainable use of aquatic ecosystems requires effective monitoring and management, which in turn requires data. Conventional means of sampling are limited geographically and temporally, and so, EO data has filled the gap with a shift to the ecosystem approach. Although EO does not observe fish stocks directly, measurements (such as SST, sea surface height, OCR, ocean winds, and sea ice) characterize the critical habitat that in turn influences marine resources. Also, Chl-a is key metric for assessing the health and productivity of marine ecosystems on a global scale. For example, it is used operationally by organizations such as the US National Oceanic and Atmospheric Administration (NOAA) to detect and monitor harmful algal blooms that impact fisheries as they can cause harvesting closures at shellfish beds and their toxicity can cause mass mortalities to fish and marine mammals.

The concept of regime shifts in biotic and abiotic oceanic conditions has been widely accepted since this term was first applied to marine ecosystems by Lluch-Belda et al. (1989). Published studies have reported a large number of abrupt and substantial changes in the state of marine ecosystems in many oceanic regions across the world, such as the North Atlantic in the early 1960s and late 1980s (Reid et al. 1998; Beaugrand 2004; Raitsos et al. 2005), North Pacific in the late 1970s and 1980s (Hare and Mantua 2000; Benson and Trites 2002; PICES 2005), and tropical Pacific in the 1970s (Lees et al. 2006). Prevalence at large spatial scales and the ability to propagate through different trophic levels are among the most important criteria of regime shifts (Collie et al. 2004; de Young et al. 2004, 2008). Therefore, characterizing the nature of these changes has important fisheries and ecological management implications.

Coastal, shelf, and semienclosed marine ecosystems are often heavily subjected to human-induced perturbations and so experience more complicated regime shifts through multiple forcings relative to their oceanic counterparts (Oguz and Gilbert 2007). The Mediterranean Sea is of special interest because of dramatic changes that took place in its southeastern part (Levantine Basin), including the Nile delta shelf (Figure 23.4). After the construction of Aswan High Dam (AHD) along the Nile River, the Egyptian fisheries off the Mediterranean showed rapid downward and upward shifts in the mid-1960s and early 1980s, respectively. A lack of long-term monitoring data, including available in situ data, has constrained research activities. Therefore, a time-series analysis of total fish catch, the pelagic to demersal (P/D) ratio, and EO-derived Chl-a concentrations from the Nile delta shelf and coastal lakes of the delta have been used to provide insight. There remain limitations due to the lack of accompanying in situ data with issues around the use of OCR in coastal waters discussed in IOCCG Report No. 3 (2000).

Chl-a concentrations in the Nile delta shelf (see transect area in Figure 23.4) were computed by applying the MedOC4 regional bio-optical algorithm of Volpe et al. (2007) to

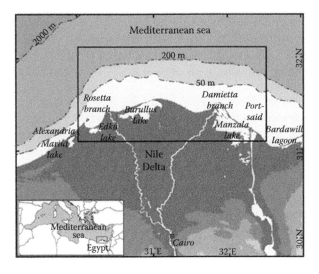

FIGURE 23.4
Location map showing the Nile delta shelf and the transect area used for time-series analysis of EO-derived Chl-a data. (From Moufaddal, W.M., and S. Lavender, Assessment of the Possible Key Factors for Fall and Rise of the Coastal Fisheries off the Nile Delta: A Remote Sensing Approach, *Proceedings of the Second Gulf Conference & Exhibition on Environment and Sustainability*, 620–635, Kuwait, 2009a.)

the 10-year GlobColour multisensor merged monthly averaged data. Among region-specific algorithms, MedOC4 has been shown to be the best algorithm for the South East Mediterranean region—in terms of unbiased EO-derived Chl-a estimates and an improved uncertainty in coastal Mediterranean waters (Volpe et al. 2007). It is a fourth-order polynomial expression based on the OC4 functional form (see Section 23.2), but the coefficients are locally determined for the Mediterranean Sea. As this study had no in situ Chl-a concentrations to validate the algorithm to the specifics of the coastal Nile River delta region, its accuracy remains questionable although pseudo-true color MODIS imagery do show a greenish coloration where the algorithm is predicting high Chl-a concentrations (Lavender et al. 2009; Moufaddal and Lavender 2009b).

The EO-derived seasonal pattern, temporal and spatial distribution, of surface Chl-a in the Nile delta shelf for the period 1997–2006 is shown in Figure 23.5. Winter (December–March) and spring (March–May) imagery show an extensive bloom off the delta coast encompassing the highest phytoplankton Chl-a concentrations, whereas in summer (May–September) and autumn (September–December) there is an apparent shrinkage in the size and magnitude of the bloom. These results confirm earlier observations by Dowidar (1984) and Halim et al. (1995) that the Nile phytoplankton bloom has been temporally and spatially shifted compared to the pre-dam conditions; the classical (pre-dam) bloom was exclusively composed of diatoms (Halim 1960; Aleem and Dowidar 1967) and the magnitude varied from year to year and location to location depending on the volume of the incoming flood, prevailing hydrodynamics, and nutrient concentration with a standing crop of the order of 1 million or more cells/L (Halim et al. 1967; Aleem 1972). Nowadays, diatoms are no longer the dominant phytoplankton species with dinoflagellates becoming equally important (Halim et al. 1976; Dowidar 1984). Moreover, the phytoplankton standing crop rarely exceeds a few thousands cells/L during the post-dam flood (Aleem 1972; Halim et al. 1976; Dowidar 1984). It is also noticeable that the bloom is stronger and more pronounced near the outlets of the Egyptian coastal lakes and urban outfalls, and a general south to north gradient in Chl-a concentration can be observed during most of the year.

Figure 23.6a demonstrates the mean monthly surface Chl-a variations (range is from ~0.4 to 1.25 mg/m^3) within the delta shelf with the bracketing dashed lines showing the 95% confidence interval. The trend has been calculated from a linear regression with the regression statistic shown in the figure ($R^2 = 0.0715$). The low correlation is not unexpected as there is a strong seasonal signal (see Figure 23.5). Monthly anomalies were computed as departures from the 9-year (1998–2006) climatological monthly values; the limited data in 2007 was not used to calculate the climatologies. Figure 23.6b shows the monthly anomalies plotted and a trend line calculated as for Figure 23.6a. The correlation coefficient has increased marginally ($R^2 = 0.1964$), but remains low. It can be seen that there are still strong variations from the linear trend with, for example, a significant peak (positive anomaly of 0.257 mg/m^3) for December 1998.

Figure 23.7a demonstrates a comparison of fish landings for three different Egyptian fishing areas throughout a 28-year period (1962–1999); Mediterranean and Red Sea off the Egyptian coast, and coastal lakes of the Nile delta. The total fish catch has varied greatly for all the areas during this period, but with an overall upward trend from the early 1980's onwards. Figure 23.7b summarizes the range of mean P/D ratios off the Nile delta before and after the construction of the AHD as well as during the early to mid-1980s (Halim et al. 1995). The P/D ratio mainly remained above average until the mid-1960s. Afterwards, in 1996, the ratio dropped abruptly below the average and remained at around 0.5 until 1979 before it returned to the pre-dam values of the early 1980s.

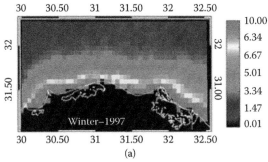

(a)

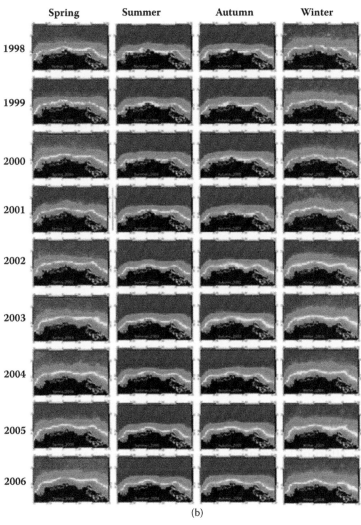

(b)

FIGURE 23.5
(**See color insert.**) (a) Enlarged view of the Winter 1997 mean seasonal Chl-a concentrations. (b) Mean seasonal Chl-a concentrations off the Nile delta during 1997–2006, as calculated using the MedOC4 algorithm of Volpe et al. (2007) applied to the GlobColour dataset. (From Moufaddal, W.M., and S. Lavender, Assessment of the Possible Key Factors for Fall and Rise of the Coastal Fisheries off the Nile Delta: A Remote Sensing Approach, *Proceedings of the Second Gulf Conference & Exhibition on Environment and Sustainability*, 620–635, Kuwait, 2009a.)

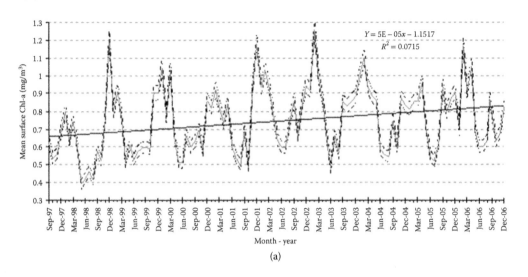

$$Y = 5E - 05x - 1.1517$$
$$R^2 = 0.0715$$

Month - year

(a)

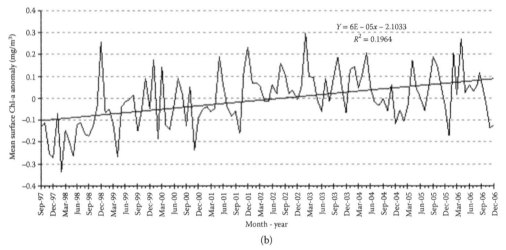

$$Y = 6E - 05x - 2.1033$$
$$R^2 = 0.1964$$

Month - year

(b)

FIGURE 23.6
(a) Mean monthly and (b) monthly anomaly variability of Chl-a for the period from 1997 to 2006, as derived by application of the MedOC4 algorithm to GlobColour merged EO-derived data.

The combination of EO-derived Chl-a and fish landings tends to support the theory of increased nutrient loading/enrichment (Caddy 1993, 2000) rather than an improvement in fishing effort or techniques. This is not to say that an increase (or decrease) in fishing intensity is without effect, but it seems unlikely, as suggested by Nixon (2003), that changes in fishing effort alone would explain the dramatic recovery of the Egyptian Mediterranean fisheries. The P/D ratio data tends to support the nutrient loading theory as the timing of the decline in the P/D ratio below the long-term average and its recovery (Figure 23.7) coincides with disappearance of the historical Nile phytoplankton bloom in the mid-1960s after the construction of the AHD and the return of a bloom (in a new modern form) in the early 1980s. The return is coincident with increasing fertilizer use, expanded agricultural drainage, increasing human population, and subsequent dramatic extensions of urban water supplies and sewage collection systems in Egypt (Nixon 2003).

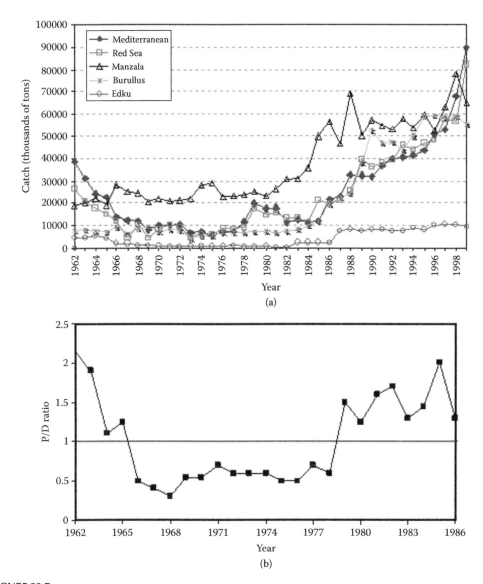

FIGURE 23.7

(a) Total fish catch for Egyptian marine fisheries yielded from the coastal lakes of the Nile delta and the Mediterranean and Red Seas during 1962–1999 (NIOF, 1962–1986, and GAFRD, 1987–1999) with each location shown by a different symbol and (b) the pelagic to demersal ratio for fish landings off the Nile delta during 1962–1986. (From Halim, Y. et al., *Effects of Riverine Inputs on Coastal Ecosystems & Fisheries Resources*, FAO Fisheries Technical Paper 349, 1995.)

The significant peak in the December 1998 monthly Chl-a anomaly (Figure 23.6a) could be linked with the 1997–1999 Pacific ENSO. In addition, during 1998–2001 period there were a series of high flood years where the AHD had to be opened because of high water levels within Lake Nasser. In addition, Barale et al. (2008) highlighted a Mediterranean-wide December 1998 positive anomaly with Bosc et al. (2004) also showing that the southern Levantine basin had raised levels. This highlights the difficulty in separating anthropogenic linked changes from natural variability (interannual to multidecadal scale).

The Societal Applications in Fisheries and Aquaculture using Remote Sensing Imagery (SAFARI) project aims to accelerate the assimilation of EO data into fisheries research and management by facilitating the application of rapidly evolving satellite technology (Wilson 2011). Of the many types of EO data, the paper concludes that OCR is the most important to fisheries because it is the only biological measurement.

23.4 Summary and Comments on the Future

In summary, OCR allows us to explore the relationships between the marine ecosystem, biogeochemical cycles, and climate change with EO showing a clear link between ocean color–derived products and changes in temperature and stratification at regional to global scales. There is a need for sensors with increased spectral resolution in the blue–green region to characterize pigment absorption amplitude and width plus lower (UV) wavelengths for the more accurate separation of pigments and CDOM.

Decadal-scale changes in the distribution and abundance of ocean phytoplankton remain uncertain, resulting from processes occurring at various timescales, including (possibly) the centennial one of anthropogenic global warming, interdecadal ones at basin-to-global scales, and interannual to seasonal ones. One of the most challenging aspects for understanding variability in biological processes is associating with detecting changes in the environmental forcings responsible. Important climate variables linked to ocean circulation and productivity are sea-level pressure, surface winds, SST, surface air temperature, and cloudiness. Interannual-to-seasonal timescales have been intensively investigated on scales ranging from regional to global thanks to the 10+ year ocean color time series. However, as stated by Henson et al. (2010), a time series of about 40 years in length would be needed to distinguish global warming trends from natural variability.

Model assimilation research has concentrated on Chl-a as an estimate of biomass, but recent trends are focused on assimilating the radiances themselves akin to assimilation within meteorological models, as discussed by Robinson in Chapter 3 "Ocean-Colour Data—An Aid to Modelling" of IOCCG (2008). Here, the application of the bio-optical algorithm transfers from an EO preprocessing step to a step that occurs within the numerical ocean model itself.

Despite the increasing use of EO and ocean models, we cannot forget in situ data that remain a key component of the three-point triangle; without in situ data, it is difficult to be sure that we fully understand the results obtained. Therefore, an integrated observing strategy consisting of EO-derived data, time series stations, gliders, floats, and moorings will be necessary to detect the full suite of biological responses to global warming (Henson et al. 2010).

References

Aleem, A. A. 1972. "Effect of River Outflow Management on Marine Life." *Marine Biology* 15:200–08.
Aleem, A. A., and N. M. Dowidar. 1967. "Phytoplankton Production in Relation to Nutrients along the Egyptian Mediterranean Coast." *Studies in Tropical Oceanography* 5:305–23.

Allen, J. I., J. Aiken, T. R. Anderson, E. Buitenhuis, S. Cornell, R. J. Geider, K. Haines, et al. 2010. "Marine Ecosystem Models for Earth Systems Applications: The MarQUEST Experience." *Journal of Marine Systems* 81:19–33.

Barale, V., J.-M. Jaquet, and M. Ndiaye. 2008. "Algal Blooming Patterns and Anomalies in the Mediterranean Sea as Derived from the SeaWiFS Data Set (1998–2003)." *Remote Sensing of Environment* 11:3300–13.

Beaugrand, G. 2004. "The North Sea Regime Shift: Evidence, Mechanisms, and Consequences." *Progress in Oceanography* 60:245–62.

Behrenfeld, M. J., and P. G. Falkowski. 1997. "Photosynthetic Rates Derived from Satellite-Based Chlorophyll Concentration." *Limnology and Oceanography* 42:1–20.

Benson, A. J., and A. W. Trites. 2002. "Ecological Effects of Regime Shifts in the Bering Sea and the Eastern North Pacific Ocean." *Fish and Fisheries* 3:95–113.

Bosc, E., A. Bricaud, and D. Antoine. 2004. "Seasonal and Interannual Variability in Algal Biomass and Primary Production in the Mediterranean Sea, as Derived from 4 Years of SeaWiFS Observations." *Global Biogeochemical Cycles* 18: GB1005. doi: 10.1029/2003GB002034.

Bricaud, A., A. Morel, and L. Prieur. 1981. "Absorption by Dissolved Organic Matter of the Sea (Yellow Substance) in the UV and Visible Domains." *Limnology and Oceanography* 26:43–53.

Caddy, J. F. 1993. "Toward a Comparative Evaluation of Human Impacts on Fishery Ecosystems of Enclosed and Semi-Enclosed Seas." *Reviews in Fisheries Science* 1:57–95.

Caddy, J. F. 2000. "Marine Catchment Basin Effects versus Impacts of Fisheries on Semi-Enclosed Seas." *ICES Journal of Marine Science* 57:628–40.

Carr, M.-E., M. A. M. Friedrichs, M. Schmeltz, M. N. Aita, D. Antoine, K. R. Arrigo, I. Asanuma, et al. 2006. "A Comparison of Global Estimates of Marine Primary Production from Ocean Color." *Deep Sea Research Part II: Topical Studies in Oceanography* 53(5–7):741–70.

Collie, J. S., K. Richardson, and J. H. Steele. 2004. "Regime Shifts: Can Ecological Theory Illuminate the Mechanisms?" *Progress in Oceanography* 60:281–302.

de Young, B., M. Barange, G. Beaugrand, R. Harris, R. Ian Perry, M. Scheffer, and F. Werner. 2008. "Regime Shifts in Marine Ecosystems: Detection, Prediction and Management." *Trends in Ecology and Evolution* 23(7):402–09.

de Young, B., R. Harris, J. Alheit, G. Beaugrand, N. J. Mantua, and L. J. Shannon. 2004. "Detecting Regime Shifts in the Ocean: Data Considerations." *Progress in Oceanography* 60:143–64.

Dowidar, N. M. 1984. "Phytoplankton Biomass and Primary Productivity of the Southeastern Mediterranean." *Deep-Sea Research-II* 31:983–1000.

Falkowski, P. G., R. J. Scholes, E. Boyle, J. Canadell, D. Canfield, J. Elser, N. Gruber, et al. 2000. "The Global Carbon Cycle: A Test of Our Knowledge of Earth as a System." *Science* 290:291–96.

Fasham, M. J. R. 1993. "Modelling the Marine Biota." In *The Global Carbon Cycle*, edited by M. Heimann, 457–504. Springer-Verlag, Heidelberg.

Franz, B. A., and P. J. Werdell. 2010. "A Generalized Framework for Modeling of Inherent Optical Properties in Remote Sensing Applications." *White Paper, Proceedings of Ocean Optics 2010*, Anchorage, Alaska.

GAFRD. 1987–1999. *Annual Fishery Statistics Reports*. General Authority for Fish Resources Development, Cairo, Egypt.

Garibaldi, L., and R. J. R. Grainger. 2002. "Chronicles of Catches from Marine Fisheries in Eastern Central Atlantic for 1950–2000." *International Symposium on Marine Fisheries, Ecosystems, and Society in Western Africa*, Dakar, Senegal.

Halim, Y. 1960. "Observations on the Nile Bloom of Phytoplankton in the Mediterranean." *ICES Journal of Marine Science* 26(1):59–67.

Halim, Y., S. K. Guergues, and H. H. Saleh. 1967. "Hydrographic Conditions and Plankton in the South East Mediterranean during the Last Normal Nile Flood (1964)." *International Review of Hydrobiology* 52(3):401–25.

Halim, Y., S. A. Morcos, S. Rizkalla, and M. Kh. El-Sayed. 1995. "The Impact of the Nile and the Suez Canal on the Living Marine Resources of the Egyptian Mediterranean Waters (1958–1986)." *Effects of Riverine Inputs on Coastal Ecosystems & Fisheries Resources*. FAO Fisheries Technical Paper 349, FAO, Rome.

Halim, Y., A. Samaan, and F. A. Zaghloul. 1976. "Estuarine Plankton of the Nile and the Effect of Fresh Water Phytoplankton." In *Fresh Water on the Sea*, edited by S. Skreslet, R. Leinbo, J. B. L. Matthews, and E. Sakshaug, April 22–25, 1974. Oslo, Geilö, Norway: The Association of Norwegian Oceanographers.

Hare, S. R., and N. J. Mantua. 2000. "Empirical Evidence for North Pacific Regime Shift in 1977 and 1989." *Progress in Oceanography* 47:103–45.

Henson, S. A., J. L. Sarmiento, J. P. Dunne, L. Bopp, I. Lima, S. C. Doney, J. John, and C. Beaulieu. 2010. "Detection of Anthropogenic Climate Change in Satellite Records of Ocean Chlorophyll and Productivity." *Biogeosciences* 7(2):621–40.

IOCCG. 2000. *Remote Sensing of Ocean Colour in Coastal, and Other Optically-Complex, Waters*, edited by S. Sathyendranath, 140. *Reports of the International Ocean-Colour Coordinating Group*, No. 3. Dartmouth, Canada: IOCCG.

IOCCG. 2006. *Remote Sensing of Inherent Optical Properties: Fundamentals, Tests of Algorithms, and Applications*, edited by Z. Lee, 126. *Reports of the International Ocean-Colour Coordinating Group*, No. 5. Dartmouth, Canada: IOCCG.

IOCCG. 2007. *Ocean-Colour Data Merging*, edited by W. Gregg. *Reports of the International Ocean-Colour Coordinating Group*, No. 6. Dartmouth, Canada: IOCCG.

IOCCG. 2008. *Why Ocean Colour? The Societal Benefits of Ocean-Colour Technology*, edited by T. Platt, N. Hoepffner, V. Stuart, and C. Brown. *Reports of the International Ocean-Colour Coordinating Group*, No. 7, Dartmouth, Canada: IOCCG.

IOCCG. 2010. *Atmospheric Correction for Remotely-Sensed Ocean-Colour Products*, edited by M. Wang. *Reports of the International Ocean-Colour Coordinating Group*, No. 10, Dartmouth, Canada: IOCCG.

Johannessen, S. C., and W. L. Miller. 2001. "Quantum Yield for the Photochemical Production of Dissolved Inorganic Carbon in Seawater." *Marine Chemistry* 76:271–83.

PICES. 2005. *PICES Advisory Report on Fisheries and Ecosystem Responses to Recent Regime Shifts*. North Pacific Marine Science Organization, Sidney, Canada. 12 p.

Lavender, S. 2012. "Ocean Measurements and Applications, Ocean Color." In *Encyclopaedia of Remote Sensing*, edited by E. Njoku. Berlin Heidelberg: Springer-Verlag. Springer Reference (www.springerreference.com). doi: 10.1007/SpringerReference_327180.

Lavender, S. J., W. M. Moufaddal, and Y. D. Pradhan. 2009. "Assessment of Temporal Shifts of Chlorophyll Levels in the Egyptian Mediterranean Shelf and Satellite Detection of the Nile Bloom." *Egyptian Journal of Aquatic Research* 35(2):121–35.

Lees, K., S. Pitois, C. Scott, C. Frid, and S. Mackinson. 2006. "Characterizing Regime Shifts in the Marine Environment." *Fish and Fisheries* 7:104–27.

Lluch-Belda, D., R. J. Crawford, T. Kawasaki, A. D. McCall, A. D. Parrish, R. A. Schwartzlose, and P. E. Smith. 1989. "Worldwide Fluctuations in Sardine and Anchovy Stocks: The Regime Problem." *South African Journal of Marine Science* 8:195–205.

Longhurst, A. 1998. *Ecological Geography of the Sea*, 398. London: Academic Press.

Longhurst, A. R., and W. G. Harrison. 1989. "The Biological Pump: Profiles of Plankton Production and Consumption in the Upper Ocean." *Progress in Oceanography* 22:47–123.

Mantoura, R. F. C., S. W. Jeffrey, C. A. Llewellyn, H. Claustre, and C. E. Morales. 1997. "Comparison between Spectrophotometric, Fluorometric and HPLC Methods for Chlorophyll Analysis." In *Phytoplankton Pigments in Oceanography: Guidelines to Modern Methods. Monographs on Oceanographic Methodology*, edited by S. W. Jeffrey, R. F. C. Mantoura, and S. W. Wright. Vol. 10, 361–80. France: UNESCO Publishing.

Maritorena, S., O. Hembise Fanton d'Andon, A. Mangin, and D. A. Siegel. 2010. "Merged Satellite Ocean Color Data Products Using a Bio-Optical Model: Characteristics, Benefits and Issues." *Remote Sensing of Environment* 114(8):1791–804. doi: 10.1016/j.rse.2010.04.002.

Mélin, F., G. Zibordi, and S. Djavidnia. 2009. "Merged Series of Normalized Water Leaving Radiances Obtained from Multiple Satellite Missions for the Mediterranean Sea." *Advances in Space Research* 43(3):423–37.

Morel, A., and J.-F. Berthon. 1989. "Surface Pigments, Algal Biomass Profiles, and Potential Production of the Euphotic Layer: Relationships Reinvestigated in View of Remote-Sensing Applications." *Limnology and Oceanography* 34:1545–62.

Morel, A., Y. Huot, B. Gentili, P. J. Werdell, S. B. Hooker, and B. A. Franz. 2007. "Examining the Consistency of Products Derived from Various Ocean Color Sensors in Open Ocean (Case 1) Waters in the Perspective of a Multi-Sensor Approach." *Remote Sensing of Environment* 111:69–88.

Moufaddal, W. M., and S. Lavender. 2009a. "Assessment of the Possible Key Factors for Fall and Rise of the Coastal Fisheries off the Nile Delta: A Remote Sensing Approach." *Proceedings of the Second Gulf Conference & Exhibition on Environment and Sustainability*, 620–35, February 16–19, 2009, Kuwait.

Moufaddal, W. M., and S. Lavender. 2009b. "Biogeochemical Response to Mesoscale Circulation Features and the High Levels of the Nile Flood in the SE Levantine Basin, as Revealed by Ocean-Color Remote Sensing." *Egyptian Journal of Aquatic Research* 35(4):431–43.

NIOF. 1962–1986. *Annual Fishery Statistics*. Alexandria, Egypt: National Institute of Oceanography and Fisheries.

Nixon, S. W. 2003. "Replacing the Nile: Are Anthropogenic Nutrients Providing the Fertility Once Brought to the Mediterranean by a Great River?" *Ambio* 32(1):30–9.

Oguz, T., and D. Gilbert. 2007. "Abrupt Transitions of the Top-Down Controlled Black Sea Pelagic Ecosystem during 1960–2000: Evidence for Regime-Shifts under Strong Fishery Exploitation and Nutrient Enrichment Modulated by Climate-Induced Variations." *Deep-Sea Research-I* 54: 220–42.

O'Reilly, J. E., S. Maritorena, D. Siegel, M. C. O'Brien, D. Toole, and B. G. Mitchell. 2000. "Ocean Color Chlorophyll a Algorithms for SeaWiFS, OC2, and OC4: Version 4." SeaWiFS Post-launch Technical Report Series, Vol. 11. SeaWiFS post-launch calibration and validation analyses: Part 3. Maryland: NASA Goddard Space Flight Center.

Pinnock, S., O. Fanton D'Andon, and S. Lavender. 2007. "GlobColour: A Precursor to the GMES Marine Core Service Ocean Colour Thematic Assembly Centre." *ESA Bulletin* 132:42–9.

Raitsos, D. E., P. C. Reid, S. Lavender, M. Edwards, and A. J. Richardson. 2005. "Extending the SeaWiFS Chlorophyll Data Set Back 50 Years in the Northeast Atlantic." *Geophysical Research Letters* 32:L06603. doi: 10.1029/2005 GL 022484.

Reid, P. C., B. Planque, and M. Edwards. 1998. "Is Observed Variability in the Long-Term Results of the Continuous Plankton Recorder Survey a Response to Climate Change?" *Fisheries Oceanography* 7:282–88.

Robinson, I. S. 2010. *Discovering the Ocean from Space*. Berlin Heidelberg: Springer. doi: 10.1007/978-3-540-68322-3_2.

Siegel, D. A., S. Maritorena, N. B. Nelson, M. J. Behrenfeld, and C. R. McClain. 2005. "Colored Dissolved Organic Matter and the Satellite-Based Characterization of the Ocean Biosphere." *Geophysical Research Letters* 32:L20605.

Siegel, D. A., S. Maritorena, N. B. Nelson, D. A. Hansell, and M. Lorenzi-Kayser. 2002. "Global Ocean Distribution and Dynamics of Colored Dissolved and Detrital Organic Materials." *Journal of Geophysical Research* 107:3228.

Swan, C. M., D. A. Siegel, N. B. Nelson, C. A. Carlson, and E. Nasir. 2009. "Biogeochemical and Hydrographic Controls on Chromophoric Dissolved Organic Matter Distribution in the Pacific Ocean." *Deep Sea Research. Part I* 56:2175–92.

Volk, T., and M. I. Hoffert. 1985. "Ocean Carbon Pumps, Analysis of Relative Strengths and Efficiencies in Ocean-Driven Atmospheric CO_2 Changes." *Geophysical Monograph* 32:99–110.

Volpe, G., R. Santoleri, V. Vellucci, M. Ribera D'Alcàla, S. Marullo, and F. D'Ortenzio. 2007. "The Colour of the Mediterranean Sea: Global versus Regional Bio-Optical Algorithms Evaluation and Implication for Satellite Chl-a Estimates." *Remote Sensing of Environment* 107:625–38.

Wang, X. J., M. Behrenfeld, R. Le Borgne, R. Murtugudde, and E. Boss. 2009. "Regulation of Phytoplankton Carbon to Chlorophyll Ratio by Light, Nutrients and Temperature in the Equatorial Pacific Ocean: A Basin-Scale Model." *Biogeosciences* 6:391–404. doi: 10.5194/bg-6-391-2009.

Wilson, C. 2011. "The Rocky Road from Research to Operations for Satellite Ocean-Colour Data in Fishery Management." *ICES Journal of Marine Science* 68:677–86.

Yoder, J., K. Moore, and R. Swift. 2001. "Putting Together the Big Picture: Remote-Sensing Observations of Ocean Color." *Oceanography* 14(4):33–40.

Zepp, R. G. 2002. "Solar Ultraviolet Radiation and Aquatic Carbon, Nitrogen, Sulfur and Metals Cycles." In *UV Effects in Aquatic Organisms and Ecosystems*, edited by E. W. Helbling, and H. Zagarese, 137–83. Cambridge: Royal Society of Chemistry.

Section VII

Wetland, Soils, and Minerals

24

Wetland Classification

Maycira Costa, Thiago S. F. Silva, and Teresa L. Evans

CONTENTS

24.1 Introduction..461
24.2 Remote Sensing Methods 462
    24.2.1 Optical Remote Sensing...462
        24.2.1.1 Application of Optical Remote Sensing...465
    24.2.2 Synthetic Aperture Radar Remote Sensing 467
        24.2.2.1 Application of Synthetic Aperture Radar Remote Sensing 470
    24.2.3 Light Detection and Ranging Remote Sensing 473
24.3 Conclusions..474
24.4 Summary...474
References...474

24.1 Introduction

Wetlands are transitional ecosystems between land and water, which provide essential environmental services including flood control, climate regulation, carbon storage, aquifer recharge, and biodiversity management. Ecologically, wetlands have high biodiversity and serve as breeding grounds and habitat for several species of plants, invertebrates, fish, and wildlife. Biogeochemically, wetlands recycle many essential nutrients and may also act as sinks for organic carbon in the form of peat (Mitsch and Gosselink 2007).

Despite the importance of these ecosystems, the global coverage of wetlands is not well known (Davidson and Finlayson 2007), with estimates ranging from 7 to 10 million km² or 4% to 8% of global land coverage, of which approximately half are located in tropic and subtropic regions (Lehner and Döll 2004; Mitsch and Gosselink 2007).

The spatial distribution of wetlands is key information for supporting applications in resource management, habitat reconstruction, species at risk recovery, and biogeochemical budgets, and it is the most relevant instrument for promoting legal protection and conservation (Mitsch and Gosselink 2007). Therefore, developing efficient techniques for mapping and deriving biophysical properties of wetland ecosystems is of critical importance for managing and understanding these ecosystems (Rebelo et al. 2009). Information is required at multiple spatial scales, from global to local (Rebelo et al. 2009) and temporally (Hamilton et al. 1996; Silva et al. 2010), to allow for guidance to policy makers. Classic methods for mapping wetland habitats have been largely based on ground surveys of soil and vegetation inventories gathered through extensive and time-consuming fieldwork requiring ancillary data analysis and visual estimations of ground cover. As a consequence, such methods are only practical on small scales and do not provide spatially continuous information over large regions (Lee and Lunetta 1996; Mitsch and Gosselink 2007). In many cases, remote sensing technology offers the only reliable method for determining ecologically valuable information regarding the characteristics of wetland habitats across a diverse range of scales (Kerr and Ostrovsky 2003; Davidson and Finlayson 2007; Rebelo et al. 2009). The objectives of this chapter are to review the various remote sensing theories and techniques for classifying wetlands.

Mapping of wetland ecosystems with imagery acquired by satellite and airborne platforms has been successful in different regions of the world. Some examples are in Table 24.1.

24.2 Remote Sensing Methods

24.2.1 Optical Remote Sensing

The earliest studies of wetlands using remote sensing were based on analog aerial photography and were mostly related with identification and delineation of wetland areas (Polis et al. 1974; Howland 1980). The U.S. Fish and Wildlife Service National Wetlands Inventory used archived color-infrared (IR) air photos extensively for identifying and delineating wetlands in the United States (Wilen et al. 1999). Although currently superseded in capabilities by digital multispectral or hyperspectral systems, analog aerial photos are still considered a relatively cheap method to obtain high-resolution coverage for mapping purposes and are still in use (Carpenter et al. 2011), often for historical wetlands mapping for which no satellite data exist.

Optical satellites have been used to study wetlands in different parts of the world for detection of open water surfaces, vegetation, inundation cycle, and submerged substrate. Depending on the study's goal, different sets of knowledge are required in regard to interaction of electromagnetic radiation and matter. Here, we summarize basic knowledge to facilitate choosing/using different satellite imagery for the mapping/management of wetlands.

The main physical principle behind the detection of open water surface using satellites operating in the optical spectra (visible [VIS] and IR) is the strong absorption of near-and shortwave-IR radiation (0.7–2.4 µm) by water and, therefore, low returns to the satellite. Pure water reflection in the VIS range is also low and depends heavily on the composition and concentration of water optical constituents, resulting in a large variety of water spectral responses (Kirk 2010).

Detection of vegetation with optical sensors takes into consideration different parameters, which will be more or less important depending on the scale of the study—leaf level (pigment type and concentration, leaf morphology, and water content), plant level (leaf density, orientation, and overall canopy structure), and community level (biomass and density) (Silva et al. 2007). As optical spectral satellites operate in a maximum resolution of approximately 0.5 m, the expected important interactions are observed at individual and community levels; still, leaf-level interactions play an important role on the final radiance reaching the satellite. The interaction at the leaf level is mostly dependent on pigment composition, cellular structure, and water content of the leaves. Pigment composition affects the VIS wavelengths (0.4–0.7 µm) because of strong absorption, especially in the blue and red wavelengths, and therefore low returns to the satellite, and high reflectance in the green wavelengths, and therefore high returns to the satellite. Phenology and

TABLE 24.1

Remote Sensing of Wetlands: Examples

Author	Sensor(s)	Spatial Resolution (m)	Classification Method	Location	Land Cover	Number of Classes	Overall Accuracy (%)
Hoekman et al. (2010)	L-band SAR	50	Pixel-based	Borneo	Mixed land cover (including wetland)	17	85
Rebelo (2010)	Optical and L-band SAR	12.5	Pixel-based	Lake Chilwa, Malawi	Wetland	6	89
Zomer et al. (2009)	Optical	12.5	Pixel-based	United States	Wetland	5	84
Costa et al. (2007)	Optical	4	Pixel-based	Fraser, Canada	Wetland	7	92
Durieux et al. (2007)	Optical and L-band SAR	100–300	Object-based	Siberia, Russia	Wetland	7	89
Grenier et al. (2008)	Optical and C-band SAR	10–20	Object-based	Quebec, Canada	Wetland	5	81
Simard et al. (2002)	L- and C-band SAR	100–400	Pixel-based	Gabon, Africa	Mixed land cover (including wetland)	10	84
Souza-Filho et al. (2011)	Airborne L-band SAR	10	Pixel-based	Bragança coastal plain, Brazil	Coastal wetland	4	83
Costa (2004)	L- and C-band SAR	12.5	Object-based image analysis	Amazon, Brazil	Wetland	3	>95
Hess et al. (2003)	L-band SAR	100	Pixel-based	Amazon basin, Brazil	Wetland	8	81
Martinez and Le Toan (2007)	L-band SAR	12.5	Pixel-based	Óbidos floodplain, Brazil	Wetland	5	81
Walker et al. (2010)	Optical and L-band SAR	12.5–30	Object-based	Amazon, Brazil	Mixed land cover (including wetland)	15	58
Evans et al. (2010)	L- and C-band SAR	50–100	Object-based	Pantanal, Brazil	Wetland	5	81
Costa and Telmer (2006)	L- and C-band SAR	12.5	Pixel-based	Pantanal, Brazil	Lakes	3	83

stress status may change the pigments and consequently the absorption and reflectance signature of the vegetation. The near-IR region of the electromagnetic spectrum (0.7–1.0 μm) generally shows the highest reflectance values for plants because of the interaction of radiation with the leaf cell structure. Very little absorption occurs in this region, with the exception of a small absorption band at approximately 0.98 μm. About 40%–50% of the radiation is being reflected on reaching a leaf, whereas another 60%–50% is transmitted. The high transmission value results in even higher overall reflectance for canopies, as the transmitted light can then be reflected by leaves located further down in the canopy. The location and magnitude of the transition between the red and near-IR spectra, defined as the red edge, is an important feature for the detection of the general vegetation health status (Thomson et al. 1998). In the shortwave-IR range (1.0–2.4 μm), water absorption features are the dominant factor, with an overall decreasing trend in reflectance, marked by well-defined absorption features at 1.4 and 1.9 μm.

A common method to delineate different vegetation types in optical imagery is the use of vegetation indexes, mathematical combinations of spectral bands that enhance vegetation response features (Table 24.2). The most commonly used index is called the Normalized Difference Vegetation Index (NDVI) (Rouse et al. [1974]), which capitalizes on the shift from strong absorption of red wavelengths to strong reflectivity at the near-IR region. A similar index named Normalized Difference Water Index (NDWI) is also of importance for wetland studies. Unlike the NDVI, the NDWI explores the relationship between the water absorption bands at 0.98 and 1.4 μm (Gao 1996). An alternative formulation of the NDWI was also suggested by McFeeters (1996), based on the difference between the green and near-IR channels, which enhances the signal of open water surfaces.

Besides considerations about pigments and water content of the vegetation, successful mapping of wetlands with optical imaging systems must consider the flooding status. The presence of water tends to reduce the reflectance signal in all wavelengths, especially in the IR spectra where the most important water absorption bands are located. Vegetation density and structure of the canopy play a role in the intensity in which the water spectrum is mixed with the vegetation spectrum. The spectra of a pixel corresponding to a surface with dense vegetation and flood underneath will have little influence from the water (Jakubauskas et al. 2000).

Successful detection of submerged targets (for instance, intertidal macroalgae, substrate) using remote sensing technology must consider (1) the spectral properties of the water column, specifically the light attenuation by water optical constituents, (2) the uniqueness of the spectral properties of the substrate, (3) the spectral and spatial resolution of the imaging sensors, and (4) the spectral attenuation properties of the atmosphere

TABLE 24.2

Normalized Vegetation Indexes

Index Name	Equation	Proponent
Normalized Difference Vegetation Index	$NDVI = \dfrac{\rho_{NIR} - \rho_{Red}}{\rho_{NIR} + \rho_{Red}}$	Rouse et al. (1974)
Normalized Difference Water Index (moisture)	$NDWI_{Gao} = \dfrac{\rho_{NIR} - \rho_{SWIR}}{\rho_{NIR} + \rho_{SWIR}}$	Gao (1996)
Normalized Difference Water Index (water surfaces)	$NDWI_{McFeeters} = \dfrac{\rho_{Green} - \rho_{NIR}}{\rho_{Green} + \rho_{NIR}}$	McFeeters (1996)

(Hedley and Mumby 2003). As the turbidity of the water increases because of particulates and colored dissolved organic matter (CDOM), so does the attenuation of light, making it difficult to detect the substrate. The attenuation of light is spectrally dependent. For instance, near-IR wavelengths are strongly absorbed by water molecules, and therefore not very useful for substrate detection. Further, the presence of phytoplankton and CDOM reduces the penetration of blue and red light in the water column because of their specific absorption characteristics at this region of the spectra (Phinn et al. 2005; Kirk 2010). The presence of epiphytes also changes the spectral properties of the substrate by reducing the typical green reflectance peak of certain species of macroalgae (Drake et al. 2003; O'Neill et al. 2011).

Studies using imagery acquired by hyperspectral airborne sensors have shown the potential to accurately map benthic substrate because of their high spectral and spatial resolutions (Dierssen et al. 2003; Dekker et al. 2005; Peneva et al. 2008; O'Neill et al. 2011; O'Neill and Costa submitted). However, the acquisition of airborne hyperspectral imagery over large areas is still not cost-effective, and the high spatial resolution imagery acquired by satellites (IKONOS, GeoEye-1, QuickBird-2, and WorldView-2) are showing promising results at the habitat scale (Mumby and Edwards 2002; Mishra et al. 2005; Fornes et al. 2006; O'Neill and Costa in press). New satellites such as WorldView-2 advanced the spectral resolution capability of these satellites by adding two new bands in the VIS spectra, a coastal blue (440–450 nm) and a yellow (585–625 nm), which potentially allow better substrate detection in tropical clear and temperate more turbid waters, respectively.

24.2.1.1 Application of Optical Remote Sensing

Eelgrass beds are found in the inter- and subtidal coastal ecosystems, are of vital importance for physical shoreline stability, provide nursery ground and food source for a variety of animals including commercial fish, and are important in the carbon balance of the marine ecosystem. Despite their importance, 2%–5% of the global seagrass habitats are lost annually as a result of human actions. With this in mind, Parks Canada, a federal government institution, requested a protocol to effectively map eelgrass beds on the British Columbia coast. To accomplish this, O'Neill, Costa, and Sharman tested the suitability of satellite-based multispectral (IKONOS) and airborne-based hyperspectral (AISA [airborne imaging spectroradiometer for applications]) high spatial resolution imagery to map eelgrass beds in coastal temperate waters of British Columbia, Canada. The authors chose high spatial resolution imagery because of the requirements to capture the spatial pattern of the eelgrass meadow. The differences in the spectral resolution of the sensors allowed the authors to test for the mapping accuracy of cheaper satellite-based multispectral versus expensive airborne-based hyperspectral imagery. IKONOS is a satellite with bands located in the blue, green, red, and near-IR spectra, a 4×4 m spatial resolution, and 11.3 km swath. The AISA sensor was mounted in an airplane that flew at 1500 m elevation, providing imagery at a 2×2 m spatial resolution, 592 m imagery swath, and at a 2 nm spectral resolution in the 408 to 2494 nm spectral range.

Because little was known about the general characteristics of the substrate and their reflectance properties in turbid waters (attenuation coefficient around 1.0 m^{-1}), the authors decided to conduct a detailed survey during the same month in which the images were acquired. The survey consisted of defining the distribution of the dominant substrate and their reflectance properties. Reflectance spectra were acquired over different substrates with a set of radiometers on the same day in which images were acquired. The authors also defined the differential global positioning system (GPS) location of dominate substrate and the percentage cover of eelgrass, and acquired above-water photographs and underwater videography of the field sites. A total of 507 field sites were surveyed and used either for training or for validation of the classification.

Both images were processed in ENVI (however, the final protocol was also written for PCI Geomatica). The following were the main imagery processing steps: geometric, atmospheric, and glint corrections; deep water and land masking; and finally the production of classified maps created using a maximum likelihood classifier (MLC). Step 1: The geometric correction procedures were different for the IKONOS and AISA imagery because of the mosaicking of flight lines required for the latter. Flight lines were first individually geocorrected according to the airplane's positional measurements, followed by manual matching of side lines. IKONOS was corrected using the satellite orbit model and ground control points acquired in the field with differential GPS. Step 2: A land mask was created for each image based on respective digital number thresholds and subtracted from each image aiming to avoid confusion in the classification step. Step 3: Atmospheric correction was conducted considering the empirical line calibration method. This method uses band-specific regression lines to derive reflectance values based on measure in situ reflectance. Step 4: Glint correction was applied for both images by assuming that the near-IR reflectance from water is null, and therefore, any measured reflectance is due to glint and is linearly related to the reflectance in the VIS wavelengths. Step 5: Masking deep water regions was done by defining a reflectance threshold for each image. This threshold was derived based on the reflectance difference between deep waters and eelgrass substrate. Step 6: MLC was applied for the classification of the different substrate. The MLC algorithm calculates the probability that a given pixel belongs to a specific class based on training sites, and each pixel is therefore assigned to the class for which it had the highest probability of affinity. MLC is a relatively simple supervised classification method that is available in most imagery processing software, which facilitates the use of this approach by agencies interested in baseline maps of wetlands distribution. For the AISA imagery, a reduced number of bands were chosen for the classification aiming to optimize the separability of eelgrass from all the other classes. For the IKONOS imagery, all the spectral bands were considered in the classification algorithm. A total of 20% and 80% of the surveyed sites were used for training of the classification algorithm and validating the classification results, respectively. Final evaluation of the classified maps (Figure 24.1) was performed based on user's and producer's accuracies of shallow and deep eelgrass occurrence, and total accuracy considering all substrates.

The final accuracy results suggested that the AISA imagery provided the best classification with 83%, 85%, and 93% for total accuracy, producer's accuracy, and user's accuracy, respectively for shallow and deep eelgrass (Figure 24.1). The classification results for IKONOS were slightly inferior at 68%, 79%, and 91%, respectively. The main source of confusion in the classification was between classes with similar spectral reflectance, for instance, green algae and eelgrass; further, errors were larger for the IKONOS imagery because of lower spatial resolution, which is prone to more spectral mixing at the pixel level. However, the final IKONOS user's and producer's accuracies for eelgrass were defined as sufficient if the purpose is a cost-effective, efficient method for monitoring of eelgrass beds. With this technique, historical eelgrass distribution can be accessed with historical imagery so that the impact of human activities can be better determined. Further, monitoring of eelgrass replantation areas and evaluation of ecosystem health status can more frequently be done and over larger areas.

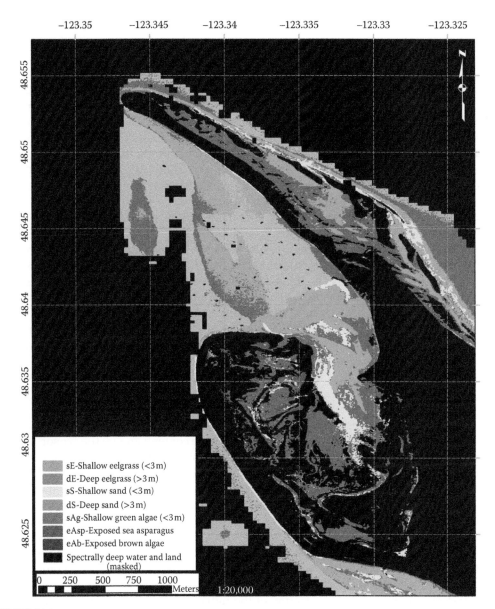

FIGURE 24.1
(See color insert.) Airborne imaging spectroradiometer for applications (AISA) classification. (From O'Neill et al. 2011.)

24.2.2 Synthetic Aperture Radar Remote Sensing

Optical imagery is a useful tool for mapping and monitoring land surfaces but is problematic for using in certain regions where cloud cover or smoke is significant. Optical wavelengths also lack the ability to penetrate forest canopy for flood detection. Synthetic aperture radar (SAR) technology is capable of overcoming these limitations and has been successfully used for mapping inundation, land cover, and biophysical properties in regions of dense vegetation or with frequent cloud cover. SAR systems can acquire images at different wavelengths, polarization, and incidence angles. As a result, different interactions or scattering mechanisms between the radiation and the target, and therefore different signal strengths, return to the satellite. For land and water applications, the most commonly used wavelengths of SAR satellite systems are in a range from 2 to 30 cm (Table 24.3). A radar antenna can transmit and receive microwave radiation in horizontal and horizontal polarization (HH), respectively, or in vertical and vertical polarization (VV), respectively. These are called like-polarized systems. Furthermore, the antenna can transmit and receive in horizontal and vertical polarization (HV), respectively, or vertical and horizontal polarization (VH), respectively. These are called cross-polarized systems. In addition to these options, the incidence angle in which radiation reaches the ground surface varies from very steep (20° off-nadir) to very shallow (60° off-nadir) (Ulaby et al. 1981). Currently, SAR satellites only operate in a particular wavelength and a combination of polarizations; however, data acquired at different angles of incidence have been collected by ERS-1, ERS-2, ENVISAT/ASAR, Radarsat-1 and Radarsat-2 (C-band), JERS and ALOS/PALSAR (L-band), and TerraSAR-X and Cosmo-Skymed (X-band), among others.

Using SAR systems to map and monitor wetlands has been effectively shown, with some notable areas of focus including monitoring the extent and timing of flooding (Hess et al. 2003; Costa 2004; Martinez and Le Toan 2007; Arnesen et al. submitted), biomass estimation (Le Toan et al. 1997; Costa et al. 2005), and mapping of wetland vegetation (Le Toan et al. 1997; Simard et al. 2002; Costa 2004; Costa and Telmer 2006; Lucas et al. 2007; Evans et al. 2010; Rebelo 2010; Silva et al. 2010; Evans and Costa 2013). Accuracy of the results for the above-mentioned studies varies, mostly because wetlands around the world have distinct characteristics in terms of vegetation type, timing of flooding, and so on.

The geometry and organization of the vegetation are important factors for characterizing the radar signal. Overall, the total backscattering from wetland herbaceous plants is dependent on the characteristics of (1) the canopy, such as density, distribution, orientation, shape of the foliage, dielectric constant, height, and components of the canopy; (2) the ground surface, such as roughness and moisture content; and (3) the sensor, such as polarization, incidence angle, and wavelength. For instance, dense, tall (1.5 m), vertically oriented herbaceous plants of marshes show double-bounce mechanism with L (HH and VV) and even C-HH at low incidence angles (Hess et al. 1995; Costa 2004; Costa and Telmer 2006; Evans et al. 2010). A double-bounce mechanism results in a high signal return to the satellite, and is caused by the interaction of the radiation with the stem/trunk, followed by a change in direction toward the surface (water) and a strong bounce back toward the radar antenna (dihedral corner reflector behavior). The characteristics of both the plants and the sensor are important to explain the double-bounce interaction. The combination of the long wavelength (L-band, 25 cm), the polarization (HH), and the steep incidence angle ($\theta < 25°$) allows higher penetration of the radiation through the canopy. At a lower incidence angle, the pathway of the incidence wave through the canopy is minimal; therefore, the radiation is less attenuated by the canopy, favoring double-bounce signal. For less dense herbaceous plants of the flooded wetlands, the backscattering values, or the

TABLE 24.3

Satellite SAR Band Designations for Land and Water Applications

Band Designation	Wavelength (cm)
L	15.0–30.0
C	3.7–7.5
X	2.4–3.7

returned signal to the satellite, are not as high as those observed for high-density stands because of the increase in the forward scattering off water patches, a scattering mechanism known as specular reflection (Pope et al. 1997; Hess et al. 2003; Costa 2004; Evans et al. 2010). Another type of interaction between plants and microwave radiation is volume scattering, which is characterized by the interaction of the radiation within the canopy of the vegetation, with the leaves and stems. The resultant backscattered radiation toward the antenna is not as strong as it is for double-bounce interaction but higher than that observed for specular reflection mechanism (Dobson et al. 1996).

In wetlands with woody vegetation, the signal returning to the satellite changes because of the presence of trunks and the higher density of the canopies as compared to nonwoody vegetation. Both modeled and measured backscattering coefficients are consistent with the interpretation that L-band imagery with HH polarization and lower incidence angles ($\theta < 30°$) produced canopy-volume scattering (medium signal return) and double-bounce (high signal return) interactions (Hess et al. 1995; Costa 2004; Evans et al. 2010; Evans and Costa 2013). At C-band, independent of the polarization and the incidence angle, canopy-volume scattering is predominant. Flood detection beneath trees is only possible with C-band, HH polarization, and low incidence angle, when trees lose their leaves, which, in the case of some species, is during the flood period in the Amazon floodplain. L-band is generally recommended for flooding detection beneath tree canopies because of the generated high signal return (Hess et al. 2003; Costa 2004; Lucas et al. 2007; Evans et al. 2010).

Beyond backscattering analysis, SAR imagery also allows the use of interferometric and polarimetric information. Radar polarimetry is based on the ability of some SAR systems to record the full polarimetric information of the returned radar beam, which can then be compared to the polarization characteristics set by the system on sending the radar beam. As different target–radiation interactions will change the original polarization characteristics of the sent radiation in different manners, information about the nature of such processes can be derived from the comparison (Henderson and Lewis 1998). Common approaches for extracting information from polarimetric SAR data are the calculation of backscattering polarimetric variables (Table 24.4), which combine the complex polarimetric information into interpretable quantities, based on the observed radar backscattering at each polarization. The simplest polarimetric attributes are the backscattering coefficients for each polarization combination themselves (HH, VV, HV, and VH). Beyond those, co- and cross-polarized ratios can be derived (e.g., HH/VV and HH/HV), as well as the total power returned. Other attributes include the indexes derived by Pope et al. (1997), including the Biomass Index (BMI = VV + HH/2), the Canopy Structure Index (CSI = VV/[HH + VV]), and the Volume Scattering Index ([VSI] considers HV, VH, VV, and HH polarizations), which can together characterize and separate different types of vegetation. BMI is mostly related to the proportion of woody versus herbaceous biomass, with lower values implying that more herbaceous biomass is present. The CSI indicates the relative presence of vertical scatterers (trunks and stems), and the VSI relates to the relative density of the canopy.

A different set of parameters can be produced by the use of target decomposition methods, which attempt to characterize the complex polarimetric information as a combination of independent measurements that have an underlying physical explanation (Table 24.4) (Cloude and Pottier 1997). For example, Freeman–Durden decomposition estimates the proportion of each of the three main radar backscattering mechanisms (volumetric, surface, and double-bounce) (Freeman and Durden 1998), whereas Cloude–Pottier decomposition characterizes the physical backscattering processes to derived parameters using second-order statistics and associates these parameters to a classification scheme.

TABLE 24.4

Polarimetric Variables Derived from Backscattering and Polarimetric Decomposition

Attributes Derived from Backscatter Data	Common Symbol	Description
Pope Parameters (Pope et al. 1994)		
Biomass Index	BMI	Relative amount of woody biomass, compared to leafy biomass
Canopy Structure Index	CSI	Relative presence of vertical scatterers (e.g., trunks and stems)
Volume Scattering Index	VSI	Relative thickness or density of the canopy
Attributes from the Decomposition Data		
Cloude–Pottier Decomposition (Cloude and Pottier 1997)		
Average alpha angle	α_m	Type of dominating scattering (surface ($\alpha_m = 0°$), volumetric ($\alpha_m = 45°$) or double-bounce ($\alpha_m = 90°$) scattering)
Entropy	H	Number of dominant scattering mechanisms. If entropy is low ($H < 0.3$), the dominant scattering mechanism can be recovered; if entropy is high, it should be considered as a mixture of possible scatterer types; if $H = 1$, polarization information becomes zero and the target scattering is truly a random noise process
Anisotropy	A	A complementary parameter to entropy, measuring the relative importance of the second and third scattering mechanisms, when $H > 0.7$. High entropy and low anisotropy correspond to random scattering. High entropy and high anisotropy correspond to the presence of two scattering mechanisms with the same probability

Source: Adapted from Sartori et al. *IEEE Transactions on Geoscience and Remote Sensing* 49, 4717–912, 2011.

Besides polarimetry, radar interferometry has also been used successfully to study wetland environments. Radar interferometry, or InSAR, is based on the study of small changes in the surface backscattering characteristics (signal decorrelation), based on small changes in sensor-viewing conditions (SAR baseline), and can be seen as an analog technique comparable to stereoscopy in aerial photography. The major problem in radar interferometry is to separate the signal decorrelation that results from the intended changes in radar baseline and from the decorrelation caused by temporal changes (e.g., moisture and phenology) (Richards 2007). This technique has been classically used to produce very accurate digital elevation models (DEMs) and to retrieve information about vegetation characteristics (Wegmüller and Werner 1997). However, interferometry can also be used successfully to detect fine changes in water surface height. For example, Alsdorf et al. (2001) used SAR interferometry to track water level changes in the Amazon floodplain, with centimeter precision. Currently, the European Space Agency offers experimental river and lake altimetric data for select regions in the world, available at http://tethys.eaprs.cse.dmu.ac.uk/RiverLake/shared/main.

24.2.2.1 Application of Synthetic Aperture Radar Remote Sensing

The Nhecolândia region of the Brazilian Pantanal is part of a large continuous tropical wetland that exhibits a high biodiversity of flora and fauna species and many threatened habitats. The spatial distribution of these habitats influences the abundance and interactions of animal species, and the change or destruction of this habitat can cause the disturbance of key biological processes. The region has a highly heterogeneous and dynamic landscape, with forest, savanna, wild grasslands, introduced pastures, seasonal waterways (locally known as *vazantes*), herbaceous vegetation, aquatic macrophytes, and tens of thousands of geochemically diverse lakes generally divided into three categories: two classes of freshwater lakes (locally known as *baías* and *salobras*) and brackish lakes (locally known as *salinas*). Despite the importance of the Nhecolândia wetland, the internal phytogeography of this region, and the spatial distribution of the habitats within it, is still inadequately documented (Pott et al. 2011). Evans and Costa used a dual-season (February and August) set of fine spatial resolution C-band (ENVISAT/ASAR 12.5 m, HH and HV polarization; RADARSAT-2, 25 m, HH and HV polarization) and L-band (ALOS/PALSAR, 12.5 m, HH and HV polarization) SAR imagery to map these habitats by using a hierarchical object-based image analysis approach. The authors chose this approach for several reasons: (1) SAR data allowed for the acquisition of cloud-free imagery during both wet and dry seasons, (2) fine spatial resolution imagery permitted the separation of the highly heterogeneous habitats, (3) dual-polarization (HH and HV) and dual-band (C and L) SAR imagery provided additional information regarding the vegetation cover not possible with a single polarization and/or band, and (4) a hierarchical image-analysis technique allowed for an examination of features (such as mean and standard deviation of image objects, object size, and shape) and spatial and hierarchical relationships of objects rather than single pixels.

Field data (geographic coordinates, vegetation description, and photographs) were acquired for 209 ground sites and 75 lakes concurrent with the dry season imagery acquisition. In addition, water geochemistry was determined for each of the lakes to confirm the relationship between lake geochemistry and vegetation assemblages.

All SAR images were processed in PCI Geomatica, and habitats were classified using Definiens eCognition software. The main image-processing steps included radiometric calibration, geometric correction and mosaicking of images, SAR speckle filtering, and classification. Step 1: Images were georeferenced and projected to Universal Transverse Mercator coordinates (zone 21, row K) using the WGS84 reference ellipsoid. Each set of images was mosaicked to form cohesive coverage of the study area. Step 2: Images were filtered to reduce the effect of speckle by using a Kuan filter with a 3×3 kernel. Step 3: Classification using Definiens eCognition software. The basic steps for image classification in eCognition were as follows: (1) segmentation into image objects: scale = 50, shape = 0.005 (heavily emphasizing radiometry over shape), compactness = 0.5 (equal emphasis on smoothness and compactness), and more heavily weighting the dry season imagery to better separate the lakes from seasonal flooding areas; (2) collection of training objects for habitat classes: objects were selected based on approximately 50% of the ground reference data; (3) backscattering analysis; (4) development of mean class thresholds: mean values for training objects were exported, and collective mean and standard deviation of the means were calculated for each class to develop thresholds; (5) development of additional hierarchical rules to further separate classes based on several features as primary inputs (mean, standard deviation, brightness, maximum difference, proximity, shape, and compactness); and (6) validation of final classification product: classified objects were compared to the remaining 50% of the field data sites for defining the accuracy of the classification.

This study resulted in a Level 1 classification (Figure 24.2) defining seven habitats (forest woodland, open wood savanna, open grass savanna, agriculture, swampy grassland, *vazantes*, and lakes), which was achieved at an overall accuracy of 83%. In general, the greatest confusion for Level 1 classification results was found with adjacent successional classes: for example, forest woodland with open wood savanna. These errors were expected given

(1) the similarity of these classes in terms of the backscattering characteristics of the vegetation structure and (2) the highly heterogeneous nature of the landscape. In many cases, these landscape units are found adjacent to one another, and clear-cut borders between the different cover types are not readily apparent, especially given the dynamic nature of the landscape in terms of inundation. A Level 2 classification defined only the "lakes" class from Level 1 and divided it into fresh water lakes with floating and emergent vegetation (*baías*), fresh lakes with the presence of *Typha* (*salobras*), and brackish lakes (*salinas*) (overall accuracy results of 81%) (Figure 24.3). For Level 2, the largest confusion was for the *Typha* class, where approximately half were misclassified. This confusion is likely a result of two main factors: the stand of *Typha* sp. observed in a lake may have been too small to be captured as a single image object; and the stand of *Typha* sp. may have been sparse and mixed with other aquatic vegetation, thereby reducing the backscattering signal considerably.

The authors found that combination of dual-season, C- and L-band, high spatial resolution imagery was essential for providing a relatively high overall accuracy of 83% for the Level 1 land cover classification, and 81% for the Level 2 lakes classification. These classification results are well within the overall accuracy ranges reported by similar wetlands studies especially given the highly heterogeneous nature of the landscape.

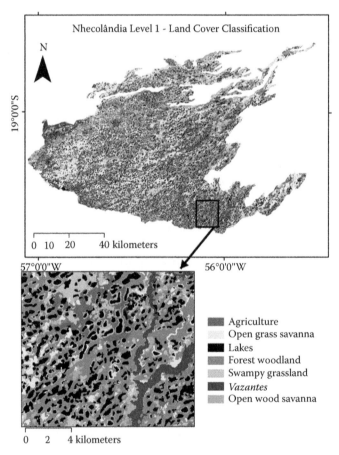

FIGURE 24.2
(See color insert.) Classification Level 1 of wetland habitats: Pantanal. (Adapted from Evans T. L. and M. Costa, *Remote Sensing of Environment*, 128, 118–37, 2013.)

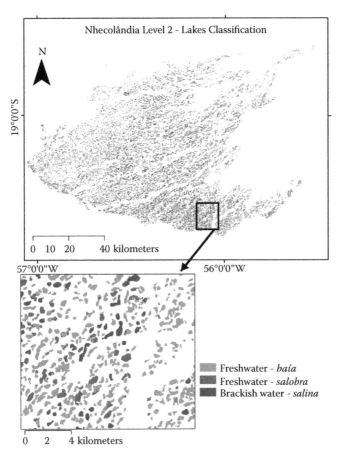

FIGURE 24.3
(See color insert.) Classification Level 2 of wetland lakes: Pantanal. (Adapted from Evans T. L. and M. Costa, *Remote Sensing of Environment*, 128, 118–37, 2013).

24.2.3 Light Detection and Ranging Remote Sensing

The technique known as light detection and ranging (LiDAR) has gained prominent status on remote sensing applications in recent times because of its remarkable capacity to reconstruct three-dimensional information in natural environments. LiDAR remote sensing is based on the same principle as radar but using high-frequency laser pulses instead of microwave beams. By measuring the return time of each laser pulse after it is reflected by an object at the surface, a point return cloud can be produced, indicating the relative three-dimensional structure of the imaged area. These returns can then be used to extract actual ground surface (DEMs), canopy surface models, and be related to canopy density and vegetation biomass (Gilmore et al. 2008). As the majority of LiDAR systems operate mostly in the IR region, saturated soils and free water surfaces may attenuate the incoming signal, thereby reducing the returning signal (Hopkinson et al. 2005). Shorter wavelengths (e.g., green LiDAR) can penetrate the water column and have been used systematically to obtain bathymetric measurements, but must be used carefully, because of potential harmful effects to living beings (Wang and Philpot 2007).

24.3 Conclusions

The objective of this section was to provide a comprehensive review of remote sensing techniques for classifying land cover of wetland ecosystems. This section also provided successful case studies to show the use of some of these techniques in regard to temperate coastal substrate and tropical floodplain mapping applications.

24.4 Summary

Information about the coverage of global wetlands is not well known, and it is required at multiple spatial scales, from global to local and temporally. Remote sensing technology offers the only reliable method for determining ecologically valuable information regarding the characteristics of wetland habitats across a diverse range of scales. The objectives of this chapter are to review the various remote sensing theories and techniques for classifying wetlands. To accomplish this, we provide a review of optical, SAR, and LiDAR remote sensing and applications of wetlands classification.

References

Alsdorf, D. E., L. C. Smith, and J. M. Melack. 2001. "Amazon Floodplain Water Level Changes Measured with Interferometric SIR-C Radar." *IEEE Transaction on Geoscience and Remote Sensing* 39:423–31.

Arnesen, A. S., T. S. F. Silva, L. L. Hess, E. M. L. M. Novo, C. M. Rudorff, B. D. Chapman, and K. C. McDonald. "Monitoring Flood Extent in the Lower Amazon River Floodplain Using ALOS/PALSAR ScanSAR Images." *Remote Sensing of Environment* 130:51–61.

Carpenter, L., J. Stone, and C. R. Griffin. 2011. "Accuracy of Aerial Photography for Locating Seasonal (Vernal) Pools in Massachusetts." *Wetlands* 31:573–81.

Cloude, S. R., and E. Pottier. 1997. "An Entropy Based Classification Scheme for Land Applications of Polarimetric SAR." *IEEE Transactions on Geoscience and Remote Sensing* 35:68–78.

Costa, M. P. F. 2004. "Use of SAR Satellites for Mapping Zonation of Vegetation Communities in the Amazon Floodplain." *International Journal of Remote Sensing* 25:1817–35.

Costa, M. 2005. "Estimate of Net Primary Productivity of Aquatic Vegetation of the Amazon Floodplain Using Radarsat and JERS-1 Imagery." *International Journal of Remote Sensing* 26:4527–36.

Costa, M., E. A. Loos, A. Shaw, C. Steckler, and P. Hill. 2007. "Hyperspectral Imagery for Mapping Intertidal Vegetation at Roberts Bank Tidal Flats, British Columbia, Canada." *Canadian Journal of Remote Sensing* 33:130–41.

Costa, M. P. F., and K. H. Telmer. 2006. "Utilizing SAR Imagery and Aquatic Vegetation to Map Fresh and Brackish Lakes in the Brazilian Pantanal Wetland." *Remote Sensing of Environment* 105:204–13.

Davidson, N. C., and C. M. Finlayson. 2007. "Earth Observation for Wetland Inventory, Assessment and Monitoring." *Aquatic Conservation: Marine and Freshwater Ecosystems* 17:219–28.

Dekker, A. G., V. E. Brando, and J. M. Anstee. 2005. "Retrospective Seagrass Change Detection in a Shallow Coastal Tidal Australian Lake." *Remote Sensing of Environment* 97:415–33.

Dierssen, H. M., R. C. Zimmerman, A. Robert, D. T. Valerie, and C. Davis. 2003. "Ocean Color Remote Sensing of Seagrass and Bathymetry in the Bahamas Banks by High-Resolution Airborne Imagery." *Limnology and Oceanography* 48(1):444–55.

Dobson, M. C., L. E. Pierce, and F. T. Ulaby. 1996. "Knowledge-Based Land-Cover Classification Using ERS-l/JERS-1 SAR Composites." *IEEE Transactions on Geoscience and Remote Sensing* 34(1):83–99.

Drake, L. A., F. C Dobbs, and R. C. Zimmerman. 2003. "Effects of Epiphyte Load on Optical Properties and Photosynthetic Potential of the Seagrasses *Thalassia testudinum* Banks Ex Konig and *Zostera marina.*" *Limnology and Oceanography* 48(1, Part 2):456–63.

Durieux, L., J. Kropáček, G. D. de Grandi, and F. Achard. 2007. "Object-Oriented and Textural Image Classification of the Siberia GBFM Radar Mosaic Combined with MERIS Imagery for Continental Scale Land Cover Mapping." *International Journal of Remote Sensing* 28(18):4175–82.

Evans, T. L., and M. Costa. 2013. "Landcover Classification of the Lower Nhecolândia Subregion of the Brazilian Pantanal Wetlands Using ALOS/PALSAR, RADARSAT-2 and ENVISAT/ASAR Imagery." *Remote Sensing of Environment* 128:118–37.

Evans, T. L., M. Costa, K. Telmer, and T. S. F. Silva. 2010. "Using ALOS/PALSAR and RADARSAT-2 to Map Land Cover and Seasonal Inundation in the Brazilian Pantanal." *IEEE Journal of Selected Topics in Applied Earth Observations and Remote Sensing* 3(4):560–75.

Fornes, A., G. Basterretxea, A. Orfila, A. Jordi, A. Alvarez, and J. Tintore. 2006. "Mapping *Posidonia Oceanica* from IKONOS." *ISPRS Journal of Photogrammetry and Remote Sensing* 60:315–22.

Freeman, A., and S. L. Durden. 1998. "A Three-Component Scattering Model for Polarimetric SAR Data." *IEEE Transactions on Geoscience and Remote Sensing* 36:963–73.

Gao, B. 1996. "NDWI—A Normalized Difference Water Index for Remote Sensing of Vegetation Liquid Water from Space." *Remote Sensing of Environment* 266:257–66.

Gilmore, M., E. Wilson, N. Barrett, D. Civco, S. Prisloe, J. Hurd, and C. Chadwick. 2008. "Integrating Multi-Temporal Spectral and Structural Information to Map Wetland Vegetation in a Lower Connecticut River Tidal Marsh." *Remote Sensing of Environment* 112:4048–60.

Grenier, M., S. Labrecque, M. Garneau, and A. Tremblay. 2008. "Object-Based Classification of a SPOT-4 Image for Mapping Wetlands in the Context of Greenhouse Gases Emissions: The Case of the Eastmain Region, Québec, Canada." *Canadian Journal of Remote Sensing* 34(2 Suppl): 398–413.

Hamilton, S. K., S. J. Sippel, and J. M. Melack. 1996. "Inundation Patterns in the Pantanal Wetlands of South America Determined from Passive Microwave Remote Sensing." *Archiv für Hydrobiologie* 137(1):1–23.

Hedley, J. D., and P. J. Mumby. 2003. "A Remote Sensing Method for Resolving Depth and Subpixel Composition of Aquatic Benthos." *Limnology and Oceanography* 48(1, Part 2):480–88.

Henderson, F. M., and A. J. Lewis. 1998. "Principles and Applications of Imaging Radar." In *Manual of Remote Sensing,* edited by F. M. Henderson and A. J. Lewis, 131–87. New York: John Wiley & Sons.

Hess, L. L., J. M. Melack, C. C. F. Barbosa, and M. Gastil. 2003. "Dual-Season Mapping of Wetland Inundation and Vegetation for the Central Amazon Basin." *Remote Sensing of Environment* 87:404–28.

Hess, L. L., J. M. Melack, and S. Filoso. 1995. "Delineation of Inundated Area and Vegetation along the Amazon Floodplain with the SIR-C Synthetic Aperture Radar." *IEEE Transactions on Geoscience and Remote Sensing* 33(4):896–904.

Hoekman, D. H., M. A. M. Vissers, and N. Wielaard. 2010. "PALSAR Wide-Area Mapping of Borneo: Methodology and Map Validation." *IEEE Journal of Selected Topics in Applied Earth Observations and Remote Sensing* 3(4):605–17.

Hopkinson, C., L. E. Chasmer, G. Sass, I. Creed, M. Sitar, W. Kalbfleisch, and P. Treitz. 2005. "Vegetation Class Dependent Errors in Lidar Ground Elevation and Canopy Height Estimates in a Boreal Wetland Environment." *Canadian Journal of Remote Sensing* 31(2):191–206.

Howland, W. 1980. "Multispectral Aerial Photography for Wetland Vegetation Mapping." *Photogrammetrical Engineering and Remote Sensing* 46:7–99.

Jakubauskas, M., K. Kindscher, A. Fraser, D. Debinski, and K. P Price. 2000. "Close-Range Remote Sensing of Aquatic Macrophyte Vegetation Cover." *International Journal of Remote Sensing* 21(8): 3533–38.

Kerr, J. T., and M. Ostrovsky. 2003. "From Space to Species: Ecological Applications for Remote Sensing." *Evolution* 18(6):299–305.

Kirk, J. T. O. 2010. *Light and Photosynthesis in Aquatic Ecosystems*. Cambridge: Cambridge University Press.

Le Toan, T., F. Ribbes, L. Wang, N. Floury, K. Ding, J. A. Kong, and M. Fujita. 1997. "Rice Crop Mapping and Monitoring using ERS-1 Data Based on Experiment and Modeling Results." *IEEE Transaction of Geoscience and Remote Sensing* 35(1):41–55.

Lee, K. H., and R. S. Lunetta. 1996. "Wetland Detection Methods." In *Wetland and Environmental Application of GIS*, edited by J. Lyon and J. G, McCarthy, 249–84. New York: Lewis.

Lehner, B., and P. Döll. 2004. "Development and Validation of a Global Database of Lakes, Reservoirs and Wetlands." *Journal of Hydrology* 296(1–4):1–22.

Lucas, R. M., A. L. Mitchell, and A. K. E. Rosenqvist. 2007. "The Potential of L-Band SAR for Quantifying Mangrove Characteristics and Change: Case Studies from the Tropics." *Aquatic Conservation: Marine and Freshwater Ecosystems* 17:245–64.

Martinez, J. M., and T. Le Toan. 2007. "Mapping of Flood Dynamics and Spatial Distribution of Vegetation in the Amazon Floodplain Using Multitemporal SAR Data." *Remote Sensing of Environment* 108:209–23.

McFeeters, S. K. 1996. "The Use of the Normalized Difference Water Index (NDWI) in the Delineation of Open Water Features." *International Journal of Remote Sensing* 17:1425–32.

Mishra, D. R., S. Narumalani, D. Rundquist, and M. Lawson. 2005. "High-Resolution Ocean Color Remote Sensing of Benthic Habitats: A Case Study at Roatan Island, Honduras." *IEEE Transactions on Geoscience and Remote Sensing* 43(7):1592–1604.

Mitsch, W. J., and J. G. Gosselink. 2007. *Wetlands*. Hoboken, NJ: John Wiley & Sons.

Mumby, P. J., and A. J. Edwards. 2002. "Mapping Marine Environments with IKONOS Imagery: Enhanced Spatial Resolution Can Deliver Greater Thematic Accuracy." *Remote Sensing of Environment* 82:248–57.

O'Neill, J. D., and M. Costa. In press. "Mapping of Eelgrass (*Zostera marina*) at Sidney Spit, Gulf Island National Park Reserve of Canada, Using High Spatial Resolution Satellite and Airborne Imagery." *Remote Sensing of Environment*. 10.1016/j.rse.2013.02.010.

O'Neill, J. D., M. Costa, and T. Sharma. 2011. "Remote Sensing of Shallow Coastal Benthic Substrates: In Situ Spectra and Mapping of Eelgrass (*Zostera marina*) in the Gulf Islands National Park Reserve of Canada." *Remote Sensing* 3:975–1005.

Peneva, E., J. A. Griffith, and G. A. Carter. 2008. "Seagrass Mapping in the Northern Gulf of Mexico Using Airborne Hyperspectral Imagery: A Comparison of Classification Methods." *Journal of Coastal Research* 24(4):850–56.

Phinn, S., A. Dekker, V. Brando, and C. Roelfsema. 2005. "Mapping Water Quality and Substrate Cover in Optically Complex Coastal and Reef Waters: An Integrated Approach." *Marine Pollution Bulletin* 51:459–69.

Polis, D. F., M. Salter, and H. Lind. 1974. "Hydrographic Verification of Wetland Delineation by Remote Sensing." *Photogrammetric Engineering & Remote Sensing* 40(1):75–78.

Pope, K. O., E. Rejmankova, and J. F. Paris. 1997. "Detecting Seasonal Flooding Cycles in Marshes of the Yucatan Peninsula with S IR-C Polarimetric Radar Imagery." *Remote Sensing of Environment* 59:157–66.

Pott, A., A. K. M. Oliveira, G. A. Damasceno-Junior, and J. S. V. Silva. 2011. "Plant Diversity of the Pantanal Wetland." *Revista Brasileira de Biologia* 71(1 Suppl):265–73.

Rebelo, L.-M. 2010. "Eco-Hydrological Characterization of Inland Wetlands in Africa Using L-Band SAR." *IEEE Journal of Selected Topics in Applied Earth Observations and Remote Sensing* 3(4):554–59.

Rebelo, L.-M., C. N. Finlayson, and N. Nagabhatla. 2009. "Remote Sensing and GIS for Wetland Inventory, Mapping and Change Analysis." *Journal of Environmental Management* 90:2144–53.

Richards, M. 2007. "A Beginner's Guide to Interferometric SAR Concepts and Signal Processing." *IEEE Aerospace and Electronic Systems Magazine* 22:5–29.

Rouse, J., R. Haas, J. Schell, and D. Deering. 1974. "Monitoring Vegetation Systems in the Great Plains with ERTS." In *Third Earth Resources Technology Satellite-1 Symposium*, December 10–14, 1973, NASA SP-351, Washington, DC, Goddard Space Flight Center. Washington, DC: National Aeronautics and Space Administration, Scientific and Technical Information Office, 309–17. Washington, DC.

Sartori, L. R., N. N. Imai, J. C. Mura, E. M. L. M. Novo, and T. S. F. Silva. 2011. "Mapping Macrophyte Species in the Amazon Floodplain Wetlands Using Fully Polarimetric ALOS/PALSAR Data." *IEEE Transactions on Geoscience and Remote Sensing* 49:4717–912.

Silva, T., M. P. F. Costa, and J. M. Melack. 2010. "Spatial and Temporal Variability of Macrophyte Cover and Productivity in the Eastern Amazon Floodplain: A Remote Sensing Approach." *Remote Sensing of Environment* 114:1998–2010.

Silva, T., M. P. F. Costa, J. M. Melack, and E. M. L. M. Novo. 2008. "Remote Sensing of Aquatic Vegetation: Theory and Applications." *Environment Monitoring Assessment* 140:131–45. doi:10.1007/s10661-007-9855-3.

Simard, M., G. de Grandi, S. Saatchi, and P. Mayaux, 2002. "Mapping Tropical Coastal Vegetation Using JERS-1 and ERS-1 Radar Data with a Decision Tree Classifier." *International Journal of Remote Sensing* 23(7):1461–74.

Souza-Filho, P. W. M., W. R. Paradella, S. W. P. Rodrigues, F. R. Costa, J. C. Mura, and F. D. Gonçalves. 2011. "Discrimination of Coastal Wetland Environments in the Amazon Region Based on Multi-Polarized L-Band Airborne Synthetic Aperture Radar Imagery." *Estuarine, Coastal and Shelf Science* 95(1):88–98.

Thomson, A., R. Fuller, T. Sparks, M. Yates, and J. Eastwood. 1998. "Ground and Airborne Radiometry over Intertidal Surfaces: Waveband Selection for Cover Classification." *International Journal of Remote Sensing* 19(6):1189–205.

Ulaby, F. T., R. K. Moore, and A. K. Fung. 1981. *Microwave Remote Sensing: Microwave Remote Sensing Fundamentals and Radiometry*. Reading, MA: Artech House.

Walker, W. S., A. Member, C. M. Stickler, J. M. Kellndorfer, S. Member, K. M. Kirsch, and D. C. Nepstad. 2010. "Large-Area Classification and Mapping of Forest and Land Cover in the Brazilian Amazon: A Comparative Analysis of ALOS/PALSAR and Landsat Data Sources." *IEEE Transactions on Geoscience and Remote Sensing* 3(4):594–604.

Wang, C.-K., and W. D. Philpot. 2007. "Using Airborne Bathymetric Lidar to Detect Bottom Type Variation in Shallow Waters." *Remote Sensing of Environment* 106:123–35.

Wegmüller, U., and C. Werner. 1997. "Retrieval of Vegetation Parameters with SAR Interferometry." *IEEE Transactions on Geoscience and Remote Sensing* 35:18–24.

Wilen, B. O., V. Carter, and R. J. Jones. 1999. *National Water Survey on Wetland Resources*. United States Geological Survey Water Supply Paper 2425. Accessed August 27, 2012. http://water.usgs .gov/nwsum/WSP2425/mapping.html.

Zomer, R. J., A. Trabucco, and S. L. Ustin. 2009. "Building Spectral Library for Wetlands Land Cover Classification and Hyperspectral Remote Sensing." *Journal of Environmental Management* 90:2170–77.

25

Remote Sensing Applications to Monitoring Wetland Dynamics: A Case Study on Qinghai Lake Ramsar Site, China

Hairui Duo, Linlu Shi, and Guangchun Lei

CONTENTS

25.1 Introduction 479
25.2 Materials and Methods 480
    25.2.1 Study Area and Datasets 480
    25.2.2 Image Preprocessing and Land Use and Land Cover Classification 481
    25.2.3 Wetland Dynamics Analysis 482
    25.2.4 Landscape Analysis 482
25.3 Results 484
    25.3.1 Land Use and Land Cover Conversion Matrix 484
        25.3.1.1 Period 1977–1987 484
        25.3.1.2 Period 1987–2000 486
        25.3.1.3 Period 2000–2005 487
        25.3.1.4 Period 2005–2011 488
    25.3.2 Landscape Pattern Dynamics 488
    25.3.3 Patch Shape Changes 490
25.4 Conclusions and Discussion 492
References 493

25.1 Introduction

Wetlands store 25%–30% of the terrestrial soil organic carbon pool and thus play a significant role in reduction of carbon concentration in the atmosphere (MA 2005; Mitsch and Gosselink 2007). Moreover, wetlands also function for flood control, biodiversity, and habitat conservation. However, human activities such as converting wetlands to agricultural lands are fragmenting wetlands' landscapes, disturbing their ecosystems, reducing their distributions, and degrading their biogeochemical and ecosystem functions (Waddington and McNeil 2002; Euliss et al. 2006). Thus, wetlands are among the most threatened ecosystems in the world (MA 2005).

To mitigate the degradation trend of wetlands' ecosystems, it is essential to provide wetland and land managers and decision makers for carbon cycle science and management with accurate information of the distribution of wetlands and their dynamics. Remotely sensed data that reveal spatial distribution of objects on the earth's surface provide great potential for this purpose. Many studies have been conducted in remote sensing–based monitoring of wetland dynamics (Prigent et al. 2001; Toyra et al. 2002; Papa et al. 2006; Zhang et al. 2009; Nie and Li 2011; Panigrahy et al. 2012; Wang et al. 2012). For example, Panigrahy et al. (2012) investigated and assessed wetlands in India using geospatial technologies including remote sensing. Nie and Li (2011) used an object-oriented image classification method to extract wetland information for Alpine Wetland Dynamics from 1976 to 2006 in the vicinity of Mount Everest and analyzed spatial and temporal variability of the wetland ecosystem. Zhang et al. (2009) explored wetland dynamics in China's Sanjiang Plain by combining maximum likelihood supervised classification and postclassification change detection techniques with Landsat multispectral scanner subsystem (MSS) and Thematic Mapper (TM) images acquired in 1976, 1986, 1995, 2000, and 2005. They concluded that human activities led to the loss of the wetland area and induced the degradation of the marsh hydrology and ecosystem. Toyra et al. (2002) proposed a multisensor approach for wetland flood monitoring. Prigent et al. (2001) showed the first global wetland dynamics monitoring program based on multiple satellite datasets. Although there are reports, the global coverage and dynamics of wetlands is still not well known.

On the other hand, the contracting parties to the Ramsar Convention have adopted a series of guidelines (Ramsar 2008) to help the managers develop wetland conservation plans (Ramsar 1971). Monitoring of ecological changes of wetlands is one of the key technical guidelines that assist wetland managers in understanding the dynamics of wetlands and therefore, to timely detect all possible risks and threats to the ecosystem components, processes, and ecosystem services of Ramsar sites (Guan et al. 2011). Considering the international importance of wetlands in China, the State Forestry Administration of China has commenced monitoring of Ramsar sites since 2005 and drafted a National Standard of Monitoring Indicators System for Ramsar sites, including traditional and specific monitoring methodologies and techniques. As a part of the program, this study focused on the application of Landsat TM images to monitor the wetland dynamics, especially the impacts of both nature- and human-induced disturbances on the land use and land cover (LULC) conversions, at the Qinghai Lake Ramsar site in China.

25.2 Materials and Methods

25.2.1 Study Area and Datasets

Qinghai Lake is located at the northeastern part of the Tibet-Qinghai Plateau (Figure 25.1) with the geographic coordinates of 97°53′–101°13′E and 36°28′–38°25′N. It is the largest inland salt water lake wetland complex in China and has an area of about 495,200 ha, which consists of water body, islands, and marshes. The lake was much larger than it is now. During the Bei Wei dynasty (AD 386–557), it possessed its maximum area of 820,000 ha and then the water level dropped down by 90 m in 1908 with the lake area decreasing to 480,000 ha. This trend continued until 2005 with a further decrease of water level by more than 10 m.

There are five islands in the lake, which are used as the breeding grounds for bar-headed geese (*Anser indicus*), black-necked cranes (*Grus nigricollis*), whooper swans (*Cygnus cygnus*), and great cormorants (*Phalacrocorax carbo*). More than 163 species of birds, with a total population of at least 160,000, have been recorded within this wetland. The wetland was designated as a provincial level nature reserve in 1975 and upgraded to a national level nature reserve in 1984. When China made its access to Ramsar Convention in 1992, it was selected as one of the six Ramsar sites.

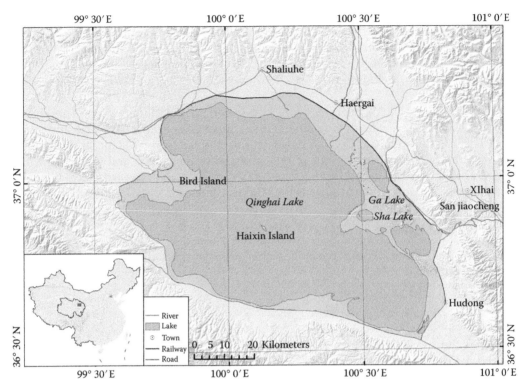

FIGURE 25.1
(See color insert.) The location of the Qinghai Lake wetland.

Landsat multispectral images were downloaded from the U.S. Geological Survey (USGS) website: http://glovis.usgs.gov. The images included MSS data for the year 1977 and TM and Enhanced TM (ETM) datasets for the years 1987, 2000, 2005, and 2011. The MSS images consisted of 79 m spatial resolution green (0.5–0.6 μm), red (0.6–0.7 μm), near-infrared (0.7–0.8 μm), and middle-infrared (0.8–1.1 μm) bands, and a 234 m spatial resolution far-infrared (10.4–12.6 μm) band. Each of the TM image datasets included six channels at a spatial resolution of 30 m × 30 m and one channel at a spatial resolution of 120 m × 120 m. The wavelengths of TM band 1 to band 5 and band 7 were 0.45–0.53, 0.52–0.60, 0.63–0.69, 0.76–0.90, 1.55–1.75, and 2.08–2.35 μm, respectively, and the wavelength of TM band 6 was 10.4–12.5 μm. The ETM images had similar characteristics to the TM except that the pixel size of band 6 was increased to 60 m × 60 m. In addition, the LULC maps of scales at both 1:100,000 and 1:25,000 were obtained for the year 1990 and were updated for the year 2000.

25.2.2 Image Preprocessing and Land Use and Land Cover Classification

All the remotely sensed data were georeferenced to the Universal Transverse Mercator coordinate system using ground control points with the ERDAS IMAGINE 9.3 software package. Moreover, the coregistration of the Landsat images was conducted. A postclassification algorithm was used to detect the LULC changes. That is, the LULC classifications were done for each of the years 1977, 1987, 2005, and 2011 and the LULC changes were then analyzed by comparing the classification maps between any two years. The LULC classification for each year was carried out using supervised classification. The training areas were selected and validated through field survey. The LULC classification and change detection maps were converted into shapefiles and put into a geographic database using ArcGIS10.0 software. The wetland LULC conversions and landscape changes over the past four periods, including 1977–1987, 1987–2000, 2000–2005, and 2005–2011, were then analyzed.

A total of eight LULC categories were considered in the classification. The classes included residential, farmland, grassland, mashes, lakes, saline land, bare land, and sandy areas. This classification was defined based on the national standards for land classification from the Chinese State Administration for Quality Inspection and Quarantine and the Chinese National Standard Commission (2007) and by combining with local land use, management, and utilization, as well as the land use change.

25.2.3 Wetland Dynamics Analysis

Wetland dynamics that account for wetland changes over a period can be expressed by the following formula:

$$K = \frac{U_b - U_a}{U_a} \times \frac{1}{T} \times 100\% \qquad (25.1)$$

where K is the rate of dynamic changes of wetland at a certain period T, U_a and U_b are the wetland areas at the start and the end of the study period T, and if the period is T years, K is the annual change rate of the wetland. Because of significant regional differences in wetland dynamics, we need to use the relative wetland change rate to reflect the regional differences in wetland change, which can be described as follows:

$$R = \frac{U_b - U_a/U_a}{C_b - C_a/C}$$

where C_a and C_b are the regional wetland areas at the start and the end of a study period T. A positive R indicates the trends of change are the same for both a subarea or an LULC category and the whole wetland region, while a negative R indicates the trend of change for a subarea or an LULC category and the whole region are not in the same direction. An absolute value $R > 1$ means that the rate of change in the study area is larger than the regional wetland change. The conversion matrix of LULC types for each of the four periods 1977–1987, 1987–2000, 2000–2005, and 2005–2011 was obtained based on Table 25.1.

25.2.4 Landscape Analysis

A consistent spatial resolution of 30 m × 30 m was used in this study. The characteristics of landscape for each of the years 1977, 1987, 2000, 2005, and 2011 were quantified using Fragstats 3.3 landscape modeling software (McGarigal et al. 2002) (http://www.umass .edu/landeco/research/fragstats/fragstats.html). The used spatial metrics included patch number (NP), patch area (CA) (ha), percentage of landscape (PLAND), mean patch area (MPA) (ha), patch edge density (ED) (m/ha), mean shape index (MSI), and mean patch fraction dimension (MPFD). The dynamics of the landscape was then analyzed by comparing the values of the spatial metrics for each of the periods from 1977 to 1987, 1987 to 2000, 2000 to 2005, and 2005 to 2011. The landscape or spatial metrics were defined and listed in Table 25.2.

TABLE 25.1

Conversion Matrix of LULC Types

k Period		Conversion Matrix	Total Area (CA1)
		k + 1 Period	
X_1	A	$A_{11}\ A_{12}\cdots\cdots A_{1j}$	
	B	$B_{11}\ B_{12}\cdots\cdots B_{1j}$	
	C	$C_{11}\ C_{12}\cdots\cdots C_{1j}$	
$\cdot$	A	$\cdots\cdots$	
	B	$\cdots\cdots$	
	C	$\cdots\cdots$	
X_i	A	$A_{i1}\ A_{i2}\cdots\cdots A_{ij}$	
	B	$B_{i1}\ B_{i2}\cdots\cdots B_{ij}$	
	C	$C_{i1}\ C_{i2}\cdots\cdots C_{ij}$	
Total Area (CA2)			
Rate of change (%)			

Note: The row represents the *i* LULC type of the *k*th period, the column represents *j* LULC of the (*k* + 1)th period. CA1 and CA2 represent the total area of LULC type in the *k*th period and (*k* + 1)th period, respectively. Conversion proportion (%) = CA1/CA2 × 100%.

TABLE 25.2

Landscape Indexes Used in This Study

Landscape Index		Abbreviations	Formula	Description
Patches	1. Number of patches	NP	$NP = n_i$	Landscape patch number, value: NP ≥ 1, no upper limit. The index describes habitat heterogeneity and a larger value implies a more fragmented landscape.
Edge	1. Total edge length	TE	$TE = \sum_{k=1}^{m} e_{ik}$	Total edge length of all patches (m) (TE > 0)
	2. Edge density	ED	$ED = \dfrac{\sum_{k=1}^{m} e_{ik}}{A}$	Total edge length (m) divided by the landscape area
Area	1. Total patch area	CA	$CA = \sum_{j=1}^{n} a_{ij}$	Total patch area of each patch type (CA > 0)
	2. Largest patch area	Max. A		Area of the largest patch
	3. Smallest patch area	Min. A		Area of the smallest patch

(Continued)

TABLE 25.2 (*Continued*)

Landscape Indexes Used in This Study

Landscape Index		Abbreviations	Formula	Description
	4. Mean patch area	MPA		Area of a particular patch type divided by total landscape area (%)
	5. Percentage of landscape	PLAND		Area of a patch type divided by the whole landscape area
Shape	1. Average patch shape index	MSI	$MSI = \dfrac{\sum\limits_{i=1}^{m}\sum\limits_{j=1}^{n}\left(0.25P/\sqrt{a_{ij}}\right)N}{N}$	Edge length of each patch type divided by root-squared area, then multiplied by quadrate shape index, and finally accumulated over all patches. MSI > 1.
	2. Mean patch fraction dimension	MPFD	$MPFD = \sum\limits_{i=1}^{m}\sum\limits_{j=1}^{n}\left[\dfrac{2\ln(0.25P_{ij})/\ln(a_{ij})}{N}\right]$	MPFD value ranges between 1 and 2 and is the average patch dimension.

25.3 Results

25.3.1 Land Use and Land Cover Conversion Matrix

The obtained area of the Qinghai Lake wetland is 571,376.9 ha out of which an area of 510,456.80 ha accounting for 89.34% of the landscape was not changed from 1977 to 2011. In 2011, it was found that the areas of lake, grassland, sandy areas, marshes, farmland, bare land, saline land, and residential area were 434,926.3, 79,298.5, 26,903.1, 20,741.3, 5,582.0, 1,681.3, 2,061.6, and 182.9 ha, respectively (Figure 25.2). However, LULC conversions took place during each of the periods.

25.3.1.1 Period 1977–1987

From Table 25.3, we can see the changes of marshes and lake areas are negative, indicating that both LULC types decreased whereas all other LULC types increased from 1977 to 1987. The largest increase of LULC types was bare land, then residential areas, saline land, farmland, sandy land, and grassland. Bare land increased by 711.35%, reflecting the environmental degradation, and at the same time, the residential area and farmland also increased significantly, indicating the human intervention with the wetland ecosystem.

With regards to the residential area, the original residential area remained unchanged, but some farmland and grassland were converted into residential area. In particular, the conversion from farmland area accounted for 74.8% of the total converted residential area. About 94% of the farmland, grassland, and marshes remained unchanged during this period. However, conversion did take place between each of these three LULC types. The farmland expansion happened mainly because of reclamation of wetlands within the area of Qinghai Lake. With the implementation of land tenure since 1978, this reclamation

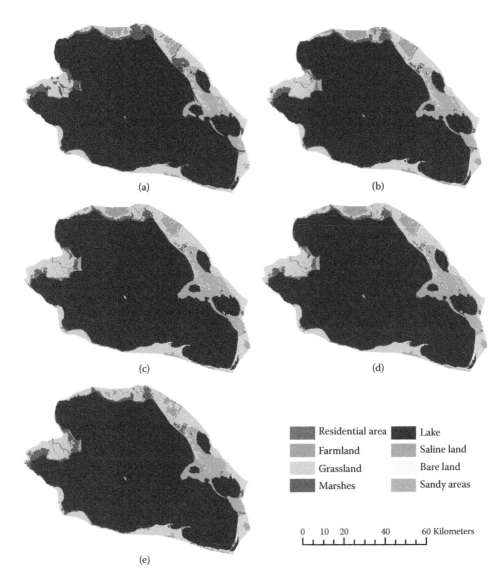

FIGURE 25.2
(See color insert.) LULC classification maps of the wetland landscape of Qinghai Lake for the years (a) 1977, (b) 1987, (c) 2000, (d) 2005, and (e) 2011.

reached its peak in 1987 with a total farmland area of 11,190.33 ha. This conversion was mainly from grassland (98.28%). On the other hand, during this period 7.54% of the newly reclaimed farmland was abandoned.

The conversion between grassland and all other LULC types also took place during this period. Grassland was mainly converted to marshes, sandy areas, and farmland. Meanwhile, the main source of land transferred to grassland was marshes because of the drying up (83.47%). Only 6.17% of the sandy area was vegetated and 3.08% of the degraded grassland was restored, indicating the trend of land degradation. With the decrease of the water table, marshes and lake areas were converted into bare land. However, very little bare land was converted to any other LULC types.

TABLE 25.3

Land Use Conversion Matrix from 1977 to 1987

1977–1987	RA	FL	GL	ML	L	S	SL	BL	Total
RA	54.5								54.5
FL	25.9	7,773.9	416.0	35.3					8,251.1
GL	8.7	3,344.3	54,647.4	6,149.9	97.5	1,690.5	3.4	200.0	66,141.6
ML		58.7	13,174.9	11,728.0	123.3	1,766.1	148.7	512.5	27,512.3
L			885.5	6,326.5	435,042	2,465.1	1757.3	141.1	446,617.7
S			547.7	195.4	134.5	20,165.7	120.2	69.0	21,232.4
Sl			91.0	161.2	129.0	659.7	128.0	236.9	1,405.8
BL			0.2	9.00	1.4			150.9	161.5
Total	89.2	11,177	69,762.6	24,605.2	435,528	26,747.2	2157.5	1310.5	571,376.9
Total (%)	+63.6	+35.5	+5.48	−10.57	−2.48	+25.97	+53.47	+711.3	
Annual change (%)	6.36	3.55	0.55	−1.06	−0.25	2.60	5.35	71.13	

Note: RA, residential area; FL, farmland; ML, marsh land; L, lakes; S, sandy area; SL, saline land; and BL, bare land.

TABLE 25.4

Land Use Conversion Matrix from 1987 to 2000

1987–2000	RA	FL	GL	ML	L	S	SL	BL	Total
RA	32.2	44.4	8.1						84.7
FL	30.0	9968.7	1,180.3	11.4					11,190.4
GL	55.0	772.0	6,6574.7	1,153.9	38.6	948.0	44.1	167.4	69,753.6
ML		90.3	12,453.6	10,336.6	124.8	429.5	814.4	356.1	24,605.2
L			241.6	2,993.7	430,372.5	488.4	1335.9	95.8	435,527.9
S			694.1	476.1	27.0	24,988.3	437.8	123.8	26747.2
SL			23.9	16.1	81.90	870.9	1157.8	6.9	2,157.5
BL			416.4	28.9	7.6	127.3	44.6	685.7	1,310.5
Total	117.2	10875.3	81,592.6	15,016.6	430,652.3	27,852.5	3834.5	1435.8	571,376.9
Total (%)	38.4	−2.82	16.97	−38.97	−1.12	4.13	77.73	9.57	
Annual change (%)	2.95	−0.22	1.31	−3.00	−0.09	0.32	5.98	0.74	

Note: RA, residential area; FL, farmland; ML, marsh land; L, lakes; S, sandy area; SL, saline land; and BL, bare land.

25.3.1.2 Period 1987–2000

From 1987 to 2000, farmland and marshes were reduced while lake area almost remained the same (Table 25.4). Compared to the period of 1977–1987, the change rate of the lake area was lower. Marshes decreased, while bare land and grassland development increased. Overall, the trend of ecological degradation slowed down (Table 25.4).

Abandoned residential area was mainly converted to farmland and grassland, while a reverse trend was also observed. Of the newly developed residential areas, 64.7% came from farmland and then grassland. During the 1980s, massive reclamation for farmland was not well planned and the fertility of the reclaimed land decreased quickly. Hence, a large amount of farmland was abandoned soon after cultivation, which led to the reduced rate of reclamation in the 1990s. From 1987 to 2000, the rate of land reclamation to abandoned area was 0.72:1, which meant that more farmland was abandoned than reclaimed. Conversion from grassland to all other LULC types was observed. Grassland was mostly converted to marshes, sandy areas, and farmland, whereas marshes were the main source of land conversion to grassland (82.92%), followed by farmland (7.86%). During this period, the increased desertification area was 1.63 times of the newly restored area and, meanwhile, the rate of grassland converted to desert was 1.37 times of the restored grassland, which implied the land degradation continued, but the degradation rate decreased compared to the period of 1977–1987.

25.3.1.3 Period 2000–2005

From 2000 to 2005, the two major LULC types showed a decreased annual change, which was characterized by the increased marsh area and the decreased saline land area (Table 25.5). Moreover, increases in the residential area and bare land and the reduction of farmland were also observed. The newly increased residential areas mainly came from grassland and farmland. On the other hand, with the implementation of the national policy on returning farmland to grassland and shifting crop cultivation to animal husbandry, most of the abandoned farmland was artificially restored into grassland. During this period, the reclamation for farmland was significantly reduced. Therefore, farmland was converted to grassland, which accounts for 63.8% of the newly restored grassland. Unfortunately, this period was also a period of accelerated land degradation because the rate of desertification was 5.45 times the rate of land restoration. In particular, the rate of grassland converted to desert was 16.69 times the rate of grassland restoration. Compared to the period of 1987–2000, the expanded desertification was observed mainly because of the degradation of grassland.

TABLE 25.5

Land Use Conversion Matrix from 2000 to 2005

2000–2005	RA	FL	GL	ML	L	S	SL	BL	Total
RA	103.8	3.4	15.4						122.6
FL	45.7	6516.7	4,306.9	6.0					10,875
GL	58.4	720.4	73,356.9	4,566.4	73.3	2,174	78.3	565	81,592
ML		0.9	2,072.8	11,768.3	570.1	313.3	163.2	129.0	15,017.6
L			60.1	339.6	42,8677	414.2	1,099	62.7	43,0652
S			130.2	252.8	20.0	27,054	394.2	1.3	27,852.4
SL			61.3	516.6	302.7	1,305.9	1,613.5	35.3	3,835.3
BL			104.2	39.1	29.0	144.6	9.5	1,109.4	1,435.8
Total	207.9	7241.4	8,0107.8	17,488.6	42,9672	31,405.8	3,357.7	1,902.7	571,384.2
Total (%)	69.51	−33.41	−1.82	16.45	−0.23	12.76	−12.45	32.52	
Annual change (%)	6.95	−3.34	−0.18	1.65	−0.02	1.28	−1.25	3.25	

Note: RA, residential area; FL, farmland; ML, marsh land; L, lakes; S, sandy area; SL, saline land; and BL, bare land.

25.3.1.4 Period 2005–2011

From 2005 to 2011, marshes and open water area increased mainly because the rising water level led to shrinking of all other LULC types (Table 25.6). With the implementation of the ecological conservation programs in the Tibet-Qinghai Plateau, the newly reclaimed farmland was regulated in this period. Compared to the period of 2000–2005, much more farmland was converted to grassland. Grassland was mostly converted to marshes. However, the source of the land that was converted most to grassland was sandy land, which accounts for 39.64% of the newly formed grassland. Moreover, there was also 34.27% of the newly formed grassland coming from farmland. Land restoration characterized this period in which the rate of desertification to restoration of desert land was 0.12:1. The corresponding rate for grassland was 0.14:1, which was mainly contributed to by the return of farmland to grassland and other ecological restorations that were implemented since 2002. Because of the rising water level, about 30% and 82% of the sandy land area and saline land, respectively, became water bodies.

25.3.2 Landscape Pattern Dynamics

In Qinghai Lake, major LULC categories were lake (water body), grassland, marshes, sandy areas, and farmland and the rest of the LULC types including saline land, bare land, and residential areas accounted for less than 1% of the whole landscape (Table 25.7). Based on the patch area percentage (PLAND), during the whole period from 1977 to 2011, the lake area varied from 78.17% to 75.20%, grassland from 11.57% to 14.02%, and marshes from 4.81% to 2.63%. For each of these three LULC types, the land percentages had a difference of more than 2%. During the period of 34 years, 50.18% of farmland and 45.39% of marshes were converted to other LULC types. Also 3.8% of the lake area was changed.

Lake dominated this wetland. Both the area and edge length of the lake continuously shrank at 1977–1987, 2000, and 2005 after which the lake area increased. From 1977 to 2005, the lake area was reduced by 16,972 ha and increased by 5,265.9 ha from 2005 to 2011.

TABLE 25.6

Land Use Conversion Matrix from 2005 to 2011

2005–2011	RA	FL	GL	ML	L	S	SL	BL	Total
RA	167.8	7.6	36.3						211.7
FL	6.0	5499.1	1,733.4						7,238.5
GL	9.0	75.4	74,240.2	5,208.1	177.4	286.1	12.5	90.7	80,099.5
ML			955.4	14,548.2	1,728.7	53.6	166.7	36.1	17,488.6
L			18.0	60.4	429,535.7	29.0	1.4	27.7	429,672.3
S			2,005.2	413.7	1,532.1	26,282.8	767.3	404.8	31,405.9
SL			8.5	130.0	1,857.2	245.5	1,105.0	11.6	3,357.7
BL			301.3	380.9	95.3	6.1	8.7	1,110.4	1,902.7
Total	182.9	5,582.0	79,298.5	20,741.3	434,926.3	26,903.1	2,061.6	1,681.3	571,376.9
Total (%)	−13.6	−22.9	−1.00	18.60	1.22	−14.34	−38.60	−11.64	
Annual change (%)	−2.72	−4.58	−0.20	3.72	0.24	−2.87	−7.72	−2.33	

Note: RA, residential area; FL, farmland; ML, marsh land; L, lakes; S, sandy area; SL, saline land; and BL, bare land.

TABLE 25.7

Landscape Indexes Used to Quantify the Landscape Dynamics of the Wetland at Qinghai Lake

Year	Type	NP	CA	PLAND	MPA	Min. A	Max. A	TE
1977	RA	21	57.06	0.01	2.72	0.26	10.44	16,320
	FL	24	8,252.37	1.44	343.85	3.88	2,827.50	228,420
	GL	56	66,134.52	11.57	1,180.97	2.76	22,416.85	116,7300
	ML	26	27,504.18	4.81	1,057.85	1.71	12,634.73	1,056,060
	L	37	446,634.72	78.17	12,071.21	0.99	439,747.26	803,100
	SL	10	1,406.52	0.25	140.65	19.50	293.69	133,980
	BL	5	162	0.03	32.40	61.37	100.16	16,920
	S	15	21,228.57	3.72	1,415.24	3.18	16,593.93	463,380
1987	RA	35	84.51	0.01	2.41	0.26	10.44	25,050
	FL	24	11,191.41	1.96	466.31	3.46	3,910.61	300,420
	GL	273	6936.68	12.21	255.45	0.99	29,677.40	2,236,170
	ML	354	24,610.05	4.31	69.52	0.99	6,602.16	2,228,880
	L	34	435,547.35	76.23	12,810.22	0.99	428,185.00	773,220
	SL	91	2,157.48	0.38	23.71	0.99	558.81	386,700
	BL	57	1,310.85	0.23	23.00	0.99	379.53	252,780
	S	73	26,741.61	4.68	366.32	0.99	20,281.60	817,740
2000	RA	19	118.8	0.02	6.25	0.53	26.28	21,240
	FL	31	10,879.74	1.90	350.96	2.91	4,159.52	250,980
	GL	301	81,592.02	14.28	271.07	0.00	32,031.00	2,301,840
	ML	316	15,019.56	2.63	47.53	0.99	2,913.68	1,796,460
	L	33	430,651.26	75.37	13,050.04	0.99	423,360.00	766,200
	SL	91	3,832.38	0.67	42.11	0.09	476.10	558,720
	BL	35	1,437.84	0.25	41.08	0.99	275.85	249,360
	S	83	27,848.34	4.87	335.52	0.99	20,897.50	847,560
2005	RA	19	209.25	0.04	11.01	1.17	41.16	33,030
	FL	40	7,239.69	1.27	180.99	2.91	2,616.58	247,620
	GL	140	80,126.46	14.02	572.33	0.54	33,635.90	1,894,230
	ML	283	17,492.94	3.06	61.81	0.99	3,746.13	1,615,470
	L	28	429,662.34	75.20	15,345.08	0.99	413,937.00	759,240
	SL	65	3,353.49	0.59	51.59	0.99	742.37	487,920
	BL	31	1,898.28	0.33	61.23	0.99	455.70	261,540
	S	60	31,397.49	5.50	523.29	0.99	24,717.10	726,210
2011	RA	13	183.15	0.03	14.09	1.97	49.18	26,910
	FL	48	5,575.32	0.98	116.15	2.91	1,505.70	239,520
	GL	142	79,329.69	13.88	558.66	0.99	24,000.50	1,943,730
	ML	207	20,737.44	3.63	100.18	0.99	4,488.25	1,741,380
	L	28	434,928.24	76.12	15,533.15	1.08	427,918.00	770,190
	SL	99	2,061.36	0.36	20.82	0.99	648.09	349,560
	BL	30	1,672.38	0.29	41.81	1.17	423.91	308,880
	S	53	26,892.27	4.71	507.40	1.08	20,617.20	732,390

Note: RA, residential area; FL, farmland; GL, grassland; ML, marshes; L, lake (water body); SL, saline land; BL, bare land; and S, sandy area.

During the whole period of 34 years, the patch number of lake also decreased, implying the drying up of small lakes. In 1977, the whole lake area was 446,634 ha; this decreased to 429,662 ha by 2005 and then increased to 434,928 ha in 2011. In 1987, a small lake in the middle area of sandy land was observed separately from the main body of the lake. In 2005, another medium-sized lake with a total area of 8937.75 ha was formed by separating from the main lake body. With the increase of water level, this small lake was reunited with the main body of the lake in 2011. From 1977 to 2005, six small lakes disappeared, but only one was restored in 2011.

From 1977 to 2000, more than 12,480 ha of marshes disappeared, which accounted for 45.5% of the whole marsh area of Qinghai Lake. MPA decreased, indicating shrinking of the marshes. Since 2000, the smallest patch area remained unchanged and the total patch area, MPA, and largest patch area increased slightly, but the number of patches decreased continuously, indicating the merging of small patches of marshes.

From 1977 to 1987, more than 2939 ha of farmland were reclaimed. The number of patches increased rapidly mainly because of the new tenure policy. Many small patches of farmland merged and newly reclaimed small patches of farmland appeared at the edges. From 1987 to 2000, the total patch area of farmland decreased slightly, whereas there was a mild increase in the largest patch area. The number of farmland patches increased by 29% and the total edge length was reduced by 16.5%. This was explained by the coexistence of both farmland reclamation and abandonment. From 2000 to 2011, the total patch area and the largest patch area of farmland decreased rapidly, the number of patches increased quickly, and other parameters remained unchanged. This showed the decrease of farmland from 11,191 ha in 1987 to 5,575 ha in 2011.

From 1977–1987, 2000, and 2005, the total patch area and the largest patch area of sandy patches increased continuously, whereas the number of sandy patches and the total edge length increased during the first two periods and then decreased in the third. The dynamic of these parameters for sandy areas implied the development of severe desertification. From 1977 to 2000, large sandy areas were expanded and at the same time, newly formed small sandy patches were added at the front edges of the expansion areas. This process was followed by the merging of small sandy patches, which led to larger sandy patches from 2000 to 2005. From 2005 to 2011, the total patch area, number of patches, and largest patch area of sandy decreased, whereas the total edge length and smallest patch area increased. This indicated the disappearance of small sandy patches and shrinking of larger sandy patches as well as forming of heterogeneity patches within the larger sandy patches.

The trend of change for bare land patches was similar to that of the sandy patches, characterized by the rapid increase from 1977 to 2005 and then slight decrease from 2005 to 2011. Saline land increased before 2000 and then decreased. Residential area increased from 57 ha in 1977 to 209 ha in 2005 and then decreased to 183 ha in 2011. Both the largest and smallest patch areas increased continuously. These implied the expansion of human activity impacts as well as the village development.

25.3.3 Patch Shape Changes

Grassland, lake, and marshes had larger values of patch EDs than other types (Figure 25.3), which reflected the heterogeneity of patch distribution. The patch ED value for residential area was the smallest. The values of MSI and MPFD for marshes, saline land, and bare land were relatively greater (Figures 25.4 and 25.5). These index values suggested that these LULC types were restricted by the geographic and topographic characteristics of

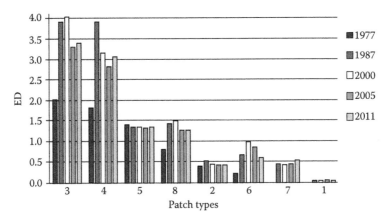

FIGURE 25.3
The edge density of each patch type and its dynamics over time: (1) residential area, (2) farmland, (3) grassland, (4) marshes, (5) lake, (6) saline land, (7) bare land, and (8) sandy area.

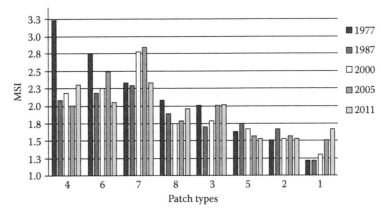

FIGURE 25.4
The average patch shape index (MSI) of each patch type and its dynamics over time: (1) residential area, (2) farmland, (3) grassland, (4) marshes, (5) lake, (6) saline land, (7) bare land, and (8) sandy area.

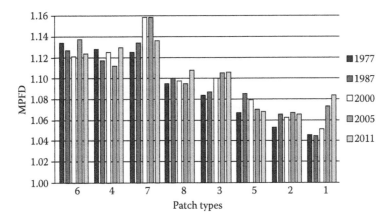

FIGURE 25.5
The mean patch dimension of each patch type and its dynamics over time: (1) residential area, (2) farmland, (3) grassland, (4) marshes, (5) lake, (6) saline land, (7) bare land, and (8) sandy area.

the study area and they were mixed with other LULC types. For instance, the marshes were distributed along the lake shore as narrow belts and intermingled with sandy areas, grassland, and bare lands. Other LULC types such as the residential area and farmland had small values of ED due to human intervention and regular shapes. The lake with a huge area had a regular shape and hence its value of ED was low.

The degradation and disappearance of marshes led to an increase of patch number from 26 to 354 from 1977 to 1987 and thus the significant increase in the edge length and patch ED, implying the fragmentation of marshes and the decrease of MSI and MPFD. Meanwhile, the adding of grassland patches within the marshes resulted in the increase of internal edge length and ED. During the period of 1987 to 2000, more than 9587 ha (38%) of the marsh wetland disappeared. From 2000 to 2005, marsh areas increased rapidly, which led to a significant decrease in the number of patches from 316 to 283. The patch ED of marshes continuously decreased due to merging of small marsh patches into larger ones and hence the values of both mean patch shape index and MPFD decreased. From 2005 to 2011, the area of marsh patches continuously increased and the patch edge length increased too, while the patch number decreased from 283 to 207. The patch ED, mean patch shape index, and MPFD of marshes increased, which implied the expansion and complexity of marshes with irregular edge shape.

25.4 Conclusions and Discussion

Remote sensing has been widely applied in wetland classification and dynamics monitoring globally (Lowry 2006). The European Space Agency, in collaboration with Ramsar Convention Secretariat, launched the GlobWetland project in 2003. More than 20 countries including China and 51 wetland managers participated in this project (ESA 2012). In 2010, China launched a wetland eco-compensation program and its aim was not only to provide local communities with sustained livelihood but also, through the eco-compensation (one form of payment for ecosystem services), to conserve and sustain the ecological characteristics of the wetlands in China. Achieving the goals requires a mechanism that can be used to timely and effectively monitor the dynamics of the wetlands and to provide accurate information. Remote sensing technologies offer this potential.

This study showed the use of Landsat satellite images in monitoring the dynamics of the wetland in Qinghai Lake. Overall, the area of the lake continuously decreased, the landscape was fragmented, and the wetland ecosystem functions were degraded during the first three periods from 1977 to 1987, 1987 to 2000, and 2000 to 2005. These results were mainly induced by increased human activities. After 2005, the lake area increased and the wetland landscape and functions started to recover. This recovery may be mainly attributed to the implementation of wetland conservation plans in China recently. The obtained results from this study will greatly help manage wildlife habitats (Forman and Godron 1986) and map ecosystem services of the whole wetland complex (Chen and Zhang 2000; Liu 2001; Feng and Feng 2004).

However, it has to be pointed out that in this study we lacked the accuracy assessment data of the obtained LULC classification maps because of time limitations. This will be conducted in the near future. Thus, readers have to take caution in using the results in this study.

References

Chen, Z. X., and X. S. Zhang. 2000. "The Value of Chinese Ecosystem Service." *Chinese Science of Bulletin* 45(1):17–22.

Chinese State Administration for Quality Inspection and Quarantine and Chinese National Standard Commission. 2007. *China Land Use Classification*. Beijing: Chinese Standard Press.

Euliss, N. H., R. A. Gleason, A. Olness, R. L. McDougal, H. R. Murkin, R. D. Robarts, R. A. Bourbonniere, and B. G. Warner. 2006. "North American Prairie Wetlands are Important Nonforested Land-Based Carbon Storage Sites." *Science of the Total Environment* 361:179–88.

Feng, Z. W., and Z. Z. Feng. 2004. "Key Ecological Challenges and Responses at Qinghai Lake Basin." *Journal of Ecological Environment* 13(4):467–69.

Forman, R. T. T., and M. Godron. 1986. *Landscape Ecology*. New York: John Wiley & Sons.

Guan, L., P. Liu, and G. Lei. 2011. "Ramsar Site Ecological Character Description and Its Monitoring Indicators." *Journal of Forest Inventory and Planning* 30(2):1–9.

Liu, X. Y. 2001. "Water Level Trend Analysis of Qinghai Lake." *Arid Zone Research* 18(3):58–62.

Lowry, J. 2006. *Low-Cost GIS Software and Data for Wetland Inventory, Assessment and Monitoring*. Ramsar Technical Report No. 2. Gland, Switzerland: Ramsar Convention Secretariat. ISBN 2-940073-30-9.

McGarigal, K., S. A. Cushman, M. C. Neel, and E. Ene. 2002. "FRAGSTATS: Spatial Pattern Analysis Program for Categorical Maps." Computer software program produced by the authors at the University of Massachusetts, Amherst.

Millennium Ecosystem Assessment (MA). 2005. *Ecosystem and Human Well-Being: Wetlands and Water Synthesis*. Washington, DC: World Resources Institute/Island Press.

Mitsch, W. J., and J. G. Gosselink. 2007. *Wetlands*. Hoboken, NJ: John Wiley.

Nie, Y., and A. Li. 2011. "Assessment of Alpine Wetland Dynamics from 1976–2006 in the Vicinity of Mount Everest." *Wetlands* 31:875–84.

Panigrahy, S., T. V. R. Murthy, J. G. Patel, and T. S. Singh. 2012. "Wetlands of India: Inventory and Assessment at 1:50,000 Scale Using Geospatial Techniques." *Current Science* 102:852–56.

Papa, F., C. Prigent, W. B. Rossow, B. Legresy, and F. Remy. 2006. "Inundated Wetland Dynamics over Boreal Regions from Remote Sensing: The Use of Topex-Poseidon Dual-Frequency Radar Altimeter Observations." *International Journal of Remote Sensing* 27:4847–66.

Prigent, C., E. Matthews, F. Aires, and W. B. Rossow. 2001. "Remote Sensing of Global Wetland Dynamics with Multiple Satellite Data Sets." *Geophysical Research Letters* 28(24):4631–34.

Ramsar. 1971. "Convention on Wetlands of International Importance Especially as the Waterfowl Habitat." Ramsar (Iran), February 2, 1971, UN Treaty Series No. 14583. As amended by the Paris Protocol, December 3, 1983, and Regina Amendments, May 28, 1987.

Ramsar. 2008. "10th Meeting of the Conference of the Parties to the Convention on Wetlands." Ramsar (Iran), 1971. http://www.ramsar.org/doc/res/key_res_x_16_e.doc.

Toyra, J., A. Pietroniro, L. W. Martz, and T. D. Prowse. 2002. "A Multi-Sensor Approach to Wetland Flood Monitoring." *Hydrological Processes* 16:1569–81.

Waddington, J., P. McNeil. 2002. "Peat Oxidation in an Abandoned Cutover Peatland." *Canadian Journal of Soil Science* 82:279–86.

Wang, L., I. Dronova, G. Peng, W. Yang, Y. Li, and Q. Liu. 2012. "A New Time Series Vegetation-Water Index of Phenological-Hydrological Trait across Species and Functional Types for Poyang Lake Wetland Ecosystem." *Remote Sensing of Environment* 125:49–63.

Zhang, S., X. D. Na, B. Kong, Z. Wang, H. Jiang, H. Yu, Z. Zhao, X. Li, C. Liu, and P. Dale. 2009. "Identifying Wetland Change in China's Sanjiang Plain Using Remote Sensing." *Wetlands* 29:302–13.

26

Hyperspectral Sensing on Acid Sulfate Soils via Mapping Iron-Bearing and Aluminum-Bearing Minerals on the Swan Coastal Plain, Western Australia

Xianzhong Shi and Mehrooz Aspandiar

CONTENTS

26.1 Introduction..496
26.2 Setting of Study Area...498
26.3 Data Acquisition and Laboratory Measurement..499
    26.3.1 HyMap Data..499
    26.3.2 Soil Samples Collection...499
    26.3.3 Proximal Hyperspectral Data...499
    26.3.4 Soil Properties Measurement..500
    26.3.5 Mineralogy Verification...500
26.4 Reflectance Spectral Characterization...500
26.5 Mapping Method...503
    26.5.1 Preprocessing..503
    26.5.2 Mapping Methods..503
26.6 Results and Discussion...504
    26.6.1 Ferric Iron Content Mapping..504
    26.6.2 Indicative Iron Secondary Minerals Mapping and Classification...............505
        26.6.2.1 End-Members Selection...505
        26.6.2.2 Mapping and Classification..506
        26.6.2.3 Iron-Bearing Minerals Distribution and Space Pattern...............507
        26.6.2.4 Soil Acidity Map Deducted from Iron-Bearing Minerals Map......507
    26.6.3 Aluminum-Bearing Minerals, Carbonates, and Sulfates Mapping and
        Classification...508
        26.6.3.1 Mapping and Classification Method..508
        26.6.3.2 Non-Iron-Bearing Minerals Distribution and Space Pattern...........508
        26.6.3.3 Toxicity Map Derived from Aluminum-Bearing Mineral and
            pH Distribution...509
26.7 Conclusion and Future Work...510
26.8 Summary...511
References..511

26.1 Introduction

Acid sulfate soils (ASSs) are widely spread in coastal wetland areas, inland lakes, and mining sites all around the world. They can be very deleterious to the environment and can result in serious issues that can constrain both environmental and economic development. ASS is a kind of soil/sediment containing sulfide minerals, which may either already contain some amounts of sulfuric acid (actual acid sulfate soils [AASSs]) or have the potential to produce sulfuric acid (potential acid sulfate soils [PASSs]) (Fitzpatrick 2003). PASSs are not harmful until they are disturbed, whereas AASSs can have very low pH values (<2.8) and high concentrations of trace metals. Evidently, sulfides are the source and cause of ASSs and are mainly composed of pyrite, although they sometimes contain some other sulfides in lesser amounts, such as pyrrhotite, chalcopyrite, greigite, and iron monosulfides, which are usually formed as monosulfidic black ooze (MBO) in coastal areas. The formation of ASSs involves several chemical, biological, and electrochemical processes, but the mechanism can be summarized as the oxidation of pyrite described by the following formula:

$$FeS_2 + \frac{15}{4O_2} + \frac{7}{2H_2O} = 2SO_4^{2-} + Fe(OH)_3 + 4H^+$$

From the formula, it is clear that the oxidation produces not only sulfuric acid but also sulfate and iron-bearing secondary minerals, which precipitate at different pH values. Generally, copiapite forms when pH < 1.5, jarosite when pH is in the range of 1.5–2.8, schwertmannite when pH is in the range of 2.8–4.5, goethite when pH < 6, hematite when pH is from 7 to 8, and ferrihydrites when pH > 5 (Anderson 1994; Alpers et al. 1994; Bigham 1994; Fitzpatrick 2003; Montero et al. 2005). Thus, these secondary iron-bearing minerals can be used as indicators of pH conditions when they are formed. However, the pH values of soils/sediments not only depend on how much acidity is generated from pyrite oxidation but also on the balance between acid productivity, acid-neutralizing capacity, and availability of water and oxygen (Harries 1997). Once the formation of ASS occurs, the products of acidity may react with surrounding minerals, such as carbonates including calcite, dolomite, and siderite, and it will lead to acid neutralization and the potential production of gypsum. Meanwhile, some aluminosilicate minerals in soils, such as k-feldspar and kaolin group minerals, may also participate in the neutralization process, consuming H^+ and releasing Al^{3+} (Blowes et al. 2003), and generate some aluminum-bearing minerals under different pH conditions, for instance, gibbsite forms in the primitive state of aluminosilicate neutralization in near neutral conditions (Bigham and Nordstrom 2000) and dissolves by reacting with sulfuric acid when pH further decreases to the range of 5 to 4. When pH drops below 4, soluble aluminum sulfate, such as halotrichite, pickeringite, and alunogen, may form (Bigham and Nordstrom 2000). The depletion of pyrite from the ASS results in the reduction of acidity production; eventually, this will become lower than the buffering capability of the soils and will result in the formation of insoluble hydroxysulfates such as alunite and basaluminite when pH increases to 5 (Bigham and Nordstrom 2000). At a mine site, Kim et al. (2003) found that aluminum sulfate was precipitated in the pH range of 4.45–5.95 and aluminum ions were mostly removed from the mining drainage when pH was >5. Like the presence of iron-bearing secondary minerals, aluminum-bearing minerals also can be regarded as good indicators of pH conditions and various concentration of Al^{3+}. Based on the possible abundant distribution of kaolin, alunite, and gibbsite minerals and their indicative nature, in addition to their apparent spectral features, these three minerals have been selected as good indicators of the potential presence of Al^{3+} toxicity in ASSs.

From the formation and evolvement of ASSs, we can see that the harmfulness of this kind of soil mainly contains three aspects: the first aspect is the acidity produced from the oxidation of pyrite, which primarily is responsible for the degradation of soil structure and consequently reduces productivity of crops and vigor of vegetation in ASSs. Moreover, it is also a contributing factor and leading cause of the increased release of trace metals with decreasing pH values. Of the heavy metals, aluminum plays the main negative role in damaging aquatic systems, killing fishes, crustaceans, oysters, and plants, and even polluting human water resources in most coastal types of ASSs, whereas pollutants such as arsenic, copper, lead, and mercury are also observed in some acid mining drainage (AMD). Their negative effects on the environment also include salinity.

A good understating and location of ASS are vitally important for subsequent land reclamation and management. Conventional methods, such as intensive field investigation and followed laboratory analysis, have been proved to be time consuming and expensive in other places and other similar applications. Therefore, there is an urgent need to explore new ways to monitor and map the spatial extent and severity of these harmful soils. With the advancement of remote sensing technologies, such as hyperspectral sensing, the identification of subtle spectral features and improved pixel resolutions of sensors make it possible to transcend previous limitations and effectively map the spread of ASSs and monitor their impacts on the environment. Numerous studies have been conducted on monitoring and mapping AMD, which is similar in formation mechanism to an ASS. Alpers et al. (1994), Bigham (1994), Crowley et al. (2003), Fitzpatrick (2003), and Montero et al. (2005) have studied the relationship between iron-bearing secondary minerals and pH values in acid mine drainage; Crowley et al. (2003) and Montero et al. (2005) conducted research into the spectral characteristics of main iron-bearing secondary minerals in AMD areas; Blowes (1997), Swayze and Smith (2000), and Lau et al. (2008) adopted different sensors and mapping methods in various applications related to AMD; and most of these works mainly focused on characterizing the reflectance spectra of associated iron-bearing minerals and tried to suggest the extent and severity of AMD via the identification and mapping of these iron-bearing minerals. However, much less research has been done on the environment affected by ASSs. Lau (2008) mapped associated iron-bearing secondary minerals in South Yunderup, Western Australia, and created a risk map of ASSs based on the distribution of ferric iron contents, carbonate buffering capacity, and comparison of non-ASSs. Based on the work of Lau (2008), this study will further the works on the mapping abundance of specific iron-bearing minerals or mineral mixtures rather than just mapping ferric iron content and meanwhile attach more importance on mapping aluminum-bearing minerals, which has been neglected in previous studies.

It is a clear fact that all environmental issues involved in ASSs are directly or indirectly related to acidity, the more acidic the soil, the worse the environmental condition would be; therefore, pH values become the primary environmental index in these areas, whereas toxicity from the high concentration of aluminum is more harmful to the environment than acidity, although it is derived from acidity. Thus, this study focuses on monitoring both pH and aluminum toxicity circumstance on Swan Coastal Plain using hyperspectral sensing and aims to explore a new, effective way of mapping the extent and severity of this kind of soil in the environment. There are three specific objectives in this study: (1) to characterize the spectral features of minerals related to ASSs, (2) to map the distribution of these related minerals, and (3) to subsequently deduce maps of soil acidity and toxicity from the minerals map and the relationship between minerals and soil properties.

26.2 Setting of Study Area

The Swan Coastal Plain lies to the west of the Darling Fault and extends westward to the Indian Ocean. It is mainly deposited by river in the east and wind in the west. Eolian deposits formed a three-dune system, namely, the Bassendean Dune System, Spearwood Dune System, and Quindalup Dune System, from east to west and also from old to young (Mcarthur and Bettenay 1974; Gozzard 2007; Rivers 2009). Between these dune systems, a series of wetlands, swamps, and lakes spreads on the depressions that occur approximately parallel to the coastline. The Swan Coastal Plain has very thick sands, but most of the concerns related to ASSs in Western Australia are found in materials formed during the Holocene geological period, which was about 10,000–6,000 years ago after the last major sea-level rise (Western Australia Planning Commission 2008). The sediments originating from seawater provide enough resources of sulfate and iron to form iron sulfides, which are the main cause of the ASSs. South Yunderup (Figure 26.1), which is in the southwestern area of the Swan Coastal Plain, was selected as a typical example of coastal ASSs in the region. It lies in the area of the Peel Inlet, which is 80 km south of Perth (the capital of Western Australia). Two main rivers flow into the Peel Inlet: one is the Serpentine River, which comes from the north, and the other is the Murray River from the east, which travels through the Pinjarra Plain and Guildford formation and forms a large delta composed of levees, lakes, and abandoned channels (Degens 2009). From the topography map of the Swan Coastal Plain area, most areas in South Yunderup are below

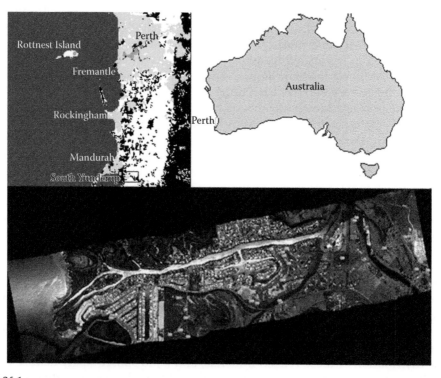

FIGURE 26.1
(See color insert.) Range of study area: red points are the sampling sites. (Modified from Lau et al. (2008).)

5 m Australian Height Datum (AHD), varying between −6 and 9 m, which means that most sediments or soils in this area are possibly related to ASSs (Degens 2009).

26.3 Data Acquisition and Laboratory Measurement

26.3.1 HyMap Data

The hyperspectral HyMap imagery was from the Commonwealth Scientific and Industrial Research Organisation (CSIRO) and acquired over South Yunderup at 12:45 pm on December 1, 2005, on a bearing of 258° and at an altitude of 1447 m ASL. This generated a pixel size of approximately 2.9 m at nadir for a swath width of 1.7 km and a flight-line length of 7 km (Lau et al. 2008). The HyMap sensor is an airborne whiskbroom sensor developed by Integrated Spectronics Pty, Ltd., Australia, providing 128 contiguous bands in the range from 450 to 2500 nm with a spectral resolution of 20 nm, pixel resolutions of 3–20 m depending on the operating height of the aircraft but usually about 5 m, and a signal-to-noise ratio greater than 400:1.

26.3.2 Soil Samples Collection

To understand seasonal environmental changes, we designed our experiments to seasonally collect samples from the study area. Sixty-one samples were collected four times from 10 different sites, of which 9 samples were collected on October 20, 2011; 11 samples on November 22, 2011; 30 samples on February 16, 2012; and 11 samples on May 24, 2012. Several field datasets from the study by Lau et al. (2008) were also used as reference. To obtain samples that were more representative of those observed by the airborne sensor, we collected the sediments that were within 5 mm thickness of the surface, but subsurface samples were also collected for the purpose of comparing the properties and mineralogy of the soil in depth. At each site, four to five samples were collected, two of which were chosen by locating materials with contrasting color, possibly representing indicative minerals, while the other two were chosen randomly. Each sample was 10 × 5 cm square and was collected carefully to keep the surface intact when transporting back to the laboratory for further measurement. Except 11 samples collected on May 24,2012, others were just measured for pH and Eh. Each sample in the other 50 was divided into two parts: one was kept intact and preserved in a sealed container at −5°C, whereas the other was dried in an oven and then milled to powder. The powder was further divided into three parts: one was used for x-ray diffraction (XRD) analysis, one was used for pH and Eh measurement, and the last was measured with an Analytical Spectral Devices (ASD) FieldSpec 3 spectrometer together with the preserved, undried samples.

26.3.3 Proximal Hyperspectral Data

In remote sensing, changes in the environment generally are found by comparing a series of multitemporal images, but in reality it usually is hard to acquire the needed images due to either the cost or the difficulty in accessing data at a suitable time. It is particularly difficult to get the data using an airborne sensor in a routine. Thus, portable spectrometers are a good alternative, allowing the identification of objects and detection of changes by comparing known reference spectra with spectra collected from target objects. In this study, an ASD FieldSpec 3 was used to collect spectra from soil samples in the laboratory. The ASD FieldSpec 3 covers a spectral range from 350 to 2500 nm, which is very beneficial as it closely matches the spectrum from HyMap, making it an ideal tool for ground truth. HyMap has 128 bands, whereas ASD data have more than 2000 bands; therefore, it is necessary to resample the ASD spectral data to the resolution of HyMap. It is noteworthy that the resampling will result in the loss of some spectral information. Four or five spectra were acquired by scanning each sample, which were then averaged to a representative spectrum.

26.3.4 Soil Properties Measurement

Soil properties such as pH and Eh values of samples have been measured in the laboratory by a TPS WP-80D dual pH-mV meter, which provides a pH resolution of 0.01 pH, millivolt resolutions of 0.15 and 1 mV, and a temperature resolution of 0.1°C. This study adopted the method of mixing soil and water with a ratio of 1:5 by weight. After mixing, the container was shaken for about 2–3 minutes to keep the soil in solution and then allowed to rest for 2 minutes. After that, the pH value was measured by putting the pH meter into the container at water above the soil for approximately 2 minutes until the values stabilized. Before using the pH meter, a calibration of the pH sensor was performed using a buffer solution of pH 7.

26.3.5 Mineralogy Verification

Samples selected for XRD analysis were dried and milled to powder, whereas for scanning electron microscope (SEM) analysis samples were kept undamaged to maintain the soil structure and grain size. XRD scanning was undertaken by a Bruker D8 Advance at the Centre for Materials Research at Curtin University, Western Australia, and SEM was done by an Eiss Evo 40XVP at the same research center. The mineral composition was identified by the analysis of the intensity and the position of material phase. Mineralogical identification was used to choose suitable end-members prior to mineral mapping and image classification and to verify the results of mineral mapping and classification. Figure 26.2 shows an example of a sample, including the form of the material at the sampling location, SEM analysis, and the spectra of the materials collected at the sample location.

26.4 Reflectance Spectral Characterization

The mineralogy related to ASSs in the study area mainly consists of pyrite, iron monosulfides, hematite, goethite, ferrihydrite, schwertmannite, jarosite, gypsum, calcite, dolomite, kaolinite, alunite, and gibbsite. To conveniently analyze the spectral features of these minerals, they were divided into two groups: (1) iron-bearing minerals and (2) non-iron-bearing minerals or background minerals (Figure 26.3).

Pyrite is the major source of ASS, but its spectrum displays very low reflectance and has almost no distinctive features in visible and near-infrared (VNIR) and short-wave infrared (SWIR). This makes it very difficult to detect by reflectance spectroscopy, whereas other iron-bearing minerals have diagnostic reflectance spectral features that enable their identification by hyperspectral sensing. The common spectral features of these iron-bearing secondary

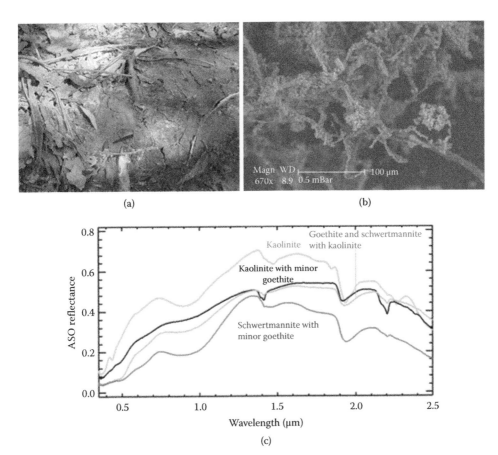

(a)   (b)

(c)

FIGURE 26.2
Jarosite with schwertmannite observed in the study area: (a) picture of the sampling site; (b) Web-like structure of jarosite and schwertmannite pin cushions in SEM image, and (c) spectral measurements of the sample.

minerals are that they all have strong characteristic absorption in the spectral range of 750–1100 nm and most of them have absorption features in the ranges of 550–650 nm and 450–550 nm (Figure 26.3). Absorptions in these three ranges result from crystal field transitions of ferric cations: absorptions in the range of 450–550 nm are caused by the $^6A_{1g}$–$^4A_{1g}$ and ^4E_g transitions, absorptions from 550 to 650 nm are related to the $^6A_{1g}$–$^4T_{2g}$ transition, and absorptions between 750 and 1100 nm are relevant to the $^6A_{1g}$–$^4T_{1g}$ transition (Hunt and Ashley 1979; Burns 1993; Crowley et al. 2003). Within these three ranges, however, different iron-bearing minerals species also have subtle differences in absorption features, which distinguish them from one another. Generally, hematite has a strong absorption characteristic near 870 nm and a strong absorption feature near 550 nm caused by Fe^{3+} electron transition and electric charge transfers (Montero et al. 2005). Goethite possesses a diagnostic broad asymmetric absorption near 940 nm, which was used to differentiate hematite's 870 nm and two other strong absorptions near 480 and 674 nm; all of the three features are related to crystal field transitions (Sherman et al. 1982; Crowley et al. 2003; Montero et al. 2005). Ferrihydrite has a diagnostic absorption feature near 500 and 950 nm, slightly different from hematite and goethite. For ferric hydroxysulfates, schwertmannite has a strong absorption feature near 900 nm and another near 500 nm with a steep edge and jarosite has strong absorption features near 430, 920, 1850, and 2265 nm due to the bond between Fe and OH– and Fe and SO_4^{2-}.

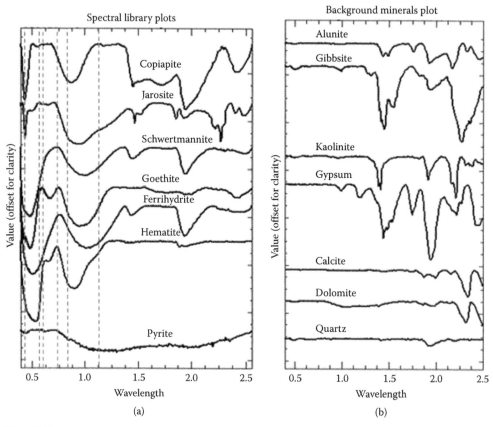

FIGURE 26.3
Reflectance spectra of main minerals related to acid sulfate soils (ASSs). (a) is the spectra of main iron-bearing minerals related to ASS and (b) is the spectra of main background minerals related to ASS. Courtesy of USGS Spectral Library.

For non-iron-bearing minerals such as aluminum-bearing minerals and carbonates, we usually use absorption features in the range of 1300–2500 nm, which are caused by molecular vibration of H_2O, CO_3^{2-}, and OH. The kaolin group, muscovite, and illite display a combination of an Al–OH bend overtone and an OH stretch (Clark et al. 1990; Montero et al. 2005), for example, kaolin group minerals, including kaolinite, dickite, nacrite, and halloysite, display strong absorption features in the range of 2160–2200 nm and less strong absorption features near 2380 nm due to the vibration of the Al–OH molecule (Montero et al. 2005). Gibbsite has a strong diagnostic absorption near 2268 nm, which persists strongly in mixtures, as well as three less strong absorptions near 1452, 1521, and 1549 nm. Alunite has one diagnostic absorption feature in the range of 2160–2170 nm, which may be modified in mixtures; one near 1760 nm, which is similar to gypsum and related to SO_4^{2-}; and a doublet absorption at the ranges of 1440–1436 nm and 1474–1480 nm, which vary with the composition of Na^+ and K^+. Carbonate minerals, including calcite and dolomite, however, have common absorption features near 2320 nm. Gypsum has a strong absorption feature at 1750 nm due to the presence of SO_4^{2-}, in addition to three diagnostic water absorption features. Quartz does not have any diagnostic features in the range of 400–2500 nm, and although it dominates the mineral composition in the sediments and soils around South Yunderup it does not cause ASS issues. Detailed spectral features of these background minerals are illustrated in Figure 26.3.

26.5 Mapping Method

26.5.1 Preprocessing

To remove atmospheric effects, hyperspectral correction (HyCorr), which was developed based on atmosphere removal algorithm (ATREM), was used to correct the at-sensor radiance to apparent ground reflectance. A further correction was applied to the apparent reflectance by using ground spectral measurements collected within the flight area. Anthropogenic objects within imagery play a negative role in the processes of target objects identification, mapping, and interpretation of ASSs; thus, it was found to be necessary to create a mask to exclude these objects. Particularly in this study, roof materials contained iron oxides that are spectrally very similar to iron-bearing minerals in soils or sediments and could be very likely confused with the mapping results of iron-bearing minerals, which we mainly want to achieve, and vegetation, which acts as a barrier to mineral identification; water bodies such as the sea, rivers, lakes, and creeks and roads in this area are also mapped out and combined together to generate a mask that excluded these pixels from further image processing and mapping.

26.5.2 Mapping Methods

The mapping methods involved in this study are mainly about spectral feature fitting (SFF), spectral indices, and linear unmixing methods. SFF is an absorption-feature-based detection algorithm to match image spectra to selected reference spectra using a least-squares technique (Clark et al. 1990; Clark and Swayze 1991). This method requires the data to be reduced to reflectance and the continuum that always corresponds to a background signal to be removed from the reflectance data prior to further analysis. Multispectral feature fitting is an advanced SFF that allows users to define multiple and specific wavelength ranges to extract absorption features for each end-member, and more importantly it also allows users to define optional weights for each spectral range to emphasize the importance of certain features; thus, it is usually believed to produce more accurate results than those produced by simple SFF.

Spectral indices are a method of using ratios of bands to map target objects. They have long been used in remote sensing from the advent of multispectral sensing, and a wide range of spectral indices, such as normalized difference vegetation and leaf area indices for vegetation vigor and chlorophyll content detection, are widely applied in various applications. To date, spectral indices are widely used as an effective mapping method in geology and mineral identification (Hewson and Cudahy 2010). The crux of this method is to find useful bands or band ratios that can reflect the physical and chemical properties of target rocks or minerals. This requires some relevant expertise and experience.

Unmixing is important for the identification of soil mineral composition and the mapping of soil types because soils usually contain several minerals or mineral compounds mixed together in a small area. Complete linear spectral unmixing is a relatively simple subpixel mapping method that is based on the hypothesis that all end-members are mixed in reflectance in each pixel, which can be explained by a linear model that is described by $Y = W_1 \times E_1 + W_2 \times E_2 + W_3 \times E_3 \ldots + W_n \times E_n$, where E_n is the reflectance of the nth end-member, W_n is the reflectance proportion of this end-member accounting for the pixel, and Y is the total reflectance in the pixel. In essence, this technique requires the solving of a set of n linear equations to get the values of W for each end-member in each pixel. The results of linear unmixing include a series of relative abundance images for all end-members. In each image, the pixel values indicate the percentage of reflectance of a certain end-member in a pixel. The problem with this method is that it requires all the end-members in an image to be identified. Matched filtering (MF) is a similar method that does not aim at finding all the end-members in the pixels but only targets specific end-members; thus, it is commonly called partial unmixing. This method aims to maximize the response of the target spectrum and suppress the response of background; thus, MF was selected for mapping iron-bearing minerals.

SFF and MF mapping methods were integrated in ENVI software as tools to specially process hyperspectral data, whereas spectral indices are one of the functions in MMTG A-list, which was developed by CSIRO and could be embedded in ENVI software; thus, ENVI 4.3 and MMTG A-list are the main tools for mapping in this study.

26.6 Results and Discussion

26.6.1 Ferric Iron Content Mapping

A map of ferric iron content abundance was produced by the spectral indices function in MMTG A-list (Figure 26.4). This method uses a depth of 900 nm absorption between 776 and 1050 nm in continuum-removed HyMap imagery using a fitted second-order polynomial. The resultant map shows high abundances of ferric iron in wetlands in the northern central part of the scene, moderate abundances in the vacant housing lots along Murray Waters Boulevard, South Yunderup, and low abundances around Batavia Quay. Forty-eight of the 51 samples collected from 10 different sites were validated by XRD, and it was shown that they contain ferric iron–bearing minerals, including jarosite, goethite, ferrihydrite, hematite, and so on. The result of ferric iron content mapping was used to act as a guide for fieldwork and also as a reference for comparing the result of further specific iron-bearing minerals mapping and classification. The area in the red-dashed quadrangle was selected as a typical area to display the result of flowing minerals mapping and classification.

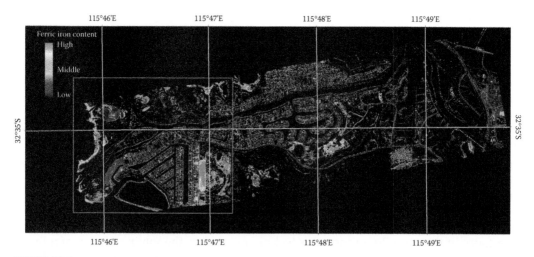

FIGURE 26.4
Map of ferric iron contents generated from HyMap imagery using spectral indices.

26.6.2 Indicative Iron Secondary Minerals Mapping and Classification

26.6.2.1 End-Members Selection

Eight end-members were selected for iron-bearing minerals mapping, five of which were from the United States Geological Survey (USGS) Spectral Library, including jarosite, goethite, ferrihydrite, schwertmannite, and hematite; the remaining three were from samples collected from the study area, and their mineral composition was verified by XRD (Figure 26.5). It is noteworthy that although copiapite is a common mineral related to AMD and was sometimes found in ASSs in other studies, because it is unstable and rarely observed in this study area copiapite was not selected as an end-member. The spectra from the USGS Spectral Library, which have 420 bands, and the spectra collected from the laboratory with an ASD were resampled to 125 HyMap bands. It is noteworthy that some spectral features were lost and some absorption positions were shifted during the process of resampling. For instance, the absorption of jarosite at 2265 nm in the spectra from the USGS Spectral Library was shifted to 2264.5 nm in the HyMap bands and the absorption features near 430, 920, and 1855 nm were lost. The mixture of jarosite and goethite exhibits the spectral superimposition of jarosite and

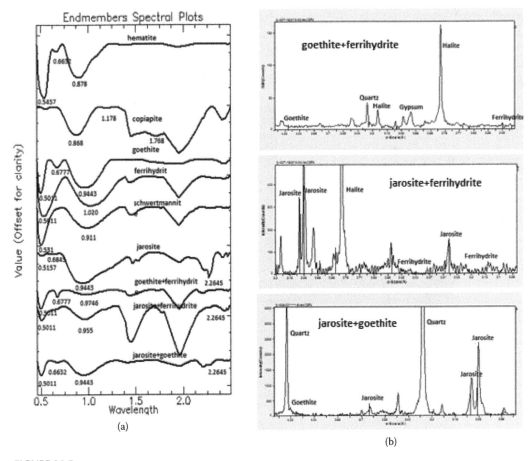

(a)

(b)

FIGURE 26.5
End members for iron bearing mineral mapping: (a) spectral characteristic of end members, among which of mineral mixture are from the study area, and the others are from USGS spectral library and (b) XRD analysis results of the mineral mixtures.

goethite, with absorption features near 501.1, 677, and 944.3 nm like goethite and absorption features at 2264.5 nm like jarosite, but the absorbing depth is less intense than that of pure jarosite. However, for the mixture of jarosite and ferrihydrite, the feature of jarosite is still clearly near 2264.5 nm and the mixture shows less depth than pure jarosite, but the absorption features of ferrihydrite are not apparent. However, goethite dominates the reflectance in the mixture of goethite and ferrihydrite in the iron absorption range. It should be noted that although the mineral mixtures selected for end-members contain abundant quartz and halite (XRD result in Figure 26.5), the reflectance spectra do not show any diagnostic features of quartz and halite , and the presence of quartz and halite does not influence the spectral features of iron-bearing minerals, thus these mineral mixtures can be selected as end-members.

26.6.2.2 Mapping and Classification

MF was used to map the relative abundance of each end-member, and the results of matching were put into a classification. The results of MF unmixing contain several MF score images for all end-members, and each score image suggests the percentage of that end-member in each pixel; thus, every pixel in a scene is relevant to several values that represent different proportions of different end-members. The classified image is pixel based, which means that one pixel can only have one value. Thus, to produce a class map we need to solve the problem of which end-member can represent a certain pixel in the scene. Undoubtedly the end-member that has the largest value in a certain pixel in MF score images should become the representative of that pixel. Thus, each pixel is classified according to its largest MF score value. The classification result is shown in Figure 26.6.

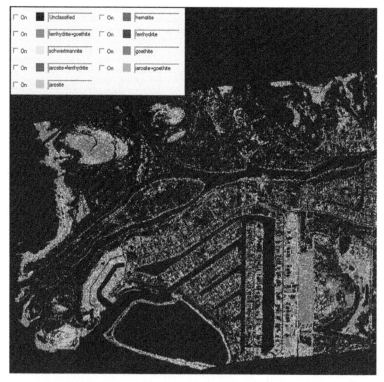

FIGURE 26.6
(**See color insert.**) Image classification of iron-bearing minerals.

26.6.2.3 Iron-Bearing Minerals Distribution and Space Pattern

The minerals classification map in Figure 26.6 shows that iron-bearing minerals occurred widely throughout the study area. The most striking feature of these minerals' distribution is that they extend linearly along watercourses and roads. This can be explained by the mechanism of formation of these minerals. First, the soils along the banks of watercourses are mainly from the sediments excavated from adjacent wetlands and constitute a high abundance of pyrite or MBO. Once these sediments were excavated from inundated reducing conditions and deposited along the banks, they were exposed to the atmosphere; they oxidized rapidly and generated iron-bearing minerals in different landform and drainage conditions. In addition, in the dry season the sediments in the riverbed were exposed to the air and subsequently iron-bearing minerals were produced parallel to the river centerline. The other interesting phenomenon is that these minerals concentrated on some excavated sediments piles, dried water pools, and uncovered soil piles from building construction.

26.6.2.4 Soil Acidity Map Deducted from Iron-Bearing Minerals Map

The relationship between pH values and indicative minerals is summarized in Table 26.1. The results are consistent with the findings of previous studies. It is important to note that jarosite was not found in any of the samples alone, but the presence of jarosite mixtures with goethite or ferrihydrite was frequently found in the samples. From the consideration of the link between minerals and pH values from previous studies and from the samples collected at our study site, we adopted the relationships as shown in Table 26.1 to deduce the acidity map (Figure 26.7). During the process of acidity mapping, the pH range <1.5 and range of 1.5–2.8 were merged into <2.8 and the range of 2.8–4.5 and range of 2.9–4.5 were merged into 2.8–4.5. The mineral compositions and pH values of most samples were consistent with the result of the classifications. But it is noted that some factors influence the effectiveness of verification, including whether we can strictly match a pixel in a classification image to the site of the sample collected, the environment alteration over time, and the method to assess the result of subpixel unmixing.

TABLE 26.1

Relationship between Mineral Species and pH Values

Minerals Verified by XRD in Samples	Sample Amount	pH Average	pH Range in Previous Studies	Reference	Applied pH Range
Jarosite	0		<2.8	Crowley et al. (2003)	<2.8
Schwertmannite	4	3.30	2.8–4.5	Crowley et al. (2003)	2.8–4.5
Ferrihydrite	6	6.7	>5	Montero et al. (2005)	6–7
Goethite	4	5.08	<6	Bigham (1994)	4.5–6
Jarosite + ferrihydrite	8	4.26			
Jarosite + goethite	6	4.28	2.9–3.5	Swayze and Smith (2000)	2.8–4.5
Goethite + ferrihydrite	3	6.38			6–7
Hematite	2	7.5	7–8	Bigham (1994)	7–8
Other iron minerals	10	5.05			

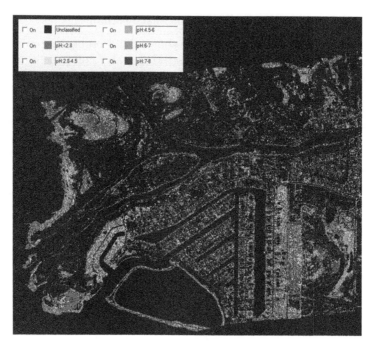

FIGURE 26.7
(See color insert.) Map of soil acidity deduced from the distribution of iron-bearing minerals.

26.6.3 Aluminum-Bearing Minerals, Carbonates, and Sulfates Mapping and Classification

26.6.3.1 Mapping and Classification Method

The end-members of background minerals (or non-iron-bearing minerals) were from the USGS Spectral Library, including kaolinite, gibbsite, alunite, gypsum, dolomite, and calcite (Figure 26.2). The spectral indices method was used to map the abundance of kaolin group minerals (including kaolinite, dickite, nacrite, and halloysite), carbonate (including dolomite and calcite), and gypsum by using the absorption features at 2200, 2320, and 1750 nm, respectively. However, the multirange SFF method was used to map the abundance of gibbsite and alunite. For gibbsite the spectral range focused on the vicinity of 2268 nm and the other three less important ranges near 1452, 1521, and 1549 nm, whereas for alunite spectral ranges were defined in the range of 2160–2170 nm and a range near 1760 nm. Afterward, each end-member's regions of interest (ROIs) were extracted by a two-dimensional scatterplot by choosing pixels with a higher matched score and a lower error, and then the selected ROIs were input into the classification using a method of image classification from ROIs. The result of classification of non-iron-bearing minerals is illustrated in Figure 26.8.

26.6.3.2 Non-Iron-Bearing Minerals Distribution and Space Pattern

From the result of non-iron-bearing minerals classification map (Figure 26.8), we can see that kaolin group minerals and gypsum are widely distributed in the study area and their distribution is relatively independent of pH conditions; the reasons for kaolin is probably its large mass of accumulation and relatively low chemical reaction rate, but for

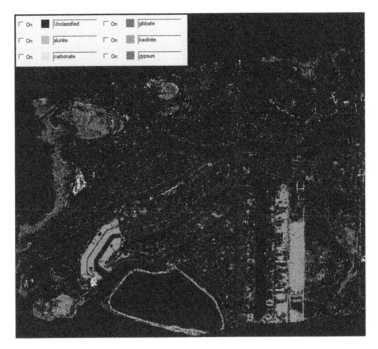

FIGURE 26.8
(See color insert.) Image classification of non-iron-bearing minerals.

gypsum the reason is that gypsum can form in a wide range of pH conditions. Alunite and gibbsite occur in a smaller area than the former two, and furthermore most alunite occurs scattered in small areas where it is in medium to near neutral pH condition; this is consistent with the mechanism of precipitation of alunite, which usually precipitates above pH 5. Gibbsite very rarely appears in areas with pH below 2.8 and is rarely seen in areas of pH below 4, but it is mostly dispersed in areas that tend to have neutral pH. This is consistent with the description given by Kittrick (1977) and Blowes et al. (2003). It is interesting to note that small amounts of carbonates are mapped as occurring in low pH condition and are sometimes distributed linearly along roads or banks of water bodies; this is very likely because the materials of road construction and reinforcing dam walls contain masses of carbonates.

26.6.3.3 Toxicity Map Derived from Aluminum-Bearing Mineral and pH Distribution

Like iron-bearing secondary minerals, aluminum-bearing minerals also can be regarded as good indicators of different pH ranges and various concentrations of Al^{3+}. When pH is more than 5, there is a low abundance of dissolved Al^{3+} in soil or sediments whether aluminum-bearing minerals exist in the environment or not, and thus it would appear to have low toxicity. When pH is less than 4, kaolin minerals and aluminum hydroxide gibbsite will dissolve and thus an environment containing kaolin or gibbsite displays high toxicity, whereas when pH is between 4 and 5 the concentration of Al^{3+} will be at a moderate level. When pH is below 2.8, strong acidity will accelerate the reaction with kaolin to release much more Al^{3+} and almost all the gibbsite will be dissolved to aluminum ions; therefore, this area will become highly toxic. Comprehensively considering both

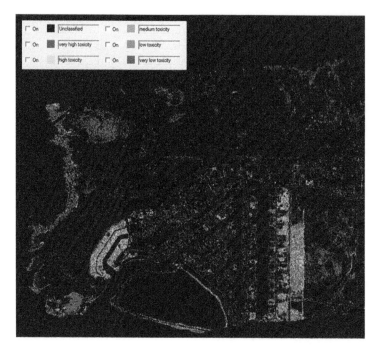

FIGURE 26.9
(See color insert.) Map of soil toxicity deduced from the distribution of aluminum-bearing minerals and pH.

pH conditions and the distribution of aluminum-bearing minerals, five grades of toxicity of the relative concentration of Al^{3+} were proposed: very high toxicity, high toxicity, medium toxicity, low toxicity, and very low toxicity. When intersections between pixels of gibbsite, kaolin, and alunite and pixels in each range of pH presented in the pH classification were operated, the results of gibbsite zone intersecting with the zone of pH < 2.8, kaolin zone intersecting with the zone of pH < 2.8, and alunite zone intersecting with the zone of pH < 2.8 were merged into a very high toxicity zone. Likewise, the intersection results of gibbsite, kaolin, and alunite with a pH zone from 2.8 to 4.5 were merged into a high toxicity zone; the intersection results with a pH zone from 4.5 to 6 were merged into a medium toxicity zone; and the intersection results with a pH zone from 6 to 7 were merged into a low toxicity zone. All the areas with pH values above 7 were defined as very low toxicity zones. The toxicity distribution is illustrated in Figure 26.9.

26.7 Conclusion and Future Work

The harmful aspects of ASS can be categorized by two main causes: one is strong acidity and the other is the consequently high concentration of Al^{3+}. Iron-bearing secondary minerals, such as jarosite, goethite, ferrihydrite, or their mixture, were reaffirmed to have a good link with the pH conditions and can be regarded as good indicators of environmental conditions. The toxicity from aluminum is mainly determined by the dissolution of aluminosilicates and aluminum hydroxides, and the presence of aluminum sulfates usually suggests Al^{3+}-releasing conditions. Thus, kaolin, gibbsite, and alunite were selected to act as indicators of potential toxicity. Soil acidity and soil toxicity maps were created from the relationships between the related minerals and soil properties.

There still remains some work for the future: the first is to establish the relationships among reflectance spectral features, pH values, and concentrations of trace metals such as Al^{3+} and apply them to directly deduce the severity and extent of soil acidity and soil toxicity on soil surface. The second is to use a novel method to detect mineral distributions in different depths under the surface (<1.5 m) by using an automated core scanning technology (such as HyLogger™) to scan soil cores to deduce soil properties under the surface and find possible changes with time. Thus, we have not only airborne and proximal sensors to detect acidity and aluminum toxicity environments on the surface but also the ability to detect soil or sediment properties below the surface. The use of these tools may assist in not only assessing current environmental conditions but also predicting future environmental changes.

26.8 Summary

ASSs can be extremely detrimental to the environment when iron sulfide minerals are exposed to the atmosphere. Iron oxides, iron hydroxides, and iron sulfates, which form from the oxidation of pyrite, were used as indicators of pH, and aluminum-bearing minerals such as kaolin, gibbsite, and alunite were also used as indicative minerals to suggest aluminum-releasing potential in ASSs. These minerals have diagnostic reflectance spectral features in the regions of VNIR and SWIR of the electromagnetic spectrum. Generally, iron oxides have strong absorptions near 900 nm due to crystal field processes and iron sulfates have diagnostic features at 2265 nm, whereas kaolin displays apparent features of Al–OH near 2200 nm. Gibbsite has a strong diagnostic absorption near 2268 nm, and alunite has absorption features in the range of 2160–2170 nm. These absorption features allow the assessment of environmental conditions by identifying and mapping these minerals by hyperspectral sensing. The mechanism of formation of ASSs was reviewed, and associated mineralogy was characterized spectrally and verified by XRD and SEM analysis. Intensive soil sampling has been undertaken on a seasonal basis, with soil property measurements and spectral analysis performed in the laboratory. The relationship between soil pH values and mineral species was revalidated after analyzing datasets from the field and the laboratory. These relationships were applied to deduce the acidity distribution by mapping iron-bearing minerals. A toxicity map of Al^{3+} release from the soil was derived from the distribution of aluminum-bearing minerals and their potential discharging capability in varying pH conditions.

References

Alpers, C. N., D. W. Blowes, D. K. Nordstrom, and J. L. Jambor. 1994. "Secondary Minerals and Acid Mine-Water Chemistry." In *Environmental Geochemistry of Sulfide Mine-Wastes*. J. L. Jambor and D. W. Blowes, eds., Mineral. Assoc. Canada Short Course Handbook, Vol. 22, 247–70.

Anderson, J. E. 1994. "Spectral Characterization of Acid-Mine and Neutral-Drainage Bacterial Precipitates and Their Relationship to Water Quality in a Piedmont Watershed." *Virginia Journal of Science* 45:175–85.

Bigham, J. M. 1994. "Mineralogy of Ochre Deposits Formed by Sulfide Oxidation." In *Environmental Geochemistry of Sulfide Mine-Wastes*. J. L. Jambor and D. W. Blowes, eds., Mineral. Assoc. Canada Short Course Handbook, Vol. 22.

Bigham, J. M., and D. K. Nordstrom. 2000. "Iron and Aluminum Hydroxysulfates from Acid Sulfate Waters." *Reviews in Mineralogy and Geochemistry* 40(1):351–403.

Blowes, D. W. 1997. "The Environmental Effects of Mine Wastes." In Proceedings of Exploration '97, Fourth Decennial International Conference on Mineral Exploration: Prospectors and Developers Association of Canada, Toronto, ON. A. G. Gubins, ed., 887–92.

Blowes, D. W., C. J. Ptacek, J. L. Jambor, and C. G. Weisener. 2003. "The Geochemistry of Acid Mine Drainage." In *Environmental Geochemistry* B. S. Lollar, ed.

Burns, R. G. 1993. "Origin of Electronic Spectral of Minerals in the Visible to Near-Infrared Region." In *Remote Geochemical Analysis: Elemental and Mineralogical Composition*, edited by C. M. Pieters C. M. Pieters and P. A. J. Englert, eds., Vol. 4, 3–29. Cambridge, UK: Cambridge University Press.

Clark, R. N., A. J. Gallagher, and G. A. Swayze. 1990. "Material Absorption Band Depth Mapping of Imaging Spectrometer Data Using the Complete Band Shape Least-Squares Algorithm Simultaneously Fit to Multiple Spectral Features from Multiple Materials." In *Proceedings of the Third Airborne Visible/Infrared Imaging Spectrometer (AVIRIS) Workshop, California*, JPL Publication 90–54,176–86.

Clark, R. N., and G. A. Swayze. 1991. "Mapping with Imaging Spectrometer Data Simultaneously Fit to Multiple Spectral Features from Multiple Materials." In *Proceedings of the Third Airborne Visible/Infrared Imaging Spectrometer (AVIRIS) Workshop*, California, JPL Publication 91–28, 2–3.

Crowley, J. K., D. E. Williams, J. M. Hammarstrom, N. Piatak, and I. M. Chou. 2003. "Spectral Reflectance Properties (0.4–2.5 μm) of Secondary Fe-Oxide, Fe-Hydroxide, and Fe-Sulfate-Hydrate Minerals Associated with Sulfide-Bearing Mine Wastes." *Geochemistry: Exploration, Environment, Analysis* 3:219–28.

Degens, B. P. 2009. "Acid Sulfate Soil Survey of Superficial Geological Sediments Adjacent to the Peel-Harvey Estuary." Department of Environment and Conservation, Government of Western Australia.

Fitzpatrick, R. W. 2003. "Overview of Acid Sulfate Soil Properties, Environmental Hazards, Risk Mapping and Policy Development in Australia." In *Advances in Regolith*, 122–25. Roach I. C., ed., CRC LEME.

Gozzard, J. R. 2007. *Geology and Landform of the Perth Region*. East Perth, WA: Geological Survey of Western Australia.

Harries, J. 1997. *Acid Mine Drainage in Australia: Its Extent and Potential Future Liability*. Supervising scientist report 125. Canberra: Supervising scientist.

Hewson, R. D., and T. J. Cudahy. 2010. "Geological Mapping Accuracy Issues Using ASTER in Australia Applications in ASTER." In *Land Remote Sensing and Global Environmental Change: NASA's Earth Observing System and the Science of ASTER and MODIS*. B. Ramachandran, C. Justice, and M. Abrams, eds., New York: Springer.

Hunt, G. R., and R. P. Ashley. 1979. "Spectral of Altered Rocks in the Visible and Near Infrared." *Economic Geology* 74:1613–29.

Kim, J. J., S. J. Kim, and C. O. Choo. 2003. "Seasonal Change of Mineral Precipitates from the Coal Mine Drainage in the Taebaek Coal Field, South Korea." *Geochemical Journal* 37:109–21.

Kittrick, J. A. 1977. "Mineral Equilibria and the Soil System." In *Minerals in Soil Environments*. J. B. Dixion and S. B. Weed, eds., 1–25. Madison, WI: Soil Science Society of America.

Lau, I. C. 2008. "Report to the DEC on Hyperspectral Imagery of South Yunderup for Mapping the Effects of Acid Sulphate Soils." *CSIRO Exploration and Mining Open File Report No: P2008/1086*, July 2008.

Lau, I. C., R. D. Hewson, C. Ong, and D. J. Tongway. 2008. "Remote Mine Site Rehabilitation Monitoring Using Airborne, Hyperspectral Imaging and Landscape Function Analysis (LFA)." *The International Archives of the Photogrammetry, Remote Sensing and Spatial Information Sciences* XXXVII, Part B7. Beijing.

Mcarthur, W. M., and E. Bettenay. 1974. *The Development and Distribution of the Soils of the Swan Coastal Plain, West Australia.* Soil publication No. 16.

Montero, I. C. S., G. H. Brimhalla, C. N. Alpers, and G. A. Swayzec. 2005. "Characterization of Waste Rock Associated with Acid Drainage at the Penn Mine, California, by Ground-Based Visible to Short-Wave Infrared Reflectance Spectroscopy Assisted by Digital Mapping." *Chemical Geology* 215:453–72.

Rivers, M. 2009. "Overview of the Peel Inlet and Harvey Estuary—Genesis to Water Quality, ARWA Centre for Ecohydrology." *Australian Journal of Earth Science (1996)* 43:31–44.

Sherman, D. M., R. G. Burns, and V. M. Burns. 1982. "Spectral Characteristics of the Iron Oxides with Applications to the Martian Bright Region Mineralogy." *Journal of Geophysical Research* 87: 10169–80.

Swayze, G. A., and K. S. Smith. 2000. "Using Imaging Spectroscopy to Map Acidic Mine Waste." *Environmental Science and Technology* 34:47–54.

West Australia Planning Commission. 2008. *Acid Sulfate Soils Planning Guidelines.* Perth, WA: Western Australia Planing Commission.

Index

A

ABA, *see* Area-based approach
Above-ground biomass (AGB), 273, 274, 402, 408
AC, *see* Agreement coefficient
Accuracy assessment, 82–84, 147
  based on matrix error, 59–60
  change detection, 60–61
  choice of sampling unit for, 60
  for classification and modeling
    change detection maps, 52
    continuous variables maps, 52–54
    error matrix, 46–49
    large-area land cover, 51
  components and issues of
    analysis, 50–51
    response design, 49–50
    sampling design, 49
  criticisms of Kappa, 61
  example of, 66–68
  measures of, 61–62
  pixel, 62–66
  real application
    accuracy measurement of, 73
    image description, 69–73
  results
    from different algorithms, 119–121
    from different datasets, 117–119
  of soft classification maps, 57–58
Acid mining drainage (AMD), 497
Acid sulfate soils (ASSs), 496–497
  aluminum-bearing minerals
    mapping and classification method, 508
    non-iron-bearing minerals, 508–509
    toxicity map derived from, 509–510
  data acquisition and laboratory
    measurement, 499–500
  ferric iron content mapping, 504
  harmful aspects of, 510
  indicative iron secondary minerals,
    see Indicative iron secondary minerals
  mapping method, 503–504
  reflectance spectral characterization, 500–502
  study area, setting of, 498–499
Acquisition satellite images, 16
Across-track scanners, 4–7
Active microwave sensors, 8
Active remote sensing technique, LiDAR, 156

Active sensors, 4, 17
Advanced remote sensing sensors/systems,
  240–242
Advance Very High Resolution Radiometer
  (AVHRR), 13
Aerial photographs, 10, 17, 181, 292–294
AGB, *see* Above-ground biomass
Aggregation index (AI), 213, 216–217
Agreement coefficient (AC), 53–54
AGRI4CAST Action, 324, 325
Agricultural drought monitoring, 318
Agricultural food production, 315
Agricultural remote sensing, 335
  applications, 347
  spectral, spatial, and temporal
    considerations, 337–338
Agricultural sustainability, 316–317
Agro-meteorological models, 326
Agronomic models, 319
Agronomic practices, uniform, 333
AI, *see* Aggregation index
Air- and spaceborne sensors, 8
Airborne color infrared (CIR), 269
Airborne Hyperspectral Scanners
  (HyMap), 239–240, 242
Airborne hyperspectral sensors, 198
Airborne imaging spectroradiometer for
  applications (AISA), 240, 241,
  253, 465–467
Airborne laser measurements, 265
Airborne laser scanning (ALS), 260
  applications of
    forest biomass and disturbance
     monitoring, 273–274
    forest growth and site type
     prediction, 272
    forest management operations
     mapping, 273
    forest stock and yield value prediction,
     274–275
    inventory information value, forest
     management planning, 276
    sampling-based forest inventories, 275–276
    tree and stand variables estimation,
     268–272
  vs. MLS and TLS, 262–265
  principle of, 261
  scanner data, 293

Airborne medium- and small-footprint
waveform LiDARs, 407–408
Airborne multispectral imagery
for agricultural remote sensing, 335
detection of crop stress, 341
Airborne scanners, 5
Airborne sensors, 240–242
Airborne small-footprint discrete-return
LiDAR
profiling system, 405–406
scanning system, 401–405
Airborne Visible/Infrared Imaging
Spectrometer (AVIRIS), 14, 198,
239–242, 339–340
for crop stress detection, 344
integrated with LVIS, 409
Along-track scanners, 4–7
ALS, *see* Airborne laser scanning
Aluminum-bearing minerals
mapping and classification method, 508
non-iron-bearing minerals, 508–509
toxicity map derived from, 509–510
Alunite, 502, 509
AMD, *see* Acid mining drainage
Angular texture signature (ATS), 162–163
ANN, *see* Artificial neural network
Anniversary images, 15
Anomaly vegetation index model, 318
Aquatic ecosystems, use of, 448
ArcGIS10.0 software, 482
Area-based approach (ABA), 262, 266, 267
estimation of timber assortments with, 280
inventory errors, 275
tree and stand variables estimation,
268–269
Area-based LiDAR methods, 402–403
Area-based statistical models, 404–405
Areal units using k-NN weights, 299–302
Artificial neural network (ANN), 142, 196, 248,
249, 251
ASSs, *see* Acid sulfate soils
ASTER sensor, 17
Atlanta, 144, 148
Atmosphere removal algorithm (ATREM), 503
Atmospheric correction, 466
Atmospheric windows, 6
Attenuated waveforms, correcting, 407–408
Automatic calibration model, 160
Average patch shape index, 491
AVHRR, *see* Advance Very High Resolution
Radiometer
AVIRIS, *see* Airborne Visible/Infrared Imaging
Spectrometer

B

Background minerals, spectral features of,
500, 502
Backscattering decomposition, polarimetric
variables derived from, 469, 470
Bacterial leaf blight, 341
Band-specific regression lines, 466
Beer's law, 408
Benchmarking process, 174
forest landscape dynamics, 178, 179
lake data for, 184–185
BEPS, *see* Boreal ecosystem productivity
simulator
Bidirectional reflectance distribution function
(BRDF), 384–385
BigFoot, 424–425
Binaghi approach, 64–67, 73, 77, 80, 82
Binary change detection, 128
Binary nonchange detection, 128
Biogeographical model, 176
Biomass, 423
estimation, 404, 409
mapping
large-footprint satellite LiDAR for, 408
profiling LiDAR for, 406
models, 410–412
remote sensing of, 400–401
Biomass-harvest index models, 319, 322
Biomass Index (BMI), 469
Biome-biogeochemical cycles (Biome-BGC)
model, 424, 428
Black spruce forest, reflectance, 381–382
Boreal ecosystem productivity simulator (BEPS)
models, 424, 428–431
NPP using, 434–436
Brazilian Amazon
LULC classification, *see* Land use/land cover
(LULC) classification
vegetation change detection in,
see Vegetation change detection, in
Brazilian Amazon
Brazilian waterweed, 198
BRDF, *see* Bidirectional reflectance distribution
function
Broadband images, 337
Brovey transformation (BT), 211

C

CAI, *see* Cellulose absorption index
Calibrated MS-NFI estimators, 300
Calibration-based methods, 160

California's mediterranean-type ecosystems, 195
Canonical correlation analysis, 246
Canonical discriminant analysis (CDA), 246, 247
Canopy architecture
  on reflectance, 380–383
  three-dimensional (3D), 376–379
Canopy Chlorophyll Content Index (CCCI), 345
Canopy cover, 404
Canopy gap fraction, 376
Canopy height (CH), 400, 423
Canopy height model (CHM), 265, 267, 403
  TCA, 271–272
Canopy-related metrics, ITD, 270–271
Canopy returns, QMH of, 406
Canopy Structure Index (CSI), 469
Canopy transparency parameters (CTPs), 273
Canopy vertical profile (CHP), 408
Canopy volume, 404
Carbon (C) cycle modeling for terrestrial ecosystems, 421–422
  case studies, 432–436
  features of, 432
  integrating, 424–425
  RS data and, 422–424
Carnegie-Ames-Stanford approach (CASA), 424–426
CART, *see* Classification and regression trees
CASA, *see* Carnegie-Ames-Stanford approach
CASI, *see* Compact Airborne Spectrographic Imager
Catastrophic landscape changes, 173
CB method, *see* Contour-based method
CCCI, *see* Canopy Chlorophyll Content Index
CCDF, *see* Conditional cumulative distribution function
CCDs, *see* Charge-coupled devices
CDA, *see* Canonical discriminant analysis
CDOM, *see* Chromophoric dissolved organic matter
Cellulose absorption index (CAI), 354
CH, *see* Canopy height
Change detection
  accuracy assessment, 60–61
  multitemporal ALS and, 281
  standard error matrix, 52
  techniques, 128
Charge-coupled devices (CCDs), 5–7
China, mapping terrestrial ecosystem of, 434–436
China's Sanjiang Plain, wetland dynamics in, 480
Chlorophyll a (Chl-a), 444, 446, 448
CHM, *see* Canopy height model

CHM-based tree detection, 270
CHP, *see* Canopy vertical profile
CHRIS-PROBA sensor, 381, 384
Chromophoric dissolved organic matter (CDOM), 446–447
Circular local density, 162
Classical classification model, 58
Classification and regression trees (CART), 245, 248–251
Classification methods, 196
Classification scheme, response design, 49–50
Classification tree analysis, 116, 119, 120, 122
Climate, 443
Cloude–Pottier decomposition, 469, 470
Clumping index
  of conifer forests, 382
  indirect measurement of, 378
  map, examples of, 390–391
Clustering methods, 160, 244
Coarse radiometric resolution, 15
Coastal Zone Color Scanner (CZCS), 444
Coefficient of variation (CV), 30, 38
Cokriging methods, 354, 355, 359
Cokriging variance, 34–35
Colored dissolved organic matter (CDOM), 465
Color infrared (CIR) image, 344
Commercial farming, 328
Commercial satellite sensors, 240–242
Commission errors, 46, 270
Commonwealth Scientific and Industrial Research Organisation (CSIRO), 499
Community Land Model (ED) model, 178, 179
Compact Airborne Spectrographic Imager (CASI), 240–242, 354
Compacted soils, 141
Complete linear spectral unmixing, 503
Conditional cumulative distribution function (CCDF), 36, 37, 359
Confidence-based quality assessment, 51
Confusion matrix, *see* Errors, matrix
Connected component analysis groups pixels, 162
Conservation tillage, 353, 370
Continental-scale estimation of terrestrial C sinks, 433
Continuous variables maps, accuracy assessment, 52–54
Contour-based (CB) method, 224
Conversion matrix of LULC types, 482, 483
Cost-efficient sampling design strategy, 23, 32, 39
Cost-plus-loss approach, 276
Crisp classification model, 63, 83
Crop Acreage and Production Estimation (CAPE) project, 325–326

Crop acreage estimation, 318–319
Crop condition monitoring, 317–318
Crop cultivation, 316
Crop growing model, 317–318
Crop-info parameters, 317
Cropland Data Layer, 319
Crop leaf water potential, 342, 343
Crop monitoring systems, 326–328
Crop phenophase monitoring, 319–320
Crop production monitoring, 316
Crop production system, 336
Crop residue cover, 353
    methods for mapping, 354
    QuickBird images with, 360–362
    regression models of, 354, 367–369
    sequential Gaussian cosimulation for
        mapping, 358–360, 363–367
    study area and datasets, 355–356
    transformations of QuickBird images,
        356–357
Crop stress detection, 341, 344–345
CropWatch system, 320–323
Crop Water Stress Index (CWSI), 318, 342, 343
Crop yield estimation, 319
Cross-correlogram spectral matching
    technique, 248
Cross-polarized systems, 468
Cross-tabulation matrix for accuracy
    assessment, 50
Cross-validation accuracy, 51
Cross-variogram, 354, 358, 359
CSIRO, *see* Commonwealth Scientific and
    Industrial Research Organisation
CTPs, *see* Canopy transparency parameters
CV, *see* Coefficient of variation
CWSI, *see* Crop Water Stress Index
CZCS, *see* Coastal Zone Color Scanner

D

Damaged crown projection area (DCPA), 274
Data acquisition
    and laboratory measurement, 499–501
    sensors, 262
Data collection
    LULC classification, 113–114
    and preprocessing, 129–130
Data fusion, 114–115
Data-specific regression models, 268–269
Data transformation techniques, 246–248
DBH, *see* Diameter at breast height
DCD, *see* Dirección de Coordinación de
    Delegaciones

DCPA, *see* Damaged crown projection area
Decision trees, 250
Deep water regions, masking, 466
Definiens eCognition software, 471
Defuzzyfication process, 73
Delaunay triangulated irregular network, 162
ΔCHM method, 274
DEMs, *see* Digital elevation models
Department of Agriculture & Cooperation
    (DAC), 325
Derivative analysis, 339
Design-based approach, 299
Design-based sampling design strategy, 24–25,
    27, 33–34
Detailed "from–to" vegetation change
    trajectories detection, 128, 133
Detailed vegetation change trajectories,
    135–138
DETER, *see* Real Time Deforestation
    Monitoring System
Deterministic approaches, 98
Diameter at breast height (DBH), 91, 267, 271,
    278–279
Different classification algorithms, comparison
    of, 116–117
Digital elevation models (DEMs), 89, 161, 265,
    470, 473
Digital image analysis methods, 18
Digital image processing
    core of, 18
    techniques, 15
Digital surface model (DSM), 265
Digital terrain model (DTM), 265
Dilation, 161
Dimension reduction methods, 246, 335–336
Dirección de Coordinación de Delegaciones
    (DCD), 326
Direct measurement of LAI, 376–379
Direct monitoring model, 317
Discrete wavelet transforms (DWT), 248
Discriminant analysis, 293
Disease detection, precision farming
    applications, 341–342
Diuraphis noxia, 341
Dominant forest types, 177
Dominant invasive plant species, 195
Double-bounce mechanism, 468
Downy brome, maps of, 198, 199
Drought monitoring, 322
DSM, *see* Digital surface model
DTM, *see* Digital terrain model
DWT, *see* Discrete wavelet transforms
Dyadic wavelet decomposition, 226–227, 232

E

Early warning system, 316
Earth observation (EO), 444, 446, 448
Earth Observation for Sustainable
  Development program, 51
Earth System Models, 176, 184–185
Ecological process, 173
Ecosystem demography (ED) model,
  178–180
Ecosystem respiration (RE), 429
Edge-based segmentation, 244
Edge-detection methods, 224–229
Edge probability, 227, 232, 233
Eelgrass beds, 465
Electromagnetic energy, 11
Electromagnetic radiation, 3
Electro-optical sensors, 4–6, 11
El Niño–Southern Oscillation (ENSO), 448
Empirical model, 32–33
End-members, 142–143
  fraction images
    derived from LSMA, 146
    derived from multilayer perceptron, 149
  LSMA, 130
  scatterplots of, 145
  selection, 505–506
  SMA, 243
Enhanced TM Plus (ETM+), 353, 354
ENVI software, 504
EO, *see* Earth observation
ERDAS Imagine 8.4, 355
ERDAS IMAGINE 9.3 software package, 481
Error-propagation algorithm, 251
Errors
  estimation, 25
    current and potential methods, 302–306
    model-based error estimation at pixel
      level, 306–308
  matrix
    accuracy assessment based on, 59
    among different datasets, 120
    hypothetical example of, 46
    Kappa analysis, 47–48
    LULC classification, 117, 118
    in mathematical terms, 47
    normalized accuracy, 48–49
    for soft classification map, 77, 79
    of TM image classification result,
      135, 136
  sources of, 88–92
  and uncertainty budgets, 93
ETM+, *see* Enhanced TM Plus

European Space Agency (ESA), 18, 445, 470
Evaluation protocol, response design, 49

F

FACET model, 175
fAPAR, *see* Fraction of absorbed
  photosynthetically active radiation
FAST method, *see* Fourier amplitude sensitivity
  test method
Feature extraction
  dimension reduction methods, 336
  hyperspectral transformation and, 246–248
Feature selection, dimension reduction
  methods, 335–336
Ferric iron content mapping, 504
Ferrihydrite, 501
FI, *see* Forest inventory
Field data, 471
Field trees, 271–272
Filtering, 157
Final field plot weights, estimation of, 301
Finland
  forest inventory, *see* Forest inventory
  k-NN estimation parameters in, 298
  MS-NFI of, 305
  post-stratified estimation in, 302
Finnish multisource national forest inventory
  (MS-NFI), 293–295, 297, 306
FIRE-BGC model, 175
Flood detection beneath trees, 469
Fluorometry, 444–445
Foliage clumping, 378, 382, 391
Forest-BGC model, 428
Forest biomass, estimation of, 91
Forest canopy, 261, 262, 273
Forest canopy reflectance and transmittance
  (FRT) model, 180
Forest carbon
  generation of, 92
  uncertainties of, 94
Forest disturbance, 16
Forest ecosystem process models, 175
Forest hazards, risk of, 274
Forest inventory (FI)
  ABA, 266, 267
  acceleration of, 281
  ALS, MLS and TLS comparison, 262–265
  ITD, 267–268
  laser scanning, 260–262
  NFIs, 260
  point cloud metrics and surface models,
    265–266

Forest landscape dynamics, 175–176
    model-data integration, 178, 180–181
    model initialization, 176–178
    model validation/benchmarking, 178, 179
Forest management operations mapping, 273
Forest management planning, 260
    acceleration of, 281
    ALS inventory information value, 276
    inventories, in Scandinavia, 269
Forest resource information, 260
Forests
    biomass and disturbance monitoring, 273–274
    growth and site type prediction, 272
    mapping and monitoring, 260, 280
    stock and yield value prediction, 274–275
Forest stand age, 424
Forest stands, 223–224
Forest stock, prediction of, 274–275
Forest variables, prediction of, 266, 268, 269
Forest yield value, prediction of, 274–275
Fourier amplitude sensitivity test (FAST)
    method, 93, 94
Fractal analysis, advantage of, 208
Fractal shape index, 213, 218–219
Fraction images
    development, with LSMA, 130
    Landsat images, comparison of, 133, 134
    LSMA
        end-members derived from, 146
        impervious surface derived from, 147
    multilayer perception
        end-members derived from, 149
        impervious surface derived from, 150
Fraction of absorbed photosynthetically active
    radiation (fAPAR), 425
Fragstats 3.3 landscape modeling software, 482
Framework classification, road identification
    using LiDAR data, 158–160
Free air CO_2 experiment (FACE), 174
Freeman–Durden decomposition, 469
Fundamental land cover types, 159
Fungal pathogen infection, 341–342
Fusion of LiDAR, 409–410
Fuzzy ARTMAP, 116, 119–122
Fuzzy classification, 143
Fuzzy error matrix, 64, 67
Fuzzy logic classifier, 160

G

GAP model, 175
Gaussian cosimulation algorithm, 36, 97, 98
Gaussian decomposition, 408

GCPs, *see* Ground control points
Generalized Inherent Optical Property (GIOP)
    algorithm, 446, 447
Genie Pro algorithm, 182
GeoEye-1, 240, 241
Geographic information system (GIS), 18, 88–90
Geometric correction procedures, 466
Geometric errors, 90
Geometric processing for NFIs, 295
GeoSafras, 325
Geoscience Laser Altimeter System (GLAS), 408
Geospatial modeling, 52–54
Geostatistical methods, 340, 354
Gibbsite, 502, 509
GIEWS, *see* Global information and early
    warning system
GIOP algorithm, *see* Generalized Inherent
    Optical Property algorithm
GIS, *see* Geographic information system
GLAS, *see* Geoscience Laser Altimeter System
Glint correction, 466
GLOBal Biophysical Products for Terrestrial
    (CARBON) LAI algorithm, 390
Global climatic changes, 173
Global clumping index map, examples of,
    390–391
Global Environmental Monitoring Index
    (GEMI), 387
Global information and early warning system
    (GIEWS), 323
Global LAI map, 390–391
Global LAI products, 389
Global Lakes and Wetlands Database
    (GLWD), 184
Global positioning system (GPS), 18, 89,
    90, 142, 466
Global Production Efficiency Model
    (GLO-PEM), 426–427
Global-scale modeling, 422
Global terrestrial observation system, 424
Global vegetation models, 176, 180
GlobColour, 445
GLO-PEM, *see* Global Production
    Efficiency Model
Glyphosate-resistant (GR) crops, 346
Goethite, 501
Gómez approach, 65–66, 73, 77, 82
GPS, *see* Global positioning system
Graphical user interfaces (GUI), 322
Grassland, 487
GR crops, *see* Glyphosate-resistant crops
Greater spectral resolution, 17
Green vegetation (GV), 130–132, 135

Grid spacing, 37
Ground-based discrete-return scanning
    LiDAR, 409
Ground-based LAI measurement, theory and
    techniques of, 376–379
Ground control points (GCPs), 337
Ground data, 295
Ground resolution cell, 5, 6
Ground survey–based methods, 353
GV, *see* Green vegetation
GV-index-based method, 137–139
Gypsum, 502

H

Healthy plant leaves, 376, 380
Height metrics, 403, 404
Hematite, 501
Hemispherical photography, techniques of, 379
Herbicide drift detection, precision farming
    applications of, 346
Hierarchical clustering (HC), 196
Hierarchical mapping system, 245
High-performance liquid chromatography
    (HPLC), 444–445
High-resolution (HR) images, 294
High spatial resolution commercial satellite
    imagery, 239
High spatial resolution sensors, 240–242
High temporal resolution, satellite
    hyperspectral image data, 253
Horizontal and horizontal (HH)
    polarization, 468
Horizontal and vertical (HV) polarization, 468
Horvits–Thompson estimator, 27
Hough transform, 163
HPLC, *see* High-performance liquid
    chromatography
HR images, *see* High-resolution images
Hughes phenomenon, 250, 335
HYDICE, 240–242
HyMap data, 198, 499
HyMap imagery, 200, 504
Hyperion, 240–242, 251
    hyperspectral data, 197
    images, 144, 147
Hyperspectral airborne sensors, 465
Hyperspectral data, 195
Hyperspectral images
    for agricultural remote sensing, 335, 337
    nitrogen stress detection, 344
    of plant canopy reflectance, 346
    soil mapping, 340

VIs, 338
    for weed sensing, 345
Hyperspectral remote sensing
    data, SVMs classification of, 250
    imagery to agriculture, 335
Hyperspectral sensors, 14–15, 192, 240–242
Hyperspectral transformation, 246–248

I

Ice, Cloud, and Land Elevation Satellite
    (ICESat), 408
Iceplant, mapping of, 195, 196
ICESat, *see* Ice, Cloud, and Land Elevation
    Satellite
IDL tools, *see* Interactive Developing
    Language tools
IDRISI's image segmentation, 211, 212
IFOV, *see* Instantaneous field of view
IFRS, *see* International Financial Reporting
    Standards
IHS transformation, *see* Intensity–hue–saturation
    transformation
IJI, *see* Interspersion and juxtaposition index
IKONOS, 240–242, 244, 245, 465, 466
Image-aided sequential Gaussian cosimulation
    algorithm, 31, 36
Image objects (IOs), 244–245, 252
Image preprocessing, 481–482
Image processing techniques, 195–198
Image segmentation, 117
    IDRISI, 211, 214
    techniques, 244–245, 252
Imaging radar sensors, 8–9
Impervious surface extraction
    case study, 144–148
    and impact, 141–142
    methods for, 142–144
Improved k-NN method (ik-NN), 297
Increased radiometric resolution, 17
India remote sensing (IRS) images, 208, 210, 216
Indicative iron secondary minerals
    end-members selection, 505–506
    iron-bearing minerals, 507
    mapping and classification, 506
    soil acidity map deducted from iron-bearing
        minerals map, 507–508
Indirect measurement of LAI, 376–379
Individual tree analysis (ITA) methods, 405
Individual tree detection (ITD), 262, 267–268
    tree and stand variables estimation, 270–271
Inherent optical property (IOP) products, 446
Instantaneous field of view (IFOV), 5, 6

Institute of Remote Sensing and Digital Earth
(RADI), 320
Intensity–hue–saturation (IHS)
transformation, 229
Interactive Developing Language (IDL)
tools, 322
Intercepting area in LAI, 375
Interferometric synthetic aperture radar
(InSAR), 400, 422, 470
International Financial Reporting Standards
(IFRS), 274
Interspersion and juxtaposition index (IJI),
212–213, 216–217
Invasion research
image processing and classification
algorithms, 195–198
imaging spectroscopy in, 194
incorporating phenology, advantages of,
198–201
IOP products, *see* Inherent optical property
products
IOs, *see* Image objects
Iron-bearing minerals
distribution and space pattern, 507
image classification of, 506
mapping, end-members for, 505
map, soil acidity map deducted from,
507–508
spectral features of, 500–502
IRS images, *see* India remote sensing images
ISODATA algorithm, 197
ITA methods, *see* Individual tree analysis
methods
ITD, *see* Individual tree detection

J

JABOWA model, 175
Jarosite minerals, 501
Joint Global Ocean Flux Study (JGOFS), 444

K

Kaolin group minerals, 502
Kappa coefficient, 48, 59, 61
Kappa statistic measures, 66
Kernels, 117, 122
KHAT value, error matrix, 48–49
k-most-similar-neighbor (*k*-MSN) imputation
method, 269, 271
k-nearest-neighbor method (k-NN), 116, 120,
122, 293
for areal units estimation using, 299–302

estimation and parameters, 296–299
uncertainty estimation using, 308
Kriging variance, 26–27, 33–35
Kuan filter, 471

L

Labeling protocol, response design, 49
Lag, separation vector, 358
LAI, *see* Leaf area index
Lake landscape dynamics
model initialization, lake data for, 184
model validation/benchmarking, lake data
for, 184–185
remote sensing images, lakes from, 181–184
Lake map, 182
Land application, satellite SAR band
designations for, 468
Land cover, 422
LANDIS model, 177
Land mask, 466
Land restoration, 488
Landsat Enhanced Thematic Mapper Plus
(ETM+) sensor, 424
Landsat images, 181, 182
ETM+, 12, 13
multispectral, 481
thematic mapper, 90, 99–100, 117, 121, 215, 353
BT of, 211
capabilities of, 112
multispectral image, 123
radiometric rectification of, 210
Landscape analysis, 482–484
Landscape indexes, 483–484, 489
Landscape metrics, 207, 208
Landscape models, 175
Landscape pattern dynamics, 488–490
Landscape segmentation, 211, 213, 216
Landscape shape index (LSI), 212–213, 216
Land Surface Water Index (LSWI), 428
Land use/cover area-frame survey
(LUCAS), 325
Land use/land cover (LULC) classification, 61,
83, 208, 209, 216
accuracy assessment results
from different algorithms, 119–121
from different datasets, 117–119
categories, 70, 71, 212
data collection and preprocessing, 113–114
different datasets for, 115–116
distribution of, 135, 137
and evaluation, vegetation change
detection, 132–133

framework of, 112–113
image preprocessing and, 481–482
maps classification, 485, 492
object-based classification for, 60
from remote sensing data, 111–112
Land use/land cover conversion matrix,
484–488
Laplacian of the Gaussian (LOG) operator, 229
Large-area land cover products, accuracy
assessment, 51
Large-footprint satellite LiDAR for biomass
mapping, 408
Large-scale models, 174
Laser-based tree height, ITD, 267
Laser point clouds
ITD, 267–268
tree variables measurement, 266
Laser pulse, forest canopy, 261, 262, 273
Laser scanning (LS)
forest inventory, instruments used for,
263–264
in measuring forests, 260–262
Laser Vegetation Imaging Sensor (LVIS), 407
L-band, 469
LDA, *see* Linear discriminant analysis
Leaf angle distribution of canopy, 377
Leaf area index (LAI), 273, 274, 375, 422–423
angles on reflectance, 384–385
canopy architecture on reflectance, 380–383
ground-based measurement of, 376–379
leaf optical properties, 380
products, global, 389–392
retrieval using remote sensing data
algorithm for, 385–390
principles of, 379–380
vegetation background, 383–384
Leaf optical properties
of LAI retrieval, 380
seasonal variations in, 392
Least-squares technique, 503
LI-COR LAI-2000 Plant Canopy Analyzer,
377–379
Light detection and ranging (LiDAR), 9–10, 90,
91, 260–262
remote sensing, 473
airborne discrete-return profiling,
405–406
airborne medium- and small-footprint
waveform, 407–408
airborne small-footprint discrete-return
scanning, 399–401
area-based methods, 402–405
biomass models, 410–412

fusion of, 409–410
ground-based discrete-return
scanning, 409
ITA methods, 405
satellite large-footprint waveform, 408
of vegetation biomass, 399–401
road extraction
applications in, 156–158
using remote sensing, 158–164
Like-polarized systems, 468
Linear discriminant analysis (LDA), 195, 196,
245, 248
Linear spectral mixture analysis (LSMA),
128, 142, 243
fraction images
development with, 130
of end-members derived from, 146
of impervious surface derived from, 147
Linear spectral unmixing method, 144–146
Line extraction, Radon analysis for, 164
Local maximum (LM)-based methods, 224
Local variability–based sampling design
strategy, 36–39
Local variance method, 30–31
Logistic regression models, 273
LOG operator, *see* Laplacian of the Gaussian
operator
LS, *see* Laser scanning
LSI, *see* Landscape shape index
LSMA, *see* Linear spectral mixture analysis
LSWI, *see* Land Surface Water Index
LUCAS, *see* Land use/cover area-frame survey
LULC classification, *see* Land use/land cover
classification
LVIS, *see* Laser Vegetation Imaging Sensor

M

Machine-learning approaches, 160–161
Macroscale crop condition information, 317
Management zone, 336
Map generation, method for, 97
Mapping accuracy, 197
Mapping camera, 10
Mapping errors, 91
Mapping invasive weed species, 345
Mapping method, ASSs, 503–504
Map production of forest variables, 302, 303
Margfit, 48
Marine ecosystems, 448–449
Marker-controlled watershed segmentation,
231–234
Marker image generation, 231

Markov chain Monte Carlo method, 165
Markov graph approach, 164
Markov model, 358
MARS, *see* Monitoring agricultural resources
Matched filtering (MF), 504, 506
Matérn class model, 307
Mathematical morphology, 161
Maximumlikelihood classification technique, 197
Maximum likelihood classifier (MLC), 112, 117, 118, 120, 248, 466
MBO, *see* Monosulfidic black ooze
Mean CH (MCH), 407, 411, 412
Mean patch dimension, 491
Measurement errors of tree variables, 91, 97
Mechanistic model, 425
MESAM classification approach, *see* Multiple-end-member spectral angle mapper classification approach
MESMA, *see* Multiple end-member spectral mixture analysis
MF, *see* Matched filtering
Micro-meteorological model, 425
Microscale crop condition information, 317
Microwave RS model, 318
Middle infrared (MIR), 380, 385, 388
Military training, 207–208
    methods of, 211–213
    study area and datasets, 208–211
Miller's theorem, 377
Mineral species
    iron-bearing, 501
    and pH values, relationship between, 507
Minimum noise fraction (MNF), 130, 195, 246, 247
MIR, *see* Middle infrared
Misclassification errors, 118, 122
MISR, *see* Multiangle Imaging SpectroRadiometer
Mixed-effects model, 411
Mixed pixels, 142, 318
Mixture tuned matched filtering, 341–342
MLC, *see* Maximum likelihood classifier
MLP, *see* Multilayer perceptron
MLS, *see* Mobile laser scanning
MNDVI, *see* Modified NDVI
MNF, *see* Minimum noise fraction
Mobile laser scanning (MLS)
    vs. ALS and TLS, 262–265
    applications, 279–280
    in Evo study area, 279, 280
MOD17A2, 432
MOD17A3, 432

MOD17 algorithm, *see* MODIS primary production product algorithm
Model-based approaches
    remotely sensed data, 28–29
    sampling design strategy, natural resources, 25–27, 33–34
Model-based error estimation at pixel level, 304, 306–308
Model-data integration, 174–175, 178–181
Modeling errors, 91
Model initialization, 174
    forest landscape dynamics, 176–178
    lake data for, 184
Model validation, 174
    forest landscape dynamics, 178, 179
    lake data for, 184–185
Moderate-Resolution Imaging Spectroradiometer (MODIS), 15, 178, 179
    image, 70, 73, 83
    sensor, 391
Modified NDVI (MNDVI), 387
Modified simple ratio (MSR), 387
MODIS, *see* Moderate Resolution Imaging Spectroradiometer
MODIS BRDF product, 390
MODIS primary production product (MOD17) algorithm, 431–433
Monitoring agricultural resources (MARS), 324–325
Monosulfidic black ooze (MBO), 496
Monte Carlo method, 94, 98
MS-NFI, *see* Multisource national forest inventory
MSR, *see* Modified simple ratio
MSS, *see* Multispectral scanner subsystem
Multiangle Imaging SpectroRadiometer (MISR), 383
Multidate remotely sensed images, 198
Multilayered feed-forward ANN algorithm, 251
Multilayer perceptron (MLP), 143, 149
Multiple-end-member spectral angle mapper (MESAM) classification approach, 248, 250
Multiple end-member spectral mixture analysis (MESMA), 243, 252
Multiple image classifications, 197
Multiscale wavelet decomposition, 233
Multiscan TLS measurements, 277
Multisensor data
    integration of, 114–115
    use of, 253

Multisource methods, NFIs
    input data, 294–295
    k-NN
        estimation and parameters, 296–299
        weights/predictions, estimating areal
            units using, 299–302
    need for, 293
    pixel-level predictions and map
        production, 302
    preprocessing, 295–296
Multisource national forest inventory
    (MS-NFI), 293, 300
Multispectral feature fitting, 503
Multispectral images, 335, 337, 346
Multispectral remote sensing, 341, 344
Multispectral scanner subsystem (MSS), 480
Multispectral scanning systems, 18
Multistep marching algorithm, 164
Multitemporal ALS, 281
Multitemporal Landsat thematic mapper (TM)
    images, 128–129
Multivariate time series model, 33

N

Nairobi Work Program, 317
NASA, 18
NASA-CASA model, 433
NASS, *see* National Agricultural Statistics
    Service
National agricultural monitoring, 325–326
National Agricultural Statistics Service
    (NASS), 319
National Cooperative Soil Survey, 90
National forest inventories (NFIs), 260
    error estimators, 302–306
    history of, 291–293
    multisource methods, *see* Multisource
        methods, NFIs
    spatial input data for, 294–295
National Oceanic and Atmospheric
    Administration (NOAA), 448
Natural resource management, products for,
    87–88
Natural resource maps, 87, 92–93
Natural resources, sampling design strategy
    of, *see* Sampling design, strategy of
    natural resources
Navigation sensors, 262
NDHD, *see* Normalized difference between
    hotspot and darkspot
NDVI, *see* Normalized difference vegetation
    index

NDWI, *see* Normalized Difference Water Index
Nearest-neighbor (NN) methods, 266, 267
Net ecosystem production (NEP), 424
Net primary production (NPP), 424,
    434–436, 447
Network training mechanism, 251
NFIs, *see* National forest inventories
Nile delta shelf, 449–450
Nile phytoplankton bloom, 450, 452
Nitrogen stress detection, precision farming
    applications, 344–345
NN methods, *see* Nearest-neighbor methods
NOAA, *see* National Oceanic and Atmospheric
    Administration
Non-iron-bearing minerals, 500, 502, 508–509
Nonlinear spectral mixture analysis, 243
Nonparametric decision tree learning
    technique, 250–251
Nonpoint source pollution, 141
Nonsite-specific assessment, 62–63, 68
Nonstatistical-based algorithms, 116, 122, 123
Nonthematic errors, response design, 50
Nordic countries, NFIs in, 292
Normalized accuracy, 48–49
Normalized difference between hotspot and
    darkspot (NDHD), 390
Normalized difference vegetation index
    (NDVI), 338, 341–342, 344, 464
    evaluation of simulated, 181
    time series, 320
    transformations of QuickBird images,
        356, 357
Normalized Difference Water Index
    (NDWI), 464
North China Plain, 69, 82, 83
NPP, *see* Net primary production

O

OBIA method, *see* Object-based image analysis
    method
Object-based approaches, 159–160
Object-based classification, 60, 116, 122, 142
Object-based image analysis (OBIA) method,
    244–245
Object-specific resampling method, 30
Ocean Biology Processing Group (OBPG), 445
Ocean color radiometry (OCR), 444, 446
Oceanic ecosystem, 443
OCR, *see* Ocean color radiometry
Office of Global Analysis (OGA) of FAS, 323
Off-nadir imaging for stereoscopic
    acquisitions, 16

OGA, *see* Office of Global Analysis
Omission error, 46
Operational forest management planning, 260
Optical high spatial/spectral resolution
    sensors/systems, 240–242
Optical imagery, 409, 467
Optical remote sensing, 462–465
   application of, 465–467
   sensors/systems, 240–242
Optical satellites, 462
Optical sensor data, 121
Optical spectral satellites, 462
Optical wavelengths, 467
Optimal hyperplane, 117
Optimal spatial resolution for sampling and
    mapping, 30–32
Optimal temporal resolution for sampling and
    mapping, 32–34
Oscillating mirrors, 5
Overall accuracy, error matrix, 46–47

P

PALS, *see* Portable airborne LiDAR system
Panchromatic image, 355
PAR, *see* Photosynthetically active radiation;
    Photosynthetically available radiation
Partial least squares regression (PLSR), 246, 247,
    339, 345
Partial unmixing, 504
Passive microwave sensors, 8–9
Passive sensors, 4, 17
PASSs, *see* Potential acid sulfate soils
Patch shape index, 490–492
PCA, *see* Principal component analysis
PCD approach, *see* Phase-coded disk approach
P/D ratio, *see* Pelagic to demersal ratio
Pearson product–moment correlation
    coefficient, 213, 217, 218
Pelagic to demersal (P/D) ratio, 450, 452
Permafrost degradation, 182
Pest detection, precision farming applications,
    341–342
Phase-coded disk (PCD) approach, 164
pH distribution, toxicity map derived from,
    509–510
Photosynthesis–transpiration coupling
    method, 430
Photosynthetically active radiation (PAR), 427
Photosynthetically available radiation (PAR),
    447–448
Phragmites australis, 197
Physics-based models, 385

Phytoplankton carbon biomass, 443, 444
Pigment composition, 462
Pinus genus, 248
Pixel accuracy assessment measures for soft
    classification map, 62–66
Pixel-based segmentation, 244
Pixel image, 506
Pixel-level predictions of forest variables, 302
Pixel reflectance, 380–381
Pixels, 11
Plant water stress detection, precision farming
    applications, 342–344
Plot expansion factors, 296, 299–300
Plot location errors, 90
PLSR, *see* Partial least squares regression
PnET-II process model, 176
Point- and pixel-based classification, 159
Point-based approaches, 270
Point cloud, 401, 403
Polarimetric decomposition, polarimetric
    variables derived from, 469, 470
PolInSAR, 400
Polynomial regression model, 98
Ponderosa pine forest stand, VHR aerial
    imagery, 225
Pope parameters, 470
Portable airborne LiDAR system (PALS), 406
Positional accuracy, 45
Position error, 89
Postclassification algorithm, 481
Post-stratified estimation of NFI, 301–302
Potential acid sulfate soils (PASSs), 496
Potential-stress models, 319
Power models, 404
PPARR3, *see* Primary production algorithm
    round robin
Precision agriculture, 333–335
   applications
     crop stress detection, 344–345
     herbicide drift detection, 346
     pest and disease detection, 341–342
     plant water stress detection, 342–344
     soil mapping, 340–341
     weed detection and mapping, 345–346
   future direction of, 346–347
   management of, 336
   remote sensing data in, 335–336
   spectral, spatial, and temporal
     considerations, 337–338
   VIs, 338–340
Predicted ecosystem status, 175
Prediction residual error sum of squares
    (PRESS) statistic, 246

Preprocessing, field data collection and, 113–114

PRESS statistic, *see* Prediction residual error sum of squares statistic

Primary production algorithm round robin (PPARR3), 448

Principal component analysis (PCA), 211, 246, 247

Probability maps of crop residue cover, 363–367, 370

Probability sampling design, 51

PRODES, *see* Program for the Estimation of Deforestation in the Brazilian Amazon

PROD operator, *see* Product operator

Producer's accuracy, 47

Product operator (PROD), 65

Program for the Estimation of Deforestation in the Brazilian Amazon (PRODES), 127

Pulse scanners, 262

Pushbroom scanners, 6

Pyrite, 500

Q

Qinghai Lake, 480, 481, 485, 488

Quadratic mean height (QMH), 406, 412

Quality assessment of natural resource maps, 93

Quartz, 502

QuickBird, 240–242, 244, 245, 251
  data, 197
  images, 113, 336, 341, 344
    with crop residue cover, 360–362
    of multispectral bands, 355, 356
    transformation of, 356–357

R

Radar altimeters, 9

Radar antenna, 468

Radar data, 114, 118, 121–123

Radar, fusion with, 410

Radar interferometry, 470

Radar polarimetry, 469

Radiation interception
  concave leaf surface, 375
  curled broadleaf area for, 376
  and sunlit leaf, 391

Radiative transfer models, LAI algorithms based on, 388–390

Radiometers, 8

Radiometric characteristics, treetop detection on, 230

Radiometric processing for NFIs, 295, 296

Radiometric resolution, 15

Radon analysis for line extraction, 164

Radon transform, 163–164

Ramsar Convention, 480

Random forest (RF), 161, 269

Random process, spatial variability of, 358

Random sampling, simple, 34

Range and Training Land Assessment (RTLA) program, 209

Range-image clustering method, 278

Rayleigh probability density function, 227

RDVI, *see* Renormalized Difference Vegetation Index

Real Time Deforestation Monitoring System (DETER), 127

REDD, *see* Reducing Emissions from Deforestation and Forest Degradation

Reduced simple ratio (RSR) for LAI retrieval, 387, 388

Reducing Emissions from Deforestation and Forest Degradation (REDD), 273

Reflectance
  canopy architecture on, 380–383
  illumination and observation angles on, 384–385
  spectra, 465
  spectral characterization, 500–502
  threshold, 466
  vegetation background on, 383–384

Regime shifts, concept of, 449

Regional variation in CDOM, 447

Region-based segmentation, 244

Regions of interest (ROIs), 508

Regression analysis, 293

Regression-based methods, 94

Regression-kriging method, 341

Regression models, 98, 269, 354, 360
  of crop residue cover, 367–369

Relative variance contribution (RVC), 98–100

Remotely sensed data, 89, 127–128, 192
  characteristics of
    photographs, 10
    radiometric resolution, 15
    satellite images, 10–11
    spatial resolution, 11–12
    spectral resolution, 12–15
    temporal resolution, 15–17
  natural resources using, 87–88
  sampling design strategy, role of, 28–29
  variables selection from, 114–116

Remote sensing (RS), 3–4, 174–175
  applications, 17–18, 316–317
    advances in, 317–320
    algorithms and methods, 444–448

crop monitoring systems, 326–328
CropWatch system, 320–323
fisheries, 448–454
GIEWS, 323
global and national operational
systems, 320
MARS, 324–325
national agricultural monitoring,
325–326
for sampling design, *see* Sampling design,
strategy of natural resources
USDA FAS, 323–324
of biomass, 400–401
data
C cycle models and, 432
cloud detection in, 296
integrating, 424–425
electro-optical sensors, 4–6
images, 177, 181–184
light detection and ranging, 9–10
methods, 463
for forests, 294
LiDAR, 473
optical remote sensing, 462, 464–467
SAR, 467–473
passive microwave and imaging radar
sensors, 8–9
potential of, 260
satellite, 239
systems/sensors, 239
technology, 14
thermal infrared sensors, 6–8
of vegetation biomass using LiDAR
systems, 401
Remote sensing–based approach, 400
Remote sensing–based mapping systems,
93, 95, 96
Remote sensing data in precision agriculture,
335–336
applications
crop stress detection, 344–345
herbicide drift detection, 346
pest and disease detection, 341–342
plant water stress detection, 342–344
soil mapping, 340–341
weed detection and mapping, 345–346
spectral, spatial, and temporal
considerations, 337
VIs, 338–340
Remote sensing data of LAI retrieval
algorithm, 385–390
principles, 379
angles on reflectance, 384–385

canopy architecture on reflectance,
380–383
leaf optical properties, 380
vegetation background, 383–384
Remote sensors, *see* Remote sensing
Renormalized Difference Vegetation Index
(RDVI), 387
Repeat cycle, temporal resolution, 15
Response design, accuracy assessment
component, 49–50
Resultant backscattered radiation, 469
Revised Universal Soil Loss Equation (RUSLE)
model, 90, 99, 370
RF, *see* Random Forest
RMSD, *see* Root-mean-square deviation
RMSEs, *see* Root-mean-square errors
Road centerline extraction, 163–164
Road classification refinement, 162–163
Road clusters, 158–161
Road extraction, 155–156
LiDAR applications in, 156–158
using LiDAR remote sensing
road clusters, 158–161
road networks, 162–164
Road identification, algorithms used for,
160–161
Road networks, 162–164
ROIs, *see* Regions of interest
Root-mean-square deviation (RMSD),
see Root-mean-square errors
Root-mean-square errors (RMSEs), 52, 61, 66,
210, 304, 369
calculation of, 360
mathematical expression of, 144
MESMA, 243
RS, *see* Remote sensing
RS-based C cycle modeling for terrestrial
ecosystems
BEPS model, 428–431
Biome-BGC model, 428
CASA model, 425–426
GLO-PEM, 426–427
MOD17 algorithm, 431–432
VPM model, 427–428
RS-based vegetation properties, 422–424
RS-based yield estimation, 326–327
RTLA program, *see* Range and Training Land
Assessment program
Rule-based approach, 159
RUSLE model, *see* Revised Universal Soil Loss
Equation model
Ruspini partition, 64, 66
RVC, *see* Relative variance contribution

S

SAAs, *see* Semianalytical algorithms
Sacramento–San Joaquin Delta, 197, 198
SAFARI project, *see* Societal Applications in Fisheries and Aquaculture using Remote Sensing Imagery project
Salinas, 471
Salobras, 471, 472
SAM, *see* Spectral angle mapper
Same period comparing model, 317
Sampling-based forest inventories, 260, 275–276
Sampling design
    accuracy assessment component, 49
    for mapping vegetation cover percentage, 34–35
    strategy of natural resources, 23–24
        case study, 37–39
        design-based, 24–25, 27
        improvement of, 34–35
        local variability–based, 36–37
        model-based, 25–27
        optimal spatial resolution, 30–32
        optimal temporal resolution, 32–34
        role of remotely sensed data, 28–29
Sampling error, 91, 92, 97
Sampling schemes, 49
Sampling units, 49
SAR, *see* Synthetic aperture radar
Satellite-based crop monitoring system, 328
Satellite-based observations of Chl-a, 445
Satellite-derived mapping algorithm, 353
Satellite Hyperion hyperspectral data, 240
Satellite image classification methods, 293
Satellite images, 10–11, 17, 336
Satellite large-footprint waveform LiDAR, 408
Satellite remote sensing, 15, 142, 239
Satellite SAR band designations for land and water applications, 468
Satellite sensors, advanced, 376
SAVI, *see* Soil Adjusted Vegetation Index
Savitzky–Golay filtering, 339
Scale-space theory, 226
Scanning electron microscope (SEM) analysis, 500, 501
Scatterometers, 9
Schwertmannite minerals, 501
Seasonal effect, tree species classification, 252
Seasonal variations
    patterns of LAI, 392
    of vegetation background, 376, 383–384
SEGCLASS, 211, 212, 216

Segmentation
    module, 211
    process, 116
Self-organizing map (SOM) neural network, 143–144
SEM analysis, *see* Scanning electron microscope analysis
Semianalytical algorithms (SAAs), 446
Sensitivity analysis, natural resource mapping
    methodological framework, 96–99
    methods for, 92–96
Sequential Gaussian cosimulation, 340, 341, 354, 355
    for mapping crop residue cover, 358–360, 363–367
SFF, *see* Spectral feature fitting
Shortwave infrared (SWIR), 197, 337
Signal-processing tool, 246–248
Simple power model, 403, 404
Simple random sampling, 34
Simple ratio (SR), for LAI retrieval, 385, 386
Simplest polarimetric attributes, 469
Single-forest variables, 268
Single-scan TLS measurements, 277
Site-specific assessment, 62–63
Site type classification for forest stands, 272
SMA, *see* Spectral mixture analysis
Small-scale process, 174
Smoothing scale, 229
SM technique, *see* Spectral matching technique
Snow-damaged crowns, detection of, 274
Sobol's method, 94, 95
Societal Applications in Fisheries and Aquaculture using Remote Sensing Imagery (SAFARI) project, 454
Soft classification maps
    accuracy assessment of, 57–58
    pixel accuracy assessment measures for, 62–66
    and reference data, 73
Soil acidity map, 507–508
Soil Adjusted Vegetation Index (SAVI), 336, 344–345
Soil erosion, 87, 90
    basic model for, 99–100
    dynamics of, 37–39
    uncertainties of, 94
Soil fertility, model factors of, 339
Soil heterotrophic respiration, 429
Soil mapping, precision farming applications, 340–341
Soil maps, 336
Soil properties, measurement of, 500

Solar zenith angles, 379, 384

SOM neural network, *see* Self-organizing map neural network

Spatial characteristics, treetop detection on, 230–231

Spatial interpolation methods, 91

Spatial metrics, 207–208, 212, 216

Spatial resolutions, 11–12, 92, 294
agricultural remote sensing, 337
results for, 213–214

Spatial sensitivity analysis methods, 95

Spatial uncertainty, procedure of, 96–99

Spatial variability–based method, 31–32

Spatial variability of random process, 358

Spectral angle mapper (SAM), 195, 196, 248, 250

Spectral feature fitting (SFF), 503, 508

Spectral indices, 503

Spectral matching (SM) technique, 195, 248–250

Spectral mixture analysis (SMA), 242–244

Spectral resolution, 12–15, 337

Spectral unmixing, 243–244, 354

Spectral values, uncertainties of, 89

Spectrophotometry, 445

SPOT-5 sensor, 16

SR, *see* Simple ratio

Stand variables estimation, 268–272

Stand-wise field inventory (SWFI), 260

Statistical-based algorithms, 116

Statistical learning theory, 117

Statistical models, 319

Stem curve estimation, 278–279

Stem location mapping with TLS, 278

Stem mass, 423

Stepwise masking system, *see* Hierarchical mapping system

Stochastic approaches, 98

Stratified k-NN, 301

Strip tillage sites, 355

Subsistence farming, 328

Superior spatial resolution, 17

Supervised learning models, 250

Support vector machines (SVMs), 117, 160–161, 248–250

Swan Coastal Plain, 498–499

Sweden, k-NN estimation parameters in, 298

SWFI, *see* Stand-wise field inventory

SWIR, *see* Shortwave infrared

Synthetic aperture radar (SAR), 423
remote sensing, 467–473
technique, 294

T

TAC, *see* Tracing Radiation and Architecture of Canopies

Tasseled cap transform, 137

Taylor series expansion–based methods, 94

TCA, *see* Tree cluster approach

Temperature condition index model, 318

Template-based methods, 161

Template matching (TM)-based methods, 224

Temporal resolution, 15–17

10 band ratios, transformations of QuickBird images, 356, 357

Terrestrial ecosystems, 421–422
carbon (C) cycle modeling for, *see* Carbon (C) cycle modeling for terrestrial ecosystems
RS-based C cycle modeling for, *see* RS-based C cycle modeling for terrestrial ecosystems

Terrestrial laser scanning (TLS)
vs. ALS and MLS, 262–265
applications, 276–279

Terrestrial primary productivity, global mapping of, 432–433

Textural images, identification of, 114

Thematic accuracy, 45

Thematic errors, 89–90

Thematic mapper (TM)
data, 73
images, 353, 480

Thermal inertia model, 318

Thermal infrared (IR) sensors, 4, 6–8

Three-dimension (3D)
canopy architecture, 376–379
model-based methods, 224
space, stem location mapping, 278

Tibet-Qinghai Plateau, 488

Timber assortments, estimation of, 280

TLS, *see* Terrestrial laser scanning

TM, *see* Thematic mapper

TM-based methods, *see* Template matching-based methods

Toxicity Map, 509–510

Tracing Radiation and Architecture of Canopies (TRAC), 378–379

Traditional design-based approaches of remotely sensed data, 28

Traditional multispectral classifiers, 248

Traditional photographic emulsions, 18

Traditional supervised classification techniques, 142

Traditional vegetation model, 176

Trajectory error matrix, 52
Tree attribute data acquisition from TLS, 279
Tree cluster approach (TCA), 262, 271–272
Tree-crown delineation, 235
    edge-detection methods and, 224–225
    treetop extraction and, 285
Tree-crown delineation algorithm
    boundaries enhancement
        dyadic wavelet decomposition, 226–227
        edge probability with magnitude
            information, 227
        scale and geometric consistency
            constraints, 228, 235–236
    edge detection, 228–229
    marker-controlled watershed segmentation,
        231–232
    treetop identification, 229–231
Tree crown dimensions, 266–268, 271
Tree detection errors, 270
Tree height, 423
Tree species classification
    advanced remote sensing sensors/systems,
        240–242
    considerations, 251–252
    future directions, 253
    ITD, 270–271
    satellite remote sensing, 239
    techniques and methods
        advanced classifiers, 248–251
        hierarchical mapping system, 245
        hyperspectral transformation and feature
            extraction, 246–248
        OBIA method, 244–245
        SMA, 242–244
Treetops
    detection, 229–231
    extraction, 235
    tree crowns and, 236
Tree variables
    estimation of, 268–272
    measurement, laser point cloud, 266
2D-layer searching technique, 278
Two-phase sampling
    procedure, 275–276
    for stratification, 28
Two-step approach, change detection maps, 52
Two-step space–time kriging procedure, 33
Two-time-point laser data, 272

U

UAV, *see* Unmanned aircraft vehicle
Unbiased cokriging estimator, 34–35

Uncertainties, natural resources
    methodological framework, 96–99
    methods for, 92–96
    sources of, 88–92
Uniform agronomic practices, 333
United States Geological Survey (USGS)
    Spectral Library, 505, 508
Universal Transverse Mercator (UTM), 89, 144,
    210, 471, 481
Unmanned aircraft vehicle (UAV), 262, 264
Unsupervised classification techniques, 142
Urban environment, tree species
    classification, 252
USDA Foreign Agricultural Service (FAS),
    323–324
User's accuracy, 47
U.S. Fish and Wildlife Service National
    Wetlands Inventory, 462
U.S. Geological Survey (USGS), 481
U.S. National Forest Inventory and Analysis
    (FIA) sampling design, 23–24
UTM, *see* Universal Transverse Mercator

V

Variance-based methods, 93–94
Variance map of crop residue cover, 363–367, 370
Variogram model, 307
Variogram of crop residue cover, 354, 358,
    360–361
Vazantes, 471
Vegetation background on reflectance, 383–384
Vegetation biomass
    estimate of, 399–400
    LiDAR remote sensing of, *see* Light detection
        and ranging, remote sensing
    remote sensing using LiDAR systems, 401
Vegetation change detection in Brazilian
    Amazon
    analysis of, 133–138
    deforestation, 127
    methods
        data collection and preprocessing,
            129–130
        detailed "from–to" vegetation change
            trajectories detection, 133
        fractional images development with
            LSMA, 130
        land use/cover classification and
            evaluation, 132–133
        multitemporal Landsat thematic mapper
            (TM) images, 128–129
        vegetation gain/loss detection, 130–132

Vegetation condition index model, 318
Vegetation gain/loss detection, 130–132
    analysis of, 133–136
Vegetation, geometry and organization of, 468
Vegetation indices (VIs), 344
    LAI algorithms based on, 385–388
    in precision agriculture, 338–340, 342
VEGETATION sensor, 391
Vegetation temperature condition index
    model, 318
Vegetation water supply index model, 318
Vertical and horizontal (VH) polarization, 468
Vertical and vertical (VV) polarization, 468
Vertical drought index model, 318
Vertical point height distributions, 266
Very-high-resolution (VHR) aerial imagery
    data preparation, 225–226
    results, 232–234
    study sites, 225
    tree-crown delineation algorithm
        boundaries enhancement, 226–228
        edge detection, 228–229
        marker-controlled watershed
            segmentation, 231–232
        treetop identification, 229–231
VIs, *see* Vegetation indices
Visible and near-infrared (VNIR), 197
Visual interpretation of medium- and
    large-scale aerial imagery, 224
VNIR, *see* Visible and near-infrared
Volume Scattering Index (VSI), 469

VPM model, 427–428
VSI, *see* Volume Scattering Index

W

Water application, satellite SAR band
    designations for, 468
Water bodies, classification of, 215
Watershed-based region detector, 267, 268
Water stress detection, plant, 342–344
Wavelet-based methods, 226
Wavelet-merging technique, 115
Wavelet transform (WT), 246–248
Weed sensing, precision farming applications,
    345–346
Wetland dynamics, remote sensing
    applications, 479–480
    analysis, 482
    materials and methods, 480–484
    results
        landscape pattern dynamics, 488–490
        LULC conversion matrix, 484–488
        patch shape changes, 490–492
Wetland ecosystems, 461–462
Wetland habitats, Level 1 classification of,
    471–472
Wetland lakes, Level 2 classification of, 472, 473
Whiskbroom scanning, 5
Wood procurement chain, 260
WorldView-2 satellite, 240–242, 244, 465
WT, *see* Wavelet transform

Printed and bound by CPI Group (UK) Ltd, Croydon, CR0 4YY

01/11/2024

01782604-0014